Fourth Edition

PROCESS GEOMORPHOLOGY

Dale F. Ritter
Desert Research Institute

R. Craig Kochel
Bucknell University

Jerry R. Miller
Western Carolina University

Boston   Burr Ridge, IL   Dubuque, IA   Madison, WI   New York   San Francisco   St. Louis
Bangkok   Bogotá   Caracas   Kuala Lumpur   Lisbon   London   Madrid   Mexico City
Milan   Montreal   New Delhi   Santiago   Seoul   Singapore   Sydney   Taipei   Toronto

McGraw-Hill Higher Education

A Division of The **McGraw-Hill** *Companies*

PROCESS GEOMORPHOLOGY, FOURTH EDITION

Published by McGraw-Hill, a business unit of The McGraw-Hill Companies, Inc., 1221 Avenue of the Americas, New York, NY 10020. Copyright © 2002, 1995, 1986, 1978 by The McGraw-Hill Companies, Inc. All rights reserved. No part of this publication may be reproduced or distributed in any form of by any means, or stored in a database or retrieval system, without the prior written consent of The McGraw-Hill Companies, Inc., including, but not limited to, in any network or other electronic storage or transmission, or broadcast for distance learning.

Some ancillaries, including electronic and print components, may not be available to customers outside the United States.

 This book is printed on recycled, acid-free paper containing 10% postconsumer waste.

2 3 4 5 6 7 8 9 0 QPD/QPD 0 9 8 7 6 5 4 3 2

ISBN 0-697-34411-8

Publisher: *Margaret J. Kemp*
Developmental editor: *Lisa Leibold*
Associate marketing manager: *Tami Petsche*
Senior project manager: *Kay J. Brimeyer*
Senior production supervisor: *Laura Fuller*
Coordinator of freelance design: *Rick D. Noel*
Cover designer: *Kay Fulton*
Cover image: *Photo by R. Craig Kochel*
Senior photo research coordinator: *Lori Hancock*
Senior supplement producer: *David A. Welsh*
Media technology producer: *Judi David*
Compositor: *Shepherd, Inc.*
Typeface: *10/12 Times Roman*
Printer: *Quebecor World Dubuque, IA*

The credits section for this book begins on page 545 and is considered an extension of the copyright page.

Library of Congress Cataloging-in-Publication Data

Ritter, Dale F.
   Process geomorphology.—4th ed./Dale F. Ritter, R. Craig Kochel, Jerry R. Miller.
     p. cm.
   Includes bibliographical references and index.
   ISBN 0-697-34411-8 (acid-free paper)
    1. Geomorphology. I. Kochel, R. Craig. II. Miller, Jerry R. (Jerry Russell), 1960–
III. Title.

GB402 .R57 2002   2001031509
551.41—dc21   CIP

www.mhhe.com

To my family
D.F.R.

To Travis and Kasei
R.C.K.

To Suzanne, Rebecca, and Mary
J.R.M.

CONTENTS

Preface   ix

CHAPTER 1

Process Geomorphology: An Introduction   1

Introduction   2
The Basics of Process Geomorphology   3
    The Delicate Balance   4
    Force, Process, and Resistance: The Ingredients
        of Balance   6
        Driving Forces   7
        The Resisting Framework   9
    Thresholds and Complex Response   14
    The Principle of Process Linkage   16
    The Time Framework   16
Summary   19
Suggested Readings   19

CHAPTER 2

Internal Forces and Climate   20

Introduction   21
The Endogenic Effect   21
    Epeirogeny   22
    Orogeny and Tectonic Geomorphology   25
        Tectonic Geomorphology   25
    Volcanism   31
Climatic Geomorphology   34
    Climate, Process, and Landforms   34
    Climate Change and Geomorphic Response   36
        Sea-Level Fluctuation   37
        Geologic and Vegetal Screens   38
Summary   40
Suggested Readings   41

CHAPTER 3

Chemical Weathering and Soils   42

Introduction   43
Decomposition   44
    Processes of Decomposition   46
        Oxidation and Reduction   47
        Solution   47
        Hydrolysis   48
        Ion Exchange   48
    Mobility   49
        Leaching   49
        pH   50
        The Eh Factor   50
        Fixation and Retardation   51
        Chelation   51
    The Degree of Decomposition   52
        Mineral Stability   52
        Secondary Minerals   53
        Estimates Based on Chemical Analyses   56
    The Rate of Chemical Weathering   56
        The Complicating Factors   57
Soils   58
    Soil Properties   58
    Soil Horizon Nomenclature and Description   61
    Soil Classification   62
    Pedogenic Controls and Regimes   65
        Podzolization   68
        Laterization   70
        Calcification   70
    The Factor of Time   71
    Geomorphic Significance of Soils   73
Summary   78
Suggested Readings   78

CHAPTER 4

Physical Weathering, Mass Movement,
and Slopes   79

Introduction   80
Physical Weathering   80
    Expansion of Rocks and Minerals   81
        Thermal Expansion   *81*
        Unloading   *82*
        Hydration and Swelling   *83*
        Salt Weathering   *87*
    Growth in Voids   89
    The Significance of Water and Interaction of Chemical
        and Physical Weathering Processes   92
Physical Properties of Unconsolidated Debris   92
    Shear Strength   95
        Internal Friction   *95*
        Effective Normal Stress   *96*
        Cohesion   *96*
    Measurement of Strength   99
Mass Movements of Slope Materials   99
    Slope Stability   100
    Mass Wasting Processes   102
        Heave and Creep   *102*
        Rockfalls   *105*
        Slides   *107*
        Flows   *113*
    Morphology of Mass Movements   123
Slope Profiles   125
    The Rock-Climate Influence   129
    Slope Evolution   131
    Relict Hillslope Forms   132
Summary   133
Suggested Readings   133

CHAPTER 5

The Drainage Basin—Development,
Morphometry, and Hydrology   134

Introduction   135
Slope Hydrology and Runoff Generation   137
    Infiltration   138
    Subsurface Stormflow and Saturated Overland Flow   139
    The Stream Hydrograph and Response to Basin
        Characteristics   141
    Effect of Physical Basin Characteristics   141
Initiation of Channels and the Drainage Network   144
    Basin Morphometry   147
        Linear Morphometric Relationships   *150*
        Areal Morphometric Relationships   *150*
        Relief Morphometric Relationships   *153*
    Basin Morphometry and the Flood Hydrograph   154
    Basin Evolution   156
Basin Hydrology   160
    Subsurface Water   160
        The Groundwater Profile   *161*
        Movement of Groundwater   *161*

Aquifers, Wells, and Groundwater Utilization
    Problems   *162*
    Surface Water   164
        Flood Frequency   *165*
        Paleoflood Hydrology   *168*
Basin Denudation   173
    Slope Erosion and Sediment Yield   173
        Wash   *174*
        Sediment Yield (Soil Loss)   *175*
        Factors Affecting Sediment Yield   *176*
    Sediment Budgets   184
    Rates of Denudation   186
Summary   188
Suggested Readings   188

CHAPTER 6

Fluvial Processes   189

Introduction   190
The River Channel   190
    Basic Mechanics   190
    Flow Equations and Resisting Factors   193
Sediment in Channels   195
    Transportation   195
    Entrainment   196
    Bank Erosion   200
    Erosion of Bedrock Channels   202
    Deposition   204
    The Frequency and Magnitude of River Work   205
The Quasi-Equilibrium Condition   206
    Hydraulic Geometry   207
    The Influence of Slope   211
    Channel Shape   213
Channel Patterns   214
    Straight Channels   215
    Meandering Channels   217
    Braided Channels   221
    Anastomosing and Anabranching Channels   223
    The Continuity of Channel Patterns   224
Rivers, Equilibrium, and Time   225
    Adjustment of Gradient   228
    Adjustment of Shape and Pattern   230
Summary   231
Suggested Readings   231

CHAPTER 7

Fluvial Landforms   232

Introduction   233
Floodplains   233
    Deposits and Topography   235
    The Origin of Floodplains   237
        The Humid-Temperate Model   *237*
        Spatial and Temporal Effects   *241*
Fluvial Terraces   242

Types and Classification   243
The Origin of Terraces   243
  Depositional Terraces   243
  Erosional Terraces   244
Terrace Origin and the Field Problem   246
Piedmont Environment: Fans and Pediments   248
Alluvial Fans   248
  Fan Morphology   251
  Fan Deposits and Origins   255
Pediments   259
  Morphology and Topography   260
  Processes   262
  Formative Models   263
Deltas   264
Delta Classification, Morphology, and Deposits   264
Delta Evolution   266
Summary   269
Suggested Readings   270

CHAPTER 8

Wind Processes and Landforms   271

Introduction   272
The Resisting Environment   272
The Driving Force   275
Entrainment and Transportation   276
Processes   276
The Effect of Saltation   277
Erosional Features   280
Deposits and Features   282
Ripples   285
Dunes   286
Fine-Grained Deposits   292
Summary   295
Suggested Readings   295

CHAPTER 9

Glaciers and Glacial Mechanics   296

Introduction   297
Glacial Origins and Types   297
The Mass Balance   301
The Movement of Glaciers   302
Internal Motion   303
  Basic Mechanics   303
  A Simple Model of Internal Flow   304
  Extending and Compressive Flow   306
Sliding   308
Motion with Deforming Subglacial Sediments   312
Velocity Variations with Time   313
Ice Structures   315
Stratification   316
Secondary Features   316
  Foliation   316
  Crevasses   316

Summary   319
Suggested Readings   320

CHAPTER 10

Glacial Erosion, Deposition,
and Landforms   321

Introduction   322
Erosional Processes and Features   322
Minor Subglacial Features   322
Cirques   328
  Cirque Morphology   328
  Cirque Formation   329
Glacial Troughs   332
Deposits and Depositional Features   335
Drift Types   335
  Nonstratified Drift   335
  Stratified Drift   338
The Depositional Framework   338
Marginal Ice-Contact Features   341
  Moraines   341
  Stratified Marginal Features   344
Interior Ice-Contact Features   349
  Interior Moraine   349
  Fluted Surfaces and Drumlins   350
Proglacial Features   354
Summary   356
Suggested Readings   357

CHAPTER 11

Periglacial Processes and Landforms   358

Introduction   359
Permafrost and Ground Ice   360
Definition and Thermal Characteristics   360
Distribution, Thickness, and Origin   361
Hydrology   366
Periglacial Processes   368
Frost Action   368
  Frost Wedging   369
  Frost Heaving and Thrusting   370
  Frost Sorting   371
  Frost Cracking   372
Nivation   373
Pedogenesis in Permafrost Terrain   373
Mass Movements   374
  Frost Creep   374
  Solifluction (Gelifluction)   374
Periglacial Landforms   377
Landforms Associated with Permafrost   377
  Ice Wedges and Ice-Wedge Polygons   377
  Pingos   379
  Thermokarst   381
Patterned Ground   381
  Classification   381
  Origin   385

Landforms Associated with Mass Movement   386
Stratified Slope Deposits   386
Gelifluction Features   386
Rock Glaciers   387
Blockfields   389
Cryoplanation Terraces   391
Relict Periglacial Features and Their Significance   395
Environmental and Engineering Considerations   396
Building Foundations   397
Roads and Airfields   397
Utilities: Water and Sewage   399
Pipelines   399
Implications for Global Warming   400
Applications to Planetary Geology   401
Summary   405
Suggested Readings   405

CHAPTER 12

Karst—Processes and Landforms   406

Introduction   407
Definitions and Characteristics   407
The Processes and Their Controls   408
Karst Rocks—The Resisting Framework   408
Lithology   408
Porosity and Permeability   409
The Driving Mechanics and Controls   410
The Solution Process   410
Solution Rates—The Controlling Factors   412
Spatial Variations in Limestone Solution   413
Karst Hydrology and Drainage Characteristics   413
Surface Flow   413
Karst Aquifers and Groundwater Flow   414
The Relations Between Surface and Groundwater   415
Morphology of Karst Drainage   418
Surficial Landforms   419
Closed Depressions   419
Dolines   419
Doline Morphometry   422
Uvalas and Poljes   423
Karst Valleys   424
Allogenic Valleys   424
Blind and Dry Valleys   424
Pocket Valleys   425
Cockpit and Tower Karst   425

Limestone Caves   428
Cave Physiography   428
Entrances and Terminations   428
Passages and Passage Morphology   428
The Origin of Limestone Caves   430
Summary   432
Suggested Readings   432

CHAPTER 13

Coastal Processes and Landforms   433

Introduction   434
Coastal Processes   436
Waves   437
Wave Generation   437
Wave Modification Near the Coast   440
Tsunamis and Seiches   445
Tides and Currents   445
Tides   445
Nearshore Currents   448
Beaches   451
The Beach Profile and Equilibrium   451
Nearshore Bars   453
Beach Morphodynamics   455
Shoreline Configurations and Beach Landforms   460
Beach Cusps   461
Large-Scale Rhythmic Topography and Capes   464
Coastal Topography   466
High-Relief, Erosional Shorelines   467
Low-Relief, Depositional Shorelines   470
Shoreline Change   473
Rates of Change   474
Causes of Shoreline Erosion   476
Rising Sea Level   476
Coastal Storms   478
Barrier Islands   481
Distribution and Characteristics   481
Geomorphic Processes and Dynamics   483
Summary   490
Suggested Readings   490

Bibliography   491

Credits   545

Index   548

PREFACE

Geomorphology has undergone a drastic change in scope and philosophy during the last several decades. In the past, the discipline was primarily concerned with the evolutionary development of landscapes under a wide variety of climatic and geologic controls. Landforms were used as indicators of relative age with little emphasis placed on how the features were formed. More recently, geomorphologists have recognized the need for an applied as well as historical focus in the discipline. The simple truth is that both elements are needed to be fully conversant with the Earth's surface.

The historical aspect of geomorphology has been revitalized by concerns about global change. It is particularly important that geomorphologists document how the physical environment responded to past climatic changes and tectonic events. Such analyses provide the basis for realistic predictions of what environmental responses might be expected during future climatic and/or tectonic fluctuations. Nonetheless, we believe that valid interpretations of geomorphic history must be based on a thorough understanding of the processes involved in landform development. Geomorphologists, therefore, must be cognizant of process mechanics prior to analyzing how landform history manifests past climatic or tectonic phenomena.

In addition, today's geomorphologist must relate to problems that face hydrologists, engineers, geologists, pedologists, foresters, and many other types of earth scientists. The bond that unites geomorphology with so many apparently diverse disciplines is the common need to understand the processes operating within the Earth's surficial systems. Thus, although the historical aspect of landscapes remains important, it is absolutely essential for earth scientists to have a basic understanding of surface mechanics. This edition of *Process Geomorphology,* like its predecessor, is an attempt to satisfy those needs. The prime purpose of the book remains as it was, to provide undergraduate students with an introductory understanding of process mechanics and how process leads to the genesis of landforms.

A wealth of new information concerning surficial process has emerged since the last edition was completed, and many new techniques to analyze process have been developed. In most chapters new data and interpretations have been assimilated within the format of the last edition. A lengthy bibliography is again presented so that students wishing to pursue a particular topic in greater depth will find a ready nucleus of source material. Most references cited were purposely selected from journals and books that will most likely be found in libraries of North American colleges and universities.

We extend our thanks to those who reviewed all or part of the manuscript for this edition, and we renew our great appreciation to the reviewers of earlier editions of *Process Geomorphology*. Special thanks are given to Dru Germanoski for his insights and discussions concerning fluvial mechanics and to Nick Lancaster for his invaluable guidance concerning new concepts in the field of aeolian geomorphology. The conscientious efforts of all who made this edition of *Process Geomorphology* happen are deeply appreciated. Shortcomings and errors in the book are, of course, ours.

D.F.R.
R.C.K.
J.R.M.

PROCESS GEOMORPHOLOGY:

An Introduction

G. W. Stose by U.S. Geological Survey

Introduction
The Basics of Process Geomorphology
   The Delicate Balance
   Force, Process, and Resistance:
      The Ingredients of Balance
      Driving Forces
      The Resisting Framework
   Thresholds and Complex Response
   The Principle of Process Linkage
   The Time Framework
Summary
Suggested Readings

INTRODUCTION

One of the remarkable aspects of planet Earth is the infinite variety of its surface forms. It is probably safe to assume that as humans became aware of their physical environment, landscape was the first geological characteristic they noted. Familiar surface features guided their travels and established their territorial boundaries. As time passed, people learned how best to utilize regional characteristics for different purposes, such as agriculture, trade, and military tactics. They also learned that some landforms possess certain peculiarities that somehow, almost imperceptibly, set them apart from others. Gradually these isolated observations grew into an organized collection of knowledge, and a separate branch of science was born.

Geomorphology is best and most simply defined as the study of landforms. Like most simplistic definitions, the actual meaning is somewhat vague and open to interpretation. For example, landforms can be analyzed in different ways depending on what information is being sought. Engineers may need to understand the physical properties of a particular landform being considered as a potential construction site. They may have little, if any, interest in how or when the landform was developed. Their analytical approach would be different from that of a geologist collecting data needed to understand the effect of climate change on the same landform. Thus, geomorphic data may be derived for different purposes in which the scientific goals are quite variable. Geomorphologists, therefore, sometimes create subdisciplines such as climatic geomorphology or historical geomorphology (or in our case, process geomorphology) to emphasize a specific focus of landform analysis. Academicians may find one approach more appealing than another and, in fact, the subject is commonly taught in both geology and geography departments.

Traditional geomorphology has been excessively descriptive. Much emphasis in the past was given to placing landforms, both regional and local, into some evolutionary model, so that the field was concerned primarily with historical interpretations. In recent years, however, the discipline has become more quantitative, and the character of geomorphic research is now more applied. Today's geomorphologist often deals with problems that link him or her directly with other professionals working at the surface. Obviously, geomorphology has greater breadth than any definition can adequately express. This lack of rigid philosophical boundaries is probably its greatest attribute because geomorphology lends itself strongly to the interdisciplinary approaches needed to solve our major scientific problems associated with global change and environmental stress.

More important than a precise definition is the fact that geomorphology is and probably always will be a field-oriented science. Map and photo analyses are necessary first steps to good geomorphic work, and laboratory data support interpretations. But the real test of geomorphic validity is outdoors, where all the evidence must be pieced together into a lucid picture showing why landforms are the way we find them and why they are located where they are. A prime requisite for a geomorphologist is to be a careful observer of relevant field relationships. This trait cannot be easily taught, and truly outstanding geomorphologists usually develop it by learning from their own mistakes. Geomorphic processes are remarkably subtle, and minor changes of basic controls can result in an infinite array of landforms. Invariably, the person with the greatest experience under varied conditions will make the most viable geomorphic interpretations. Thus, a geomorphologist, like any other scientist, must learn the trade. There are no shortcuts that produce geomorphic insight. It must be acquired gradually through long field experience.

This book will concentrate on processes that create the features we see at the surface of Earth. **Process** can be defined as the action involved when a force induces a change, either chemical or physical, in the materials or forms at Earth's surface. In simpler terms, process may be thought of as the method by which one thing is produced from something else. It may not be clear why this approach is more beneficial than using some other criterion, such as climate or time, as a central theme. Geomorphology stands at the interface between geology and many other disciplines that deal with surficial phenomena. Today, geomorphologists must be aware of the problems facing hydrologists, civil engineers, pedologists, foresters, urban planners, and other specialists. And because those scientists are working in an environment underlain and partly controlled by the geologic fabric, they must be concerned with geologic concepts and problems. It follows that a common interest must unite these apparently diverse fields, because they all function in the same place at the same time. The universal need to understand processes is basic to all surficial disciplines.

Most important, an understanding of process is critical in geoscience itself. In fact, it seems certain that for geomorphology to remain a viable part of geoscience, future research concerning surficial processes must provide geologists and geophysicists with useful information (Ritter 1988). This linkage is already emerging. For example, sedimentologists and geomorphologists are joining ranks to provide a more solid basis for interpretation of the stratigraphic record (Schumm 1981; Kraus and Middleton 1987). The meaning of abrupt changes in a depositional sequence may well be reflecting geomorphic response to external factors such as climate change or tectonics in the source area. It is also possible, however, that sequence breaks are caused by unique events within the geomorphic system itself (Smith et al. 1989; Brizga and Finlayson 1990; Kraus 1987). Exactly how any disruptive event is transferred through a geomorphic system and manifested in the depositional framework is an extremely important research thrust that will require input from both disciplines.

Geomorphologists can also provide insight into the study of important geoscience questions such as: (1) the analysis of plate boundary conditions (Tosdal et al. 1984; Taylor et al. 1985), (2) determination of groundwater flow systems (Larkin and Sharp 1992), (3) landform stability analysis to determine safety of hazardous waste storage (Wells and Gardner 1985), and (4) geomorphic input related to the development of new dating techniques such as cosmogenic radionuclide dating (Pavich et al. 1985; Cerling 1990) and luminescence dating (Berger 1992).

The research directions cited above are still hampered by the lack of explicit understanding of cause and effect in geological phenomena. We know, for example, that Holocene climate changes were severe enough to upset the delicate balance at the surface. What is confusing is that diverse geomorphic responses often result from the same climatic trends. Thus, even for the very recent past we commonly are unsure of cause and effect.

Finally, we should recognize that human activities are also critical ingredients in the nature of cause and effect (for a classic example see Gilbert 1917). This is especially significant in analyses of environmental problems and geological hazards. It follows then that geomorphology should play a major role in environmental science. Unfortunately, environmental science has historically been equated with biochemical systems, and the important physical component of our environment has been overlooked or referred to engineers for mitigation rather than study. We believe that understanding problems involving our physical environment is an important endeavor that must be accomplished before a mitigation plan is proposed. Furthermore, we suggest that the scientific understanding required stems logically from the application of concepts rooted in process geomorphology. Engineering and environmental geology, therefore, can benefit greatly from contributions made by geomorphologists.

A good example of human influence on geomorphic processes and environmental problems occurred in southern California in the 1970s and 1980s (Kuhn and Shepherd 1983; Kochel and Ritter 1990). Here one effect of climate flux was to produce accelerated erosion of the bluffs overlooking the Pacific Ocean. In the preceding several decades, wave action had not been severe because the prevailing dry climate during that interval created very few major storms. Beaches, shorelines, and sea cliffs were relatively stable. This led to large-scale urbanization along the coast, and with it excessive watering of lawns, irrigation, and the use of septic tanks, leach lines, and cesspools. The extensive use of water caused a steady rise in the water table, which is a prime culprit in slope failure (see chapter 4). As the minor climate fluctuation produced more precipitation and more erosive storm waves, the sea cliffs were primed for failure by the preceding human activities. Landslides and other mass movements became more numerous, and blocks of the coastal bluffs (often supporting homes) slipped downward into the ocean (fig. 1.1). Clearly, humans are geomorphic catalysts.

THE BASICS OF PROCESS GEOMORPHOLOGY

Assuming that our focus on process is a viable way to examine geomorphology, we must identify those concepts that, when integrated, constitute the basic principles of process geomorphology. They are listed here and discussed in detail on the following pages.

1. A delicate balance or equilibrium exists between landforms and processes. The character of this balance is revealed by considering both factors as systems or parts of systems.
2. The perceived balance between process and form is created by the interaction of force and resistance.
3. Changes in driving force and/or resistance may stress the system beyond the defined limits of stability. When these limits of equilibrium, or **thresholds,** are exceeded, the system is temporarily in disequilibrium and a major response may occur. The system will develop a different equilibrium condition adjusted to the new force or resistance controls, but it may establish the new balance in a complex manner.
4. Various processes are linked in such a way that the effect of one process may initiate the action of another.
5. Geomorphic analyses can be made over a variety of time intervals. In process studies the time framework utilized has a direct bearing on what conclusions can be made regarding the relationship between process and form. Therefore, the time framework should be determined by what type of geomorphic analysis is desired.

Figure 1.1 Large landslide and earthflow along the California coast.
(© John Shelton)

Figure 1.2 Interpretation of slope
adjustment to geology by
G. K. Gilbert.
Equilibrium slope developed at *a* is
maintained at times *b* and *c*.

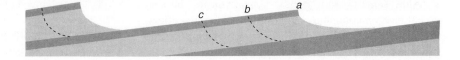

THE DELICATE BALANCE

The idea that some form of balance or equilibrium exists between landforms and the processes that create them is not new. It was clearly expressed by G. K. Gilbert during the latter part of the nineteenth century in his classic reports on the geology in the western United States (see Gilbert 1877). Essentially, Gilbert believed that under any given climate and tectonic setting, landforms reflect some unique accommodation between the dominant processes and the local geology. He often used the terms "dynamic adjustment" and "balanced condition" to describe this relationship. An example of Gilbert's perception of equilibrium is shown in figure 1.2. Here we see a series of slopes that are adjusted to alternating weak and resistant rock layers. The slopes developed on the different units are produced and maintained by the interaction of geology and processes such as mass movement, sheetwash, and river flow. Importantly, Gilbert believed that continuous erosion would not change the slope angles as long as the processes and their climatic and tectonic controls remained constant. Thus, the slopes at times *b* and *c* will be a mirror image of the slopes at time *a* because the process types and rates have not changed through time. If tectonic or climatic controls change, processes will also change, and new slope characteristics will develop in an adjustment to the altered processes.

In the first half of the twentieth century, Gilbert's ideas were pushed aside when geomorphologists espoused the concept developed by W. M. Davis that landscapes change continuously with time and progress through distinct stages that can be identified by regional geomorphic characteristics. It was not until after World War II that the equilibrium approach was revitalized in a number of papers reemphasizing the importance of the adjustment between process and form (Horton 1945; Mackin 1948; Strahler 1950, 1952a; Leopold and Maddock 1953). This shift in emphasis resulted in the **dynamic equilibrium concept** in which J. T. Hack (1960b) essentially brought back Gilbert's approach as a philosophical framework for geomorphic analyses. Dynamic equilibrium suggests that elements of landscape rapidly adjust to the processes operating on the geology, and thus process and form reveal a cause and effect relationship. The forms within a landscape maintain their character as long as the fundamental controls do not change.

Hack's model was presented as an alternative to Davis's perception of landscape development over long time intervals. For our process emphasis, this long-term evolution of landscapes may be less important than the episodic changes in climate or tectonics that occur more frequently (Ritter 1988). These changes tend to disrupt equilibrium and often require that a different relationship be established between landform character and the processes altered by climatic or tectonic events. How rapidly this new equilibrium relationship is developed is a critical question that depends primarily on the magnitude of the climatic or tectonic change. Clearly, some landforms (e.g., moraines) are related to processes and climates that no longer exist and, therefore, are still readjusting to modern conditions.

Many geomorphologists believe that the balance between form and process is best demonstrated by considering both factors as systems or parts of systems. A *system* is simply a collection of related components. For example, suppose we define a drainage basin as a system and consider its measurable parts to be basin area, valley-side slopes, floodplains, and stream channels. The balance or equilibrium condition within our system is revealed by statistical relationships between the various parameters; for example, basin area may be directly related to total channel length, and so on.

The systems approach has become highly sophisticated (Chorley 1962; Chorley and Kennedy 1971), and different types of systems have been identified and used in geomorphology (Schumm 1977). For our purposes, it is best to consider landforms and processes as part of the same open system in which energy and/or mass are continually added or removed. Any flux in energy or mass requires that the processes and their statistically related landforms adjust to maintain balance in the system.

The systems approach has these advantages:

1. It emphasizes the intimate relationship between process and form.
2. It stresses the multivariate nature of geomorphology.

3. It reveals that some forms may not be in balance because they owe their character to relict conditions. Some glaciated regions, for example, may have landforms that were adjusted to geomorphic controls different from those of the present.

Although some geomorphologists feel that equilibrium is difficult to define (Thorn and Welford 1994) or is too imprecise to serve as a geomorphic paradigm (see Phillips and Renwick 1992), we believe that the concept has merit if applied carefully. Equilibrium implies that landforms (and presumably processes) exist in some type of unchanging condition. In theory this requires that factors which ultimately control landforms and process (such as climate and tectonics) must also remain unchanged. In reality changes do occur in the controlling factors with time. Thus, the true meaning of equilibrium depends on the time interval over which our balance is being considered. Schumm and Lichty (1965) argued that different time intervals, which they called *cyclic, graded,* and *steady,* are critical to our understanding of the process and landform relationship, and the distinction of these is extremely important in our perception of equilibrium. Indeed this insight was followed by the further suggestion (Chorley and Kennedy 1971) that different kinds of equilibrium are related to each particular interval of time (fig. 1.3). **Static equilibrium** is that which exists over the short steady-time interval (days or months). In this framework of time, landforms do not change and therefore they are truly time-independent. In **steady-state equilibrium,** landforms and/or processes are considered over graded time, perhaps 100 to 1000 years (Schumm 1977). The equilibrium demonstrated in this interval is one in which changes do occur, but their offsetting effects tend to maintain the system in a constant average condition (fig. 1.3). In contrast, **dynamic equilibrium** must be considered over cyclic time, perhaps millions of years (Schumm 1977). In this case, even though fluctuations of variables occur, they are not offsetting and the average condition of the system is progressively changing (fig. 1.3).

Other geomorphologists considered equilibrium to be definable in only one distinct time interval. For example, Wolman and Gerson (1978, 196) state:

> All definitions recognize that both the processes and specific forms represent averages and that the characteristics which define equilibrium must be measured "over a period of years," in Mackin's (1948) phrase, to allow for short-term variations. The "equilibrium" of a stream at a given point will, over time, fluctuate around a mean. The effect of climate change is discernible only when a trend in the width, or any other descriptive parameter can be detected, or when a "new" form is maintained for a sufficient period of time to permit distinction between the previous and the newly established values.

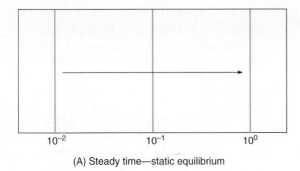

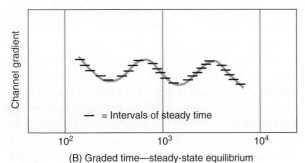

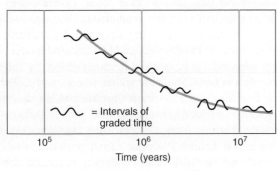

Figure 1.3 Different time intervals and associated equilibrium in geomorphic analyses.
(A) Steady time (static equilibrium). No change in channel gradient over short periods. (B) Graded time (steady-state equilibrium). Constant average channel gradient with periodic fluctuations above and below the average condition. Measurements made during intervals of steady time within the graded time period may show no change in channel gradient. (C) Cyclic time (dynamic equilibrium). Gradual lowering of the average channel gradient over long time intervals. Intervals of graded time and steady-state equilibrium exist within the cyclic time scale.

(Schumm 1977; Chorley and Kennedy 1971)

Clearly, Wolman and Gerson suggested that equilibrium can be defined only as a graded-time phenomenon, a very important difference from the analyses discussed above.

With the foregoing perspective of equilibrium, it is apparent why the concept has been difficult to define or understand. Time is a major factor in understanding equilibrium, and effective use of the concept in geomorphology demands that the time framework be specified.

TABLE 1.1	Common Systems of Units Used in Mechanical Analyses.

Units[a]

Systems	Length	Mass	Force	Time
cgs	centimeter	gram	dyne	second
fps	foot	slug	pound	second
mks	meter	kilogram	newton	second

[a]1 slug = 1 lb sec^2 ft^{-1}; 1 dyne = 1 g cm sec^{-2}; 1 newton (N) = 10^5 dynes.

We will examine the time factor in process geomorphology later in the chapter.

Force, Process, and Resistance: The Ingredients of Balance

In this book, we loosely apply Newtonian physics as the underlying basis for the analysis of process. Our approach may at times seem imprecise and theoretical, but we believe that the concepts presented are all supported by fundamental physical laws. In fact, it is possible to analyze Earth processes in very quantitative terms using units, or their derivatives, shown in table 1.1 to define the pertinent mechanics (for example see Carson 1971; Selby 1982).

One of the mechanical properties identified by Newton is *force,* which, on the basis of his laws of motion, can generally be regarded as anything that changes or tends to change the state of motion in a body. In our perception of surficial mechanics, we refer to a **driving force** as simply the application of energy on Earth materials. A factor commonly used in mechanical analyses is *stress,* which is nothing more than a measure of a driving force applied per unit area of any material. In fact, in a conceptual sense, landforms can be thought of as manifesting the interaction between driving forces (stresses) and the resistance (strength) of Earth materials.

Energy used to produce driving forces in geomorphology is derived directly from climate, gravity, and heat generated inside Earth. Resistance is provided by lithology and structure of the geologic framework. The link between these two components is process. Thus, as stated earlier, process may be considered as the method by which one thing is produced from something else, and as the vehicle by which a quantity of one system is transferred into, and participates in, the mechanics of another system.

In general, processes are either exogenic or endogenic. *Exogenic* processes operate at or near Earth's surface and are normally driven by gravity and atmospheric forces. *Endogenic* processes are different because the energy that initiates the action is located inside Earth. The processes themselves may operate at the surface,

| TABLE 1.2 | Annual Heat Balance and the Transfer of Heat in Different Latitude Zones. |

Zones of Latitude (degrees)	Fraction of Total Area	Short-Wave Radiation Absorbed (cal/cm²/min)	Long-Wave Radiation Emitted (cal/cm²/min)	Poleward Transport of Heat Across Latitude Parallels (cal/min)
0–20	0.34	0.39	0.30	57×10^{15} (20°)
21–40	0.30	0.34	0.30	77×10^{15} (40°)
41–60	0.22	0.23	0.30	50×10^{15} (60°)
61–90	0.14	0.13	0.30	
Weighted mean		.30	.30	

Source: Handbook of Applied Hydrology, ed. by Ven T. Chow, copyright © 1964. Reprinted by permission of The McGraw-Hill Publishing Co., Inc.

but their energy source is usually well below the surface. Both types of processes may sometimes be involved in the development of the same landform. For example, the shape of a volcanic cone is the product of both endogenic volcanism and normal exogenic slope processes.

Thus, we suggest that geomorphology can be examined by using physical concepts that revolve around the application of force on surface materials. In our model the effect of processes depends on how vigorously the forces drive them and how strongly their action is resisted by the geological framework. Process, in this sense, allows us to explain the incredible variety of landforms at Earth's surface.

Driving Forces   Having suggested that energy is exerted on Earth materials as a driving force, we should briefly examine the major forces in our systems. Although each of these has been detailed after long and careful study, we will treat them only briefly to fit our specific needs.

Climate   Radiation emitted from the sun is the major source of energy needed to drive exogenic processes. Radiation is expressed in terms of heat, a form of energy possessed by molecules of matter because of their motion. Heat could be expressed in normal units of energy, but it has historically been measured in the special, more convenient units of calories or British thermal units (Btu). A calorie, for example, is simply the amount of heat required to raise the temperature of a specified mass or weight of water one degree centigrade.

If an imaginary plane were placed at the outer limit of the atmosphere, perpendicular to the incoming rays of sunlight, it would receive 2.0 cal/cm²/min of radiant energy over its entire surface, a parameter known as the *solar constant*. This energy value is significantly lowered before it can be utilized geomorphically because much of the radiation is reflected back into space by clouds, particulate matter in the atmosphere, and Earth's surface. Therefore, the amount of energy absorbed in the system (*insolation*) and actually available for work averages about 0.30 to 0.35 cal/cm²/min over the entire globe.

The average insolation value varies greatly with latitude (table 1.2) and with the seasons. Because the total heat budget does not change, the Earth-atmosphere system must return to space as much heat as it receives, which it does in the form of long-wave, blackbody radiation. It is significant that although absorbed radiation decreases with increasing latitude, heat loss is fairly constant (table 1.2). This produces an obvious temperature differential between the equator and the poles, which drives a poleward transfer of heat in the oceans and, even more, in the atmosphere. The transfer of solar energy demands a series of complex processes that generate the various components of our weather. These processes tend to establish reasonably well-defined temperature and precipitation patterns for all portions of Earth's surface. The average of weather conditions at any place, considered over a long period of time, is called *climate*. Climate represents the net result of how solar energy is distributed or transferred in the Earth-atmosphere system.

On a local or regional scale, the mechanics of heat transfer help explain the great differences between oceanic and continental climates. Continents heat and cool faster than oceans, with more extreme variations in temperature. Because the thermal character of the surface controls the heating of the adjacent air, we can expect temperatures over land and water to function in the same way as the surface itself. The seasonality of midlatitude continental climates can also be understood in terms of the relative rates of heating and cooling and the associated pressure changes of the land-water settings.

On a larger scale, the inequality of heat with latitude (table 1.2) requires a transfer of heat from the equatorial region to the poles. The precise mechanics of this transfer involves a series of complexly interrelated processes that are a basic concern of meteorology. Variables include world circulation patterns of air masses, vertical and horizontal pressure distribution, fronts, rotation of Earth, and the distribution of landmasses and oceans. In general, heat is transferred poleward by air motions, controlled by the average air pressure and wind patterns of Earth (fig. 1.4).

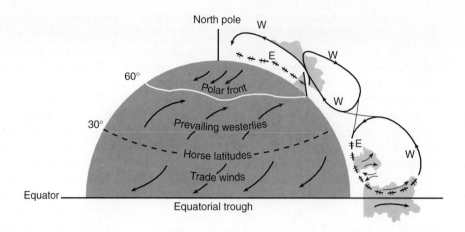

Figure 1.4 Prevailing wind patterns and cellular air motions in the Northern Hemisphere.
(Source: Adapted from C. G. Rossby, 1941)

Ocean currents also are important dispensers of heat. Warm ocean currents moving toward the poles flow under cold air and give off energy as latent heat that ultimately warms the overlying atmosphere. The distribution of heat is not complete in the sense that all regions attain equivalent temperatures. It is effective, however, in stabilizing temperature conditions on a regional basis, providing Earth with a rather well-defined temperature pattern.

Precipitation is also controlled by the same processes that distribute temperature, and the pattern of precipitation generally follows the major zones of atmospheric circulation. Precipitation is greatest near the equatorial trough and least near the 30° sinking limb of the low-latitude convection cell (fig. 1.5). Secondary peaks of precipitation occur in the middle latitudes with the unsteady pressure cells of that region. Precipitation occurs when large air masses are cooled or burdened with added moisture. Commonly precipitation is triggered when moisture-laden air rises to higher and cooler elevations. The principal mechanisms that lift air masses and thereby initiate precipitation are convection, orographic effects, and frontal mechanics. In convection, warm air, being lighter than surrounding air, rises until condensation forms the familiar bulging shape of cumulus clouds. The orographic effect occurs when air masses are forced to rise over high mountain ranges. Frequent rain occurs on the windward side of the mountains, with a characteristic rain shadow on the leeward side. Frontal activity involves the interaction of low-pressure and high-pressure zones. A *front* is simply the line of contact between the moist air of the low-pressure cell and the dry air associated with a high-pressure cell. At the frontal contact, moist air is forced to rise up and over the dry air and so is cooled; the result is precipitation.

Gravity   The second major driving force, gravity, manifests itself in a myriad of endogenic and exogenic geomorphic processes. Combined with the climatic engine, gravity determines the rigor of fluvial power, mass

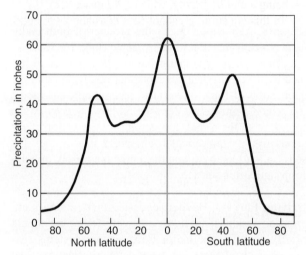

Figure 1.5 Generalized world distribution of average precipitation according to latitude.
(Adapted from *Handbook of Applied Hydrology*, ed. by Ven T. Chow. Copyright © 1964. Reprinted by permission of The McGraw-Hill Companies.)

wasting, glaciation, tidal effects on coastal processes, and the movement of groundwater. Internally, gravity bears directly on the process of isostasy, which tends to control the distribution of Earth materials of different densities, ultimately powering regional uplift. Gravity is ubiquitous, affecting all substances. The force of gravity is applied continuously in every system at, above, and beneath the surface and so can never be completely ignored in any consideration of process.

Sir Isaac Newton's classic work on the force of gravity was published in 1687, introducing his law of universal gravitation. Simply stated, the law says that there exists between any two objects a mutual attractive force that is a function of the two masses (m_1 and m_2), the distance separating them (r), and the universal gravity constant (G):

$$F = G \frac{m_1 m_2}{r^2}$$

Thus, gravity attraction between two objects is an action-reaction phenomenon. Each body exerts a force on the other that is equal in magnitude but oppositely directed along a straight line joining the two bodies. Our main interest in gravity is how it affects geomorphology, especially surficial matter. The gravitational force exerted on surface materials is measured in terms of the amount of acceleration that the force imparts to any freely falling particle having mass. It is normally expressed by the equation

$$g = \frac{GM}{r^2}$$

where M is the mass of Earth. In most scientific work g is assumed to be constant, having a value of 980 gals (a *gal* being a unit of gravity, which is 1 cm/sec/sec).

In this equation, g cannot be a constant, as we normally assume, because it depends on several variable factors. The distance (r) changes because of topographic irregularities and because Earth is not a perfect sphere. The density of Earth materials is not evenly distributed and so may vary along the line connecting the masses. In addition, the rotation of Earth introduces a counteracting force and causes a distinct latitudinal variation in gravity. Therefore, g is not distributed regularly at the surface. This fact is a justifiable concern of geophysicists because slight variations in gravity have real significance, especially as an exploration tool. However, the variation in gravity at Earth's surface is so small compared with the total magnitude that for most exogenic analyses g can reasonably be considered to be constant, and this is normal practice in process analyses. On the other hand, the minor variations that reflect internal density or mass differences are extremely important in endogenic processes. We will discuss gravity again in chapter 2 when considering isostatic adjustments as a geomorphic process.

Internal Heat Thermal energy is generated inside Earth, primarily by radioactive decay and secondarily by friction caused by Earth tides and rock deformation. Earth transmits to the surface about 2.4×10^{20} cal each year of its internal heat. The total amount of heat is minor compared with the heat received at the surface from solar radiation, but it does indicate that heat, no matter what its origin or gradient, is being transferred from place to place within Earth. The mechanics of heat transfer is significant since the energy distributed drives internal geological processes.

Measurements of heat reaching the surface are difficult and costly, and often they are affected by secondary factors such as groundwater, variations in conductivity, and recent volcanism. In addition, measurements are not randomly spaced but tend to be concentrated in areas of some specific interest so that large regions exist for which little or no data are available. Nonetheless, the de-

velopment of sophisticated instrumentation and interest in ocean tectonics have produced a storehouse of information that is beginning to yield a reasonable picture of surface heat flow. Except for local abnormalities, heat emerges from all parts of Earth in amazingly equal amounts, with average continental and oceanic values differing by only 0.2 μ cal/cm^2/sec (Wyllie 1971). If radioactivity is the major source of internal heat, the equality of heat flow from continents and ocean floors would require an unusual distribution of radioactive minerals beneath the two environments unless the thermal condition were balanced by heat transfer processes such as conduction and convection. These processes may be demonstrated by examining heat flow for major physiographic regions of Earth, as shown in table 1.3. Note that heat flow from ocean ridges and trenches differs considerably from average values for entire ocean basins; ridge crests are abnormally high and trenches notably low. Heat may be actively rising under ocean ridges as part of a convective overturn, while the low heat values beneath the trenches presumably represent the descending limbs of the overturning cells. On continents, as one would expect, the lowest heat flow values occur in the very stable shield areas and the highest in the most recent orogenic belts and their associated regions of Cenozoic volcanism.

The transfer of internal heat plays a significant role in determining the major topographic framework of Earth. Heat transfer drives processes beneath the surface causing uplift and deformation, distributes rock masses of varying resistance, and controls the volume of ocean basins, thereby influencing the position of sea level. Precisely how or if heat flow relates to gravity distribution is debatable, but certainly the two forces combined represent a major geomorphic element.

The Resisting Framework Landforms reflect a balance between the application of driving forces and the resistance of the material being worked on. Having reviewed the salient features of the driving forces in our systems, we should now examine the resisting elements, but exactly how to do so is rather perplexing. It is tempting simply to state that the resistance in geomorphic systems is geology; the geologic affect on geomorphology is so pervasive and so varied that any brief review of its role in determining process and form must be inadequate. A complete discussion of the geological control of geomorphology would require an analysis of every possible geologic framework in every possible climatic and tectonic regime. Although such an effort is impossible here, some general examples will show how geological resistance manifests itself in landforms.

Lithology The resisting force in geomorphology is implemented through the two major geologic variables, lithology and structure. The diverse origins of rocks

exogenic gravity forces approx. constant while endogenic gravity forces aren't constant but dynamic

TABLE 1.3	Heat Flow Values from Major Geologic Features.

Geologic Feature	Average Heat Flow ($\mu cal/cm^2/sec$)
Land Features	
1. Precambrian shields	0.92
Australian shield	1.02
Ukrainian shield	0.69
Canadian shield	0.88
S. African shield	1.03
Indian shield	0.66
2. Post-Precambrian	
Nonorogenic areas	1.54
Europe	1.67
Interior Lowlands, Australia	2.04
Interior Lowlands, N. America	1.25
S. Africa	1.36
3. Post-Precambrian	
Orogenic areas[a]	1.48
Appalachian area	1.04
E. Australian highlands	2.03
Great Britain	1.31
Alpine system	2.09
Cordilleran system	1.73
Island arcs	1.36
4. Cenozoic volcanic areas[b]	2.46
Ocean Features	
1. Ocean basins	1.28
Atlantic	1.13
Indian	1.34
Pacific	1.18
Mediterranean seas	1.20
Marginal seas	1.83
2. Ocean ridges	1.82
Atlantic	1.48
Indian	1.57
Pacific	2.13
3. Ocean trenches	0.99
4. Other ocean areas	1.71

Adapted from W. H. K. Lee and S. Uyeda, *Geophysical Monograph 8* , p. 147, 1965, copyright by the American Geophysical Union.

[a]Excluding Cenozoic volcanic areas.
[b]Excluding geothermal areas.

create lithologies at the surface that differ vastly in their chemical and mineralogic compositions, textures, and internal strengths. In geomorphology we are concerned with the modern resisting framework, regardless of its history. It is important to gain an overall picture of the crustal and surface rock distributions as they presently exist.

Table 1.4 synthesizes several estimates of the bulk chemical composition of the lithosphere. As expected, the chemistry of continental crust is higher in silica and

K_2O than that of oceanic crust, and lower in CaO, MgO, and total iron. Such a chemical distribution can be converted into reasonable estimates of the volume-percentage of common rock types and their modal mineral composition (table 1.5). The significance of these analyses is to emphasize that the resisting framework in geomorphology basically entails only two igneous and metamorphic rock suites and approximately ten mineral varieties. The crust consists primarily of a silicic assemblage (granites, gneisses, schists, granodiorites, and diorites) that makes up 48 percent of the crustal volume and a mafic association that constitutes about 43 percent. Obviously the silicic group is plutonic or metamorphic in origin and is dominantly continental; the mafic types are overwhelmingly volcanic and rooted beneath the oceans.

Although these estimates were made decades ago, more recent data based on modern seismic techniques and numerous geological approaches have not resulted in significant changes in the compositional values presented in table 1.4 (see Percival and Berry 1987 for references). What has become clear, however, is that the bulk composition of the lower crust seems to vary depending on the tectonic environment during formation of the rocks (Percival and Berry 1987). In addition, Holbrook et al. (1992) suggest that the lower part of the continental crust may be more mafic than the data show because of a bias in the location of seismic studies. This premise is consistent with analyses of lower-crust xenoliths derived from many regions of the world (Rudnick 1992). Similarly, some data suggest that the crust beneath the conterminous United States is more mafic than one might guess (table 1.6). Pakiser and Robinson (1966) point out that, based on seismic velocities, the total U.S. crust is 54 percent mafic by volume (55 percent by weight). In addition, they show that the mafic content is considerably greater in the provinces of the eastern United States. In general, the eastern regions have a crust that is predominantly mafic, and the western provinces a crust that is mostly silicic.

If Pakiser and Robinson are correct, it is even more interesting to examine the igneous rocks exposed at the surface in the Appalachian and Cordilleran regions (table 1.7). In the Appalachians, where the crust is predominantly mafic (as noted in table 1.6), the surface igneous rocks are overwhelmingly calc-alkalic, plutonic rocks. Of the rocks of this type indicated in table 1.7 (84.5 percent of the total), 96 percent of the plutons are granites. In contrast, the igneous rocks exposed in the Cordilleran system are mainly extrusive (63.6 percent), and of these 77 percent are basaltic or andesitic in composition. The thick mafic crust in the eastern United States supports a surface rock assemblage that is dominantly granitic, in contrast to a silicic crust supporting mafic surface rocks in the west.

In North America, sedimentary rocks make up most of the exposed materials (table 1.8) even though they are

TABLE 1.4	Weight Percent of Common Elements in Earth's Lithosphere.

	Continental Crust			Oceanic Crust		Total Lithosphere	
	Poldervaart 1955	Pakiser and Robinson 1966	Ronov and Yaroshevsky 1969	Poldervaart 1955	Ronov and Yaroshevsky 1969	Poldervaart 1955	Ronov and Yaroshevsky 1969
SiO$_2$	59.4	57.8	61.9	46.6	48.7	55.2	59.3
TiO$_2$	1.2	1.2	0.8	2.9	1.4	1.6	0.9
Al$_2$O$_3$	15.5	15.2	15.6	15.0	16.5	15.3	15.9
Fe$_2$O$_3$	2.3	2.3	2.6	3.8	2.3	2.8	2.5
FeO	5.0	5.5	3.9	8.0	6.2	5.8	4.5
MgO	4.2	5.6	3.1	7.8	6.8	5.2	4.0
CaO	6.7	7.5	5.7	11.9	12.3	8.8	7.2
Na$_2$O	3.1	3.0	3.1	2.9	2.6	2.9	3.0
K$_2$O	2.3	2.0	2.9	1.0	0.4	1.9	2.4

TABLE 1.5	Abundance of Rock and Mineral Types in Earth's Crust.

Rocks	% of Crustal Volume	Minerals	Modal %
		Quartz	12
Sands	1.7	K-feldspar	12
Clays and shales	4.2	Plagioclase	39
Carbonates	2.0	Micas	5
Granites, gneiss, and crystalline schist	36.9	Amphibole	5
Granodiorite and diorite	11.2	Pyroxene	11
Syenite	0.4	Olivine	3
Basalt, gabbro, amphibolite, eclogite	42.5	Clay	4.6
Peridotite, dunite	0.2	Calcite and dolomite	2.0
		Magnetite	1.5
		Others	4.9
Total	99.1	*Total*	100.0

From A. Ronov and A. Yaroshevsky, *Geophysical Monograph 18*, ed. by P. Hart, pp. 37–57, 1969, copyright by the American Geophysical Union.

only a minor constituent of the total crustal volume. Their ultimate source, however, is older igneous, metamorphic, and sedimentary rocks, and so their chemistry and mineralogy reflect changes induced by exogenic geomorphic processes. Geomorphology, therefore, becomes an important link in the rock cycle.

The wide areal distribution of sedimentary rocks undoubtedly causes a surface mineral composition different from that shown in table 1.5. At the surface, quartz and feldspars are dominant and probably exist in equal amounts (feldspar 30 percent, quartz 28 percent); calcite and dolomite increase to about 9 percent; and clay minerals and micas become much more significant, rising to approximately 18 percent of the surface material (Leopold et al. 1964).

In any given climate each rock type will respond to the processes of weathering and erosion in a different manner and at a different rate. With time and tectonic stability, high-standing landmasses commonly will be underlain by resistant rocks, and low-standing regions will be formed from rocks that are more susceptible to weathering and erosional attacks. These effects of differential weathering and erosion in landscape development are stressed in every introductory course in the basics of geology. In fact, we are conditioned early in our geological training to view regional topography as a mirror of gross lithology, tectonics, and geologic history. For example, the concept of physiographic provinces stresses this approach, causing us to think of geological controls in geomorphology as regional phenomena. It is worthwhile to emphasize, however, that geomorphic processes will accentuate lithologic differences on many scales. Mega-scaled differentiation produces regional features such as mountains and plains (fig. 1.6), and can be utilized in an erosional topography as a first approximation of the gross lithologic distribution. Within any large

TABLE 1.6	Volume Percentage and Chemical Composition of Silicic and Mafic Crust in the United States.

Volume Percentage

Western Provinces	Silicic	Mafic
California coastal region	75	25
Sierra Nevada	50	50
Pacific NW (coastal)	28.6	71.4
Columbia Plateau	22.3	77.7
Basin and Range	66.7	33.3
Colorado Plateau	62.5	37.5
Rocky Mountains	62.5	37.5
Average	56.4	43.6
Eastern Provinces		
Interior Plains and Highlands	40.0	60.0
Coastal Plain	57.2	42.8
Appalachian Highlands and Superior Upland	37.5	62.5
Average	42.6	57.4
Total United States	46.3	53.6

Chemical Composition

	Western	Eastern
SiO_2	60.0	57.1
TiO_2	1.1	1.3
Al_2O_3	15.1	15.2
Fe_2O_3	2.3	2.3
FeO	4.9	5.7
MgO	4.5	5.6
CaO	6.3	7.5
Na_2O	3.0	3.0
K_2O	2.0	2.1
Total	99.2	99.8

From L. C. Pakiser and R. Robinson, *Geophysical Monograph 10*, pp. 620–26, 1966, copyright by the American Geophysical Union.

TABLE 1.7	Area of Different Igneous Rocks Exposed in the Appalachian and Cordilleran Regions of the United States (Percent).

	Cordilleran	Appalachians
Plutonic Rocks		
Calc-alkalic rocks (granite, granodiorite, quartz monzonite, quartz diorite, diorite, gabbro, anorthosite)	33.6	84.5
Alkalic rocks (syenite, monzonite, others)	0.4	neg.
Ultramafic rocks (periodotite, pyroxenite)	0.5	neg.
Hypabyssal Intrusives		
Calc-alkalic rocks (porphyries, quartz diabase, diabase)	1.5	7.4
Alkalic rocks (porphyries)	0.3	0.2
Extrusive Rocks		
Calc-alkalic rocks (basalt, dacite, andesite, rhyolite)	63.4	7.9
Alkalic rocks (trachyte, latite, phonolite, others)	0.2	neg.
Total	99.9	100.0

Source: E. Daly, ed., *Igneous Rocks and the Depths of the Earth.* Copyright © 1933 by McGraw-Hill Publishing Co., Inc.

TABLE 1.8	Rocks Exposed at the Surface of the North American Continent (expressed as % of area).

	Gilluly 1969	Blatt and Jones 1975
Sedimentary	61.5	52
Volcanic	8.2	11
Plutonic	3.8	6
Metamorphic and total Pє	26.5	31

region of similar rock type, small lithologic discrepancies will also surrender to geomorphic processes and appear at the surface as minor landform deviations. These tiny blips in the general landscape provide critical information about geological history and exert important controls on subsequent geomorphic developments (figs. 1.7 and 1.8).

Lithologic diversity must be considered on a variety of levels. Large areas underlain by crystalline rocks or sedimentary rocks may develop a distinct regional character, but smaller variations within the region are revealed in subtle topographic changes that often provide significant geologic and geomorphic information. The geomorphologist must be able to read these subtle topo-

graphic modifications in order to present a coherent interpretation of history and process.

Structure   Geologic structures that influence landforms also range in magnitude from large, areawide tectonic styles to minor features that exert only local control. Structural influence is readily apparent only when the rocks and climate involved are conducive to differential weathering and erosion. In depositional environments,

Figure 1.6
Mountains and surrounding plains, looking west-southwest from a point about 1.6 km northeast of Boulder at an elevation of 2152 meters. Boulder County, Colo., ca. 1934.

(T. S. Lovering, U.S. Geological Survey)

Figure 1.7
A topographic irregularity caused by differences in lithology—West Spanish Peak, from the northwest. Dikes cutting flat-lying Eocene strata. Spanish Peaks quadrangle, Huerfano County, Colo.

(G. W. Stose by U.S. Geological Survey)

Figure 1.8
Variations in lithology as evidenced in cuestas formed by hard sandstones north of Galisteo
Creek, N.M. The rocks in succession from left to right are Mansamo red beds, Morrison,
Dakota, and Mancos (Galisteo Creek and the Santa Fe Railroad in foreground).
(W. T. Lee, U.S. Geological Survey)

structures may be buried by thick accumulations of sediment that mask the surface expression of the underlying structure. Comparably, the internal structure may not be immediately evident in erosional topography formed in areas with distinctly similar lithology, such as shields or crystalline mountain cores, but minor structures still may produce a discernible topographic control. Spacing of joints, for example, is recognized as a prime factor in the development of the longitudinal "staircase" profiles that characterize glaciated valleys in mountains held up by rocks of uniform lithology. The most likely lithologic environment to display structural control is a sedimentary sequence with alternating resistant and nonresistant units, such as the Valley and Ridge province of the Appalachian Mountains. There resistant sandstone and conglomerate layers form ridges that are separated by intervening valleys underlain by easily eroded shales and limestones. The regional topography reveals the pervading structure of plunging anticlines and synclines because the ridges cross the countryside in a sinuous pattern that shows the character of the underlying folds. We will examine the relationship between tectonics and geomorphology in greater detail in chapter 2.

Thresholds and Complex Response

The third basic principle of process geomorphology involves the **threshold concept.** Any concept proposing equilibrium inherently implies a contrasting state of disequilibrium. If variations in controlling factors demand a response within the system, there must be a period of readjustment during which process and form are out of equilibrium. Landslides, subsidence, and gulley erosion are examples of disequilibrium generated when the variables of force and/or resistance are altered so they can no longer maintain a balanced relationship (fig. 1.9). Schumm (1973) recognized that if a system in equilibrium can be defined by real parameters, it follows that there must be parameter values that represent the limits of the balanced condition. If these limiting values are exceeded, the system enters a condition of disequilibrium. The limits of equilibrium are critical conditions called thresholds (Schumm 1973).

If parameter values are pushed to the limiting condition by variations of external controlling factors, the threshold is known as an *extrinsic threshold*. Examples are numerous in nature; geologists will be most familiar with thresholds related to the fluctuating climate that characterized much of the Pleistocene epoch.

A more subtle type of threshold, however, is the *intrinsic threshold,* where instability and failure of a system occur even though external variables remain relatively constant. The threshold conditions develop in response to gradual, often imperceptible, changes within the system. In many cases the threshold represents a deterioration of resistance rather than an increase in driving forces. For example, a region characterized by periodic heavy rains may have stable slopes for a long time,

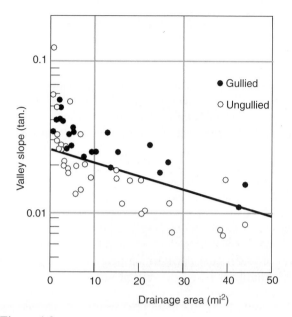

Figure 1.9
Threshold relationship between gullied and ungullied valley
floors in several drainage basins of northwest Colorado.
(Patton and Schumm 1975)

but continuous freeze and thaw or other soil-forming
processes gradually reduce the cohesion of slope material. Eventually one storm, no more severe than thousands that have preceded it, triggers slope failure.

Disruptions of equilibrium are common enough to
be considered as a fundamental characteristic of geomorphic systems. This perception has led many geomorphologists to consider the threshold concept as the primary working model of geomorphology (Coates and
Vitek 1980). What geomorphologists have found, however, is that applying the threshold concept in real situations is difficult and is often complicated by failure to
consider the factor of time. All disrupting events cause a
response in geomorphic systems, but designating each
response as being associated with a threshold event has
created some confusion. For example, major floods and
hurricanes usually cause fluvial and beach systems to be
thrown into a state of disarray. However, unless the frequency of these disturbances has been increased by
changes in the fundamental climatic controls, the systems will return to their pre-event equilibrium conditions
before the advent of the next disturbance. Therefore, we
suggest that the crossing of a threshold boundary *requires* that the system develop a new equilibrium condition defined by average parameter values that are significantly different from those prevailing before the
disrupting event occurred. If the system returns to its
original state before another disrupting event of the
same magnitude is repeated, a threshold has not been
crossed even though the system is in temporary disequilibrium. The real question, then, is whether the disrup-

tion of equilibrium produces landform or process
changes that are meaningful in geomorphic analyses
(Ritter 1988); in other words, what is temporary disequilibrium and what is geomorphically significant disequilibrium? We will return to these questions later in the
chapter when we discuss time.

Threshold-crossing events often initiate a series of
reactions, which are known as a **complex response.** The
sequence is complex because all processes and system
components may not reach the threshold conditions at
the same time, and at any given location the initial response may be different from the final adjustment that
produces a new equilibrium condition.

Complex response was first demonstrated experimentally by Schumm and Parker (1973) in a study of an
artificial drainage basin. An induced base-level decline
at the mouth of the basin caused downcutting of the
trunk river at that point and the formation of terraces. At
the same time, however, tributary channels were unaffected and remained in their equilibrium state. With
time, the site of channel incision migrated progressively
upstream until the base level of each tributary was lowered and channel entrenchment ensued. The tributary inclusion, however, provided so much sediment to the
trunk river that aggradation began at the basin mouth because the stream was incapable of transporting the increased load derived from entrenchment of the tributary
channels. Clearly, the processes functioning in different
parts of the systemic basin were out of phase.

The same sequence of events can occur during
major glacial stages when sea level declines dramatically. The effect of that base-level decline will be initially felt at the mouths of major rivers such as the Mississippi. Tributaries in that huge basin may not
experience the expected incision until long after the initiating event.

It is somewhat difficult to document complex response in natural settings because the sequence of
events commonly requires long periods of time to reach
a final adjustment. For example, studies of drainage
basins in central Nevada have shown that responses to a
severe drought occurring about 2000 years ago are still
exerting controls on the modern fluvial system, which
has yet to adjust completely to the threshold-crossing
event (Miller et al. 1999). Nonetheless, some very in-
formative studies can be found in earlier work. Gilbert
(1917), for example, was able to show that hydraulic
mining for gold on the west flank of the Sierra Nevada
produced abnormally high sediment loads to rivers
draining the mined areas. The coarse fractions of the
mine tailings gradually invaded the channels of the
major downstream rivers as sand and gravel bedload.
The rivers, unable to transport such an overwhelming
load, adjusted by drastically raising their channel bottoms as the material was deposited. As each segment
filled, the gradient increased so that the river acquired

the capacity needed to transport the sediment farther down the valley. The rise in channel level stopped at different times in each segment of each river, depending on the distance from the source and the amount of load. For example, the channel of the Yuba River at Marysville (Calif.) rose about 6 m (19.1 ft) between 1849 and 1905, when it reached its highest level. The Sacramento River at Sacramento (Calif.) elevated about 3 m (10.8 ft), attaining its highest level in 1897. The Gilbert study also revealed a significant lesson for environmental managers. This follows because both rivers continued to aggrade even though legal action, intended to remedy the environmental damage, required mining to stop in 1884. The channel filling at the two stations continued because upstream reaches, no longer receiving great volumes of sediment, had excess energy on their steepened gradients and therefore entrenched the channel bottom. Sediment from the entrenchment was transported to the downstream river segments where it contributed to the aggradation at Marysville and Sacramento. Thus, for a period of time, part of the fluvial system was filling and part was entrenching. Note that the upstream entrenchment was a response to a second disrupting action caused when cessation of mining eliminated the major source of load. Therefore, failure to understand fluvial response and the time needed to restore equilibrium resulted in a presumed remedial action becoming part of the problem because it exaggerated the initial aggradational response at the Marysville and Sacramento reaches.

The Principle of Process Linkage

Complex adjustments to altered conditions often involve a chain reaction of responses that we will call process linkage. **Process linkage** essentially operates on the domino principle; it means that the changes that occur in one process or landform during an adjustment period often initiate subsequent responses in totally different processes and/or landforms. Linkage works because a driving force can transfer from one process type to another as its effect filters through a system, or it can even shift to processes operating in totally different systems. Thus, a myriad of different processes can be involved in the response to a single threshold-inducing force.

A case history exemplifies how process linkage works. On May 18, 1980, Mount St. Helens in southwestern Washington experienced a violent volcanic eruption. The widespread effects of the eruption have been documented in a series of short papers published as U.S. Geological Survey Circular 850, and a major Professional Paper (Lipman and Mullineaux 1981). The initial process response occurred during the eruption as a massive debris avalanche that deposited enormous volumes of rock, ice, and other debris in the upper 17 miles of the North Fork Toutle River valley (fig. 1.10). The deposits are up to 600 feet thick at places. Physical,

chemical, and biological characteristics of lakes close to the eruption were drastically altered, and benthic faunas in the adjacent rivers were destroyed.

Immediately following the avalanche, snow and ice that had melted during the eruption provided enough water to generate a mudflow in the same valley. In addition to environmental damage, the mudflow deposited about 25,000 acre feet of sediment in the Cowlitz River channel (fig. 1.10). This elevated the channel floor and decreased its cross-sectional area, making the valley bottom prone to more frequent flooding (fig. 1.11). Furthermore, a significant volume of sediment reached the Columbia River, where it created a shoal area that blocked the channel used for shipping (see Lombard et al. 1981).

The Mount St. Helens catastrophe involved a number of process links that demonstrate that the location of the dominant response shifted progressively downstream. In addition, the single driving force in this case was produced inside Earth. The results of volcanic action, however, were quickly transferred into slope processes (debris avalanche and mudflows) and from there into fluvial, hydrologic, glacial, and lake systems and processes. Physical changes even altered the biological balance.

The remarkable ability of process action to shift from one form to another in response to a single impetus is a critical ingredient in process geomorphology. It often provides the only explanation for isolated or apparently unprovoked geomorphic actions.

The Time Framework

Earlier in the chapter we saw that geomorphology can be considered over different time intervals, which Schumm and Lichty (1965) called cylic, graded, and steady. The question is, which of these time spans is most conducive to demonstrating the mechanics of process geomorphology?

To visualize the time factor more clearly, consider a hypothetical drainage basin and the component subsystems (rivers) within it. If we observe a single cross-section of any river, we can describe the channel morphology at that point by a group of parameters such as width, depth, slope, and shape. Any single flow event in the river may bring about temporary scouring or filling, which causes immediate changes in the channel parameters but does not affect them permanently; they will return to their previous state when the flow event passes. Measurements taken an hour or day apart will show different values for the variables, but they will always be internally consistent and apparently adjusted to their external controls. Significantly, observed sediment and water discharges through the cross-section are dependent variables when viewed in this time sense because they are modified by changes in the channel configuration. For example, if the river scours its bed, the moving load will increase and the channel area will expand, al-

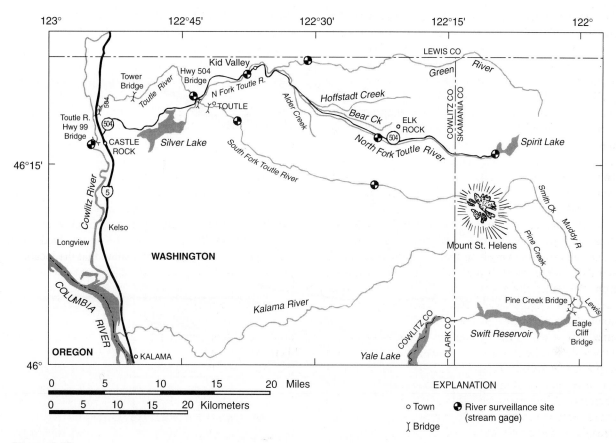

Figure 1.10

Map showing location of river surveillance sites in the lower Toutle and Cowlitz river systems.

(Source: R. E. Lombard, et al., U.S. Geological Survey Circular 850-C, 1981.)

lowing a greater volume of water to be held within the banks. Thus both water and sediment discharge are temporarily affected by the channel changes. However, we are observing the river for only a very short period of time, and any equilibrium that we define is almost instantaneous (or "steady" in the terminology of Schumm and Lichty). Since no permanent changes can be expected on this temporal scale, we can justifiably say that processes and landforms are time-independent when considered in this sense (see fig. 1.3A).

Over longer periods, the steady-time channel measurements will vary with mean values of sediment and water discharge; the statistical relationship between the external and internal variables defines the equilibrium state for graded time (see fig. 1.3B). In fact, as quoted earlier, Wolman and Gerson suggest that equilibrium in geomorphic systems can only be defined on a graded-time interval. On this time scale (unlike steady time), sediment and water discharge function as independent variables to determine the morphologic and flow characteristics of the stream reach. In graded time, changes in external variables due to modifications of climate or base level may require a new set of equilibrium conditions within the channel. If and when thresholds are exceeded, some significant change in the river is necessary and (again in contrast to steady time) the adjustment will be permanent unless still another change occurs in the external controls. Commonly the response takes the form of channel incision or aggradation, pattern adjustments, or modification of sinuosity.

The point is that following a threshold crossing, the system is temporarily out of balance, and there exists a certain time interval during which the river is approaching a new equilibrium state. The time involved is intermediate between long-term cyclic time and instantaneous steady time. Absolute time in years is not the important element in recognizing graded time as a valid geomorphic concept. Its significance lies in the fact that, once equilibrium is achieved, disruption of the balance will be counteracted by each subsystem's ability to reestablish a new equilibrium quickly (in the geologic sense); graded time thus does not involve continuous and progressive change in the landscape. In addition, all parts of the regional system may not be affected simultaneously or in precisely the same way.

If we consider our hypothetical basin over cyclic time, the balance (or temporary imbalance) seen in steady or graded time becomes irrelevant. The inexorable

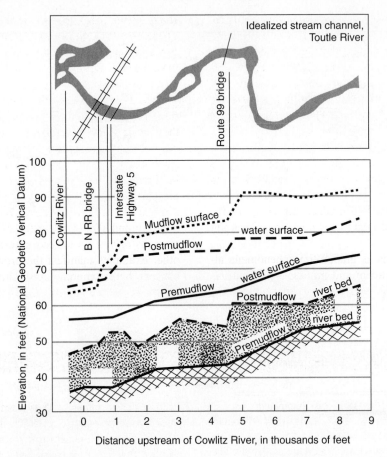

Figure 1.11

Channel bottom and surface elevation of the lower Toutle River prior to and
after the mudflows of May 18, 1980. Pre- and post-eruption water surfaces
are based upon a flow of 38,000 cubic feet per second.

(Source: R.E. Lombard, et al., U.S. Geological Survey Circular 850-C, 1981.)

loss of sediment and energy from the basin suggests that
the system must be continuously approaching, but perhaps never attaining, an ultimate equilibrium condition.
In this time framework, landforms within the basin
should be progressively losing relief in phase with the
deteriorating systemic energy.

Clearly a hierarchy of time exists, and how we interpret geomorphology depends, in a real sense, on which
time scale is used. In process geomorphology, analyses
are best made on a steady- or graded-time scale, assuming landforms and processes to be time-independent or
only temporarily time-dependent phenomena.

We now return to the question raised earlier concerning the meaning of thresholds. Let's examine the
question by considering a catastrophic storm that results in massive changes in beach and/or stream-
channel morphology caused by severe erosion and
deposition. The geomorphic significance of the event
depends on whether the beaches or stream channels
return to their prestorm morphology before another
storm of similar magnitude occurs. The time required
for the system to revert to its original character is

known as the *recovery time* (Wolman and Gerson
1978). The point here is that a true threshold crossing
requires a new set of equilibrium conditions, and these
will not be established if the time between disruptive
storms is longer than the recovery time; that is, another
storm of equal magnitude will not hit before the systems return to their original conditions.

Given that climate represents long-term averages of
weather, catastrophic events should be considered as normal, albeit rare, components of the equilibrium condition,
and recovery time will be shorter than the frequency of
such major events. An unusual storm, therefore, places a
system in temporary disequilibrium, but a rapid return to
its original condition suggests that a threshold has not been
crossed. This type of phenomenon is common over steady
time intervals (days to tens of years), in which disruptive
events can be annoying or even dangerous. (Ritter et al.
1999). In fact, it is during these temporary episodes that
most natural hazards are activated. To lessen the impact of
such events we should strive to recognize the causes of instability that result in these disasters. Such catastrophes,
however, are not necessarily threshold crossings.

Definition of Thresholds

In light of the above, we believe that thresholds are best defined as follows (see Ritter et al. 1999): "Thresholds are values of parameters that define the limits of equilibrium in geomorphic systems. When thresholds are exceeded, changes in process and/or form are irreversible, and a new equilibrium condition, adjusted to altered controlling factors, must be developed over a graded time interval."

This definition will not satisfy all geomorphologists. However, we stress that the threshold concept is inextricably linked to how one defines equilibrium. In our approach, we accept the Wolman and Gerson perception that equilibrium is a graded-time phenomenon and, therefore, we are driven to a similar time constraint for thresholds.

Defining thresholds as graded-time phenomena allows us to equate landform and process modifications with changes in the controlling factors of geomorphology. It also provides us with the link needed to understand events that occurred during Quaternary history. These concepts are not easily comprehended, and our perceptions are still evolving, but geomorphologists must provide useful information to other scientists if the discipline is to be considered as a viable part of geoscience. For example, in this model, steady time disruptions may or may not be preserved in landforms, but they are probably recorded as isolated units in a stratigraphic sequence. It is important for geologists to understand that such units do not necessarily signify a change in climate or a tectonic event. However, a change from predominantly fine-grained to predominantly coarse-grained deposition probably does reflect a threshold crossing and should relate to landform and/or process changes driven by altered external factors. This and other examples manifest the promise of the threshold concept, but geomorphologists must agree on its meaning before its true power will be realized.

SUMMARY

In this chapter we have suggested that the process approach in geomorphology is more useful than other approaches because it relates better to a variety of disciplines that examine phenomena occurring at the surface of Earth. A set of basic principles constitute the framework of process geomorphology. Processes and the resulting landforms will be analyzed as balances between the driving forces (e.g., climate, gravity) and the resistance offered by the geologic framework that makes up Earth's surface. Processes will be considered on time scales that are geologically significant.

SUGGESTED READINGS

The following references will help you understand the approach used in this book.

Baker, V., and Twidale, C. 1991. The reenchantment of geomorphology. *Geomorphology* 4:73–100.

Coates, D. R., and Vitek, J. D., eds. 1980. *Thresholds in geomorphology.* London: Allen and Unwin.

Hack, J. T. 1960. Interpretation of erosional topography in humid temperate regions. *Amer. Jour. Sci.* Bradley vol. 258-A:80–97.

Ritter, D. F. 1988. Landscape analysis and the search for geomorphic unity. *Geol. Soc. Amer. Bull.* 100:160–71.

Schumm, S. A. 1973. Geomorphic thresholds and complex response in drainage systems. In Morisawa, M., ed., *Fluvial geomorphology,* pp. 299–310. Pubs. in Geomorphology, 4th Ann. Mtg. S.U.N.Y., Binghamton.

———. 1977. *The fluvial system.* New York: John Wiley and Sons.

Schumm, S. A., and Lichty, R. W. 1965. Time, space, and causality in geomorphology. *Amer. Jour. Sci.* 263:110–19.

Vitek, J. 1989. A perspective on geomorphology in the twentieth century: Links to the past and future. In Tinkler, K., ed., *History of geomorphology from Hutton to Hack,* pp. 293–324. London: Unwin Hyman.

Wolman, M. G., and Gerson, R. 1978. Relative scales of time and effectiveness of climate in watershed geomorphology. *Earth Surf. Proc. and Landforms* 3:189–208.

2

INTERNAL FORCES
AND CLIMATE

© John Shelton

Introduction
The Endogenic Effect
   Epeirogeny
   Orogeny and Tectonic Geomorphology
     Tectonic Geomorphology
   Volcanism
Climatic Geomorphology
   Climate, Process, and Landforms
   Climate Change and Geomorphic Response
     Sea-Level Fluctuation
     Geologic and Vegetal Screens
Summary
Suggested Readings

INTRODUCTION

One of the major principles of process geomorphology is that spasmodic disruptions of equilibrium are significant components of surficial mechanics. In many cases the cause of instability is external to the geomorphic system being affected. In this chapter we will examine the primary external controls on geomorphic systems.

In theory, exogenic processes unimpeded by opposing forces will gradually reduce the landscape to a rather dull, featureless surface with only minor topographic irregularities to interrupt its sameness. However, because the surface of Earth does possess relief, and has done so throughout geologic time, Earth must be constructed in such a way that exogenic processes are not always preeminent. The surface is not a static environment but a locale where denudational processes have been repeatedly counteracted as new mass is elevated or created by endogenic forces. Each influx of mass not only brings with it escalated relief but also new potential energy that is available to accelerate or change the character of the exogenic processes. The introduction of new mass and energy occurs primarily through volcanic activity and vertical uplift of the land surface.

In addition to pulses of endogenic input, we know from the glacial and interglacial episodes of Quaternary history that the climatic regime is not a static phenomenon. Although the equilibrium condition in geomorphic systems is adjusted to climate, changes in climate are equally significant because they serve as spark plugs of temporary instability.

The importance of major climate change and significant endogenic input is that they tend to occur as irregular events separated by long periods of relative constancy. These factors create conditions whereby threshold responses in geomorphic systems are required. Furthermore, periods of climate change and endogenic action are the ingredients needed to decipher the sequential nature of geomorphology. Therefore, the major external controls are the basis for *Quaternary geomorphology,* another subarea of the discipline, in which the determination of history rather than process is a common goal of geomorphic investigation. We will occasionally examine various aspects of how geomorphology may be used in Quaternary studies. However, we will take these excursions into Quaternary geomorphology cautiously so that we do not subvert our primary process theme.

THE ENDOGENIC EFFECT

Endogenic forces are an integral part of what we observe at Earth's surface. Volcanic activity and seismicity are unmistakable manifestations of how internal energy and force can affect landforms and processes. Often, however, exertion of internal force results in equally important though less dramatic geomorphic effects. For example, figure 2.1 shows that most land surface exists at two persistent levels, which correspond to the continents and the ocean floor. The hypsometric levels (fig. 2.1) for Earth also indicate the physiographic importance of continents and the ocean floor; surfaces at elevations between sea level and 1 km account for 20.8 percent of the

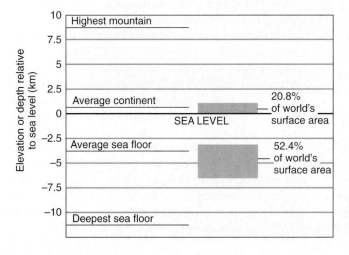

Figure 2.1
Features and areas of Earth standing at different elevations.
(Source: Data from P. Wyllie, *The Dynamic Earth,* copyright 1971 John Wiley & Sons.)

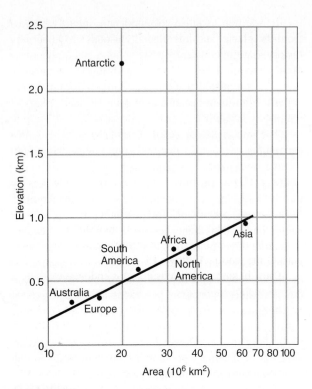

Figure 2.2
Relationship between area and elevation of the continents.

average height of land and depth of sea floor

world's total area, and ocean bottoms between 3 and 6 km in depth comprise about 52.4 percent of the total surface area. The average elevation of continents is about 0.875 km, and the average depth of the ocean floor is approximately 3.729 km.

The rocks that make up continents are drastically different in mineralogic and chemical composition from those that form the ocean basins. They are richer in silica and less dense than the ocean rocks, so that continental crust is thicker and stands higher than the ocean floor. Exactly how this monumental difference between oceanic and continental crust developed is not certain because theories concerning the origin of continents lack hard data. The reconstruction of events that led to the formation of protocontinents depends entirely on indirect evidence of the early history of Earth. Nonetheless, the maintenance of high continents and low ocean basins represents a balance perpetuated by endogenic mechanics. In addition, as figure 2.2 shows, with the exception of Antarctica, larger continents stand at higher elevations, a fact that must reflect some primary systemic operation. Our task, then, is to decipher what internal process is responsible for this topographic phenomenon.

Epeirogeny

The processes by which the crust is deformed are collectively known as *diastrophism* or *tectonism* and are commonly divided into two distinct tectonic styles. *Orogenic*

processes culminate in the formation of structural mountains. Such mountain systems are typified by intense disruption of the included rocks due to folding and overthrusting; the effects are usually localized in narrow, elongated belts. In contrast, *epeirogenic* processes cause uplift or depression on a regional scale and proceed without internal disruption of original rock structures. Response to driving forces is rather passive in this type of deformation. Although gentle tilting of strata may accompany the vertical displacements, folding and thrusting are absent during the movement. Mountain systems, however, are also affected by epeirogeny *after* the orogeny that formed them has ended. Such vertical displacement of rocks and surfaces is driven by the fundamental gravitational force.

In chapter 1 we briefly examined the force of gravity and the various factors that determine its effect on a body at any location. The net force on any mass is the vector sum of all gravitational attractions acting on it. Each body, therefore, possesses a discrete amount of potential energy because mutual attractions can be transferred into a kinetic form that is capable of doing work. In geomorphology, we can simplify the perception of gravity by considering it to be an energy field consisting of horizons of equal potential energy. In this model, sea level becomes an extremely important equipotential surface in Earth's gravitational field. Even though the surface may be slightly distorted because of local factors, the inflections are small in amplitude compared with the radius of Earth and limited in areal extent.

The sea-level equipotential surface is called the **geoid,** which on land is usually defined by the water level in a series of imaginary canals cut through the solid mass. Surface topography is referred to the geoid because any elevation is determined by extending upward a succession of planes that are parallel to sea level. At any point to be measured, a surveying instrument is set tangent to the geoid, with the tangential line being the perpendicular to a vertical plumb line over the site. As with any body, however, the direction of the plumb bob itself is the vector sum of all the gravitational forces acting on it and so may not be perfectly normal to Earth's center of gravity. To resolve this complication, geodesists utilize a second surface called the **spheroid,** which is a mathematical representation of sea level with all irregular influences removed. Essentially, the spheroid is the *hypothetical* sea-level surface of Earth with no lateral variations in density or topography and with a vertical change in density that is uniform from the center of gravity to the surface. With such a mass distribution, gravity would vary consistently from pole to equator, and its theoretical sea-level values can be easily calculated. The differences between the predicted values of gravity, calculated under the above assumptions, and the actual measured values are called **gravity anomalies;** they indicate the departure of the geoid from the spheroid.

Because few gravity measurements are made at sea level, most observations must be reduced into separate components indicating the portion of the measurement produced by mass and the portion due to distance. Corrections for each must then be made before a gravity measurement can be compared with the spheroidal value. A general tendency exists for anomalies produced by mass, called *Bouguer anomalies,* to be strongly negative in mountains and increasingly more negative with higher elevations, demonstrating a most important principle in geomorphology—that surface topography is somehow related to the internal distribution of mass.

The idea that topography is influenced by the distribution of mass within Earth is not new. It was expressed by Leonardo da Vinci, and its concrete formulation as a hypothesis arose from analyses of data obtained in the mid-nineteenth-century land surveys in India. C.E. Dutton, working in the Colorado Plateau, introduced the term **isostasy** to define the internal process involved. In essence, the result of isostatic adjustment is a condition in which large, elevated regions such as continents or mountain ranges are compensated by a mass deficiency in the crustal rocks beneath them. The process of isostasy requires that at some depth beneath sea level the pressure exerted by overlying columns of rock will be the same, regardless of how high the various columns stand above sea level. Mountains, ocean basins, shields, and other large topographic or physiographic regions are balanced with regard to the total mass overlying each area at some internal level called the *depth of compensation.*

Isostatic compensation is usually produced by changes in the thickness of the crust. Thus, high-standing mountains are normally supported by a thick wedge of low-density crustal rocks extending downward into the surrounding mantle. However, it is not true that every isostatic adjustment is accomplished by variations of crustal thickness. For example, the elevated surface of the Colorado Plateau is underlain by a relatively thin crust. The uplift and subsequent support of the high plateau were caused by a mass deficiency in mantle rocks that make up the lower reaches of the lithosphere (Morgan and Swanberg 1985; Parsons and McCarthy 1995).

Gravity measurements indicate that this isostatic balancing act is rarely completed and certainly not constant; that is, internal mass distribution and topography are not in perfect equilibrium. The portion of the mass imbalance that has not been corrected by the process of isostasy is called the *isostatic anomaly.* If the isostatic anomaly is zero, the system is in perfect balance. We know, however, that the equilibrium condition is easily upset so that negative or positive anomalies are not unusual. A negative isostatic anomaly indicates deficiency of mass at the locality, and the surface should have a tendency to rise because more matter must be added at depth to establish the equilibrium state. Similarly, positive isostatic anomalies should portend sinking because they indicate an excess of mass beneath the surface.

Because most topographic blocks, local or regional, are not perfectly equalized, vertical movements of crustal segments are inherent in the attempt to establish equilibrium. When isostatic balance is disrupted by erosion, thick sediment deposition (at Lake Mead, for example), or tectonics, a counteraction by isostasy is required to restore equilibrium. It is also known that the accumulation of massive glaciers is accompanied by depression of the surface; conversely, when the weight is removed as the ice disappears, the surface will rise to reestablish the isostatic balance. This response to glacial and interglacial conditions is called *glacio-isostasy* and represents one of the most important geomorphic processes in high latitudes of the world (Andrews 1974, 1991).

Thus, in sum, isostasy is a major endogenic process that causes epeirogenic diastrophism and, in general, is responsible for maintaining the topographic relationship between large blocks of the crust. The relief between ocean basins and continents probably reflects the isostatic balance established because the crustal thickness and or density of lithospheric rocks underlying the two areas are different. The same process, however, functions on scales smaller than continents or ocean basins.

Keep in mind that isostasy is not the only process that produces vertical movement of rock surfaces. Increased understanding of the mantle and athenosphere suggests that migrating mantle plumes can also result in surface uplift. Excess heat developed over hot spots in the upper athenosphere is transferred into the mantle where thermal expansion and increased buoyancy associated with the plumes initiates surface uplift. Uplifts caused by mantle plumes normally are regional in extent (1000–2000 km diameters), have vertical displacements of 1–2 km, and are closely associated with extensive volcanism (Sleep 1990). The plume concept has been used to explain elevated surfaces in volcano-rich areas, which are underlain by a crust that is too thin to provide the necessary isostatic support (for example the western Cordillera of the United States, see Parsons et al. 1994). Nonetheless, isostasy is by far the dominant process involved in vertical movements of Earth rocks and surfaces.

The fact that isostasy works is uniquely important in geomorphology because it requires vertical motion of the surface. When the movement is upward, isostasy produces potential energy that is available for use in exogenic processes. Geologists often interpret uplift characteristics from the thickness and texture of sediment deposited during and following major orogenic events (Burbank and Beck 1991). Estimates of uplift magnitudes and rates are also sometimes based on the position and age of the youngest raised strata (Lundberg and Dorsey 1990) or exposure of intrusive igneous rocks or changes in cooling rates of those rocks (Harayama 1992; Benjamin et al. 1987).

The critical question surrounding any of these techniques is how much of the upward displacement of rocks or the volume of accumulated sediment is manifesting an actual uplift of the surface rather than upward displacement of underlying rocks (England and Molnar 1990). This follows because significant erosion may occur as uplift is progressing, thereby retarding the elevation of the surface. The problem is partly solved by using only well-dated landform surfaces or shoreline phenomena as the basis for determining uplift rates and magnitudes (Stewart and Vita-Finzi 1998). Even this leaves serious questions concerning the viability of uplift interpretations. For example, rates are not constant but change significantly during different intervals of the same uplift event (Gardner et al. 1987). This fact has been clearly documented in areas of isostatic rebound caused by removal of ice or water that had previously depressed the surface (Gutenberg 1941; Crittenden 1963). In these cases, initial uplift rates in response to unloading are very rapid but decrease gradually as the system returns to its former equilibrium. An example of how this works was demonstrated by Ten Brink (1974), who derived a detailed uplift curve (fig. 2.3) that shows how shorelines along the coast of Greenland were progressively elevated during and after deglaciation. The initial response to dissipation of the ice cover was extremely rapid, perhaps up to 100 m/ky. The

curve indicates that the 9000 yr. B.P. shoreline is now about 150 m above sea level, and the 8000 yr. surface now stands at about 80 m. Thus, at least 70 m of uplift occurred during the first 1000 years of the isostatic recovery. In contrast, only 20 m of the total uplift occurred during the 4000-year interval between 6000 and 2000 years B.P. when isostatic equilibrium was once again established. The average uplift rate for the total recovery period was about 20 m/ky, but clearly this rate is misleading because it includes an early-adjustment rate of 70 m/ky and a late-adjustment rate of only 5 m/ky. The significance of this and other studies is that rates of isostatically driven uplifts are not constant but change exponentially during the recovery period.

Uplift rates also vary as the site of tectonic activity migrates, or where the style of local tectonism differs from that which characterizes an entire region. This phenomenon is apparent along nonsubducting plate margins, such as the Pacific coastal zone of the United States, where uplift rates tend to vary from 0.1 to 0.3 m/ky (Muhs et al. 1992; Kelsey and Bockheim 1994, Hanson et al. 1994; Kelsey et al. 1996). Within the region, however, rates may be an order of magnitude higher along active faults (Anderson 1990; Kelsey and Bockheim 1994), or where the plate-margin tectonic style reflects greater activity (West and McCrumb 1988; Merritts and Bull 1989; Orme 1998). In general, uplift rates in areas next to subducting plates are 1–10 m/ky, but even in these environments the uplift rate is drastically altered by variations of tectonic style (Marshall and Anderson 1995; Fisher et al. 1998).

Meaningful uplift rates are important in geomorphology because they are one of the ingredients that determine whether a surficial system can remain in equilibrium during the uplift event. This can occur only if denudation rates and uplift rates are essentially equal. Denudation (erosional lowering of the surface) will be discussed later in chapter 5 when we consider the development of drainage basins. However, we should point out here that, except in rare situations (Formento-Trigilio and Pazzaglia 1998), denudation occurs at much lower rates than uplift (see tables 5.9 and 5.10). Thus, equilibrium between uplift and denudation seems unlikely except at very active plate margins where deformation is continuous for extended time intervals (Liu and Yu 1989; Lundberg and Dorsey 1990; Burbank and Beck 1991). If rates of uplift exceed by far the prevailing rates of denudation, the system will cross a threshold and enter into disequilibrium. How long it will take to establish a new balance depends on how radical the difference is between the rates of uplift and denudation. Schumm (1963c) discusses the complications involved in making such an analysis; although uplift rates are significantly higher than denudation rates, the uplifts occur in short, spasmodic bursts rather than as long, continuous events.

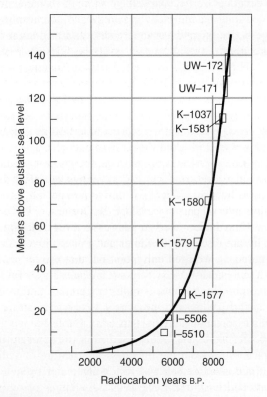

Figure 2.3

Uplift curve along the Greenland coast based on ^{14}C dates of marine fossils in emerged strandlines.

(Adapted from N.W. Ten Brink, "Glacio-isotasy: New Data from West Greenland and Geophysical Implications," *Geological Society of America Bulletin* 85:219–228, 1974. Used with permission of the author.)

Orogeny and Tectonic Geomorphology

Scientists have pondered the origin of mountains ever since they first recognized that rocks in mountain belts were structurally different from those in other areas. The intense folding and overthrusting displayed within mountainous regions led geologists to realize that significant crustal shortening was involved in their formation, but learning what caused the deformation was hindered by ignorance of the interior of Earth. Initially, a progressively cooling and shrinking Earth was suggested to explain the needed compressional stress, but this idea was rejected after scientists recognized the continuous addition of internal heat by radioactivity and accepted a cold origin for Earth. Other proposals met an equally unsatisfactory fate. The advent of the plate tectonic theory, however, forced geologists to reevaluate mountain building in light of the new global model. A detailed discussion of plate tectonics is beyond the scope of this book, and the basic concepts of the plate model can be found in any textbook of physical geology. For our purposes it is sufficient to say that mountains and ocean features such as island arcs and trenches are intimately associated with the seismicity and volcanism found at plate margins where the lithosphere is being actively consumed.

The effect of orogeny on geomorphic process is less tangible than that produced by epeirogeny because the actual rock deformation occurs well below the surface. Thus, geomorphic response does not happen during the application of orogenic forces. Instead, these regions are usually affected after orogeny is over, when isostasy raises the thick pile of low-density and highly deformed rocks. Exceptions to this general premise can be found in areas of active tectonism where vertical uplift is linked to a deforming plate boundary. For example, foreland zones of Costa Rica are experiencing rapid uplift adjacent to a subducting plate boundary (Fisher et al. 1998; Marshall and Anderson 1995).

Tectonic Geomorphology   The fact that deformation leaves an imprint on landscapes is one of the oldest tenets in geomorphology (see Everett et al. 1986). Initially, regional tectonics were used to explain diversity of character in large-scale topography. For example, geomorphologists recognized that the block mountains and intervening basins of the Basin and Range province in the western United States and the sinuous valleys and ridges in the Folded Appalachians of Pennsylvania were reflections of significantly different tectonic styles. Such observations are interesting in themselves, but the influence of tectonic style is primarily physiographic and does little to further our understanding of process. This follows because many of these landforms are reflecting tectonic events that occurred in the distant past. Their surface expressions are created primarily by denudational processes applied over a long period of time.

Thus, the resulting features tell us little about the formative tectonics.

More recently, the relationship between form and tectonics has become the basis for much more detailed and sophisticated geomorphic studies and has led to the development of a special branch of the discipline known as tectonic geomorphology (for reviews see Summerfield 1991; Whitehouse 1992). Tectonic geomorphology as now practiced deals with how tectonic activity affects processes and morphology in geomorphic systems and, conversely, how landforms can be used to assess tectonic activity. It works because the emphasis is on *neotectonic* events, that is, those occurring during the Quaternary and/or continuing to the present. In contrast to the landforms discussed above, these features are directly linked to the tectonic activity that produced them. Significantly, in many studies landforms are used to provide insight into the style and rate of tectonic processes, and reveal a shift in emphasis from the explanation of form to the analysis of what form tells us about process.

Tectonic geomorphology has been used for a variety of scientific purposes (Morisawa and Hack 1985), and studies range in scale from contrasting the tectonic style of Earth and Mars (Arvidson and Guinness 1982) to the analysis of a single landform such as an alluvial fan (Keller et al. 1982). The approach has tremendous implications in environmental planning (Schowengerdt and Glass 1983), especially with regard to landform stability, seismicity and earthquake risk, and prediction of fault characteristics and movement periodicity (Bull and McFadden 1977; Wallace 1977, 1978; Buckman and Anderson 1979; Nash 1980; Colman and Watson 1983; Wesnousky 1986; DePolo and Slemmons 1990; Menges 1990; Wesnousky et al. 1991). Helpful reviews of basic concepts and uses of tectonic geomorphology can be found in Bull (1984) and Keller and Pinter (1996).

The power of tectonic geomorphology rests in the ability of measurable landform properties to reflect the time and intensity of tectonic activity, especially over the time interval encompassed by the Quaternary Period. The various indices used to gauge the level of tectonic activity are discussed in Keller and Pinter (1996). For our purposes, we need only point out that certain landforms and geomorphic settings are particularly useful in these analyses: (1) fault-bounded mountains and their associated piedmont zones, (2) geomorphic surfaces such as marine and fluvial terraces, alluvial fans and stream channels, (3) fault scarps, and (4) plate margins.

Fault-bounded mountains   Tectonic activity along fault-bounded mountains can be estimated by two parameters: (1) *mountain-front sinuosity,* which is the ratio of the length measured along the junction of the mountain and piedmont to the total straight-line length of the mountain front, and (2) the ratio between valley floor width and valley height in major river valleys measured

[margin note: 1st suggestion of how mountains formed]

at a given distance upstream from the mountain front (for example, see Bull and McFadden 1977; Bull 1984).

Theoretically, a tectonically inactive region shows mountain fronts with high sinuosity because continuing erosion and deposition would progressively shift the position of the mountain front away from the range-bounding fault. Valleys in the mountains should be wide and shallow under tectonic stability because lateral erosion controlled by a stable base level would dominate the system. Conversely, in tectonically active areas the mountain-front sinuosities are low and the mountain valleys are narrow and deep (fig. 2.4). This follows because repeated vertical movements preserve the coincidence between the bounding fault and the mountain front, and the rivers upstream from the front are spurred into downcutting on the upthrown side of the fault. The basis for this expected response has been established by countless observations that river gorges are related to enhanced stream power brought on by rapid, fault-related uplift (for example, see Jackson et al. 1982).

We should caution here that every straight mountain front is not the result of recent uplift along a range-bounding fault (see Formento-Trigilio and Pazzaglia 1998), and every situation should be examined for the influence of other geomorphic factors. Nonetheless, a classification of relative tectonic activity has been created using the characteristics of processes and features in the piedmont adjacent to fault-bounded mountains and in the mountain block itself (table 2.1). Such a classification is useful for regional reconnaissance or for detailed studies of individual mountain fronts that may be bounded by a variety of fault types. The utility is enhanced by the fact that much of the analytical data are easily derived from topographic maps and aerial photographs.

Geomorphic surfaces and features The use of geomorphic surfaces as indicators of tectonic activity is readily apparent where fans or fluvial terraces are offset by faults, or where marine terraces and wave-cut platforms stand well above sea level (fig. 2.5; see Bull and Cooper 1986; Pillans 1990; Merritts and Bull 1989). Horizontal displacement on faults can also be documented by variations in geomorphic landforms. For example, detailed studies (Wesson et al. 1975; fig. 2.6) recognize numerous tectonically produced landforms along active strike-slip faults, and especially those of the San Andreas fault system (see Keller et al. 1982, fig. 2.7). Offset drainages, stream diversions, and horizontal displacements of fan-pediment complexes are common results of lateral movements, and additionally creation of unique evolutionary histories of entire drainage systems (Matmon et al. 1999). Other features were recognized as

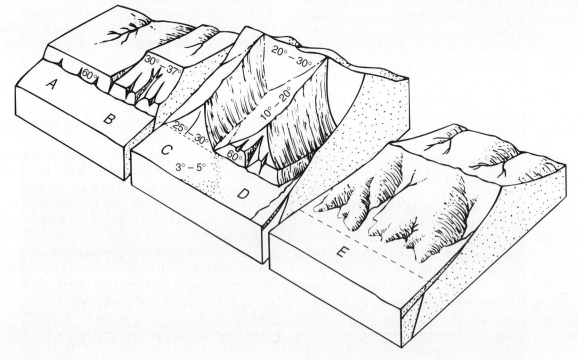

Figure 2.4A

Sequence of development of fault-generated mountain landscapes. (A) Block diagrams showing initial uplift stages (A, B), maximum relief (C, D), and dominance of fluvial erosion during subsequent stage of tectonic quiescence (E).

(Redrawn with permission from W. B. Bull, "Tectonic Geomorphology" in *Journal of Geological Education*, 32:310–324. Copyright © 1984 National Association of Geology Teachers.)

Figure 2.4 B-1

(B) Photos showing fault-generating landscapes depicted in block diagrams of figure 2.4(A). Photo 1 is equivalent to B and C and represents maximum relief of fault-bounded mountains. Looking at active front of Black Mountains on east side of Death Valley, Calif.

(Photo courtesy of Steve Wesnousky.)

Figure 2.4 B-2

Photo 2 illustrates stage E in figure 2.4(A). Stable boundary of Sonoma Range, north-central Nevada.

(Photo courtesy of Steve Wesnousky.)

TABLE 2.1	Classification of Relative Tectonic Activity of Mountain Fronts in the Quaternary Period.

Typical Landforms at Mountain Fronts

Relative Tectonic Activity	Rate of Activity	Piedmont	Mountain
Highly Active			
Class 1			
A	$\Delta u/\Delta t \geq \Delta cd/\Delta t + \Delta pa/\Delta t$	Unentrenched fan	V-shaped cross-valley profile in bedrock
B	Same as above	Unentrenched fan	U-shaped cross-valley profile in alluvium
Moderately Active			
Class 2	$\Delta u/\Delta t < \Delta cd/\Delta t > \Delta pd/\Delta t$	Entrenched fan	V-shaped valley
Class 3	Same	Same	U-shaped valley
Class 4	Same	Same	Embayed mountain front
Inactive			
Class 5			
A	$\Delta u/\Delta t \ll \Delta cd/\Delta t > \Delta pd/\Delta t$	Dissected pediment	Dissected pediment embayment
B	$\Delta u/\Delta t \ll \Delta cd/\Delta t = \Delta pd/\Delta t$	Undissected pediment	Pediment embayment
C	$\Delta u/\Delta t \ll \Delta cd/\Delta t < \Delta pd/\Delta t$	Undissected pediment	May have characteristics of active mountain front

Source: W. B. Bull, "Tectonic Geomorphology," *Journal of Geologic Education* 32:310–24, 1984.

t = time, u = uplift, cd = channel downcutting, pa = piedmont aggradation, pd = piedmont degradation.

Figure 2.5

Uplifted marine terraces (wave-cut platforms) along the central California coast.

(Raymond Guillemette)

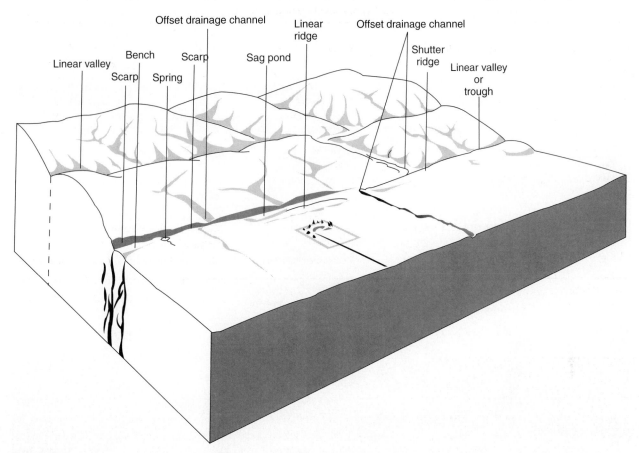

Figure 2.6
Block diagram showing landforms developed along recently active strike-slip faults.
(Wesson et al. 1975, U.S. Geol. Surv. Prof. Paper 941A, Fig. 11, p. A21.)

products of vertical displacement along the fault. Perhaps of greatest importance, Keller and his colleagues were able to explain almost all of the fault-related features by simple shearing associated with a bend in the trace of the fault. Therefore, landforms here were used as primary evidence to explain the mechanics of local deformation.

Longitudinal profiles of river terraces, alluvial valley floors, and stream channels are also sensitive indicators of regional tectonism. For example, repeated geodetic surveys have shown clearly that two areas in Louisiana and Mississippi are sites of rapid modern uplift. Burnett and Schumm (1983) examined the characteristics of fluvial features crossing these uplifts (fig. 2.8) and found considerable geomorphic evidence to substantiate the geodetic findings. Terrace and valley floor profiles show pronounced convexities where they cross the axes of uplifts or anticlines, being abnormally gentle upstream from the axes and oversteepened downstream from the axes (King and Stein 1983). Where tectonic activity is modern, the stream channel itself may be deformed.

Considerable attention is now being given to the use of river channel properties and vagaries in their slopes in a broad spectrum of tectonic analyses (Seeber and Gornitz 1983; Ouchi 1985; Bull and Knuepfer 1987; Harvey

and Wells 1987; Rhea 1989; Merritts and Vincent 1989). For example, significant work has been accomplished in the Mississippi River Valley where river anomalies and geomorphic features have been cited as evidence of tectonic activity (Russ 1982; Saucier 1987; Guccione et al. 1994; Cox 1994; Merritts and Hesterberg 1994; Boyd and Schumm 1995). The tectonism revealed by geomorphic analyses is often linked directly to modern seismicity in the New Madrid zone, but aseismic areas in the Mississippi embayment may also reveal the tectonic/fluvial relationship. In addition, surface abnormalities often reflect past or ongoing activity along fault systems that are buried under a thick accumulation of river alluvium (Spitz and Schumm 1997).

We believe that this type of analysis will become more prevalent in the future. Precisely how much response will occur in modern rivers experiencing neotectonism depends on the river size. Larger rivers may possess enough energy to keep degradational pace with the rate of uplift, and therefore, they may experience little change. The valley floors and terraces, however, may show considerable change in character at the uplift axes. Nonetheless, fluvial features clearly reflect neotectonism and should be considered as prime criteria in tectonic geomorphology (Gomez and Marron 1991).

Figure 2.7
Physiographic features of recent faulting along a segment of the San Andreas fault near the Carrizo Plain, southern California.
(Wesson et al. 1975, U.S. Geol. Surv. Prof. Paper 941A, Fig. 12, p. A22.)

Fault-scarp analyses In addition to analyzing fault-bounded mountains and deformation of surfaces, geomorphologists are providing important insights concerning the ages of fault scarps and the seismic implications associated with the recurrence interval of fault movements. Any scarp face (e.g., terrace, fault) tends to change its slope with time. The basic assumption in dating scarps is that the upper part of a scarp will have its slope lowered with time by erosion. Concurrently, the basal zone adjacent to the fault will rise with time because of the accumulation of the debris being transported down the scarp face. Therefore, slope change can be modeled by a transport law. The law utilized in scarp dating is based on the assumption that mass transport on slopes is analogous to the movement of gases and liquids in the physical process

known as diffusion. The equation utilized includes a factor of slope angle, which presumably declines in a linear manner with time (for derivation, discussions, and applications see Wallace 1977; Nash 1980; Colman and Watson 1983; Hanks et al. 1984; Koons 1989; Willgoose et al. 1990; Turko and Knuepfer 1991). Because the rate of down-slope transport of unconsolidated debris is affected by many factors (climate, aspect, sediment texture, and lithology), the dating of fault scarps has numerous uncertainties (Andrews and Hanks 1985; Pierce and Colman 1986; Mayer 1987; Colman 1987; Nash 1987) and the relationship may, in fact, be nonlinear (Andrews and Buckman 1987). Nonetheless, the approach does hold great promise especially when combined with other dating techniques.

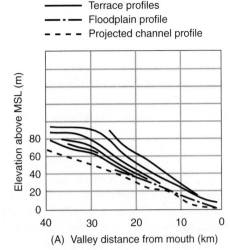

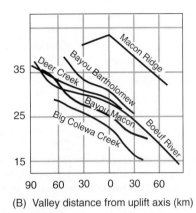

Legend:
——— Terrace profiles
—·— Floodplain profile
- - - Projected channel profile

(A) Valley distance from mouth (km)

(B) Valley distance from uplift axis (km)

Figure 2.8
(A) Longitudinal profiles of terraces, floodplain, and a projected channel profile of the Pearl River, a major river crossing the Wiggins uplift.
(B) Longitudinal valley profiles of streams crossing the Monroe uplift. Macon Ridge, a remnant of a deformed Pleistocene Mississippi River terrace, is also shown. MSL = mean sea level.
(From A. W. Burnett and S. A. Schumm, "Alluvial River Response to Neotectonic Deformation in Louisiana and Mississippi," *Science* 222:49–50. Copyright 1983 by the AAAS. Reprinted by permission.)

Dating involved in any tectonic geomorphology study, including the models used to estimate ages on declining scarp slopes, should be bolstered by other geochronological data. Absolute ages derived from standard procedures such as radiocarbon dating are preferable, but other techniques such as soil chronosequences (Harden 1982) provide relative time ranges that add credence to ages derived from diffusion models.

Plate-margin analyses   Tectonic geomorphology at plate margins is most clearly preserved in the physiography of coastal zones, and documented primarily by the same landforms that we examined above. The type of plate margin determines which geomorphic features are most diagnostic of the tectonic activity involved. For example, margins that are marked by transform faults, such as those in California and southern Alaska, are characterized by features associated with strike-slip faulting (see fig. 2.6). However, shear forces oblique to the transform margin may alter the tectonic style and produce vertical movements that uplift and deform a variety of geomorphic surfaces (for example see Orme 1998). Margins indicated by rift valleys result from the extension produced as plates move in opposite directions. Landforms here are similar to those along fault-bounded mountains.

Converging plate margins are the progenitors of orogenic belts and the mountain ranges contained within. As discussed earlier, ancient orogenic belts display landforms depicting the tectonic activity only after denudation and epeirogeny enhance the deep-seated structures. Nonetheless, current plate motions do produce landforms at converging margins, which reflect neotectonic activity. In most cases, the features produced manifest vertical rather than lateral movements, and consist primarily of uplifted and deformed terraces, raised shoreline deposits, incised rivers, and so on. These margins are prevalent in modern island arc settings of the circum-Pacific and in the mountainous coasts found from western South America to Mexico. Geomorphic evidence in these environments has been used to explain variable rates of uplift, and in some cases, the physical dynamics found in the subduction zone (Gardner et al. 1992). It is also possible to identify variations of tectonism and seismicity parallel to the trend of the subducting margin (Ramirez-Herrera and Urrutia-Fucugauchi 1999). For greater detail concerning tectonic geomorphology in coastal zones having different plate margins, see Stewart and Vita-Finzi (1998).

The selected cases just presented are only a few examples of how the relationships among landform, process, and tectonics can be employed to gain insight in numerous geological and environmental investigations. The practice of tectonic geomorphology will certainly increase in the future, and every geomorphologist should become familiar with its basic premises and uses.

Volcanism

Although some volcanic activity is associated with orogenic events, volcanism is such a major input of endogenic force that it deserves separate treatment. Volcanism is nothing more than a surface manifestation of the internal processes that create and mobilize magma. Although volcanoes are spectacular in eruption and unique in topographic form, describing examples of active volcanoes is not necessary here. Most physical geology texts do this, and excellent treatises on volcanoes and volcanic landforms are available (Bullard 1962; Ollier 1969; Green and Short 1971; Macdonald 1972; Short 1986; Ollier 1988; Francis 1993). It may be pertinent, however, to examine briefly how internal variables control the type of volcanic eruption and the ensuing topographic form.

The violence of a volcanic eruption is determined mainly by the composition of the magma and the amount of gas in it. By affecting the viscosity of the magma, these factors influence the observed differences in volcanic activity. Highly fluid, basaltic magmas are produced in both continental and oceanic environments. The more viscous, high-silica lavas that crystallize as andesites, dacites, and rhyolites are generally restricted to continents or marginal island arcs. In orogenic zones, where most viscous lava occurs, the erupting magma also contains more gas than the true oceanic types. The combined effect of higher gas content and greater viscosity creates a tendency toward more explosive eruptions. This trend is magnified in continental volcanoes where violent explosions are commonplace, and solid volcanic ejecta, called *tephra,* is the dominant extrusive material rather than flowing lava. Although we suggest that some relationship exists between magma composition, gas content, and geographic location, these links are very general because magma composition can be changed in any area by a variety of mixing processes (Rhodes et al. 1989; Geist et al. 1998, 1999), and gas content can vary significantly even within a single volcano (Ida 1995).

In most classifications, major volcanic landforms consist of three types: plateaus or plains, cones, and calderas. *Lava plains* and *plateaus* are extremely flat surfaces, both continental and oceanic, that have been aggraded by overlapping flows of fluid lava with a mafic composition. Although the nature of the vents is not clear, they probably are fissure types distributed over wide areas or a series of unconnected pipe vents. The dominant characteristic of lava plains and plateaus is the enormous volume of lava extruded and spread over a vast surface area. For example, in the Columbia River Plain of Washington and Oregon, areas greater than $200{,}000$ km² are known. Indeed, a part of the total mass includes mappable flow units combined as the Teepee Butte Member, which may cover an area of more than $52{,}000$ km² and have a volume over 5000 km³. These flow units are most likely generated from linear vent systems (Reidel and Tolan 1992). In the Deccan Plain of India, more than $500{,}000$ km² are covered, and oceanic plains can be even more extensive (Kuno 1969). Commonly the total thickness of the extrusive rocks exceeds 2000 m—great enough to bury a mountainous terrain. In fact, in the Columbia River Plain peaks of buried mountains (called steptoes) project through the flat surfaces as isolated "islands" of older rock.

The surfaces of lava plains and plateaus show minor perturbations where broad, low cones rise 30–60 m above the general level. However, individual flows, 2 to 50 m thick, can extend for hundreds of kilometers from the cones, blending imperceptibly into lava issued from fissured vents to create a surface that normally slopes less than 1°. Lava plains are almost always composed of basaltic rocks because silicic magma tends to be too viscous for the long distance of flow necessary to create a planar topography.

Some silica-rich volcanics do underlie flat plains, but these rocks did not crystallize from lava and the surface extends over much smaller areas. Silicic plains develop when incandescent volcanic glass is erupted within dense clouds of gas capable of flowing for considerable distances. The welded tuffs (ignimbrites) resulting from the ash flows are common in the Cordilleran region of the western United States. Field evidence there suggests that the ash was vented from both linear and arcuate fissures and central pipes.

The second major volcanic landform occurs as the topographic expression surrounding a single vent. These surfaces, called *cones,* have a variety of shapes depending on the type of volcano and the predominant mode of eruption. *Shield volcanoes* develop cones that are typical of those seen in Hawaii and Iceland. Shield volcanoes are always built up from fluid, mafic magma; tephra is only a minor part of the erupted material, and explosive eruptions are rare. Subaerially, slopes on shield cones are usually between 2° and 10° (fig. 2.9). During the shield-building phase, however, a distinct nearshore platform is developed below sea level because lava flowing into the ocean is rapidly chilled. Because subaqueously chilled volcanic rock has different properties, the cone slope steepens dramatically below sea level. For example, the Hawaiian volcanoes Kilauea and Mauna Loa have flank slopes that average about 4° above sea level and 13° below sea level (Mark and Moore 1987). The massive Hawaiian shields rise at least 4800 meters above the ocean floor, and some (Mauna Loa, Mauna Kea) have an absolute relief greater than 9 km, making them among the largest topographic mountains of the world. The magma in the Hawaiian chain moves surfaceward in rift zones that generally parallel the Hawaiian Ridge (topographic high in the central Pacific). The rift zones usually contain hundreds of fissures that the ascending magma uses as vents for eruption. The Hawaiian rifts extend discontinuously for 3500 km across the center of the Pacific plate to a point where they abruptly bend to join the Emperor Chain. This linear group of seamounts, similar in all aspects to the Hawaiian volcanoes, continues northward for another 2500 km.

The Hawaiian eruption rates, estimated between 0.05 and 0.1 km³ a year (Moore 1970; Swanson 1972), are the greatest known on Earth. The primary Hawaiian magma originates in the upper mantle, and no genetic relationship exists between the volcanoes and the age or structure of the adjacent sea floor. The plate-interior location of the island chain precludes an orogenic origin for the volcanic activity. This apparent lack of coincidence between plate tectonics and volcanism has prompted the hypothesis that volcanic action here occurs over a roughly circular hot spot in the upper mantle, about 300 km in diameter. Volcanism occurs as the Pacific plate rides over the heated region, and successive

Figure 2.9
Gentle slopes on cone of shield volcano in Hawaii. View of Mauna Loa in distance taken from crater area of Kilauea.
(R. Craig Kochel)

generations of volcanoes are carried away from the heat source. (For detailed syntheses of all aspects of the Hawaiian Islands see Decker et al. 1987 and Moore and Clague 1992.)

The Icelandic shields commonly are smaller in all dimensions than the Hawaiian volcanoes. Although they originate in a plate-margin environment, they are not related to orogenic events. The magma probably rises along the mid-Atlantic ridge in tensional fractures associated with the forces that move the oceanic plates apart.

Most other cone-type volcanoes take the form of a **composite cone,** so named because the cone is constructed of interbedded lava flows and layers of tephra. The flows are usually blocky and the tephra mostly cinder or ash, but the characteristics of all components may vary with the viscosity of the ascending magma. Composite cones, like the one in figure 2.10, are more peaked than shield volcanoes, commonly rising several thousand meters above a narrow circular base. Most develop from a single pipe vent. Side-slope angles are typically high, ranging from 10° to 35°, because the lack of fluidity in the magma leads to a more localized deposition. Most continental volcanoes are of this type; in North America, famous examples are Mount Rainier (Washington), Mount Shasta (California), and Mount St. Helens (Washington).

The third major landform associated with volcanic activity is known as a caldera. **Calderas** are large depressions in volcanic regions that result where eruption spews forth large quantities of material, creating an empty space in the underlying magma chamber. This results in an inward collapse of the upper part of the volcanic cone, often along ring fractures that develop during the eruption (Druitt and Sparks 1985), although different mechanics may prevail in marine environments (Brown et al. 1991). The caldera depression, therefore, has a larger diameter than the original crater and, in fact, the primary distinction between a normal volcanic crater and a caldera is size. Depressions larger than 1.6 km (1 mi) in diameter are usually accepted as being calderas (Macdonald 1972) because they almost certainly result from subsidence rather than incremental construction of a cone.

The largest calderas are associated with volcanoes that produce tephra sheets. For example, the caldera basin buried under the thick ignimbrite sheets of the Yellowstone Plateau is 70 km by 45 km. In addition, some of the most well-developed calderas, such as the one occupied by Crater Lake, Oregon (fig. 2.11), occur on the summits of composite cones, and their origin is directly associated with the occurrence of widespread ash falls.

Figure 2.10
Typical slope topography of composite volcanoes, Mount Shasta, northern California.
(Dale Ritter)

This classification of volcanic landforms is clearly based genetically on magma type, erupted material, and the character of volcanic activity. It may be, however, that this approach adds little to our understanding of volcanic geomorphology. Thouret (1999) argues that classical classifications should be revised to include additional features that reveal the mode of geomorphic processes in volcanic terrains and the fact that volcanic landforms result from both constructive and destructive processes.

CLIMATIC GEOMORPHOLOGY

Climate, Process, and Landforms

Because of climate's vast effect on many surface phenomena, it was inevitable that scientists would attempt to group climates into some useful classification. The most successful classifications, such as the widely used Köppen system or more simplified approaches (Bull 1991), are those in which climate groups are distinguished on the basis of observed temperature and precipitation values. Geomorphologists, however, should be more concerned with how energy in any climatic regime is utilized in geomorphic work than with how the climate is classified. In other words, we need to know why certain landforms develop most efficiently under a given prevailing climate.

The relationship between landforms and climate is the basis for a major philosophic approach in geomorphology known as **climatic geomorphology,** which has been most forcefully championed by European scientists (Tricart and Cailleux 1972; Büdel 1982). Essentially, the underlying premise of this approach is that geomorphic mechanics vary in type and rate according to the particular climatic zone in which they function. If that assumption is correct, landforms produced from these mechanics will be different from region to region and will reflect the dominant climate.

There is little doubt that a relationship exists between the type of dominant processes in a region and the prevailing climate. Wilson (1968) suggests that the relationship between climate and process be called a **climate-process system,** and the relationship between climate, process, and landforms be called a **morphogenetic system.** Figure 2.12 is a diagrammatic representation of six climate-process systems that ideally relate climate to a set of paramount processes. The diagram was derived by combining the relationships of several specific processes and climates; the arbitrary system names are meant to be descriptive of the climatic type. For instance (other factors being equal), we can expect that a region with a mean annual temperature of 0° C and mean annual precipitation of 500 mm will function as a periglacial climate-process system, and those processes that function most efficiently in such a climate will prevail. Each climate-process system in the figure can be converted into a morphogenetic system by defining the landforms that most commonly result from the processes involved (table 2.2).

Figure 2.11
Aerial photo of Crater Lake, southern Oregon.

(© John S. Shelton)

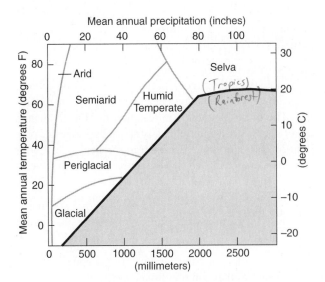

Figure 2.12
Six possible climate-process systems as suggested by Wilson (1968). Each set of temperature-precipitation values tends to drive processes that function most efficiently under those climatic conditions.

(From L. Wilson in *Encyclopedia of Geomorphology,* ed. by R.W. Fairbridge, copyright 1968 by Dowden, Hutchinson & Ross, Inc. Used by permission of Professor Rhodes W. Fairbridge.)

TABLE 2.2	Simple Morphogenetic Systems and Their Landscape Characteristics.	
System Name	**Dominant Geomorphic Processes**[a]	**Landscape Characteristics**[b]
Glacial	Glaciation Nivation Wind action (freeze-thaw)	Glacial scour Alpine topography Moraines, kames, eskers
Periglacial	Frost action Solifluction Running water	Patterned ground Solifluction slopes, lobes, terraces Outwash plains
Arid	Desiccation Wind action Running water	Dunes, salt pans (playas) Deflation basins Cavernous weathering Angular slopes, arroyos
Semiarid (subhumid)	Running water Weathering (especially mechanical) Rapid mass movements	Pediments, fans Angular slopes with coarse debris Badlands
Humid temperate	Running water Weathering (especially chemical) Creep (and other mass movements)	Smooth slopes, soil covered Ridges and valleys Stream deposits extensive
Selva	Chemical weathering Mass movements Running water	Steep slopes, knife-edge ridges Deep soils (laterites included) Reefs

Source: L. Wilson, "Morphogenetic Classification" in *Encyclopedia of Geomorphology,* ed. by R. W. Fairbridge, copyright 1968 Dowden, Hutchinson & Ross, Inc.
[a]Processes are listed in order of relative importance to landscape (not in order of absolute magnitude). List is abbreviated.
[b]Both erosional and depositional forms are included. List is neither comprehensive nor definitive, merely suggestive.

In reality, minor heterogeneity of geology may produce subtle variations in the topographic outcome. In addition, relict landforms developed under a former climate, and seasonally changing climates may complicate the expected form. Nonetheless, the morphogenetic approach does provide a reasonable framework for the analysis of landforms. Absolute servility to this approach, however, leads us away from the study of processes, since the method passes quickly from climate to form. This nonchalant leap from climate to landform overlooks the intermediate phase of process and does not indicate the utility of the morphogenetic approach in geomorphology. We believe that one of the premier scientific contributions that can be made by geomorphologists is to explain the effects of climate *change* on process; thus climate-process systems take on greater significance than their related morphogenetic systems. Having reasonably well-defined boundaries between different climate-process systems that are based on real temperature and precipitation values allows geomorphologists to become major players in understanding the magnitude of past climate changes and in predicting responses that will occur in surficial environments in future climatic changes.

Climate Change and Geomorphic Response

Climate in and of itself is important because it is one of the fundamental controls on process mechanics. Changes in climate are equally, if not more, significant (see Bull 1991) because they lead to responses (threshold crossings) in geomorphic systems that we can use to interpret past geomorphic events and predict environmental scenarios that will develop as a result of future climate changes. Overwhelming evidence of past climate change has been documented in such diverse fields as pedology (soil science), palynology (study of pollen), archeology, paleontology, oceanography, and many others. The most dramatic manifestation of climate change is, of course, that several widely spaced episodes of prolonged glaciation have occurred in the history of Earth. The most recent of these, encompassed by the Pleistocene epoch, was characterized by distinct alternations of glacial and interglacial stages when climate was patently different from the present. Thus the ephemeral nature of climate is beyond question. The reasons for climate change, however, are open to considerable debate because they are tucked away in the complex maze of interactions that control climate itself.

Detailed examination of the dynamics of climate change are well beyond the purposes of this book and are best left to the climatologists. In fact, some climatic variations are presently inexplicable. In general, however, the variables that trigger climate change exist in several major groups; significantly, the variables in each category function over different time intervals.

First, variations in the composition of the atmosphere may alter the amounts of solar radiation reaching or escaping Earth's surface. This is especially true when changes in the CO_2 content alter the greenhouse effect or when recurring volcanism introduces large volumes of fine-grained ejecta into the atmosphere. Normally the volcanic effect on climate is short-lived. It has been suggested, however, that a major volcanic event or episode occurring during a pronounced climatic transition may be a contributing factor in global cooling and perhaps in the onset of glaciation (Rampino and Self 1993). Variations in atmospheric factors can produce tangible climate change in a matter of years or decades.

Second, astronomical motions may produce changes in the pattern and intensity of solar radiation. For example, parameters such as the tilt of Earth's axis and the orbital path around the sun vary from maximum to minimum values in specific intervals of time. Periodically, a number of such factors coincide to produce minimum solar radiation, enough decrease in radiation to induce an episode of glaciation. The astronomical motions have a periodicity such that their effect on climate would occur in a time span ranging from 20,000 years to 100,000 years (see Hays et al. 1976; Imbrie et al. 1984).

Third, many scientists believe that a cause and effect relationship exists between climate change and variations in the elevation and distribution of continents. The concept implies that the major cooling needed to produce glaciation occurred when landmasses were high. This effect is most clearly understood in the analyses of Cenozoic tectonism and its influence on worldwide climate (Ruddiman 1997). It also has been suggested that the distribution of landmass and ocean basins associated with the mechanics of polar wandering and plate tectonics is important in controlling climate. Landmass movements probably influence climate over cyclic time intervals, perhaps millions of years.

In geomorphology the cause of climate change is less important than the adjustments demanded within surficial systems in response to the change. However, the required adjustments are often conditioned by temporal and/or spatial factors. Analyses of ice cores taken in the Greenland Ice Sheet Program (GISP) suggest that significant climatic changes can occur very quickly. Oxygen isotope data, dust concentrations, and snow accumulation rates suggest that oscillations of glacial and nonglacial conditions at the end of the last major Pleistocene glaciation occurred rapidly, within periods of 1–3 years and certainly no longer than 50 years (Dansgaard

et al. 1989; Taylor et al. 1993; Alley et al. 1993). Similar changes are recorded in Antartica cores (for discussion see Bloom 1997, p. 397).

The rapidity of significant climate change adds a complicating factor in geomorphic interpretation because we do not know if geomorphic response can keep pace with the climatic shift or whether a distinct lag exists between cause and effect. Bull (1991) refers to this lag as response time, which, in his analysis, includes a phase before an action begins and a phase during which a new equilibrium is established. Assuming that a lag exists, it may exceed the time between successive climatic shifts, and thus it is possible that responses are manifesting an altered climate that has already been changed again. The possibility of a lag in response is increased because climate does not directly influence process. Instead, climate affects certain environmental factors which, in turn, exert control over process. Here we will briefly examine several of these intermediary links between climate change and geomorphic response.

Sea-Level Fluctuation   One major intermediary factor related to climate change that is responsible for generating geomorphic adjustments is the phenomenon of sea-level change. Sea level is a most important horizon in geomorphology. In addition to serving as the datum for relative gravity analyses, sea level represents the ultimate base level for rivers draining the land and the theoretical end point of continental erosion. We know from geologic history that sea level is not constant relative to the adjacent land masses. Its relative position has shifted through time because (1) the continents are subject to isostatic uplift or depression and (2) the water level itself may rise or fall, an adjustment known as *eustatic change*. Eustatic sea-level change occurs on a worldwide basis in contrast to isostatic movements, which are determined by the local or regional gravity environment.

The concept of eustatic sea-level change (eustasy) is not new nor has it been readily embraced by the geological community (for an excellent review see Dott 1992). Perhaps the most acceptable argument for eustasy stems from its apparent link to the accumulation and dissipation of glaciers. Glaciers are interruptions of the hydrologic cycle because water locked up in ice represents precipitation that is not being returned to the ocean. As a result, when the volume of ice held on the land increases during a period of glaciation, the ocean volume shrinks and its surface is lowered. Conversely, if all modern glaciers were melted, sea level would rise by an amount commensurate with the volume of water released from storage. Clearly, the growth and wastage of glaciers attendant to glacial and interglacial stages of the Quaternary are major progenitors of eustatic sea-level change.

It is ironic that the most widely accepted evidence of eustasy is also the evidence used to document isostasy. This fact presents geoscientists with a nasty interpretive problem because vertical change in relative sea level is produced by the mechanics of two different processes that are both based on the same evidence (see Morner 1980 for discussion). Therefore, sea-level change attributable to eustasy cannot be accurately determined without knowledge of isostatic movement during the same interval. For example, the growth and demise of the ice sheets that result in changes of ocean volumes are also responsible for crustal depression and isostatic uplift in response to glacial loading and unloading. For this reason many studies of Quaternary eustatic change are conducted around tropical islands that are far removed from the complicating effects of glaciation. Nonethelss, accepting glacial eustasy as fact, the most important question for us is how variations in sea level effect geomorphic systems.

Sea-level fluctuations have a direct influence on a number of shoreline processes and features (e.g., beaches, barrier islands). These will be discussed in chapter 13. Sea-level change also permeates the fluvial system and through it other systems. For example, in coastal regions river terraces commonly result when alternating cutting and filling are initiated by fluctuating sea level. Theoretically, entrenchment should accompany glacial expansion (when sea level is decreasing), whereas filling would take place during the waning phase of the glacial cycle (when sea level is rising). Fisk (1944) interpreted the depositional terrace sequence in the lower Mississippi Valley as being related directly to waxing and waning glaciations and their effects on eustatic sea-level change. Although the details will continue to change as new data are collected, such depositional terraces represent a good example of the connection between eustatic change and fluvial response.

The downcutting produced during glaciation may be gradually propagated upstream, causing a similar erosional response in each tributary basin. At that time, slopes may be regraded to new levels and groundwater tables may be lowered. Thus, by process linkage, one external change in climate begins a chain reaction of adjustments throughout the geomorphic regime. How far responses will be propagated upstream in a system as large as the Mississippi basin depends on the erosive power of the river and how long conditions remain stable in the interior reaches. We know, for example, that some tributaries in central Missouri were apparently not affected by the lower Mississippi sequence recognized by Fisk (Brakenridge 1981). In fact, Brakenridge (1981) suggests that the terrace and floodplain sequence along the Pomme de Terre River shows no clear correlation with either sea-level history or glacial chronology in the upper Missouri-Mississippi basin. Aggradation occurred during glacial, nonglacial, and interstadial periods. Terrace formation by river downcutting proceeded in the late Pleistocene, as well as in the Holocene when major glaciation and eustatic base-level changes were absent.

Presumably this indicates that alternating filling and cutting can be directly induced by climate change without glaciation or base-level change. In the following section we examine how such responses might occur and whether the relationship between climate and threshold response is as straightforward as it appears.

Geologic and Vegetal Screens   The preceding discussion reveals that it may be more difficult to explain climatic effect on landforms located far from the site of sea-level change. River terraces in the interior of continents have historically been attributed to the fluctuating climate accompanying glacial and interglacial conditions. The swing from arid to humid climate as glaciation begins (and vice versa) affects not only the prevailing discharge but also the amount and type of sediment delivered to the rivers. Such changes in fundamental river controls logically produce the trenching and filling.

Pleistocene and Holocene climates in North America are reasonably well known. In general terms, average temperatures in the Pleistocene were probably about 7° C lower than present during glacial stages and 3° C higher than present during interglacials. Average annual precipitation was as much as 25 cm greater during glaciation than now and 12 cm lower during the warmer periods. These values vary depending on latitude and the technique used to make the estimate (Flint 1971; Budyko 1977).

Given this information, it is ironic that little agreement exists as to which climate produced the filling and which caused the trenching in the process of terrace formation. To complicate matters, different interpretations of the cause and effect relationship in nonglaciated regions of the United States seem to be supported by field evidence (for a discussion see Flint 1971, pp. 304–307). The ambiguity of cause and effect in terrace development relates to the second intermediary link between climate change and response, or how climatic ingredients are filtered into geomorphic systems through geologic and vegetal screens.

A major climatic influence in geomorphic systems occurs because temperature and precipitation are fundamental controls on mean annual runoff and the magnitude of erosion. As expected, runoff increases with higher annual precipitation, but at constant precipitation the runoff will decrease with higher temperature because evapotranspiration is known to be greater in warmer regions (fig. 2.13). The amount of sediment yielded from basin slopes is an indication of erosion and is also a function of temperature and precipitation (fig. 2.14).

The curves shown in figure 2.14 were derived using data from basins in the western United States averaging 3900 km^2 in area and may not be suitable for other settings. Significantly, however, they show that the influence of climate on erosion is filtered through a vegetal screen. For example, figure 2.15 represents the 50° F

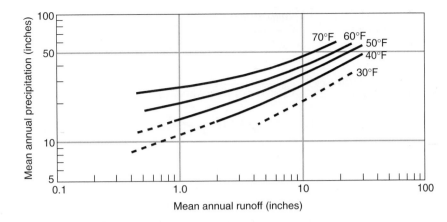

Figure 2.13
Curves illustrating the effect of temperature on the relation between mean annual runoff and mean annual precipitation.
(Source: After W. B. Langbein, et al., U.S. Geological Circular 52, 1949)

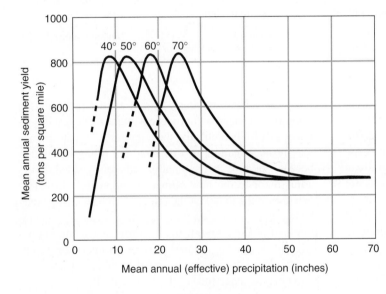

Figure 2.14
Curves illustrating the effect of temperature on the relation between mean annual sediment yield and mean annual precipitation.
(From S. A. Schumm, "Quaternary Paleohydrology" in *The Quaternary of the United States*, ed. by H. E. Wright and D. G. Frey, pp. 783–784, Princeton University Press, 1965. Used with permssion of the author.)

(10° C) curve, where effective precipitation is the mean annual precipitation adjusted for that prevailing temperature (Langbein and Schumm 1958). Under the stated conditions, maximum sediment yield occurs at approximately 30 cm of effective precipitation because the density and type of vegetation developed at that precipitation and temperature is most conducive to water-related erosion. Where precipitation is lower than 30 cm, not enough water is available to erode the slopes. Above 30 cm of precipitation, the vegetal cover becomes more dense and changes from desert brush to grasses. The ubiquitous root systems tend to fix sediment on the slopes, and therefore, with increased precipitation sediment yield becomes progressively lower until it nearly stabilizes under a forest cover.

The Langbein-Schumm analysis will be discussed further in chapter 5. For now, it is important for you to recognize that climatic change does not affect geomorphic systems directly but is passed through geological screens (rock type, soils) and vegetal screens that determine how

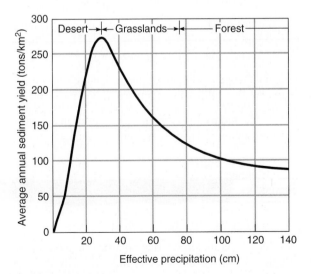

Figure 2.15
Average annual sediment yield as it varies with effective precipitation and vegetation.
(Langbein and S. A. Schumm, *Transactions of the American Geophysical Union*, vol. 39, p. 1077, 1958, copyright by the American Geophysical Union.)

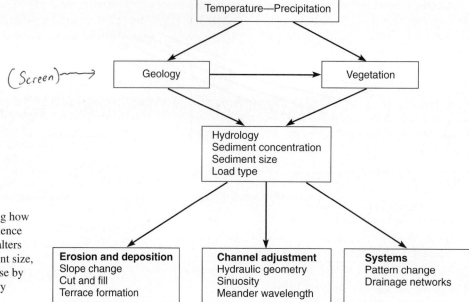

Figure 2.16

Hypothetical flow chart showing how climatic variables exert an influence on rivers. A change in climate alters sediment concentration, sediment size, or load type, requiring a response by the river system. Responses vary depending on local conditions.

much water and sediment get to a river channel. Variations in these two factors require threshold adjustments that can occur in a number of different ways (fig. 2.16). In addition, the lag in response is controlled by how rapidly a new vegetal screen and its characteristics can develop under a new climatic regime. This factor is not as straightforward as we would like. Predictions of response to possible future climate changes are clouded because vegetation biomes before the Holocene (ca. 10,000 years B.C.) have no modern analogs (Overpeck et al. 1992). This fact suggests that scientists cannot predict with certainty what types of vegetation communities will develop if our climate reverts back to conditions that were prevalent during the Pleistocene.

Another problem is that the same climatic change may prompt entirely different sediment yields and therefore different geomorphic responses. For example, considering figure 2.15 again, assume that a 15-cm decrease in precipitation occurs in a particular drainage basin that had an effective precipitation of 45 cm prior to the change. The reduced precipitation will result in a greatly increased sediment yield. However, the same 15-cm de-

crease in a basin having a 35-cm annual precipitation prior to the change will produce a major reduction in sediment yield. Theoretically, then, the same climate change may result in cutting by one river and filling by another because the type and amount of sediment yielded during adjustment to the new climate is oppositely affected. What this tells us is that the effect of climate change may be highly dependent on the antecedent values of temperature and precipitation. If that is a true statement, knowledge of preexisting climate may be as important in understanding how systems respond to climate changes as knowing the magnitude of the change itself.

In sum, the relationships among climate, process, and landform are not easily determined because the effect of change is sidetracked into ancillary factors. The adjustments of these factors in the new climate provide variable conditions of load and water that spur responses that are not predictable under the present state of our knowledge. Thus, we have not been able to describe clearcut relationships between climate and landforms because we are far from understanding the climatic geomorphology scheme.

SUMMARY

In this chapter we briefly examined climate and endogenic factors as major external controls on geomorphic systems. Endogenic influence occurs primarily through the addition of mass and energy by volcanism and tectonic activity. The most important tectonic processes are those producing vertical movements of the surface. One of these is the isostatic adjustment required when the in-

ternal mass balance is upset. Other vertical movements, associated with faulting and warping, are integral parts of a subdiscipline known as tectonic geomorphology, in which the relationships among tectonics, processes, and landforms are utilized in a variety of geologic and environmental studies. In most cases, vertical displacements induce threshold conditions and responses in the affected surficial systems.

Climatic geomorphology explores the relationship between prevailing climates and the landforms expected to form under those conditions. Of greater significance is the fact that climate change is a threshold-producing phenomenon, although the cause and effect relationship may pass through intermediary factors such as eustatic sea-level fluctuations or vegetal and geological screens.

Suggested Readings

The following references provide greater detail concerning the concepts discussed in this chapter.

Bull, W. B. 1984. Tectonic geomorphology. *Jour. Geol. Educ.* 32:310–24.

———. 1991. *Geomorphic responses to climate change.* New York: Oxford Univ. Press.

Decker, R., Wright, T., and Stauffer, P., eds. 1987. *Volcanism in Hawaii.* U.S. Geol. Survey Prof. Paper 1350, vols. 1 and 2.

Keller, E., and Pinter, N. 1996. *Active tectonics: Earthquakes, uplift, and landscape.* New Jersey: Prentice-Hall.

Morisawa, M., and Hack, J., eds. 1985. *Tectonic geomorphology.* Proc. 15th Binghamton Geomorph. Symposium. Boston: Allen and Unwin.

Short, N. 1986. Volcanic landforms. In Short, N., and Blair, R., eds., *Geomorphology from space,* pp. 185–253. Washington, D.C., NASA.

Stewart, I., and Vita-Finzi, C., eds. 1998. *Coastal tectonics.* Geological Society, London, Spec. Publ. 146.

CHEMICAL WEATHERING AND SOILS

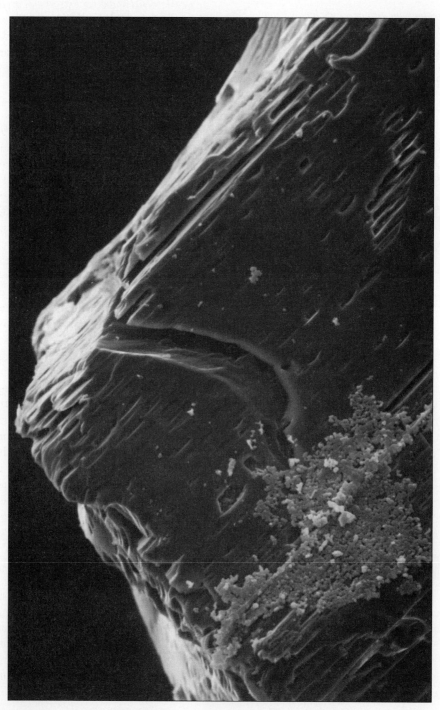

Dale Ritter

Introduction
Decomposition
    Processes of Decomposition
        Oxidation and Reduction
        Solution
        Hydrolysis
        Ion Exchange
    Mobility
        Leaching
        pH
        The Eh Factor
        Fixation and Retardation
        Chelation
    The Degree of Decomposition
        Mineral Stability
        Secondary Minerals
        Estimates Based on Chemical Analyses
    The Rate of Chemical Weathering
        The Complicating Factors
Soils
    Soil Properties
    Soil Horizon Nomenclature and Description
    Soil Classification
    Pedogenic Controls and Regimes
        Podzolization
        Laterization
        Calcification
    The Factor of Time
    Geomorphic Significance of Soils
Summary
Suggested Readings

INTRODUCTION

Most of Earth's surface is not composed of solid rock but is underlain by the unconsolidated remains of thoroughly altered rock. The fresh rocks and minerals that once occupied the outermost position reached their present condition of decay through a complex of interacting physical, chemical, and biological processes, collectively called _weathering_. Weathering progressively alters the original lithologic character until what finally remains in the space of the former rock is an unconsolidated mass consisting of (1) new minerals created by the weathering processes, (2) minerals that resisted destruction, and (3) organic debris added to the weathered zone.

Because every mineral species has, by definition, a unique chemical composition or atomic arrangement, it is not surprising that each type resists or responds to weathering in a special way. Although there is a wide variety of climates driving the processes and an almost endless array of rock structures and mineral types, the bulk of Earth's crust is composed of a surprisingly small number of mineral varieties with an equally limited chemistry. With this advantage, we can explore the

mechanics of weathering efficiently even though the systems involved are exceedingly complex.

Weathering is usually divided into separate domains of chemical processes (_decomposition_) and physical processes (_disintegration_). The processes are quite distinct; disintegration involves no chemical reactions but simply produces smaller particles from larger ones. Nonetheless, the two types of weathering operate simultaneously and, in fact, each may directly affect the character and rate of the other. Breaking a large rock into smaller particles increases the total surface area and thereby accelerates chemical attack on the material. For example, Hoch et al. (1999) showed that continuous disaggregation by freezing of welded tuff at high elevations produced smaller and fresher surfaces and led to abnormally high amounts of chemical weathering. It is now clear, however, that increase in mineral surface area is not totally due to a decrease in particle size, but involves a number of controlling factors. The effect of these controls on the rate of chemical weathering will be discussed later in the chapter. Conversely, expansion of minerals by chemical processes may exert enough internal stress to hasten the disintegration of the rock. Realistically, then, the physical and chemical functions of weathering may be so intimately intertwined that to consider them as unique processes is mostly a matter of convenience. In this chapter we will deal with only the chemical and biological aspects of weathering, which are the dominant factors in the development of soils. Disintegration will be examined in chapter 4 when we consider the stability of slopes, because many of the processes that break rocks apart are also important in the erosion of unconsolidated slope material by mass wasting.

A _soil_ is the residuum that results from weathering over an extended period of time. Given the proper conditions, a distinct layering, called the _soil profile_, will develop in the residual material, and its characteristics will directly reflect the weathering and soil-forming processes. The geologist utilizes soils and other weathering phenomena as clues to the intricacies of geologic history and the relative age of unconsolidated deposits. Although most soils are young in the geological sense, some profiles have been preserved in the older record, and these provide critical evidence about environmental conditions at the time of formation. Mineral compositions of many sedimentary rocks reflect the combined tectonic and climatic conditions present during their creation, and a correct interpretation of such rocks requires a knowledge of weathering and soil-forming processes. Landforms are often recognized as relict features because their soil character is inconsistent with the prevailing modern climate. These are examples of how we utilize an understanding of soils in many aspects of geology. The study of chemical weathering and soils is not simply an adventure into esoteric geomorphology; a working knowledge of weathering processes is essential to any scientist interested in the surface environment.

DECOMPOSITION

Rocks and minerals are usually not in equilibrium with conditions that exist at or near Earth's surface. **Decomposition** refers to the processes that create substances that are more nearly stable in that environment. This march toward equilibrium is accomplished by alteration of original materials and/or production of new mineral types.

The most important agent of the weathering regime is water introduced as rain. Because rainwater is usually mildly acidic, a significant part of chemical weathering can be visualized as a process whereby minerals assimilate hydrogen ions and/or water and release cations to the soil liquid. Complicating this simple model are two facts: (1) water entering the ground is not chemically pure but contains a variety of ions captured from the atmosphere and introduced from surface materials; and (2) organic processes, involving metabolism of microorganisms and decay of vegetal matter, add gases and organic acids to the system. These organic functions are of such importance that some authors suggest that chemical weathering proceeds in two stages. The first stage, driven primarily by inorganic processes, is called *geochemi-*

cal weathering and produces rotten rocks or *saprolites.* A saprolite is a soft, typically clay-rich, thoroughly decomposed rock that often has structures of the unweathered parent rock preserved in the unconsolidated mass (fig. 3.1). The depth of chemical alteration in the development of a saprolite can be significant. For example, in the Piedmont region of Virginia weathering depths on granite rocks were noted to be as much as 30 m (for excellent discussions of saprolite development, see Pavich 1986, 1989; Pavich et al. 1989). The second stage of decomposition, called *pedochemical weathering,* leads to the formation of soils from the saprolitic material; it is chiefly a biologically controlled phenomenon.

The processes involved in geochemical weathering are associated with two sequential responses. As rainwater percolates into and through exposed rock material, the first important response is the breaking apart of the structures of the parent minerals. As water surrounds a mineral or penetrates its structure through micro-openings and cleavages, chemical reactions occur that tend to disrupt the mineral's orderly atomic arrangement. Atoms along exposed mineral surfaces are not satisfied electrically and so may attract the dipolar water molecules. If

Figure 3.1
Saprolite developed on carbonate rocks in the Shenandoah Valley, central Virginia. Note original rock structures are preserved but hammer easily penetrates the highly weathered mass.
(R. Craig Kochel)

the attraction causes water to dissociate into H^+ and OH^-, these will bond to the exposed ions of the mineral, as shown in figure 3.2. Hydrogen is then in the proper position to replace mineral cations, thereby releasing them to the surrounding fluid. The process involves a number of phenomena that collectively are referred to as dissolution or solution. Release of ions from a mineral surface may have a profound effect on the pH of the liquid, as hydrogen is progressively depleted and hydroxyls are concentrated.

The actual mechanics involved in removing cations from a mineral surface is much more complex than figure 3.2 implies. Two mechanisms have been suggested as controlling factors in models treating the rate and style of dissolution (for discussion see Brantley et al. 1986). In one model a thin residual layer is formed on the mineral surface as reactions occur. Analyses regarding the chemical weathering of albite (Chou and Wollast 1984, 1985) provided the first compelling evidence that a thin (20–30 Å) layer does develop during the progress of chemical weathering. Initially, the alkali ion (Na^+) is replaced at the mineral surface by H^+ similar to the exchange shown in figure 3.2. Continued exchange initiates the gradual buildup of a residual layer that is depleted in sodium and enriched in silica and aluminum. Eventually, the system reaches a steady state controlled by the rate at which sodium and hydrogen ions can be diffused through the layer. Other researchers agree that at least a partial layer is probable, but suggest that the ultimate rate of chemical reactions is a function of factors other than the diffusion process such as pH of the fluids (Holdren and Speyer 1985; Brantley and Stillings 1996; Dahlgren et al. 1999).

In the second model, the disruption of mineral structures is produced by reactions occurring at localized areas of the mineral surface. It has been suggested for years that mineral surfaces do not retreat uniformly as the diffusion-layer model predicts, but rather are characterized by the development of numerous, widely spaced holes called etch pits (fig. 3.3), which expand and produce a nonuniform retreat of the original surface (Lagache 1976; Berner and Holdren 1977, 1979: Helgeson

et al. 1984; Berner et al. 1985). Etch pits were presumed to form at sites of low resistance to weathering determined by internal properties of the mineral. More recent studies of feldspars in soils (Lee et al. 1998) demonstrate clearly that their weathering behavior is significantly influenced by intragranular microtextures (e.g., exsolution lamellae) and microstructures (e.g., dislocations) in the mineral. In fact, solution at the position of dislocations may be the major source of solutes during various stages of decomposition in the field.

It is clear that some confusion still exists concerning the chemical processes of dissolution, and indeed some researchers believe that both processes may be simultaneously at work in the chemical weathering of minerals (see fig. 3.4, and discussion in Nahon 1991). In light of these models it also seems clear that the end products of weathering may be directly influenced by the manner in which the minerals are destroyed. The process may proceed layer by layer from the external surface, or it may break the mineral into many small pieces, each retaining the structure of the original material. Some silicate minerals, for example, break into molecular chains that are easily recombined with available cations to form layered clay minerals.

The second important response in chemical weathering relates to how mobile or immobile ions are after they are liberated from their parent mineral. Considerations here are whether released ions or groups of ions are readily fabricated into new, stable minerals, or whether they can be removed in solution by percolating soil fluids. The ultimate fate of released particles depends on what they are and on the characteristics of the fluid into which they are released. The equilibrium state, for example, can only be attained in a closed system, for continuous addition and removal of water surrounding the decomposing mineral will alter the pH, carry some of the released ions away, or provide new elements that can combine chemically with those escaping the mineral. The final result is determined by a myriad of interreactions that are influenced by how the original structure breaks apart and how mobile the ionic or molecular particles are under the physical and chemical constraints

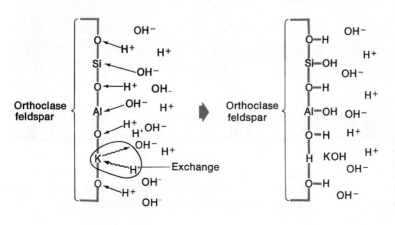

Figure 3.2
Ion exchange and chemical bonding take place as surface of orthoclase feldspar comes in contact with a solution containing dissociated H^+ and OH^- ions.

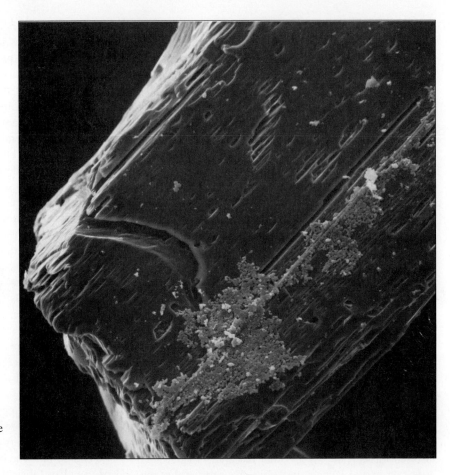

Figure 3.3
Scanning electron microscope photograph of hornblende grain showing weathering of etch pits that are elongated along cleavage planes.
(Dale Ritter)

Figure 3.4
Schematic representation of an albite grain after 500 hours of dissolution (modified from Chou and Wollas 1984). (A) The shaded area of 5 Å is where the three main components of albite have been dissolved; the heavy line is the surface of the remaining solid after dissolution; the dashed line represents the boundary between the residual layer and the fresh albite. (B) Localization of the residual layer on the walls of an etch pit.

(From *Introduction to the Petrology of Soils and Chemical Weathering* by Daniel B. Nahon, Figure 1.4, p. 13. Copyright © 1991 by John Wiley & Sons, Inc. Reprinted by permission.)

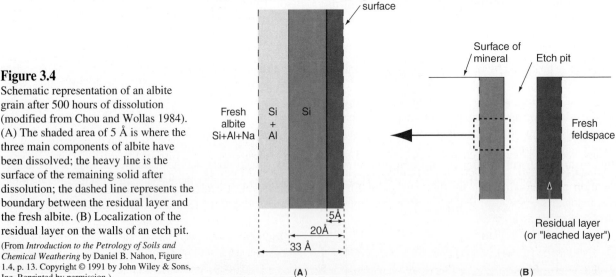

within the weathering zone. Clearly, various combinations of mineral types and chemical processes can result in a variety of final products. We refer you to Loughnan (1969) for basic concepts, and White and Brantley (1995) for a more recent treatment of processes and problems associated with chemical weathering.

Processes of Decomposition

The common chemical reactions involved in decomposition are oxidation and reduction, solution, hydrolysis, and ion exchange. Each process plays a particular role in the overall scheme of chemical alteration, although all function simultaneously in the weathering zone.

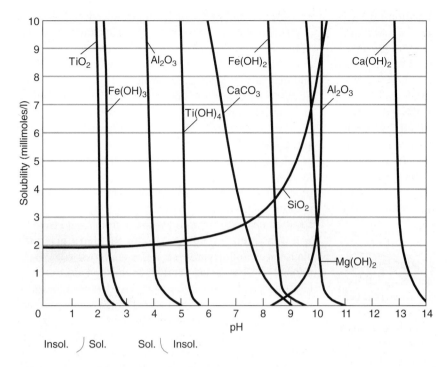

Figure 3.5
The relationship of solubility of common substances to various pH conditions.

(From F. C. Loughnan, *Chemical Weathering of the Silicate Minerals.* Copyright 1969 Elsevier Publishing Co., New York. Reprinted by permission of the author.)

Oxidation and Reduction Oxidation occurs when an element loses electrons to an oxygen ion. The process tends to occur spontaneously above the water table where atmospheric oxygen is readily available; therefore, most elements at Earth's surface exist in an oxidized state. Below the water table the environment is generally reducing; however, high concentrations of organic matter may cause local reducing conditions to occur above the water table.

The capacity for oxidation depends on the redox potential (Eh), the magnitude of which is controlled by the abundance of organic matter and the accessibility of free oxygen. In most soils, Eh values range from −350 to +700 millivolts, sufficient to keep the majority of common elements in their oxidized state. However, some elements, such as aluminum, change from a reduced to an oxidized form with considerable difficulty. The most common elements affected by fluctuations of Eh are iron, manganese, titanium, and sulfur. Iron is easily oxidized to the ferric (Fe^{+3}) state:

$$2Fe^{+2} + 4HCO_3^- + 1/2 \ O_2 + 2H_2O \rightarrow Fe_2O_3 + 4H_2CO_3$$

Its appearance as crustations on grains or as reddish-brown stains along fractures usually signifies the first tangible evidence of decomposition. The redox phenomenon exerts a significant control on the mobility of certain elements as they are affected by decomposition. This role of the Eh factor will be discussed later in the chapter.

Solution The process of solution is critical in chemical weathering because when atoms are removed from a mineral, the structure becomes unstable. However, the precise way in which a mineral collapses varies with its crystalline structure and the mobility of its constituent ions. When an atom is removed from its parent mineral, it may remain in solution and be taken completely out of the system by the downward moving fluids. This depends on the concentrations of reactable ions in the fluid and on other chemical characteristics of the medium.

Most common elements and minerals are soluble to some degree in normal groundwater, where pH values usually range from 4 to 9 (fig. 3.5). In the weathering of silicate materials, note that silica is soluble under all normal groundwater conditions. Solubility is low when the silica is contained in quartz and much greater (~ 115 ppm) as amorphous silica. The dissolution of quartz, estimated in numerous experimental studies, ranges from 6 ppm (Morey et al. 1962) to 11 ppm (Rimstidt 1997). In general, it is difficult to to determine quartz dissolution in the field (Schulz and White 1999) because (1) quartz has such a slow rate of dissolution, (2) silica contribution from quartz is not easily distinguished from that released from other silicate minerals, and (3) a portion of the dissolved silica can be taken into the structures of secondary minerals. In addition, the surface area of grains contributing to quartz dissolution varies with the density of etch pits and fractures. In cases where these variables can be controlled, the rate of quartz dissolution determined in field studies and long-term analyses seems to be similar to rates derived from experimental work (White et al. 1998; Schulz and White 1999; Stonestrom et al. 1999), but this conclusion does not necessarily hold for other silicate minerals. In fact, a number of researchers suggest that dissolution rates of most silicate minerals determined in the laboratory are simply not applicable as estimates of natural chemical weathering (White et al. 1996; Lee et al. 1998; Murphy et al. 1998; Dahlgren et al. 1999).

Aluminum oxides are virtually insoluble under normal groundwater conditions, and iron in the ferric state can be dissolved only by rather acidic fluids. It is not surprising, therefore, that mature weathering profiles in humid climates should be characterized by the presence of abundant ferric iron and aluminum. Other cations are readily soluble, and their removal to the water table concentrates the least soluble constituents in the weathered zone.

Hydrolysis The reaction between mineral elements and the hydrogen ion of dissociated water is called hydrolysis. Chemically, hydrolysis involves a reaction between a salt and water to produce an acid and a base; it is probably the most important mechanism in breaking apart structures of the silicate minerals. During the process, metallic cations are separated from the mineral structure and replaced by H^+, which is held in the original aluminosilicate complex (see fig. 3.2). As discussed before, most of the replaced cations are soluble in natural waters, and, therefore, the processes of hydrolysis and solution are linked.

In addition to freeing cations, hydrolysis usually produces H_4SiO_4, HCO_3^-, and OH^-, all of which can be considered to be in solution. In the common case where water and acid attack an aluminosilicate mineral, the reaction leaves the H^+ embraced in segments of the original structure, and these are subsequently recombined into clay minerals. The hydrolysis of orthoclase feldspar shown here and in figure 3.2 demonstrates this response:

$$\text{(orthoclase)} \qquad \text{(kaolinite)}$$

$$2KAlSi_3O_8 + 2H^+ + 9H_2O \rightarrow H_4Al_2Si_2O_9 + 4H_4SiO_4 + 2K^+$$

The removal of the metallic cation will proceed as long as free hydrogen ions are available, easily replaced cations are present, and the solvent has not reached saturation with respect to the ion being liberated. The continuous introduction of fresh water and the development of organic acids will assure a ready source of H^+. Since OH^- is carried downward to the groundwater table by percolating water, it seems unlikely that highly alkaline water can ever be produced in open systems with constantly moving water. Therefore, the normal product of a continuously leached system is a residue in which all mobile cations have been freed from the original mineral. After the cations are released, they either remain in solution or attach to the surfaces of minerals held in colloidal suspension. Eventually they are carried to the water table and delivered to the regional streams.

Clearly, the effect of hydrolysis decreases as clays depleted of cations become the dominant aluminosilicate in the weathering zone. In fact, as more clays form, they commonly become colloidally suspended in the fluid and may adsorb H^+ to their surfaces. This may cause the liquid to become more acidic rather than more alkaline as hydrolysis proceeds.

Ion Exchange Ion exchange is the substitution of ions in solution for those held by mineral grains. Although all minerals possess some capability for ion exchange, the process is most effective in clay minerals. The ions to be exchanged are held on the surfaces of clays because unsatisfied charges, exposed hydroxyl groups, and isomorphic substitutions such as Al^{+3} for Si^{+4} have given the clays an overall negative charge. Cations are held at the mineral surfaces by adsorption. In soils, *adsorption* usually refers to the attraction of ions and water molecules to the surfaces of colloidal particles in an attempt to neutralize their negative charges. The adsorbed ions are not held too tightly and therefore are susceptible to replacement by or exchange with other cations.

Each clay species has a different propensity for adsorbing cations, called its *cation exchange capacity* (c.e.c.), which is expressed as the number of milliequivalents per 100 grams of clay. When the adsorbed ion is hydrogen, the c.e.c. directly influences the pH value that the clay assumes. Kaolinite, for example, takes on a pH of 4 to 5 under complete adsorption, while H^+-montmorillonite attains pH value as low as 3. Colloidal suspensions of these clays create acids that are capable of attacking other minerals. In soils, colloids of organic compounds produce similar acids because the c.e.c. of organic matter is usually rather high, ranging from 150 to 500 (Birkeland 1974).

It should also be noted that cations other than H^+ may be adsorbed by clays and that the prevailing environment controls which cation types are more likely to be adsorbed. For example, in humid regions colloidal clays will adsorb H^+ and Ca^{++} more readily than Mg^+, Na^+, or K^+. In well-drained arid soils, Ca^{++} and Mg^{++} are usually the most prominent exchangeable ions, and H^+ is the least common. In poorly drained arid soils, the number of adsorbed sodium ions often equals or exceeds the calcium (Buckman and Brady 1960).

Ion exchange is governed by the composition and pH of the interstitial water as well as the type of ion in the exchangeable position. In general, strongly acidic waters allow H^+ to replace metal cations of the parent minerals, but this tendency changes as the water becomes neutral. At higher pH values, the mineral cations may remain in the exchangeable position or, in fact, H^+ may be replaced by metallic cations. In soils, the pH of the soil-water mixture is an indicator of the number of cations held in the exchange position by clays. This characteristic is commonly referred to as a *percentage of base saturation,* meaning the percentage of exchange sites occupied by cations other than hydrogen. The higher the percentage of base saturation, the higher the pH of the soil-water complex because less H^+ is held by the clays.

Mobility

The extent to which chemical weathering will alter the parent mineralogy depends largely on the relative mobilities of the constituent ions. Some ions are easily removed (high mobility) from the weathering system under normal groundwater conditions, while others are relatively difficult to remove (high immobility). The presence of highly mobile ions in a mature weathering zone indicates that some factor is impeding the transfer of the ion from the system. Orthoclase feldspar, for instance, will hydrolyze to kaolinite (as shown earlier) if all the potassium is lost in the process. If some potassium is retained, the clay product will be illite rather than kaolinite, as shown in the following reaction:

(orthoclase)   (illite)

$$3KAlSi_3O_8 + 2H^+ + 12H_2O \rightarrow KAl_3Si_3O_{10}(OH)_2$$
$$+ 6H_4SiO_4 + 2K^+$$

Accordingly, then, since K^+ is a mobile ion, its immobile behavior in the formation and preservation of illite must be linked to the prevailing character of the fluid or to an incomplete breakdown of the orthoclase that traps the K^+ in all molecules of the original structure. In either case, the easily removed potassium is rendered immobile and remains in the weathered zone.

The relative mobilities of common cations are as follows, in order of decreasing mobility:

$$(Ca^{+2}, Mg^{+2}, Na^+) > K^+ > Fe^{+2} > Si^{+4} > Ti^{+4}$$
$$> Fe^{+3} > Al^{+3}$$

The exact order of very mobile ions (Ca^{+2}, Mg^{+2}, Na^+, K^+), however, seems to be at least partly dependent on the type of mineral involved (Siegel and Pfannkuch 1984).

The mobility distribution is probably related to a parameter known as the *ionic potential,* which is expressed as the ratio of the valence (Z) to the ionic radius (r). In general, Z/r is a useful first approximation of mobility because very mobile ions have Z/r values less than 3; those forming immobile precipitates have values between 3 and 9.5; and those forming soluble, complex anions are usually greater than 9.5 (fig. 3.6). However, as each cation can be immobilized by external factors, those factors in a sense greatly influence how effectively the major processes of chemical weathering will work. For example, Shoji and his colleagues (1981) found a different mobility sequence than the normal one just shown. They concluded that volcanic glass parent rock and the secondary minerals derived from its weathering exert a strong control on the ion mobilities. Nonetheless, if all processes do function without hindrance, the mobile ions will be depleted from the system and the immobile ions will be progressively concentrated. The major external factors that control mobility include leaching, pH, Eh, fixation and retardation, and chelation.

Leaching   The most important factor influencing ion mobility is the amount of water leaching through the

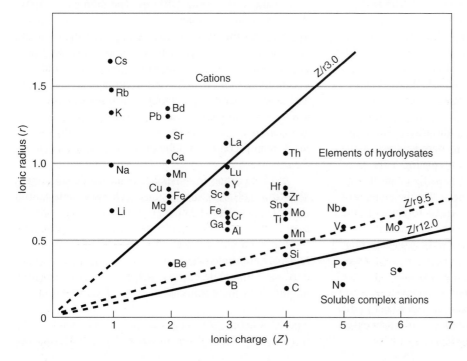

Figure 3.6
Grouping of some common elements according to their ionic radius (*r*), ionic charge (*Z*), and ionic potential (*Zr*).

weathering zone. How frequently rainwater moves through the system, and in what quantity, depends mostly on climate but also on the permeability of the material. The significance of leaching is severalfold. First, it removes in solution the constituents that have been separated from minerals by hydrolysis and ion exchange and, by providing new hydrogen, allows these processes continuously to alter the original material toward an ultimate degraded condition. Second, it directly affects the pH of fluids surrounding minerals and thereby helps determine which elements remain in solution. Third, it provides the mechanism by which dissolved ions and clays are transferred from higher levels to lower levels in the weathering zone; this transfer introduces new components to each level and so may engender the chemical environment needed to precipitate new minerals.

In general, the absence of continual leaching makes the weathering zone function like a closed system. Because ions removed from the parent minerals are held in the surrounding fluid, the exchange processes will continue only until an equilibrium condition is established between the ions in the fluid and those in the mineral. For all practical purposes, further chemical weathering is impossible beyond this stage, and mobile ions and easily decomposed minerals will remain in the system. It seems clear that the retardation of leaching due to insufficient rainfall is a prime factor in causing immobility and explains how very mobile ions are abundant in arid-climate soils. Conversely, in humid regions where leaching is continuous, the same mobile ions are removed completely from the residual soils. Thus, it seems reasonable that the end result of weathering should reflect a climatic control. However, this perception has been rigorously challenged, especially for weathering in arid climates (Pope et al. 1995). In addition, the climatic effect on the *rate* at which soils reach this ultimate condition is even less certain (White and Blum 1995), and its consideration requires an understanding of how long-term versus short-term weathering rates are calculated. We will visit this question later in the chapter.

pH   In addition to its previously discussed influences, H^+ concentration (pH) also plays an important role in the solubility of most common elements (see fig. 3.5). Stated another way, the mobility of ions is partly controlled by pH. In fact, some researchers believe that the amount of acids introduced by rainwater-derived solutions is the primary factor determining the extent of decomposition (Singer 1980; Nesbitt and Young 1984). Although rainwater reaching the surface is slightly acidic, processes operating beneath the surface can considerably alter the original pH value. Thus, pH is not an independent variable in weathering but may, in fact, be dependent on the type and degree of inorganic and organic processes within the weathering zone (e.g., Viers et al. 1997).

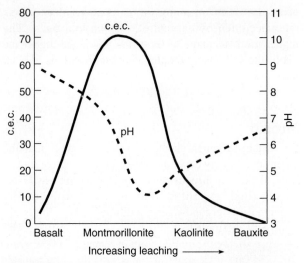

Figure 3.7
Relationship of cation exchange capacity (c.e.c.) and pH with continuous weathering of a basalt. Transition from basalt to bauxite represents gradual change in mineralogy with increased leaching.
(Loughnan 1969)

In geochemical weathering the pH is influenced by leaching, by the c.e.c. of residual minerals, and by the composition and structure of the parent minerals. The effect of leaching and cation exchange capacity can be shown diagrammatically by considering the progressive weathering of a basalt (fig. 3.7). With continued leaching, the pH is related almost inversely to the type of clay produced during various stages of the weathering. Assuming complete mobility of Ca, Mg, and Na, the first clay developed is an H^+-montmorillonite with a high c.e.c. value (70–100 meq/100 g). The high c.e.c. allows the clay to adsorb many of the hydrogen ions made available by the continuous leaching, and the colloidal acid formed causes a pronounced drop in pH. As leaching continues, however, montmorillonite is degraded to kaolinite, which becomes the dominant clay mineral. The much lower c.e.c. of kaolinite permits less hydrogen adsorption, and the pH rises. The final product, bauxite, can adsorb virtually no H^+, and the pH climbs to an almost neutral condition. Obviously, pH is dependent on the other factors.

It is important to remember that organic substances also form colloidal suspensions with very high cation-exchange capacities. Although the values vary with climate and the type of organic matter, it is not uncommon for these humic acids to lower the pH below 4. All feldspars are subject to dissolution by organic acids, and even very immobile cations such as Al^{+3} can be readily mobilized by these acids (Manley and Evans 1986).

The Eh Factor   As suggested earlier, the redox potential exerts a control on ion mobility. Eh affects ion mobility because oxidation-reduction reactions are reversible. When soils become waterlogged, free oxygen is excluded and strongly reducing anaerobic bacterial action begins. This generates lowered Eh values that, in turn, cause some elements to revert to their reduced forms. In some cases, ease of reduction is also dependent on pH (Connell and Patrick 1968). Such transitions have a direct influence on the mobility of certain elements that may be relatively insoluble in one form and easily dissolved in the other. Removal of iron oxides attached to clays, for example, can result from only a minor lowering of Eh, which transforms ferric iron to the more soluble ferrous type. Apparently, a fluctuating water table can have important ramifications on the redox potential (Eh) and with it the relative mobility of important ions.

The significance of Eh in decomposition and ion mobility can be briefly stated as follows:

1. Ferrous iron (Fe^{+2}) commonly binds silica tetrahedra in the structures of silicate minerals. The oxidation of the iron to the ferric state requires impossible internal adjustments, and the lattice structure is destroyed (Carroll 1970).
2. The by-products of the oxidation process may facilitate the decomposition of other, more stable minerals. An example is the oxidation of pyrite produces sulfuric acid:

$$4FeS_2 + 14H_2O + 15O_2 \leftrightharpoons 4Fe(OH)_3 + 8H_2SO_4$$

This acid lowers pH and will react with any nearby mineral susceptible to attack by acid. Groundwater in regions near zones of sulfide mineralization is usually very acidic.

3. The solubility or insolubility of some elements in groundwater having a normal pH is directly controlled by whether the elements exist in oxidized or reduced form. This is especially true for iron and titanium.

Fixation and Retardation   Some cations, especially potassium, have a tendency to be retained or fixed in the weathered zone, so that their mobility is considerably lower than might be expected. The fixation of potassium may account for its low concentration in seawater and the widespread distribution of illite in sedimentary rocks. The precise mechanism involved is poorly understood, but it is probably related to the fact that potassium silicates (muscovite, K-feldspars) are more resistant to chemical weathering than are other minerals with similar lattice structures. Potassium is most readily fixed in clay minerals with expandable lattices such as chlorite, illite, micas, vermiculite, and montmorillonite. For example, studies of deeply weathered, subtropical soils (Gardner and Walsh 1996) show that Ca and Na were almost totally lost from the system, but the concentration of K

and Mg experienced almost no change from the base to the top of the soils. This clearly indicates that potassium and magnesium are preferentially retained (fixed) in the structures of clay minerals (illite, chloite, hydrobiotite, etc.) while calcium and sodium are readily leached. Why this should be so, and why potassium is so much more susceptible to this phenomenon than other ions, is not clear, but Wear and White (1951) suggest that the unique size of the K^+ ion provides the most stable structure when combined with oxygen in layered silicates.

Chelation   The process of chelation represents one of the most dramatic effects on the mobility of ions. Lehman (1963) defines chelation as "the equilibrium reaction between a metal ion and a complexing agent, characterized by the formation of more than one bond between the metal and a molecule of the complexing agent and resulting in the formation of a ring structure incorporating the metal ion." This means that metallic ions that are extremely immobile under normal conditions can be mobilized by reacting with complexing agents and be vertically transported as part of the compound.

Most complexing agents involved in chelation are organic compounds nurtured in soils by alteration of humus into a plant acid called fulvic acid (Wright and Schnitzer 1963), although other organic processes also produce chelating complexes. Lichens, for example, secrete such materials (Schatz 1963). The complexing agent ethylenediaminetetraacetate (EDTA) is by far the most completely understood chelator, and its structure (fig. 3.8) shows how the metallic ion is bonded and held.

Although the chemistry of the chelating process is far from clear, its importance in weathering has been recognized for decades (Schatz et al. 1954). During this time several excellent studies have demonstrated that EDTA and other chelators may play a dominant role in mobilizing iron and aluminum under pH conditions that are not conducive to Fe and Al solution (Atkinson and

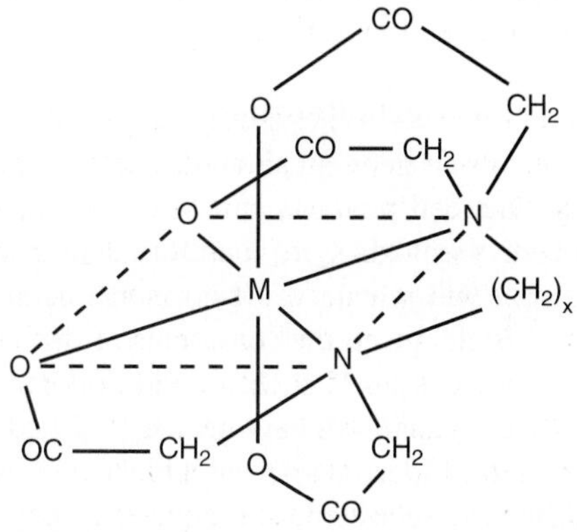

Figure 3.8
Structure of complexing agent EDTA. Metallic ion (M) is bonded within structure.
(Source: *Soil Science Society of America Proceedings* 27:169, 1963.)

Wright 1957; Wright and Schnitzer 1963; Schalscha et al. 1967). Using several different chelating agents Schalscha and his co-workers (1967) were able to extract iron from a variety of minerals. Significantly, their study revealed no correlation between pH and the amount of liberated iron. EDTA released more iron from magnetite and hematite, for example, than did hydrochloric acid, even though the HCl solution was more acidic. It is also known that quartz and aluminosilicates can rapidly dissolve even under neutral pH conditions where the environment is anoxic and organic-rich, such as in peat bogs (Bennett et al. 1991). The increased solubility is caused by the action of organic-acid-silica complexes (Bennett and Siegel 1987; Bennett et al. 1988), which are stable at neutral pH conditions.

In soils, chelating agents become soluble in water as they are oxidized (Wright and Schnitzer 1963). Iron and aluminum are locked in the ring structure and carried downward with the percolating water. The downward movement continues until the entire mass is flocculated because of small changes in ionic content of the soil water or until the complex is broken by microbial action. In either case further downward movement is curtailed, and the iron and aluminum are redeposited.

Even before soils begin to develop, chelation can be important in the alteration of the bare rock. Jackson and Keller (1970) showed that the presence of the lichen *Stereocaulon vulcani* accelerated the chemical weathering of volcanic rocks in Hawaii. Rocks covered by lichen growth had a thicker weathering crust than lichen-free rocks, and the crust was enriched in iron and depleted in titanium, silicon, and calcium. It should be noted, however, that the weathering rate of Hawaiian basalts is much higher beneath vascular plants than lichens. In fact, even though dissolution does occur under a lichen cover, it is significantly enhanced by the presence of vascular plants, and in contrast to areas of surface-covering lichens, even proceeds rapidly along vesicles, joints, and fractures when they are invaded by plant roots (Cochran and Berner 1996).

The Degree of Decomposition

To utilize soils in geological studies, it is important to know what the end products will be in a completely altered system. Geologists are therefore concerned about what material will remain when no more chemical reactions are possible, given the constraints of climate, vegetation, and rock type. If such a condition can ever be reached, a steady state will have been established between the driving forces (decomposition) and the resisting materials. Although the weathering zone may get progressively thicker, the upper parts of the profile are degraded to their ultimate form. In thoroughly leached soils, all minerals remaining presumably will be stable and all mobile ions will be gone. Thus, if we have some idea as to

what final products are expectable in any climate, we can estimate the degree of weathering (how far the material has progressed toward the steady state) by observing the mineral and chemical composition of the soil.

Mineral Stability   In considering the degree of weathering, a logical first question to ask is what minerals will be most rapidly destroyed in a thoroughly leached open system and, conversely, which types will remain if a steady state is attained. Perhaps the first analytical answer to this question was provided by Goldich in 1938. In a detailed study of the weathering of several varieties of igneous and metamorphic rocks, he suggested that the mineral stability of the rock-forming silicates is directly related to their order of crystallization. Quartz, being the last to crystallize, forms under the lowest temperature conditions and therefore should be most stable in the surface environment. High-temperature minerals such as olivine and pyroxene are least stable and weather most rapidly. With the exception of muscovite and the plagioclase feldspars, the Goldich stability series (table 3.1) reflects the number of oxygen atoms shared with adjacent silica tetrahedra in the structure. Quartz and orthoclase, for example, have four shared oxygens in their structure, making their internal bonding strength much higher than that of a mineral like olivine, which shares no oxygens. Although the Goldich series is not invariable, and certain minerals will shift position according to microenvironments found in specific study areas, it still remains a viable guide to the realative stability of silicate minerals.

Since Goldich's work, other guides to mineral stability have been developed, and studies using the various techniques have led to the significant conclusion that the extent of chemical weathering can generally be estimated from the mineral assemblage contained in a soil as compared to the minerals of the parent material. Furthermore, this conclusion provides a basis for the determination of relative ages of different soils, which is a major concern of most Quaternary geomorphologists.

In detail, the correlation between the different methods used to determine mineral stability is far from perfect, and the factors that control mineral stability are not clearly defined. For example, the relationship between

TABLE 3.1	Weathering Stability of the Common Silicate Minerals.
Olivine, Anorthite	Least Stable
Pyroxenes, Ca-Na Plagioclase	
Amphiboles, Na-Ca Plagioclase	
Biotite, Albite	
K-Feldspars	
Muscovite	
Quartz	Most Stable

[handwritten annotations: "most soluble" at top right; "least soluble" at bottom right]

After Goldich 1938.

stability and silicate structure recognized by Goldich is not as strong as we often think. Plagioclase feldspars show considerable variation in stability depending on the mineral variety, even though they all possess the same framework structure; micas show the same type of discrepancy. Zircon appears to be an extremely persistent mineral under weathering, yet it is structurally the same as olivine, a notably unstable mineral species. Despite these apparent difficulties, the use of mineral stability estimates is still an acceptable technique for gaging degree of decomposition, but the analyses must be made with care.

Secondary Minerals The minerals just discussed are components of the original rock, commonly referred to as *primary minerals.* The processes of chemical weathering, however, play a dual role. In addition to destroying primary minerals, they also create new minerals by recombining or reprecipitating materials liberated from the parent rocks. These *secondary minerals,* born within the weathered zone, are distinctly more stable than their primary ancestors because they reach equilibrium in the temperature-pressure environment of the soil rather than in the magmatic or metamorphic conditions that created the original crystals. In fact, dissolution rates of the primary minerals of the Goldich series and secondary minerals (table 3.2) demonstrate this stability because dissolution of secondary minerals such as kaolinite and gibbsite is comparable or slower than some of the most important primary silicates (e.g., feldspars). The most common secondary products of weathering are clay minerals and amorphous hydrous oxides of iron, aluminum, silica, and titanium.

TABLE 3.2	Mean Lifetime of a 1 mm Crystal at 25° C and Ph-5.

Mineral	Log Rate (mol/m²/s)	Lifetime (y)
Quartz	−13.39	34,000,000
Kaolinite	−13.28	6,000,000
Muscovite	−13.07	2,600,000
Epidote	−12.61	923,000
Microcline	−12.50	921,000
Prehnite	−12.41	579,000
Albite	−12.26	575,000
Sanidine	−12.00	291,000
Gibbsite	−11.45	276,000
Enstatite	−10.00	10,100
Diopside	−10.15	6,800
Forsterite	−9.50	2,300
Nepheline	−8.55	211
Anorthite	−8.55	112
Wollastonite	−8.00	79

Redrawn from Chemical Weathering and Global Chemical Cycles, by Lasaga et al. *Geochimica et Cosmochimica Acta* 58 Table 1, p. 2362 Copyright © 1994.

Clay Minerals The advent of X-ray diffraction and electron microscopy provided scientists with the capability to examine the actual lattice structures of clay minerals. As a result, clay mineralogy has developed into a separate scientific discipline, the details of which are well beyond the purposes of this book. Nonetheless, clay minerals are important indicators of the degree and character of weathering, and they are significant components in the physical and chemical attributes of soil. As such, they deserve brief mention.

Clay minerals are aluminum silicates in which silica tetrahedra and aluminum octahedra are bonded together in a layered atomic structure. A silica tetrahedron has one silicon atom surrounded by four oxygen atoms; an aluminum octahedron consists of a single aluminum atom bonded to six oxygen atoms. In clays the individual tetrahedrons and octahedrons are linked together in planes, forming distinct sheets or layers typified by either a tetrahedral structure (silica sheet) or an octahedral structure (alumina sheet).

Most clays are either 1:1 layer silicate or a 2:1 layer silicate (table 3.3). A 1:1 structure has as its fundamental building block one layer of silica tetrahedra and one layer of aluminum octahedra. In 2:1 mineral structures, one aluminum octrahedral layer is positioned between two layers of silica tetrahedra. Other clay varieties do form, but they either represent combinations of the basic types (mixed-layer) or originate under very special weathering conditions (chain silicates).

The mineral *kaolinite* represents the 1:1 clay mineral group that has the greatest importance in soils. As shown in table 3.3, kaolinite is constructed of crystal units consisting of one tetrahedral sheet (silica sheet) and one octahedral sheet (alumina sheet). These two layers are held together by ions that are mutually shared by aluminum and silicon atoms in the separate sheets. Importantly, the bonding strength between adjacent, two-sheet crystal units is also very strong. The bonding strength between the units is significant because it keeps the lattice spacing fixed, thereby preventing expansion of the structure when the clay is wetted; as a result, cations and water do not penetrate between the kaolinite crystal units. This factor explains the relatively low cation exchange capacity of kaolinite because the ion exchange process is restricted to the external surfaces of the mineral. The fixed structure also accounts for the low plasticity (the capacity to be molded) and swelling characteristics of the mineral.

Of the minerals possessing a 2:1 structure, the smectite group and the mica group have the greatest significance in weathering and soil development. The *smectite* group is characterized by ion substitutions that occur primarily in the octahedral layer. This leads to wide variations in chemical compositions of the group minerals. *Montmorillonite,* the most common type of the smectite group, has the lattice structure shown in

TABLE 3.3	Classification of the Clay Minerals.		

		Structure	Common minerals
		1:1	Kaolinite
			Dickite
			Nacrite
			Halloysite ($4H_2O$)
			Allophane

Lattice structure of kaolinite

		2:1	*Mica group*
			Muscovite, 2M
			Illite
			Glauconite
			Biotite
			Vermiculite
			Smectite group (Expanding)
			Montmorillonite
			Many other types
			Chlorite group　2:1:1
			Chlorite
			Chamosite

Lattice structure of montmorillonite

Redrawn from *The Nature and Properties of Soils*, 6th ed. by Harry O. Buckman and Nyle C. Brady. Copyright © 1960 by Macmillan Publishing Company.

table 3.3. The obvious difference from kaolinite is that montmorillonite has a three-sheet crystal unit. Like kaolinite, the three layers comprising the unit (two tetrahedral sheets and one octahedral sheet) are bound tightly together by shared atoms. In contrast to kaolinite, however, the bonding between adjacent crystal units is notably weak, and the mineral lattice is capable of expansion upon wetting. Therefore, cations and water molecules easily penetrate the mineral interior, where cation exchange takes place along the surfaces of the crystal units. The result is a clay with greater plasticity and swelling and a cation exchange capacity that is much higher than can be generated when only external exchange sites are available.

Illite is the most common mica-group mineral found in soils. Because its lattice is 2:1, illite is structurally similar to montmorillonite. It differs from montmorillonite, however, in that some of the silicon in the tetrahedral sheets has been replaced by aluminum. This condition creates an unbalanced charge in the silica sheets, a charge that is primarily satisfied by potassium ions placed between the crystal units. As a result, the bonding between the units is stronger in illite than in montmoril-

lonite. Therefore, the values of c.e.c., swelling, and plasticity in illite are lower than those found in smectite clays, but they still exceed those found in kaolinite.

The mineral *vermiculite,* also part of the mica group, is found in many soils. It differs from illite because it normally has little K^+ in the interlayer zone. Instead, that position is more likely to be occupied by Ca^{++} or Mg^{++}. These cations do not provide the strong bonding noted in illite and mica; therefore, vermiculite is prone to some expansion.

Other clays found in soils have variations on the structures we have been discussing. *Chlorite,* for example, has a crystal unit consisting of alternating 2:1 layers and octahedral layers. Therefore, it is often designated as a 2:1:1 clay. Bonding in chlorite is strong because ion substitutions create opposite charges in the layers, making the lattice nonexpanding. In addition, *mixed-layer* clays can form when different crystal units interstratify with one another or with hydroxides of magnesium or aluminum. As you might expect, the physical and chemical properties of mixed-layer clays are quite variable.

Chemical weathering of common silicate minerals produces only a few abundant groups of clay minerals;

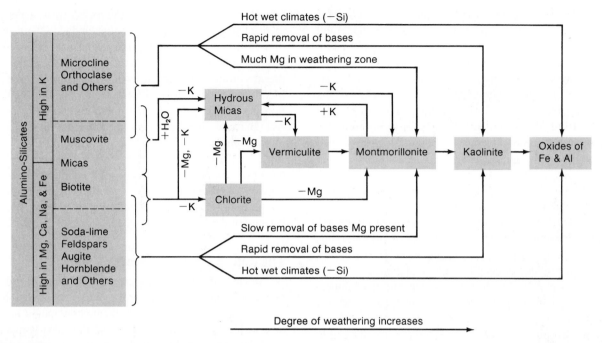

Figure 3.9
Diagram showing the general conditions for the formation of the various silicate clays and the oxides of iron and aluminum. In each case, genesis is accompanied by the removal of soluble elements such as K, Na, Ca, and Mg.

(From *Elements of the Nature and Property of Soils,* 12th Edition by Nyle C. Brady and Ray R. Weil. Copyright © 2000. Reprinted by permission of Prentice-Hall, Inc., Upper Saddle River, NJ.)

kaolinites, smectites, and micas such as illite (Nesbitt and Young 1989). However, differences in stability are present even within the realm of clay minerals. Jackson and his colleagues (1952) proposed a distinct sequence of clay mineral development in which muscovite or illite is the initial product of weathering and, assuming effective leaching, progressively degrades through montmorillonite to a final kaolinite clay (fig. 3.9).

The evolutionary origin of clay minerals should be applied with caution because some species have a variety of origins. Kaolinite, for example, is a very complex mineral. It may derive directly from feldspar without the development of an intervening secondary mineral (Keller 1978, 1982), and it seems certain that a phase of solution can be involved in the transformation. In fact, electron microprobe analyses of phyllosilicates (Murphy et al. 1998) revealed grains having a biotite core surrounded by a fringe of kaolinite. This suggests that the two micas might be linked; the precipitation of kaolinite being preceded by a partial dissolution of the host biotite. This transition is apparently accomplished without the formation of an intermediate chlorite or montmorillonite as suggested in figure 3.9. Thus, the origin of kaolinite may involve many precursors and processes, and does not necessarily follow the sequential pathways that have been suggested (e.g., Jeong 1998a,b). Nonetheless, assuming that a sequential development of clay types is possible, the type and amount of clay

should provide a reasonable basis for estimating the degree of weathering in a soil. It should also be apparent that specific clay minerals will tend to be dominant in mature soils developed under a particular set of climate conditions. For example, halloysite seems to form best in a stagnant moisture regime where there is a depositional overburden acting as a source of silica (Violante and Wilson 1983).

The apparent sequential development of clays emphasizes the important point that clay minerals themselves are highly reactive and readily alter from one form to another. In most weathering situations, the composition of the initial clay developed is usually controlled by the ions available in the parent rock, and many of these ions may be highly mobile. With time, however, climate and other factors operating with the weathering zone exert a greater influence, and clays will alter to forms that are more nearly in equilibrium with the chemical environment. Therefore, the total assemblage of clay minerals in soils developed under a particular climate should indicate how far the processes of chemical weathering have proceeded toward this equilibrium state.

A note of caution is perhaps needed at this point. In some situations all clay minerals in a soil are not necessarily derived directly from weathering of the parent materials. Colman (1982) examined the clay-sized material in the weathered outer fringe (called a *rind*) of basalt and andesite boulders contained in clay-rich soil horizons at least 100,000 years old. The material in the rinds is predominantly disorganized or amorphous allophane and iron oxides or hydroxides. However, in the surrounding soil itself the clay fraction consists of highly crystalline, well-developed clay minerals. Because of the rind composition, it appears likely that the clay minerals in the soil horizon are not the result of simple transitions from weathering of the boulders. Instead, the clay minerals were probably introduced to the soil as fully developed crystalline forms by some external process.

Hydrous Oxides In addition to clay minerals, a number of oxide and hydroxide compounds are formed in the weathering zone as important secondary minerals. The oxides are normally sesquioxides of iron and aluminum occurring mainly as *hematite,* Fe_2O_3, or its hydrated form *limonite,* $Fe_2O_3 \cdot H_2O$, and *gibbsite,* $Al_2O_3 \cdot 3H_2O$ [or preferably $Al(OH)_3$]. In general, these minerals are crystalline, but when a poorly ordered ion arrangement exists in the structure, they are considered to be amorphous. The sesquioxides are usually stable end products of weathering unless the chemical environment becomes strongly acidic.

Estimates Based on Chemical Analyses The degree of chemical weathering has also been gaged on the basis of total chemical analyses. The most common approach utilizing such data is a direct comparison of the fresh parent rock with the saprolite or soil derived in situ from it. To make an analytical comparison of the two, Al_2O_3 is usually considered to be constant since its mobility is extremely low under normal conditions. The weight percent values of oxides in the soil are recalculated by using the conversion factor $\%Al_2O_3$ (fresh rock)/$\%Al_2O_3$ (weathered material) as a multiplier of each constituent. Table 3.4 demonstrates this method by showing a hypothetical

comparison of a saprolite developed from a parent metagabbro. The recalculated values indicate the degree of weathering, and relative gains and losses from the original composition can be used to compare nearby exposures or different horizons within the same soil profile. Maximum errors in this method occur when the assumption of a completely constant aluminum value is incorrect. Because aluminum may be mobilized by chelators or humic acids, it is likely that some aluminum will be lost from the upper weathering zones and some gained in the lower horizons. Nonetheless, the method provides a good first approximation of the relative degree of weathering, even though absolute gains and losses are suspect because of the aluminum problem.

The Rate of Chemical Weathering

If estimates of the degree of chemical weathering tell us how far decomposition has proceeded, rates of chemical weathering differ in that they indicate how long it might take to reach the steady state. If certain environmental conditions are met, the parameters used to estimate the degree of weathering can also serve as the basis for calculating the rate of weathering. The most important requirement is a sample base consisting of a group of soils that differ in age but have the same parent material and have formed under similar climates and vegetation. Soil assemblages of this type, called chronosequences, have been used to quantify changes in weathering properties with age (e.g., Birkeland 1999; Harden 1987; Markewich et al. 1990; Merritts et al. 1992; White et al. 1996). Weathering rates are calculated from chronosequences by measuring the change that occurs in some weathering parameter during the time elapsed between soils having a different age. For example, figure 3.10 shows the change in weight percentage of key minerals in different soils of the Merced chronosequence of central California (Harden 1987; White et al. 1996). The relative wt % of plagioclase and hornblende decreases consistently with age and can be used to calculate a weathering rate. Other mineral species are more variable with time and may actually increase in wt % because they are concentrated by their resistance (e.g., quartz) or by development as secondary products (e.g., clay minerals). Results of studies using this approach provide reasonable estimates of long-term, average rates of weathering, but they are probably not constant during the entire episode of soil development, and they cannot be used to calculate the age of a soil in regions having different environmental settings.

Other approaches used to estimate rates of chemical weathering are based on short-term criteria and primarily involve the analyses of solute fluxes in laboratory and natural waters. In laboratory studies, specific mineral species are crushed and exposed to pristine water that serves as the solvent. Some studies will alter the

TABLE 3.4	Gains or Losses of Chemical Constituents in a Hypothetical Example of Weathering.			
	Original Rock	Saprolite	Adjusted %[a]	Loss or Gain[a]
SiO_2	50.3	41.3	24.78	−25.52
Al_2O_3[b]	18.3	30.6	18.3	0
Fe_2O_3	2.5	11.3	6.78	+4.28
FeO	9.4	1.2	0.72	−8.68
MgO	5.5	0	0	−5.5
CaO	12.6	0	0	−12.6
TiO_2	1.1	0.1	0.06	−1.04
H_2O	0.2	14.6	8.76	+8.56
Total	99.9	99.1	59.40	−40.5

[a]To obtain loss or gain, multiply each value in the saprolite column by 0.6. Subtract the adjusted percentage from the original percentage in the fresh rock.
[b]Al_2O_3 (fresh)/Al_2O_3 (saprolite) = 18.3/30.6 = 0.6.

composition of the solvent if their emphasis is to ascertain what effect a chemical change in the fluid would have on the dissolution process (e.g., variation in pH, see fig. 3.5). Ions removed from the mineral surface by processes of decomposition are contained in the fluids moving through the experimental system. The flux in solute concentrations, measured periodically, serves as the data base for calculations of ion loss per time, and therefore, the rate of dissolution. Dissolution rate constants have been determined in the laboratory for all major silicate minerals (for references to specific minerals see Lasaga et al. 1994; White and Brantley 1995).

The rate of chemical weathering can also be estimated in field studies where solute fluxes are periodically measured in the main stream leaving a particular catchment. The assumption here is that rocks and soils within the watershed are the only sources of dissolved constituents in river water leaving the basin. We know, however, that this is not true because rainwater is not pure, and a significant portion of the solutes measured in the river water can be introduced as part of precipitation (Cleaves et al. 1974). Even when the chemistry of precipitation is known and is included as part of systemic input, other factors may introduce serious errors into the calculation of a weathering rate. Drever and Clow (1995) provide an excellent discussion of the theoretical basis for catchment analyses and the problems associated with rate calculations.

The Complicating Factors   One of the major problems in catchment determinations of weathering rates identified by Drever and Clow (1995) is the selection of sampling interval. The usual interval (weekly sampling) may miss significant temporary events such as floods, and in times of little precipitation, may tend to overemphasize the solute concentrations of base flow. Because stream chemistry tends to vary with discharge, the sampling factor could introduce systematic errors in estimating the annual solute flux. This problem is exacerbated in regions where seasonal shifts in precipitation and temperature dominate the system. For example, solute production and discharge in alpine areas are controlled by the presence of snowpacks, which tend to hold dissolved constituents in place until they can be released in greater amounts during the melt season (Campbell et al. 1995; Horton et al. 1999). The entire problem of solute concentrations was revealed in a different manner by White and Blum (1995) who argue that there is no climatic control on the derivation of catchment solutes. This follows because the correlations between yearly variations in precipitation and solute fluxes within individual watersheds are stronger than the correlations of these factors between watersheds having different climatic regimes.

Another major complication of weathering rates is the application of rates derived in the laboratory. Although experimental dissolution rates can be presented as mean lifetimes of any minerals species (see table 3.2), it is doubtful that application of laboratory rates to long-term field analyses is valid. In fact, White et al. (1996) compared mineral abundances predicted from an experimental rate constant to the actual residual abundances found in the Merced soil chronosequence. They found that the experimental rate for albite predicted a residence time of 575 ky for that mineral, but greater than 50 percent of the original plagioclase in the same size fraction remains in the Turlock Lake soil after 600 ky of weathering. The same type of discrepancy was found in the consideration of K-feldspar. It seems clear that the prediction of long-term weathering trends based on short-term rate analyses is fraught with difficulties, even though as pointed out earlier some evidence exists suggesting that quartz dissolution rate might be applicable over a wide range of timescales. In general, long-term field rate for silicates are often one to three orders of magnitude slower than experimental rates. The reason for some of this disparity in rate values is found in the amazingly complex interactions of factors that control decomposition in natural settings. However, much of the

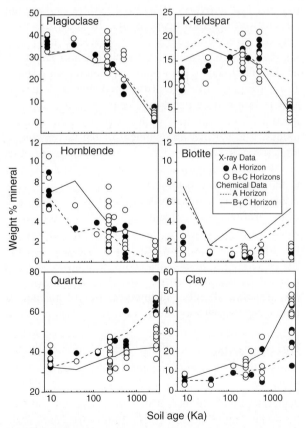

Figure 3.10
Change in weight % of minerals in Merced soil chronosequence with age.

(Figure 2, p. 2535 by White et al. from *Geochimica et Cosmochimica ACTA.* Copyright © 1996. Reprinted with permission from Elsevier Science.)

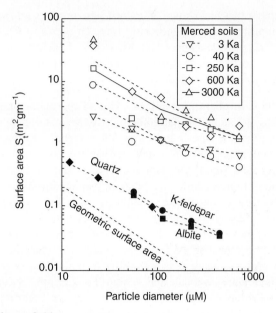

Figure 3.11

Relationship of surface area and particle size in Merced soil chronosequence. Geometric areas are calculated for smooth spheres of equivalent diameter (dashed line). Quartz, K-feldspar, and albite data are BET measurements of freshly crushed material used in dissolution experiments.

(Figure 6, p. 2539 by White et al from *Geochimica et Cosmochimica ACTA.* Copyright © 1996. Reprinted with permission from Elsevier Science.)

discrepancy seems to arise from the diverse methods used to estimate the surface areas of mineral grains under chemical attack (for detailed discussions see White 1995; White et al. 1996).

Total surface area of minerals exposed to fluids producing dissolution is a fundamental variant in the calculation of weathering rates. It is normally estimated in two ways. One approach employs the grain-size distribution in sample materials, combined with assumptions concerning the shape (e.g., sphere, cube) of specific mineral types. This approach provides an estimate known as the *geometric surface area*. The procedure, commonly used in field analyses of weathering, implies that the surface area is totally dependent on particle size. In contrast, experimental analyses of surface areas often include a procedure that measures the amount of gas (usually nitrogen or krypton) required to attach a layer of their molecules to a mineral surface. This procedure allows calculation of a *BET surface area,* so named in honor of the experimental chemists (Brunauer-Emmett-Teller) who originally devised the procedure (Brunauer et al. 1938). BET surface areas are always greater than geometric surface areas because they incorporate all irregularities that exist on the grain surface, including etch pits and microfractures (fig. 3.11). The ratio of BET areas to geometric areas is a factor known as surface roughness, which is a fundamental parameter used to calculate dissolution rate constants. Surface roughness

tends to increase with time because etch pits continue to develop, and grains begin to disintegrate along zones of internal weakness (Lee et al. 1998). Clearly, the time-dependent nature of roughness and surface area causes dissolution rates to be variable with time.

The use of weathering rates must be conditioned by some degree of caution. The use of experimental rate constants to predict absolute weathering rates of natural systems is probably futile without a consistent and accurate determination of mineral surface areas and a detailed understanding of processes that drive the weathering machine in natural settings. It should also be clear that weathering rates determined in any field situation, regardless of the timescale involved, cannot be applied to another region having a different climatic and geologic setting.

SOILS

Weathering processes that continue over an extended period of time result in a mass of soil that is measurably different from the original rock in its morphological, physical, chemical, and mineralogical properties. Soils also contain a significant amount of organic material that is added progressively to the system as decaying vegetation and microorganisms. In addition, pronounced layering develops in the weathered mass during the transition from being simply decomposing rock or alluvium to being a true soil. The soil layers, or horizons, are generally unconsolidated, but in some instances, the soil particles may become irreversibly cemented to one another with $CaCO_3$, silica, or iron oxides. The vertical arrangement of the layers constitutes a diagnostic property of soils known as the soil profile. The profile extends from the surface downward to the fresh parent material. The time in absolute years needed to form a soil profile, as well as the perfection of the profile's development, varies widely with the intensity of the weathering processes and the character of the original material. Nevertheless, where the soil profile is well developed, its character reflects the environment under which it formed and serves as the basis for classification of soils and interpretation of paleoenvironmental conditions.

Soil Properties

The horizons or layers within the soil profile can be defined using a variety of properties that distinguish the degree to which the parent materials have been altered by weathering processes. In addition to the characteristics discussed earlier in the chapter, such as pH, Eh, and c.e.c., the most important criteria are color, texture, structure, organic matter content, and moisture retention. *Color* in soils is often an indicator of the constituents within the horizon. For example, layers with high or-

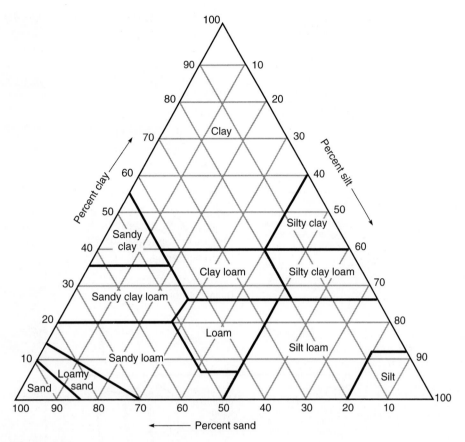

Figure 3.12
Percentages of clay (< 0.002 mm), silt (0.002–0.05 mm), and sand (0.05–2.0 mm) in basic soil textural classes as defined by the U.S. Department of Agriculture.
(Soil Survey Staff, 1951)

ganic content are typically black to dark brown, whereas the presence of ferric iron is indicated by yellow-brown to red colors. Light gray to white colors are associated with a concentration of SiO_2 or $CaCO_3$. It should be recognized, however, that small amounts of a pigmentor can cause rather intense discoloration, and so color alone may be a poor index of the total quantity of the pigmenting substance. *Texture* is simply the relative proportions of different particle sizes in a soil horizon, and is somewhat analogous to the property of sorting as used by geologists. However, it is based only on the particles that are less than 2 mm in diameter (fig. 3.12). *Structure* in soils is a unique characteristic in that it designates the shape developed when individual particles cluster together into aggregates called *peds* (fig. 3.13). In clay-rich soils, the openings between peds may play an extremely important geomorphic role by providing the primary avenues for downward percolation through an otherwise impermeable soil. *Organic matter* in soils consists mainly of dead leaves, branches, and the like, called *litter,* and the amorphous residue, called *humus,* that develops when litter is decomposed. Litter may form at mean annual temperatures as low as freezing,

but its optimum production occurs at about 25° to 30° C and decreases rapidly above those levels. Microorganisms that convert litter to humus begin to function at temperatures slightly above freezing (5° C), but the optimum temperature for their life activities may be as high as 40° C (fig. 3.14). It is significant that at temperatures between 0° and 25° C, humus is produced in abundance, but above 25° C little if any humus is accumulated. Humus has an important effect on soil formation because it includes chelators that promote the leaching of iron and aluminum, and increases water absorption. In addition, the development of humus releases CO_2 in high concentrations, leading to unusual amounts of carbonic acid within the humic zone and an associated lowering of the pH.

The total quantity of water that can be held in a soil is the *available water capacity* (AWC). By combining this parameter with the bulk density (dry weight of soil/unit volume), an estimate of the depth of wetting can be made (see Birkeland 1999). Such information is significant in that it relates to many soil properties, especially those affected by the redistribution of material during the downward percolation of water.

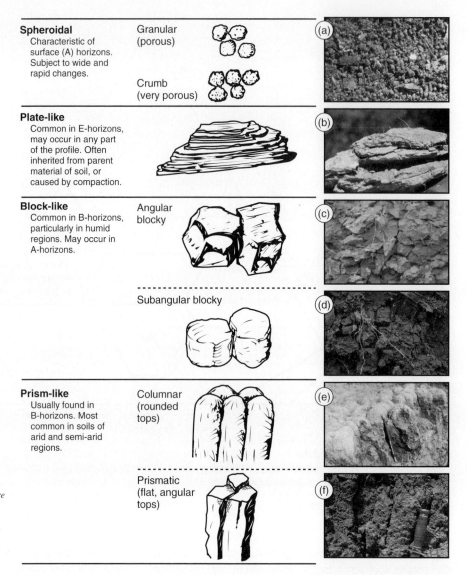

Spheroidal
Characteristic of surface (A) horizons. Subject to wide and rapid changes.
Granular (porous)
Crumb (very porous)
(a)
Plate-like
Common in E-horizons, may occur in any part of the profile. Often inherited from parent material of soil, or caused by compaction.
(b)
Block-like
Common in B-horizons, particularly in humid regions. May occur in A-horizons.
Angular blocky
(c)
Subangular blocky
(d)
Prism-like
Usually found in B-horizons. Most common in soils of arid and semi-arid regions.
Columnar (rounded tops)
(e)
Prismatic (flat, angular tops)
(f)

Figure 3.13
Major types of soil structure.
Figure 4.10, p. 105 from *Elements of the Nature and Property of Soils,* 12th Edition by Nyle C. Brady and Ray R. Weil. Copyright © 2000. Reprinted by permission of Prentice-Hall, Inc., Upper Saddle River, NJ.
(Photos: a–d, f: Courtesy of Raymond R. Weil e: Courtesy of James Arndt)

90
80
70
60
50
40
30
20
10
Quantity
A
B
0° 30°
5° 40°
10° 50°
15° 60°
20° 70°
25° 80°
30° 90°
35° 100°
40° 110°
45° C 120° F
A = organic raw material production in humid climate
B = destruction of organic matter in an aerated environment (aerobic)

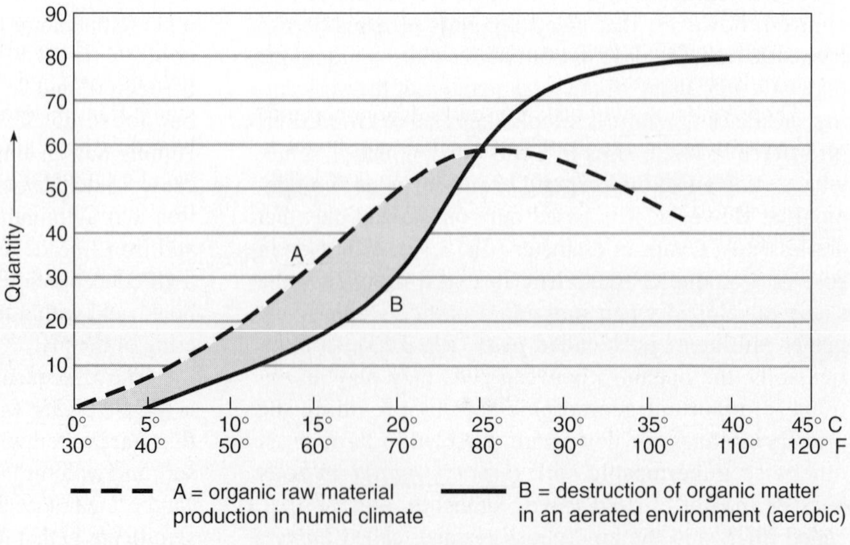

Figure 3.14
Production and destruction of organic matter in humid climates. The difference between the amount produced and the amount destroyed by microorganisms controls the accumulation of humus.
(Senstius 1958)

Available water capacity can be calculated if the moisture content at an upper limit (*field capacity*) and a lower limit (*permanent wilting point*) are known. Field capacity is determined by allowing a saturated sample to drain by gravity for at least 48 hours, by which time the remaining water content is held by adhesion to mineral and organic particles. After field capacity is reached, water can still be taken from the soil by plants until the tensional stresses holding the water in place become too great for the plants to break. At that point, the vegetation wilts. The water remaining in the soil is defined as the permanent wilting point. Both field capacity and permanent wilting point are expressed as a weight percentage according to the following equation:

$$P_w = \frac{W_s - W_d}{W_d} \times 100$$

where P_w is moisture percentage, W_s is total soil weight, and W_d is weight of soil after drying at 105° C. The available water capacity is simply the difference between the moisture content at field capacity and that at the permanent wilting point.

Soil Horizon Nomenclature and Description

Assuming that the vertical arrangement of the properties described above are distinct enough to identify a soil horizon, the pertinent consideration then becomes what nomenclature should be used to convey that information. In the United States, two systems of soil nomenclature are now in use. One is outlined in the *Soil Survey Manual* and is used in field descriptions of soil profiles (Soil Survey Division Staff 1993). The second is designed for the systematic classification of soils, and is based on the definition of diagnostic horizons, which, in many cases, can only be delineated following detailed laboratory analyses. The soil classification system used in the United States will be discussed in the next section. We will concentrate here on the nomenclature used to describe soils in the field.

Three kinds of symbols are used to denote horizons and layers in a soil profile. Capital letters, as shown in table 3.5, designate master horizons. Lowercase letters are used as suffixes to indicate specific characteristics of layers in the master horizon (table 3.6), and numbers are used as suffixes to connote vertical subdivision within a horizon or layer. In addition, numbers are prefixed to the master horizon designations to indicate a significant change in particle size or mineralogy within the soil. These signify a difference in the material from which the horizons have formed. In 1975 the Soil Conservation Service (S.C.S.) used Roman numerals as the prefix but have since changed to Arabic numerals (Soil Survey Division Staff 1993) (Note that the S.C.S. is now referred to as the National Resources Conservation Service). The number 1 is never used because it is implied to represent

Horizon[a]	Characteristics
O	Upper layers dominated by organic material above mineral soil horizons. Must have > 30% organic content if mineral fraction contains > 50% clay minerals, or > 20% organics if no clay minerals.
A	Mineral horizons formed at the surface or below an O horizon. Contains humic organic material mixed with mineral fraction. Properties may result from cultivation or other similar disturbances.
E	Mineral horizons in which main characteristic is loss of silicate clay, iron, or aluminum, leaving a concentration of sand and silt particles of resistant minerals.
B	Dominated by obliteration of original rock structure and by illuvial concentration of various materials including clay minerals, carbonates, sesquioxides of iron and aluminum. Often has distinct color and soil structure.
C	Horizons, excluding hard bedrock, that are less affected by pedogenesis and lack properties of O, A, E, B horizons. Material may be either like or unlike that from which the solum presumably formed.
R	Hard bedrock underlying a soil.

TABLE 3.5   Nomenclature of Soil Horizons.

Adapted from the Soil Survey Staff, 1960, 1975, 1981.

[a]Horizons can be divided into subhorizons by adding Arabic numbers.

the material in the surface zones. Therefore, if no changes occur downward into the profile, prefix numbers are not needed.

The master horizons are designated by the capital letters O, A, E, B, C, and R (table 3.5). Horizons at the ground surface are called either O or A depending on the nature and amount of organic constituents they contain. The O horizon is dominated by undecomposed or partially decomposed materials, such as leaves, needles, lichens, or fungi. Mineral fragments represent only a small fraction (generally less than 50 percent) of the horizon by weight. In contrast, the A horizon is dominated by mineral grains and is normally considered to be the thin, dark-colored surface layer where decomposed organic matter is concentrated and where clays and mobile components are continuously leached downward, or *eluviated.* The E horizon underlies the O or A horizon. It is characterized by intense leaching that removes Fe^{+3} or organic coatings from the mineral grains, a process that usually imparts a bleached gray color to the horizon. The C and R horizons exist at the base of the profile. The C horizon is usually thought of as the underlying, unconsolidated parent materials that have been unmodified, or only very slightly modified, by soil-forming processes. The R horizon is simply consolidated bedrock beneath the soil.

TABLE 3.6	**Some Common Descriptive Symbols to Be Used in Conjunction with Master Soil Horizons.**			
Symbol[a]	Meaning		Symbol[a]	Meaning
a	Highly decomposed organic matter		o	Residual accumulation of sesquioxides
b	Buried genetic horizon		p	Tillage or other disturbance
c	Concretions or nodules		q	Accumulation of silica
f	Frozen soil		r	Weathered or soft bedrock
g	Strong gleying		s	Illuvial accumulation of sesquioxides or organic matter
h	Illuvial accumulation of organic matter		t	Accumulation of silicate clay
i	Slightly decomposed organic matter		w	Development of color or structure
k[b]	Accumulations of carbonates		x	Fragipan character
m	Cementation or induration		y	Accumulation of gypsum
n	Accumulation of sodium		z	Accumulation of salts more soluble than gypsum

From Soil Survey Division Staff 1993, *Soil Survey Manual.*

[a]Symbol used with other profile designations (e.g., Ap, Bt, Btk).

[b]Formerly designated as ca.

The B horizon represents the transitional zone between the A and C horizons and historically has been considered as the *illuviated* zone: that is, a zone of accumulation and concentration of the material brought down from the A horizon. Some illuviation occurs because solid particles, mostly clays, are transported in suspension by waters percolating downward from the A horizon to the B horizon. This physical transfer of solids in the soil profile is known as translocation (see Birkeland 1999 for discussion). The nature of the B horizon is highly variable, and it may reflect any or all of the weathering processes discussed earlier. Nonetheless, it commonly has reddish hues, iron and aluminum concentrations as sesquioxides, stable primary minerals, and a high clay content due to illuviation or to in situ mineral growth.

The K horizon is no longer used by the Soil Survey Division Staff (1993). However, it has been retained by a number of pedologists and geomorphologists working in desert environments to designate a subsurface horizon characterized by extreme carbonate accumulation (see, for example, Birkeland 1999). The significance of $CaCO_3$ in soils of arid lands is that it commonly accumulates semisystematically within the profile, providing insights into soil age and the stability of geomorphic surfaces. The progressive buildup of $CaCO_3$ was initially classified by Gile and his coworkers into four stages on the basis of its morphology within the soil horizon (Gile et al. 1965, 1981). Machette (1985) added two additional classes beyond stage IV for a total of six morphologic stages of development (fig. 3.15). Horizons characterized by stages III–VI of Machette (1985) are generally designated as K horizons, whereas layers with stage I or II morphology are delineated as Bk horizons.

Transitional zones between the master horizons are indicated by the use of both capital letters. For example, a zone transitional between the A and B horizons may have characteristics typical of both. It is designated as AB or BA, the first letter indicating which master horizon it most clearly resembles. Alternatively, a transitional horizon may contain distinct parts, each characterized by the properties of a separate master horizon. In this case, the capital letters are separated by a virgule (/). The first letter represents the component that makes up the greatest volume of the transitional layer. For example, A/B may symbolize a layer in which materials with B horizon characteristics are surrounded by those with the properties of an A horizon.

Figure 3.16 represents a hypothetical soil profile using the nomenclature outline above. It should be clear that the nomenclature used for the field description of designated horizons provides a wealth of information about the soil's properties. For instance, to those familiar with the system, it is immediately known that the Btk horizon contains accumulations of both silicate clay and $CaCO_3$. An unfortunate trait of the S.C.S. system, however, is that it appears to be in a perpetual state of flux, making it difficult to interpret past field descriptions.

Soil Classification

Soils classification is intended to group and name soils that exhibit pronounced similarities. The trait or traits being classified serve as the philosophic basis for the groupings. The choice of diagnostic traits depends on what the classifier deems important, and because individuals disagree on that point, no classification will satisfy all members of the pedologic community. In fact, Birkeland (1999) points out that there is no worldwide agreement on which classification system to use and, in general, each country has developed its own set of nomenclature.

The first soil classification came from Russian soil scientists in the late nineteenth century, specifically from Dokuchaiev and his students. Although fraught

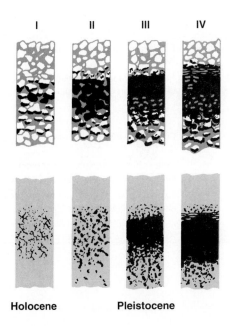

Stage	Gravelly sequence	Nongravelly sequence
I	Thin, partial or complete carbonate coatings.	Carbonate filaments and/or faint coatings on grains.
II	Carbonate coatings are thicker and there are some fillings in interstices.	Carbonate nodules separated by low-carbonate material.
III	Carbonate occurs essentially throughout the horizon, which is *plugged* in the last part of the stage.	
IV	Upper part of horizon nearly pure $CaCO_3$ exhibiting weakly platy structure; remainder of horizon plugged with carbonate.	
V	Laminar layering and strongly expressed platy structure; incipient brecciation and pisolith formation.	
VI	Brecciation and recementation of $CaCO_3$ layers are common.	

Figure 3.15
Schematic diagram of the stages of carbonate horizon formation in gravelly and nongravelly parent materials. Carbonate accumulations are indicated in black for clarity. Morphologic descriptions include stage additions by Machette (1985).
(Modified from Gile 1975)

with inconsistencies, Dokuchaiev's scheme had a great influence on American pedologists. His groups were defined in part by climate and vegetation, and until recently these genetic criteria were accepted as the basis of all soil classifications used in the United States. In fact, many Russian terms are still integrated in American soil nomenclature.

In the United States C. F. Marbut was the leading force in efforts to systematize soils. Marbut's classification, developed in the 1920s and 1930s, was based on characteristics that are present only in "mature" soils. The system, therefore, had no place for soils that were not fully developed. A more refined classification (Baldwin et al. 1938; Thorp and Smith 1949), designed to rectify many of the problems inherent in Marbut's system became the most extensively used soil classification in

the United States until a fundamentally new system was introduced. In 1960 the S.C.S. completely revised the descriptive nomenclature and the classification of soils (Soil Survey Staff 1960). This revision is not simply a formulation of new class names (although that occurred) but represents a fundamental departure from the philosophical basis of earlier classifications. Until the new system was devised, all classifications were essentially genetic in scope; that is, the major soil classes were based on climatic and vegetal factors. Although the subdivisions were linked to observable aspects of the profile, these were not explicitly defined, and soil scientists inescapably allowed their knowledge of climate and vegetation to influence decisions about placing a soil in a particular group. In many cases, pedologists were classifying the genetic factors and not the tangible resulting

genetic criteria/factors are climate and vegetation

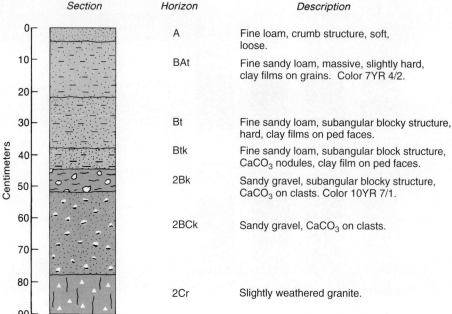

Figure 3.16
Hypothetical soil profile.
Source: Modified from Soil Survey Staff,
Soil Taxonomy, USDA Handbook 436,
Soil Conservation Service, 1975.

(Color from Munsell color charts) (see
Survey Division Staff 1993)

Horizon	Description
A	Fine loam, crumb structure, soft, loose.
BAt	Fine sandy loam, massive, slightly hard, clay films on grains. Color 7YR 4/2.
Bt	Fine sandy loam, subangular blocky structure, hard, clay films on ped faces.
Btk	Fine sandy loam, subangular block structure, $CaCO_3$ nodules, clay film on ped faces.
2Bk	Sandy gravel, subangular blocky structure, $CaCO_3$ on clasts. Color 10YR 7/1.
2BCk	Sandy gravel, $CaCO_3$ on clasts.
2Cr	Slightly weathered granite.

TABLE 3.7 | **Formative Elements in Names of Soil Orders in Soil Taxonomy.**

Name of Order	Formative Element in Name of Order	Derivation of Formative Element	Mnemonicon and Pronunciation of Formative Elements
Afisol	alf	—	pedalfer
Andisol	and	Modified from ando	ando; volcanic
Aridisol	id	L. *aridus,* dry	arid
Entisol	ent	—	recent
Gelisol	el	—	jell
Histosol	ist	G. *histos,* tissue	histology
Inceptisol	ept	L. *inceptum,* beginning	inception
Mollisol	oll	L. *mollis,* soft	mollify
Oxisol	Ox	F. *oxide,* oxide	oxide
Spondosol	od	Gk. *spodos,* wood ash	podzol; odd
Ultisol	ult	L. *ultimus,* last	ultimate
Vertisol	ert	L. *verto,* turn	invert

From Soil Survey Staff 1999, *Soil Taxonomy* p. 126.

---

soil. Although an understanding of some climatic parameters are required at certain steps in the classification of soils by the S.C.S. system, it is nongenetic; that is, it is based on very specific, often quantitative, soil properties. We will briefly describe the most recent version, referred to as Soil Taxonomy, without becoming mired in its details. For a more in-depth discussion, consult Soil Survey Staff (1999).

Soils in the S.C.S. system are grouped into 12 *orders* distinguished by the diagnostic horizons in their profiles (table 3.7; fig. 3.17). The orders are subdivided into *suborders,* which are defined by a physical or chemical soil property that in some cases requires quantitative laboratory data to be recognized (table 3.8). The terms used to designate a particular suborder represent the combination of two syllables; the prefix indicates the diagnostic property of the suborder and the suffix reveals the order, since it is composed of several key letters of the order name (tables 3.7, 3.8). An Argid, for instance, is an Aridisol with an argillic horizon, and an Aquent is a Entisol developed in wet (or saturated) sediments. Further subdivision into *great groups* and *subgroups* is made on the basis of even more detailed properties than those differentiating the suborders. For

nongenetic criteria/factors are specific, often quantitative soil properties

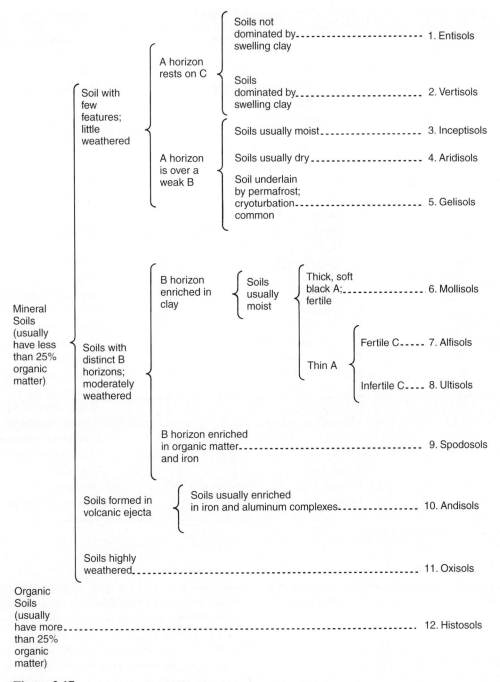

Figure 3.17
Simple classification system of mineral and organic soils of the world based on *Soil Taxonomy* by the S.C.S.

(Modified from Harpstead and Hole 1980)

example, within the Aquent suborder, a great group characterized by a low temperature is a Cryaquent (the prefix *cry-* indicates coldness). Subgroup nomenclature requires an additional descriptive word placed before the great group name. A Cryaquent with a layer enriched in pyroclastic debris becomes an Aquandic Cryaquent.

It is obvious even from a brief description that the S.C.S system allows for extremely detailed classification of soils. However, with 12 orders, 64 common suborders, and countless great groups and subgroups, the number of possible combinations of terms is enormous. It seems fair to say that the system requires more than a casual understanding of pedology for its successful use.

Pedogenic Controls and Regimes

In the previous section, it was shown that soils can be systematically grouped on the basis of profile similarities. In general, these result from a unique combination of certain factors that control the magnitude and types of soil-forming processes. The external factors of climate,

TABLE **3.8**	Formative Elements in Names of Suborders in Soil Taxonomy.

Formative Element	Derivation of Formative Element	Connotation of Formative Element
Alb	L. *albus*, white	Presence of an albic horizon
Anthr	Modified from G. Anthropos, human	Modified by humans
Aqu	L. *aqua*, water	Aquic conditions
Ar	L. *arare*, to plow	Mixed horizon
Arg	L. *argilla*, white clay	Presence of an argillic horizon
Calc	L. *calcis*, lime	Presence of a calcic horizon
Camb	L. *cambiare*, to exchange	Presence of a cambic horizon
Cry	Gr. *kryos*, icy cold	Cold
Dur	L. *durus*, hard	Presence of a duripan
Fibr	L. *fibra*, fiber	Least decomposed stage
Fluv	L. *fluvius*, river	Floodplain
Fol	L. *folia*, leaf	Mass of leaves
Gyps	L. *gyspum*, gypsum	Presence of a gypsic horizon
Hem	Gr. *hemi*, half	Intermediate stage of decomposition
Hist	Gr. *histos*, tissue	Presence of organic materials
Hum	L. *humus*, earth	Presence of organic matter
Orth	Gr. *orthos*, true	The common ones
Per	L. *per*, throughout in time	Perudic moisture regime
Psamm	Gr. *psammos*, sand	Sandy texture
Rend	Modified from Rendzina	High carbonate content
Sal	L. base of *sal*, salt	Presence of a salic horizon
Sapr	Gr. *saprose*, rotten	Most decomposed stage
Torr	L. *torridus*, hot and dry	Torric moisture regime
Turb	L. *turbidis*, disturbed	Presence of cryoturbation
Ud	L. *udus*, humid	Udic moisture regime
Ust	L. *ustus*, burnt	Ustic moisture regime
Vitr	L. *vitrum*, glass	Presence of glass
Xer	Gr. *xeros*, dry	Xeric moisture regime

From Soil Survey Staff 1999, *Soil Taxonomy* p. 127.

biota, topography, parent material, and time have long been recognized as the prime controls, and they are commonly expressed as an equation (Jenny 1941):

$$S \text{ or } s = f(cl, o, r, p, t \ldots)$$

where S = soil, s = any soil property, cl = climate, o = biota, r = topography, p = parent material, and t = time. The dots following the t represent external factors that have a significant local influence on soil development, such as the influx of eolian dust in many arid regions. Although the soil equation cannot be solved in quantitative terms, the relative effect of each factor has been determined, with some difficulty, by analyzing field sites where four factors are held essentially constant and one is allowed to vary (Jenny 1941).

The implication here is that soil-forming factors are independent variables, a perception that is probably invalid except for the factor of time. In fact, many researchers now suggest that time is the most important factor controlling differences in soils (Birkeland 1984b; McFadden and Weldon 1987). Other variables seem to be interdependent; climate directly controls vegetation and animal form, rock type (parent material) influences relief, and relief often affects climate. In addition, other difficulties arise because many soils are polygenetic; that is, they developed under more than one set of controlling factors. Some scientists would argue that all soils are polygenetic because processes and conditions that are involved in the development of any soil are never constant (Johnson et al. 1990). As a result soils achieve polygenetic complexity with time because of fluctuations in climate and vegetation and because the internal soil properties and conditions are continually changing. Johnson and his colleagues also suggest that the development of a soil profile does not always progress smoothly to an equilibrium state. Instead, a myriad of regressive external or internal factors may slow, alter, or reverse the expected soil evolution with time (Johnson and Watson-Stegner, 1987).

A good example supporting the above argument is found in the analysis of the Yarmouth-Sangamon Paleosol (YSP) in the midwestern United States (Woida and

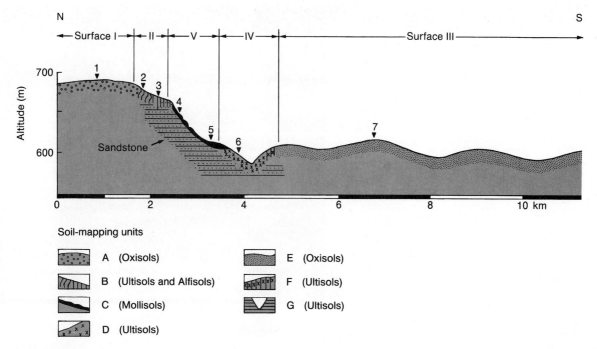

Figure 3.18
Soil catena showing relationship between slope and soil type.
(Lepsch et al. 1977)

Thompson 1993). Here the YSP is a composite profile of at least two soils that formed under paleoenvironmental conditions ranging from periglacial to intense seasonality. The expected differences in profiles stemming from development under vastly different conditions were masked by homogenizing factors such as shrink and swell processes and postburial diagenesis. These factors along with an erosional episode are clear examples of regressive processes and conditions. The resulting soil (the YSP) has an isotropic profile that does not reveal the complex polygenetic history involved in its development.

Parent material is usually considered to be rocks that are weathering in situ and unconsolidated sediment that was transported from its place of origin by various surficial processes and deposited at a different locality. It is generally conceded that parent material exerts its greatest control in the early stages of soil development or in very dry regions. However, the influence of parent material may be significant in mature soils. This is demonstrated best where textural and/or mineralogical variations of parent rock exist in the same climatic and topographic setting. For example, in the piedmont zone of North Carolina, soils developed from siliceous mica gneiss and granite gneiss are considerably different from those developed on diorites and gabbros. The greater content of ferromagnesian minerals in the mafic rocks led to formation of thick, clay-loam A horizons and reddish B horizons. These are notably absent in soils over-

lying the siliceous gneisses (see Buol et al. 1973 for discussion and references).

The relief factor essentially refers to the effect of local topography on soil development. Therefore, consideration of this factor is really an attempt to analyze soil changes brought about by the influence of slope and the position of the profile on a sloping surface. In this type of analysis, geomorphologists have made use of the concept of a soil catena. A **soil catena** consists of a group of soil profiles whose characteristics change gradually beneath a sloping surface. The profile changes result because variations in soil-forming factors are produced by differences in geomorphic processes acting on the slope materials, especially drainage of groundwater, transport of surface sediment, and removal of mobile chemical elements. The steepness of the slope is important because it tends to control the magnitude of sediment transport by wash, creep, and mass movements, and it also allows underground water to move more rapidly through the soil. As a result, soils in the steep upper-slope location tend to be freely drained, oxidized to red-brown colors, and coarse-grained. Lower-slope soils, where gradients are less steep, commonly have higher moisture contents because the material is poorly drained. They tend to be clay-rich and blue-gray to gray in color. Midslope profiles are often transitional in color and texture between the upper- and lower-slope characteristics (fig. 3.18). Therefore, a catena manifests the interaction of soil processes and slope processes. For this

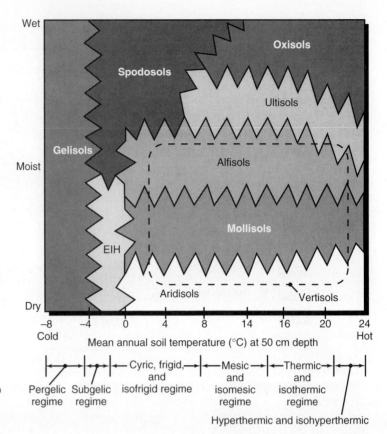

Figure 3.19

Approximate temperature and moisture regimes for which eight of the soil orders are typically found. The remaining four soil orders (including Andisols, Entisols, Inceptisols, and Histosols) may form under any temperature and moisture regime. Vertisols require clay-rich sediments and are most common in regions with temperature and moisture conditions shown inside the dash-lined box.

(Figure 3.5, p. 68 *Elements of the Nature and Property of Soils,* 12th Edition by Nyle C. Brady and Ray R. Weil. Copyright © 2000. Reprinted by permission of Prentice-Hall, Inc., Upper Saddle River, NJ.)

reason, the concept is significant in geomorphology and, when applied judiciously, soil catena can be used to estimate relative ages of geomorphic landforms (Swanson, 1985; Berry 1987; Birkeland and Burke 1988; Birkeland et al. 1991).

The facility of a catena to reflect process is complicated because slopes are not always underlain by a single rock type. In addition, variations in climate, duration of soil development, and landscape evolution will make the meaning of catena profile changes more difficult to interpret. Nonetheless, the catena concept can be successfully applied in any area regardless of geologic complexity or climate if sufficient analytical care is exercised (for more complete discussions, see Young 1972; Ollier 1976; Gerrard 1981; Birkeland 1999).

In addition to steepness, slopes also affect soil development because their orientation often promotes differences in microclimate and vegetation even where the regional climate is the same. Because of lower direct insolation and temperature, north-facing slopes tend to have greater moisture contents and vegetal cover than south-facing slopes. This leads to thicker A horizons with a higher organic content.

Climate and vegetation have been recognized historically as the most dominant controls on soil formation (*cl* and *o* in the Jenny equation). In fact, most soil orders are restricted to specific climatic regimes (fig. 3.19) and, as demonstrated for the United States in figure 3.20,

their distribution is clearly related to climatic zonation. This follows because temperature is a primary control on the rates at which chemical processes occur within a soil, and water is involved in most chemical and physical weathering processes. In addition, moisture and the availability of chelating agents associated with certain plant communities dictate the depth and magnitude of leaching within a profile. As a result, certain combinations of climate and vegetation (called pedogenic regimes) produce distinct trends in soil-forming processes. Several examples of regimes can be used to illustrate this point.

Podzolization The processes that result in the removal of iron, aluminum, and organic matter from the A horizon and their accumulation in the B horizon are referred to as *podzolization.* Podzolization is common in humid-temperature climates, especially in soils that develop under forest vegetation. For example, most soils in the eastern half of the United States where precipitation exceeds 70 cm annually are podzolized to some extent. To the west where precipitation is less than 56 cm, another pedogenic regime (calcification) becomes dominant. Podzolization produces a continuous translocation of mobile substances as they are released from parent minerals. Because forest vegetation uses less alkali and alkaline earth elements than do grasses or shrubs, the surface litter is normally quite acidic. As a result, most soluble

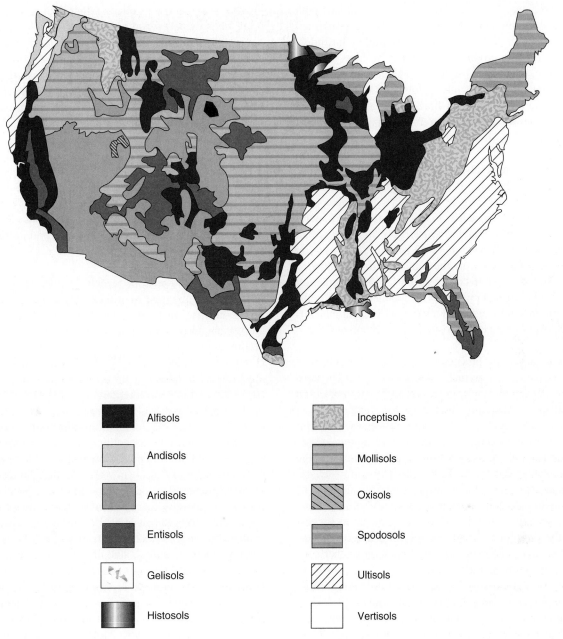

Figure 3.20
The major occurrences of the soil orders in the United States. Significant occurrences of Gelisols and Oxisols are
not found in the contiguous 48 states.
(Modified from Soil Survey Staff 1998)

ions (Ca, Na, Mg) in the upper soil are leached, and colloids of all types adsorb H^+. Therefore, Spodosols, Alfisols, and Ultisols, the soils most affected by podzolization, are notably acidic.

Where podzolization is very active, an E horizon may be found overlying a spodic B horizon. This bleached zone represents the ultimate form of podzolization because almost all the Al^{+3}, Fe^{+3}, and organic matter are removed and the residuum is highly concentrated in silica. Although the soil in a true podzol is acidic, the E horizon can be developed where the pH of the zone is

still within the insolubility range of Al^{+3} and Fe^{+3}. Thus, most of the translocation of these insoluble ions involves chelation rather than dissolution by colloidal acids. Precipitation of the chelated complexes at depth is generally controlled by changes in the chemical environment or by the decomposition of the complexes by microorganisms (see Berner and Berner 1996; Birkeland 1999).

Beneath the zone of maximum eluviation, Spodosols and Alfisols are usually brown or grey-brown, but as the temperature increases they grade progressively into Ultisols having yellow and red colors. The greater humus content in cooler regions, combined with the translocation and redeposition of organic matter, accounts for the drabber color in the higher latitudes. However, as the production of humus declines with increased temperatures, the ferric content rather than the organic matter becomes the predominant factor in determining the color of the B horizon. The characteristic color of podzolic profiles, therefore, changes from north to south in the eastern United States.

The end products of podzolic weathering include resistant primary minerals, iron and aluminum sesquioxides or hydrates, and kaolinitic clays. Most of the mobile ions have been removed from the system in solution, but some cations may be retained by fixation in the clay mineral structures. Clays accumulate in the B horizon in any or all of three ways: (1) the component polymers of clays can be carried in solution from the A horizon and recombined in the B horizon; (2) clays can be leached as already formed minerals and physically accumulated; or (3) clays can form in situ by chemical alteration of parent minerals that were originally in the position of the B horizon.

Laterization The processes resulting in Oxisols constitute a pedogenic regime called *laterization*. The conditions needed for these processes to function are high rainfall and temperature, intense leaching, and oxidation. Thus, Oxisols, and the laterization processes that form them, most commonly exist in tropical regions, especially where wet and dry seasons alternate and drainage is good. Most Oxisols are associated with relatively stable landscapes whose ages are measured in millions of years. Climatic conditions over these time frames have varied greatly, prompting some investigators to suggest that the development of Oxisols is a discontinuous process, proceeding only when climatic and other soil-forming conditions allow significant laterization to occur (see, for example, Valeton 1994; Achyuthan 1996). In fact, Oxisols have been found in a wide variety of climatic regimes, including the subarctic, but these profiles are likely to have formed during warmer, more humid conditions than characterize the region today. Nonetheless, to consider Oxisols the zonal soil of the tropics may not be absolutely correct.

Laterization is marked by two essential differences from podzolization: (1) organic accumulation is inhibited; and (2) silica is leached from the system in addition to the common mobile ions, leaving abnormally high concentrations of hydrated oxides of iron and aluminum in the soil. In some cases silica combines with available alumina to form kaolinitic clays, but where silica is more extensively leached the excess alumina is present as gibbsite $Al(OH)_3$ or, more rarely, boehmite $(AlO(OH))$. The type of parent material and the organic content also exert a direct influence on whether the clays will be kaolinitic or gibbsitic. Where granites are the parent rocks, less silica will be removed during lateriza-

tion because it exists as quartz rather than as amorphous SiO_2. Oxisols developed on granites, therefore, tend to have more kaolinite in their profiles; quartz may be a notable end product. Where basalts are the parent rocks, gibbsite tends to be more dominant because the more soluble amorphous silica is thoroughly leached, leaving little available to form the kaolinite. When limestone is the parent material, the main secondary mineral seems to be boehmite.

Laterization is perhaps the most poorly understood of the pedogenic regimes. We know that iron and aluminum can be dissolved in acidic fluids, yet the environments of their accumulation are very acidic. Precisely how these substances are concentrated is not clear, and much controversy has arisen over interpretations concerning the environment of laterization.

Calcification A third pedogenic regime, *calcification*, functions in subhumid to arid climates where precipitation is insufficient to drive the soil water downward to the water table. Ions mobilized in the A horizon are reprecipitated in the B horizon, where zones of $CaCO_3$ commonly develop (fig. 3.21). The depth of the carbonate zone is a function of the annual precipitation (fig. 3.22) and in general represents a first approximation of the vertical extent of leaching. Arkley (1963), however, showed that this relationship holds only in regions without pronounced seasonality of rainfall or orographic effects. In addition, we now know that windblown dust is a major source of $CaCO_3$ in soils, and that the character of parent material can determine the water-holding capacity and thereby the depth of carbonate deposition. (Gile et al. 1981; Reheis et al. 1995). There is also some evidence to suggest that $CaCO_3$ in soil can be precipitated by microorganisms; and, thus, the depth is affected by biotic activity (Wright and Tucker 1991; Amit and Harrison 1995). Nonetheless, reasonable estimates of carbonate distribution can be made if the mitigating factors are considered (Marion et al. 1985; Mayer et al. 1988; McFadden et al. 1991; McDonald et al. 1996).

Regardless of the inherent difficulties in evaluating calcification, for our purposes it is useful to visualize the depth of the carbonate zone as a gross index of the magnitude of calcification. Soils derived from those processes all possess a K horizon or a calcium carbonate zone, but its position within the soil profile rises proportionately with decreasing rainfall and/or the efficiency of the calcifying processes.

Calcification takes place most efficiently under grass or brush vegetation because such plants utilize large quantities of alkali and alkaline earths in their life processes. When the plants die, the mobile ions are returned to the soil as the litter is decomposed. The amount escaping this recycling process depends on precipitation, temperature, and related bacterial activity. It is also known that root activities control the local geo-

Figure 3.21
Soil profile in an area of Wyoming
characterized by the accumulation of
$CaCO_3$. Note continuous nature of
horizons beneath the surface.
(Photograph courtesy of Soil Survey Staff 1999)

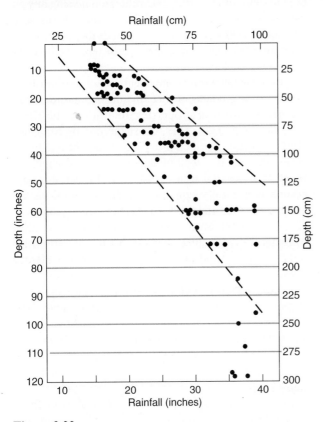

Figure 3.22
The depth of carbonate zones in soils as related to mean annual
rainfall.
(Jenny and Leonard 1939)

chemical environment. These tend to lower soil pore
P_{CO_2} during seasonal droughts and result in the precipitation of caliche (Schlesinger 1985).

The two S.C.S. soil orders formed primarily by calcification are the Mollisols and Aridisols. Mollisols
occur mainly in the temperate zones of the United

States, especially in the northern Great Plains states (the
Dakotas and Nebraska). In these soils the profile
(fig. 3.23) usually has an undifferentiated, black A horizon ranging from 30 to 120 cm in thickness, which reflects the high accumulation of humus in cooler temperature zones. The B horizon is usually light-yellow to
brown, and carbonate may exist as disseminated material or in a distinct layer at the base of the horizon. In
more arid areas the Aridisols become dominant. In these
environments, soils are often immaturely developed and
$CaCO_3$, if present as a secondary accumulation, will be
near or at the surface. Generally, then, in cooler areas
with less than 64 cm of annual precipitation, increasing
aridity is shown in the profiles by a progressively less
developed A horizon and a carbonate zone rising closer
to the surface. Where the temperature is higher, humus
production is less and a red hue becomes more prominent in the calcified soils. This gives rise to the reddish-
colored Mollisols and Aridisols.

Because of the immobility of the soluble ions, most
clay minerals formed in the calcification regime are
montmorillonite or illite types. In some cases kaolinite is
abundant near the top of the profile where slightly acidic
soil water is present, especially where humus is abundant. The retention of the alkaline earths, however, is
not favorable for the development of kaolinite in the
deeper zones.

types of clays formed in the calcification regime

The Factor of Time

The general perception that climate is the dominant control of soil formation seems to be changing. Many soil
experts now believe that time may be the most significant factor in explaining differences between soils of
any given landscape and that the climatic effect may be
negligible (Birkeland 1984; Claridge and Campbell
1984; Eggleton et al. 1987; Nesbitt and Young 1989) or

indirect (McFadden 1988). There is no question that the development of diagnostic soil properties is a time-dependent phenomenon. Therefore, properties such as horizon development, clay mineralogy, total soil morphology, iron oxide composition, and $CaCO_3$ accumulation can be expected to change during soil formation (Birkeland 1999; Arduino et al. 1986; Harden 1982, 1987; McFadden and Hendricks 1985; McFadden and Weldon 1987). Recognize, however, that the value of each property changes at a different rate and that, given enough time, each will eventually reach a condition where the property no longer changes or its rate of change becomes negligible. At that time, the property has attained a steady-state relationship with the soil-forming environment. When all properties of a soil profile reach this condition, the soil itself is said to be in a steady state. Birkeland (1999) presents an excellent discussion on the amount of time needed for various diagnostic properties to culminate in a steady state. As shown in figure 3.24, an A horizon with pronounced organic accumulation will generally reach a steady-state condition before a B horizon can be recognized or while it is patently immature. Diagnostic B horizons (Bt, Bk) require more time. A true oxic B horizon presumably attains steady state only when all the weatherable minerals have been altered to stable forms and, therefore, is the diagnostic horizon requiring the most time for its complete development.

Because diagnostic components of soil profiles form over variable amounts of time, the major soil orders that are based on those properties should also be time-dependent (fig. 3.24). Thus, the suggestion by Birkeland (1999) that the distribution of major soil orders should correlate with the ages of deposits or landscapes on which they develop seems reasonable. In fact, that suggestion is generally supported by the relationships between soil orders and deposit ages in the United States.

The recognition of time as an important control of soil character prompted the search for ways to utilize soil properties as indicators of age. For example, Harden (1982) provided the methodology needed to derive a *profile development index* which reflects the total time elapsed during the development of any given soil. Thus, soils may represent a powerful dating technique if rates of change in their properties can be determined in ab-

Figure 3.23
Mollisols usually have dark-colored A horizons due to the accumulation of humus, and light-yellow to brown B horizons. Greater aridity will decrease the thickness of the A horizon and may lead to the accumulation of carbonate within the B horizon.

(Photograph courtesy of Soil Survey Staff 1999)

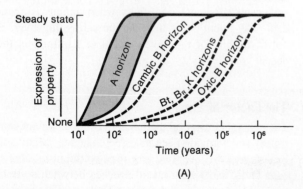

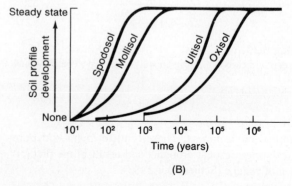

Figure 3.24
Diagram showing the variations in time to attain the steady state for (A) various soil properties and (B) various soil orders.
(After Birkeland 1984)

solute terms. To meet this requirement, numerous attempts have been made to establish soil chronosequences whose tenure of development could be limited by radiometric and other numeric dating techniques. A *soil chronosequence* is a genetically related suite of soils in which vegetation, topography, and climate are similar (Harden 1982). The data obtained from studies of chronosequences (constructed using data from individual soils of differing age) are generally transformed into *chronofunctions,* which represent the continuous change in soil properties with time. It is now clear that the development of chronofunctions is plagued by a variety of difficulties (for an excellent discussion, see Harrison et al. 1990). In fact, some investigators have argued that univariant chronofunctions that accurately portray the development of soils with time cannot be adequately derived (Yaalon 1975; Harrison et al. 1990). There is no question, however, that chronosequence studies have led to significant advances in our understanding of geomorphic processes, landscape evolution, and soils development.

Geomorphic Significance of Soils

Our discussion thus far has been a brief treatment of the basic processes involved in weathering and soil development. It should be clear from these discussions that the rate and nature of weathering and pedogenesis are intimately related through the influences of relief, parent material, and time to the local geomorphic setting and the sequence of geomorphic events that are responsible for creating the existing landscape. What may be less obvious is the importance of soils data in the study of geomorphology. However, soil characteristics affect geomorphic processes in a number of significant ways; in fact, a unique soil property may dictate the mechanics of a surficial system. Once such a property is identified, it provides critical information needed for environmental control or for regional and local planning (McComas et al. 1969). For example, in certain situations soil properties are known to control the stability of building foundations (Baker 1975), hydrologic response to precipitation (Cooley et al. 1973), and the perma-

nence of road construction (Weinert 1961, 1965). Moreover, because soil properties are altered by time and climate change, soil formation has even greater importance in deciphering the sequence of events in Quaternary history. Given the above, there has been an increasing trend in recent years for researchers to focus on the genetic relationships between soils and landscapes, a field of study that McFadden and Knuepfer (1990) refer to as *Soil Geomorphology.*

One of the most common uses of weathering and soils is to establish the relative ages of Quaternary deposits and, by inference, the sequence of geomorphic events. For example, soil-geomorphic studies are commonly applied to the dating of Quaternary faulting, particularly normal faults that allow unconsolidated deposits to be offset vertically (Machette 1978, 1988; Amit et al. 1995). In this case, soil data may be used to approximately determine the time of faulting by estimating the age of the faulted and unfaulted deposits, or by examining soil-catena relations along the surface of the fault scarp (fig. 3.25) (Birkeland 1999; McCalpin and Berry 1996). In addition, tectonic events can lead to the episodic deposition of sediment at the base of the fault scarp, creating sequences of soils buried by wedge-shaped deposits of colluvium (fig. 3.26). By characterizing the degree of soil development within the colluvial wedges, it is possible to estimate the duration between the tectonic events.

The use of soils in Quaternary geomorphology is a dual-edged sword. The fact that soil-forming processes change with time is a basic ingredient of historical interpretation. However, because soil-forming factors are not constant, the record preserved in profiles is more difficult to ascertain. This is especially true in areas that have been affected by the pronounced climatic fluctuations of alternating glacial and interglacial episodes. As the climate changes, the dominant factors of soil formation in any given area must also change accordingly. Thus, many soils preserve in their profiles characteristics that reflect more than one set of soil-forming factors. As discussed earlier, these *polygenetic soils* (sometimes called *complex soils*) complicate the record because thickness

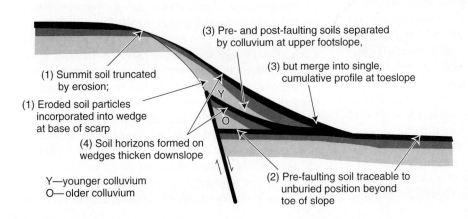

(3) Pre- and post-faulting soils separated by colluvium at upper footslope,

(3) but merge into single, cumulative profile at toeslope

(1) Summit soil truncated by erosion;

(1) Eroded soil particles incorporated into wedge at base of scarp

(4) Soil horizons formed on wedges thicken downslope

Y—younger colluvium
O—older colluvium

(2) Pre-faulting soil traceable to unburied position beyond toe of slope

Figure 3.25
Possible soil-catena relations on a fault scarp characterized by several episodes of movement.

(Figure 5 by McCalpin and Berry from *Catena,* Vol. 27. Copyright © 1996. Reprinted by permission from Elsevier Science.)

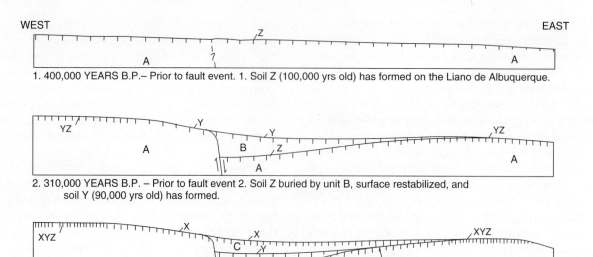

WEST   EAST

1. 400,000 YEARS B.P.– Prior to fault event. 1. Soil Z (100,000 yrs old) has formed on the Liano de Albuquerque.

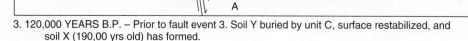

2. 310,000 YEARS B.P. – Prior to fault event 2. Soil Z buried by unit B, surface restabilized, and
   soil Y (90,000 yrs old) has formed.

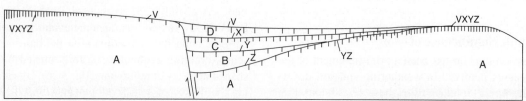

3. 120,000 YEARS B.P. – Prior to fault event 3. Soil Y buried by unit C, surface restabilized, and
   soil X (190,00 yrs old) has formed.

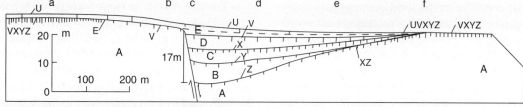

4. 20,000 YEARS B.P. – Prior to fault event 4. Soil X buried by unit D, surface restabilized, and soil
   V (100,000 yrs old) has formed.

5. PRESENT DAY– Soil V buried by unit E, surface restabilized, and soil U (20,000 yrs old) has formed.
   Soil U presently covered by small coppice dunes and thin eolian sand sheet.

Figure 3.26

Cross-sections showing sequence of fault movements for the past 500 ka in the vicinity of the Dump Fault, New Mexico.

(From Machette 1978)

and maturity of soils are reliable indices of age only when the conditions developing the soils have been maintained continuously. When conditions change, the properties of the initial soil are supplanted by new characteristics. For example, Reheis (1987) described alternating layers of carbonate and illuvial clay in the B horizon of soils in south-central Montana. She attributes the alterations to local glacial-interglacial climatic fluctuations. Because such alterations vary in the degree of completeness, complex soils are very difficult to correlate with soils in other areas.

The interpretive problems are even greater when climate change is not the cause of variations in soil properties. McFadden and Weldon (1987), for example,

show that soil-forming processes are drastically altered by incorporation of eolian dust and organic matter into interstices of permeable gravel. As the fine material continues to accumulate, it eventually lowers permeability enough to change the system from one incapable of holding water to one dominated by chemical weathering. This transition from a noncolloidal to colloidal system represents what the authors refer to as a pedogenic threshold. Importantly, the threshold condition was attained during the Holocene and, therefore, is a function of age rather than the Pleistocene-Holocene climate change. The threshold depends on the influx rate of dust and on initial permeability. Thus, the time needed to produce this threshold and the resulting diagnostic prop-

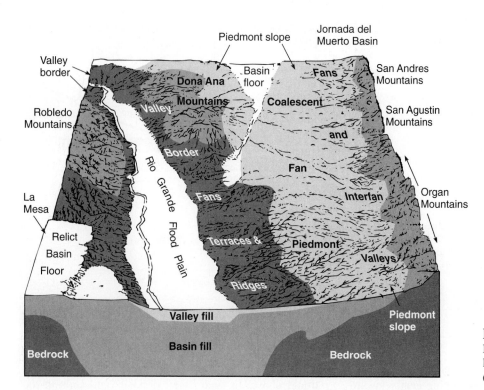

Piedmont slope

Jornada del
Muerto Basin

Valley
border

Basin
floor

Dona Ana
Mountains

Fans

San Andres
Mountains

Robledo
Mountains

Valley

Coalescent

San Agustin
Mountains

Border

and

Fan

Rio Grande Flood Plain

La
Mesa

Fans

Interfan

Organ
Mountains

Relict
Basin
Floor

Terraces &

Piedmont

Ridges

Valleys

Valley fill

Piedmont
slope

Basin fill

Bedrock

Bedrock

Figure 3.27
Block diagram showing major
landforms of the Desert Project.
(Gile 1975)

erties may vary significantly where the physical controls are different.

Soils that form on a landscape of the past are *paleosols*. They can be of three types: (1) *buried* soils are developed on a former landscape and subsequently covered by younger alluvium or rock; (2) *relict* soils were not subsequently buried but still exist at the surface; and (3) *exhumed* soils were at one time buried but have been reexposed when their cover was stripped by erosion (Ruhe 1965). Buried soils are immediately recognized as paleosols. Relict and exhumed soils are much more difficult to identify, because it must be proved that their properties are inconsistent with the modern environment or that their age is the same as the paleosurface on which they rest. Some additional geomorphic or stratigraphic data are usually needed to substantiate those requirements.

Paleosols are important features in the older rock record as well as the Quaternary and have now been recognized in most Phanerozoic time intervals (Reinhardt and Sigleo 1988). However, identifying paleosols in preQuaternary rock sequences requires careful field study and the existence of diagnostic properties (Retallack 1983, 1988). Nonetheless, the recognition and interpretation of paleosols has become so common (for example, Retallack 1990; Vanstone 1991; Gustavson 1991; Mack and James 1992) that a separate classification system has been suggested (Mack et al. 1993). The paleosol classification emphasizes morphological and mineralogical features that are easily recognized. It differs from the classification of modern soils by excluding those properties that do not survive the changes associated

with the transition of unconsolidated material into solid rock, such as compaction and chemical diagenesis.

Perhaps the most persuasive use of soils in Quaternary geomorphology arises when soil characteristics are painstakingly integrated with sequences of landform development and associated sedimentary deposits. An excellent example of this approach is documented in detailed studies of the Rio Grande valley and the adjacent slopes and intermontane basins near Las Cruces, New Mexico. These studies, sponsored by the U.S. Soil Conservation Service and known as the Desert Soil Geomorphology Project, began with the work of R. V. Ruhe in the 1950s (Ruhe 1964, 1967) and culminated approximately 20 years later in thorough syntheses of the soil geomorphology (Gile and Grossman 1979; Gile et al. 1981). A brief summary of the project's procedures and results will demonstrate how soils can be used in the analysis of Quaternary history.

The Desert Project area is physiographically divisible into distinct subareas (fig. 3.27). The east and northwest margins consist of semiconnected mountains that rise up to 2750 meters in elevation. The valley border zone is located in the valley of the Rio Grande and is characterized by deposits and surfaces formed by the river and tributaries during and after the Pleistocene epoch. The floodplain of the modern river stands at approximately 1200 meters. Between the mountains and the valley (the piedmont slope) exist a number of intermontane basins that have been affected by tributary and slope processes that function on the valley sides. The basin or piedmont areas are composed mainly of alluvial

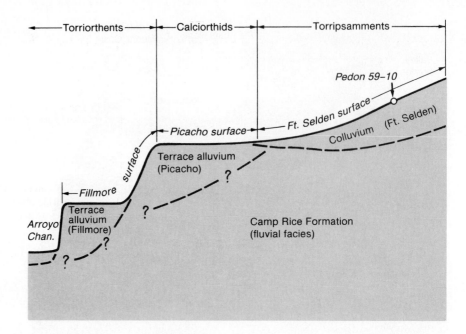

Figure 3.28
Terraces and river alluvium in the valley border subarea near Las Cruces, N.M.
(Gile et al. 1981)

fan deposits that emerge from the mountains and coalesce into smooth alluvial plains sloping gently toward the valley bottom. In some cases the piedmont deposits are graded to the valley-bottom deposits, but more often they and their related surfaces are not physically connected to the valley-border sequence.

The regional history was controlled by repeated climatic changes that produced alternating periods of deposition and erosion. In the valley zone, these alternations resulted in a series of river terraces, the surfaces of which are underlain by alluvium of the Rio Grande (fig. 3.28). Episodes of downcutting by the Rio Grande initiated trenching in the tributary arroyos of the piedmont and valley-border areas. This isolated many of the geomorphic surfaces that formed as the upper level of a depositional event. Since the middle Pleistocene, five depositional episodes, separated by intervening downcutting, have created a sequence of surfaces that document their relative ages. Soil formation on the various deposits and surfaces began at different times.

Knowing the geomorphic setting and realizing that climate change was synchronous in the valley and piedmont areas, it might be easy to assume that soil development was similar throughout the region. Actually, the meaning of the preserved soils is very confusing because their developmental histories vary on a local basis. In some zones, erosion has truncated the diagnostic horizons of a profile. At other localities, soil profiles of one age were buried by material of a younger depositional event. For example, arroyo trenching initiated in the Rio Grande valley did not proceed entirely to the mountain front prior to the incidence of the next depositional phase. In such cases, deposits and soils in the upper piedmont zone were buried by subsequent fan development (fig. 3.29). Additionally, alluvial plain deposits may not extend all the way to the axial valley. Furthermore, precise dating by ^{14}C provided only a broad age framework for events in the area because the amount of datable material was limited. What results from these complications is that the history recognized in deposits and surfaces of the valley border cannot be directly correlated with those in the valley-side sequence, and soils developed in the two areas are different because of vagaries in local history, parent material, and microclimate.

The crux of this is that Quaternary history in the Desert Project area is unintelligible without the integration of stratigraphy, soils, and geomorphology. The histories of the valley-border and piedmont areas were finally linked by relating the *degree* of soil development in each subarea to the relative age of the deposits and geomorphic surfaces within that subarea (Gile et al. 1981). This could be accomplished because at stable sites (where there is no erosive disruption of profiles) age becomes the key soil-forming factor. This was revealed especially in the carbonate horizons, which become thicker and more indurated with age (fig. 3.15). Thus, even though soils of the two subareas have different characteristics and classifications, the relative degree of their development allowed them to be correlated and the surfaces associated with the soils could be placed into a developmental sequence (table 3.9).

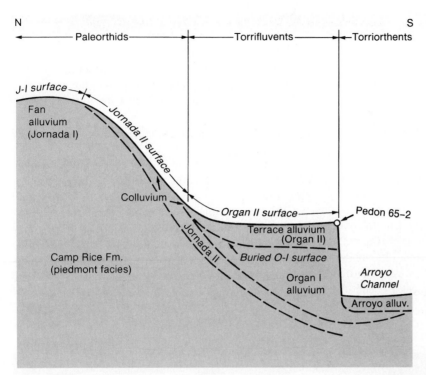

Figure 3.29
Cross-section showing burial of deposits and surfaces by younger fan and terrace alluvium near Las Cruces, N.M.

TABLE 3.9	General Relationship Between the Degree of Development in Carbonate Soils and Geomorphic Surfaces in Subareas of the Desert Soils Geomorphology Project Area Near Las Cruces, New Mexico.

	Geomorphic Surface		Stage of Carbonate Accumulation (see fig. 3.15)	
Age (yr B. P. or epoch)	Valley Border	Piedmont Slope	Nongravelly Soils	Gravelly Soils
Holocene				
0–100	(Dune)[a]	(Dune)[a]	—	—
100–7500	Fillmore[b]	Organ	I	I
> 7500 (latest Pleistocene)	Leasburg[b]	Isaacks Ranch	II	II, III
Late Pleistocene	Picacho	Jornada II	III	III, IV
Late mid-Pleistocene	Jornada I	Jornada I	III	IV (multiple laminar zones)
Early mid-Pleistocene	La Mesa		IV	

Source: Gile 1975.

[a]The dunes are not formally designated by a geomorphic surface name.

[b]Where Fillmore and Leasburg cannot be distinguished they are grouped into the Fort Selden surface.

SUMMARY

The processes of chemical weathering—hydrolysis, oxidation-reduction, solution, ion exchange—alter the exposed portion of the geologic framework and, combined with organic processes, produce soils. The degree of chemical change depends on how mobile the ions of the parent minerals are under the external and internal controls on the weathering mechanics. In regions with abundant precipitation, highly mobile ions are usually removed from the weathered zone unless the original mineral structure is incompletely broken down and elements such as potassium are fixed in the system. In contrast, where leaching is minimal, mobile ions (Ca, Na, Mg) are concentrated in the weathered zone. Immobile ions (Fe^{+3}, Al^{+3}) may be transposed in very acidic groundwaters or by special organic processes such as chelation. The mobility of most substances is also dependent on the pH and Eh of the weathering environment. The type of clay mineral formed in the weathering zone is usually a good indicator of the intensity of decomposition.

Soils are described and classified according to the soil profile. The character of the soil profile varies with parent material, climate, biota, topography, and the length of time involved in its formation. Three major pedogenic regimes—podzolization, laterization, and calcification—produce the dominant soil groups; however, changes in the controlling pedogenic factors may result in complex soils that show evidence of forming under more than one pedogenic regime. Soils are important elements in reconstructing geomorphic history, and they directly influence other surficial processes.

SUGGESTED READINGS

The following references provide greater detail concerning the concepts discussed in this chapter.

Birkeland, P. W. 1999. *Soils and geomorphology.* 2nd ed. New York: Oxford University Press.

Gile, L. H., Hawley, J. W., and Grossman, R. B. 1981. *Soils and geomorphology in the Basin and Range area of southern New Mexico: Guidebook to the Desert Project.* New Mexico Bur. Mines Min. Resources, Memoir 39.

Harden, J. W. 1982. A quantitative index of soil development from field descriptions: Examples from a chronosequence in central California. *Geoderma* 28:1–28.

Knuepfer, P., and McFadden, L., eds. 1990. *Soils and landscape evolution.* Proc. 21st Binghamton Symp. on Geomorphology. *Geomorphology* 3(3, 4).

Loughnan, F. 1969. *Chemical weathering of the silicate minerals.* New York: American Elsevier.

Machette, M. N. 1985. Calcic soils and calcretes of the southwestern United States. Geol. Soc. Am. Spec. Paper 203: 1–21.

McFadden, L. D., and Weldon, R. J. 1987. Rates and processes of soil development on Quaternary terraces in Cajon Pass, California. *Geol. Soc. Amer. Bull.* 98:280–93.

Soil Survey Staff. 1960. *Soil classification, a comprehensive system—7th approximation.* Washington, D.C.: U.S. Dept. of Agriculture, Soil Conservation Service.

———. 1999. *Soil taxonomy.* 2nd edition. U.S. Dept. of Agriculture, Soil Conservation Service, Agriculture Handbook 436.

PHYSICAL WEATHERING, MASS MOVEMENT, AND SLOPES

R. Craig Kochel

Introduction
Physical Weathering
    Expansion of Rocks and Minerals
        Thermal Expansion
        Unloading
        Hydration and Swelling
        Salt Weathering
    Growth in Voids
    The Significance of Water and Interaction of
           Chemical and Physical Weathering Processes
Physical Properties of Unconsolidated Debris
    Shear Strength
        Internal Friction
        Effective Normal Stress
        Cohesion
    Measurement of Strength
Mass Movements of Slope Materials
    Slope Stability
    Mass Wasting Processes
        Heave and Creep
        Rockfalls
        Slides
        Flows
    Morphology of Mass Movements
Slope Profiles
    The Rock-Climate Influence
    Slope Evolution
    Relict Hillslope Forms
Summary
Suggested Readings

INTRODUCTION

The transformation of rocks into unconsolidated debris is the prime geomorphic contribution of weathering and soil-forming processes. Whether the debris produced by weathering will resist erosion and become part of the regolith depends on the balance between the internal resistance of the materials and the magnitude of the external forces acting on them. The relative resistance of Earth materials influences the character of the slope that will develop on it. Extremely steep slopes, for example, can be maintained for long periods only if the underlying rock and soil is so tightly bound together that the forces and agents of erosion cannot lower the slope angle. On the other hand, gentle slopes in regions of low relief may be stable for long time spans even if the underlying material is very friable. Slope morphology reflects the properties of underlying materials as well as the environmental forces acting to modify them. Useful information can be extracted from studies of slope characteristics only when we understand the erosive processes attacking them.

The evolution of landscapes is the history of regional slope development. These slopes form by a multitude of geomorphic processes, and the properties of slopes reflect in subtle ways the temporal effect of these processes on the resisting framework. Interest in slopes and slope-forming processes is evident across the entire range of geomorphic thinking, from the analysis of a modern stability problem in slope hazard assessments to investigations of geologic history.

Mass movements are one of the most universal groups of geomorphic processes operating on all solid planetary surfaces. Because of the differences in gravity, atmosphere, and volatiles in the regolith, the predominant type of mass movement process will vary from planet to planet. However, the slope processes recognized on Earth have been recognized on the other solid planets and their moons (fig. 4.1).

Numerous links exist between geomorphic processes operating on hillslopes and hillslope form. These processes regulate the rate of production and character of **colluvium,** the products of gravity-driven mass movement, shed to downslope accumulation sites. In turn, mounting evidence exists that hillslope form affects the processes themselves. The mechanics of slope erosion are closely related to the processes of physical weathering because the forces that disintegrate rocks and minerals simultaneously lower the internal strength of the unconsolidated cover, ultimately leading to slope failure. We will begin, then, with a brief discussion of physical weathering.

PHYSICAL WEATHERING

Physical weathering culminates in the collapse of parent material and the diminution of its grain size. Rock breaks down continuously as stress is exerted along zones of weakness within the original material. These zones may be planar structures such as bedding or fractures that, upon rupture, produce fragments whose size and shape are controlled by the spacing of the planes. In other cases, failure may occur along mineral boundaries, resulting in an accumulation of particles similar in size and shape to the original rock. Although stresses are generated in different ways, in all processes of disintegration a force within the material itself is responsible for its destruction.

The stress field involved in disintegration results from either expansion of rocks or minerals themselves or pressure generated by growth of a foreign substance in the voids within the lithologic host fabric. In each case the direction of the principal stress may change according to the process involved, but the most pronounced disintegration invariably occurs where the adjacent rocks exert the least confining pressure. Intuitively, then, we should expect disintegration to be most pernicious near the surface, where static load from overburden is minimal and fractures are abundant and closely

(A)

Figure 4.1

Examples of mass movement phenomena on Mars. (A) Large slump along the walls of the equatorial canyon Valles Marineris. Note the flow features extending from the base of the large slump and eminating from the opposite canyon wall in the foreground. The canyon is 140 km wide; walls are about 2 km high. (B) Rockfall and talus deposits along steep walls of Kasei Vallis paleochannel. Main canyon is about 40 km wide with walls around 3 km high. Upper smooth surface is Lunae Planum plateau. (C) Debris flow along walls of upper Kasei Vallis. Channel floor is about 20 km wide. (NASA)

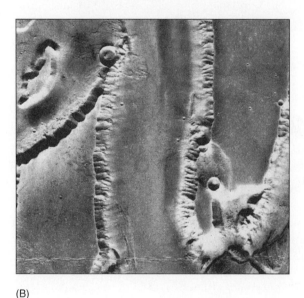

(B)

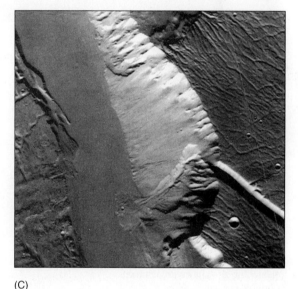

(C)

spaced. With increasing depth, confining pressure increases, fractures are less common, and the disintegrating processes become less effective.

Similar to chemical weathering phenomena, physical disintegration processes are influenced by endogenetic and exogenetic factors. The rates and relative role of individual disintegration processes vary with climate and vegetation (exogenetic factors) as well as endogenetic factors related to the structure and composition of the rock itself.

Expansion of Rocks and Minerals

Thermal Expansion   Rocks and minerals expand in response to several phenomena that can rightfully be considered agents of physical weathering. The application of intense heat cause physical disruption of rocks. The low thermal conductivity of rocks prevents the inward transfer of heat, and little, if any, change oc-

curs below the outer few centimeters. Differential stresses produced by this thermal constraint cause the rock exterior to spall off in plates or wedges 1 to 5 cm thick (fig. 4.2). Heat expansion occurs during forest fires, and in semiarid forested mountains of the western United States it may be the dominant process of physical weathering (Blackwelder 1927), although its effectiveness varies with rock composition (Ollier and Ash 1983). Whether or not insolation can drive the process has been debated for many years. Many geologists have gradually, if not grudgingly, accepted the premise that diurnal temperature fluctuations are not severe enough to produce thermal spalling (Twidale 1968) because experimental studies (Griggs 1936a, 1936b) suggested that the process is not viable.

Gray (1965), however, demonstrated that thermal spalling is indeed possible, and geomorphologists have reaffirmed thermal expansion as a method of rock disintegration (Ollier 1963, 1969; Rice 1976). Winkler

Figure 4.2
Spalling of granitic boulder caused by heat expansion during forest fire. Beartooth Mountains, Mont.
(R. R. Dutcher)

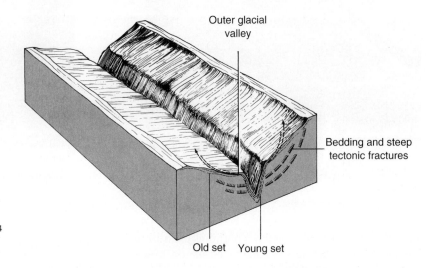

Figure 4.3
Expansion joints produced by pressure release during valley entrenchment. Vaiont River valley, Italy.

(After G. A. Kiersch, "Vaiont Reservoir Disaster,"
Civil Engineering, vol. 34, no. 3, p. 35, copyright 1964
American Society of Civil Engineers. Used by
permission of ASCE.)

(1975) pointed out that when rocks such as granite contain entrapped water, tensile stresses of approximately 250 atmospheres can develop upon heating from 10° to 50° C, which is significant enough to disrupt the rock.

Unloading   Expansion of large segments of rock mass occurs when confining pressure is released by erosion. As denudation removes overburden, the stress squeezing the underlying framework is lowered, and rocks tend to split into widely spaced sheets, 1 to 10 m thick, that are oriented perpendicular to the direction of pressure release. This sheeting tends to mirror the surface topography. Because outer sheets are relatively easy to erode, the process helps perpetuate the surficial configuration as subsequent sheets develop with a similar orientation. Although other processes aid in the removal of the

sheets, it can be readily documented that the original formation of the fractures is a pressure-release phenomenon. Rock bursts in deep mines, for example, are explicit proof that something as simple as excavation of tunnels can trigger a rapid expansion of surrounding rocks. In the natural setting, postglacial entrenchment of the Vaiont River in Italy permitted valleyward expansion of rocks and produced a joint system parallel to the valley sides (fig. 4.3). Hack (1966) demonstrated that arcuate patterns of streams, ridges, and vegetal types in the eastern United States are probably controlled by the position of curved sheets that dilated during erosion of crystalline rocks.

Although most geomorphologists favor unloading or pressure-release mechanisms to explain sheeting joints like those visible in figure 4.4, considerable evi-

Figure 4.4
Extensive sheeting joints on Enchanted Rock, a granitic inselberg in the Llano Uplift region of central Texas. Live oak trees average 10–12 m in height. Granite sheets on slopes are 1–3 m thick.
(R. Craig Kochel)

dence exists that this process may not be universally applicable. Twidale (1971, 1973, 1982) presented evidence showing that sheeting in several granitic landscapes unaffected by glacial loading may be controlled by endogenic characteristics of the bedrock and/or regionally imposed tectonic stresses. Alternative explanations for sheeting in granitic rocks reviewed by Twidale include (1) deep-seated roofing fractures, (2) petrologic heterogeneity from concentric crystallization, (3) regional compressive stresses during emplacement or subsequent metamorphism, and (4) faulting because of local tectonic stresses.

Hydration and Swelling   Expansion also occurs when minerals are formed or when they are altered by the addition of water to the structure. Although the process begins as a chemical process called **hydration,** its physical effects are particularly obvious when the clay minerals containing layers of OH⁻ or H_2O are formed. The creation of the layered structure expands the minerals and propagates stress outward from the clay particle. Clays such as bentonite (Na-montmorillonite), which do not have a fixed OH⁻ or water layer in their structure, have the capacity to absorb water into the mineral during periods of wetting. The swelling produced by wetting exerts the same outward stress as during clay formation. Most clays show this trait to some extent, but the percentage of expansion depends on the mineral species plus a myr-

TABLE 4.1	Expansion of Common Clay Minerals by Hydration.
Clay Mineral	**% Expansion**
Ca-Montmorillonite	
Forest, Miss.	145
Wilson Cr. Dam, Colo.	95
Davis Dam, Ariz.	45–85
Na-Montmorillonite	
Osage, Wyo.	1400–1600
Illite	
Fithian, Ill.	115–120
Morris, Ill.	60
Tazewell, Va.	15
Kaolinite	
Macon, Ga.	60
Langley, N.C.	20
Mesa Alta, N.M.	5

Adapted from Mielenz and King (1955).

iad of other factors (table 4.1). Montmorillonite clays, for example, drastically lose their swelling capacity when sodium is replaced by another cation (Mielenz and King 1955; Gromko 1974). Upon drying, the expanded

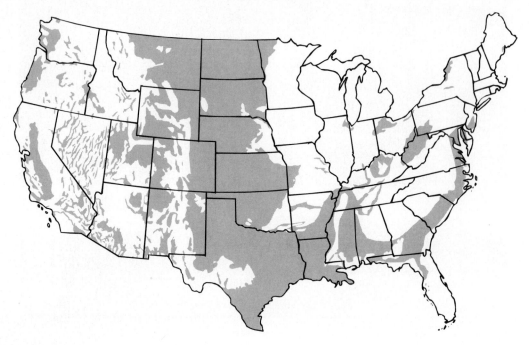

Figure 4.5

Distribution of soils with expandable clay minerals within the United States.

(From Hays, 1981 *Facing Geologic and Hydrologic Hazards,* U.S. Geological Survey Professional Paper 1240-B)

clays lose part or all of the absorbed water, initiating an alternating swelling and shrinking sequence with episodes of wetting and drying.

Soils that change volume significantly in response to changes in soil moisture are referred to as **expansive soils.** Volume changes associated with expansive soils cause major engineering problems that in the United States alone account for over two billion dollars in annual damages (Jones and Holtz, 1973). Figure 4.5 shows that expandable clays are widespread in the western, south, and southeastern United States. Figure 4.6 shows examples of the cracks and surface morphology common in expandable montmorillonite-rich soils upon desiccation. Evidence of repeated opening and closing of these cracks is recorded in the soil profiles. Soil horizons typically appear convoluted due to the collapse of material from the surface and high levels of the profile into deep cracks. These soils also may exhibit slickensides when observed in their desiccated state due to differential shear stresses generated during expansion. When wet, expandable soils are especially sticky and plastic, in marked contrast to the rigid state when they are dry.

Expansion pressures caused by these soils can range from 0.16 to 0.6 MN/m^2 (1.5 to 5.6 tons/ft^2), which can readily exceed loads imposed by small structures, causing structural damage to roads, houses, small schools, and small commercial buildings. Engineers can avoid regions likely to be plagued by expansive soils by learning

to recognize the diagnostic geomorphic characteristics of the soil group known as Vertisols (see chapter 3). Vertisols generally display unhorizonated and convoluted profiles, exhibit large cracks when dry, become very sticky when moist, and may have a peculiar micro-furrowed surface called gilgai (Gustavson 1975). Reviews of expansive soils and engineering strategies designed to minimize disruptive activity to structures can be found in Gromko (1974), Mathewson et al. (1975), and Costa and Baker (1981).

In contrast to Vertisols, the well-ordered hydrated clays have a stable structure, and destruction of the OH or water layer occurs only when the mineral is heated to at least 300° C. The disintegrating effect of these clays, therefore, occurs during their formation but, in contrast to the swelling clays, is exerted continuously until relieved. Once released, there is no way to reinstate the internal stress in these clays.

The effect of mineral expansion has been clearly demonstrated in the physical breakdown of granite in arid and semiarid regions (Wahrhaftig 1965; Eggler et al. 1969; Isherwood and Street 1976). In these settings, the major product of granite disintegration is a coarse, angular mass of rock and mineral fragments called **grus** (fig. 4.7), in which feldspars are often unaffected by decomposition. In the Laramie Range of Colorado and Wyoming, the sequence of grus development started in the Precambrian with formation of hematite by high-temperature

important to learn about the soil Vertisols when dealing w/ expansive soils.

(A)

(B)

(C)

(D)

Figure 4.6
Morphology of expandable soils (vertisols). (A) Polygonal cracks formed during desiccation (Badlands, SD). (B) Popcorn texture of a surface of dry soils rich in montmorillonite (Anza Borrego Desert, Calif.). (C) Sticky character of expansive clay soils when wet (Badlands, SD). (D) Example of structural damage to parking lot structure caused by expandable soils (San Antonio, TX).
(R. Craig Kochel)

Figure 4.7
Roadcut on slope of Mount Palomar, Calif., showing extensive grusification of biotite-rich granitic rocks of the Peninsular Ranges. Note the resistant quartz veins standing in relief from the crumbling granite. The footprints are in an apron of grus composed largely of single feldspar crystals released upon hydration of the biotites.
(R. Craig Kochel)

(A)

(B)

(C)

(D)

Figure 4.8
Granite weathering in southern California. (A) Boulder slope near Escondido showing granitic corestones left stranded between joints as grus was stripped off. (B) Corestone exhumed from grus during highway construction. (C) Low-relief bench formed by progressive downwasting of granites by grusification processes. (D) Gabbro boulder showing spheroidal weathering. Himalaya Mine, San Diego County.

(*A–C:* R. Craig Kochel; *D:* W. T. Schauer, U.S. Geological Survey)

oxidation along cleavage planes in the biotite (Eggler et al. 1969). Although this process expanded the biotite in the direction of the *c*-axis, the stress was not sufficient to cause disintegration. It did, however, weaken the biotite's ability to resist further geomorphic attack. Subsequent near-surface weathering produced clays from the biotite with as much as 40 percent increase in volume, and the stress generated by this expansion shattered the granite into grus.

Degradation of granitic landscapes is greatly facilitated by grusification driven by hydration processes. Geomorphic features produced in this manner include extensive boulder slopes (fig. 4.8A), downslope alluvial fans composed largely of grus, and unusual level but undrained benches formed in high-relief topography (fig. 4.8C). The level, boulder-strewn landscape in figure 4.8C was created by gradual downwasting of the area in which vertical migration of a subsurface weath-

ering front was followed by fluvial removal of residual grus surrounding boulder corestones (fig. 4.8B, D).

In other situations, the granite bedrock may be weakened in different ways prior to the grus development. For example, Folk and Patton (1982) showed that the first stage of grus formation in central Texas granites is the development of microsheet joints parallel to the weathering surface (fig. 4.9). Because these joints cut indiscriminately across mineral grains, they cannot be the result of biotite expansion. Instead, they precede and ultimately enhance grus development.

The process of mineral expansion manifests itself in different end products. Rocks are peeled off to produce curved surfaces; the process on a large scale is called **exfoliation** and on a smaller scale **spheroidal weathering.** Even though the resulting large domelike masses (fig. 4.10) or rounded boulders are probably a function of pressure release, it seems certain that water and mineral

Figure 4.9
Microsheet joints (Folk and Patton 1982) in central Texas granite that precede and enhance grus development.
(R. Craig Kochel)

Figure 4.10
Northeast side of Half Dome taken from the subsidiary dome at the northeast end of the rock mass, revealing exfoliation on a gigantic scale. In the foreground is an old shell disintegrating into undecomposed granite sand. Yosemite National Park, Mariposa County, Calif.
(F. C. Calkins, U.S. Geological Survey)

alteration are intimately involved (Gentilli 1968). Spheroidal boulders are formed as edges and corners of lithologic blocks are weathered more rapidly than flat surfaces, a phenomenon especially apparent where the parent rock has been fractured into a blocky framework by perpendicular joint sets. The relatively fresh spheroidal cores are usually surrounded by a zone of disintegrated flakes and spalls that is enriched in secondary clay minerals. Simpson (1964), for example, found that the clay matrix in weathered graywacke increased by 5 to 10 percent in the spalled zone and also contained abundant vermiculite, an expandable clay not present in the fresh rock. Evidence such as this seems to indicate that outward expansion caused by the development of clay minerals peels off the fresh rock layer by layer, working progressively inward from the surrounding joint openings.

Exfoliation processes also play a major role in the production of arches and natural bridges like those occurring in sandstones of the Colorado Plateau. Here, exfoliation, in concert with other weathering processes such as hydration, may produce arches where rocks are competent enough to maintain roof support. Arch production is also promoted in the presence of closely spaced joints and where cliff headwall erosion processes are active (Blair 1986).

Salt Weathering   Hydration of salts within pores of building stones and concrete develops sufficient stress to cause extensive spalling (Winkler and Wilhelm 1970). According to Winkler (1965), a similar process almost destroyed Cleopatra's Needle, an obelisk that was brought from Egypt to New York City in 1880. Salts

(A)

Figure 4.11 Tafoni
(A) Large-scale tafoni and alveolar
weathering on the face of Ayers Rock,
Australia, in sandstone. (B) Small-scale
tafoni in limestone, Southwest Texas.
(*A* by C. R. Twidale; *B* by R. C. Kochel)

(B)

trapped in spaces within the red-granite monument did not hydrate until they were placed in the humid climate of the eastern United States. Recent experimental work by Sperling and Cooke (1985) has shown that hydration of salts alone can be effective in promoting rock disintegration. Laboratory experiments indicate that sodium carbonate is most effective in granular disintegration, followed by magnesium sulfate and sodium sulfate (Goudie 1986). Volume expansion of over 300 percent can occur from absorbtion of water during hydration of these salts (Goudie 1989).

Salt weathering has been increasingly suggested as a significant component in the physical breakdown of rocks (Goudie 1997), especially as a process to explain the formation of minor weathering features such as tafoni (Rodriguez-Navarro et al. 1999). **Tafoni** are holes or de-

pressions, usually less than a few meters in width and depth, that commonly form on the undersides of rock masses or on steep rock faces (fig. 4.11). They often develop on granitic rocks in arid climates. Salt weathering by crystallization has been suggested as the genetic cause of tafoni (Evans 1969; Winkler 1975; Bradley et al. 1978; Mittershead and Pye 1994). The exact origin of tafoni, however, remains a mystery (Evans 1969; Selby 1982) because the features can form in a variety of nonarid climates (Calkin and Cailleaux 1962; Martini 1978; Watts 1979) and on rock types other than granite. Alternatively, dissolution and precipitation related to groundwater exiting along the face of an outcrop has been emphasized as an important component in the formation of tafoni or *alveolar weathering* (fig. 4.12) in permeable sandstones of the Colorado Plateau (Howard and Kochel 1988).

Tafoni – cause still not completely understood
– possible causes explained above.

Figure 4.12
Tafoni or alveolar weathering in sandstone on the Colorado Plateau, northern Arizona.
(R. Craig Kochel)

Figure 4.13
Rock disintegration products, salts, and duricrust at the *Viking 1* lander site on Mars attributed to salt-weathering processes.
(NASA)

Sources for salts are varied (Goudie 1989). Proximate sources include sea spray, seawater and relict seawater, volcanic gases, and rock weathering (particularly from evaporite beds). Nonpoint sources are precipitation and eolian dust. Examples of the geomorphic effects of salt weathering range from cliff sapping (Laity 1983) to differential ground heaving (Horta 1985) and are summarized in table 4.2. Salt weathering has been shown to be an important weathering process in the Dry Valley areas of Antarctica (Selby and Wilson 1971; Matsuoka 1995) and in the most arid of warm deserts (Goudie et al. 1997), and has been postulated to have been an extensive agent of weathering on Mars (Malin 1974). Observations made by the Viking landers (fig. 4.13)

showed clay minerals and salts thought to be products of igneous rock alteration as well as a duricrust of soil cemented by sulfate salts (Toulmin et al. 1977). On Mars salt weathering may be an effective mechanism in reducing rocks to silt or sand sizes suitable for dispersal by the wind.

Growth in Voids

A second group of processes generate stress when some substance grows within void spaces in the rock. The pressure gradient differs from that in the processes explained above because it is the openings that are expanded, not the structure of the parent minerals or rocks.

TABLE 4.2	Geomorphic Effects of Salt-Weathering Processes.

Example	References
Disintegration of rocks by expansion	Winkler and Wilhelm 1970
Tafoni formation	Bradley et al. 1978; Twidale 1978
Alcove formation and cliff sapping	Laity 1983; Laity and Malin 1985
Disintegration and preparation for deflation	Malin 1974
Loess formation	Goudie 1989
Tor formation	Watts 1981
Ground heave	Horta 1985
Contributions to duricrust formation processes	Implied in several studies

Microcracks in rocks can be produced by processes acting inside Earth (Simmons and Richter 1976; Whalley et al. 1982) and therefore may already be present before rocks are exposed at the surface. Because these spaces are not expanded simultaneously or with equal magnitude or direction, the resultant pressures differ locally and the entire system is burdened with a differential stress field. Such uneven pressure distribution is conducive to fracturing or granulation; the processes responsible for its development are probably the dominant agents of disintegration.

Plants and organisms can aid in the disintegrating processes (fig. 4.14), but they usually have the greatest effect after the parent rock has already been converted into soil. Plant roots commonly grow in fractures of the parent rock and physically pry the soil material apart. Nonetheless, compared with other processes, rootlet growth is of minor consequence.

The most significant processes of physical weathering involve forces generated by crystallization of ice (frost action) or other minerals in rock spaces (fig. 4.15). In a perfectly closed system, water experiences a 9 percent increase in volume upon freezing and almost certainly produces hydrostatic pressures that exceed the tensile strength of all common rocks. **Frost action** is most effective when the rock is saturated prior to the freezing event. Simple alternations of wetting and drying will sometimes fracture rocks, but the process is accelerated by freezing (Muridge and Young 1983). Working in Antarctica, Matsuoka (1995) has shown that frequent diurnal freeze-thaw cycles in the summer result in rapid disintegration of rocks, when saturation is common. Recent work indicates that frost-shattering may take the form of explosive rock bursting similar to that observed in mines. Rocks susceptible to violent frost-bursting phenomena are those strong enough to release large amounts of stored strain energy suddenly rather than gradually dissipate the stress caused by ice pressures (Michaud et al. 1989).

If more than 20 percent of the available pore space is empty, the expansion pressure of the freezing water may be less than the rock strength and shattering will

Figure 4.14
Tree trunk splitting a large granite sheet block along a joint, central Texas.
(R. Craig Kochel)

not occur (Cooke and Doornkamp 1974). Some evidence suggests that the intensity of frost action is related to the structure of the pores rather than simply the percentage of pore space. In a system containing a variety of pore sizes, ice crystals will preferentially grow in large pores rather than smaller ones (Everett 1961). *[handwritten margin note: ice crystals grow in larger pores before growing in smaller pores]*

Minerals can also grow in rock spaces, with results similar to those of frost action (fig. 4.16). Most commonly such fracturing occurs when percolating fluids evaporate within the pores, giving rise to supersaturated conditions and eventual precipitation of minerals. The

Figure 4.15
Frost-shattered jointed metamorphic rocks in a cold alpine environment near Powder River Pass, Bighorn Mountains, Wyo.
(R. Craig Kochel)

Figure 4.16
Pebble fractured by growth of calcite along planes of weakness, near Roberts, Mont.
(Dale Ritter)

pressures exerted in crystallization are probably greater than those produced by ice, but their absolute values depend on the concentration of the ionic constituents in the solution. The most common precipitates are sulfates, carbonates, and chlorides of very mobile cations (Ca, Na, Mg, K), and the process is therefore more prone to operate in arid and semiarid regions where the ions are rendered immobile by insufficient leaching. The growth of salts from solution within rock fractures and pores is recognized as another form of salt-weathering phenomenon. Crystal growth can occur for many reasons including (1) decreasing solubility in some salts with declining temperature, (2) common ion effects upon mixing of different salt-rich solutions, and (3) evaporation (Goudie 1989).

The Significance of Water and Interaction of Chemical and Physical Weathering Processes

[handwritten margin note: H₂O is very important in the process of disintegration]

Even a short review of physical weathering makes clear the importance of water in the disintegrating processes. Hydration, frost action, crystal growth, and swelling all require water as the basic component of the system. The amount of water need not be great. Many believe, for example, that even thin films of condensing dew in desert regions may be infinitely more destructive than insolation (Twidale 1968). Therefore, a direct relationship between climate and the prevalence of disintegration seems likely.

Peltier (1950) utilized mean annual temperature and precipitation to predict relative intensities of physical and chemical weathering (fig. 4.17). This figure broadly generalizes the relationships. For example, recent work by Caine (1992) shows that chemical weathering also plays a significant role in the removal of materials in alpine catchments by solution. Physical weathering is generally dominant where precipitation is readily available and the mean annual temperatures are near or below freezing. This analysis seems to correlate with the importance of frost action as a mechanical tool and with the fact that frost action is preeminent in those areas having the most freeze-thaw cycles during the year. The frequency of freeze-thaw events has been detailed for the United States by Russell (1943) and Williams (1964).

Where unusual local problems exist, the regional climatic characteristics may have little significance. In those cases it may be extremely important to understand in detail the climate-lithologic-weathering system, and a more sophisticated approach than those just reviewed will be necessary. An excellent example of this point was provided by Weinert (1961, 1965). In the eastern part of South Africa, the parent Karoo dolerite has been altered into a mature soil that is unsatisfactory for maintaining road foundations. In the western part of the region, this mature soil is not present. Instead, hydration of the micas has apparently disintegrated the dolerite into a grus that has considerable internal strength and is

quite sound from an engineering perspective. Weinert found that the boundary between the sound and unsound surface materials could be mapped by the distribution of evaporation and precipitation in the area.

Weathering of granitic terrain, such as the inselberg in figure 4.4, involves an intimate blend of physical and chemical processes. The large-scale weathering features such as sheeting fractures are due to physical stresses associated with the release of confining stress during unroofing of overlying sedimentary cover and possibly to some combination of mineralogical and tectonic stresses inherited during the emplacement of the pluton. The emergence of isolated remnants like Enchanted Rock above the surrounding granitic terrain involves the spatially nonuniform activity of weathering processes. The granite surrounding the inselberg has been lowered at a more rapid rate because of more efficient chemical and physical attack by water-related processes where joints are more closely spaced compared to the lower joint density within the inselberg itself (Twidale 1971, 1978).

Smaller-scale granite weathering features on the inselberg also originated from physical and chemical interactions. Runoff is concentrated along major joints, where on low-sloping surfaces it may remain long enough for dissolution to occur. Repeated dissolution along the joints eventually produces elongated potholes or gnammas (fig. 4.18A), which then collect and retain water, thereby further enlarging through a positive feedback process. Some of these potholes become sites for sediment storage and plant growth, both of which continue to promote their enlargement. Over extended periods of time, surface runoff succeeds in creating an articulated network of rillen in the granite, and other unusual phenomena occur through a combination of solutional and abrasional processes (fig. 4.18B).

PHYSICAL PROPERTIES OF UNCONSOLIDATED DEBRIS

The resistance of unconsolidated debris to the forces of erosion depends on the physical properties of the material. These properties determine whether a slope developed on any substance will be stable or, if the physical properties help determine the shape of the slope profile, when and if it attains an equilibrium condition. Given that the slope material itself directly influences the resulting process and landform, it is rather disconcerting to find that most geologists have only a vague knowledge of the basic physical properties of soil, which have been identified through years of study by engineers. These characteristics directly control the mechanics of many geologic hazards.

Before we examine these properties, however, it may be helpful to look again at the concept of driving force, which was briefly discussed in chapter 1. Figure 4.19A

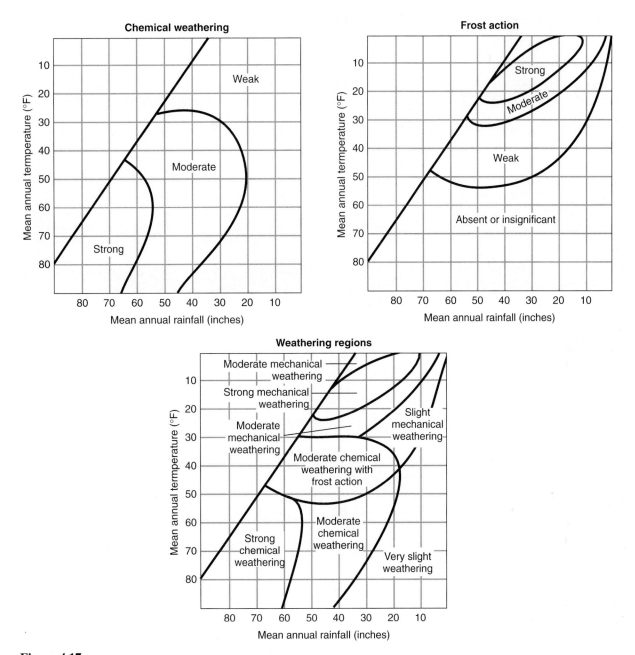

Figure 4.17
Relative intensities of physical and chemical weathering generalized for different temperature-precipitation conditions.

(From L. C. Peltier, *Annals of the Association of American Geographers,* vol. 40, p. 219, 1950. Reprinted by permission.)

depicts a boulder resting on a sloping surface. The force of gravity acts vertically on the particle, and the force magnitude stems from the weight of the particle *mg* (mass times the acceleration of gravity). Actually, the weight may be resolved into two components, one acting perpendicular to the sloping surface and one acting parallel to it. The component acting parallel to the surface tends to promote downslope movement and is measured as $W \sin \theta$, where W is the weight in pounds or kilograms. The perpendicular component tends to keep the boulder in place by pushing it into the surface and

thereby resisting the downslope motion; its magnitude is determined as $W \cos \theta$. Clearly, downslope movement of the particle is enhanced on steeper slopes because the sine value increases and the cosine value decreases as the angle θ is increased.

In the analysis of slope processes, however, engineers and scientists are usually concerned with the force acting on some potential plane of failure existing below the ground surface along which movement of a block of overlying material takes place (fig. 4.19B). In this case the exerted force derives primarily from the weight of

Figure 4.18
Physical-chemical weathering features on granite surfaces in central Texas. (A) Solution pans, or gnammas, aligned along joints.
(B) Rillen and major surface microrelief formed by a combination of solution and abrasion.

(R. Craig Kochel)

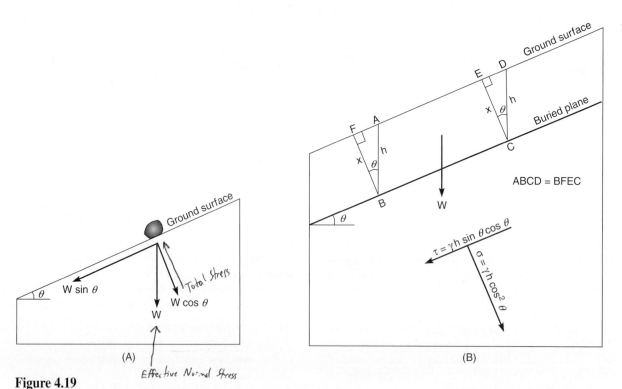

Figure 4.19
Analyses of slope stability. (A) Forces acting on a particle resting on a slope surface. (B) Stresses acting on a planar surface
covered by unconsolidated material.

the debris overlying the plane. Because the total weight of this material cannot be determined like that of a single, discrete boulder, it is calculated indirectly by multiplying the unit specific weight (γ) of the material (lb/ft³ or kg/m³) times the vertical distance (h) from the plane to the ground surface. The resolved components are now $\gamma h \sin \theta \cos \theta$ in the downslope direction and $\gamma h \cos^2\theta$ perpendicular to the plane. The reason for the change in equations is that the block of soil, represented by the parallelogram ABCD, is equal to the rectangle BFEC, and the angles FBA and DCE are equal to θ. Therefore,

$$\cos \theta = \frac{x}{h} \text{ and } x = \cos \theta h$$

Because $W = \gamma x$, substituting from above gives us $W = \gamma h \cos \theta$. Thus, the pressure acting perpendicular to the plane is

$$\sigma = W \cos \theta = \gamma h \cos \theta \cos \theta = \gamma h \cos^2\theta$$

and the shear is

$$\tau = W \sin \theta = \gamma h \cos \theta \sin \theta$$

Notice from the above that in the case of the buried plane, we are no longer talking about force because the value γh is given in units of *stress*, which by definition is the force acting on a specific area. Because γ is in lb/ft³ and h is in ft, $\gamma h = \text{lb/ft}^3 \times \text{ft} = \text{lb/ft}^2$.

When used in this sense, stress and pressure are synonymous. The perpendicular component of the total stress is called *normal stress* and is usually indicated by the symbol σ. The downslope component is called *shear stress* and is denoted in mechanical analyses by the symbol τ.

Shear Strength

The properties of matter that resist the stresses generated by gravitational force are collectively known as the **shear strength.** The detailed analysis of internal strength is an extremely complex procedure, well beyond the scope of this book. For our purposes, the shear strength of any material is derived from three components: (1) its overall frictional characteristic, usually expressed as the angle of internal friction; (2) the effective normal stress; and (3) cohesion. These factors determine shear strength by the Coulomb equation:

Coulomb equation: $S = c + \sigma' \tan\phi$

where S is shear strength (in units of stress), c is cohesion, σ' is effective normal stress, and ϕ is the angle of internal friction.

Internal Friction   Internal friction is composed of two separate types: *plane friction,* produced when one grain slides past another along a well-defined planar surface, and *interlocking friction,* which originates when particles are required to move upward and over one another.

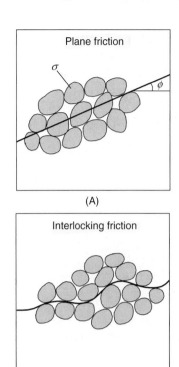

σ = normal stress
ϕ = angle of internal friction

Figure 4.20

Two types of internal friction. (A) Plane friction, in which resistance occurs along a well-defined plane that may cut through individual grains. Angle of plane friction is approximated by angle at which sliding begins.
(B) Interlocking friction, in which resistance occurs around particle boundaries.

These are shown graphically in figure 4.20. The angle of plane friction is approximated by the angle at which a block will begin to slide over another along the plane separating them. Once sliding begins, the frictional angle actually decreases slightly, requiring that a distinction be made between a static and a dynamic angle. In addition, the plane friction angle varies considerably with smoothness of the plane surface, moisture at the contact, mineralogy, and other factors.

Interlocking friction is usually greater than plane friction because extra energy must be used to move interlocked grains in an upward direction. The angle of interlocking friction also varies with mineralogy and moisture and, it would seem, must be affected by the density of packing within the mass. In loose particulate matter of any size, the angle of repose should approximate the angle of internal friction, but Carson and Kirkby (1972) point out that a platform holding a cone of debris at its angle of repose can be tilted up to 10° before the slope fails. This suggests that the angle of repose may be somewhat less than the maximum angle at which a slope can stand.

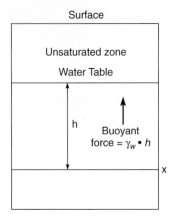

Figure 4.21
Buoyant stress produced by hydrostatic pressure beneath a water table.

Effective Normal Stress   The importance of normal stress is its capacity to hold material together, thereby increasing the internal resistance to shear. In theory, normal stress acting perpendicular to a shear surface (fig. 4.20) is absorbed by the underlying slab at the point of contact between grains. In reality, some of the shear surface is occupied by openings filled with air or water. Thus, the applied forces are resisted by solid particles and by fluids present within the void spaces. Since pore pressure exists in these interstitial spaces, it tends to support part of the normal stress. The total normal stress (σ) therefore includes two elements, effective normal stress (σ') and pore pressure (μ), such that

$$\sigma = \sigma' + \mu$$

When considering internal friction, effective normal stress (stress exerted at the solid-to-solid contacts across the shear surface) is the critical parameter rather than total normal stress.

The value of pore pressure (μ) has a direct bearing on the effective normal stress because it can add to or detract from the total stress value. For example, figure 4.21 shows a distinct water table in unconsolidated material. At some level x below the water table, a hydrostatic pressure is exerted that is equal to the specific weight of the water (γ_w) times the vertical distance (h) between the water table and the level of x. Although the pore pressure beneath the water table is acting in all directions, Terzaghi (1936) points out that a portion of the pressure will be exerted in a direction opposite to the normal stress and therefore will provide some relief from the overburden weight. In contrast, in the unsaturated zone above the water table, some of the water will be prevented from moving downward because it is attached to particles by capillarity. Simply stated, this attached moisture increases the weight of the soil. Relating water content to the effective normal stress, three possible situations can be envisioned.

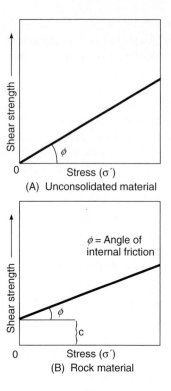

Figure 4.22
Relationship between shear strength and effective normal stress. (A) Unconsolidated material has no shear strength when effective normal stress is zero. (B) Rock material has shear strength (c) from cohesion even where no effective normal stress is present.
(From M. A. Carson and M. Kirkby, *Hillslope Form and Process,* copyright 1972 Cambridge University Press, Cambridge.)

1. In a completely dry soil, the pore pressure is atmospheric and μ is zero. Therefore, the effective normal stress and the total normal stress are the same since

$$\sigma' = \sigma - 0$$

2. Below the water table, pore pressure is positive (greater than atmospheric pressure), causing the effective normal stress to be lower because

$$\sigma' = \sigma - \mu$$

3. Above the water table, μ is negative and the effective normal stress is higher:

$$\sigma' = \sigma - (-\mu)$$

Because the effective normal stress directly influences internal friction, it is clear that dry or partially saturated soils, especially those with a high clay content, should have greater shear strength and stand at higher slope angles than equivalent materials that are thoroughly saturated.

Cohesion   Figure 4.22 demonstrates the relationship between shear strength and effective normal stress. The

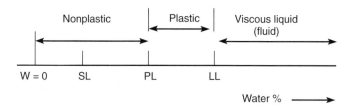

Figure 4.23

Atterberg limits and their relationship to water content (W).

TABLE **4.3**	Average Atterberg Limits in Clays Saturated with Common Cations.							
	Montmorillonite[a]		Illite[b]		Kaolinite[c]		Halloysite[d]	
	Plastic Limit	Liquid Limit	Plastic Limit	Liquid Limit	Plastic Limit	Liquid Limit	Plastic Limit	Liquid Limit
Na^+	91	442	36	65	27	41	42	46
K^+	63	173	41	75	33	52	45	48
Mg^{++}	59	164	39	84	29	50	54	60
Ca^{++}	68	155	39	86	31	54	48	60

Adapted from Grim (1962).

[a]4 samples

[b]3 samples

[c]2 samples

[d]1 sample $2H_2O$, 1 sample $4H_2O$

graph indicates that as the effective normal stress increases, the values of shear strength also rise. The relationship between the two variables defines a straight line that passes through the origin of both axes. The angle between the line and the abscissa represents the angle of internal friction. The material represented in figure 4.22A has no discernible strength when the effective normal stress decreases to zero, a condition that is common in coarse, unconsolidated detritus. Thus, movement begins immediately upon the application of stress. Solid rocks, however, possess shear strength even when σ' is removed (fig. 4.22B) because the constituent particles are bonded or cemented together. The strength revealed here is *cohesion,* a factor that is unaffected by normal stress.

Clay-rich soils also have some cohesion, presumably because adsorption of ions and water by clay minerals creates a binding structure among the particles. The cohesive strength depends on the attractive force between the particles and the lubricating action of the interstitial liquid. As Grim (1962) points out, the molecules of the inner layers of water adsorbed to the surfaces of clays are oriented by the electrical charge of the minerals; because of this, fluidity and lubrication are not possible when moisture content is low, even though water is present. For clays to become plastic and exert a lubricating action, the adsorption of water layers must continue until the outermost ones can be held but are no longer fixed in a rigid, oriented position. Fortunately, simple tests can indicate how much water a soil can absorb before it begins to behave like a plastic substance

or, if more water is added, when the substance will lose all its cohesion and become a muddy fluid.

If water is gradually added to a dry, pulverized soil, the voids fill and the mixture becomes increasingly more plastic. As more water is added, however, the cohesion decreases, and when all the pores are filled, any further input of water completely destroys the internal fabric and produces a fluid. Atterberg (1911) suggested two simple tests to indicate the transition from the solid to the plastic state and from the plastic to the liquid state; they are, respectively, the *plastic limit* and the *liquid limit.* The two limits are expressed as moisture content, determined by the weight of contained water, divided by the weight of the dry soil. The range of water content between the two limits is the *plasticity index* (LL-PL in fig. 4.23).

The values of Atterberg limits are affected by a number of auxiliary factors. First, they are related to the types of clay minerals included in the soil, although the precise limits for any particular clay species vary appreciably (see table 4.3). In general, plastic and liquid limits are higher for montmorillonite clays than for illites or kaolinites, mainly because the montmorillonites are able to disperse into very small particles with an enormous total water-absorbing area. The type of exchangeable cation in the montmorillonite, however, may engender wide variations in the limit values, especially in the case of Na^+ and Li^+. Plasticity indices vary in a similar manner. Second, limit values tend to increase when the particle size is smaller. The *activity,* defined as the ratio of

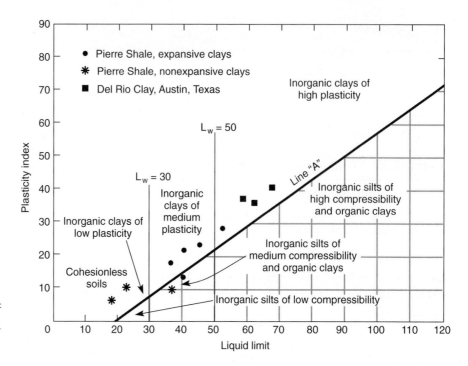

Figure 4.24

Plasticity chart showing example plots of several problem soils susceptible to expansion.

(From "Urban Geology of Boulder, Colorado: A Progress Report" by Victor R. Baker 1975 in *Environmental Geology,* Vol. 1, pp. 75–88. Reprinted by permission of Springer-Verlag GmbH & Co. and the author.)

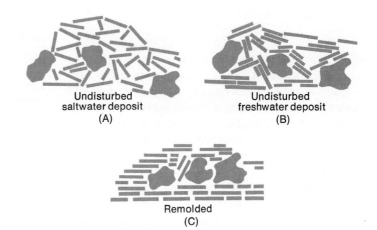

Figure 4.25

Open soil structures in sensitive clays. Structures in (A) and (B) allow water content to exceed liquid limit. Disturbance of this structure causes temporary fluidity until substance is remolded (C).

(From "Urban Geology of Boulder, Colorado: A Progress Report" by Victor R. Baker 1975 in *Environmental Geology,* Vol. 1, pp. 75–88. Reprinted by permission of Springer-Verlag GmbH & Co. and the author.)

the plasticity index to the abundance of clays (Skempton 1953), shows that plasticity increases proportionately with the percentage increase of clay-sized particles. Activity values give a cursory indication of the major clay minerals present in the soil. As the activity increases, soils tend to have a lower resistance to shear, and so the parameter has some engineering significance. Finally, drying of a soil reduces its plasticity because shrinkage during the dehydration process brings particles closer together. The attraction between particles is so strengthened that penetration by water is difficult. Conversely, repeated wetting combined with only partial drying may increase plasticity (Grim 1962).

Although the tests for Atterberg limits are rather unsophisticated, the results are consistent when determined by more than one analyst, and they are geologically significant (Casagrande 1948; Seed et al. 1964). Fig-

ure 4.24 shows the results of Atterberg limits analyses for several very problematic swelling clays and their associated nonswelling units within the same formation. The "A" line on the plasticity chart, from the Unified Soil Classification System developed by Casagrande, separates more plastic claylike soils above the line from nonplastic silt-like materials below the line. Thus, this chart can be used as a quick guide to the anticipated behavior of soil materials at a construction site.

The moisture content in fine-grained material is also a guide to its internal strength. Some soils, called **sensitive soils,** exist with natural water contents above their liquid limits (fig. 4.25), an apparent inconsistency with the limit concept. The fact is, however, that some soils develop an open honeycomb structure that is capable of holding water in excess of its liquid limit. Although the structure is unstable, in an undisturbed soil the material will be solid and possess some strength. The disruption of the internal structure by erosion, earthquake shocks, or other phenomena will cause the excess water to be released, and the solid material will become a fluid. A simple test for the liquid limit can reveal the presence of sensitivity and the potential for dangerous mass wasting.

Loess soils are an example of a widespread surficial material in glaciated regions and along rivers that drained these areas (see chapter 10). Loess is silt-sized, wind-blown sediment that exhibits dramatically different structural characteristics depending on its moisture content. Loess exhibits pronounced vertical anisotropy with respect to permeability (permeability varies depending on direction in the unit), has relatively high activity, and has plasticity indices ranging from 5 to 22 depending on its clay mineralogy (Sweeney and Smalley 1988). Although loess can maintain high, near-vertical slopes when dry, it is well-known for its sudden collapsibility when saturated and loaded (Feda 1988). Lutenegger and Hallberg (1988) have shown that the collapsibility of loess tends to be greater near its source proximal to the major river valleys. Collapse becomes imminent when the in situ soil moisture content of loess nears the liquid limit. Because of these characteristics, engineers regard loess as one of the most problematic geologic deposits worldwide.

Measurement of Strength

Although some field tests for strength have been developed, such as the Vane Shear Test and the Iowa Borehole Shear Test (Luttenegger and Hallberg 1988), most detailed analyses of shear strength in soil are done in a laboratory by means of the direct shear test, the triaxial compression test, or the uniaxial (unconfined) compression test. (The details of these tests are provided in most textbooks of soil mechanics.)

Measurement of shear strength is a complicated task, made more difficult because some of the techniques do not yield certain critical data. In the direct shear test, for example, pore pressure cannot be determined. Furthermore, laboratory tests are designed to measure stress at the moment of failure, which is analogous to shear strength. Natural materials, however, especially soils, begin to deform well below the stress level that causes rupture (fig. 4.26) and may continue to deform at a rate that increases steadily with additional stress. Ultimately the material ruptures at a critical stress called the *breaking strength.*

Determining the strength of rocks is also difficult because tests on small samples may not be reliable indicators of the overall strength of a large rock mass. Structures within the mass such as joints and other fractures may make the total mass relatively nonresistant to erosion. However, analyses of rock samples taken in zones between the fractures may indicate that the rock is quite strong. Obviously methods for estimating rock strength

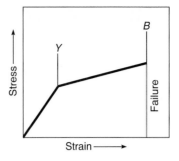

Figure 4.26

Relationship between stress and strain. Permanent deformation begins at *Y* (yield stress), but rupture of the substance does not occur until the stress value reaches *B* (breaking strength). Material behaves like a plastic substance between *Y* and *B*.

in the field are needed. One method devised by Selby (1980, 1982) is presented in table 4.4. The Selby classification allows any rock mass to be placed into one of five categories representing overall rock mass strength. Placement in a specific group is based on numerical ratings (*r*) given for various strength parameters. All parameters are not considered to be of equal significance in determining strength, so each is assigned a percentage value representing its importance. The sum of the weighted values is an estimate of the rock mass strength.

The appeal of an approach such as Selby's is that a rapid estimate of rock mass strength can be made with normal geological field equipment except for the Schmidt hammer, which is a portable engineering tool designed to measure hardness (intact strength) of materials. In addition to providing information about stability, information gained from strength analyses may provide important clues in understanding slope profiles and evolution.

MASS MOVEMENTS OF SLOPE MATERIALS

Three primary types of mass movements are slides, flows, and heaves; each has distinctive characteristics (Carson and Kirkby 1972). In **slides,** cohesive blocks of material move on a well-defined surface of sliding, and no internal shearing takes place concurrently within the sliding block. In contrast, **flows** move entirely by differential shearing within the transported mass, and no clear plane can be defined at the base of the moving debris. The velocity in flows tends to decrease from the surface downward. In **heave,** the disrupting forces act perpendicular to the ground surface by expansion of the material. This movement does not in itself provide a lateral component of transport, but it facilitates slow, downslope movement by gravity and is often an important forerunner of more rapid mass movements such as rockfalls.

Although each of these movements could theoretically function alone, all seem to be involved to some extent in most natural slope failures. Many processes can

TABLE 4.4	Field Classification of Rock Strength.				
	1	**2**	**3**	**4**	**5**
Parameter	**Very strong**	**Strong**	**Moderate**	**Weak**	**Very weak**
Intact rock strength (N-type Schmidt hammer "R")	100–60	60–50	50–40	40–35	35–10
	$r = 20$	$r = 18$	$r = 14$	$r = 10$	$r = 5$
Weathering	unweathered	slightly weathered	moderately weathered	highly weathered	completely weathered
	$r = 10$	$r = 9$	$r = 7$	$r = 5$	$r = 3$
Spacing of joints	>3 m	3–1 m	1–0.3 m	300–50 mm	<50 mm
	$r = 30$	$r = 28$	$r = 21$	$r = 15$	$r = 8$
Joint orientations	Very favorable Steep dips into slope, cross joints interlock	Favorable Moderate dips into slope	Fair Horizontal dips, or nearly vertical (hard rocks only)	Unfavorable Moderate dips out of slope	Very unfavorable Steep dips out of slope
	$r = 20$	$r = 18$	$r = 14$	$r = 9$	$r = 5$
Width of joints	<0.1 mm	0.1–1 mm	1–5 mm	5–20 mm	>20 mm
	$r = 7$	$r = 6$	$r = 5$	$r = 4$	$r = 2$
Continuity of joints	none continuous	few continuous	continuous, no infill	continuous, thin infill	continuous, thick infill
	$r = 7$	$r = 6$	$r = 5$	$r = 4$	$r = 1$
Outflow of groundwater	none	trace	slight <25 l/min/10/ m^2	moderate 25–125 l/min/10 m^2	great >125 l/min/10 m^2
	$r = 6$	$r = 5$	$r = 4$	$r = 3$	$r = 1$
Total rating	100–91	90–71	70–51	50–26	<26

From M. Selby (1980) in *Zeitschrift für Geomorphologie.* Used by permission of Gebrüder Borntraeger Verlagbuchhandlung, Stuttgart.

in fact only be explained by some combination of the primary types of movement. As figure 4.27 shows, mass movements are multifarious events. The location of a particular process near a corner of the triangle in the figure indicates the dominance of a particular primary type of movement; the closer a process is to the corner, the more dominant is that type of movement. Superimposed on the diagram are lines that show relative transport velocity for each process and the typical water content within the moving debris.

Slope Stability

when will a slope move? answered here →

A body of material on a slope will remain in equilibrium (stable) as long as the sum of the applied shear stresses does not exceed the sum of the shear strength of the slope materials. The various mass movements are alike in that all begin when the shear stress tending to displace material exceeds the resisting strength. Stability, therefore, represents some balance between driving forces (shear stress) and resisting forces (shear strength), and can be expressed as a safety ratio:

$$F = \frac{\text{resisting force (shear strength)}}{\text{driving force (shear stress)}}$$

F values greater than 1 connote slope stability, but as the ratio approaches unity a critical condition evolves and failure becomes imminent. Clearly, any factor that lowers the safety ratio (table 4.5) can trigger mass movement, and this tendency can be produced by increasing the driving force, lowering the resistance, or both. Theoretically, failure occurs when $F = 1$; this value is an excellent example of a *geomorphic threshold.* In reality, however, because of our inability to accurately measure and model all important aspects of the parameters affecting stability, slope failures sometimes occur when the F value is slightly positive. Likewise, some slopes have remained stable in spite of small, calculated negative F values. Once failure does occur, the type of movement depends on precisely how the forces interact with one another.

The stability of slope material above a suspected plane of failure can be estimated if the components of stress and strength can be measured. On shallow planar surfaces, like that shown in figure 4.19B, the driving stresses are numerically equal to $\gamma h \sin \theta \cos \theta$. Resistance is equivalent to shear strength, which is derived mathematically as $s = c + (\gamma h \cos^2\theta - \mu) \tan \phi$. Therefore,

$$F = \frac{c + (\gamma h \cos^2 \theta - \mu) \tan \phi}{\gamma h \sin \theta \cos \theta}$$

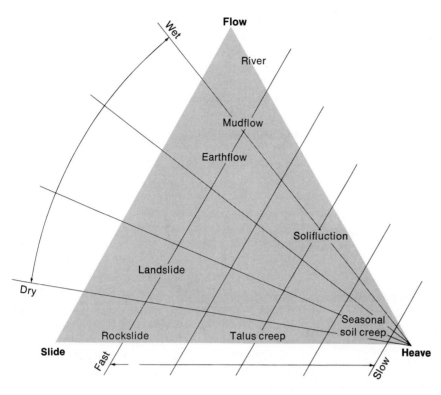

Figure 4.27
Classification of mass movement processes.
(From M. A. Carson and M. Kirkby, *Hillslope Form and Process,* copyright 1972 Cambridge University Press, Cambridge.)

TABLE 4.5	Factors that Influence Stress and Resistance in Slope Materials.

Factors that Increase Shear Stress	Factors that Decrease Shear Strength
Removal of lateral support	Weathering and other physicochemical reactions
Erosion (rivers, ice, waves)	Disintegration (lowers cohesion)
Human activity (e.g., quarries, road cuts)	Hydration (lowers cohesion)
Addition of mass	Base exchange
Natural (e.g., rain, talus)	Solution
Human (e.g., fills, ore stockpiles, buildings)	Drying
Earthquakes	Pore water
Regional tilting	Buoyancy
Removal of underlying support	Capillary tension
Natural (e.g., undercutting, solution, weathering)	Structural changes
Human activity (mining)	Remolding
Lateral pressure	Fracturing
Natural (swelling, expansion by freezing, water addition)	

After Varnes (1958).

In most analyses the vertical height of the water table above the slide plane is expressed as a fraction of the soil thickness above the plane (m), where $m = 1.0$ if the water table is at the surface, and $m = 0$ if it is at or below the sliding plane. Thus, the pore pressure can be expressed as

$$\mu = \gamma_w \, mh \cos^2\theta$$

and

$$F = \frac{c + (\gamma - m\gamma_w)\, h \cos^2\theta \tan\phi}{\gamma h \sin\theta \cos\theta}$$

The following hypothetical example will show how to determine whether the slope is stable or close to failure. If laboratory tests tell us that $\phi = 10°$, $c = 45$ lb/ft^2,

$\gamma = 165$ lb/ft^3, $\theta = 8°$, $h = 12$ ft, $m = 0.6$, and $\gamma_w = 62.4$ lb/ft^3, then

$$F = \frac{45 + (165 - 0.6 \times 62.4)\, 12 \times .98 \times .18}{165 \times 12 \times .14 \times .99}$$

$$= \frac{45 + (165 - 37.44)\, 2.12}{274.43}$$

$$F = \frac{315.42}{274.43} = 1.15$$

The slope in our hypothetical case is stable, but changes in the controlling factors could easily lead to failure. For example, a rise in the water table to $m = 0.9$ would decrease the F value to 1.0 and presumably cause slope failure.

The analysis of slope stability for curved sliding surfaces such as those found in rotational slips is more complicated because it involves balancing moments (force times distance) rather than simply the forces acting on the surface. Rotational failures are usually analyzed by dividing the material above the slip plane into vertical slices and treating each slice by a modified form of the planar analysis discussed above. The values of all slices are then summed to determine the total resistance and shear stress and an estimate of the slope stability.

Excellent discussions of slope stability, instrumentation for measuring strength parameters, and analyses of stability are found in Selby (1982), Franklin (1984), and Graham (1984). Several computer programs are now often used in assessing slope stability; they employ a variety of field data related to slope geometry and geotechnical properties of the bedrock and soil. An example of one of these programs is PC-STABL4 developed by Purdue University for the Federal Highway Administration (Federal Highway Administration 1985). This program can use several of the methods for slices and has considerable flexibility for inputting field data such as heterogeneous and anisotropic soil properties, pore water pressure, and seismic loading.

Recent approaches to assessing slope hazards have used multivariate statistical analyses to delineate areas with the highest potential for slope failure. Parameters in these analyses include a variety of geomorphic indices (e.g., slope angle, basin hydrology, soil thickness), bedrock structural features, soil geotechnical indices such as Atterberg limits, moisture, shear strength (Kenney 1984), and indices relating to vegetal cover. Crozier (1984) provides a good review of field criteria useful in assessing slope stability. The most common statistical approaches have employed stepwise multiple regression, principal components analysis, and discriminant analysis (Carrara 1983; Moser and Hohensinn 1983; Terranova and Kochel 1987; Bernknopf et al. 1988; Crozier 1984). (A good review of the various approaches used in landslide hazard analysis can be found in Hansen 1984.) Typically, these studies are based on comparative analy-

ses of stable and recently failed slopes within a selected region, in an attempt to determine which of the interdependent factors is most diagnostic for instability and also mappable over the area for use in prediction and planning.

A growing number of studies in recent years have been aimed at defining threshold limits of stability for mass movements such as landslides and debris flows. Caine (1980) surveyed debris flows worldwide to establish a relationship between precipitation intensity and duration. Numerous studies in California have demonstrated that two kinds of triggering thresholds are important in slope failures. Minimum probability thresholds (Crozier 1986), discovered through correlations between landslides and precipitation records, indicate that no failures occur before a minimum cumulative rainfall has occurred over an extended period of time (Campbell 1975; Wieczorek and Sarmiento 1983; Schrott and Pasuto 1999). These studies (and others) have also uncovered the existence of maximum stability thresholds where failures always occurred when critical rainfall intensities were exceeded. These relationships, and related issues of antecedence and cyclic instability related to recovery and climate fluctuations, are complex but show that thresholds for slope stability can be established on a regional basis.

Mass Wasting Processes

Heave and Creep   The almost imperceptibly slow movement of material in response to gravity is called **creep.** *Seasonal creep,* or soil creep, is the downslope movement of regolith that is aided periodically by the heave mechanism. *Heave* involves vertical movements of unconsolidated particles in response to expansion and contraction, resulting in a net downslope movement when these occur on even the slightest slopes. No continuous external stress is placed on the mass; it moves under gravity when its cohesion and frictional resistance are spasmodically lowered. This process occurs in the upper several feet of the soil, and its effect decreases rapidly with depth. For example, Godfrey (1997) showed that creep involved only the upper few cm in soils formed on the badlands of the Mancos Shale in Utah. The phenomenon of soil creep was first recognized in the latter part of the nineteenth century, and its ubiquity was gradually accepted as its effects were observed in the field. Historically, evidence suggesting the influence of soil creep on slope materials has included downslope curvature of bedding (fig. 4.28), stone lines, downslope growth of trees or tilting of structures (fig. 4.29), and accumulations of soil upslope from a fixed obstruction (see Young 1972). In recent decades, however, observations and measurements of seasonal creep have become more sophisticated. Precise surveying methods and the use of trenches such as Young pits

Figure 4.28
Creep in vertical Romney shale. Western Maryland Railroad cut one mile west of Great Cacapon, Washington County, Md.

(G. W. Stose, U.S. Geological Survey.)

(A)   (B)

Figure 4.29
Examples of disturbances due to creep. (A) Trees tilted during their lifetime by creep of the substrate. (B) Telephone poles tilted to nearly horizontal and turf lobes mobilized by creep processes.

(R. Craig Kochel)

(Young 1960) have become fairly standard techniques for measuring soil creep (Selby 1966, *Revue de Geomorphologie Dynamique* 1967), and new techniques are continually being developed (Finlayson 1981).

Although burrowing animals and vegetation may cause random disturbances in soils, their effects are minor compared with the heave produced by swelling or freezing and thawing. In the heave mechanism, expansion disturbs soil particles perpendicular to the ground surfaces; when the soil contracts, the vertical attraction of gravity acts on the particles. The expansion-contraction cycle thus adds a lateral component to particle motion in any soil having an inclined surface. Because gravity is

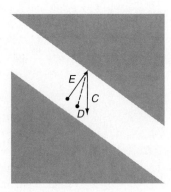

Figure 4.30
Movement of near-surface material by heaving. During expansion (*E*), particle is displaced perpendicular to the surface. During contraction (*C*), particle settles in a vertically downward direction under influence of gravity. Actual movement is shown by line (*D*).

reasonably uniform over the surface, the distance of transport in each heave event, and presumably the rate of creep in any climate, should vary with the slope angle and should decrease with depth beneath the surface. Schumm (1967a) has demonstrated a significant correlation between the rate of surficial rock creep and the sine of the slope angle, but documentation of this relationship for fine soils or below the surface is lacking. Actually, as figure 4.30 shows, the contraction event is never perfectly downward but usually moves in a direction about midway between the normal and the vertical. Because the lateral distance traveled in each heave is less than would be theoretically predicted, a clear relationship between slope angle and creep rate may be difficult to demonstrate. Considerable evidence suggests that even the direction of creep movement is often random on a short-term basis, complicating the presumed relationship even more (Fleming and Johnson 1975; Finlayson 1981). In addition, detailed measurements (Kirkby 1967) have shown the creep rate to decrease with depth (fig. 4.31). Presumably this relates to the lower frequency of heaving at depth and to the greater difficulty of expansion with increasing overburden. In any case, below a depth of 20 cm movement ceases or becomes drastically smaller (Young 1960; Kirkby 1967).

Many studies have assumed that creep movements at various depths below the surface are subparallel. Recent observations in Africa (Moyersons 1988) have detailed the complexity of creep movements, showing convergence and divergence of creep lines to be most common. These irregular movements can produce convoluted soil structures which, in paleoclimatic reconstructions, may

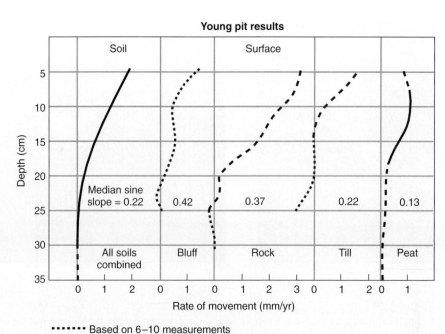

Figure 4.31
Rate of creep as it relates to type and depth of material. Rate in all soils decreases with depth.

(From M. J. Kirkby, "Measurement and Theory of Soil Creep," *Journal of Geology,* vol. 75, pp. 359–378. Copyright © 1967 University of Chicago Press. Used by permission.)

be mistaken for cryoturbation features that are found in frozen soil environments (see chapter 11).

Downslope creep rates are extremely variable because of differences in slope angle, moisture content, and measuring technique (Caine 1981). Particle size also introduces variability in the rate because fine-grained clays tend to swell more upon wetting, and frost heaving is greatest in silty material. Nevertheless, available measurements indicate that downslope rates range from 0.1 to 15 mm/yr where the soil is vegetated. Rates may increase to 0.5 m/yr or more on uncovered colluvial slopes where frost action is prevalent (Gardner 1979; Selby 1982). Volumetric rate can also be calculated; this is the volume of soil moved annually across a plane set perpendicular to the surface and parallel to the contour of the slope for a unit of horizontal distance along the plane. In semiarid regions the rate seems to be somewhat higher. Not surprisingly, slope angle can have a significant effect on creep rates. In a 16-year study of the relative importance of creep versus overland hydraulic flow processes on hillslopes in the Mojave Desert (California), Abrahams and others (1984) found that creep was only dominant at high elevations. On slopes less than 24°, distances of stone movement appear to be dominated by hydraulic activity. In arctic climates a special kind of creep process called **solifluction** is extremely important in the geomorphic scheme; it is discussed in chapter 11.

A second type of creep, *continuous creep* (Terzaghi 1950), is fundamentally different from seasonal creep in that (1) it is driven by gravity alone, (2) it may affect consolidated rock, and (3) it can function at levels well below the surface. Continuous creep is the strain response to stress and continues even though no additional stress is placed on the material. Continuous creep is especially pronounced where rocks or semi-consolidated materials with low yield stress (see fig. 4.26) are overlain by stronger substances. For example, a weak clay unit sandwiched between resistant strata is prone to deform by continuous creep. Excavations of any kind through that rock sequence will reduce the lateral confining pressure on the clay unit, and it will begin to flow. Creep of this type is not important in terms of the volume of material it moves or the distance of transport, but it is very significant as a precursor of rapid, sometimes catastrophic, mass movements. In many cases landslides are immediately preceded by accelerated creep (i.e., Furuya et al. 1999); human agents who intervene in the natural setting can trigger these events by not recognizing the potential for continuous creep (Kiersch 1964). Furuya et al. (1999) demonstrated that creep can be accelerated by groundwater seepage along shear zones.

Thick creep mantles of colluvium (1–2 m) can be produced on convex hills and may be important in engineering considerations (Moyersons 1988). Figure 4.32 shows a typical exposure of colluvium emplaced by creep.

Rockfalls   The heave mechanism is also an essential element in some rapid mass movements, especially falls. **Falls** in both rock and soils involve a single mass that travels as a freely falling body with little or no interaction with other solids (fig. 4.33A). Movement is usually through the air, although occasional bouncing or rolling may be considered as part of the motion. Rockfalls are most common where the parent material is well jointed and a steep slope is developed on the rock face. The fractures become enlarged progressively by heaving, mainly in the form of freezing and thawing, until the gravitational force exceeds the internal resistance. Undercutting of the rock or soil face by erosive agents acting at the base of the material accelerates the process. The removal of the subjacent support tends to increase tension in the overhang and so helps to create and expand incipient cracks.

Figure 4.32

Two layers of Quaternary colluvium emplaced by distinctly different episodes of creep are visible in this central Pennsylvania hillside exposure. Note the imbrication of the poorly sorted, angular colluvium. The colluvium overlies stratified glacial outwash, visible in the lower third of this view.

(R. Craig Kochel)

(A)

Figure 4.33
Types of mass movements. (A) Rockfall on
Interstate 70 west of Denver, May 5–6, 1973.
(B) Rock topple leading to a rockfall in Morgan
County, Ky.
(A): (U.S. Geological Survey)
(B): (Dale Ritter)

(B)

Along with snow avalanches and debris flows, rock-falls are one of the most significant geological hazards in alpine regions. Encounters with falling rocks are not the only hazards associated with the rockfalls. Wieczorek et al. (2000) documented more than 1000 trees snapped by a powerful air blast of up to 110 m/s associated with a rockfall in Yosemite National Park. Planners in these regions not only face the problem of delineating the runout zones for rockfalls (for example in the Colorado Rocky Mountains, Ives and Krebs 1978) but must also determine whether rockfalls have occurred under modern climatic regimes and make assessments regard-ing their frequency and magnitude. Parsons (1988) points out that assessments of future probability of mass movements are not as straightforward as for river floods (discussed in chapter 5). In a comprehensive study of recent large rockfalls in Yosemite National Park, California, Wieczorek and Snyder (1999) demonstrated that the role of structure and propagation of cracks is very complex. No reliable predictions of rockfall events could be made, but they were able to document that falls and crack development were variable over time. Studies by Matsuoka and Sakai (1999) in the Japanese Alps indicated a lack of correlation of rockfall activity with rainfall events. Rather, they observed maximal rockfall activity about 10 days after the thaw of cirque headwalls related to the movement of a thaw front into the cliff. A survey of slope failures in Vermont by Lee et al. (1997) indicated that conditions favoring failure included: (1) intense rainfall, (2) steep hillslope angle, (3) debris in shallow hillslope hollows or chutes, and (4) intersecting joints with orientations parallel to valley strike. Most Vermont failures seem to occur in late spring to early summer (May–early August). The occurrence of a slope failure affects the future probability of similar events because it removes the unstable debris from the slopes, which will require a certain recovery period before a similar triggering event, such as a given amount of rainfall, can result in movement.

Successful studies dating past rockfall events have used combinations of botanical and geomorphic data. Porter and Orombelli (1981) mapped and dated rockfalls in the Italian Alps using observations of lichens growing on rockfall deposits. Because lichens grow primarily on the upper surfaces of boulders, uniform lichen cover indicates the boulder has moved or rolled within the life span of the lichens. Additionally, once regional growth curves can be established for specific lichens (Benedict 1970), measurements of their diameters can be used to date rockfall deposits. Bull and Brandon (1998) were successful in using lichenometric dating to correlate New Zealand rockfalls with earthquakes. Hupp (1983) used observations of damage to trees on steep blocky slope sites like the one shown in figure 4.35 in Virginia to document rockfall activity within the past century.

Establishing the process responsible for the emplacement of coarse rock debris proximal to steep slopes is not always an easy task. Rockfall talus typically displays a size gradation with a coarsening of clasts downslope (fig. 4.34). Studies of boulder *fabric,* the preferred alignment of clasts, have been useful in distinguishing mass movement deposits. Flows tend to produce a longitudinal fabric parallel to flow axes (Shreve 1968), while simple rockfall deposits are less organized. Perez (1998) found distinct differences in fabric with respect to position on the talus slope. Upper talus clasts had their long axes oriented parallel to slope, suggestive of emplacement by sliding. Basal slopes fabrics were less organized indicating deposition by free fall. Considerable success in dating rockfall deposits has also been achieved by using indices of weathering measured by response to hammer impact (i.e., Nesje et al. 1994). However, Gates (1987) observed a crude fabric aligned parallel to transport in an ancient rock avalanche in the northern Black Hills of South Dakota. Recently, workers have begun to quantify block fabric associated with a variety of mass movement processes (for example, Mills et al. 1987; Bertran et al. 1997).

Topples (see fig. 4.33B) are similar to falls except that forward movement of a material block is produced by slow rotation around a fixed hinge located at the base of the block, followed by a fall. Although the process is not commonly cited, in some cases toppling may be the most important factor in cliff retreat (Caine 1982).

Slides   Rapid mass movements are grouped according to which aspects of the phenomena are most important. Because most rapid movements are not observed as they occur, the fundamental properties of motion must be interpreted after the event, based on analyses of the sedimentology and geometry of deposits or the effects of impact on permanent objects. The lack of samples and observations of these phenomena while in motion has resulted in a number of viable classifications of rapid mass movements (Sharpe 1938; Ward 1945; Varnes 1978; Hutchinson 1968). Clear-cut distinctions between the primary modes of transport are difficult to assess. The classification prepared by Varnes (1978) is adopted here because it suits our process orientation (fig. 4.36). The classification is based on the type of material being moved and the primary type of movement. Additions to the classification (Varnes 1978) are shown in table 4.6.

[margin handwritten note: usually analysis occurs after the slide]

As defined earlier, slides are slope failures that are initiated by slippage along a well-defined planar surface. The sliding mass is essentially undeformed; however, it may partially disintegrate during the sliding motion, giving rise to flow movement in the latter phase of the event. The plane of sliding may be shallow and approximately parallel to the ground surface as in the case of rockslides and debris slides, or it may penetrate to some depth as a concave surface along which rotational slip or slumping may occur (fig. 4.37).

Slides on shallow planar surfaces, called **translational slides,** are the most common of sliding phenomena. Like all forms of mass movements, the initiation of translational slides occurs when the material's resisting strength is exceeded by the shear stress. As long as F is greater than 1.0 at the potential surface of sliding, the slope will be stable. If an increase in the driving force or a decrease in resistance brings the ratio to unity, sliding will ensue. Commonly the driving force is increased by an addition of mass to the sliding block, but other factors can produce the same effect. Earthquakes, for example, generate a horizontal mass force that passes through the center of gravity of the slope material and adds an extra driving element (i.e., Shuster et al. 1996; Keefer 1999). Much of the loss of life in the January 2001 earthquake in El Salvador was caused by mass movements triggered by earthquake-induced mechanisms. Steepening of the slope also adds driving force because shear stress increases as the slope angle (θ) increases. Thus, human activity (such as undercutting) or erosion that increases slope may trigger mass movement. Actually this is a rather simplistic view because steepening also reduces the shear strength by complicated changes in cohesion, pore pressure, and effective normal stress (Terzaghi 1950; Carson and Kirkby 1972).

[margin handwritten note: $F > 1.0$ slope is stable; $F \leq 1.0$ sliding occurs]

Figure 4.34
Rockfall talus in the Colorado Rocky
Mountains. Note the general gradation of
angular talus, coarsening downslope.
(R. Craig Kochel)

Figure 4.35
Scree slope in the Virginia Blue Ridge near Waynesboro.
Largest blocks in the foreground are about 1 m long.
(R. Craig Kochel)

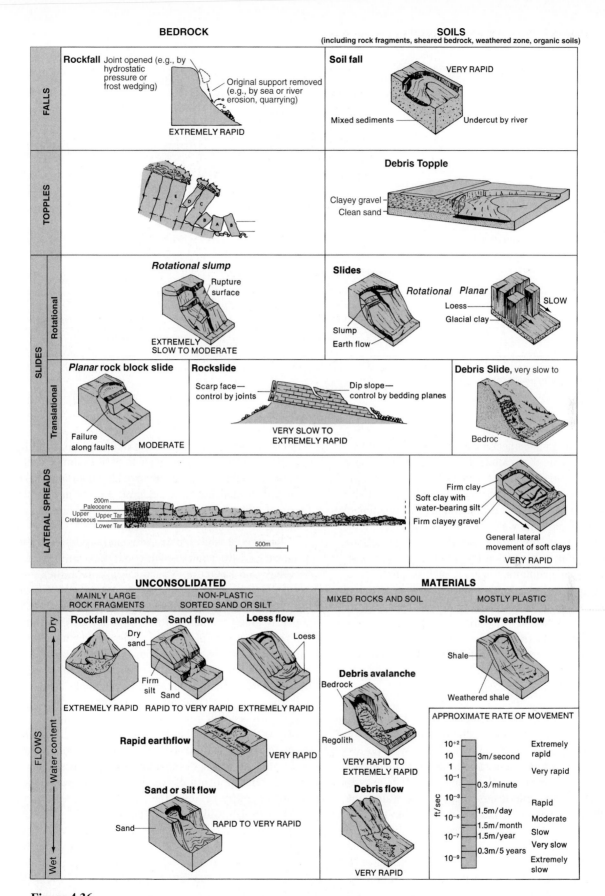

Figure 4.36

Classification of landslides.

(From D. J. Varnes, 1978, "Landslides: Analysis and Control," *TRB Special Report 176,* Transportation Research Board, National Research Council, Washington, D.C. Used by permission.)

TABLE 4.6	Classification of Mass Movement Types in Different Parent Materials.

Type of Movement			Type of Material		
			Bedrock	Engineering soils	
				Predominantly coarse	Predominantly fine
Falls			Rockfall	Debris fall	Earth fall
Topples			Rock topple	Debris topple	Earth topple
Slides	*Rotational*	*Few units*	Rock slump	Debris slump	Earth slump
	Translational		Rock block slide	Debris block slide	Earth block slide
		Many units	Rockslide	Debris slide	Earth slide
Lateral spreads			Rock spread	Debris spread	Earth spread
Flows			Rock flow (deep creep)	Debris flow	Earthflow (soil creep)
Complex			Combination of two or more principal types of movement		

From D. J. Varnes 1978 "Landslides: Analysis and Control," *TRB Special Reports 176: Landslides,* Transportation Research Board, National Research Council, Washington, D.C. Used by permission.

The sliding phenomenon can also be produced by a variety of events that reduce the internal resistance of the debris. From observation, sliding usually occurs after prolonged or exceptionally heavy rainfall, indicating that the lowering of resistance is predominantly a function of water. In the past, the water effect was interpreted to be lubrication along the sliding surface. Terzaghi (1950), however, refuted this notion by pointing out that water applied to many common minerals, such as quartz, is actually an antilubricant. Furthermore, most soils in humid regions contain more than enough water to cause lubrication at all times, yet they also fail after rainstorms. Clearly water affects strength in other ways. You will recall that shear strength is a function of cohesion (c), effective normal stress (σ'), and friction (ϕ) such that

$$S = c + (\sigma') \tan \phi$$

The response of these factors (c, σ', ϕ) to wetting is significantly more important in the initiation of slippage than is lubrication (Sidle and Swanston 1982). For example, the rise of the water table or the piezometric surface, which accompanies all prolonged rainfalls, may be the most common culprit in sliding. As the water table rises, the pore pressure (μ) at any point within the saturated mass increases, ultimately resulting in a decrease in effective normal stress (σ') and a concomitant reduction in shear strength. Numerous recent studies have focused upon linkages between landslide initiation and shallow groundwater dynamics (i.e., Miller and Sias 1998; Van Asch et al. 1999; Matsuoka 1996). Miller and Sias (1998) showed that the spatial variability in landsliding was controlled by groundwater flux. Figure 4.38 shows the close relationship between rainfall, ground-

water flux, and landsliding (Matsukura 1996). Significant progress has also been made to define rainfall intensity and cumulative rainfall thresholds important in the initiation of landslides (i.e., Schrott and Pasuto 1999). Iverson (2000) provides a detailed theoretical analysis of landslide initiation by rainfall initiation.

The type of landslide that ultimately occurs at a site is influenced by a range of factors, most importantly lithology and structure. Structural discontinuities, because they control material strength limits, play a large role in the morphology of landslides. Rocks with pronounced structure, such as foliation or inclined bedding, and prominent joint planes will tend to yield planar landslides and rockfalls, respectively. In contrast, failures in homogeneous materials and horizontally stratified units tend to be rotational in character. For example, Jacobsen and Pomeroy (1987) report that slumps and rotational earthflows account for most of the slope failures reported in the Appalachian Plateau, which occur primarily in colluvium and flat-lying fine-grained sedimentary rocks. In some cases, the style of movement is regulated primarily by the depth of weathered residuum. Crozier (1986) cogently reviews the effects of material and structure on landslide form. Often overlooked is the influence of relict bedrock structures on slope stability. Irfan (1998) clearly showed the control of landslides by relict joints in saprolitic soils. This study argues for detailed analyses of stability for development in regions characterized by saprolite at the surface.

Rockslides are usually associated with major structural features within the rock such as the stratigraphy of the rock sequence, joint patterns, and orientation of the foliation in metamorphic rocks. Massive rock units normally

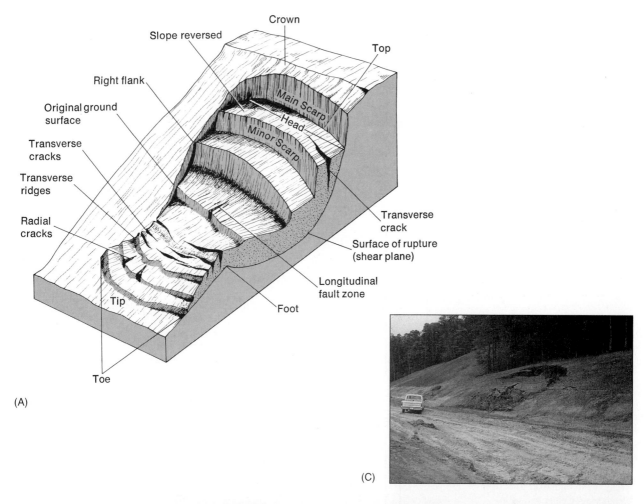

(A)

(C)

(B)

Figure 4.37

(A) Features of a rotational slide.

(B) Example of rotational slump in which toe area has disintegrated into an earthflow.

(C) Small rotational slide (slump) in colluvium along a new highway cut in the Virginia Blue Ridge

(A): (From D. J. Varnes, 1978, "Landslides: Analysis and Control," *TRB Special Report 176,* Transportation Research Board, National Research Council, Washington, D.C. Used by permission)

(B): (U.S. Geological Survey)

(C): (R. Craig Kochel)

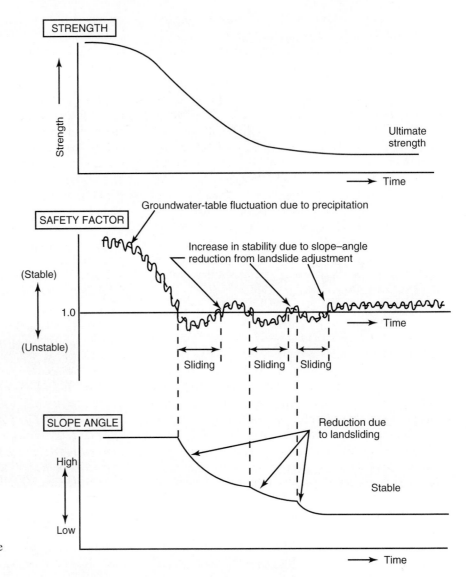

Figure 4.38
Schematic showing the relationship between weathering, slope evolution, and soil creep over time (from Matsukura 1996). Top shows decay of strength over time from weathering. Center shows fluctuations in stability with groundwater flux. Bottom shows slope evolution with respect to slope angle.

have several prominent joint sets as well as a superimposed network of randomly spaced and oriented fractures. Prior to the creation of the joints these rocks have great shear strength associated with the high cohesion in lithified materials. Once joints begin to form, however, the process becomes self-generating because shear stresses are concentrated preferentially on the unfractured zones of solid rock. With time, more and more of the original rock is consumed by jointing, and eventually the near surface becomes a cohesionless mass of densely packed angular blocks (Terzaghi 1962). At this stage, except for particle size, the rock mass resembles an aggregate of dry sand in which the cohesion within the individual particles has little bearing on slope stability. The shear strength of the total mass stems entirely from internal friction because the cohesion across the joint openings is zero. In a reconnaissance of several hundred rockslides in Alberta, Cruden and Eaton (1987) related the probability of rockslides to the geometry of slope angle and dip. Rockslides were found to be most frequent where dip was nearly parallel to slope and in overdip slopes. Rockslides were less common in reverse-dip slopes and underdip slopes.

Rockslides can be divided into two types: rock avalanche and slab failure (Carson and Kirkby 1972). Both obey the same mechanics, and they differ only in the amount of fracturing within the rock and the angle on the potential surface of sliding. In **slab failure,** cracks develop where a rock mass expands because the horizontal confining pressure is removed, allowing strain to proceed outward in the direction of the pressure release. This process is similar to pressure-release sheeting. As lateral stress is removed, a tensional zone develops in the upper part of the mass, and cracks form with the zone. The tensional fractures penetrate to a depth that is controlled by the strength of the material. Because rocks usually have high tensile strength, fracturing does not penetrate the total depth of the tensional zone. This is not always the case in unconsolidated substances.

The stability of the outer slab depends on the depth of the fracture relative to the height of the unconfined surface that is undergoing expansion. Equations have been derived to predict the maximum height attainable before slab failure occurs (Terzaghi 1943), but predictive models are not always completely reliable. For example, in one study (Lohnes and Handy 1968), tension cracks probably did not appear until slab failure was imminent. As a result, the unsupported face was considerably higher than would have been possible had the cracks developed at an earlier stage. In addition, Terzaghi (1962) considered the worst possible case, the weakest rocks, and found that most rock types could probably stand vertically up to heights of 1300 m. The observation that few vertical cliffs stand at this height even in stronger rocks indicates that fracturing drastically reduces the height that can be maintained on an unsupported rock face.

Rock avalanches occur when the joint network becomes essentially continuous down to the potential surface of sliding. An avalanche differs from slab failure in that it involves the entire mass above the sliding surface, whereas slab failure includes only the material outside the outermost continuous joint. Both types of rockslide differ from rockfalls in that the rocks fail only when fractures intersect the potential slide plane. Both heaving and sliding processes may be involved in some natural mass movements (Schumm and Chorley 1964). In fact, the incidence of sliding is so complicated that Terzaghi (1950) was able to list 19 possible causes.

Flows In true **flows,** the movement within the displaced mass closely resembles that of a viscous fluid, in which the velocity is greatest at the surface and decreases downward in the flowing mass. In many cases, flows are the final event in a movement begun as a slide, and the distinction between the two is sometimes unclear. For most types of flow, abundant water is a necessary component, but dry flows, called rock fragment flows by Varnes (1958), do occur when rockslides or falls increase drastically in velocity and lose their identity as a unitized mass. When rocks slide or fall down a steep slope, the material disintegrates as it crashes into the relatively flat surface at the base of the slope. From there the rocks travel as masses of broken debris (called *sturzstroms* in German), moving with enormous velocities over the gentler slopes in the piedmont or valley bottom. For example, the wet, mud-soaked sturzstrom at Mount Huascaran, Peru, sped at approximately 400 km an hour over a distance of 14.5 km, wreaking destruction along its path and killing some 80,000 persons (Eriksen et al. 1970; Browning 1973).

The exact mechanism required to move these large bouldery masses (usually greater than 1 million m³) for such long distances is the subject of some controversy. Shreve (1966b, 1968) proposed that the material slides continuously on a layer of compressed air that is trapped beneath the debris as it comes to the bottom of the steep slope. This air-lubrication hypothesis has been challenged by Hsu (1975) who suggests that sturzstrom movement is primarily a flow phenomenon, reviving a conclusion made earlier by Heim (1932). In most cases the highest strata involved in the fall are contained in the rear portion of the deposited debris. Shreve correctly indicates that this distribution negates viscous flow as the transporting mechanism, because in that process the uppermost layers are transported at higher velocities and so would be farther downstream in the blocky debris. The deposits, however, also have geometric features similar to those formed by lava flows and glaciers, and care must be used to distinguish between the different types of deposits (Porter and Orombelli 1980). Hsu therefore feels that the debris moves by flow, but that the mechanism differs from viscous transport in that individual particles are dispersed in a dust-laden cloud, and the kinetic energy driving the flow is transferred from grain to grain as particles collide and push one another forward. The entire mass moves simultaneously until all the original energy is dissipated by friction from the particle collisions.

Flow by this process would explain the distribution of the source rocks within the deposits. It would also permit great distances of transport because the frictional resistance decreases when grains are immersed in a buoyant interstitial fluid that reduces the effective normal stress. Variations in flow distance and velocity in different events probably depend on the properties of the interstitial substance.

Debris flows are a complex group of gravity-induced rapid mass movements intermediate between landslides and water flooding (Johnson 1970). Debris flows include a variety of grain sizes from boulders to clay mixed with varying amounts of water (table 4.7). Because of the high velocity and impact force of many debris flows, they represent significant geological hazards (i.e., fig 4.39). Costa (1988) notes that sediment concentration in debris flows ranges from 70 to 90 percent by weight (47 to 77 percent by volume) and both water and solids move together as a unit at the same velocity. **Mudflows** contain mostly fine-grained material, such as sand, silt, and clay, admixed with water. **Debris avalanches** are an extremely rapid form of debris flow typically generated on steep bedrock slopes with thin colluvial and soil cover. We will use the term debris flow to include this entire range of related processes and features as well as others such as volcanic lahar (Johnson and Rodine 1984) because of their mechanical similarities and the difficulty in distinguishing between their rates of movement and grain size following the event.

Debris flows may originate and become mobilized in a variety of ways. The prerequisite conditions for debris flows include (1) an abundant source of moisture, (2) an abundant supply of fine-grained sediment, and

TABLE 4.7	**Characteristics of Debris Flows and Comparisons with Other Materials.**							
Material	Velocity (m/s)	Slope (%)	Bulk Density (g/cm³)	Newtonian Viscosity (poise)	Depth (m)	Solids (% by weight)	Shear Strength (dn/cm²)	Flow Character
Debris flow	0.6–31	5.8–47	1.8–2.6	200–60,000	0.5–12	70–90	>>400	Laminar
Hyperconcentrated flow			1.33–1.8	20–200		40–70	100–400	Laminar and turbulent
Water flood	1–15		1.01–1.33	0.01	0.5–20	<<40	0–100	Turbulent
Cement				24			480	
Honey				115			580	

Modified from Costa (1984); Costa (1988).

Figure 4.39
Impact forces from debris flow boulders undermined part of this high-rise apartment in Venezuela. Several major cities were seriously impacted like this by flows in December 1999. The cities are situated on major debris fans. Note the huge boulder lodged on the third floor.
(Photo by L. S. Eaton)

(3) relatively steep slopes (Costa 1984). Although numerous ideas exist concerning how debris flows are initially mobilized (Johnson and Rodine 1984), most researchers feel that they begin either as a debris-laden slurry that erodes its own channel and thus increases its sediment concentration, or as a shallow landslide that provides a high concentration of unconsolidated debris and mobilizes as a flow when runoff is mixed with the debris. Sasaki (2000) showed that soil creep, initiated by rainfall, triggered debris flows in Japan. Numerous studies have shown that debris flows were initiated by shallow planar slides that evolved into flows with the addition of water downslope (i.e., Shimokawa and Jitousono 1999). The abundant moisture necessary for debris flows is commonly provided by intense rainfall or rapid snowmelt, or in some cases by snow and ice melt during

volcanic eruptions such as Mount St. Helens in 1980. Flows are typically sourced in small drainage basins where slopes are steep, where runoff can be concentrated, and where sediment supply is likely to be high. Many studies suggest that debris flows may be responsible for the bulk of the sediment eroded from a basin, although these movements occur episodically during infrequent but catastrophic events (for example, Zicheng and Jing 1987).

Debris flows can vary greatly in their physical properties between sites and even during an event at one location, but they often have many commonalities which produce distinctive erosional and depositional effects on the landscape. Figure 4.40 shows a typical debris flow channel, lateral levee deposits of coarse, unsorted debris, and an abrupt lobate terminus. Debris flows do not behave mechanically as Newtonian fluids such as water. Instead, they possess a significant internal shear strength which enables the flow to behave as a relatively rigid plug moving over a laminar boundary zone; as such, a debris flow is referred to as a Bingham material. Debris flows have a high bulk density and high viscosity compared to water flows. This shear strength, in concert with other processes operating within the flow because of its excessive sediment concentration, enables debris flows to transport exceedingly large boulders (up to tens of meters) and allows them to flow with far less turbulence than water floods. Figure 4.41A shows examples of the extreme erosion incurred by Blue Ridge debris flows in Madison County, Virginia, in June 1995. The shear strength of these Virginia debris flows is impressively illustrated by the size of the clasts transported in figure 4.41B. A survey of the impacts of the Madison flood event can be found in Morgan et al. (1999). The special rheological properties of debris flow, approximated by a Coulomb-viscous model (Johnson 1970), also account for (1) their ability to construct extensive channels with boulder levees, (2) their typical inverse grading, (3) their ability to stop abruptly, and (4) their damaging impact forces. Iverson (1997) and Major and Iverson (1998) provide good recent discussions of debris flow dynamics.

Figure 4.40
Two debris flows on a talus slope in southern Iceland. Talus is forming from the erosion of the basaltic cliff. Note the bouldery levees lining the twin debris-flow channels and the abrupt cessation of flow once slope angles were encountered that were too low to permit internal shear within the flow.
(R. Craig Kochel)

(A)

(B)

Figure 4.41
Debris flows in the Virginia Blue Ridge in June 1995. Catastrophic rainfall, centered on Graves Mill, along the Rapidan River triggered hundreds of debris flows, landslides, and other mass movements. (A) View from Kinsey Run (foreground) of a major flow deposited on the margin of an older debris fan. (B) Example of the scour along the debris-flow track and transport of large clasts in Kinsey Run.
(R. Craig Kochel)

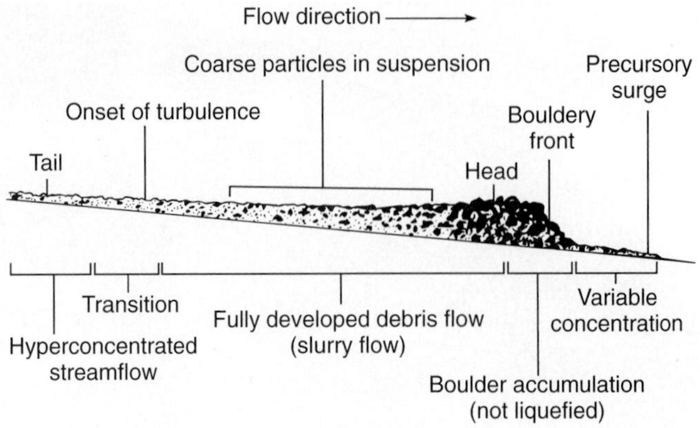

Figure 4.42

Components of a typical debris flow observed near Mount St. Helens.

(Source: T. C. Pierson, in *Hillslope Processes*, ed. by A. D. Abrahams, copyright 1986 Allen & Unwin.)

Although debris flows typically follow preexisting channels, they can flow across unchanneled alluvial fan surfaces because of their ability to construct leveed channels as the flow proceeds. Viscosities and other characteristics of debris flows can vary greatly at a single point of observation during the course of a flow. Changes in the character of the moving debris are generally regulated by additions of debris and water from lateral sources along the flow path. Debris flows have been observed to move in pulses or surges accompanied by changes in the fluid characteristics (Sharp and Nobles 1953; Pierson 1980; Costa 1984). Figure 4.42 illustrates the major components of a typical debris flow observed along streams draining the Mount St. Helens volcano in Washington (Pierson 1986). From this view, it is easy to see why the flow characteristics change with time at a given observational point. Following the passage of the viscous and coarse-grained flow front, the flow is commonly transitional to a more turbulent muddy slurry and fully turbulent streamflow. A series of gradations from landslide through debris flow is common from the source to downstream depositional areas. In companion studies of debris flows in the Mount St. Helens area, Scott (1988) showed that flow transformations are also common along the flow path as a function of changing hydraulic conditions with distance from their source, both in recent flows and in paleodebris flow deposits recognized in the stratigraphy of the flow channels. These transformations occur because of constant increases and decreases in sediment concentration within the flow, referred to as bulking and debulking, respectively. Flows may transform along their paths from debris flow to streamflow, by passing through a transitional state known as hyperconcentrated flow (Beverage and Culbertson 1964; Pierson and Costa 1987). Figure 4.43 shows an example of these flow transformations.

Because of the catastrophic nature of debris flows, direct sampling and observations of the movements are rare. Considerable progress has been made in recent studies in developing methods for estimating the physical properties and character of the flowing debris based on analyses of the remaining deposits, erosional effects along the channels, debris lines, and channel geometry (Costa 1984; Johnson and Rodine 1984; Brunsden 1984; Berti et al. 1999).

Debris flows can flow for many kilometers from source regions, but they typically come to rest on relatively low-gradient areas lacking channel confinement on alluvial fans (fans are discussed in detail in chapter 7). Despite the exceedingly complex controls on flowing debris and its deposition, progress is being made in determining threshold slope angles for debris flow deposition (Van Dine 1985; Rickenmann and Zimmermann 1993). Major and Iverson (1998) suggest that past notions of the importance of excess pore-fluid pressure may not be as important in the control of debris-flow runout. Instead, they suggest that depth and runout are more affected by grain-contact and bed friction along the flow margins. In like manner they warn that deposit thickness may not be useful in estimating the shear strength of debris flow. Webb and Fielding (1999) demonstrated that the high mobility of some debris flows can result from high percentages of clay in the flow matrix. Thus, debris flows can continue to flow on far gentler slopes than expected. Such studies, in combination with studies of the characteristics of debris flow source areas, are useful in planning for construction and preventative design in regions prone to debris flow (Hungr et al. 1987; Reneau and Dietrich 1987). Mitigation of debris flow hazards is an issue of increasing concern in mountainous regions (U.S. Geological Survey 1982) because debris fans (the depositional zones) tend to be favored for development because of their location above floodplains. Figure 4.44 shows that the three regional areas having high potential for mass movements are the Appalachian Mountains, the Pacific Coast ranges, and selected regions of high relief in the central Rocky Mountains.

A good example of the serious magnitude of the debris flow problem is the frequent and damaging flows that occurred along the Wasatch Front in Utah during 1983. Damages from these flows exceeded $250 million along this zone where most of the urban centers of the state are located. Most notable, the Thistle landslide-flow

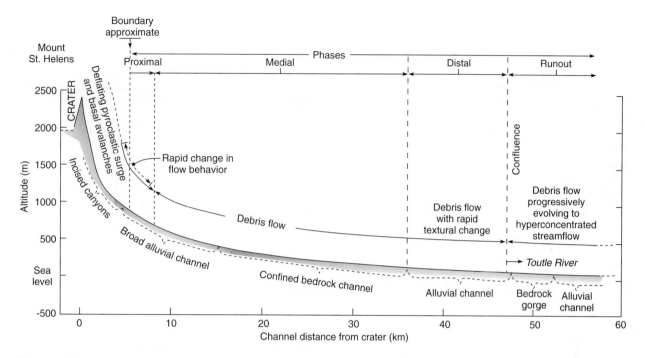

Figure 4.43
Transformations of debris flow along its path originating at Mount St. Helens.
(After K. M. Scott, U.S. Geological Survey Professional Paper 1447-A 1988.)

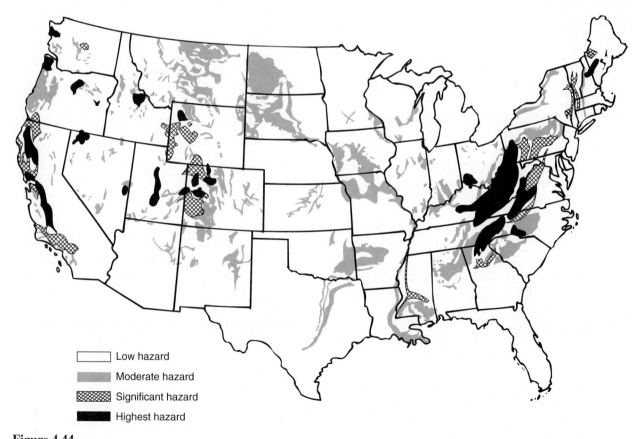

Figure 4.44
Major slope hazard regions in the United States. Darkest areas represent greatest severity. Details can be found in U.S. Geological
Survey Professional Paper 1188.
(From Hays, 1981 *Facing Geologic and Hydrologic Hazards,* U.S. Geological Survey Professional Paper 1240-B.)

(A)   (B)

Figure 4.45
Debris flow and avalanches in Nelson County, Va., during 1969. (A) Flows deposited debris on small fans at the base of first-order hillslope channels (near Lovingston). (B) Catastrophic erosion and impact forces from these flows removed some structures and devastated others (view from Davis Creek).

(A, B): (Courtesy Virginia Division of Mineral Resources, Charlottesville, VA.)

may have been the most expensive mass movement in United States history. It severed major east-west transportation arteries, U.S. Highway 6, and the main line of the Denver and Rio Grande Railroad, dammed the Spanish Fork River, and flooded about 15 homes in the temporary lake that resulted (National Research Council 1984). Some of these slides and flows occurred at sites of prehistoric mass movements as the colluvium was remobilized (Fleming et al. 1987). Activity had been observed for several years prior to the major Thistle event (Witkind 1988). Over 150 major debris flows, hyperconcentrated floods, and landslides occurred during 1983–84 following the culmination of several anomalously wet years. A combined heavy winter snowpack, followed by a cool spring and rapid late-spring melt produced most of the flows. Many of the debris flows occurred on slopes with shallow colluvial cover over weathered metamorphic rocks and were initiated by shallow soil slides caused by elevated pore water pressures at the soil-bedrock boundary (Mathewson and Keaton 1986). Detailed geomorphic studies have led to techniques for evaluating the potential for debris flows and hyperconcentrated floods along the Wasatch Front (Wieczorek et al. 1989).

Generally associated with semiarid regions having sparse vegetation, debris flows are also common in densely vegetated chaparral in California and forested mountain slopes in humid regions (Hack and Goodlett 1960; Williams and Guy 1973; Kochel 1987, 1990). Nelson County, Virginia, was the site of one of the most cataclysmic debris flow and flooding episodes in United States history on the night of August 19, 1969 (Williams and Guy 1973). Up to 30 inches of rain fell in less than

eight hours on a small, rugged area, resulting in hundreds of debris flows and avalanches, the loss of 150 lives, and massive road and bridge destruction in this rural area (fig. 4.45). The combined effect of excessive moisture from the inland remnants of Hurricane Camille moving northeastward from the Gulf of Mexico, an easterly advancing cold front, and orographic effects of the Blue Ridge Mountains produced intense local rainfall. Debris flows were mobilized on the slopes as intense rains elevated pore water pressures at the soil-bedrock interface, typically about 1 m below the surface on the steep, colluvial forested slopes. Debris flows scoured out shallow hillslope hollows. Subsequent observation of weathering on the bedrock that was under the colluvium suggests that these sites normally concentrated shallow subsurface flow (Kochel 1987). Reneau and Dietrich (1987) and Reneau et al. (1990) emphasized the importance of identifying colluvium-filled hillslope hollows in the anticipation of future debris flow events in northern California (fig. 4.46).

The linkage between climate change and debris flow activity is being investigated in many areas. Reneau et al. (1989, 1990) observed patterns in the chronology of hollow evacuation episodes in California that they attributed to changes in climate. Bovis and Jones (1993) have linked episodic activity among several forms of mass movement in British Columbia with Holocene hydroclimatic variations. Kochel (1987) observed a coincidence between the initiation of debris flow activity on alluvial fans in Virginia and the time when climate may have ameliorated enough to permit frequent invasion of tropical air masses into the central Appalachian Mountains. Debris flow activity may be a

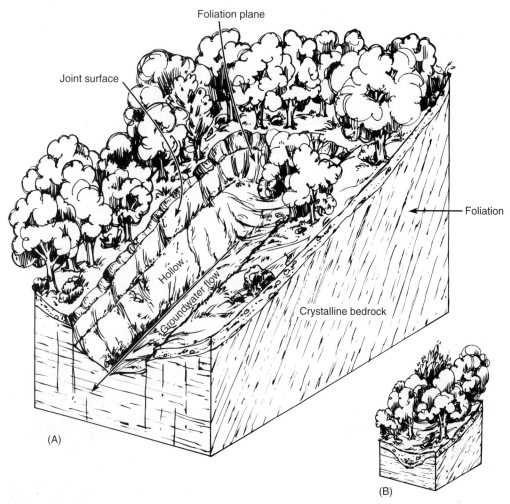

Figure 4.46
Schematic model of a colluvium-filled hollow. (A) Hollow excavated to bedrock by a recent debris avalanche/debris flow. (B) Hollow has been refilled with colluvium and slope wash after a few hundred years. The site is now primed for another debris-flow event once again.

sensitive indicator of paleoclimatic variability. Likewise, changes in the frequency and occurrence of debris flows could result from future climate variations.

The erosive potential of debris flows like those in Virginia is enormous (see fig. 4.45B). Geomorphologists must be able to recognize and assess the potential for debris flows in a specific region. Are the fans active or relict from some ancient environmental regime? What is the anticipated frequency of deposition? Are the fans and channels subject to debris flow or water flood or both? What is the magnitude and extent of anticipated future flows? Determination of process, whether debris flow or water flood, requires detailed analysis of the morphology and sedimentology of the deposits. Costa and Jarrett (1981) and Costa (1984) show how debris flow sediments can be distinguished from waterlaid deposits and discuss the likely environmental consequences of misidentification. Determination of fan activity and anticipated debris flow frequency is difficult. A

first approximation of flow activity on debris fans can be accomplished by detailed studies of the stratigraphy, sedimentology, and soils on the fans. Based on stratigraphic studies from excavations of debris fans where debris flows were deposited, Kochel (1987) showed that catastrophic flows like the 1969 Hurricane Camille flows in central Virginia occurred at least three times within the past 11,000 years (fig. 4.47). In an extensive study using more than 45 radiocarbon-dated flows in Madison County, Virginia, Eaton (1999) found similar recurrence intervals of 2000 to 3000 years for Quaternary Blue Ridge debris flow events (fig. 4.48). Similar events are known to have occurred periodically in the central and southern Appalachian Mountains in historic times. In this region these types of terrain-locked rainfall events associated with tropical moisture may be rather common. Climatologists point out that such localized intense storms often go undetected because of the low density of weather observation stations and their localization in

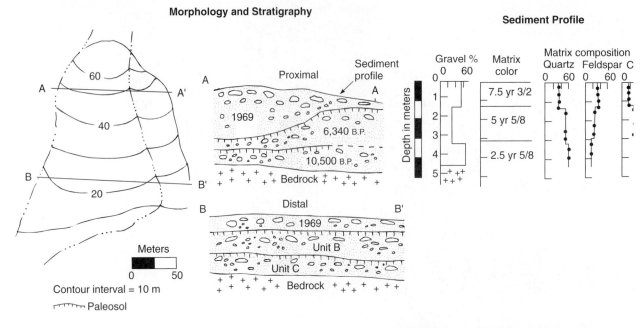

Morphology and Stratigraphy

Sediment Profile

(A)

(B)

Figure 4.47

Stratified debris fan sediments in Nelson County, Va. (A) Radiocarbon-dated organic debris shows that events like the Camille flows have occurred there at least three times in the last 11,000 years. (B) Stratified floodplain sediments in Davis Creek, Va. The top layer above the glove was deposited by the Camille flood in 1969. The two lower units have paleosols developed on them and erosional upper boundaries.

(From R. C. Kochel, Holocene Debris Flows in Central Virginia, in "Debris Flows/Avalanches: Process, Recognition, Mitigation," *Reviews in Engineering Geology,* vol. 7, p. 148; 1987.)

(B): (R. Craig Kochel)

remote regions (Michaels 1985). As population continues to expand into more remote regions of the Appalachians, damage like that caused by the Camille storm in 1969 may become increasingly common. Detailed histories of debris flow frequency and magnitude have also been reconstructed by studying debris flow-induced damage in tree rings. Hupp (1984) used a dendrogeomorphic approach at Mount Shasta, California, to elucidate a 300-year record of debris flow activity. Bowers et al. (1997) were able to determine debris flow

history in the Grand Canyon using ecological aspects of plant colonization on debris flow surfaces.

Progress is being made in identifying threshold intensities required to destabilize slopes in selected regions of rainfall and snowmelt (Campbell 1975; Caine 1980; Church and Miles 1987; Wieczorek 1987; Takabatake et al. 1998). Recent studies have clarified the relationships between hillslope fires and subsequent debris flows (Parrett 1985) and indentified a wide range of mechanisms that may ultimately produce debris flows

Figure 4.48
Poorly sorted debris-flow deposits from June 1995 overlying buried soil developed on older debris-flow deposits, Madison County, Va.
(R. Craig Kochel)

(Clague et al. 1985). Although many studies have shown that debris flows often occur in steep basins shortly after a fire, Florsheim and Keller (1987) have shown that debris flow frequency may be much lower than the frequency of fires in a given region (Meyer and Wells 1997). These kinds of studies, in concert with laboratory investigations, are beginning to uncover some of the complex threshold relationships that control debris flow initiation, transportation, and deposition (see Wieczorek et al. 1997).

The impact of catastrophic inputs of coarse sediment by debris flows has been shown to have a marked effect on the character of downstream alluvial fan and fluvial systems (fig. 4.49). Common impacts include reduction of channel flow capacity, pulses of migrating,

coarse sediment, changes in channel pattern, and development of alluvial fan and terrace surfaces (Cu 1999; Kochel et al. 1997; Eaton 1999; Cenderelli and Kite 1998).

Earthflows are one of the most common mass movement phenomena (Keefer and Johnson 1983) involving rates ranging from slow to rapid. Earthflow mechanics are complex. Some earthflows have basal lobes and other features suggestive of fluid-like flow, while others exhibit slickensides and other properties suggestive of rigid blocks moving on distinct shear surfaces. In slow earthflows the original failure of the slope usually occurs as a slump, often when the mass becomes saturated with groundwater. A rising water table and increasing pore pressure tend to lower shear resistance, and slippage results (Wells et al. 1980). If the slumped

(A)

Figure 4.49
Debris flows in Madison County, Va., 1995.
(A) Flow tracks and scour along the east side
of Kirtley Mountain. Flows were common on
this slope where foliation and joints aligned
nearly parallel to slope. Note the forest-
covered depositional area and coalescing
debris fans. (B) Kinsey Run flow (far left) and
neighboring debris flows. Note the small,
steep source basins. Debris-flow sediments
were deposited on alluvial/debris fans
throughout the area here at Graves Mill.
(L. S. Eaton)

(B)

mass is relatively wet, it may slowly bulge forward at its
front by viscous flow and take the form of tongues, superimposed piles of rolled mud, or bulbous toes
(fig. 4.50). This movement may continue at a slow pace
for many years until stability is finally reached.

Keefer and Johnson (1983), working with earthflows in California, developed a model for earthflow behavior that incorporates elevated pore water pressure
and loading from deposits of other mass movements as
the predominant mobilization processes. Earthflows
have been observed on a variety of slope angles. Their
mobility appears to depend on a combination of factors,
including unit weight of the material, shearing resistance
(shear strength) of the soil, maximum pore water pressures, and geometry of the slope mass (Keefer and Johnson 1983). Unlike most other forms of mass movement,
earthflows are commonly characterized by slow movements spread over long periods of months to years.
Episodic surges have also been observed in earthflows

commonly slow in movement and can last for months or years

(Keefer and Johnson 1983; Grainger and Kalaugher
1987), but correlations between environmental variables
and activity appear complex.

To summarize, the distinction between slides and
flows is often rather nebulous. Several generalizations
can be proposed, however, to help put mass movements
in some reasonable perspective.

1. Slides are characterized by distinct masses or blocks
   of earth materials that move along distinctive planar
   or curviplanar failure surfaces by shearing
   mechanisms.
2. Flows possess no discrete shear planes but
   experience deformation throughout their thickness.
3. Mass movement events are commonly transitional
   between several different forms both temporally at a
   site and spatially along their path. Many flows
   observed in downslope areas are actually initiated
   by slides and other phenomena, acquiring attributes

Figure 4.50
Earthflow along Mississippi River bluffs, southern Illinois.
(Courtesy of Illinois Department of Transportation)

of flow later as water and sediment are contributed to them during the course of their travel. Slope stability analyses, therefore, must consider the mechanics of sliding and the factors that might produce failure as well as the downslope products of these events.

4. The mobility of mass movements depends to a large degree on the amount of water and sediment within the displaced material, particularly in flow movements.

Morphology of Mass Movements

It is appropriate to ask how we can reconstruct the mode of mass transfer, especially since subtle transitions from one mechanism to another are common and most interpretations of movement characteristics are made after the event is over. The distinction between mass movement processes is exceedingly important from the perspective of environmental engineering because of the great differences in volume of water runoff, duration, and impact forces associated with the various mass wasting processes. Unless some concrete relationship can be established between the surface configuration of the displaced material and the genetic process, we face an insoluble problem.

Fortunately, there is some evidence to suggest a morphologic relationship that discriminates between mass movement processes. In a study of 66 landslips in New Zealand, Crozier (1973) arranged the common types of mass movements into five primary process groups: ① fluid flow (mudflows, debris flows, debris avalanches), ② viscous flow (earthflows, bouldery earthflows), ③ slide-flow (slump/flow), ④ planar slides (turf glide, debris slides, rockslides), and ⑤ rotational slides (earth and rock slumps). The aim of the study was to characterize the slope movement process using simple morphometric indices (table 4.8), assuming that the degree of flow deformation of the displaced mass could be correlated to water content (fig. 4.51). The relationship between each process group and the index values was tested statistically to ascertain whether the correlation was significant.

Crozier found that the classification index relating depth to length (D/L) was the best indicator of the process group, reaffirming Skempton's (1953) assertion about the importance of this parameter. As one would expect, the D/L value decreases markedly with greater flow (table 4.9) because the displaced material will extend farther downvalley than it would if moving as a sliding block. Some uncertainty will remain, however, unless the classification index is used in conjunction with other indices. Importantly, a definite inverse relationship was found between D/L and three other morphometric indices (flowage, dilation, fluidity). Each of these is presumably controlled by the water content of the material during its movement. Although more work

TABLE **4.8**	Morphometric Indices Used to Determine Process of Mass Movement.
Index	**Description**
Classification	. D/L Maximum depth of displaced mass prior to its displacement over maximum length.
Dilation	Wx/Wc Width of convex part of displaced mass over width of concave part; indicates lateral spreading.
Flowage	$(Wx/Wc - 1) \times Lm/Lc \times 100$ Lm is length of displaced mass; Lc is length of concave segment.
Displacement	Lr/Lc Lr is length of the surface of rupture exposed in concave segment. Low value indicates instability.
Viscous Flow	Lf/Dc Lf is length of bare surface on displaced material; Dc is depth of the concave segment.
Tenuity	Lm/Lc Indicates how dispersed or cohesive the material is during displacement.
Fluidity	Amount of flowage expected from particular type of material on distinct slope. Varies with water content.

After Crozier 1973.

Note: Compare figure 4.51.

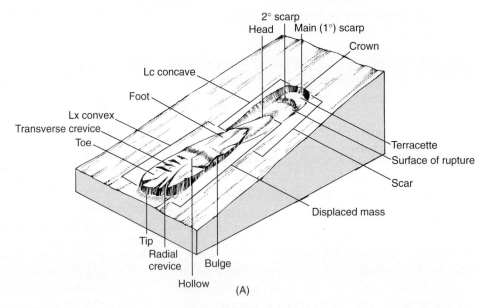

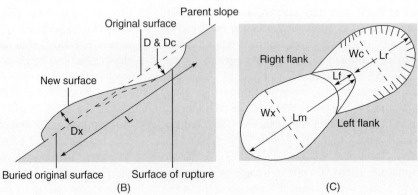

Figure 4.51

Different morphometric indices shown in diagram can be used to describe and distinguish processes of movement:
(A) landslip terminology;
(B) longitudinal section;
(C) plan view.

(From: M. J. Crozier. 1973, "Techniques for the morphometric analysis of landslips," Zeitschrift for Geomorphologie 17: 78–101. Gebrüder Borntraeger Verlagsbuchhandlung, Stuttgart. Used by permission.)

is needed before a clear relationship between morphology and process can be defined, the morphometric approach exemplified by studies like Crozier's holds real promise. Predictions of slope stability, possible modes of failure, and areas that might be affected are potential benefits if we can understand how previous movements occurred in any given region.

Landslides are ubiquitous in most all lithologies and climates. One of the primary goals of slope hazard analyses is the determination of whether landslides are active. Table 4.10 (from Crozier 1984) summarizes field criteria useful in distinguishing between active and inactive landslides. An even more fundamental concern of the geomorphologist is to be able to distinguish landslide deposits from deposits left by other geomorphic activity such as fluvial and glacial processes. Table 4.11 presents a summary of characteristics useful in making such determinations in the field. Brabb and Harrod (1989) have compiled an extensive collection of encapsulated descriptions of modern landslides and accounts of their mitigation and economic implications in virtually every country in the world. Extensive relict Quaternary landslides have been recognized in most regions as well (for example, Southworth 1988).

Research on mass movements and their effect on environmental planning is continuing at a high level.

New techniques are being applied to real-time monitoring of hillslope processes, including the installation of shallow piezometers (wells) and global positioning systems (GPS), which permit communication between parts of the hillslope not within line of sight (Gili et al. 2000). Brunsden (1993) provides an assessment of the current state of knowledge in the field and the anticipated direction of future research in each of the following areas related to mass movements: description and morphology, mapping and inventory, recognition, process monitoring, and mechanisms and causes.

SLOPE PROFILES

The profiles of natural slopes formed primarily by erosional processes are generally regarded as reflections of the major geomorphic factors—climate, rock type and structure, time, and process. The relationship between slope form and geomorphic factors, however, is not straightforward. There is no simple method for deciphering which factors will determine the precise characteristics of a slope profile, and all of the factors are involved in some way to produce slopes. Therefore, many debates about the development of slopes revolve around *to what degree* a factor is involved in slope formation rather than *whether* a factor is involved. Invariably individual value judgments are involved since very few, if any, studies can absolutely isolate the effect of one factor by keeping all others constant.

Geomorphologists have paid considerable attention to the geometry of slopes and the angles developed on different parts of the profile. Ideally, slope profiles can be divided into four general components (fig. 4.52): an ① upper *convex segment,* a ② *cliff face* (or free face), a ③ *straight segment,* having a constant slope angle, and a ④ *concave segment* at the hillslope base (Wood 1942; King 1953; Carson and Kirkby 1972). More detailed distinctions of components have been suggested. For example,

TABLE 4.9	Average Values of the Depth/Length Ratio in Different Types of Landslips as Calculated for Different Areas.
Type of Landslip	**Average D/L ratio**
Flows	1.58
Planar slides	6.33
Rotational slides	20.84

TABLE 4.10	Field Criteria for Distinguishing between Active and Inactive Landslides.	
Feature	**Active**	**Inactive**
Scarp area	Sharp breaks, unweathered	Weathered, rounded scarps
	Fresh tension cracks common	No fresh tension cracks
	Unfilled grabens, depressions	Colluvial-filled grabens, depressions
Vegetation	Tilted trees	Unaffected trees
	Vegetation differences and disruptions between slide and adjacent area	Lack of vegetative differences
	Rapid colonizers only	Mixed, with old growth vegetation
Drainage	Deranged, disrupted with sag ponds	Integrated drainage system, few ponds
Slide margins	Fresh shear planes, slickensides	No fresh planes or slickensides
Soils	Lack of soil development on exposures	Soil development on exposures
Toe area	Active lobes rolling over vegetation	Inactive lobes, no vegetative disturbance
	Deposits may block slope-base drainage	Toe deposits eroded by basal drainage

Adapted from Crozier (1984).

TABLE 4.11	Distinguishing Landslide Deposits from Coarse-Grained Sediments Deposited by Glacial and Fluvial Processes.		
Characteristic	**Landslide Deposits**	**Glacial Deposits (till)**	**Fluvial Deposits (coarse)**
Sedimentology			
Sorting	Very poor	Very poor	Variable, generally good
Roundness	Angular	Moderately rounded	Rounded
Grain size	Variable, may be large	Variable, may be large	Generally finer-grained
Fabric	Generally lacking	Generally lacking	Imbricated
Composition	Local, monolithologic	Variable, extrabasinal	Variable, within basin
Stratification	None	None to poor	Well-layered
Transport process	Creep, sliding	Ice	Tractive-bedload
Morphology			
Internal forms	Transverse ridges	Sinuous moraines	Lacking ridges
Surface relief	Hummocky	Hummocky	Organized bedforms
Drainage	Poor, undrained sags	Poor, undrained sags	Well-drained
Head region	Scarp, local	None or cirque	None
Profile	Convex-up surface	Irregular	Concave-up longitudinal
Valley form	None specific	U-shaped	None specific
Lateral associations	Source region upslope	Downstream outwash and loess	Downstream-fining fluvial deposits

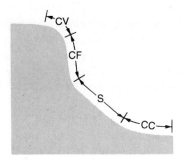

Figure 4.52
Major components of slope profiles. CC = concave segment, S = straight segment, CF = cliff race, CV = convex segment.

the classification shown in figure 4.53 recognizes nine slope components and additionally suggests that each is associated with certain dominant processes. In an actual situation, all of the components may not be present in the profile or they may have negligible significance. The upper convexity, for instance, is usually more prevalent in humid-temperature regions than in semiarid or arid climatic zones because soil creep is known to be more important in the humid environment.

Measurements in a variety of climatic zones have revealed the interesting fact that slope angles appear to be concentrated in groups with rather small ranges of values (see Carson and Kirkby 1972; Young 1972). Most pronounced are those that cluster at 43°–45°, 30°–38°, 25°–29°, 19°–21°, 5°–11°, and 1°–4°. Although any slope angle is possible, the frequency of these recurring groups is tantalizing to geomorphologists

because it probably reflects the underlying control of the great geomorphic variables.

Each of these groups has well-defined maximum and minimum values, which have been termed limiting angles by Young (1972) and threshold angles by Carson and Kirkby (1972). The general interpretation of the angular distribution is that angles within any group represent a stability regime for slopes formed in a particular climatic and lithologic setting. Under those conditions, threshold values can be exceeded if the intrinsic properties of the parent material are altered or if the climate changes. When threshold values are reached, any further change requires a fundamental response in the system that adjusts the slope angle and places it within a different stability group. Exactly why and how slopes adjust, however, is a debatable question. According to one hypothesis, groups and their limiting angles represent characteristic angles for the processes that are working on the slopes. Thus, geomorphologists recognize process as an important ingredient in the development of slope components and slope angles.

The significance of process was understood years ago by Gilbert (1877), who believed that the development of any slope was controlled by either weathering or sediment transport. Since then, the terms *weathering-limited* and *transport-limited* have been generally accepted to describe slopes formed under each process control. **Weathering-limited slopes** are created where the rate of soil or regolith production is lower than the rate of its removal by erosion. As a result, most of these profiles are determined by the character of the parent rock. Such profiles seem to prevail in dry climates or in

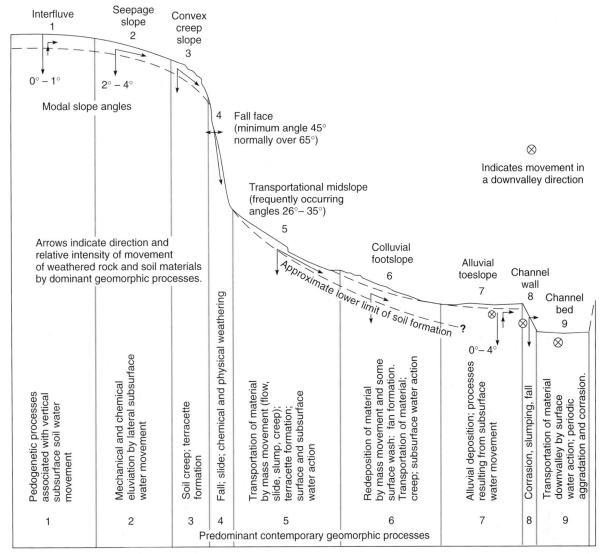

Figure 4.53

Diagrammatic representation of the hypothetical nine-unit landsurface model.

(Redrawn from J. B. Dalrymple, et al., "A Hypothetical Nine-Unit Landsurface Model," *Zeitschrift für Geomorphologie* 12:60–76, 1968. Used by permission of Gebrüder Borntraeger Verlagsbuchhandlung, Stuttgart.)

mountainous terrain where erosion is rapid, and are normally characterized by thin, weakly developed rocky soils. The rate of physical weathering tends to be at a maximum when the thickness of the residuum (the soil and colluvium) is minimal (fig. 4.54). Chemical weathering, which proceeds most efficiently under a significant cover of residuum, will be slowed, however, when the residuum becomes so thick that it interrupts the movement of water to the bedrock weathering front (an example of negative feedback). Numerous examples of weathering-limited slopes can be seen on slick-rock slopes developed in sandstones of the Colorado Plateau (Oberlander 1977; Howard and Kochel 1988). In contrast, **transport-limited slopes** are formed where the rate of weathering is more rapid than erosion. Slopes

produced under this regime normally develop on any unconsolidated parent material regardless of environment, but they are typically dominant in humid-temperate zones where vegetation cover is continuous. These profiles are less affected by parent rock and more dependent on the type and rate of slope processes.

Selby (1982) has made a cogent argument that weathering-limited slopes are directly dependent on the relative resistance of the underlying parent rocks. As evidence, he has demonstrated a high correlation between rock mass strength (see table 4.4) and the angle developed on various slope segments (fig. 4.55). A line drawn around the data points shown in figure 4.55 creates what Selby calls the *strength equilibrium envelope,* and the slopes represented by points within that envelope are

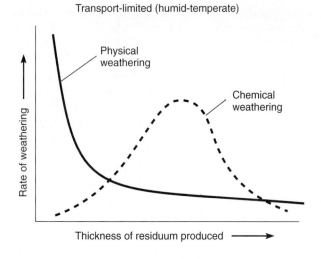

Transport-limited (humid-temperate)

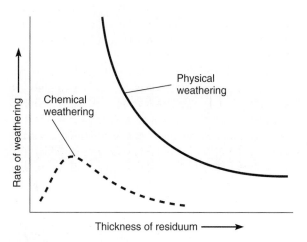

Weathering-limited (semiarid)

Figure 4.54
Schematic showing relative rates of production of weathering products on transport-limited and weathering-limited slopes.

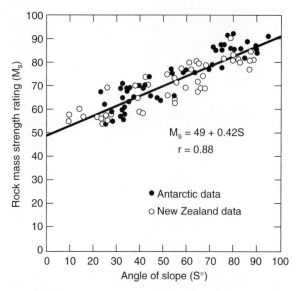

Figure 4.55
The relationship between mass strength and profile angle for all rock units studied in Antarctica and New Zealand.

(Redrawn from M. Selby, "A Rock Mass Strength Classification for Geomorphic Purposes: With Tests from Antarctica and New Zealand," *Zeitschrift für Geomorphologie* 24:31–51, 1980. Used by permission of Gebrüder Borntraeger Verlagsbuchhandlung, Stuttgart.)

referred to as *strength equilibrium slopes*. Presumably, as long as the mass strength is maintained, strength equilibrium slopes will keep a constant angle and the slope surface will retreat parallel to itself.

Whether we can expect rock strength to remain constant is debatable, however, because once joints form in bedrock, the process becomes self-generating and more of the rock will become fractured. This, of course, decreases rock mass strength and requires an adjustment in the slope angle. In addition, during the process of slope retreat, material eroded from the cliff face by rockfalls accumulates lower in the profile as talus. Technically, **talus** refers to the slope formed from the accumulation of debris eroded from a cliff face, although many geomorphologists use the term *talus* or *scree* in reference to the debris itself. Any model proposed to explain the evo-

lution of talus or scree slopes must explain the straight slope profile and basal concavity along with a gradual fine to coarse particle sorting in the downslope direction (Statham and Francis 1986). The supply of talus to the slope is controlled by weathering along mechanical discontinuities in the cliff rock, predominantly features such as jointing, bedding, and foliation. Therefore, the size of talus blocks is largely inherited from the structural characteristics of the parent bedrock. Talus extends upslope with time and eventually may bury the original rock face (fig. 4.56). If and when this occurs, a transition has been made from a weathering-limited slope to a transport-limited slope, and the strength of the parent rock is no longer significant. Instead, it is the relationship between resistance of the talus debris and the erosional forces that should determine the slope angles.

Theoretically it can be shown that in cohesionless material the angle of internal friction ϕ is equal to the slope angle θ (Carson and Kirkby 1972). Therefore, a strong correlation should exist between slope angles and the physical strength properties of unconsolidated material. In talus accumulates, angles are uncommonly high (43°–45°) because the mass is densely packed and the rock fragments are interlocked (Carson and Kirkby 1972). If the void percentage is large, little pore pressure will develop and the stable angle of slope will correspond to the ϕ angle. Continuous breakdown of the talus deposits without much clay formation will produce a sandy matrix that should stand near the angle of repose for cohesionless sands, approximately 35°. Statham and

Figure 4.56
Upslope extension of talus slopes.

Francis (1986) point out that with enough time for continued weathering, disintegration products will eventually plug pore spaces within the blocky colluvium, reducing hydraulic conductivity. The accumulation of horizons of concentrated fine-grained sediments facilitate increases in pore water pressure which eventually acts to destabilize the coarse hillslope mantles—a geomorphic illustration of a negative feedback mechanism.

Ample evidence from a wide variety of climates suggests that a weathered mixture of rock rubble and soil underlies slopes between 25° and 29° (Young 1961; Melton 1965b; Robinson 1966). As the original talus deposits are progressively broken down by weathering, the mass gradually loses its openpore framework. During times of abundant water and high water tables, the material attains positive pore pressures that reduce the effective normal stress by buoyancy. This obviously lowers shear strength and changes the relationship between internal friction and the potential failure surface. Thus, the recurrence of slope angles at 25° to 29° may be associated with saturated soils. As summarized by Skempton (1964), cohesionless materials subjected to pore pressures are likely to experience shallow landsliding along failure planes that approximate

$$\tan \theta \; 1/2 \tan \phi$$

Assuming an original ϕ angle of 45° for coarse talus deposits and 35° for a sandy mantle, the stable slope developed on these materials when they become saturated would be about 26° and 19°, respectively. In clay-rich soils the ϕ angle is much lower, and stable slope angles are considerably less.

Carson (1969) proposed that instability in slopes requires the progressive replacement of steep slopes by gentler ones. In this model, many landscapes should go through more than one phase of instability, but the exact number depends on the characteristics of the rocks and how they ultimately break down. In the initial stage, a steep rock cliff is replaced by talus or slopes developed on thoroughly fractured rocks. This phase might be followed by a change to lower slopes and eventually to the gentle slopes formed on clay-rich soils. Each slope is only temporarily stable, for as weathering changes the mantle's properties and pore pressures vary, the mass reaches its slope threshold value. Further change causes the slope to adjust rapidly into a new stability range consistent with the revised properties of the mantle. Because of the variability of soil properties and pore pressures, any limiting angle values are possible, even though they apparently cluster in recurring groups. The types of material, the number of instability phases, and the threshold values combine in any area to control the progression of slope development. The net effect of the variables is eventually to form slopes that have long-term stability with respect to rapid mass movements; at that point, creep and surface water erosion become much more significant as slope processes.

The salient point of this discussion is that recurring angles measured on slopes can be easily explained by the relationship between erosive process and the different strengths of unconsolidated materials caused by textural variations. However, whether all slope materials experience an evolution in texture as envisioned by Carson is debatable, and perhaps unnecessary to explain slope angle and form.

Processes of weathering and erosion are intimately involved in slope development. Process, however, is not an independent variable because it is directly controlled by climate and geology. Of the many variables cited as being responsible for hillslope form, only geology and climate can be considered independent variables.

The Rock-Climate Influence

As we have seen, slopes in weathering-limited situations are controlled by the mass strength of the parent rock. This is especially significant in maintenance of a cliff face. The lithologic influence on slopes is shown in both declivity and profile shape. Coherent rocks tend to support steeper slope angles, and with equal cohesion, the more massive the bedding, the steeper the slopes. Where strata contain alternately weak and resistant rocks, an irregular profile may develop, and resistant units will assume higher than normal angles where they overlie weaker rocks.

In regions where a cliff face is not present, lithology may still exert a control on slopes. Topography generally reflects lithology and the fact that "resistant" rocks underlie hills and "nonresistant" rocks become the valleys. Resistance is not defined by the intrinsic properties of a particular rock type but is a relative feature determined by how rapidly slopes developed on the rock retreat and whether the rock stands relatively high in the local topography (Young 1972). In other words, it is not so much the rock itself that determines resistance, but whether the slopes formed over the rock are controlled by processes of weathering or processes of removal. In weathering-controlled slopes, resistance is related to how rapidly the rock is weathered and is a direct function of the rock properties. In transport-limited slopes, the resistance is attributable to the rate at which regolith can be eroded, and the properties of the weathered mass and the type and magnitude of the erosional processes become important in slope development. The downslope-grading of transport-limited slopes can be explained by the direct relationship between slope gradient and the rate of downslope material transport. For these reasons, the resistance of a particular rock type and its influence on slopes can be reversed if the rock is located in different climates. For example, the characteristics of slopes formed on limestones in humid climates contrast markedly with those developed in arid climates.

With regard to climatic influence, geomorphologists have long recognized that the most common slope profile in humid-temperate regions has a distinct convex upper slope and concave lower slope. Contrary to some beliefs, straight slope segments do occur in regions with humid-temperate climate, and some profiles do contain steep cliff faces. Most cliff faces, however, are ephemeral in the sense that as soon as undercutting ceases a talus slope forms and will extend upslope until it covers the original cliff wall (fig. 4.56). If the lithology of the rock sequence underlying the slope is not uniform, cliff faces may persist because resistant units are maintained as caprocks where the weaker underlying strata retreat faster, essentially undercutting the stronger rocks.

Convex upper slopes are usually attributed to soil creep; the lower concavity probably results from soil wash, although not all slopes have this segment, particularly when there is active erosion at the slope base (Strahler 1950). The convex-concave profile is most likely to be attained after mass movements have produced a long-term angular stability. At this stage, creep and wash become the dominant slope processes; the straight segment, representing stability of slope material, is gradually diminished in size. The processes of water erosion on slopes will be discussed in the next chapter. Recognize here, however, that water flowing over and through slope material combines with mass movement to mold slope profiles, and in some cases water erosion may be the dominant process involved.

Semiarid and arid climates tend to engender slope profiles that are more angular than those found in humid-temperate regions, even though the same convex, straight, and concave segments may be present (fig. 4.57). Steep cliffs usually are present above a

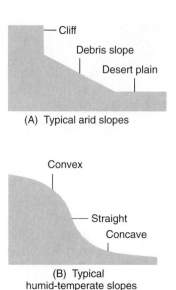

Figure 4.57
Typical slope profiles in (A) arid regions and (B) humid-temperate regions.

straight, debris-covered segment that normally stands at angles between 25° and 35°. At the base of the straight segment a pronounced change in slope occurs, and angles decrease over a short distance to less than 5°, a normal slope for most desert plains. The limited vegetal cover and low precipitation in arid zones assure that mass movements occur at higher angles and that creep is subordinated to wash. As a result, the upper slope convexity, so prominent in humid regions, is much less pronounced. However, convex bedrock slopes are common in selected semiarid regions where jointing characteristics of the rock promote development of extensive exfoliation (Bradley 1963).

Straight segments are maintained by the wash process, which is accelerated on the sparsely vegetated surfaces. Unlike similar segments in humid climates, these usually have only a thin veneer of rock debris. Thus they are not like slopes of accumulation, talus slopes, but probably represent true slopes of transportation, on which the amount of debris supplied to the straight segment from the cliff face or from weathering of the underlying rocks is removed in equal quantities to the desert plain. The angle of slope represents a balance between the processes that break debris down and the actual transporting mechanism (Schumm and Chorley 1966). Most geomorphologists feel that a general relationship between particle size and slope angle can be demonstrated. Our treatment of semiarid slopes has been greatly oversimplified. A lengthy overview of slope evolution in the Colorado Plateau by Howard and Kochel (1988) highlights the complex interactions between chemical and physical weathering processes, mass wasting, and groundwater-related processes as they work on sandstone.

Although other climatic regimes have characteristic slope forms, in most cases they are produced by the same mechanics that operate in the humid-temperate or arid zones. In the periglacial environment a special influence is exerted by magnified frost activity; a more extensive treatment of that environment is presented in chapter 11.

Very little research has focused on what aspects of hillslope profiles are most closely related to climate. A study by Toy (1977), however, utilized a rigorous statistical analysis to compare slope properties within two extended traverses in the United States (Kentucky to Nevada and Montana to New Mexico) along which considerable climatic variation occurs. The selection of sampling localities was stringent. Parent rock at each measuring site was restricted to shales dipping at less than 5°. Each slope analyzed was south-facing, within 5 miles of a weather station having records for the same 21-year period used as the climatic base, and had no effects of human activity. Toy found that climate could account for 59 percent of the variability in the upper convex segments and 43 percent of the variability in the slope of the straight segments. Arid slopes in this study were shorter, had steeper straight segments, and had shorter radii of curvatures developed at the convex crests than slopes in humid regions. In addition, of the climatic variables used in the study, those most closely associated with slope variations were spring and summer precipitation, potential evapotranspiration, and water availability (total precipitation minus total potential evapotranspiration during the 21-year period).

Toy's findings cannot be used to make sweeping generalizations about climatic effects on slope profiles because they apply only to one type of parent rock. However, the study demonstrates the type of research design needed to estimate the influence of one geomorphic factor by reducing or eliminating the effects of others.

Slope Evolution

In addition to geology and climate, the factor of time can also be considered as an independent variable. Its effect, however, is difficult to determine, especially when the time interval involved is very long. As we saw in chapter 1, some of the great debates in geomorphology revolve around the question of how slopes respond to continued erosion. Do slopes progressively flatten through time in steps or stages? Or do slopes reach an equilibrium between form and geomorphic factors that is maintained through time as slopes retreat in a parallel manner? These questions are not easily answered.

Three main types of slope evolution have been suggested: slope decline, slope replacement, and parallel retreat (fig. 4.58). In *slope decline,* the steep upper slope erodes more rapidly than the basal zone, causing a flattening of the overall angle. It is usually accompanied by

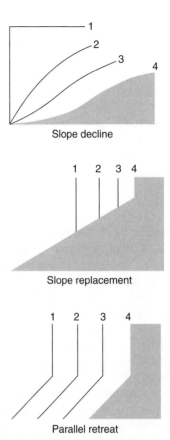

Figure 4.58

Three hypotheses of slope evolution. Higher numbers indicate increasing age of the slope.

(Adapted from A. Young, *Slopes.* fig. 14., 1972, Oliver and Boyd Publishers. Used by permission of A. Young.)

a developing convexity on the upper slope and concavity near the base. Slope decline alone cannot explain a concave profile on the lower slope unless some deposition occurs at the base. In *slope replacement,* the steepest angle is progressively replaced by the upward expansion of a gentler slope near the base. This process tends to enlarge the overall concavity of the profile, which may be in either a segmented or a smoothly curved form. Slopes evolving by *parallel retreat* are characterized by the maintenance of constant angles on the steepest part of the slope. Absolute lengths of slope parts do not change except in the concave zone, which gets longer with time.

Studies of hillslope evolution have also documented adjustments in the location of lateral convexities and concavities. Some of these studies highlight the concept of *gully gravure* (Bryan 1940), which describes how concave slope drainages (hollows) become armored with coarse colluvium shed off of neighboring convex slopes (noses), shifting drainage laterally in a manner such that the unarmored noses are preferentially eroded. In this manner, the former noses swap geomorphic roles with

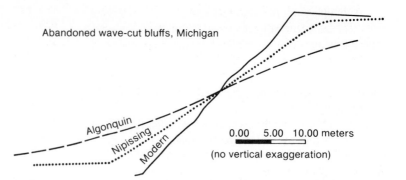

Figure 4.59
Profiles of bluffs of three different ages
developed in the same material.
(From: D. Nash. 1980."Forms of bluffs degraded for different
lengths of time in Emmet Co., Michigan." Earth Surface
Processes and Landforms 5:331–45. Used by permission of
John Wiley & Sons.)

the former hollows and become the sites of new hollows. At the same time, the former hollows become the convex noses. Recent studies of slope evolution in the southern Appalachians (Mills 1987; 1988) and in Australia (Twidale and Campbell 1986) have strengthened support for this model of gully gravure.

All three types of slope evolution can be demonstrated in actual situations (Savigear 1952; Brunsden and Kesel 1973; Cunningham and Griba 1973; Haig 1979; Nash 1980; Selby 1980, 1982; Colman 1983). In general, slope decline is most notable in humid regions, and parallel retreat seems to be more prevalent in arid regions.

In most of the foregoing cases, demonstration of slope evolution is based on the *ergodic hypothesis* (Chorley and Kennedy 1971), which states that, in the proper setting, spatial elements can be considered equivalent to time elements, and space-time transformations are therefore acceptable. For example, Nash (1980) suggests that profile variations in a series of lake bluffs in Michigan represent slope changes that have occurred during intervals between the development of the different bluffs. Because each bluff was formed in the same morainic material, it is assumed that the oldest bluff originally had a profile similar to that of the modern bluff. Therefore, the observed differences between the two profiles represent changes that have occurred on the older slope since the time of its formation (fig. 4.59).

The type of studies just mentioned have inherent value because they are based on real observations. However, they are probably valid only if significant changes in other geomorphic variables, such as climate, do not occur during the period of slope development. Thus they are restricted to geologically short time spans. Because of this, the possibility remains that the observed changes are merely transitions toward an equilibrium form that, when attained after a greater period of time, would experience no further profile alteration.

We should note that numerous attempts have been made to characterize slope evolution by employing theoretical techniques such as numerical and simulation models (for references, see Carson and Kirkby 1972; Young 1972; Selby 1982; Fernandes and Dietrich 1997). Although modeling can suggest routes that slope evolu-

tion may follow, this approach is deficient because the assumed character of the original slope profile is pure conjecture; that is, there is no sure way to know the form of the initial profile. As a result, some geomorphologists believe that models add little to our comprehension of slopes unless they are based on long-term and detailed field measurements (Dunkerley 1980; Selby 1982). Recent studies are beginning to show the importance of the magnitude and frequency of mass movement events on the evolution of hillslopes and predmont zones (Kochel et al. 1997; Eaton 1999; Wainwright 1996).

Relict Hillslope Forms

Although talus, or scree, processes are major controls in the evolution of rock slopes along resistant rock outcrops, talus can also play a major role in controlling the development of gentler slopes in temperate and alpine regions. Large quantities of talus commonly produced during glacial epochs can be found mantling the slopes of areas where the present rate of talus production is nominal. The talus cover acts to armor underlying colluvium, precluding further slope development. Relict hillslope forms tend to be common where significant changes in dominance or rates of hillslope processes have occurred within the past few tens of thousands of years (Parsons 1988). Much of the colluvium presently on Appalachian slopes, for example, was inherited from Pleistocene periglacial activity (Jacobsen et al. 1989; Clark and Ciolkolz 1988) and appears to have undergone little change during the moderate Holocene climate. Caine (1986), summarizing 20 years of data on sediment movement and storage on alpine slopes in the southern Rocky Mountains, showed that modern processes have contributed little to hillslope evolution, and he concluded that colluvium develops in episodes of increased activity during climates different from the present.

Considerable evidence indicates that present-day hillslopes may owe as much of their form to paleoslope processes as to currently operating ones. A complication arises, however, when we consider the contribution of infrequent catastrophic mass movements in long-term slope evolution. Iida and Okunish (1983) argue that although landslides are sporadic and localized events, they represent a continuous process over the long term and play a major role in hillslope development. Debris fans are a major hillslope landform within the Virginia Blue Ridge province. Excavations of these debris fans reveal that they are constructed predominantly of debris flow deposits like those produced in 1969 by Hurricane Camille rains, which recur at intervals of about 3000 to 4000 years (Kochel and Johnson 1984). Because of the infrequency of large mass movements, their role in geomorphic evolution remains difficult to assess. Some of the features that are attributed to processes that operated in formerly periglacial climates may be the result of rare, large-magnitude events functioning in the modern climatic regime.

Considerable recent work has demonstrated the influence of climatic variations on the temporal variability of mass wasting. Mason and Knox (1997) found evidence of multiple episodes of colluviation in the Upper Mississippi River Valley linked to periods when permafrost was present. Other studies show that periods of increased landsliding correlate to humid climatic intervals (Trauth et al. 2000; Alexandrowicz and Alexandrowicz 1999).

SUMMARY

The processes of physical weathering tend to break rocks and unconsolidated debris into smaller particles. The force needed to accomplish this disintegration is provided by expansion resulting from unloading, hydration of minerals, or growth of foreign substances in spaces within the parent material. Many important processes of disintegration require the presence of water. Physical weathering, combined with gravity, is instrumental in determining the type and rate of mass movements; ultimately it has a direct bearing on the slopes developed in any region.

Mass movements occur as slides, flows, and heaves, or by water-induced transport of surface debris. The magnitude and type of mass movement are partly dependent on the physical properties of the parent material. Shear strength (a function of internal friction, effective normal stress, and cohesion) determines how vigorously any substance will resist the force attempting to produce mass movement. Thus, slope failure or other mass movements can result from an increase in shear stress (driving force), a lowering of shear strength (resistance), or both. Physical weathering tends to decrease the shear strength of materials and thereby helps to initiate mass movements and control the form of the resulting slopes. Climate and lithology interact to influence slope profiles. The effect of time is shown by the manner in which slopes evolve.

SUGGESTED READINGS

The following references provide greater detail concerning the concepts discussed in this chapter.

Abrahams, A. D., ed. 1986. *Hillslope processes.* Boston: Allen and Unwin.

Brabb, E. E., and Harrod, B. L., eds. 1989. *Landslides: Extent and economic significance.* Proc. 28th Int'l. Geol. Cong. Symposium on Landslides. Rotterdam: A. A. Balkema.

Brunsden, D. 1993. Mass movement—The research frontier and beyond: A geomorphological approach. *Geomorphology* 7:85–128.

Brunsden, D., and Prior, D. B., eds. 1984. *Slope instability.* New York: John Wiley and Sons.

Carson, M., and Kirkby, M. 1972. *Hillslope form and process.* London: Cambridge Univ. Press.

Chandler, R. J., ed. 1991. *Slope stability engineering: Developments and applications.* London: Telford.

Costa, J. E. 1984. Physical geomorphology of debris flows. In Costa, J. E., and Fleisher, J. P., eds., *Developments and applications of geomorphology,* pp. 268–317. New York: Springer-Verlag.

Costa, J. E., and Wieczorek, G. F., eds. 1987. *Debris flows/avalanches: Process, recognition, and mitigation.* Geol. Soc. Amer. Reviews in Engin. Geol. 7.

Costa, J. E., and Williams, G. P. 1984. *Debris flow dynamics.* U.S. Geological Survey. Videotape.

Crozier, M. J. 1973. Techniques for the morphometric analysis of landslips. *Zeitschrift für Geomorphologie* 17:78–101.

Grim, R. 1962. *Applied clay mineralogy.* New York: McGraw-Hill.

Ollier, C. D. 1969. *Weathering.* Edinburgh: Oliver and Boyd.

Parsons, A. J. 1988. *Hillslope form.* London: Routledge.

Selby, M. J. 1982. *Hillslope materials and processes.* Oxford: Oxford Univ. Press.

Terzaghi, K. 1950. Mechanism of landslides. In Paige, S., ed., *Application of geology to engineering practice,* Berkey Vol., pp. 83–123. Boulder, Colo.: Geol. Soc. America.

Varnes, D. J. 1978. Slope movement types and processes. In Schuster, R., and Krizek, R., eds., *Landslides: Analysis and control,* pp. 12–33. Washington, D.C.: Nat. Academy of Science.

Young, A. 1972. *Slopes.* Edinburgh: Oliver and Boyd.

5

THE DRAINAGE BASIN—
DEVELOPMENT, MORPHOMETRY,
AND HYDROLOGY

R. Craig Kochel

Introduction
Slope Hydrology and Runoff
Generation
    Infiltration
    Subsurface Stormflow and Saturated Overland Flow
    The Stream Hydrograph and Response to Basin
        Characteristics
    Effect of Physical Basin Characteristics
Initiation of Channels and the Drainage Network
    Basin Morphometry
        Linear Morphometric Relationships
        Areal Morphometric Relationships
        Relief Morphometric Relationships
    Basin Morphometry and the Flood Hydrograph
    Basin Evolution
Basin Hydrology
    Subsurface Water
        The Groundwater Profile
        Movement of Groundwater
        Aquifers, Wells, and Groundwater Utilization
            Problems
    Surface Water
        Flood Frequency
        Paleoflood Hydrology
Basin Denudation
    Slope Erosion and Sediment Yield
        Wash
        Sediment Yield (Soil Loss)
        Factors Affecting Sediment Yield
    Sediment Budgets
    Rates of Denudation
Summary
Suggested Readings

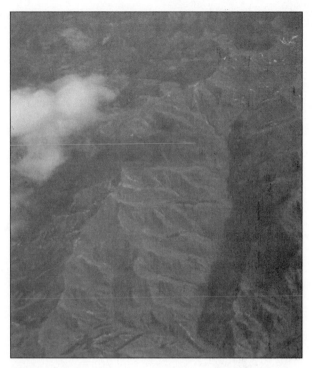

Figure 5.1A
An oblique aerial photo of drainage basins formed in volcanic rocks of the Oregon Cascade Range.
(Photo by R. C. Kochel)

INTRODUCTION

Two basic generalizations about rivers were realized long before geomorphology emerged as an organized science: (1) streams form the valleys in which they flow, and (2) every river consists of a major trunk segment fed by a number of mutually adjusted branches that diminish in size away from the main stem. The many tributaries define a network of channels that drain water from a discernible, finite area which is the **drainage basin, or watershed,** of the trunk river.

The drainage basin is the fundamental landscape unit concerned with the collection and distribution of water and sediment. Each basin is separated from its neighbor by a **divide,** or **interfluve.** (fig. 5.1A). Thus, the basin can be viewed as a geomorphic system or unit. As we will soon see, the basin is inexorably linked with hillslope processes that contribute water and sediment to the channel network in accord with the regional climate, underlying bedrock and tectonic regime, and land use by humans (fig. 5.1B). Any feature or portion of the basin can be considered a subsystem having its own unique set of processes, geology, and energy gains and losses. Furthermore, because it is possible to measure the amount of water entering the basin as precipitation and the volume leaving the basin as stream discharge, hydrologic events can be readily analyzed on a basinal scale. Likewise, much of the sediment produced within the basin is ultimately exported from the basin through the trunk river. Thus, considered on a long temporal scale, the rate of lowering of the basin surface can be estimated.

The output from a given basin compartment serves as input to the master channel and influences downstream channel characteristics and hydrologic processes in rivers. The mechanics of fluvial processes usually reflect some balance between the amount of sediment supplied for transport and the water available to accomplish this task. Throughout the discussion of drainage basins and fluvial systems, we will frequently refer to the concepts illustrated in figure 5.1B as we describe the interrelationships between various components of the fluvial system and the regulatory influence of the external variables of water and sediment in the adjustment and evolution of basins and channels.

Most Earth scientists are introduced to watersheds when they learn that drainage patterns or individual stream patterns commonly mirror certain traits of the

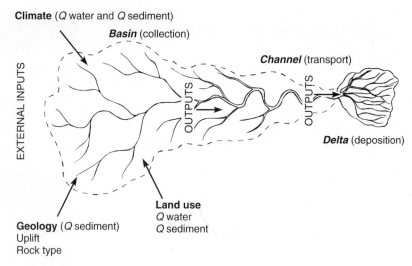

Figure 5.1B
Schematic surface components of the fluvial system. The tributaries provide links between lithology and climate and are adjusted to both. Channel characteristics vary in response to the external variables of sediment and water discharge (Q), which are influenced naturally from climate, tectonic, and lithologic factors. Human influence also modifies these variables through land use alterations.

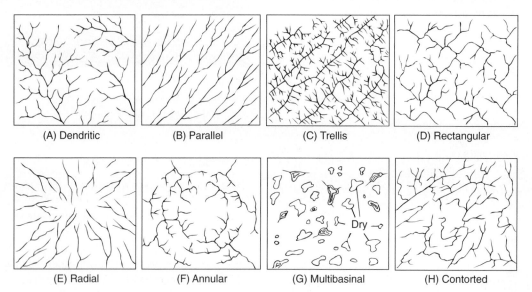

Figure 5.2A
Basic drainage patterns. Descriptions are given in table 5.1.
(Howard 1967, reprinted by permission)

underlying geology, described in figure 5.2A and table 5.1. Because the gross character of these patterns is evident on topographic maps and aerial photos, the patterns are useful for structural interpretation (fig. 5.2B) (Howard 1967) and for approximating lithology in a study of regional geology.

In a hydrologic sense, however, prior to World War II most basins were described in qualitative terms such as well-drained or poorly drained, or they were connoted descriptively in the Davisian scheme as youthful, mature, or old. The mechanics of how river channels or networks actually form and how water gets into a channel was poorly understood by geolo-

gists and hydrologists alike. This early twentieth-century view of streams and drainages contrasts markedly with the avant-garde approach presented by R. E. Horton during the latter part of this period (Horton 1933, 1945). His attempt to explain stream origins in mathematical terms and to describe basin hydrology as a function of statistical laws marked the birth of quantitative geomorphology. We now know that many of Horton's original ideas are only partially correct. Still, modern geomorphic analysis of drainage basins has its roots in Horton's original work, and his thinking has been instrumental in the development of modern geomorphology.

TABLE 5.1	Descriptions and Characteristics of Basic Drainage Patterns Illustrated in Figure 5.2A.
Pattern	**Geological Significance**
Dendritic	Horizontal sediments or beveled, uniformly resistant, crystalline rocks. Gentle regional slope at present or at time of drainage inception. Type pattern resembles spreading oak or chestnut tree.
Parallel	Generally indicates moderate to steep slopes but also found in areas of parallel, elongate landforms. All transitions possible between this pattern and dendritic and trellis patterns.
Trellis	Dipping or folded sedimentary, volcanic, or low-grade metasedimentary rocks; areas of parallel fractures; exposed lake or seafloors ribbed by beach ridges. All transitions to parallel pattern. Pattern is regarded here as one in which small tributaries are essentially same size on opposite sides of long parallel subsequent streams.
Rectangular	Joints and/or faults at right angles. Lacks orderly repetitive quality of trellis pattern; streams and divides lack regional continuity.
Radial	Volcanoes, domes, and erosion residuals. A complex of radial patterns in a volcanic field might be called multiradial.
Annular	Structural domes and basins, diatremes, and possibly stocks.
Multibasinal	Hummocky surficial deposits; differentially scoured or deflated bedrock; areas of recent volcanism, limestone solution, and permafrost. This descriptive term is suggested for all multiple-depression patterns whose exact origins are unknown.
Contorted	Contorted, coarsely layered metamorphic rocks. Dikes, veins, and migmatized bands provide the resistant layers in some areas. Pattern differs from recurved trellis in lack of regional orderliness, discontinuity of ridges and valleys, and generally smaller scale.

From Howard 1967, reprinted by permission.

Figure 5.2B
Trellis drainage formed as a result of the alternation of resistant rocks in the central Pennsylvania Valley and Ridge province. View is along the Susquehanna River north of Harrisburg, PA and shows sandstone ridges (resistant) and intervening valleys underlain by shale and limestone (less resistant) in this plunging syncline.
(Photo by R. Craig Kochel)

SLOPE HYDROLOGY AND RUNOFF GENERATION

The ultimate source of river flow is, of course, precipitation, which represents the major influx of water to any drainage basin. Precisely how much of that precipitation actually becomes part of the streamflow and what route a particular drop of water follows to reach a channel are topics of great concern to hydrologists. The components of the slope hydrological cycle (fig. 5.3) are the pathways of water to the streams.

Rainfall seldom makes direct contact with the bedrock or soil surface except in arid regions characterized by sparse vegetal cover. Most raindrops are impeded by leaves and trunks of the vegetal cover, in a process known as **interception.** Interception substantially reduces not only the erosive potential of raindrop strike but also the volume of water reaching the surface. Interception losses are quite variable because they depend on numerous hydrometeorological factors, vegetation type, land use, and seasonality. In addition, the loss is dependent on storm duration, being initially high and decreasing as the storm continues. During long storms, the vegetation may become saturated and all additional rainfall is passed on to the surface. Nonetheless, interception typically removes 10 to 20 percent of precipitation where grasses and crops are the dominant vegetation and up to 50 percent under a forest canopy (Selby 1982). Rainwater that does reach

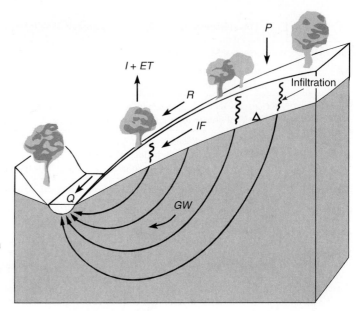

Figure 5.3

Slope hydrologic cycle. Some of the precipitation (*P*) is intercepted by vegetation (*I*) or lost by evapotranspiration (*ET*). Upon reaching the ground surface, it becomes part of stream discharge (*Q*) by direct runoff (*R*), interflow (*IF*), or groundwater flow (*GW*) after it reaches the water table (Δ).

the ground may still be prevented from contributing to streamflow by vegetation because some of it is consumed and lost by **evapotranspiration.**

Infiltration

Water on the ground surface may follow different routes to a stream channel. Water flowing into a channel in direct response to a precipitation event is called **storm runoff** or **direct runoff,** whereas the process of water entering the soil is considered **infiltration.** In 1933 Horton suggested that the rate of infiltration into soil on a sloping surface actually governs the amount of storm runoff generated by rainwater. This infiltration theory of runoff can be described as follows. Imagine a hypothetical slope with a surficial mantle of regolith formed by weathering and mass wasting processes. As precipitation falls on the slope surface, water is absorbed into the ground at a rate called the **infiltration capacity.** Infiltration capacity, typically measured in mm/hr is a function of regolith or soil thickness, soil texture, soil structure, vegetation, and the antecedent condition of the soil moisture (which depends on the recent history of precipitation). The rate of infiltration depends primarily on the interaction between three processes: (1) absorption or entry of water into the soil, (2) storage of water in the pore spaces, and (3) transmission of water downward through the soil. Infiltration rates are generally maximized when slopes are gentle with permeable soils and when rainfall intensity is low.

Recent studies of basins within the same climatic setting have shown that runoff processes, and thereby the surface hydrologic parameters such as discharge of water and sediment discharge by channels, are significantly impacted by variations in basin lithology. Kelson and Wells (1989) observed that unit runoff was consis-

tently higher from basins underlain by resistant crystalline rocks than from neighboring watersheds underlain by less resistant sedimentary rocks in the mountains of northern New Mexico. Differences in regolith production from weathering of the parent lithology and from earlier geomorphic activity (i.e., glaciation) contributed to differences in the timing of water inputs into the basin, variations in subsurface slope, infiltration capacity, and the sediment yield and erosive power of the basin streams. Sala (1988), in a comparative study of watersheds underlain by slate and crystalline rocks in Spain, found that regolith variations attributed to the lithology not only resulted in notable differences in runoff but also produced general differences in process dominance on basin hill-slopes. Slate watersheds generated large amounts of coarse debris and were dominated by creep, leading to high erosion rates. On the other hand, crystalline basins were characterized by well-integrated runoff systems with perennial flow due to thicker regolith composed of grus. Therefore, infiltration capacity varies spatially on a large scale with regional geology, but may also change locally (even along the same slope) if the controlling factors vary.

Infiltration capacity normally changes at a site during any precipitation event (fig. 5.4). It usually starts with a high value that decreases rapidly during the first few hours of the storm and then more slowly as rainfall continues, until it finally attains a reasonably constant minimum value. Typically, the ultimate infiltration rate is established by a limiting subsurface horizon with a low water transmission rate, such as a zone of textural B horizon, caliche, or the bedrock surface. The infiltration capacity changes because surface conditions change during a storm, especially as aggregated soil clumps are broken apart into smaller particles which clog some of the pores. In the time interval between rains, the infiltration

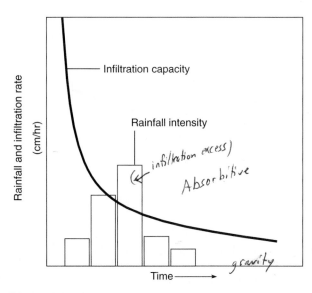

Figure 5.4
Infiltration capacity and rainfall intensity plotted against time. Infiltration capacity decreases with duration of storm. Runoff occurs only when rainfall intensity is greater than infiltration capacity.

capacity rises again as the surface dries and reestablishes its aggregated structure, and storage space within the soil increases again as soil water gradually drains downward. The frequency of rainfall, expressed as time between successive rains, can also become a significant factor affecting infiltration. Rains of relatively minor intensity may trigger disastrous floods if the infiltration capacity does not have enough time to recover between storms.

In Horton's model, as long as the infiltration capacity exceeds the rate at which rainfall strikes the surface (known as *rainfall intensity*), all incoming water will be infiltrated and none will run off. However, in those periods when rainfall intensity exceeds the infiltration capacity, runoff will occur as water moving down the slope surface; this runoff is known as **Hortonian overland flow.** Hortonian overland flow has been considered to be the primary determinant of peak flow and total direct runoff in a stream channel from a storm because its flow velocity generally ranges between 10 and 500 m/hr (Dunne 1978). The basic assumption here is that all infiltrated water is delayed greatly in its procession to the stream channels because it must percolate downward to the groundwater table and then move toward the stream by comparably slow groundwater flow velocities (see fig. 5.3). Overland flow, which occurs as broad shallow sheets or in linear depressions called *rills* is uncommon, however, except in sparsely vegetated semiarid and arid regions. Hortonian flow is virtually nonexistent in humid-temperate regions characterized by dense vegetation and thick soils (Kirkby and Chorley 1967; Dunne and Black 1970a, 1970b).

In light of the above, we must ask the logical question as to where the water comes from to produce the rapid peak flows observed in small basins of humid regions if all rainwater reaching the surface is subject to infiltration.

Subsurface Stormflow and Saturated Overland Flow

A major flaw in Horton's original model was the belief that infiltrated water moved directly downward under the influence of gravity. Research subsequent to Horton's analyses has clearly demonstrated a variety of flow within the unsaturated zone, due to the anisotropic character of regolith in most areas with respect to its ability to transmit water. Thus, significant lateral and subhorizontal flow can occur in some areas. Infiltrating water moves downward from the surface as a wetting front to the water table. In doing so, the front often encounters a soil horizon or other zone of low permeability that not only limits infiltration capacity but also tends to divert water downslope (often laterally) along its surface. This occurs because a local saturated condition develops above the low permeability zone, and lateral flow is initiated parallel to the barrier. This type of movement, called **throughflow** or **interflow** (Kirkby and Chorley 1967), can occur above the water table and permits water to take a more direct path to the stream channel than normal groundwater does. Throughflow is probably most common in the more permeable A horizon.

Where no barrier exists, infiltrated water will reach the water table and elevate its position. The water table rises rapidly adjacent to the channel, where antecedent soil moisture is greatest, and slowly in the upper-slope zone. As a result, the water table steepens immediately next to the channel and generates accelerated groundwater flow in that area. The combination of throughflow and accelerated groundwater movement is called **subsurface stormflow.** It was first recognized by Hursh (1936; Hursh and Brater 1941), but the significance of the concept was not fully realized until much later.

Subsurface stormflow from the lower parts of slopes may actually produce runoff in the form of bank seepage early in a storm. Therefore, it is possible for some of the subsurface stormflow to contribute to peak discharge. In other situations, storm runoff is generated from the movement of subsurface water through **macropores,** which are linear openings, having a much greater permeability than the surrounding regolith (Mosley 1979; Bevan and Germann 1982). Macropores can originate along root channels by **piping** (flow concentration and erosion along the rigid root) or in voids left by decayed roots, burrows, and other cavities made by plants and animals. Flow through such pipes can be rapid and turbulent, accounting for as much as one-fifth of the

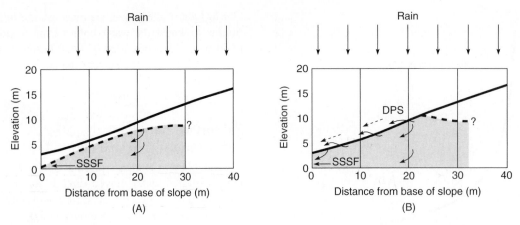

Figure 5.5
Runoff and interflow in a steep, well-drained Vermont hillslope. (A) Early in a storm the saturated zone (shaded) yields a small amount of subsurface stormflow (SSSF). (B) Late in the storm the water table rises to the surface as return flow (RF). Precipitation on the saturated area (DPS) adds to the return flow.
(Dunne and Leopold 1978)

annual contribution to runoff from hillslopes (Roberge and Plamondon 1987). Although some water may enter channels rapidly by these processes, most studies show subsurface flow to be a more important contributor to streamflow after the discharge peak, because measured velocities of subsurface flow range between 0.003 and 1.0 cm/hr (Dunne 1978), too slow to effectively contribute to most peak flows in small basins. This slow response requires yet another source of runoff to help explain the rapid production of peak flood flow. A field experimental study in a forested Appalachian watershed reported distinguishable inputs into surface runoff from throughflow sources (Hornberger et al. 1991). Significant contributions to storm runoff may occur where soil conductivity (with respect to its ability to transmit water) is high due to coarse texture and/or the presence of large macropores.

Detailed studies in Vermont (Ragan 1968; Dunne and Black 1970a, 1970b) showed that in many storms the level of the water table rises until it actually intersects the ground surface near the stream channel (fig. 5.5). In addition, zones of throughflow partially fed by laterally moving, infiltrated water may become saturated to the surface (Kirkby and Chorley 1967). In this situation some of the infiltrated water is returned once again to the surface and begins to flow downslope toward the channel; this flow is aptly called return flow. **Return flow** is obviously a form of overland flow, but its origin and location is distinctly different from Hortonian overland flow. The velocity of return flow is much greater than subsurface stormflow, attaining speeds of 3 to 15 cm/sec (Dunne 1978). In addition, direct precipitation on the saturated area marked by return flow or zones of saturated throughflow adds significant amounts of water to the volumes of direct runoff.

The combination of return flow and direct precipitation on saturated areas, called **saturation overland flow,** is now documented in many humid areas as the major contributor to direct runoff to stream channels. For example, Ragan (1968) estimated that on the average it supplied 55 to 62 percent of the total storm runoff in a small watershed near Burlington, Vermont, and was predominant in determining peak discharge. Subsurface storm flow then provided 36 to 43 percent of the total flow but exerted its greatest influence after the peak during the recessional phase of the runoff.

There are many sources of direct runoff other than Hortonian overland flow. Moreover, it is becoming apparent that the area of a watershed actually providing runoff during a storm is not temporally constant. This perception, known as the *variable source concept* or *partial area concept,* indicates that the area over which quick runoff occurs varies seasonally and during any given storm (fig. 5.6). This change is fundamentally controlled by topography, soil characteristics, antecedent moisture, and rainfall properties, and essentially means that there will be seasonal changes in the extent and location of channel networks contributing to surface flow (de Vries 1995). Areas with moderate to poorly drained soils, gentle slopes, and concave recessions along the valley walls are prone to have the greatest expansion of contributing areas during storms and on a seasonal basis.

Although the generalizations discussed above seem reasonable, they are based on a limited number of studies, and much more work is needed to refine the model (Selby 1982). Nonetheless, the observations about runoff generation have an important bearing on the geomorphic processes operating within a drainage basin, and further research in this field should be vigorously pursued (Freeze 1980).

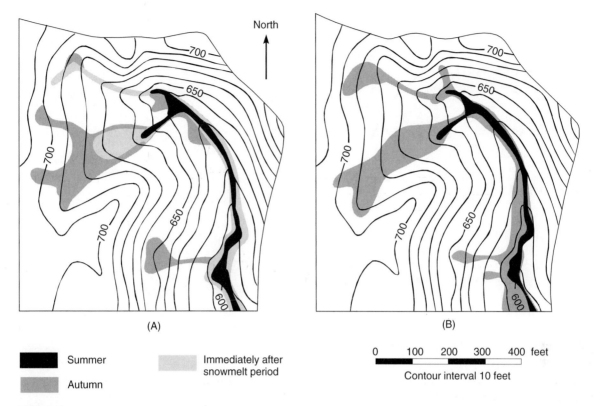

North

(A)   (B)

Summer

Autumn

Immediately after snowmelt period

0   100   200   300   400 feet

Contour interval 10 feet

Figure 5.6
Variations in saturated areas on well-drained hillslopes near Danville, Vt. (A) Seasonal changes of prestorm saturated area. (B) Expansion of saturated area during a single 46 mm rainstorm. Solid black represents beginning of storm. Light shade represents saturated area at end of storm where water table has risen to the surface.
(Dunne and Leopold 1978)

The Stream Hydrograph and Response to Basin Characteristics

Hydrologists have always been concerned about how much runoff will issue from any precipitation event and how quickly the runoff will enter the stream channels. These factors determine if and when a flood will occur and what height the river will reach at peak flow. Figure 5.7 compartmentalizes the hydrologic processes that contribute to flood runoff in a stream channel. The response of the stream to a storm is depicted graphically as a **flood hydrograph** (or storm hydrograph), which shows the passage of flood flow volume, or *discharge,* with time (fig. 5.8). Discharge can be separated, using one of numerous separation techniques (see Chow 1964), into **direct runoff** and **base flow.** Runoff represents the sum of numerous types of overland flow, interflow, and stormflow, whereas base flow arises from contributions of groundwater spread out over longer periods of time. Whereas discharge may contain elements of both if measured shortly after a storm, base flow is the sole source of water sustaining river flow in dry periods between storms.

The maximum or peak flow usually develops soon after the precipitation ends, separating the hydrograph into two distinct segments, the *rising limb* and the *recession limb.* **Lag time** is the time between the center of mass of the rainfall and the center of mass of the runoff. The rising limb generally reflects the input from direct runoff; hence, the increase in discharge with time is relatively rapid. The recessional phase, however, is controlled more by the gradual depletion of water temporarily stored in the shallow subsurface system, and thus the decrease is less pronounced with time. This phenomenon normally results in an asymmetrical hydrograph, unlike the one shown in figure 5.8. In addition, the shape of the recessional limb is not influenced by the properties of the storm causing the increased discharge but is more closely related to the physical character of the basin.

Effect of Physical Basin Characteristics

The physical characteristics of a basin contribute to the magnitude of the flood peak. The shape of the flood hydrograph is influenced not only by the temporal and spatial distribution of the rainfall, but in large manner by the physical character of the drainage basin. The flood hydrograph represents the integrated effects of basin area, channel density and geometry (basin morphometry), soils, and land use and thus is the summation

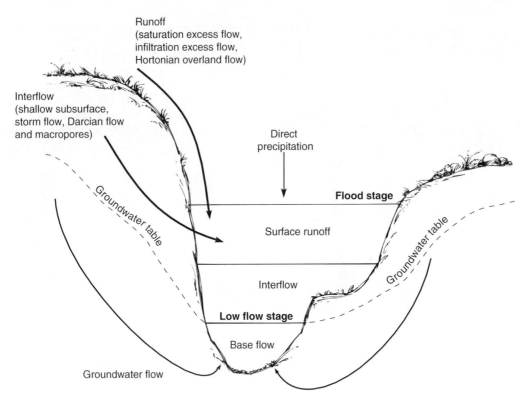

Figure 5.7
Schematic of stream channel showing major kinds of water contributions. Arrival of water from any given precipitation event is progressively delayed from runoff to interflow to groundwater flow.
(Based on Kochel 1992)

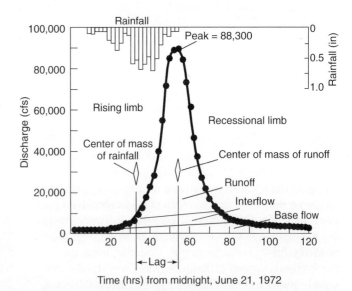

Figure 5.8
Flood hydrograph of the Hurricane Agnes flood of June 1972 on the Conestoga River at Lancaster, Pa. Although the curve is rather symmetrical, most hydrographs show significant skewness with a broader recessional limb reflecting interflow and groundwater inputs after a storm.

Time (hrs) from midnight, June 21, 1972

of the physical processing of precipitation from the divides to the site of measurement. Sophisticated models have been developed to predict how water is collected, stored, routed, and summed from all parts of a basin to achieve a final output hydrograph (for example, U.S. Army Corps Engineers 1985). If geology and topography are alike throughout an area, then rainfalls having similar properties should generate hydrographs with the same shape. On this premise, a *type* hydrograph for a

basin, called a **unit hydrograph,** has been developed, in which the runoff volume is adjusted to the same unit value (i.e., one inch of rainfall spread evenly over the basin over one day). The unit hydrograph has been used as a connecting link in many studies attempting to relate basin morphometry to hydrology. By comparing the shapes of unit hydrographs from different basins, we can see the effects of differences in physical attributes of the basin.

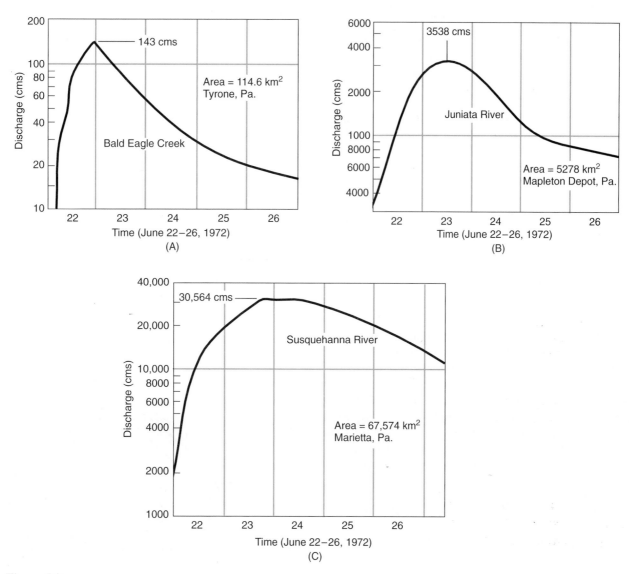

Figure 5.9

Progress of flood crests caused by Hurricane Agnes in the Susquehanna River basin, June 1972. Basins of increasing size: (A) Bald Eagle Creek; (B) Juniata River; (C) Susquehanna River.

Basin lag is the time needed for a unit mass of rain falling on the basin to be discharged through the measuring point (or basin outlet) as streamflow. Basin lag is normally estimated as the time interval between the centroid of rainfall and the centroid of the flood hydrograph (fig. 5.8) and is comprised of two distinct parts, the time involved in overland flow and the channel-transit time. Lag time for any basin is fairly consistent, with minor variations sometimes caused by the position and movement of the storm center relative to the gaging station. It seems intuitive that lag should increase with the size of the basin, but comparisons of similar-sized basins can show lag times up to three times as *sluggish* as basins with short lag times, which are referred to as *flashy* streams. Lag, therefore, is influenced by parameters other than drainage area alone, such as basin slope, the density of channels in the basin, and soils.

The time elements of basin hydrology can have a monumental influence on the magnitude of peak discharge during floods. Take, for example, the progression of flood crests recorded at several gaging stations in the Susquehanna River basin during the Hurricane Agnes flood in June 1972 (fig. 5.9). Most of the precipitation entered the basin during the period between June 19 and June 22. In the minor tributary Bald Eagle Creek, the flow peaked at 143 cms (5050 cfs) on the night of June 22, indicating a relatively quick response to the storm. In the larger Juniata River, the flood crested about 12 hours later with a considerably higher discharge value of 3538 cms (125,000 cfs). Far downstream on the main Susquehanna River, the flood peak did not occur until 12 hours after the Juniata peak, when discharge rose to 30,564 cms (1,080,000 cfs). Discharge peaked later on the main stream, and at a significantly higher magnitude than in the tributaries.

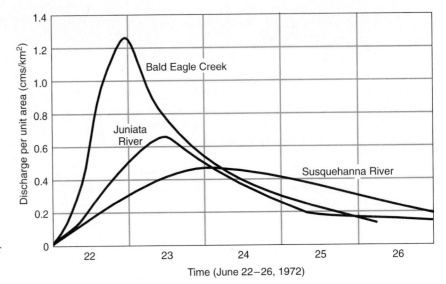

Figure 5.10

Discharge per unit area during flood caused by Hurricane Agnes in the Susquehanna River basin, June 1972. Note that the main river has considerably less discharge/area than tributaries. Main stem also peaks later and discharges floodwater over a longer time period, showing the effect of storing water on floodplains.

Lag time and peak discharge are positively correlated with basin size. However, simple expansion in basin size cannot fully explain the observed flow characteristics during a flood. Using data from figure 5.9 and replotting the peak discharge as Q/km^2, figure 5.10 reveals the interesting hydrologic property that discharge per unit area is much higher for the smallest tributary than for the massive basin of the main stem. Figure 5.11 shows how a similar relationship between basin area and discharge can be useful as a guide for anticipating the maximum flow likely for a basin of a given size. Had the increase in discharge during the 1972 Susquehanna flood been only a function of increased basin area, the peak flow at the Marietta station should have been a simple product of the Q/km^2 value of Bald Eagle Creek times the drainage area above Marietta, or 1.25 cms/km^2 × 67,574 km^2 = 84,468 cms (2,978,454 cfs), but this is almost three times the actual measured value. Somewhere within the basin, a built-in flood control mechanism exists that not only holds down the magnitude of the peak Q but simultaneously maintains abnormally high flow over a longer period of time. Most of this ability to lower peak flow can be attributed to the storage of water on floodplains and riverine wetlands. Floodplains store or retard large volumes of water until the flood crest passes a given channel locality. Additionally, differences in flood-crest travel time on the numerous tributaries in the basin and vagaries introduced by variable sources of runoff may help retard the downstream peak flow. Thus, a portion of runoff never contributes to the increasing of peak flow downstream but shows up in the record after the crest has passed as the extended part of the recessional limb of the flood hydrograph.

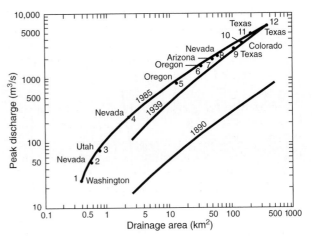

Figure 5.11

Peak discharge per unit from the largest flash floods in U.S. history. Note how the relationship indicates an increase in discharge coincident with development and land use alterations over the past 100 years.

(From Costa 1987)

INITIATION OF CHANNELS AND THE DRAINAGE NETWORK

In the Horton model, when rainfall intensity exceeds the infiltration capacity, overland flow occurs and only then does erosion become possible (see fig. 5.4). For a stream channel to develop, the erosive force (F) of the overland flow must surpass the resistance (R) of the surface being eroded. According to Horton (1945), as overland flow begins to traverse the slope, the force it exerts on a soil particle depends on the slope angle, the depth of water, and the specific weight of the water. Actually, F represents a shear stress exerted parallel to the surface by the water. The stress progressively increases downslope because the depth rises as more and more water is added to the volume of overland flow. Depending on the size of the slope material, a threshold stress is eventually reached at some point on the slope where $F > R$, and particles are dislodged or entrained. The processes and magnitude of slope erosion by water will be discussed later in the chapter when we examine sediment yield from drainage basins. Here we will look at the initiation of channels on sloping surfaces.

Surface resistance (R) is affected by the type and density of the vegetal cover (referred to in a combined factor called biomass by Graf (1979)). When vegetation intercepts raindrops before they can strike the surface, it preserves cohesion in the aggregated-clay soil structures. In addition, rootlets tend to bind soil particles, and litter often serves as a protective mat above the surface material. Vegetation also inhibits the free flow of water and retards its velocity. In areas devoid of vegetal cover, soil surfaces typically develop a hard crust upon desiccation, which provides high resistance during the initial portion of a precipitation event. This resistance is often destroyed as the storm progresses, and thus is apt to vary with time during rainfall events. The factors affecting resistance are similar to those controlling infiltration capacity.

Erosion by overland flow can, therefore, be considered a threshold process, occurring only after resistance forces are exceeded. Flowing water produces erosional forces commonly referred to as *tractive force* (τ_o), which is dependent on the character of the water, the flow depth, and the slope or gradient of the channel (Graf 1979):

$$\tau_o = \gamma_f D\,\theta c$$

where, γ_f is the specific weight of the fluid, D is flow depth, and θc is the channel gradient. Erosion typically begins as a series of subparallel rills parallel to the slope gradient. Slight variations in surface topography produce greater depth of flow, resulting in increased erosive forces in low spots. Erosion is accelerated at those points. Once water is focused into channels, this positive feedback process promotes continued evolution of channel networks at the expense of unconfined sheet flow. The actual point where rill formation begins depends on how efficiently the force of overland flow increases as the water moves downhill. This ultimately relates to infiltration capacity, rainfall intensity, and the resulting rate of runoff, or *runoff intensity.*

Assuming a constant slope and runoff intensity, the distance between the watershed divide and the upper boundary of the rills is a measurable segment called the critical length (X_c) (fig. 5.12). Horton considered the critical length to be the most important single factor in the development of stream networks, but its significance

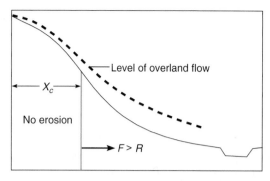

Figure 5.12
Hypothetical slope showing overland flow. No erosion occurs until the force of overland flow (F) exceeds the resistance of the surface material (R). Upslope from that point no erosion occurs. X_c is the distance from the divide to point where erosion begins.
(After Horton 1945)

can perhaps be appreciated more fully on a local scale because it is highly sensitive to changes in the factors that control erosion. Assume, for example, that a farmer wishes to plant additional row crops in an area near the divide. As the land is cleared of its original vegetal cover and replaced by the crops, resistance and infiltration capacity are drastically lowered. Under prevailing precipitation more runoff occurs, the force of overland flow increases, X_c is shortened, and rills and gullies form in the newly cropped region. Depending on their depth, these gullies may make the area impassable for the heavy equipment needed for plowing and harvesting. A useful review of gully erosion processes and models is found in Bocco (1991).

The analysis just presented is probably applicable in areas that are prone to the development of Hortonian overland flow. In humid-temperate regions, however, overland flow emanating from upper slope areas is rare, and therefore the concept of a critical length has no particular significance. In cases where saturated overland flow dominates, the erosive action probably begins closer to the slope base (Kirkby and Chorley 1967; Kirkby 1969), and rills are gradually extended upslope. This headward growth most likely occurs because the headcut in any rill exposes a substratum that has a lower resistance to erosion than the slope surface. The formation of rills does not explain stream channels because a rilled surface is still part of a slope system. Rill development, however, is the first necessary step toward a true river. It is now generally recognized that rills initiated on a slope cannot long remain as parallel unconnected channels. Deeper and wider rills develop. These master rills carry more water, and because of their greater depth, undergo downcutting until all the flow is contained within the channel and the rill becomes a tiny

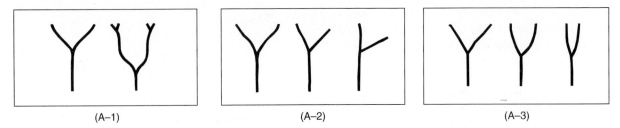

(A–1)   (A–2)   (A–3)

Figure 5.13A
Development of bifurcation angles. (A-1) The original angle is preserved. (A-2) One branch becomes dominant. (A-3) Angle decreases and branches merge into one channel; occurs on steep slopes.
(After Schumm 1956)

Figure 5.13B
View of small rilles developing a drainage network on a Texas hillslope. Note the cross-grading and micropiracy that is occuring on the slope to the right.
(Photo by R. Craig Kochel)

stream. Because they become slightly entrenched, master rills capture adjacent rills when bank caving or overtopping during high flow destroys the narrow divides between them. The repeated diversion of rills, a process called **micropiracy,** tends to obliterate the original rill distribution, and gradually the initial slope parallel to the master channel is replaced by slopes on each side that slant toward the main drainage line.

The development of new slope direction in accordance with the master channel was called **cross-grading** by Horton (1945). In the final stage of micropiracy, only one stream, confined in the master rill channel, crosses the slope. The side slopes presumably develop a new rill system sloping to the position of the initial stream, and the process repeats itself, culminating in a secondary master rill serving as an incipient tributary. Each smaller tributary evolves in a similar way until the network of streams takes form.

The network pattern develops by repeated division of single channel segments into two branches, a process known as **bifurcation.** Schumm (1956) suggests that the angle between the limbs of a bifurcated channel probably evolves in one of three ways (fig. 5.13A): (1) both limbs grow headward while preserving the original angle at their juncture; (2) one branch straightens its course and becomes dominant; or (3) the angle on steep slopes progressively decreases until the branches reunite into a single channel (fig. 5.13B). Any or all of these procedures might occur in the evolution of a network, constrained only by the fundamental erosive controls and the geologic framework. Divides between adjacent basins are predetermined by the extent to which streams can expand headward. Because critical length varies with resistance, the areal extent of the uneroded uplands partially reflects the geology and its history. Within the basin itself, smaller interfluves may be present where

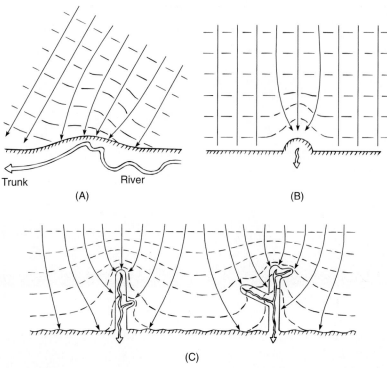

Figure 5.14

Evolution of groundwater-sapping system. Seepage begins along a stream bank at a local irregularity. Groundwater flow lines (solid) converge at the point of emergence and progressively accelerate seepage erosion. Dashed lines are lines of equal potential. This positive feedback system extends headwardly until bifurcation takes place, which is followed by development of a network system. The amphitheater-headed valley in (C) is similar to the forms shown on Mars in figure 5.15 and on Earth in figure 5.16.

(Dunne 1990)

cross-grading has not operated. These small areas parallel the main stream and preserve the slope of the original surface.

Although the discussion above demonstrates the importance of overland flow in channel initiation, many recent studies have highlighted the additional role of subsurface water as an agent of surface erosion and channel initiation (for example, Kochel et al. 1985; Howard et al. 1988; Dunne 1990). Groundwater seepage zones along hillslopes and major escarpments commonly results in groundwater flow convergence, which produces forces capable of eroding materials in the vicinity of the groundwater emergence (fig. 5.14). This process, commonly referred to as *groundwater sapping,* is important in the formation and maintenance of valleys in regions as lithologically diverse as the Colorado Plateau (Howard et al. 1988), Hawaii (Kochel and Piper 1986) and Cape Cod (Uchupi and Oldale 1994). It has also been proposed as a major valley-forming process on Mars (fig. 5.15). Valleys strongly influenced by sapping processes tend to have distinctive morphological characteristics such as amphitheater valley heads, stubby tributaries, and low channel densities (fig. 5.16). It is unlikely

in a terrestrial environment that any valleys are initiated or maintained solely by runoff or groundwater seepage processes. Most channels are probably composite forms (Schumm and Phillips 1986), but the dominance of surface or subsurface processes may be reflected in the evolution of channel and basin morphology.

Basin Morphometry

The drainage basin, the fundamental unit of the fluvial landscape, has been the focus of research aimed at understanding the geometric characteristics of the master channel and its tributary network. This geometry is referred to as the **basin morphometry** and is nicely reviewed by Abrahams (1984). Increasingly, studies have used the patterns of basin morphometry to predict or describe geomorphic processes; for example, it has been used to predict flood peaks, to assess sediment yield, and to estimate erosion rates (for example, Baumgardner 1987; Gardiner 1990). Some researchers believe that basin morphometric studies may ultimately be extended to show the influence of basin characteristics on channel cross-sections and channel attributes.

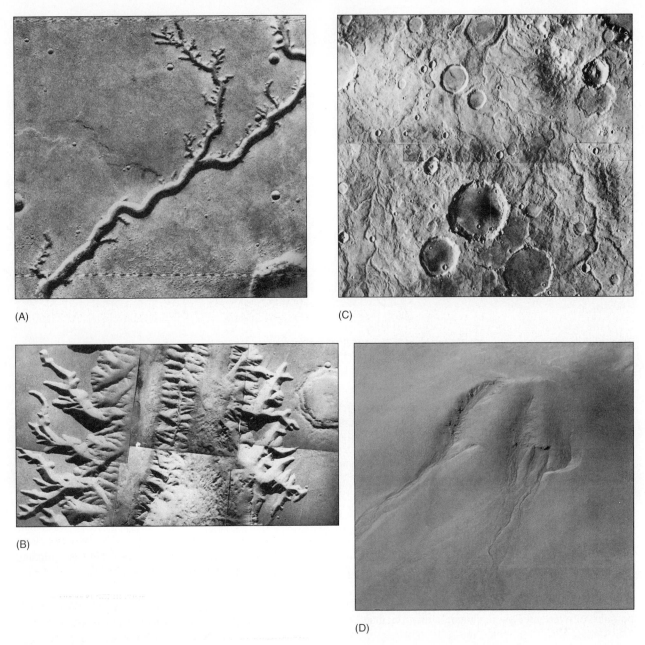

Figure 5.15
Drainage networks on Mars where groundwater-sapping processes are believed to have been the major formative process. (A) Nirgal Vallis, a longitudinal valley system with short, amphitheater-headed tributaries; channel is about 2–3 km wide. (B) Portion of Valles Marineris chasm showing stubby tributary complex (image is about 80 km across). Note how many tributaries have advanced headward along grabens and fractures. (C) Complex of smaller channel networks formed in the ancient hilly and cratered terrain. (D) June 2000 photo of the wall of Noachis Terra on Mars by Mars Global Surveyor. High resolution images from MGS seem to confirm recently active groundwater sapping processes on Mars.
(Source: NASA)

One of Horton's greatest contributions was to demonstrate that stream networks have a distinct fabric, called the *drainage composition,* in which the relationship between streams of different magnitude can be expressed in mathematical terms. Each stream within a basin is assigned to a particular order indicating its relative importance in the network, the lowest order streams being the most minor tributaries and the highest order, the main trunk river.

Figure 5.17 shows several methods of ordering steams. Horton's cumbersome method was refined by Strahler (1952b) so that stream segments rather than entire streams become the ordered units. In Strahler's sys-

(A)

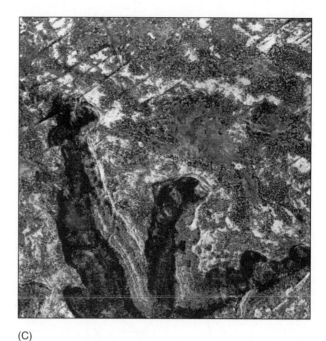

(C)

(B)

Figure 5.16
Earth examples where groundwater seepage and sapping processes have played a major formative role in valley development. (A) Northeast Kohala coast of Hawaii. The large, amphitheater-headed valleys have major springs at their head, fed from high-level aquifers (see Kochel and Piper 1986). The small, less-incised valleys in between are fed only by runoff. (B) Tributaries up-dip from the Colorado River have been significantly enlarged by groundwater-sapping processes in the permeable Navajo Sandstone. Note the lack of tributaries down-dip (to the bottom left). Runoff-dominated drainage systems typically show less influence on structural control. (C) Headward end of tributaries in the Navajo Sandstone of the Colorado Plateau in northern Arizona. Note the extension of valley heads along major joints where groundwater flow is enhanced. Compare to the right-angle junctions of valley heads on Mars in figure 5.15B.

tem, a segment with no tributaries is designated as a first-order stream. Where two first-order segments join they form a second-order segment; two second-order segments join to form a third-order segment, and so forth. Any segment may be joined by a channel of lower order without causing an increase in its order. Only where two segments of equal magnitude join is an increase in order required. The Strahler method created an apparent omission in accounting of low-order tributaries that was later accommodated in another network ordering scheme proposed by Shreve (1966a, 1967). The Shreve Magnitude, as it is called, considers streams as links within the network, with the magnitude of each link representing the sum of the link numbers of all the tributaries that feed it; that is, networks in which the downstream segments are of the same magnitude have equal numbers of links within the basins. Shreve's designations thereby express the number of first-order streams upstream from a given point. Geomorphologists investigating relationships between rainfall and runoff find the Shreve Magnitude system useful. Because the first-order streams serve as the primary collectors of rainfall within a basin, they are better flood flow predictors than the Strahler ordering system (Patton and Baker 1976).

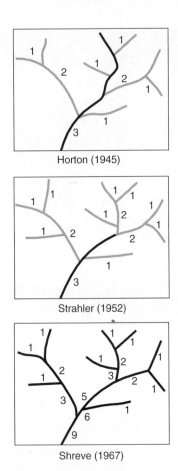

Figure 5.17
Methods of ordering streams within a drainage basin.

Shreve's system appears in many of the sophisticated runoff modeling packages which are beyond the scope of this discussion (for example, see Smart and Wallis 1971; Abrahams 1980; Abrahams and Miller 1982).

Every basin possesses a quantifiable set of geometric properties that define the linear, areal, and relief characteristics of the watershed (table 5.2), known as the basin morphometry. These variables correlate with stream order, and various combinations of the parameters obey statistical relationships that hold for a large number of basins. Two general types of numbers have been used to describe basin morphometry or network characteristics (Strahler 1957, 1964, 1968). *Linear scale* measurements allow size comparisons of topographic units. The parameters may include the length of streams of any order, the relief, the length of basin perimeter, and other measurements. The second type of measurement consists of *dimensionless numbers*, often derived as ratios of length parameters, that permit comparisons of basins or networks. Length ratios, bifurcation ratios, and relief ratios are common examples. Table 5.2 shows the most commonly used linear, areal, and relief equations, but numerous others have been derived from these.

Linear Morphometric Relationships   The establishment of stream ordering led Horton to realize that certain linear parameters of the basin are proportionately related to the stream order and that these could be expressed as basic relationships of the drainage composition. Much of linear morphometry is a function of the *bifurcation ratio* (R_b), which is defined as the ratio of the number of streams of a given order to the number in the next higher order (using Strahler ordering). The bifurcation ratio allows rapid estimates of the number of streams of any given order and the total number of streams within the basin. Although the ratio value will not be constant between each set of adjacent orders, its variation from order to order will be small, and a mean value can be used. Also, as Horton pointed out, the number of streams in the second highest order is a good approximation of R_b. When geology is reasonably homogeneous throughout a basin, R_b values usually range from 3.0 to 5.0.

The *length ratio* (R_L), similar in context to the bifurcation ratio, is the ratio of the average length of streams of a given order to those of the next higher order. The length ratio can be used to determine the average length of streams in an unmeasured given order (L_O) and their total length. The combined length of all streams in a given basin is simply the sum of the lengths in each order. For most basin networks, stream lengths of different orders plot as a straight line on semilogarithmic paper (fig. 5.18), as do stream numbers. The relationships between stream order and the number and length of segments in that order have been repeatedly verified and are now firmly established (Schumm 1956; Chorley 1957; Morisawa 1962; and many others).

Areal Morphometric Relationships   The equity among linear elements within a drainage system suggests that areal components should also possess a consistent morphometry, because dimensional area is simply the product of linear factors. The fundamental unit of areal elements is the area contained within the basin of any given order (A_O). It encompasses all the area that provides runoff to streams of the given order, including all the areas of tributary basins of a lower order as well as interfluve regions. Schumm (1956) demonstrated (fig. 5.19) that basin areas, like stream numbers and lengths, are related to stream order in a geometric series.

Although area by itself is an important independent variable (Murphey et al. 1977), it has also been employed to manifest a variety of other parameters (see table 5.2), each of which has a particular significance in basin geomorphology, especially in regard to the collection of rainfall and concentration of runoff. Numerous studies have been successful in formulating relationships between basin area and discharge. One of the more important areal factors is *drainage density* (D), which is

TABLE 5.2	Common Morphometric Relationship.

Linear Morphometry

Stream number in each order (N_o)	$No = R_b{}^{s-o}$
Total stream numbers in basin (N)	$N = \dfrac{R_b{}^s - 1}{R_b - 1}$
Average stream length	$\overline{L}_o = \overline{L}_1 R_L{}^{o-1}$
Total stream length	$L_o = \overline{L}_1 R_b{}^{s-1}\left(\dfrac{u^s - 1}{u - 1}\right)$ where $u = R_L/R_B$
Bifurcation ratio	$R_b = N_o/N_{o+1}$
Length ratio	$R_L = \overline{L}_o/L_{o+1}$
Length of overland flow	$\ell_o = \dfrac{1}{2D}$

Areal Morphometry

Stream areas in each order	$\overline{A}_o = \overline{A}_1 R_a{}^{o-1}$
Length-area	$L = 1.4A^{0.6}$
Basin shape	$R_F = \dfrac{A_o}{L_b{}^2}$
Drainage density	$D = \dfrac{\Sigma L}{A}$
Stream frequency	$F_s = \dfrac{N}{A}$
Constant of channel maintenance	$C = \dfrac{1}{D}$

Relief Morphometry

Relief ratio	$R_h = H/L_o$
Relative relief	$R_{hp} = H/p$
Relative basin height	$y = h/H$
Relative basin area	$x = a/A$
Ruggedness number (Melton 1957)	$R = DH$

Adapted from Strahler (1958).

s = order of master stream, o = any given stream order, H = basin relief, P = basin perimeter.

essentially the average length of streams per unit area and as such reflects the spacing of the drainageways. Drainage density reflects the interaction between geology and climate. As these two factors vary from region to region, large variations in D can be expected (table 5.3). In general, resistant surface materials and those with high infiltration capacities exhibit widely spaced streams, consequently yielding low D. As resistance or surface permeability decreases, runoff is usually accentuated by the development of a greater number of more closely spaced channels, and thus D tends to be higher. As a rule of thumb, where geology and slope angles are the same, humid regions develop thick vegetal cover that increases resistance and infiltration, thereby perpetuating drainage density lower than would otherwise be expected in more arid basins. Thus, drainage density not only reflects the geologic framework, but it

may serve as a useful parameter in climatic geomorphology (Daniel 1981). Methods for rapid estimation of drainage density have been devised (McCoy 1971; Mark 1974; Richards 1979; Bauer 1980).

Drainage density has also been used as an independent variable in the framing of other morphometric parameters. For example, the *constant of channel maintenance* and the length of overland flow (see table 5.2) both utilize a reciprocal relationship with density to demonstrate the link between factors that control surface erosion and those that describe the drainage net (Schumm 1956). The constant of channel maintenance indicates the minimum area required for the development and maintenance of a channel; that is, the ratio represents the amount of basin area needed to maintain one linear unit of channel length. As Schumm points out (1956, p. 607) this relationship requires that drainage networks

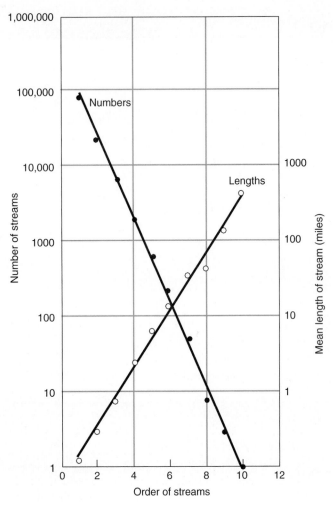

Figure 5.18
Relation of stream order to the number and mean lengths of streams in the Susquehanna River basin.
(After Brush 1961)

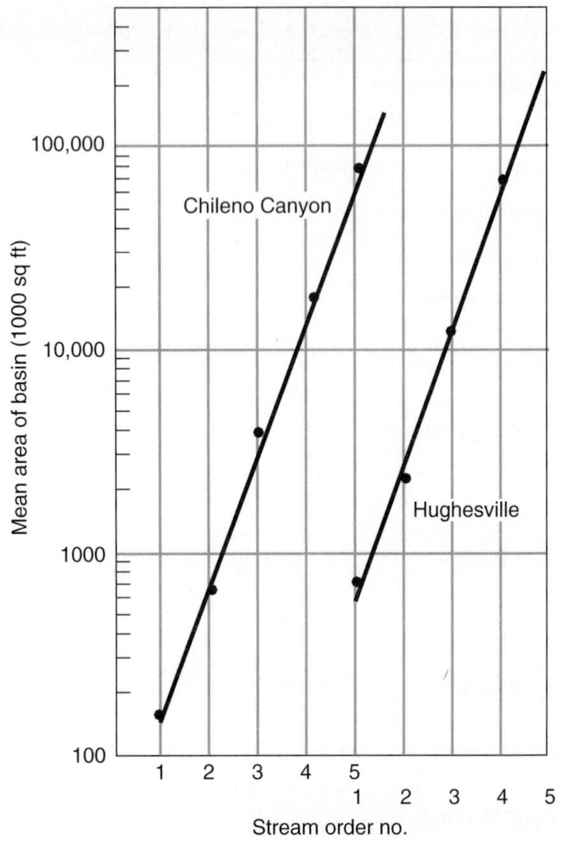

Figure 5.19
Relationship between stream order and mean basin area in two drainage basins.
(After Schumm 1956)

	TABLE 5.3	Summary of Sample Morphometric Data for Drainage Basins in Various Regions.					
Parameters[a]	Central Texas	Utah Wasatch	South California	Indiana	West Pennsylvania	Virginia	West Texas
	(19)[b]	(11)	(12)	(10)	(12)	(1)	(25)
A (km²)	12.4	29.7	2.3	156.8	122.0	34.1	32.8
L (km)	9.2	9.8	2.0	26.8	26.8	8.3	9.1
W (km)						4.7	3.5
R (km)	0.11	1.24	0.44	0.05	0.28	0.50	0.71
S	4.5	4.5	4.6	5.9	5.4	4.0	5.0
D (km/km²)	4.05	5.58	13.7	3.83	2.31	2.3	4.9
M						80.0	253.0
R	0.55	6.25	5.78	0.25	0.59	1.1	3.5
F_s (per km²)	28.7	12.4	133.3	11.0	8.4	13.8	44.2
R_b						4.4	4.5
K						1.6	2.0
Geology	Carbonate	Mixed sedimentary	Mixed metamorphic and igneous	Sandstone and shale	Sandstone and shale	Metamorphic	Carbonate

From Patton and Baker (1976) except for Virginia and West Texas, which are unpublished data from Kochel (1980).

[a]S = average Strahler order, K = shape based on lemniscate, M = Shreve magnitude.

[b]Number of basins.

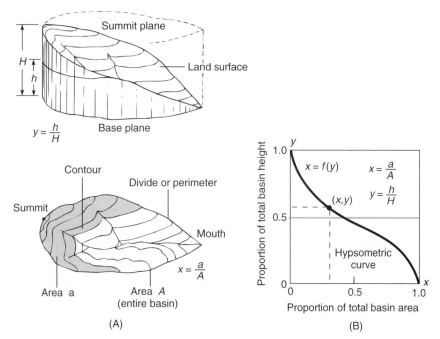

Figure 5.20

Ingredients of a hyposometric analysis. (A) Diagram showing how dimensionless parameters used in analysis are derived. (B) Plot of the parameters to produce the hyposometric curve.

(Strahler 1952b)

develop in an orderly way because the meter-by-meter growth of a drainage system is possible only if sufficient area is available to maintain the expanding channels.

Relief Morphometric Relationships   A third group of parameters shown in table 5.2 indicates the vertical dimension of a drainage basin; it includes factors of gradient and elevation. Like stream numbers, length, and area, the average slope of stream segments in any order approximates a geometric series in which the first term is the mean slope of the first-order streams. This relationship is reasonably valid as long as the geologic framework is homogeneous. Channel slopes and surface slopes are closely akin to the parameters for length. Horton suggested, for example, that the length of overland flow as a function of only the drainage density is at best an approximation because overland flow also depends on slope parameters.

As relief refers to elevation differences between two points, slopes that connect the points are the integral factors affecting the flow of runoff. The most useful relief parameters are the *maximum basin relief* (highest elevation on the basin divide minus the elevation of the mouth of the trunk river) and the *divide-averaged relief* (the average divide elevation minus the mouth elevation). The *relief ratio* (Schumm 1956), the maximum basin relief divided by the longest horizontal distance of the basin measured parallel to the major stream, indicates the overall steepness of the basin.

A different relief relationship is found by *hypsometric analysis* (Strahler 1952b), which relates elevation and basin area. As figure 5.20A shows, the basin is assumed to have vertical sides rising from a horizontal plane passing through the basin mouth and under the entire basin. Essentially, a hypsometric analysis reveals how much of the basin occurs within cross-sectional segments bounded by specified elevations. The relative height (y) is the ratio of the height (h) of a given contour above the horizontal datum plane to the total relief (H). The relative area (x) equals the ratio a/A, where a is the area of the basin above the given contour and A is the total basin area. The hypsometric curve (fig. 5.20B) represents the plot of the relationship between y and x and simply indicates the distribution of mass above the datum. The form of the curve is produced by the *hypsometric integral (HI)*, which expresses, as a percentage, the volume of the original basin that remains. In natural basins most *HI* values range from 20 to 80 percent, higher values indicating that large areas of the original basin have not been altered into slopes. Although computing the hypsometric integral can be tedious, methods have been introduced which streamline the procedure (Chorley and Morley 1959; Haan and Johnson 1966; Pike and Wilson 1971). Some researchers have found it to be an effective means of describing successive phases of landscape evolution (for example, Miller et al. 1990).

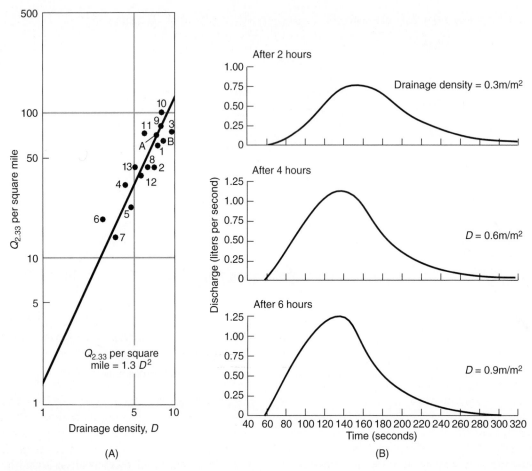

Figure 5.21
(A) Discharge (mean annual flood, $Q_{2.33}$) controlled by drainage density in 13 basins. (B) Effect of increasing drainage density on flood hydrograph in an experimental drainage system.

(A): (Carlston 1963), (B): (Zimpfer 1982)

Basin Morphometry and the Flood Hydrograph

The application of geomorphic principles to environmental hazards, such as flood potential, has led to a significant amount of research attempting to identify relationships between basin morphometry and stream flooding (see review in Patton 1988). Clearly, the shape and character of a stream flood hydrograph should be affected greatly by the manner in which a basin collects and routes water through its network. Stream hydrology, as defined by the flood hydrograph and by time elements such as flood frequency and lag, is significantly related to many components of basin and network morphometry. The interdependence of morphometry and hydrology is statistically real but does not necessarily indicate a cause and effect relationship; given two apparently related factors, one factor is not necessarily the cause of changes in the other. The high correlation probably exists because both factors vary in a consistent way with the same underlying climatic and geologic controls. In general, area and relief factors are closely related to flow magnitude, and length elements to the timing of hydrologic events. All morphometric types, however, are themselves so complexly woven together that no single factor can be isolated as a completely independent variable (Murphey et al. 1977).

Because basin area and peak discharge are highly correlative, we could expect that many other areal parameters will be similarly related to discharge. Every factor involving area differs in its success as a predictor of discharge, but one parameter, drainage density, seems to have considerable value as a gage of peak flow. In a study of 15 small basins in the southern and central Appalachians and the Interior Lowland Plateau region, Carlston (1963) demonstrated a very close relationship between drainage density and mean annual flood (fig. 5.21A). Notably, the basins in his sample have wide variations in relief, valley-side and channel slopes, and precipitation characteristics; yet none of these factors disrupts the relationship between flood magnitude and drainage density. Similar relationships have been observed in experimental studies (fig. 5.21B). Carlston suggests that the general capacity of a terrain to infiltrate

| TABLE 5.4 | Regression Forumulas for Predicting Flood Magnitudes from Drainage Basin Morphometry in Diverse Hydrogeomorphic Regions. |

Region	Equation	R^2	Probability
Central Texas	$Q_{max} = 17,369M^{0.43}(R)^{0.54}F^{-0.96}$	0.85	0.001
	$Q_{max} = 36,650M^{0.64}(R_h)^{0.54}(D)^{-1.68}$	0.74	0.01
Southern California	$Q_{max} = 155M^{1.04}(R)^{-0.83}F^{-0.73}$	0.85	0.001
	$Q_{max} = 380M^{0.89}(D)^{-1.87}$	0.86	0.0001
North-Central Utah	$Q_{max} = 23M^{0.90}(R)^{1.19}F^{-1.58}$	0.72	0.005
	$Q_{max} = 38,618M^{2.20}(R_h)^{2.51}F_1^{-3.73}$	0.83	0.005
Indiana	$Q_{max} = 424M^{0.46}(R)^{0.73}F^{0.21}$	0.67	0.01
	$Q_{max} = 424M^{0.82}(R_h)^{0.67}(D)^{0.56}$	0.66	0.05
Appalachian Plateau	$Q_{max} = 100M^{0.79}(R)^{0.19}F^{-0.29}$	0.92	0.0001
	$Q_{max} = 38M^{0.89}(D)^{-0.50}$	0.91	0.0001

Source: Patton and Baker (1976).

M = basin magnitude, R = ruggedness number, F_1 = first-order channel frequency; D = drainage density, R_h = relief ratio, Q_{max} = maximum peak discharge.

precipitated water and transmit it through the underground system is the prime controlling factor of the relationship between drainage density and near annual flood discharge in basins up to 260 km² in area. In larger basins, channel transit time plays the dominant role in the flow character. The rate of base flow, found to be inversely related to drainage density, is also dependent on terrain transmissibility. Thus, as Horton suspected earlier, high transmissibility (as evidenced by infiltration capacity) spawns low drainage density, high base flow, and a resultant low-magnitude peak flood. In contrast, an impermeable surface will generate high drainage density and efficiently carry away the abundant runoff; base flow will be low and peak discharge high.

Patton and Baker (1976) demonstrated predictive relationships between several morphometric parameters and peak flood discharges for streams in several physiographic regions of the United States (tables 5.3, 5.4). They found that areal morphometric parameters such as drainage density and stream frequency accounted for much of a model's ability to predict peak discharge, along with the relief measure known as *ruggedness number* (R) which is the product of relief and drainage density. These data were used to develop an index of flash flood potential (Beard 1975). Patton and Baker found that basins with high flash flood potential had greater ruggedness numbers than low-potential watersheds. Dingman (1978), however, warned that the relationship between drainage density and flow can be overridden by other effects in the basin such as floodplain or channel storage. In addition, where saturated overland flow is the major source of runoff, drainage density may not be related to the efficiency at which a basin is drained. Costa (1987) investigated the morphometry of basins associated with the largest historic floods in the United States. Although these flash flood basins did not uniformly possess the basin attributes expected from

studies like that of Patton and Baker (1976), Costa was able to find some commonalties. Basins with flashy or peaked flood hydrographs generally contained significant area of exposed bedrock, occurred in semiarid to arid climates, were short, and had high relief.

In recent years a large amount of research has focused on the development of more sophisticated models of runoff that are linked closely with geomorphic attributes of the basin and their impact on the production of floods. One of the predominant models is the **geomorphic unit hydrograph** (Rodriquez-Iturbe and Valdez 1979). The success of modeling efforts have been mixed (Patton 1988), partly because of our incomplete understanding of the complex interrelationships between rainfall-runoff events and the contributing basin networks. Further work using small, instrumented watersheds, as well as numerical analytical approaches that explore relationships between the geomorphic unit hydrograph and basin parameters (Chutha and Doodge 1990), will refine our understanding and perhaps lead to more reliable models for predicting floods using basin parameters.

Abrahams (1984) aptly summed up the difficulties in elucidating quantitative relationships in basin networks by noting that the apparent randomness arises largely from independent variation of a large number of factors such as lithology and microclimate. The possible interrelationships between hydrology and morphometry are seemingly infinite, and the parameters are so complexly related that equations will not explain all the variability. Still, the hydrogeomorphic approach has some validity and should not be abandoned in future research. The hydrogeomorphic approach is especially applicable in determining regional flood hazards (Baker 1976). Figure 5.22 is a schematic model showing the expected influence of variations in basin morphometry on the flood hydrograph based on generalizations from a large number of studies. In each case, the influence of a specified

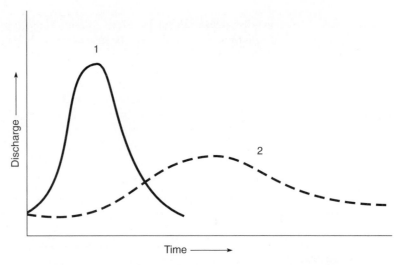

Characteristic	Flashy (hydrograph 1)	Sluggish (hydrograph 2)
Basin area	Small	Large
Drainage density	High	Low
Basin magnitude	High	Low
Relief	High	Low
Ruggedness number	High	Low
Basin shape	Equidimensional	Elongate
Soils	Thin	Thick
Vegetation	Sparse	Dense
Storm track	Down the basin	Up the basin

Figure 5.22

Idealized flood hydrograph and generalized responses to drainage basin characteristics. The effect of an individual characteristic is shown assuming the other characteristics are held constant.

morphometric variable is displayed assuming that all other morphometric, geologic, and climatic variables remain constant. Although the relationships summarized here are somewhat qualitative until more conclusive research is completed, they offer a general guide for planners making an initial assessment of expected flood character in ungaged basins.

Basin Evolution

Although morphometric values differ from basin to basin, each network still obeys the statistical relationships discussed above. Many authors have suggested that morphometry reflects an adjustment of geomorphic variables that is established under the constraints of the prevailing climate and geology (for example, Chorley 1962; Leopold and Langbein 1962; Strahler 1964; Doornkamp and King 1971; Woldenberg 1969). Essentially, once a network is established, the basinal characteristics can be defined by the same quantitative terms at any time during the drainage growth. As the basins and networks evolve, an equilibrium is eventually produced by the interplay of climate and geology and maintained as a time-independent phenomenon. Once the components within a basin become balanced, any changes in climate or geology will be compensated for by adjustments of the basin parameters in such a way that the relationships of drainage composition will be preserved. As originally

conceived, however, these relationships issued from well-developed stream systems, and the measurements needed to derive the equations were made on topographic maps of these basins. Such an approach provides no insight as to how quickly morphometric balance is attained or what changes in its character occur as the basin ages. It seems appropriate, therefore, to consider the influence of time on the morphometry of a basin.

Some studies have touched on the question of how rapidly morphometry is established and what changes occur in its nature as the basin evolves. These studies found that a quantitatively balanced drainage net forms rapidly in erodible material. This was clearly demonstrated by Schumm (1956) in the Perth Amboy, N.J., badlands, by Morisawa (1964) on the uplifted floor of Hebgen Lake, Mont., and by Kirkby and Kirkby (1969) on the raised beach and harbor floor around Montague Island, Ak. These studies showed that a balanced drainage composition evolved in a period no longer than a few years (for the Alaska case, in only a few days). Certainly, the time needed for drainage development to occur in erodible material is insignificant in terms of geological time. However, the amount of time needed in absolute terms probably varies according to the resistance of the material, the climate, and the initial slope angles. Unfortunately, examples of drainage establishment are documented only in areas where the least resistant materials underlie the system. Precisely how long it

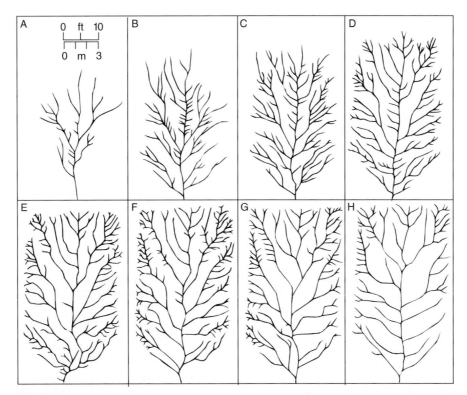

Figure 5.23
Chronologic development of an experimental drainage system in response to simulated rainfall.
(Parker 1977)

takes to form a balanced network in regions underlain by resistant crystalline rocks is rather conjectural.

Drainage evolution has been examined extensively by experimental means at the Rainfall Erosion Facility at Colorado State University by S. A. Schumm and his students and colleagues. Although experimental studies suffer some drawbacks in comparison to natural systems, such as measures of scale, they are exceedingly useful in providing conceptual models for processes that cannot be monitored in the field over the life span of an observer (Schumm et al. 1987).

Parker (discussed in Schumm 1977 and Schumm et al. 1987) found that patterns grow by headward extension until dissection reaches the watershed margin. The total drainage density and the mode of growth depended on the slope of the basin. Phillips and Schumm (1987) demonstrated a significant influence of slope on the form of drainage networks. They found that as slopes steepen the pattern changed from dendritic to parallel. Dendritic patterns predominated on slopes less than 1 percent, semiparallel patterns prevailed on 3 percent slopes, and parallel patterns formed once slopes attained 5 percent— clearly demonstrating the influence of slope angle on junction angle. The basin with a lower slope had a higher density because tributaries formed inside the basin during early extension growth, whereas in the steeper basin streams extended to the margins before interior links

were added (Parker 1976). In general, Parker's work showed that the network will develop as many streams and as much length as is needed to efficiently drain the basin. However, distinctive changes in drainage density were commonly observed in different parts of the basin during its network evolution (fig. 5.23).

The experiments summarized by Schumm et al. (1987) outline predictable sequences of drainage network evolution which highlight processes such as stream piracy in facilitating the growth process. Similarly, this experimental work was successful in establishing strong linkages between variations in sediment yield and activity in the evolving channel network. One significant cautionary note arising from their work is that different portions of the basin typically evolve disharmoniously, reemphasizing the concept of complex response often cited in Schumm's field studies. Thus, sediment yield and storage will operate in a complex manner, and making correlations between valleys and even between different sites along the same valley is difficult.

Once basin elements attain a statistical balance, further changes in morphometry are usually revealed in the shape, drainage texture, or hypsometry of the basin. The area limits of any basin presumably are determined by the hydrophysical controls denoted by Horton and by the competition for space between adjacent basins. During the period of expansion to its peripheral limits, however,

Figure 5.24
Relation of stream order to stream length and number in drainage basins of the Ontonagon Plain, Mich. A = streams between Mineral and Cranberry rivers; data from maps; stream orders not known. B = Little Cranberry River, C = Weigel Creek, D = Mill Creek.
(Hack 1965)

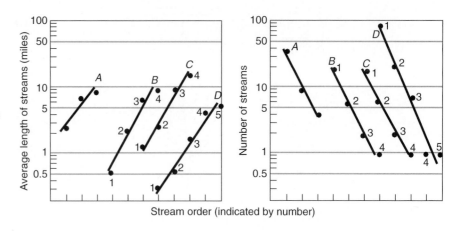

each parameter should change in such a manner as to maintain the original quantitative relationships among the factors. Hack (1957) showed that for a large number of basins, the stream length and basin area are related by the simple power function

$$L = 1.4\,A^{0.6}$$

where L is the distance from any locality on a stream to the divide at the head of the longest segment above the locality, and A is the basin area above the locality. Many investigations of basin growth indicate a tendency toward elongation (for example, Hack 1965; Miller 1958; Jarvis and Sham 1981), while others suggest that widening occurs as rapidly as elongation, particularly in large basins (Mueller 1972; Shreve 1974; Moon 1980). Figure 5.24 demonstrates that although many local variables influence basin shape and development, distinct relationships exist between stream order and morphometric parameters such as stream length and stream number.

Part of the controversy regarding trends in basin evolution through time relates to a concept suggested by Schumm (1956), that the locus of intense erosion migrates with time from the basin mouth vicinity early in basin development to the headward region in later stages. In addition, Ruhe (1952) demonstrated that both drainage density and stream frequency increase systematically with time in areas underlain by glacial deposits of known age. The rate of textural change is not constant; it was probably greatest during the first 20,000 years and then decelerated. The marked transition in the rate of textural evolution perhaps represents the time at which complete equilibrium was established and the basins were filled with as many streams as possible (Leopold et al. 1964).

Throughout the period of growth in Ruhe's study, the channel lengths and numbers seem to obey Horton's geometric laws, suggesting that texture may be a surrogate for time in the analysis of basin evolution. This proposition was given added credence when Melton

(1958) found stream frequency (F_1) and drainage density (D) to be related by a simple equation:

$$F_1 = 0.694\,D^2$$

In deriving this equation, Melton used as his sample 156 basins with widely different geology, erosional history, and probable age. Because all stages of drainage development were thrown into the statistical pot, the significant empirical relationship between the variables most likely reflects a general trend followed by basins as they grow. The dimensionless ratio F_1/D^2 called *relative density,* should indicate how completely the stream network fills the basin (Melton 1958). High relative-density values suggest that stream lengths are short and the basin outline is not yet completely filled with the stream network. As the drainage evolves and expands into each basin niche, the ratio decreases until equilibrium is established.

Melton's growth law is one in which spatial parameters can be substituted for time. However, the universal applicability of the growth equation has been questioned. Abrahams (1972) suggested that time and space are interchangeable in morphometry only if the basins analyzed are environmentally similar and are of the same order. Further studies, however, did not completely substantiate this criticism (Wilcock 1975), and a reasonable argument can be made that linear, areal, and relief factors are probably involved in the statistical relationships affected by time. An interesting addendum to the controversy over basin adjustments is provided by a simulation study modeling long-term basin morphologic development (Wilgoose et al. 1991). This study suggests that once a basin attains dynamic equilibrium in a setting where fluvial sediment transport is dominant, all basins will be similar except for variations in drainage density, in spite of differences in tectonics and lithologic resistance.

Alterations of basin morphometry do occur with time. These changes do not violate the steady-state concept because parameters vary with one another in a sys-

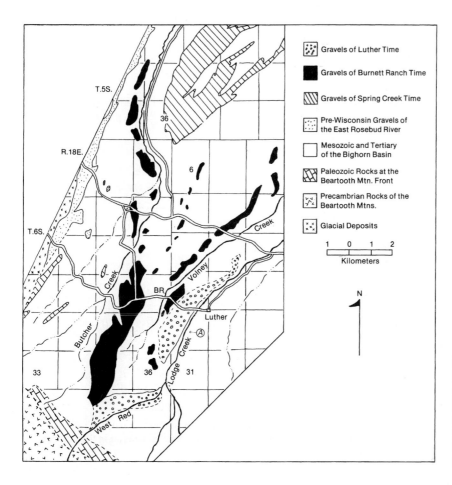

Figure 5.25
Map of terrace deposits in a small piedmont area of the Beartooth Mountains, Mont. Major stream piracy occurred near *A* between deposition of Burnett Ranch gravels and deposition of Luther gravels.
(Ritter 1972)

tematic way. It is tempting to assume, therefore, that a well-balanced network with discernible morphometry evolved in an orderly manner from infancy to its present state. Such an assumption, however, may be totally erroneous because basin morphometry may tell us little, if anything, about *basin history.*

A cogent example of this is the drainage system of Volney Creek, a small basin in southern Montana within the watershed of the Yellowstone River. Volney Creek heads in the piedmont region of the Beartooth Mountains, rising approximately 3 km from the mountain front. This basin, covering an area of 60 km² is underlain entirely by Mesozoic and Tertiary clastic sedimentary rocks. The basin exhibits all the normal morphometric characteristics of neighboring piedmont basins, and it is presumed to be in equilibrium because its morphometry is balanced statistically, as discussed above. The geomorphic history of the basin, however, shows that its evolution was anything but orderly (Ritter 1972). Two terraces standing well above the present level of Volney Creek are capped by gravel containing crystalline clasts that had to be derived from the Beartooth Mountains. In fact, upstream tracing of the terraces demonstrates that they cross through the present basin divide and continue mountainward in the neighboring valley of West Red

Lodge Creek (fig. 5.25). The significance of this is that prior to early Wisconsinan glaciation (Burnett Ranch time in fig. 5.25), the valley now occupied by Volney Creek was the drainage avenue for the master stream of the area. The crystalline-rich gravel capping the terraces indicates that the paleoriver drained from the mountains, and its basinal area must have been vastly greater than that of the modern Volney Creek. Immediately preceding the glaciation, the river was diverted into its present position in the valley of West Red Lodge Creek, leaving the Volney segment abandoned until it was occupied by the very small modern stream.

Diversions of the Volney Creek type are common in piedmont regions and may be important phenomena in the expansion of any drainage system, especially in the early stages (Howard 1971). Such changes, nonetheless, are catastrophic events in basin development because they drastically alter basin properties such as area, relief, and stream length. The internal adjustment to changes spurred by piracy must occur rapidly, for most basins possess a balanced morphometry even though such spasmodic events must be commonplace.

Finally, detailed studies of the spatial characteristics of basin networks can reveal adjustments to regionally active tectonics. For example, Pubellier et al. (1994) were

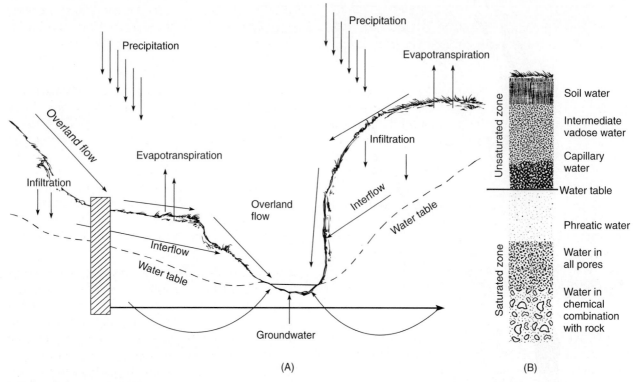

Figure 5.26
Surface and shallow subsurface hydrology of a watershed. (A) Schematic of basin showing directions and relative water movements measured in discharge (volume per time). (B) The groundwater profile as it would appear in homogenous material in the vertical slice on the left.

able to delineate blind subsurface faults and faults using abnormalities of basin networks in the Philippine Islands.

BASIN HYDROLOGY

For planning purposes, estimates are needed of how much water exists in a drainage basin and whether it is available for use. Hydrologists usually employ a concept known as the **water balance** or **hydrologic budget** to make such estimates. The water balance simply refers to the balance that must exist between all water entering the basin (input), all water leaving the basin (output), and any changes in the amount of water being stored.

Figure 5.26 illustrates the hydrologic cycle with respect to a drainage basin. The major inputs are rainfall and snow, and the major outputs are evapotranspiration and streamflow. Water is stored in lakes, soil moisture, and groundwater, and changes in these values actually represent losses or gains of available water. Although the basin cannot actually be divorced from the global hydrologic system, we can view the basin hydrologic budget simplistically by the following equation:

$$\Delta S = I_P - O_E - O_T - O_R$$

A change in storage, measured as recharge to the groundwater system (ΔS), is equal to the difference be-

tween input as infiltration from precipitation (I_P) and losses or outputs from evaporation (O_E), transpiration (O_T), and runoff (O_R). Although runoff provides the predominant connection between our basin and the global hydrologic system, some groundwater may exit the basin by regional subsurface flow pathways. For example, a positive change in groundwater storage indicates that the underground reservoir is being recharged, but in the budget it represents a loss because the availability of the water is being lowered. On the other hand, groundwater runoff (base flow) represents an input because water availability is increased as it is released from storage.

Subsurface Water

The increased demand for water that has accompanied population growth, industrial expansion, and extended irrigation in the United States has brought with it a marked increase in the utilization of groundwater. Groundwater as a resource is highly sought after because it typically maintains chemical and physical characteristics that fluctuate far less radically than surface water supplies. For example, groundwater maintains a constant year-round temperature, a more constant chemical composition year-round, and is less affected by periodic droughts than surface reservoirs. The volume of water contained in fractures and pores in Earth's underground reservoir is enormous; it is estimated at almost 8 million km³ in the outer 5 km of the crust (Todd 1970). Unfortunately, this vast resource is not evenly distributed. Some regions are blessed with abundant groundwater whereas others are seriously deficient. Ironically, many of the most rapidly expanding areas of the United States are in regions with low reserves of surface and groundwater. The lack of available water, coupled with the growing demand and ever-present threat of contamination, poses a major challenge to geologists and requires that we continue to expand our knowledge concerning the distribution, movement, and utilization of groundwater. For example, some regions possess groundwater with exceptionally high natural concentrations of dissolved minerals, whereas others are plagued by anthropogenic contaminants. Groundwater geology, or **hydrogeology,** has evolved into a separate discipline in itself; however, because it is an important part of basin hydrology, we will briefly examine how the groundwater system works.

The Groundwater Profile Groundwater in porous and permeable rocks or unconsolidated debris usually has a rather distinct distribution that can be visualized as a vertical zonation known as the *groundwater profile* (fig. 5.26B). The *unsaturated zone* above the water table is variable in water content within its pores and fractures. Air occurring in pores that are only partly water-filled is physically connected to the atmosphere. Slugs of infiltrating water move vertically down, under the influence of gravity, through the unsaturated zone in response to episodic precipitation events at the surface. Flow paths include movement through the interstices of porous sediment and bedrock, flow through bedrock fractures, and accelerated flow along zones of increased permeability referred to as macropores such as root channels and burrows. Interruptions in the complete downward transfer of water through the unsaturated zone commonly result from water being held by surface tension or that which is extracted by plants during evapotranspiration.

The uppermost portion of the unsaturated zone is the *zone of soil moisture,* which experiences large fluctuations in water quantity and quality because of variations in infiltration, evapotranspiration, and atmospheric pressure. The *intermediate zone* below the soil moisture zone experiences downward flow toward the water table. This zone, commonly thick in semiarid regions, may be totally absent in many humid areas.

The base of the unsaturated zone is marked by a zone of variable thickness where all available pore space is saturated with water. This area of apparent saturation is known as the *capillary fringe.* Capillary action draws water upward through the pores from the zone of saturation and holds this water in tension. Even though this zone is saturated, water will not drain freely into a well placed within the capillary fringe because the hydrostatic pressure remains less than atmospheric pressure, as it does through-

out the unsaturated zone. The thickness of the capillary fringe depends on pore diameter, its thickness being inversely proportional to the diameter of the pores.

All pore spaces below the capillary fringe are filled with water; hence this area is called the *saturated* or *phreatic zone.* Phreatic water will drain freely into a well because the hydrostatic pressure in this zone exceeds that of atmospheric pressure. The *water table* marks the level where the hydrostatic pressure is equal to atmospheric pressure and defines the upper boundary of the saturated zone. This zone extends downward until rocks become dense enough that small occurrences of interstitial water are no longer interconnected. The saturated zone may extend for tens of thousands of feet in sedimentary basins, being less extensive in crystalline rocks.

Movement of Groundwater Water in the saturated zone is not static but can move in any direction depending on the distribution of potential energy fields within the zone. Groundwater flow in the saturated zone is not solely regulated by gravity as it is in the unsaturated zone. Each unit volume of water contained in the saturated (phreatic) zone possesses a certain amount of potential energy, called its *potential* or *head.* The amount of potential varies from droplet to droplet, being dependent on the pressure of each drop and its elevation above some datum. When pressure and elevation are known, the potential for each unit volume of water can be calculated, and water particles having the same potential can be contoured along surfaces known as *equipotential surfaces* (fig. 5.27). Although some diffusion occurs, groundwater particles move along paths that are perpendicular to equipotential surfaces.

The movement of groundwater is a mechanical process whereby some of the initial potential energy of the water is lost to friction generated as the water moves. It follows that water moving from one point to another must have more potential energy at the beginning of the transport route than at the end. Thus, groundwater always moves according to the following rules: (1) it moves from zones of higher potential toward zones of lower potential, and (2) it flows perpendicular to the equipotential surfaces, at least in simple homogeneous and isotropic materials (Hubbert 1940).

The velocity and discharge of groundwater flow is directly proportional to the loss of potential (head) that occurs as water moves from one point to another (fig. 5.28). This concept was demonstrated in 1856 by H. Darcy in his famous experiments on flow through porous media. He also observed that flow velocity was inversely proportional to the length along the flow path. Darcy's experiments led to the fundamental law of groundwater flow, expressed as

$$V = K \frac{h_1 - h_2}{L}$$

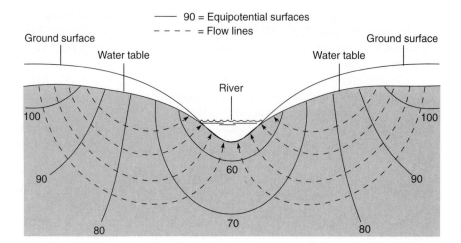

Figure 5.27
Movement of groundwater according to distribution of potential in the underground system. Water moves from high to low potential and perpendicular to the equipotential surfaces.

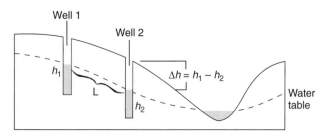

Figure 5.28
Water table and loss in head as water moves from well 1 to well 2 in unconfined aquifer.

where V is a volumetric flow rate similar to velocity, h is head, L is the distance between the two points of measurement (i.e., between monitoring wells), and K is a constant of proportionality representing the permeability of the medium, known as the *hydraulic conductivity*. In figure 5.28, the velocity of flow can be calculated from the difference in hydrostatic level (head) between two wells ($h_1 - h_2$) if the permeability of the material is known or can be estimated. Discharge can also be calculated by including the cross-sectional area of the aquifer to the equation so that

$$Q = K I A$$

where Q is discharge, A is the cross-sectional area of an aquifer perpendicular to flow ($w \times d$), and I, called the

hydraulic gradient, equals $\dfrac{h_1 - h_2}{L}$.

Darcy's Law is applicable to groundwater flow through porous materials, but adjustments are often made to accommodate the effects of variable media and fluid properties on the actual values of K. More complex relationships must be used to estimate flow characteristics through fractures in relatively impermeable host materials.

The water table, at the top of the phreatic zone, is an important hydrologic feature (see fig. 5.26). Because hydrostatic pressure everywhere along the surface of the water table is equal to atmospheric pressure, the potential of the water table is equal to atmospheric pressure. The potential of water there is completely a function of elevation, and water at this surface will always move from higher to lower elevations. This partially explains why the water table is generally a mirror image of surface topography. Whereas surface water and groundwater systems are physically connected, the levels of rivers, lakes, swamps, and other wetlands are merely surface extensions of the underground water table (see fig. 5.27).

Aquifers, Wells, and Groundwater Utilization Problems *Aquifers* are lithologic bodies that store, transmit, and yield water in economic amounts. The most common aquifer is called an *unconfined aquifer* because it is open to the atmosphere and its hydrostatic level (shown by the level at which water will stand in a well) is within the water-bearing unit itself. The hydrostatic level in an unconfined aquifer is the water table. In some other aquifers the water is held in a porous and permeable unit that is not connected vertically to the atmosphere but instead is overlain and underlain by less permeable layers called confining layers (or aquitards). Water in these *confined aquifers* (fig. 5.29) will rise above the top of the aquifer when it is penetrated by a well or open hole. The level to which water will rise is called the *potentiometric surface*. The height of the potentiometric surface above the aquifer depends on the difference in potential at the point where precipitation or infiltration enters the aquifer, called the recharge zone, and the position of the screen at the base of the well or hole (fig. 5.29). If the potentiometric level is above the elevation of the ground, surface water will flow freely out of the well without pumping as *artesian flow*.

The development of an aquifer for water supply requires wells; the larger the demand, the larger and more numerous the wells. As a well is pumped, the hydrostatic level (water table or potentiometric surface) sur-

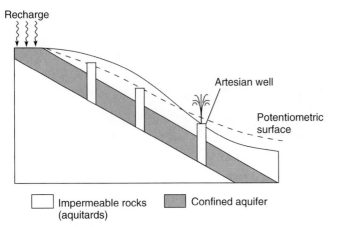

Figure 5.29
Confined aquifer and potentiometric surface.

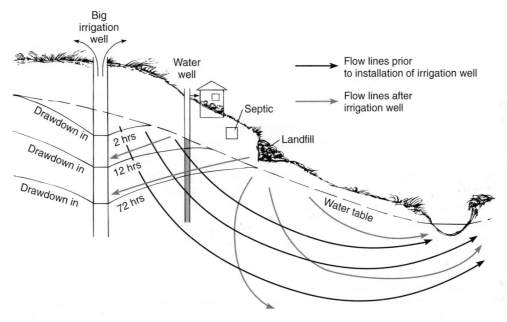

Figure 5.30
Development of cone of depression drawing down the water table due to the pumping of a major irrigation well. The cone of depression can alter groundwater flow patterns significantly as indicated here where the leachate from a landfill, which normally would move to the right, would now flow toward the water supply well for the home. This example assumes a simple homogenous system.

rounding the well declines and forms an inverted cone known as the *cone of depression* (fig. 5.30). The cone develops because the release of water (or of pressure in a confined aquifer) is greatest near the well. This causes pronounced lowering of the water table or potentiometric surface, called *drawdown,* adjacent to the well. The effect of pumping decreases away from the well, so drawdown is less toward the perimeter of the cone. In the initial phase of pumping, the rate of drawdown is normally high, but as pumping continues the rate gradually decreases until the cone attains a nearly constant form. The dimensions of the quasi-equilibrium cone depend on the rate at which the well is pumped and the hydrologic properties of the aquifer.

Utilizing groundwater can cause a number of environmental problems if the system is not studied carefully before development. For example, excessive drawdown can occur locally if wells are placed too close to one another and the radius of influence (the maximum diameter of the cone of depression) of adjacent wells overlap. This produces abnormally high drawdown in the zone of overlap because the drawdowns from all interfering wells are cumulative at any point. On a regional scale, most problems arise when more water is pumped from the aquifer over a period of years than is returned to the aquifer by natural or artificial recharge, a practice known as *overdraft.* In such a case the aquifer is actually being mined of its water, and on a long-term basis the

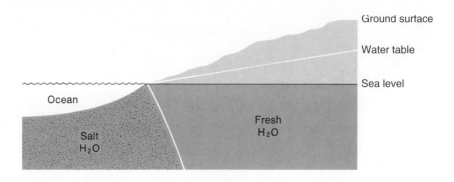

Figure 5.31
Relationship of salty and fresh groundwater in an aquifer of a coastal region. Lowering of water table by overdraft requires a 40-fold rise in the boundary between the fresh and salt water and increases the possibility of pollution of the aquifer by salt water.

normal hydrostatic level may be drastically lowered. The socioeconomic importance of these issues is well illustrated by an article in *National Geographic* (Zwingle 1993) that treats the concerns over depletion of groundwater in the major hydrostratigraphic unit of the U.S. High Plains region, the Ogallala Aquifer.

The effects of continued overdraft vary, but two types of responses will demonstrate the problems that can result from the misuse of the groundwater system. First, in confined aquifers, the drawdown of the potentiometric surface reflects the decrease of pressure within the aquifer. Because the pressure is lowered, water from the overlying confining layer seeps downward into the aquifer. As the confining layer drains, the normal load exerted by the weight of the overlying rocks compacts the confining layer and decreases its thickness. Ultimately, the process culminates in measurable subsidence of the ground surface. Some areas, such as Mexico City, Las Vegas, the Central Valley of California, and the Houston-Galveston area, have experienced 1 to 5 meters of overdraft subsidence, creating a variety of annoying and hazardous conditions such as cracking of buildings, strain on buried pipelines, highway subsidence, and destruction of well casings.

A second major effect of overdraft usually occurs in coastal regions where an aquifer is physically connected to the ocean. There, two fluids (ocean water and fresh water) having different densities are separated by a sharp boundary, as shown in figure 5.31. The location of the interface depends on the hydrodynamic balance between fresh water (density = 1.00 g/cm^3) and salt water (density = 1.025 g/cm^3). In general, the depth below sea level to the saltwater boundary is about 40 times the height of the water table above sea level. In such a situation, continued overdraft can cause pollution of the aquifer because minor lowering of the water table necessitates a much greater rise of the saltwater-freshwater interface, a phenomenon known as *saltwater intrusion*. For example, a 2-meter drawdown of the water table requires a concomitant 80-meter rise of the salt water, and any well extending to a depth greater than the new interface level will be polluted with nonpotable water. Calvache and Pulidd-Bosch (1991), using a numerical modeling approach, eloquently demonstrated how expanding

groundwater withdrawal is responsible for the gradual encroachment of saline water into important regional aquifers in coastal Spain.

Many cities along the coasts of California, Texas, Florida, New York, and New Jersey have been affected by saltwater intrusion. This phenomenon, however, can occur wherever two fluids of different density exist within the same aquifer. For example, the water supply of Las Vegas, Nevada, is in jeopardy from pollution by high-magnesium groundwater located about 24 km south of the city. The intrusion is a response to the large overdraft from the aquifer beneath Las Vegas.

Linkages between groundwater and surface water systems have been the focus of many studies since the late 1970s. These have highlighted interactions related to channel initiation (Dunne 1990; Kochel et al. 1985), particle entrainment, floodplain aquifers and river-stage fluctuations (e.g., Sophocleous 1991), and the effects of shallow groundwater circulation on soil development and chemistry and of soil properties on groundwater (e.g., Richardson et al. 1992). In light of our ever-increasing concern about the environment, increased knowledge of the interactions between subsurface and surface hydrologic systems is becoming very important.

Surface Water

The major export from drainage basins to the global hydrologic system occurs as stream discharge. The volume of water passing a given channel cross-section during a specific time interval, is expressed as

$$Q = wdv = Av$$

where Q is discharge in m^3/s (cms) or ft^3/s (cfs), w is width, d is depth, A is area (wd), and v is velocity. Measurement of discharge is relatively simple by dividing the channel cross-section into segments of even width (fig. 5.32A). Depth and velocity are measured in each segment, and then total discharge is computed by summing the discharges of all of the sections across the channel. The most difficult measurement to obtain is velocity. Velocity varies with depth in open channels (fig. 5.32C), thus surface velocity is not very representative of the mean velocity. Average velocity is obtained

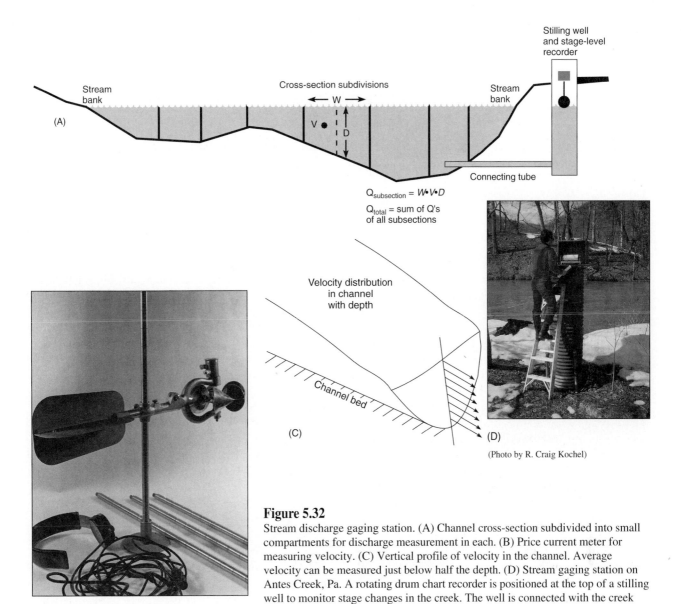

$$Q_{subsection} = W \cdot V \cdot D$$

Q_{total} = sum of Q's of all subsections

(Photo by R. Craig Kochel)

Figure 5.32
Stream discharge gaging station. (A) Channel cross-section subdivided into small compartments for discharge measurement in each. (B) Price current meter for measuring velocity. (C) Vertical profile of velocity in the channel. Average velocity can be measured just below half the depth. (D) Stream gaging station on Antes Creek, Pa. A rotating drum chart recorder is positioned at the top of a stilling well to monitor stage changes in the creek. The well is connected with the creek via lateral pipes below the floodplain.

with a current meter (fig. 5.32B) adjusted to measure the flow rate just below the halfway mark on the depth in each channel subsection.

To be of any scientific value, discharge must be measured repeatedly at the same locality; this gives hydrologists a better understanding of how flow varies with time. In the United States these sampling localities, or *gaging stations,* have been in operation at some sites since the late 1800s. Thus, a wealth of flow data for a large number of streams is available from the U.S. Geological Survey and numerous state water surveys. Common measurement frequencies vary from continuous recording to intervals of one hour. The average of these data are published annually as *mean daily discharge.* The *mean annual discharge* is the average of the daily values over the entire period of record, assuming that the station has been maintained for longer than one year.

Actually, discharge is not frequently measured by the procedure described above because it is very time-consuming. Instead, discharge is estimated from a *rating curve* (fig. 5.33), or *rating table,* which relates a range of discharge values to the elevation of the river above some datum, called the *river stage* or *gage height.* Once the rating curve for a specific station has been constructed from a series of actual discharge measurements (the data points on fig. 5.33), only the gage height must be observed directly to be able to predict discharge.

Flood Frequency   Geomorphologists are interested in the frequency and magnitude of flow events because each has an important bearing on how basins and channels evolve and function. Mayer and Nash (1987) compiled a collection of papers focusing on the geomorphic effects and frequency of catastrophic floods. The

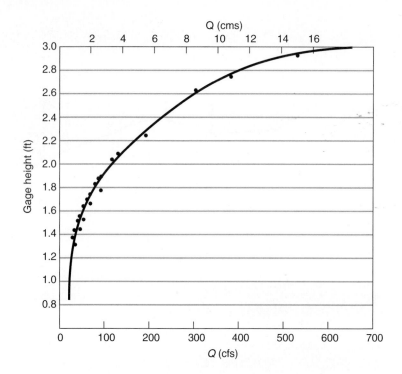

Figure 5.33
Rating curve for low flow. Rock Creek near Red Lodge, Mont.

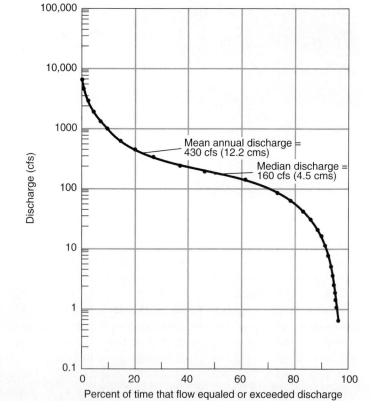

Figure 5.34
Flow duration curve for the Powder River near Arvada, Wyo., 1917–1950.
(Leopold and Maddock 1953)

frequency of a given discharge is graphed as a **flow duration curve** (fig. 5.34), which relates any discharge value to the percentage of time that it is equaled or exceeded. At any station, then, the lowest daily discharge in the period of record will be equaled or exceeded 100 percent of the time. The largest flow will be equaled or

exceeded only once out of the entire number of days in the sample, giving it a percentage value slightly greater than zero.

The most common approach to finding the frequency-magnitude relationship (especially for floodplain planning) is to consider only the peak discharge during

TABLE 5.5	Annual Discharge Extremes and Averages for the Pecos River near Langtry, Texas, 1900–1977.

Year	Discharge (m3/s)			Year	Discharge (m3/s)		
	Minimum	Average	Maximum		Minimum	Average	Maximum
1900	7.3	—	2991	1939	3.9	8.1	162
1901	4.5	17.8	310	1940	4.6	8.8	157
1902	5.8	19.6	936	1941	5.7	51.4	523
1903	5.0	12.9	363	1942	7.4	21.8	264
1904	3.0	13.8	2013	1943	5.4	9.9	313
1905	14.2	40.8	1314	1944	4.4	8.1	250
1906	11.6	27.4	2516	1945	3.9	9.9	774
1907	5.0	12.9	327	1946	4.2	10.3	1817
1908	5.9	20.4	1901	1947	4.2	7.2	171
1909	4.2	10.1	727	1948	3.1	6.4	1434
1910	4.3	10.2	2851	1949	4.0	10.9	2753
1911	5.3	11.6	755	1950	2.8	7.2	1255
1912	3.4	6.4	512	1951	2.6	5.1	229
1913	3.8	15.4	1761	1952	2.3	3.9	100
1914	6.4	10.9	1873	1953	2.0	4.5	414
1915	10.2	32.9	1453	1954	3.0	78.3	27,392
1916	5.7	14.9	2711	1955	2.8	8.0	758
1917	4.1	8.3	176	1956	2.1	4.4	208
1918	2.9	6.1	1448	1957	3.3	15.1	1073
1919	3.5	30.8	243	1958	3.6	10.4	1075
1920	6.3	20.9	1219	1959	3.9	11.3	1328
1921	7.4	17.8	517	1960	2.9	5.7	134
1922	6.2	15.5	2152	1961	2.4	6.6	411
1923	4.2	6.9	358	1962	1.7	5.3	503
1924	3.6	10.0	1816	1963	2.2	4.1	88
1925	3.8	14.5	1705	1964	1.8	12.3	1037
1926	5.0	20.5	553	1965	2.9	6.5	436
1927	3.2	10.0	408	1966	2.9	7.4	433
1928	3.9	11.4	464	1967	2.6	—	60
1929	3.8	8.9	176	1968	2.6	5.0	626
1930	2.7	6.1	785	1969	2.3	5.5	172
1931	5.1	10.3	240	1970	1.7	3.7	62
1932	5.3	32.0	324	1971	2.0	13.6	2490
1933	6.7	11.9	126	1972	3.3	5.6	123
1934	3.8	6.8	229	1973	2.4	4.4	93
1935	3.6	4.4	2359	1974	1.6	41.9	16,128
1936	4.9	11.4	559	1975	6.0	9.3	274
1937	5.0	10.2	78	1976	3.9	10.8	1660
1938	5.3	12.1	881	1977	4.5	6.8	170

Source: Data from International Boundary and Water Commission.

each year (*annual series*) or only the discharges above some predetermined censoring value (*partial duration series*). The statistical samples in these time series approaches are much smaller than in the analyses utilizing daily records; however, they are useful in studies of major flow events, provided the gaging station has a record of considerable length. The annual discharges shown in table 5.5 have been ranked according to the magnitude during the years of record displayed. The **flood recurrence interval** can be calculated from an an-nual series like this one using one of a number of formu-las. The simplest plotting formula, the Weibull Method, calculates the recurrence interval by taking the average time between two floods of equal or greater magnitude:

$$R = \frac{n+1}{m}$$

where R is the recurrence interval in years, n is the num-ber of discharge values (the number of years in an an-nual series), and m is the magnitude rank of a given

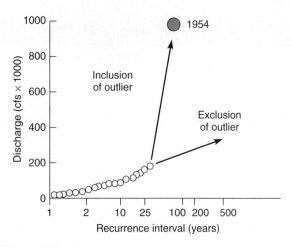

Figure 5.35
Flood frequency curve for the Pecos River near Langtry,
Texas. Note the extreme outlier representing the 1954 flood
and the problem for environmental planners of how to interpret
this point.

flood. The results of this type of analysis are then plotted
on probability graph paper to show the relationship of
discharge to recurrence interval (fig. 5.35), and the curve
is used by hydrologists to estimate the magnitude of a
flood that can be expected *within a specified period of
time.* U.S. governmental agencies have adopted another
method called the Log Pearson Type III (U.S. Water Resources Council 1981), which utilizes three parameters
from the annual flood series distribution—the mean, the
standard deviation, and the skewness. There is little
agreement among the experts as to which of the numerous analysis procedures works best (for a review and
discussion of procedures for analyzing flood frequency
see Dunne and Leopold 1978; Singh 1987; Klemes
1987; Kumar and Chandler 1987; Cunnane 1987).

The Pecos River flood frequency curve (fig. 5.35)
shows that on the average the flow is expected to equal
or exceed 2800 cms (100,000 cfs) once during a 25-year
interval. Also from this curve, the 100-year flood would
be estimated to have a discharge of about 27,400 cms
(970,000 cfs) if the outlier is included. Bear in mind,
however, that a flood recurrence interval estimate represents a probability statement and does not indicate that a
flood of given size can only occur in time at regularly
spaced intervals specified by its frequency. A river, unfortunately, does not understand statistical theory. Thus,
there is no reason to believe that 25-year floods or
100-year floods will be distributed evenly over time. In
the next 75 years, the Pecos River could experience
three successive 25-year floods during the first 5 years,
or there could be one now and several toward the end of
the time period. Similarly, a flow equaling or exceeding
the 25-year magnitude may fail to occur at all. The *prob-*

ability that a flow of a given magnitude will occur during any year is the reciprocal of the recurrence interval:

$$P = \frac{1}{R}$$

where P is the probability of flow being equaled or exceeded in any one year, and R is the recurrence interval.
Thus, a 50-year flood has a 2 percent chance of occurring in any given year, and a 10-year flood has a 10 percent chance. Over a 100-year period, the 50-year flood
has an 86 percent chance of occurring while the 10-year
flood has virtually a 100 percent chance of occurrence,
and so on. These probabilities are computed using the
probability equation

$$q = 1 - \left(1 - \frac{1}{T}\right)^n$$

where q is the probability of a flood with a recurrence
interval (T) occurring in the specified number of years
(n) (Costa and Baker 1981). One should realize, however, that the chance of experiencing a flood of any
specified recurrence interval is never zero; there is always some risk, and predicting the year in which it will
happen is virtually impossible.

Even though it may be impossible to predict when a
given flood will occur, magnitude-frequency analyses
have considerable practical value in river management,
especially for higher-frequency floods such as the
20-year (q_{20}), 10-year (q_{10}), or 5-year (q_5) floods. Using
the annual series, the *mean annual flood,* which is the
arithmetic mean of all the maximum yearly discharges
in the sample, is the flow that should recur once every
2.33 years ($q_{2.33}$). In our Pecos River example the mean
annual flood has a discharge of 790 cms (28,000 cfs)
and plots at a recurrence interval of approximately
2.33 on the flood frequency curve (fig. 5.35).

Paleoflood Hydrology   Perhaps the single most difficult problem faced by hydrologic planners is to estimate
the magnitude of the maximum probable flood that can
be expected to occur within any given basin. Such estimates usually require extension of the flood frequency
curve far beyond the limits provided by historically
measured flow events. In other cases, estimates are required for basins that have very short gaging records or
lack any kind of historical record of flow events. This is
very risky business because critical decisions about the
type and cost of flood control and protection are based
on these estimates. Most hydrologists agree that simple
extension of statistical flood frequency relationships beyond about 1.5 times the period of historical record are
invalid. In the United States, gaged observations rarely
extend beyond 100 years, thus making assessments of
flood magnitudes beyond the 100- to 150-year recur-

rence interval very uncertain. Standard hydrologic methods of estimating peak flood flows from basins lacking flood series data are similarly suspect.

Recently, geomorphologists have developed a suite of techniques that rely on the sedimentary and botanical record for information about floods that occurred along a given stream prior to direct observation or measurement. This approach is called *paleoflood hydrology* (Kochel and Baker 1982). Damage caused by floods to trees growing along the channel and floodplain can be used to establish the time of specific flood events, often down to a single year (Sigafoos 1964; Phipps 1985; Hupp 1988). Sometimes, the height of flood-induced damages on trees, such as corrasion scars, can be used to estimate a minimum flood stage (Yanosky 1982), thereby permitting estimates of paleoflood magnitude using standard indirect techniques to reconstruct flood flows (see Dalrymple 1960; Benson 1971). Costa (1983) reviews a number of flow regime-based techniques for reconstructing paleoflood depth and velocity that use the dimensions of flood-transported boulders in combination with the characteristics of channel cross-section and channel slope. A number of techniques focusing on analysis of Holocene floodplain stratigraphy have also been used to reconstruct paleoflood history. These techniques commonly interpret the stratigraphic sequence of floodplain sediments and the relationships between tributary fan sediments and mainstream sedimentation to bracket time intervals of major flow events (for example, see Costa 1978; Patton, Baker, and Kochel 1979; Blum and Valastro 1989).

One of the most successful and widely used stratigraphic approaches is based on the analysis of *slackwater deposits* (Kochel and Baker 1982; Kochel, Baker, and Patton 1982; Baker et al. 1983), which are relatively fine-grained sediments deposited in areas of backflow or flow separation from the main current. Application of the slackwater method works best in narrow bedrock reaches along streams where large stage increases result from small increases in discharge and where channel cross-sections are not subject to major change during floods. During high-flow conditions, water is backflooded into tributary mouths, shallow caves, eddies in areas of flow expansion, and in the lee of flow obstacles (fig. 5.36). These areas are characterized by relatively low-velocity conditions where sediment is deposited rapidly from suspension, leaving a record of the flood event in an ever-growing stratigraphic section of flood sedimentation units. The time of deposition of individual paleoflood slackwater units can be derived by radiocarbon dating of organic matter deposited with the sediment. Fine-grained organics, commonly found at the top of a unit, provide the most accurate estimate of the flood because larger material such as a log may have been eroded from older flood sediments and redeposited by

younger events (Kochel and Baker 1982, 1988). Waythomas and Jarrett (1994) also discuss other criteria for distinguishing between coarse-grained paleoflood sedimentation units, including lichen cover, position, clast weathering, and cover by colluvium.

Slackwater sediments can be used to estimate paleoflood stage by tracing a given unit up a tributary canyon and projecting its highest or terminal elevation back to the main valley (fig. 5.37). Experiments by Kochel and Ritter (1987) and field observations during historical floods indicate that these deposits provide a close approximation of the true paleoflood stage. Paleoflood stage information can be used to compute paleoflood discharge using a number of methods. The most common is combining the paleoflood stage information with channel cross-sections and profiles to model the paleoflood flow characteristics using the step-backwater method (Hydrologic Engineering Center 1982). For examples of this see O'Connor and Webb (1988); Baker (1987); Baker and Pickup (1987); and Jarrett (1991). New applications of statistical methods have made the use of discharge-censored data, such as minimum paleoflood stages, even more attractive (Stedinger and Cohn 1986; Stedinger and Baker 1987).

Slackwater deposits have been used to extend flood frequency curves over periods of 2000 to 10,000 years and have been used in a broad range of climatic zones (Moss and Kochel 1978; Patton and Dibble 1982; Kochel et al. 1982; Yang and Xie 1997; Kochel and Parris 2000). When used in conjunction with the historical record, paleoflood data can provide realistic estimates of major floods that plot as *outliers* on flood frequency plots (see fig. 5.35) like that for the Pecos River (Kochel 1988). In this case, conventional statistical techniques for estimating flood frequency applied in 1980 provided estimates of the recurrence interval of the 1954 flood on the Pecos River ranging from 81 years to approximately 10 million years (Kochel and Baker 1982). With the inclusion of slackwater paleoflood data (fig. 5.38) in the analysis the estimated recurrence interval is revised to approximately 2000 years, which is a more reasonable value utilizing both the historical and stratigraphic record of flooding (Kochel et al. 1982).

Although a certain amount of error is associated with all of the paleoflood hydrologic methods (errors can be minimized through careful selection of field sites), they enable hydrologists to make reasonable estimates of the recurrence intervals for large, infrequent floods and are now frequently used as resources in planning decisions. Baker et al. (1987) used slackwater deposits to conclude that overdesign of flood control projects in Arizona's Salt River would be unwise given a record of Holocene flooding spanning thousands of years. Cooley (1990) applied paleoflood techniques to alluvial streams in Wyoming to improve flood frequency

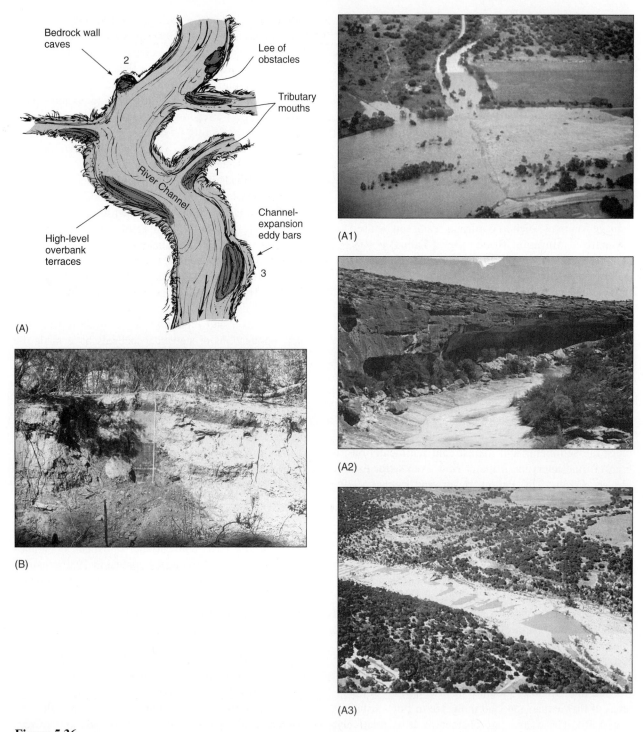

Figure 5.36

Slackwater deposits. (A) Location of common sites for slackwater sediment accumulation and preservation. Locations 1, 2, and 3 are illustrated by photos from the Pecos River in southwest Texas. (B) Stratigraphic package of over 2000 years slackwater sediment in a tributary mouth site along the Pecos River. Some of the flood units contain logs and other organic material useful in dating paleoflood events.

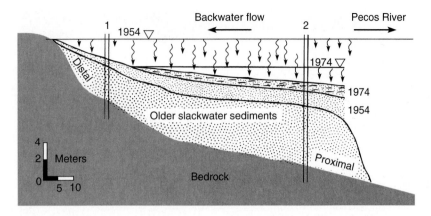

Figure 5.37
Schematic of on- and off-lap sequences and peak flood stage in a tributary valley for the 1954 and 1974 floods on the Pecos River, Texas. Sections in the proximal region (area 2) contain both floods, while distal regions (area 1) farther up the tributary record only the larger 1954 flood. Paleostage reconstructions are based on the elevation of the most distal sediments of each flood unit.
Kochel et al. 1982

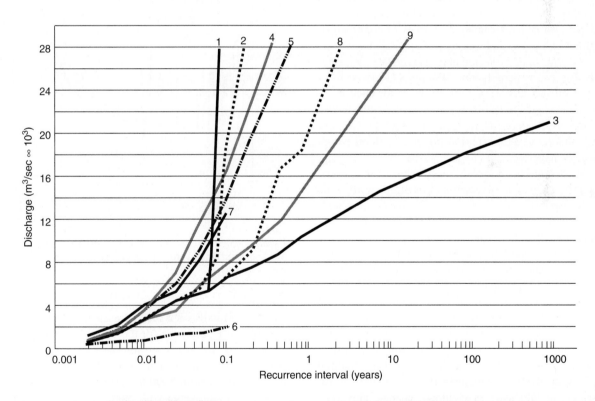

1	(n+1)/m with outliers	6	U.S.G.S. 1977 (1954 contributing area)
2	(n+1)/m historical data	7	U.S.G.S. 1977 (entire Pecos drainage)
3	(n+1)/m without outliers	8	slackwater sediments
4	log Pearson III (outliers, calculated skew)	9	modified log Pearson (slackwater)
5	log Pearson III (outliers, regional skew)		

Figure 5.38
Range of potential flood frequency curves calculated using a variety of common standard techniques applied to the Pecos River flow data. Estimates of flood frequency for the 1954 flood outlier range from less than 100 years to more than 20 million years. Slackwater paleoflood deposits were used to provide a more realistic estimate based on physical flood evidence of around 2000 years.
Kochel and Baker 1982

| TABLE 5.6 | Common Methods for Estimating Paleofloods. |

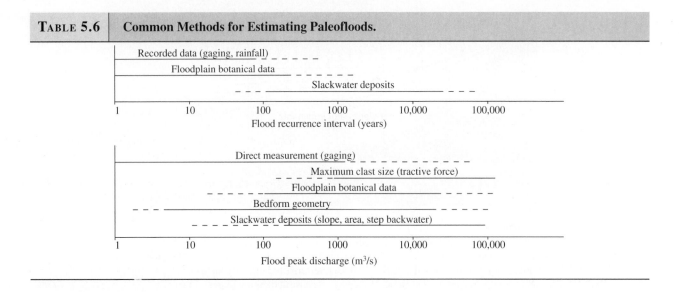

curves used in highway design. Successful techniques used in the Wyoming study included study of terraces, slackwater deposits, debris lines, dendrogeomorphology, and soils. Table 5.6 provides a summary of common paleoflood techniques and the range of applications.

A tremendous explosion in research related to paleohydrology has occurred since Schumm's (1965) seminal paper focusing attention on the prospects for Quaternary paleohydrologic studies (see, for example, entire books dealing with paleohydrologic studies such as Gregory 1983; Starkel et al. 1991). Paleohydraulic flood reconstructions have been used to gain perspective on the magnitude of some of the most catastrophic flows experienced in the Quaternary record such as the great Missoula floods responsible for carving the Channeled Scablands of eastern Washington (Baker 1973; Baker and Nummedal 1978; O'Connor and Baker 1992) and similar glacial lake-related floods in Siberia (Baker et al. 1993). Similar techniques have been used to detail the hydrology of Pleistocene lakes and breakout floods associated with the midcontinent portion of the Laurentide ice sheet (Kehew and Lord 1986; Lord and Kehew 1987).

Paleohydrological techniques also promise to play a major role in assessing the impact of human modifications on global climate, by facilitating the reconstruction of Holocene hydrologic regimes. Recent research has focused on fluvial responses to climatic change. This work is being done in arid bedrock channels (O'Connor et al. 1994) as well as in alluvial channels in semiarid regions (McQueen et al. 1993) and humid climates (Knox 1993; Patton 1988; Martin 1992). Paleoflood studies have been able to elucidate connections between climate and hydrology (Hirschboeck 1987) as well as the spatial variations in flooding between small basins (Martinez-Goytre et al. 1994). Ely (1997) found significant correlations in the frequency of Holocene paleofloods with climate fluctuations in the American southwest. Knox (1993;

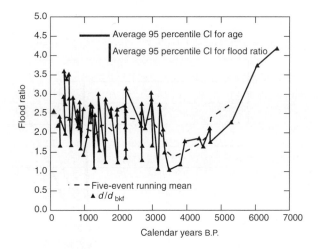

Figure 5.39
Holocene flood variations in the upper Mississippi River basin in response to small variations in climate.
(From Knox 1993)

Knox and Kundzewicz 1997) presented a detailed reconstruction of paleofloods for the upper Mississippi River showing distinctive hydrologic variations during the Holocene. His data indicate that a drier and warmer climate prevailed from 5000 to 3300 B.P. Since then it has been cooler and wetter with more frequent large floods, perhaps similar to the devastating high water experienced in the summer of 1993 (fig. 5.39). Regional and even global correlations are now beginning to appear in paleoflood syntheses (Ely and Baker 1990; O'Connor et al. 1994; Smith 1992; Baker et al. 1995 Gregory et al. 1995; Kale et al. 1997) indicating that these methods may be able to function as useful tools for reconstructing hydroclimatic variations during the Holocene. The apparent correlation of regional paleoflood events with climatic variations should be anticipated because of the

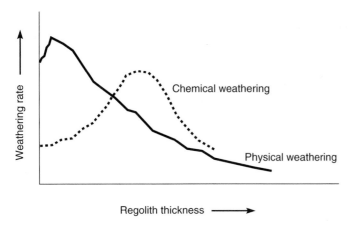

Figure 5.40
Variation in weathering rates as a function of regolith thickness over time. Physical weathering is most effective early when cover is minimal. Chemical rates increase as cover thickens, primarily due to increased time for reaction with regolith water, eventually slowing as thickening cover retards downward movement of water.

high sensitivity of the hydrologic system to climate change (Barron et al. 1989).

BASIN DENUDATION

In addition to being fundamental hydrogeomorphic entities, basins are also geographic compartments where sediment is manufactured, eroded, and deposited, and from which, given sufficient time, the debris will ultimately be removed. The amount of sediment leaving a basin can be readily converted to an estimate of lowering of the basin surface, called **denudation,** which is usually expressed as a time parameter or rate. Denudation seems to have no rigorous definition, but because it implies removal of basin material, it is commonly used as a synonym for erosion. The two processes differ, however, in that denudation refers only to those eroded products that are removed completely from the basin. Considerations of denudation also assume that the sediment is derived in equal portions from all subareas of the watershed and thus, that there is an equal lowering of the surface over the entire basin. Denudation rates tell us little about those erosive processes that simply redistribute sediment *within* the basin, and they do not account for the fact that at any given time some parts of the basin are probably aggrading rather than eroding. Denudation is the long-term sum of the overall erosive process, and even though it is analogous to erosion, it is not precisely the same. Furthermore, the basin surface may not actually lower even with a high denudation rate if active uplift of the basin is proceeding at a greater rate.

Estimates of modern denudation are usually based on measurements of stream load made at gaging stations or on the volume of reservoir space lost when sediment accumulates behind a dam. All types of load (suspended, bed, and dissolved) are included in the analyses at gaging stations, which require that the weight values of the load be converted to volumetric terms. Once the volume of sediment and chemical load leaving the basin is known, it is divided by the area of the watershed above

the gaging station to find the vertical dimension, which represents the amount of surface lowering (for details see Ritter 1967). Rates are commonly expressed in inches or centimeters per 1000 years. The type and volume of solid load has a tremendous influence on river behavior. Thus, we must examine the mechanism by which slopes are eroded and what factors affect the mode and rate of the erosive processes.

Slope Erosion and Sediment Yield

The linkages between erosion-sediment yield and the weathering processes (discussed in chapters 3 and 4) are many. Gilbert (1877) introduced a method of analyzing hillslope denudation using mass balance upon which many later investigators have built (e.g., Ahnert 1987). Weathering rates (fig. 5.40) are obviously influenced by regolith thickness because cover thickness regulates water flux to the parent material (important in chemical weathering reactions) and determines the efficacy of frost action on fresh bedrock. Regolith thickness in turn is regulated by denudation processes responsible for removing weathering products, thus illustrating the complex system of feedbacks operating to ultimately determine regolith thickness (fig. 5.41).

a complex system of feedbacks exists b/w weathering rates and regolith thickness

Hillslope morphology is a function of the complex interaction of relief-gravity processes, solar radiation, precipitation, kinetic energy, and the resistance properties of the surface, including vegetation cover and shear strength of the regolith (Brandt and Thomas/Thornes 1987). A raindrop possesses a considerable amount of kinetic energy, derived from its mass and the velocity it attains during its fall. Under the influence of gravity, a raindrop accelerates until its force is equal to the frictional resistance of the air, the speed at that point being the *terminal velocity.* As the distance needed to attain this condition is very short, most rain strikes the surface at its terminal velocity, although its absolute speed varies with wind, turbulence, drop size, and so on. In high-intensity rains, drops usually reach a maximum size of approximately 6 mm and a terminal velocity of

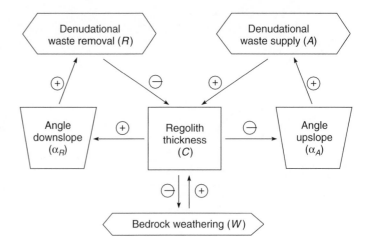

Figure 5.41
Schematic mass balance for denudation in a watershed.
(Ahnert 1987)

about 9 m/s. The impact of such rain can directly displace into the air particles as large as 10 mm in diameter and, by undermining downslope support, can indirectly set even larger pebbles in motion.

The amount of soil moved by splash depends on several interrelated factors. First, the kinetic energy of raindrops is directly related to splash movement (Kneale 1982). However, the kinetic energy of raindrops sometimes varies in unexpected ways. For example, Mosley (1982) found that rain passing through a forest canopy had greater total kinetic energy than normal rainfall and that rainsplash was three times greater under the canopy than in open areas. Second, the type of soil being struck is extremely important in determining the magnitude of splash movement. Free (1960), for example, found that splash loss varied as $E^{0.9}$ for a silt loam and $E^{1.46}$ for a sand, where E is the kinetic energy. Over a five-year period the total splash loss from the sandy surface was calculated at 1600 tons/acre, an amount three times greater than the loss from the loam, probably because the fine-grained soil had greater cohesion. Actually, the manner in which the soil particles aggregate (Luk 1979) and the dispersive properties of the surface material (Yair et al. 1980; Rendell 1982) are more significant controls than simple textural composition. Distinguishing between the energy required for initial detachment of particles and that required to deposit and remobilize them is also important (Hairsine and Rose 1991). Third, the rate and amount of splash transport appear to be a function of slope angle (Savat 1981; Reeve 1982), but the precise relationship is quite variable and not easily determined (Bryan 1979). The effectiveness of rainsplash may also be influenced by the antecedent soil moisture between storms. Rewetting experiments greatly reduced the amount of soil splash by increasing soil shear strength (Truman and Bradford 1990).

In addition to direct transportation, splash has several other erosion-inducing effects on the soil. By detaching particles, it destroys the structure of the soil and breaks apart resistant aggregates of clays. These physi-

cal processes make the soil much more susceptible to erosion by surface flow. Furthermore, as splash disperses the clays, they tend to form a fine-grained crust as they settle back on the surface. This crust forms a semipermeable barrier that reduces infiltration and promotes runoff, thereby increasing soil loss by overland flow. Research has shown that aggregate stability may play a central role in regulating rainsplash erosion (Farres 1987). Breakdown of individual soil aggregate appears to be the primary limiting factor in the system. Detailed fieldwork indicates that removal of particles early in a rain event is slow but gradually increase as aggregates begin to break up. Finally, the rate of splash erosion declines again as new crust begins to form (fig. 5.42).

Although the importance of rainsplash in total sediment transport is debated, some studies indicate that the process may be a major component of hillslope denudation. Morris (1986) found that rainsplash accounted for up to 88 percent of the fine-sediment flux during 1982 in a Colorado Front Range drainage basin.

Wash   Most natural slopes are too irregular to permit a uniform flow of water, or *wash*, over the entire surface; flow is deeper over depressions and shallower over flat reaches or high spots. The variable depth of flow produces differences in the eroding and transporting capabilities of the water. Wash does not imply that a regular sheet of debris is being carried continuously down the slope surface. In areas where *sheet flow* might be possible, only fine-grained particles can actually be moved efficiently, and those only if the surface has been prepared for erosion by rainsplash or weathering processes that reduce cohesion. In areas of concentrated flow, larger sediment can be moved, but the ability to erode depends more on the hydraulic force of the water and less on the condition of the surface.

When rainfall and flow become intense, small shallow channels may be formed which periodically shift position so that in the long run erosion is more or less even across the slope. In fine-grained soils, a set of well-

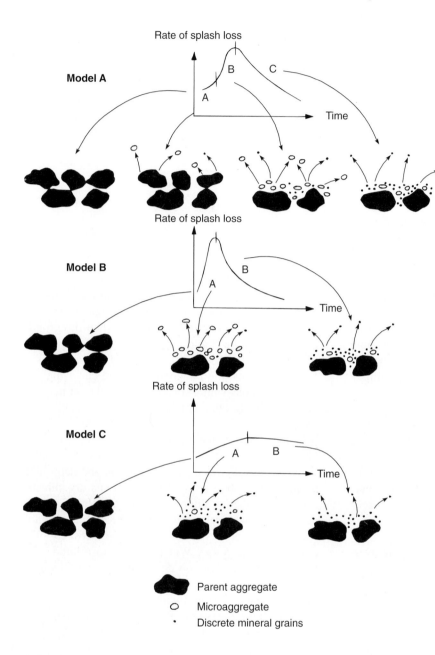

Figure 5.42
Schematic showing ideas proposed for losses due to soil splash over time. Model A represents an integration of research by many workers and shows the general changes in the rate of soil loss as a rainfall event proceeds. Models B and C illustrate two very different concepts of what might actually occur at different locations, which when averaged yield the curve in model A. In B, the erosion is high early in a storm as soil breaks up from larger aggregates formed between rains. In C, the soil breaks into fine fragments and is mobilized throughout the storm.
(Farres 1987)

defined subparallel *rills* is usually formed. Rills vary in size with the erodibility of the soil, but normally they are only several centimeters wide and deep. Heaving and other processes can obliterate these tiny channels in periods between heavy rains, especially in highly seasonal climates where rain may be lacking for months at a time. The periodic destruction of rills allows new channels to form in an entirely different location and ensures less than equal lowering of the entire slope surface. Some rills escape this spasmodic destruction by entrenching to greater depths, a difficult task that only a few of the largest rills accomplish. These "master" rills become relatively permanent and eventually evolve into true rivers.

In soils that are sandy or coarser, the channels are usually braided because the material, although easily eroded, is transported with difficulty. Sediment commonly accumulates as temporary bars within the channel,

and these subdivide the channel and the flow into a multitude of small passageways. Most braided channels are wider (up to 5 m) and deeper (1–10 cm) than rills, but they also change their position regularly because the bar deposits require a continuously shifting channel environment (see discussion of braided streams in chapter 6).

Sediment Yield (Soil Loss)   Assuming that all factors can be assessed within reasonable limits of certainty, the soil loss by water erosion should be able to be approximated by the **Universal Soil Loss Equation:**

$$A = RK(LS)CP$$

where A is the average annual soil loss, R is the rainfall factor, K is the erodibility factor, LS is the slope steepness factor, C is the cropping and management factor, and P is the conservation factor (for details see Smith

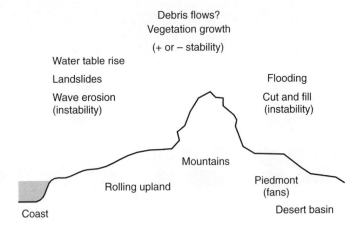

Figure 5.43
Schematic generalization showing the variation in landform response to increased precipitation episode between 1978 and 1983 in San Diego County, California.

and Wischmeier 1962). The precision of such an equation is uncertain because in reality it predicts total soil *movement* rather than soil loss. Haigh and Wallace (1982), for example, showed that ground losses from strip-mine dumps in Illinois were twice as great as those estimated by the U.S.L.E. Nonetheless, the analysis is widely used, and it probably provides useful ballpark data for predicting sediment loss from slopes.

Factors Affecting Sediment Yield A number of interrelated geologic, hydrologic, and topographic factors cause the magnitude of sediment yield to vary widely from region to region. The most important of these are (1) climate and its regulation of precipitation and vegetation, (2) basin size, (3) elevation and relief, (4) rock type, and (5) human activity (land use).

Climate: Precipitation and Vegetation Intuitively we would expect the amount of sediment yielded from any basin to be related in some systematic fashion to the amount of incoming precipitation. Actually, the correlation between the two is not that direct because precipitation is complexly interrelated with other factors that influence erosive capability. The amount of runoff from any given precipitation, for example, varies with temperature. Vegetation is a function of both precipitation and temperature and serves as a protective screen against erosion of surface material. Precipitation, then, cannot be considered as a completely independent variable with regard to denudation, even though it may be a dominant factor.

In an important paper, Langbein and Schumm (1958) documented the relationship between sediment yielded from basins averaging 3900 km² in area and *effective precipitation*, a parameter derived by adjusting the magnitude of precipitation to values expected at a mean temperature of 10° C. Under those conditions, they were able to show that as precipitation rises from zero, sediment yield increases rapidly to maximum yield value at about 30 cm of effective precipitation (see fig. 2.15). Any increase in precipitation above 30 cm

promotes a decline in sediment yield because the density and type of vegetation begin to play an active role in protecting the slopes from erosion. Vegetation generally begins to exert a control on erosion when the ground cover is between 8 and 60 percent, values typical in semiarid and subhumid climate. Thus, as L. Wilson (1973) suggests, the Langbein-Schumm curve may be valid for regions with a continental climate but may not be applicable to other climatic regimes, especially nonseasonal types. In any case, the demonstration by Langbein and Schumm that the relationship between precipitation and sediment yield is nonlinear and very complex seems to be valid. In a study of erosion and deposition resulting from short-term alternate wet and dry climatic cycles in southern California, Kochel and Ritter (1990a) described variable response across the region to a significant episode of increased precipitation between 1978 and 1983 (fig. 5.43). Wet intervals correlated with phases of aggradation on desert alluvial fan channels (Kochel et al. 1997). They observed insignificant geomorphic change in areas with antecedent precipitation already above the 30 cm threshold on the Langbein-Schumm curve. In contrast, arid regions were characterized by significant erosion and deposition as the wet episode elevated annual precipitation values toward the peak of the curve from much lower values. Although these observations are not direct sediment yield measurements, the geomorphic changes probably correlate well with sediment yield and provide field support for the Langbein-Schumm model.

Precisely at what precipitation the maximum values of sediment yield will occur depends on the specific climatic setting (for example, see Mourner 1960). Total values may relate more to the seasonality of the climate than to the mean annual precipitation (L. Wilson 1973), some studies tend to support this contention (Corbel 1959; L. Wilson 1972; Jansen and Painter 1974). Variations in sediment yield under different climates may be partly offset by an increase or decrease in dissolved load. Normally, dissolved load will increase regularly with precipitation, but maximum solution is probably

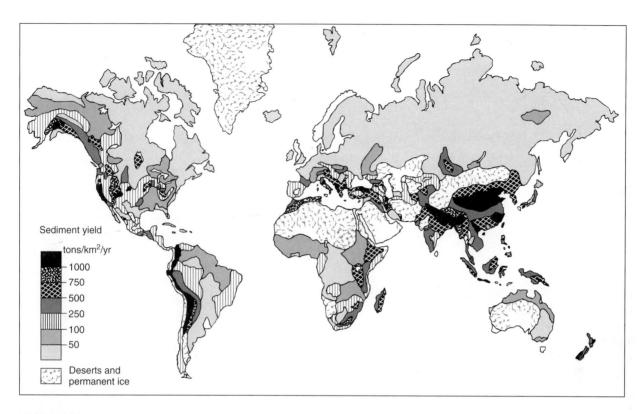

Figure 5.44
Generalized global sediment yield.
(After Walling 1987)

reached at about 63 cm of annual precipitation, with little additional dissolution resulting from higher precipitation (Leopold et al. 1964). Regardless, the amount of dissolved load usually does not exceed solid load even in regions of high precipitation, where chemical loss should be great (Li 1976; Leigh 1982). This probably occurs because the magnitude of solution loss is dominated by rock type rather than climate or vegetation (Garrels and Mackenzie 1971).

Generalizations about sediment yield patterns due to the influence of climate are difficult to make. Many studies exist that show direct correlations between major climatic shifts and regional sediment yields during the Quaternary Period (for example, Fuller et al. 1998; Zolitschka 1998). Figure 5.44 (Walling 1987) indicates that in general the sediment yield from mountain rivers is three times that of plains rivers, with glaciated mountain watersheds having the highest yields. Plains rivers typically have their highest sediment yields in subtropical-tropical areas. Generalizations regarding dissolved loads are even more difficult to make because of the lack of data.

Basin Size   A number of studies have suggested that sediment yield decreases markedly as the size of the drainage basin increases. This hypothesis is supported by remarkably high yields in very small basins of the

midwestern United States (Schumm 1963c) and also by data from the major sediment-discharging rivers of the world (Milliman and Meade 1983; fig. 5.45). Ichim (1990) showed that sediment yields increase with decreasing stream order. The explanation for this phenomenon seems to lie in several topographic realities: (1) small basins generally have steep valley-side slopes and high-gradient stream channels that efficiently transport sediment; (2) small basins with significant relief transport much of the sediment by mass movement and debris flow processes; (3) in basins filled to capacity with streams, the drainage density always remains high near the basin divide, but it may decrease with time in the central part of the basin; (4) floodplain area increases as the basin expands, especially in the central and lower reaches of the basin (Hadley and Schumm 1961), leading to lower basin-average slopes and greater opportunity for storage of sediment along the valley floor.

Taken together, these factors lead us to realize that in natural basins most sediment is produced in the small headward subareas, but during its downstream transit a significant portion may be stored in the floodplain system. How long it will remain in storage depends on the rigor of the geomorphic and hydrologic processes. It has been assumed that sampling over a long enough time would show that sediment stored within the basin is flushed rapidly from the system during episodes of rejuvenation.

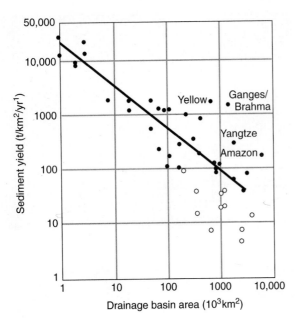

Figure 5.45

Comparison of sediment yields and drainage basin areas for all major sediment-discharging rivers (greater than 10×10^6 t yr^{-1}). Open circles represent low-yield rivers draining Africa and the Eurasian Arctic. Smaller basins have larger yields, although the largest rivers (Amazon, Yangtze, Ganges/Brahmaputra, and Yellow) all have greater loads than their basin areas would predict.

(Milliman and Meade 1983)

Inclusion of such spasmodic bursts of erosion in a long-term sample would tend to temper the variations in yield that appear to be related to basin size.

The relationship between sediment yield and basin size is less real than it appears because we tend to emphasize the extreme erosion in small upland basins and we do not fully comprehend the equilibrium state of the master streams. For example, Jansson (1988) noted that considerable scatter exists in sediment yields from small basins in semiarid areas, making the sediment yield–runoff relationship less clear. Also, ample evidence exists that downstream decrease in sediment yield occurs during episodes of accelerated erosion in small basins because storage in stream channels and valleys is increased (Trimble 1977). This occurs because the main channel is unable to transport the additional load. Thus, sediment yield from the basin mouth decreases during times of high erosion in the upland basins, indicating that rivers are in disequilibrium conditions throughout the basin. The reverse can also happen; that is, a reduction of erosion in the headland basins by sediment control measures starves the main channel load and creates erosion of the previously stored sediment, leading to increased yield measured at the basin mouth (Trimble 1977). Thus, the apparent relationships between basin size and sediment yield may simply stem from the possi-

bility that different parts of the drainage basin are responding in opposite directions to external erosional stimuli in the small marginal zones. These kinds of perturbations in the sediment yield–basin area relationship were also noted in Romania by Ichim (1990), who ascribed the differences to variations in bedrock resistance and spatial variability in the magnitude-frequency relationship within the basin.

Waythomas and Williams (1988) suggest that the correlation between sediment yield per area and basin area is spurious and may be meaningless because the basin area term occurs as a component of both axes. They note that the relationship may be falsely strengthened because the plot is largely one of drainage basin area versus basin area. They recommend that plots use annual sediment load, rather than sediment yield, as the dependent variable and either basin area or distance downstream as the independent variable.

Modern sediment yield values, therefore, may be related to recent changes in land use practice rather than an inherent areal control. For example, Kochel and Ritter (1990b) showed that dramatic channel and floodplain scour during floods since the mid-1980s along Sexton Creek, a Mississippi River tributary in southern Illinois, was due to recent changes in land use within the basin (fig. 5.46). Clearing of upland forests in southern Illinois in the late 1800s, followed by intensive agriculture on steep slopes, promoted regional denudation of loess-covered slopes and uplands, resulting in extensive aggradation of valley floors (Orbock-Miller et al. 1993). Starting in the early 1930s, the upland sections of these watersheds were reforested, when the area became part of the Shawnee National Forest, thus sharply reducing the delivery of sediment to the newly aggraded valley floors. Meanwhile, farming continued on the valley flood-plains formed of newly deposited fine-grained sediment. During this transition, streams such as Sexton Creek evolved from low-sinuosity gravel-bed braided systems to sinuous meandering channels. Beginning with a series of large floods in 1986, Sexton Creek and others in the area have experienced wholesale destruction of the fine-grained floodplain and a shift in channel form back to less sinuous gravel-bed systems in response to the decline in delivery of fine-grained sediment from the uplands (Kochel and Ritter 1990b). These floods also caused a loss of agricultural property and some bridge damage along the valley floors.

Elevation and Relief   Mountainous terrains with excessive elevation and relief are known to produce abnormally high sediment yields, particularly where rocks are nonresistant (Corbel 1959; Hadley and Schumm 1961; Schumm 1963c; Ahnert 1970) or affected by recent or current tectonism (Li 1976). For basins at least 3900 km^2 in area, the greatest denudation rates average about 0.9 m/1000 years and probably occur in mountain

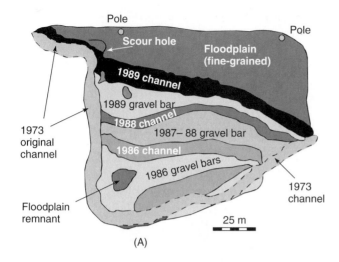

(A)

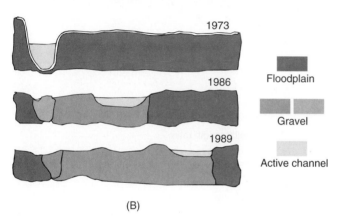

(B)

(C)

Figure 5.46

Channel changes and disequilibrium response triggered by land use changes upstream, Sexton Creek, southern Illinois. (A) Plan view map of channel adjustments between 1973 and 1989. Note the narrow, sinuous, fine-grained channel mapped in 1973. Major adjustments began suddenly with a flood in 1986, replacing the sinuous channel with a straight, wide, gravel-bed channel concurrent with the erosion of fine-grained floodplain alluvium. (B) Change in channel and floodplain cross-section taken normal to the axis of the 1989 channel in (A). (C) Sexton Creek channel in 1988 after a significant flood. Mud from the former valley fill sediment is visible along the cut bank on the left. The channel at this location increased width by over 500 percent since 1986, replacing a sinuous, fine-grained channel with a coarse-grained, cobbly straight reach.

(C) (Photo by R. Craig Kochel)

belts where relief and elevation are greatest and rocks are erodible (Schumm 1963c).

In the arid climate of the western United States, sediment yields are a function of the relief-length ratio (fig. 5.47). Utilizing this fact and holding area constant, Schumm (1963c) showed that the relationship between relief and denudation is definable in quantitative terms. It has also been shown that elevation alone produces disparate denudation rates because low-lying areas in any climatic regime yield less sediment than higher basins of comparable size (Corbel 1959).

Significantly, relief and elevation analyses demonstrate clearly that denudation rates are not constant through time. As relief and elevation of a basin gradually diminish during its evolution, the rate of surface lowering decreases proportionally, and each successive interval of stripping requires a longer period of time (fig. 5.48).

Rock Type   With similar climate and topography, basins underlain by more soluble clastic sedimentary rocks and low-rank metamorphics usually produce abundant suspended loads and so are characterized by higher rates of denudation than regions of crystalline rocks (Corbel 1964). The relationship of lithology and denudation is poorly understood in a quantitative sense and may be obscured by other rock characteristics, such as fracturing, which cause different lithologic units to behave similarly with respect to denudation. Even so, properties that commonly reflect lithology, such as the infiltration capacity, seem to be systematically related to sediment yield (fig. 5.49), indicating that the lithologic influence is real. At this time, however, geomorphologists have not been able to adequately reveal its fundamental character.

Kelson and Wells (1989) observed that both runoff and sediment yield were strongly influenced by lithology in New Mexico. Unit runoff and sediment yield were consistently higher for basins on resistant rocks such as crystallines compared to basins developed on weaker sedimentary lithologies. Increased runoff also increased stream erosive power in the resistant watersheds. Differences in sediment storage and the timing of water and sediment movement through the basins were also attributed to lithologic influence on regolith production and basin hydrology. A study of sediment influx into the Pacific Ocean from southern and central California rivers showed distinct variations with structure and lithology (Inman and Jenkins 1999). Sediment yields

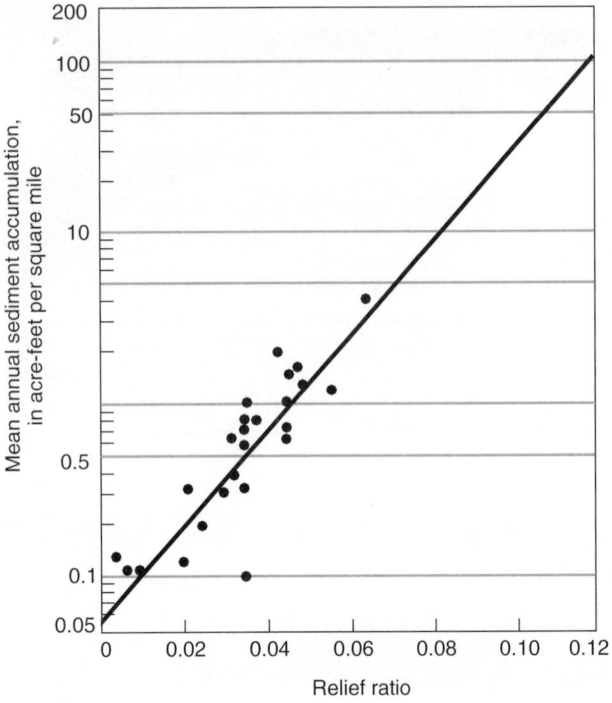

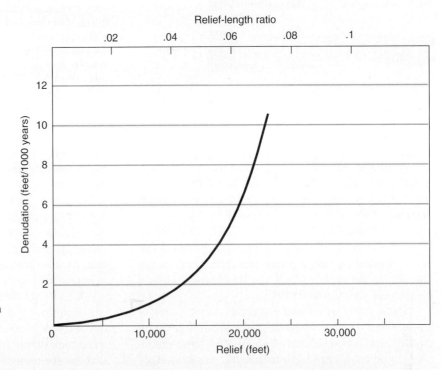

Figure 5.47

Relation between mean annual sediment accumulation in reservoirs and the relief ratio for basins located on the Fort Union formation in the upper Cheyenne River basin.

(Hadley and Schumm 1961)

Figure 5.48

Relation of denudation rates to relief-length ratio and drainage-basin relief. Denudation rates are adjusted to drainage areas of 1500 square miles. Curve is based in the average maximum denudation rate of 3 feet per 1000 years when relief-length ratio is 0.05.

(Schumm 1963c)

were highest from faulted Cenozoic sediments, followed by the Coast Ranges, and then the Peninsular Ranges.

The Human Factor George Perkins Marsh was the first scientist to present a brilliant and interdisciplinary discussion of the magnitude of human impact on physical and biological systems (Marsh 1885). In his treatise,

Marsh frequently made reference to the impact of human activity on sediment yield and denudation.

Most estimates of denudation are significantly higher where human activity disturbs the natural setting (fig. 5.50). Exactly how much humans accelerate erosion varies with the type of land use and the particular environment. Anthropogenic interference has the potential to

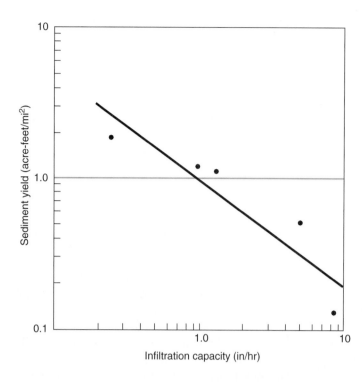

Figure 5.49
Relation between infiltration capacity and sediment yield, indicating lithologic influence.
(Hadley and Schumm 1961)

(A)

(B)

Figure 5.50
(A) Excessive sedimentation from a poorly managed agricultural field in Lancaster County, Pa., during the Tropical Storm Agnes flood of 1972. (B) Disturbed landscape in the anthracite coal mining district of central Pennsylvania. Mining disrupts the ground and destroys vegetation, and the waste piles like this one near Trevorton are sources of increased sediment yield.

(Photos by R. Craig Kochel)

alter drastically the natural sediment yield (Moore 1979; Dunne 1979; Toy 1982). Projections of changes in world land use between 1700 and 2000 are shown in table 5.7 (Richards 1990). The magnitude of change between 1950 and 1980 exceeded that between 1700 and 1850.

Evidence indicates that human activities may increase detrital loads by at least an order of magnitude (Judson 1968a; Meade 1969). Chemical loads are also expanded by pollutants introduced into the streams and the atmosphere (Meade 1969). One important contributor to accelerated erosion is the replacement of mature forest cover by intensely cultivated land (Toy 1982; Walling and Quinne 1990; Foster et al. 1990). Studies in the United States show an increase in sediment yield of one to three orders of magnitude when crops are substituted for natural vegetation (Ursic and Dendy 1965; Wolman 1967). With proper soil conservation techniques, however, the effect can be reversed and sediment yield values will decrease dramatically (Trimble and Lund 1982).

Construction associated with urbanization causes an even more dramatic rise in the sediment yield (e.g., Guy 1965), but after construction is completed, the values decrease rapidly (fig. 5.51) because much of the surface is protected from erosion by concrete citadels (Wolman 1967). Urbanization also affects the runoff characteristics within a drainage basin (McPherson 1974); consequently, the concentration of sediment in streams, as influenced by humans, depends on variations in hydrology and sediment yield. In a global review of sediment yields, Walling and Webb (1996) showed that sediment yields have increased since 1964 in most areas. Renwick (1996) suggested that human impact significantly overshadows the natural spatial variation in sediment yields.

Several workers have suggested that the abnormalities induced by human activities may be great enough to invalidate the use of sediment yields to calculate rates of denudation in drainage basins (Douglas 1967; Trimble 1977). Trimble (1977) was able to show that in large drainage basins in the southeastern United States, lowering of upland surfaces was proceeding at a rate of 95 mm/100 yr; in contrast, the denudation rate calculated by sediment in the streams was only 5.3 mm/100 yr. Thus, the *sediment delivery ratio* (sediment yield as a proportion of upland erosion) was only 6 percent. Sediment yield, therefore, may be a poor indicator of how rapidly the upland area of a basin is being lowered. The major portion of sediment eroded from the uplands is, of course, being stored within the basin by deposition. A similar example can be cited from central Virginia, where much of the denudation in small upland watersheds is accomplished by a small number of catastrophic debris avalanche and debris flow episodes occurring at return intervals on the order of thousands of years (Kochel 1987). Much of the sediment denuded from the Virginia hillslopes during these events is temporarily stored in debris fans within the basin.

Meade (1982) has suggested that the accelerated erosion produced by early settlers in the eastern United States while clearing land for cultivation has been largely arrested by soil conservation and reduced acreage of agricultural lands. However, sediment stored in the basins in response to the earlier settlement is now being eroded and is augmenting the present river loads and will continue to do so for decades or centuries to come. Modern river loads may be reflecting erosional events that occurred in the distant past. Considerable literature exists focused on determining the cause of wide-

Table 5.7	Global Land Use Changes from 1700 to 2000.	
Land Use	**Percent Change**	**Area Change (ha)**
Forest/woodlands	−19	−1.2 billion
Grassland/pasture	−8	−560 million
Cropland	+21	+1.2 billion

Source: Richards (1990).

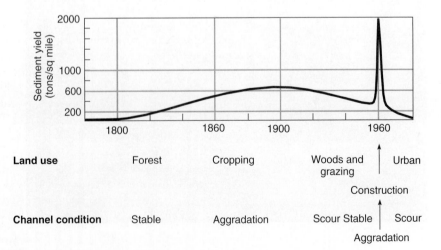

Figure 5.51

Changes in sediment yield and channel behavior in one area under various types of land use.

(Wolman 1967)

spread incision of channels (arroyos) in the American Southwest beginning in the late 1800s. Most arguments fall into two camps—climatic change (precipitation amount and intensity) and overgrazing. Working in the Southern Colorado Plateau, McFadden and McAuliffe (1997) were convinced that arroyo incision would be correlated with climate change. They believe that increased precipitation triggered accelerated erosion of sparsely vegetated slopes.

Characterizing human impact on the landscape is not simple, for the relationships between land use and sediment yield are complex (Foster et al. 1990). As shown in figure 5.52, impacts on hydrology and vegetation can influence geomorphologic and pedologic processes on varying scales of time and space (McDowell et al. 1990). Adding to the difficulty of understanding these human-induced changes is the fact that many adjustments occur in episodes that are regulated by discrete geomorphic thresholds, resulting in complex response phenomena. Before the human impacts can be assessed, the influence of geomorphic phenomena known to operate and influence systems over time scales similar to the period of significant human activity (10–1000 years) must be accounted for. McDowell et al. (1990) suggest that short-term climate change and tectonics may be the most influential of these factors and can sometimes overshadow the impact of human activity. Recent research is beginning to shed some light on the likely responses of geomorphic systems to short-term climatic fluctuations (Kochel and Miller 1997). Studies of lake sediments hold great promise of providing high-resolution records of basinal changes in sediment yield and runoff (Halfman and Johnson 1988; Foster et al. 1990). Recent work by Brown et al. (2000) in Vermont has documented a 10,000-year record of storm-induced sedimentation into northern Appalachian lakes, discernible at reasonably fine resolution.

On the other hand, numerous studies document landscape alterations, such as alluviation and erosion, that are occurring over just several decades (Knox 1977; Goudie 1986; Harvey and Schumm 1987). The complexity, frequency, and magnitude of human impacts are increasing (Goudie 1986). Arguments have long been waged regarding the relative influence of climate change versus land use in the dynamics of arroyo development in the American Southwest. Recent investigations have demonstrated significant climate changes coincident with the timing of arroyo incision and alluviation. The magnitude of these climate changes could impact basin hydrology and sediment dynamics significantly. The argument for climatic causes is strengthened because response has generally been regionally consistent rather than locally variable (for example, see Hereford 1986; Graf et al. 1991; Wells and Meyer 1993; McFadden and McAuliffe 1993).

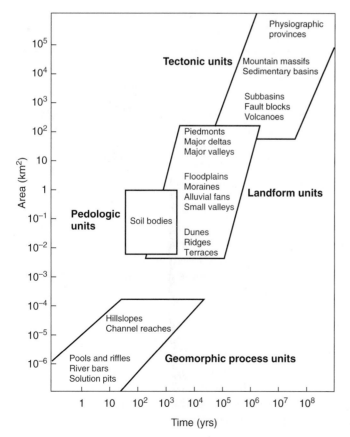

Figure 5.52

Spatial and temporal scales of human impact on geomorphic processes and landforms.

(McDowell et al. 1990)

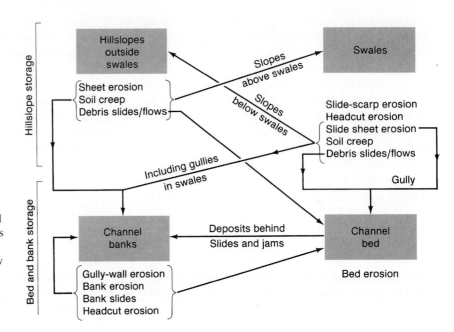

Figure 5.53
General linkages between sediment storage sites and erosional processes. Boxes indicate storage elements; listed below each box are erosional processes mainly responsible for mobilizing sediment in that element. Arrows show transfers between elements. Labels on arrows qualify or restrict location of transfers.
(Lehre 1982)

			Sediment (metric tons/km2)				
Year	Rainfall (mm)	Recurrence Interval of Peak Flow (yrs)	Mobilization on Slopes (1)	Production to Channels (2)	Redistribution on Slopes (3)	Yield: Bed + Susp. Load (4)	Bed + Bank Storage (5 = 2 − 4)
1971–72	602	1.5	86	148	−71	24	+124
1972–73	1184	15–20	1219	985	+317	1420	−435
1973–74	1046	3–5	1960	1575	+389	630	+945
1971–74	2832	—	3265	2708	+635	2074	+634

TABLE 5.8 | **Sediment Budget for a Small Drainage in Northern California for the Years 1971 to 1974.**

From Lehre (1982).

Sediment Budgets

Storage of sediment within a basin is a significant geomorphic phenomenon. Its importance is underscored by the concept of a sediment budget (Dietrich and Dunne 1978; Kelsey 1980; Lehre 1982). A **sediment budget** is a quantitative analysis of a drainage basin that shows the relationship among erosion of basin materials, discharge of sediment from a basin, and the associated changes in sediment storage. In essence, it is an accounting sheet of ins and outs of sediment and as such is similar to the hydrologic budget discussed earlier. Of greater importance, however, is its significance in planning and land management because it distinguishes the dominant erosional processes in the basin and the circumstances under which storage of sediment may change. Additionally, the sediment budget demonstrates the process linkage that we discussed in chapter 1 (fig. 5.53).

To make a complete sediment budget analysis one must identify and quantify *sediment mobilization* (processes that initiate motion and move sediment any

distance), *sediment production* (the amount of sediment reaching or given access to a channel), and *sediment yield* (sediment actually discharged from the basin). For example, table 5.8 shows the sediment budget for the years 1971 to 1974 in a small drainage basin northwest of San Francisco (Lehre 1982). In dry years or years without extreme flow events (1971–1972, 1973, 1974), most mobilized and produced sediment was stored within the basin. In contrast, during the year having a flow event with a 15- to 20-year recurrence interval (1972–1973), the amount of sediment mobilized and produced was less than the amount discharged from the basin. This indicates that sediment was taken out of storage during that year, presumably by erosion of channel banks and beds. By continuous observation, Lehre (1982) was able to demonstrate that the variations in annual sediment budgets shown in table 5.8 were accompanied by different erosive processes. In the years of low rainfall and/or peak flow, sediment was mobilized by spalling, rainbeat, and minor sliding and was moved

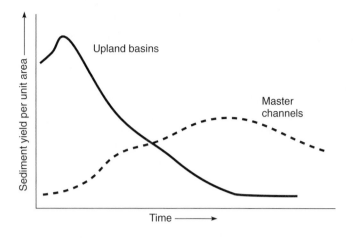

Figure 5.54
Relative timing of sediment movements in drainage basins since the end of Pleistocene glaciation. Note the shift in activity progressively downstream through the system with time.
(Church and Slaymaker 1989)

toward the channels by sheetwash. The lack of water, however, assured a minimum transport distance and most sediment went into storage. In the year of high peak flow (1972–1973), sediment was mobilized and produced mainly by debris slides and flows. It was quickly delivered to the channels and, because of the co-incidence of high channel discharge, was removed from the basin.

Sediment budget analysis is a relatively new way of looking at the inner workings of a drainage basin, and we do not yet know what its applications will be. However, it seems to hold considerable promise for land management, especially in small basins that are unstable geomorphically and subject to a variety of interrelated processes.

Recently, there has been a significant upsurge of interest in the dynamics of sediment transfer, such as sediment storage and delivery, through the drainage basin. In particular, studies have focused on the spatial distribution and residence times of sediments because of concern over the location and mobility of contaminants introduced by human activities within the watershed. These studies are beginning to reveal distinctive patterns in sediment storage and distribution along the fluvial system. For example, Florsheim and Keller (1987) observed an inverse relationship between unit stream power (which reflects the ability of a channel to generate stresses necessary to entrain sediments) and the volume of sediment stored along a specific channel reach. Sediments are generally stored upstream and downstream of channel and valley constrictions; they pass through the constricted reaches because of high unit stream power generated there. Many studies have emphasized the importance of episodic erosional events in sediment movement, particularly in small, upland watersheds (Lisle 1987; Grant and Wolff 1991; Jacobsen et al. 1989). Miller et al. (1999) discuss the significance of episodic events in mobilizing flood plain sediment containing heavy metal contaminants from 19th-century mining during a major flood on the Carson River in Nevada in

1997. Church and Slaymaker (1989) have shown that many basins in areas that experienced Quaternary glaciation may still be attempting to adjust to dramatic increases in sediment input delivered immediately after a glacial interval (sometimes referred to as paraglacial), some 10,000 years ago. Figure 5.54 demonstrates the delay in sediment transfer in response to this paraglacial sedimentation from upland basins to downstream areas along the master channels envisioned in montane areas of the Canadian Rockies.

Another generalization from studies of sediment yield and delivery indicates that the sediment delivery ratio varies systematically through a basin from headwater on downstream regions. Sediment delivery ratios are generally high, in excess of 90 percent (Trimble 1977), for upland, low-order watersheds, while they may be as low as 4 percent by the time coastal plain areas are reached (Phillips 1991). Recent advances in the use of cosmogenic isotopes such as ^{10}Be, ^{137}Cs, and Cl for dating are beginning to improve efforts to quantify sediment transfers. For example, Brown et al. (1988) showed that only a small portion of the volume of sediment mobilized from hillslopes ever leaves the basin. Considerable perturbation of expected natural patterns and rates of sediment storage and transport can occur due to the impact of artificial channelization along alluvial reaches (Simon 1989).

Sediment yield has long been known to be an important signal of tectonic, base-level climatic change in fluvial systems (i.e., Schumm 1977; Bull 1991). Chumm and Rea (1995) viewed sediment-yield patterns from areas disturbed by changes in the above parameters and found that they all result in increases in sediment yield at all scales from small experimental to the Himalayan Basin. However, they also found that if no further disturbance occurs, sediment yields will lower rapidly as the systems reestablish equilibrium. Interpretations of sediment yield promise to continue to be an important tool for interpreting the record of past adjustments of climate, base level, and tectonics.

Rates of Denudation

In spite of the problems inherent in analyzing denudation rate, various estimates have been made (fig 5.55). Table 5.9 presents a random sample of modern denudation rates for basins of varying size in the United States. The values are imprecise and probably high because in most cases they do not include an adjustment for human impact. Nonetheless, they do indicate a general tendency for denudation rates to fall between 2.5 and 15 cm per 1000 years when considered on a regional scale. Judson (1968b) recalculated the denudation rate for the entire continental United States by subtracting the effect of human occupancy from earlier estimates. His figure of

3 cm per 1000 years agrees rather closely with the denudation in very large drainage basins that are mostly unaffected by anthropogenic disturbance (Gibbs 1967) and perhaps represents a reasonable approximation for denudation on a continental scale.

Although methods other than analyses of sediment wedges have been employed (Eardly 1967; Ruxton and McDougall 1967; Clark and Jager 1969), most rates of denudation have been estimated from measurements of sediment accumulation in depositional basins. A valid estimate can be made only if (1) the volume of sediment derived by erosive processes can be accurately determined, (2) the boundaries of the source area are defin-

Figure 5.55
Junction of the main branches of the Susquehanna River at Sunbury, PA. Note the plume of suspended sediment emanating from the main branch, which recently experienced a rainfall. Water from the west branch (left) is relatively clear.

(Photo by R. Craig Kochel)

TABLE 5.9	The Influence of Geology and Climate on Suspended-Load Denudation in Basins of Different Size in the United States.			
Basin	**Location**	**Area (mi²)**	**Average Annual Suspended Load (tons × 10³)**	**Denudation (in/1000 yr)**
Mississippi	Baton Rouge, La.	1,243,500	305,000	1.3
Colorado	Grand Canyon, Ariz.	137,800	149,000	5.6
Columbia	Pasco, Wash.	102,600	10,300	0.5
Rio Grande	San Acacia, N.M.	26,770	9,420	1.8
Sacramento	Sacramento, Calif.	27,500	2,580	0.5
Alabama	Claiborne, Ala.	22,000	2,130	0.5
Delaware	Trenton, N.J.	6,780	998	0.8
Yadkin	Yadkin College, N.C.	2,280	808	1.8
Eel	Scotia, Calif.	3,113	18,200	30.4
Rio Hondo	Roswell, N.M.	947	545	3.0
Green	Palmer, Wash.	230	71	1.6
Alameda	Niles, Calif.	633	221	1.8
Scantic	Broad Brook, Conn.	98	7	0.4
Napa	St. Helena, Calif.	81	63	4.1

Data from Judson and Ritter (1964).

able, and (3) the time interval of sediment accumulation can be ascertained within reasonable limits. It is very difficult to meet all these requirements. Noneroded matter such as pelagic and volcanic rocks add to the depositional volume, material eroded from the basin as dissolved load may not be returned to the deposit by chemical precipitation, and absolute dates that bracket the time of deposition are necessarily imprecise. Nonetheless, some estimates of past rates have been made for large portions of North America (Gilluly 1949, 1955, 1964; Menard 1961). It is interesting to note that the rates found in these studies of large regions are within the same order of magnitude as those based on modern stream data. This similarity prompted the hypothesis (Ritter 1967) that when viewed on a large enough area or over a long enough time interval, denudation rates will probably be about the same. Based on current evidence, the average value will probably fall somewhere between 2.5 and 15 cm per 1000 years. That range is generally supported by estimates of solid and dissolved loads being delivered to the oceans from the world's continents (table 5.10). Estimates of total denudation must include chemical as well as solid loads if they are to be reliable over the long term. For example, Sevon (1989) found that denudation rates ranged only from 10 m per million years from historical stream sampling to 27 m per million years based on analyses of the long-term stratigraphic record for the Juniata River (a major tributary to the Susquehanna River in central Pennsylvania). These estimates, although different, are certainly in the same ballpark. The advent of new Quaternary dating technologies such as ^{137}Cs will undoubtedly yield improved estimates of continental erosion rates (Walling and Quine 1990).

The use of denudation rates for continent-sized areas (table 5.10) can lead to erroneous conclusions. For example, a 3 cm/1000 yr rate suggests that 300 m of surface lowering will be accomplished over an entire continent in a 10-million-year period. Such a rate might be used as evidence to support the generally accepted

canon of geomorphology that most of the topography of Earth is no older than Pleistocene, meaning that almost all landscapes formed in the last 2 million years. Therein lies the fallacy of denudation rates, because we know that large regions of Tertiary and older landforms do exist, especially in the Southern Hemisphere. For example, radiometric dates and geologic evidence in southeastern Australia show that much of that landscape was in its present form by the mid-Miocene, and some upland surfaces originated in the Mesozoic (Young 1983). The inconsistency of these observations with denudation analyses from river sediment arises because denudation rates are unrealistically spread over entire continents. Actually, the interiors of continental plates probably experience extremely slow denudation and may easily preserve old landscapes (Young 1983). In contrast, continental plate margins where active tectonism is occurring probably have enormously high denudation rates. Combining the two subareas provides an average rate for the continent that is indicative of neither. The point here is that a denudation rate calculated for a large basin or region tells us nothing about the tenor of erosion occurring in any component part of that basin, and the overall rate should never be used in that sense. Romero-Diaz et al. (1988) illustrated how great the spatial variation in denudation rates can be between local areas.

McLennan (1993) maintains that a general relationship exists between sediment yield and major denudational regions of the world when considered in regard to weathering history (table 5.10). Weathering history is recorded in the major element composition of suspended sediments. In this scheme, equilibrium denudational regions are areas that have a sediment yield that is in equilibrium with existing dominant weathering conditions. On the other hand, nonequilibrium denudational region areas have sediment yields generally much lower than predicted. Often, such anomalies are related to sediments trapped in lakes such as the Great Lakes system (McLennan 1993).

TABLE 5.10	Denudation Estimates Based on Solid and Dissolved Loads Delivered to the Ocean by Major Rivers of the Continents.			
Continents	Solid Load (t/km²/yr)[a]	Dissolved Load (t/km²/yr)[b]	Total Load	Denudation Rate (cm/1000 yr)
North and Central America	84	33	117	4.00
South America	97	28	125	4.28
Europe	50	42	92	3.15
Asia	380	32	412	14.10
Africa	35	24	59	2.02
Australia	28	2	30	1.03

[a]Solid load from Milliman and Meade (1983).

[b]Dissolved load from Garrels and Mackenzie (1971).

SUMMARY

In this chapter we examined a remarkable statistical balance between the spatial characteristics of river networks and the watersheds that contain them. Because the parameters of this morphometry also relate in a significant way to the hydrologic and erosional properties of most watersheds, drainage basins serve as primary units for systematic analyses of geomorphology. Drainage basins and river networks probably evolve according to fundamental hydrophysical laws, but their ultimate character is conditioned by the geological framework and the external constraints of climate. An equilibrium condition, defined in terms of mathematical balance, is probably attained early in the growth history of most basins. This does not mean, however, that basins evolve in an orderly way with time. Geologic catastrophes that upset equilibrium tend to be filtered out in a morphometric sense because basinal parameters apparently adjust to changes rather quickly.

Water flowing in basin rivers is derived from variable sources, and sediment reaching stream channels is produced and delivered by numerous erosive processes operating on basin slopes. Both water and sediment are amenable to budget analyses, which provide basic data for watershed planning. The amounts of water and sediment entering stream channels are functions of the physical properties of the basin and the effect produced by human activities. Estimates of basin denudation can be made on the basis of total sediment yield, but difficulties are created by sediment storage within the basin. The following references provide greater detail concerning the concepts discussed in this chapter.

SUGGESTED READINGS

Baker, V. R., Kochel, R. C., and Patton, P. C., eds. 1988. *Flood geomorphology.* New York: John Wiley and Sons.

Dunne, T., and Leopold, L. B. 1978. *Water in environmental planning.* New York: W. H. Freeman.

Fetter, C. W. 1994. *Applied hydrogeology.* Columbus, Ohio: Merrill.

Gregory, K. J., and Walling, D. E. 1978. *Drainage basin form and process.* New York: Halsted/John Wiley.

Kochel, R. C., and Miller, J. R. 1997. Geomorphic responses to short-term climate change. Special Issue, *Geomorphology* 19:170–368.

Mayer, L., and Nash, D. 1987. *Catastrophic flooding.* Boston: Allen and Unwin.

Rhodes, B. 1992. Fluvial geomorphology. *Progress in Physical Geography* 16:489–96.

Schumm, S. A. 1977. *The fluvial system.* New York: John Wiley.

Schumm, S. A., Mosley, M. P., and Weaver, W. E. 1987. *Experimental fluvial geomorphology.* New York: John Wiley.

Singh, V. P. 1987. *Regional flood frequency analysis.* Dordrecht: D. Reidel.

Turner, B. L., Clark, W. C., Kates, R. W., Richards, J. F., Mathews, J. T., and Meyer, W. B., eds. 1990. *The earth transformed by human action.* Cambridge: Cambridge Univ. Press.

FLUVIAL PROCESSES

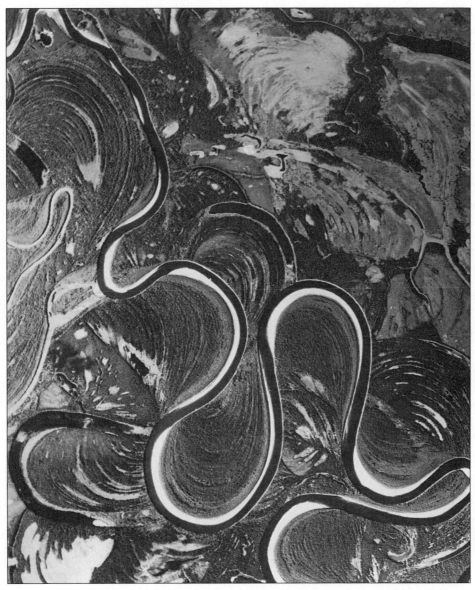

U. S. Geological Survey

Introduction
The River Channel
    Basic Mechanics
    Flow Equations and Resisting Factors
Sediment in Channels
    Transportation
    Entrainment
    Bank Erosion
    Erosion of Bedrock Channels
    Deposition
    The Frequency and Magnitude of River Work
The Quasi-Equilibrium Condition
    Hydraulic Geometry
    The Influence of Slope
    Channel Shape
Channel Patterns
    Straight Channels
    Meandering Channels
    Braided Channels
    Anastomosing and Anabranching Channels
    The Continuity of Channel Patterns
Rivers, Equilibrium, and Time
    Adjustment of Gradient
    Adjustment of Shape and Pattern
Summary
Suggested Readings

INTRODUCTION

It seems fair to say that fluvial action is the single most important geomorphic agent. Although other surficial processes are significant, streams are so ubiquitous that their influence in geomorphology can hardly be overestimated. As discussed in chapter 1, most geomorphic analyses of stream processes in the early twentieth century were based primarily on logic and qualitative observation. The few quantitative studies of fluvial mechanics were somehow lost in the wave of geomorphology that used fluvially produced landforms to reconstruct history. Our understanding of river mechanics is hampered by the difficulty of obtaining measurements in natural channels. To overcome these difficulties, investigators have extensively used laboratory flumes, where slope and velocity can be changed, and selected variables can be held constant (Schumm et al. 1987). While the results obtained from these laboratory studies have been invaluable in furthering our understanding of river systems, flumes have a limited range of discharge and depth, and they cannot take into account the many ever-changing and interdependent variables present in natural streams (Maddock 1969). As a result, we still need much more data about natural rivers before we can hope to utilize our knowledge in a predictive way. Nonetheless, the need to understand the dynamics of fluvial systems for practical purposes, such as the restoration of aquatic habitats, the remediation of contaminated rivers, and the design of flood control structures, has led to a rebirth in the study of fluvial mechanics, and data are now accumulating faster than their meaning can be assimilated. This explosion of information is undoubtedly the key to the advancement of geomorphic thinking, for without it our basic tenets will never be critically evaluated. In this chapter we will examine briefly much of the current thinking about streams; those interested in greater detail than space allows here are referred to a number of excellent books about the fluvial realm: Schumm et al. (1987), Baker et al. (1988), Bull (1991), Klingeman et al. (1998), Knighton (1998), Tinkler and Wohl (1998), and Darby and Simon (1999).

THE RIVER CHANNEL

Basic Mechanics

The ability of a river to erode and transport debris represents a balance between driving and resisting forces. It depends on how much potential energy is provided to produce flow and how much of that energy is consumed in the system by the resistance to flow. To illustrate, we can examine what happens to a constant discharge of water flowing in a long, steeply inclined channel that has no tributaries. Gravity tends to continuously accelerate the flow downstream. However, the increase in velocity is moderated by friction and turbulence generated within the water and along the channel perimeter. If these resisting elements were constant and less than the gravitational force, acceleration would continue along the entire length of the channel. In natural rivers, however, the intensity of resistance is not constant but increases with the flow velocity. Thus, at any point in a river the velocity represents the balance between the energy causing flow and the energy consumed by the resistance to flow.

A number of parameters have been developed to describe the nature of flow in open channels (table 6.1). In the case of steady flow, for example, the velocity of the water is constant at a given location and the forces causing and resisting the movement of the water through the channel are balanced. Uniform flow describes the condition in which the velocity is constant along the channel. Flow in natural channels is invariably unsteady and nonuniform, although many of the numerical equations constructed to describe flow in rivers require the simplifying assumptions of steady and uniform conditions.

One of the most important differences in the flow of water in open channels occurs between laminar and turbulent flow (table 6.1). In **laminar flow,** particles of water move in straight paths that are not disrupted by the movement of neighboring particles. In this type of flow regime, internal friction is the dominant resisting force as layers or particles of the fluid slide smoothly past one

TABLE 6.1	Flow Characterization in Open Channels.

Type of flow	Flow character
Spatial variations in velocity	
Uniform flow	Velocity is constant along the channel
Nonuniform (varied)	Velocity changes with distance along the channel
Temporal variations in velocity	
Steady flow	Velocity does not change in magnitude or direction with time
Unsteady flow	Velocity fluctuates in magnitude or direction with time
Degree of particle mixing	
Laminar flow	Fluid elements move along specific paths with no significant mixing among the adjacent layers; Re < 500
Turbulent flow	Fluid elements do not flow along parallel paths, but repeatedly move between adjacent layers; involve large-scale transfer of momentum across layer boundaries; Re > 2000

another (Leopold et al. 1964). The intensity of the resistance is related to the *molecular viscosity* of the fluid, where viscosity is governed by internal characteristics of the fluid such as temperature and the concentration of suspended sediment.

In **turbulent flow,** the water does not move in parallel layers; its velocity fluctuates continuously in all directions within the fluid. Water repeatedly interchanges between neighboring zones of flow, and shear stress is transmitted across layer boundaries in another form of viscosity, called *eddy viscosity*. Eddy viscosity greatly increases the flow resistance and thus the dissipation of energy. Because turbulence is generated along the channel boundaries, most resistance in this type of flow results from external factors such as the channel configuration and the size of the bed material.

As depth and velocity increase, the conditions at which laminar flow changes to turbulent can be predicted by a dimensionless parameter called the **Reynolds number (Re):**

$$Re = VR\rho/\mu$$

where V is the mean velocity, R the hydraulic radius, ρ the density, and μ the molecular viscosity. The hydraulic radius is determined by the relationship

$$R = A/P$$

where A is the cross-sectional area of the channel and P is the wetted perimeter (fig. 6.1). In wide, shallow channels the hydraulic radius closely approximates the mean depth.

Because the factor μ/ρ defines the fluid property called *kinematic viscosity* (υ), the Reynolds number represents a ratio between driving and resisting forces:

$$Re = VR\rho/\mu = VR/\upsilon = \frac{\text{driving forces}}{\text{resisting forces}}$$

In normal situations true laminar flow occurs where *Re* values are less than 500, and well-defined turbulent flow when *Re* is greater than about 2000.

Another dimensionless number used to describe the conditions of flow is the **Froude number (Fr):**

$$Fr = V/\sqrt{dg}$$

where d is depth and g is gravity. The Froude number is important because it can be used to distinguish subtypes of turbulent flow called *tranquil flow* (*Fr* < 1), *critical flow* (*Fr* = 1), and *rapid flow* (*Fr* > 1). The energy that is expended by these flow types differs considerably. In addition, within sand bed channels, tranquil, critical, and rapid flow have been related to the development of distinct sedimentary bedforms (fig. 6.2), which also exert an important influence on the resistance to flow in open channels (as will be discussed in more detail in the next section).

Flow within natural channels is invariably turbulent, although a very thin layer of quasi-laminar flow may be present along the channel boundaries. Most of the turbulence is generated along the water and sediment interface, causing an increase in resistance and a decrease in velocity toward the channel perimeter (fig. 6.3). Thus, across a channel the highest velocities occur near the center of the flow. The location of highest velocities may vary significantly, however, as a function of channel alignment and cross-sectional shape (fig. 6.3B), becoming more asymmetrical in meander bends (Knighton 1998).

In rivers formed in sand or finer-grained sediments with smooth channel beds, the vertical velocity profile is typically characterized by two zones of flow in addition to the laminar sublayer (fig. 6.3A). The lower zone encompasses about 20 percent of the total flow depth, and exhibits a quasi-logarithmic decrease in velocity toward the channel floor. The overlying upper zone is less affected by flow resistance along the channel bed, and vertical velocity profiles are more nearly parabolic in

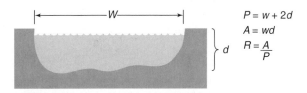

Figure 6.1
Cross-sectional measurements of a stream channel: w = width, d = depth, A = area, R = hydraulic radius, P = distance along wetted perimeter.

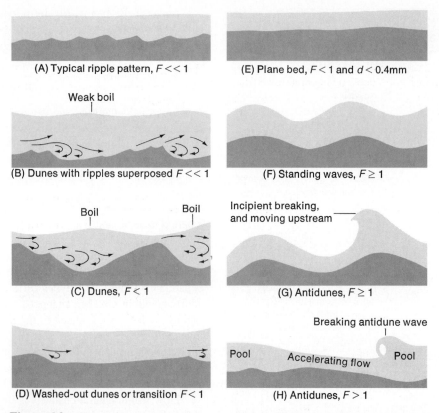

Figure 6.2

Bed forms in alluvial channels and their relation to flow conditions. F = Froude number, d = depth.

(Simmons and Richardson 1963)

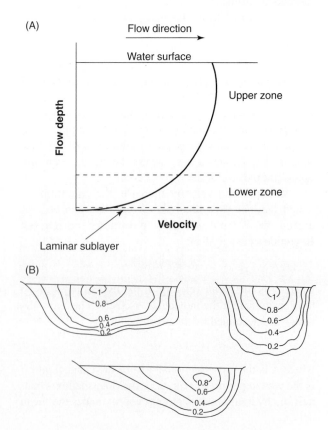

Figure 6.3

Variations in flow velocity as a function of water depth. The lower zone exhibits a quasi-logarithmic form induced by resistance along the channel bed. The upper zone is less affected by the bed roughness, and is more nearly parabolic in shape. The laminar sublayer may be absent or discontinuous in coarse-grained channels. (B) Typical variations in velocity across the channel. Isovels (lines of equal velocity) are in m/s.

(Modified from Wolman 1955)

shape (Wiberg and Smith 1991). For the purposes of calculating discharge, it is often assumed that the velocity profile exhibits a logarithmic form through the entire depth of flow (Wiberg and Smith 1991; Pitlick 1992) and that the average velocities are found at 0.6 of the depth from the water surface (Byrd et al. 2000). The highest velocities exist at, or immediately below, the surface (fig. 6.3). In rough, coarse-grained rivers where the fluid is forced around large clasts, the vertical velocity profile may become irregular or distorted (Bathurst 1988; Wiberg and Smith 1991). In fact, the lower zone of flow may be completely absent (Wiberg and Smith 1991). These effects appear most pronounced where flow depths are small compared to the size of the roughness elements (including large clasts and other topographic features) on the channel floor. Byrd et al. (2000) suggest that for these rivers, the average velocities used in the calculation of discharge may be obtained most effectively by averaging two or three measurements obtained at differing depths.

Flow Equations and Resisting Factors

Flow and resistance have been the concern of hydraulic engineers for centuries, and a number of equations have been derived to express the relationships between the two factors. Two equations of great importance to students of rivers are the **Chezy equation** and the **Manning equation.** Both were derived from equating driving and resisting forces in nonaccelerating flow, and both have been employed in a variety of fluvial investigations. Derived in 1769, the Chezy equation

$$V = C\sqrt{RS}$$

shows that velocity is directly proportional to the square root of the RS product, where S is slope of the channel. The Chezy coefficient (C) is a constant of proportionality that is related to resisting factors in the system.

The Manning equation originated in 1889 from an attempt by Manning to systematize the existing data into a useful form. The equation, when utilizing English units, is expressed as

$$V = \frac{1.49}{n} R^{2/3}S^{1/2}$$

and is similar to the Chezy formula in that velocity is proportional to R and S. In addition, the factor n, called the Manning roughness coefficient, is also a resisting element that is closely related to the Chezy coefficient because as

$$C(RS)^{1/2} = \left(\frac{1.49}{n}\right)R^{2/3}S^{1/2}$$

then

$$C = \frac{1.49R^{1/6}}{n}$$

TABLE 6.2	Manning Roughness Coefficients (n) for Different Boundary Types.
Boundary	**Manning n (ft$^{1/6}$)**
Very smooth surfaces such as glass, plastic, or brass	0.010
Very smooth concrete and planed timber	0.011
Smooth concrete	0.012
Ordinary concrete lining	0.013
Good wood	0.014
Vitrified clay	0.015
Shot concrete, untroweled, and earth channels in best condition	0.017
Straight unlined earth canals in good condition	0.020
Rivers and earth canals in fair condition; some growth	0.025
Winding natural streams and canals in poor condition; considerable moss growth	0.035
Mountain streams with rocky beds and rivers with variable sections and some vegetation along banks	0.041–0.050

Source: *Handbook of Applied Hydrology,* ed. by Ven T. Chow, copyright 1964 McGraw-Hill Publishing Co., Inc.

Manning's n is presumed to be a constant for any particular channel framework; consequently, it has been used extensively in analyses of river mechanics (table 6.2). The U.S. Geological Survey, for example, has developed a visual guide for rapid estimation of Manning's n along any given stream reach (Barnes 1968). In reality, however, resistance coefficients vary with flow stage, channels typically becoming more hydraulically efficient as discharge increases (Bathurst 1982; Knighton 1998).

It is important to recognize that resistance is not directly measured from the flow. Rather, the resistance coefficients are defined by hydraulic characteristics (e.g., S, R, V) and must be indirectly estimated from measurements of the defining parameters. Moreover, they are dependent on other factors (not included in the defining equations) because in alluvial channels their values vary with particle size, sediment concentration, and bottom configuration. The coefficients, then, represent the total resistance to flow that originates from a variety of sources. Attempts to separate the total resistance into specific source types have met with only limited success. Bathhurst (1993) suggests, however, that total resistance can be subdivided into three major components: free surface, channel, and boundary resistance. *Free surface resistance* represents the loss of energy resulting from the disruption of the water by surface waves and abrupt changes in water surface gradients (hydraulic jumps). *Channel resistance* is that which is associated with undulations in the channel bed and banks as well as alterations in channel plan form and cross-sectional shape. Most studies to date have focused on *boundary resistance,* which results from the movement of the water over either individual clasts (referred to as *grain roughness*) or microtopographic features (e.g., dunes, and ripples) that have been molded in the bed material. These microtopographic features are said to produce *form drag.*

The effect of particle size on roughness values has been evaluated in numerous studies and is perhaps the best understood of the resistance elements (Wolman 1955; Limerinos 1970). Grain roughness is primarily dependent on the depth of flow relative to the size of the particles that comprise the channel bed. This relationship, called the relative roughness, can be expressed numerically as d/D_i, where d is the water depth and D_i is a measure of the particle sizes on the channel floor. Wolman (1955), for example, demonstrated that in Brandywine Creek, Pennsylvania, the resistance-particle-size relationship is defined by the equation

$$1/\sqrt{F} = 2 \ \log d/D_{84} + 1.0$$

where F is another resistance parameter called the *Darcy-Weisbach resistance coefficient, d* is water depth, and D_{84} is the particle diameter that is equal to or larger than 84 percent of the clasts on the channel bottom. The above expression implies that as flow depth increases the influence of particle size on the resistance to flow decreases. Larger particle sizes will be associated with greater roughness for any given flow depth.

In the above equation, Wolman (1955) used D_{84} in the description of relative roughness. The use of D_{84} is somewhat arbitrary (Knighton 1998), but it is consistent with the argument by Leopold et al. (1964) that the largest particles in the bed play an important role in controlling flow resistance. Thus, it has become one of the most widely used parameters to describe grain roughness.

Microtopographic features developed in the channel bed have long been recognized as an important control on the resistance to flow in sand bed channels. For example, in the same reach of a river, different flow conditions (as determined by the Froude number) may mold bottom sediment into a variety of bed forms (fig. 6.2), which in turn have a decided influence on the values of Manning's *n* (table 6.3). Bedforms may also be developed in gravel bed rivers (Bluck 1987; Brayshaw 1985), particularly where a wide range of particle sizes exist on the channel floor. For example, pebble clusters, which may be the most prevalent type of coarse-grained bedform, consist of an exceptionally large clast that is surrounded upstream and downstream by finer-grained deposits (Brayshaw 1985)(fig. 6.4). In this case, it appears that the intensity of the resistance to flow over the microtopographic features (form drag) is related to the spacing of the bedforms on the channel bottom. The greatest resistance occurs where turbulence generated downstream of one bedform is not dissipated before the next bedform is encountered (Hassan and Reid 1990).

TABLE 6.3	Variation of Manning *n* Values with Changes in Bed form Occurring under Different Flow Conditions.
Bed Form	**Manning *n* (ft$^{1/6}$)**
Lower regime	
Ripples	0.017–0.028
Dunes	0.018–0.035
Washed-out dunes or transition	0.014–0.024
Upper regime	
Plane bed	0.011–0.015
Standing waves	0.012–0.016
Antidunes	0.012–0.020

Source: *Handbook of Applied Hydrology,* ed. by Ven T. Chow, copyright 1964 McGraw-Hill Publishing Co., Inc.

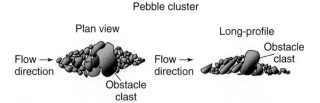

Figure 6.4
Sketch of the morphology of a pebble cluster in plan and long view. Obstacle clast is surrounded up and downtream by smaller particles.

At a larger spatial scale, irregularities of the channel bottom resulting from pools, riffles, and bars may have an equally important influence on roughness (Leopold et al. 1964; Simons and Richardson 1996). The predominant effect of pool-riffle sequences is the ponding of water upstream of the riffles. Ponding is most pronounced during low flow and, thus, resistance related to pool-riffle sequences tends to decrease with flow depth (Bathurst 1993). Nonetheless, Prestegaard (1983a) demonstrated that in 12 gravel-bed rivers of the western United States, channel bars accounted for 50 to 75 percent of the total resistance at high flow.

Aquatic and stream-side vegetation is also known to influence channel roughness. Bathurst (1993) notes that the effects on roughness may be significant enough to alter the stage-discharge relations at a given site between the summer and winter months. However, quantification of the influences of vegetation on flow resistance has proven problematic, in part because the height, density, age, and strength of the plants found within or along the channel vary significantly with both time and location.

In light of the above, it is clear that the external boundary conditions such as particle size and bed configuration generate a large amount of resistance. Some of these factors produce turbulence in the form of eddies and secondary circulation, which leads to a loss in the

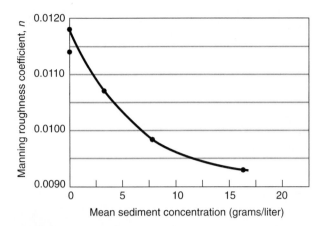

Figure 6.5
Effect of suspended load concentration on the Manning roughness coefficient *n*.
(Vanoni 1946)

energy causing flow. In contrast, sediment concentration (the amount of sediment per unit volume of water) internally affects resistance. This modification was first detailed by Vanoni (1941, 1946), who showed that an increase in the concentration of suspended sediment tends to lower resistance (fig. 6.5). As the concentration increases, the turbulent effect presumably is reduced because the mixing process within the fluid is dampened. All other factors being equal, sediment-laden water should flow at a higher velocity than clear water.

SEDIMENT IN CHANNELS

Most energy in a stream is dissipated by the many factors resisting flow in open channels. The remainder, although commonly small, is used in the important task of eroding and transporting sediment. These processes, often taken for granted, are extremely complex and poorly understood, yet they underlie some of our most basic concepts of river mechanics. We will briefly review the more significant ideas about the movement of sediment in rivers.

Transportation

In general, fine-grained sediment (silt and clay) is transported within the water column by the supporting action of turbulence. *Suspended load* usually moves at a velocity slightly lower than that of the water and may travel directly from the place of erosion to points far downstream without intermittent stages of deposition. Coarse particles may also travel in true suspension, but they are likely to be deposited more quickly and stored temporarily or semipermanently within the channel. Except for short spasms of suspension, coarse sediment usually travels as bedload. *Bedload* refers to sediment transported close to or at the channel bottom by rolling, slid-

ing, or bouncing. How long coarse debris remains stationary within a channel depends on a large number of parameters including the nature of the debris (e.g., its size, shape, and density), the interlocking relationships between the particles, the exposure to flow, and the flow characteristics of the river; such debris probably is immobile more than it is in motion. Bradley (1970), for example, showed that gravel can be stored in channel bars long enough for weathering to drastically weaken its resistance to abrasion.

Because of fluctuating discharge, at any given time a single particle may be part of either the bedload or the suspended load. As this makes the distinction between the two load types unclear, other terms have been devised to relate sediment more appropriately to river flow. *Wash load* consists of particles so small that they are essentially absent on the streambed. In contrast, *bed material load* is composed of particle sizes that are found in abundant amounts on the streambed (Colby 1963). While most, if not all, bedload is bed material load, most bed material load is transported as suspended load.

The relationship between wash load and discharge is poorly defined because most streams at any given flow can carry more fine-grained sediment than they actually do. The concentration of fines is a function of supply rather than transporting power; therefore, it is relatively independent of flow characteristics. Coarse sediment, on the other hand, is usually available in amounts greater than a stream can carry, and so its concentration should correlate more significantly with the parameters of flow such as depth and velocity. The problem, however, is that direct measurement of bedload is extremely difficult because handheld instruments can sample for only short periods, and when they are placed on the channel bottom the flow regime is disrupted. In addition, where bedload has been continuously measured, the amount of sediment passing a given channel cross-section varies significantly with time (see, for example, Leopold and Emmett 1977; Hoey 1992; Carling et al. 1998). Furthermore, the amount of bedload at any given time varies drastically in different subwidths of the channel cross-section.

Because of the difficulties surrounding direct measurement, most estimates of bedload discharge are made by means of empirical equations that attempt to determine the maximum amount of sediment that a stream can carry (its capacity) for a given set of channel, sediment, and flow conditions (Meyer-Peter and Muller 1948; Einstein 1950; Bagnold 1980; Parker et al. 1982; Williams and Julien 1989). These equations, however, are themselves problematical; their accuracy is difficult to assess because reliable measurements of bedload discharge are scarce, and variations in bedload transport for any given set of hydraulic conditions can be large. In fact, Gomez and Church (1989) compared 10 transport formulas and concluded that for coarse-grained streams,

Figure 6.6
U.S.G.S. bedload sampling station, East Fork River, Wyo. View across river shows the suspension bridge and drive mechanism of a conveyor-belt bedload sampler.
(W.W. Emmett, U.S. Geological Survey)

none of the equations are entirely adequate for predicting bedload transport.

The problems inherent in deciphering the relationship between parameters of flow and bedload transport are difficult but not insurmountable. For example, some very good bedload measurements have been made on the East Fork River, Wyoming (Leopold and Emmett 1976, 1977), where the U.S. Geological Survey installed a concrete trough across the channel floor. Sediment moving along the river bottom fell onto conveyor belts rotating within the trough and was carried to the channel side where it was weighed (fig. 6.6). Other sites have been similarly instrumented, and new techniques to assess bedload transport are continually being devised. For instance, once an understanding of the natural magnetic properties of the bed material had been determined, Carling et al. (1998) were able to use magnetic detectors to determine the rate at which individual particles moved downstream in Squaw Creek, Montana.

Entrainment

The temporal and spatial variations in bedload transport rates, noted, for example, at the East Fork River station (Leopold and Emmett 1976), are related to the mechanics involved in moving coarse-grained sediment. Most large particles do not move great distances during any transporting event. Instead, their downstream migration is characterized by spasmodic bursts of short-distance movement separated by periods during which they come to and remain at rest. The processes that initiate the bursts of motion experienced by any particle are collectively known as **entrainment.** The amount of sediment entrained depends directly on the erosive power of the flow and on the nature of the particles on the bed surface that are in the proper position to be eroded. Two streams with identical flow conditions may have different bedload or bed material discharge if one flows across a fine-sand bottom and the other over a cobble bed.

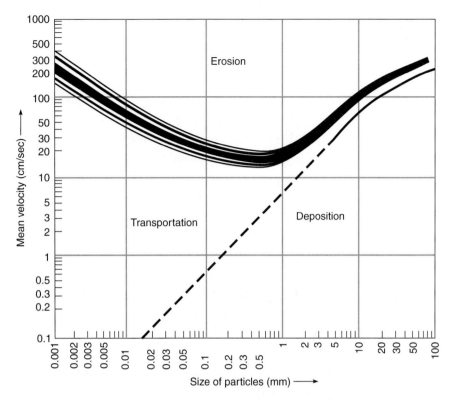

Figure 6.7
Mean velocity at which uniformly sorted particles of various size are eroded, transported, and deposited.
(Hjulström 1939, reprinted by permission)

The term *competence* refers to the size of the largest particle a stream can entrain under any given set of hydraulic conditions. The value of competence to geomorphologists depends on how we measure the sediment being moved and, more importantly, on how accurately we can determine the flow conditions. Although ascertaining competence may seem simple enough, in practice it is an excruciating problem for several reasons: (1) particles are entrained by a combination of fluvial forces including the direct impact of the water, drag, and hydraulic lift, and each of these may be best defined by a different parameter of flow; (2) flow velocity is neither constant nor easily measured, especially during high discharge; and (3) sediment of the same size may be packed together differently, be surrounded by particles of different sizes, or have shape properties that cause abnormal responses to the same flow conditions. Thus, any investigation into the mechanics of competence must settle for only partial success until we can eliminate or inhibit some of the inherent variability.

Historically, two hydraulic factors have been utilized to represent the flow condition in the competence relationship. The first, *critical bed velocity*, demonstrates the relationship between velocity and entrainment. It has been known for some time that the volume or weight of the largest particle moved in a stream varies as approximately the sixth power of the velocity (Rubey 1938). The *sixth-power law* provides a sound theoretical basis for competence studies, but it is less satisfying in practice because accurate measurement of bed velocity

is exasperating, if not impossible, in high-energy streams. A less reliable predictor of particle entrainment, mean velocity is more easily determined and has been extensively used in competence studies. Figure 6.7, for example, shows the curves produced by Hjulström (1939) that relate current velocity, particle size, and process.

The second factor, *critical shear stress* signifies the downslope component of the fluid weight exerted on a particle as motion begins (fig. 6.8). This dragging force can be expressed by:

$$\tau_c = \gamma RS$$

where τ_c is the critical shear stress, γ the specific weight of the water, R the hydraulic radius, and S the slope. In most streams transporting coarse bedload, R is closely approximated by mean depth, and, thus, critical shear stress is proportional to the depth-slope product. However, care should be taken before substituting R for d (Tinkler 1982).

The Shields diagram, shown in figure 6.9, relates a descriptor of the threshold for particle motion, called the dimensionless critical shear stress (θ), to grain Reynolds numbers. The grain Reynolds number describes the extent to which an individual particle projects above the laminar sublayer into the zone of turbulent flow, and is expressed as D/δ_o, where D is grain diameter and δ_o is the thickness of the laminar sublayer. The Shields diagram is similar to Hjulström's curve (fig. 6.7) in that both describe the flow conditions at the time of erosion.

(A)

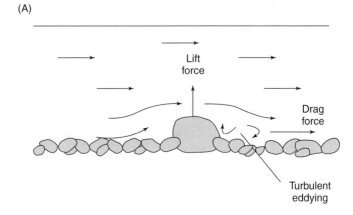

(B)

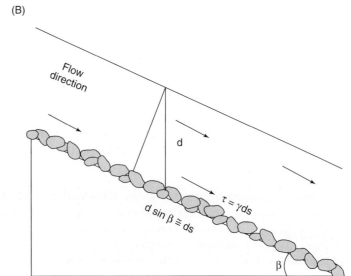

Figure 6.8

(A) Orientation of lift and drag forces acting on submerged channel bed sediment. Lift forces are due to variations in flow velocity over the top and bottom of the particle. Turbulent eddying may also create upward directed forces that act on the particles. (B) Component of flow weight exerted as shear stress on the channel bottom. The critical shear stress is equal to the depth-slope product (dS) multiplied by the specific weight of the water γ and β is the angle of slope.

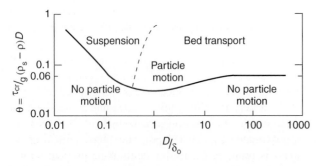

Figure 6.9

Shield curve for the entrainment of bed particles where D is grain diameter, τ_{cr} is critical shear stress, ρ_s is sediment density, ρ is fluid density, and δ_o is thickness of laminar sublayer.

The Shields diagram illustrates that within hydraulically smooth channels characterized by silt and clay, dimensionless shear stress (θ) varies with grain Reynolds numbers (D/δ_o), reaching a minimum at a value of D/δ_o of approximately 0.03 (fig. 6.9). Dimensionless shear stress increases for smaller values of D/δ_o (fig. 6.9). Given that grain Reynolds number is related to particle

size, it follows that more shear stress is required to entrain fine-grained sediments that reside below the surface of the laminar sublayer and that are not subjected to the effects of turbulent flow. Cohesion, generally associated with smaller particles, may also play a role in increasing the shear stress required for entrainment. For hydraulically rough channel beds (in which the particles are relatively large in comparison to the thickness of the laminar sublayer), motion is initiated predominantly by turbulent action (Morisawa 1985), and θ obtains a constant value of approximately 0.06 (although constant values as low as 0.03 have been reported in some studies).

Knighton (1998) notes that a disadvantage of critical shear stress formulas is that they ignore the effects of lift that may promote particle entrainment. Lift is primarily generated by differences in the velocity of the flow over the top and bottom of an individual particle, a process that creates a vertical pressure gradient leading to the upward motion of the grain (fig. 6.8). Lift may also be created by turbulent eddying generated downstream of the particle that produces locally upward directed flow. The use of critical shear stress in competence studies has been criticized for other reasons as well (Yang 1973), but the simple reality that depth and

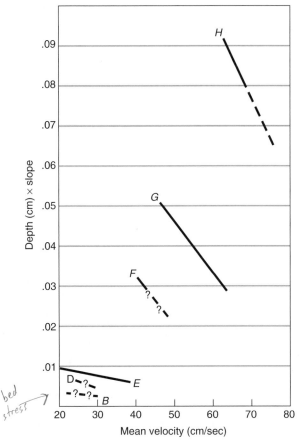

critical bed
shear stress →

Figure 6.10
Sediment particles of different sizes begin to move on the
streambed at different values of mean velocity and depth-
slope. Smallest particles (*B*) move mainly as a function of *dS*
while largest particles (*H*) move primarily as a function of the
mean velocity.
(Rubey 1938)

slope in a river are easier to measure than bed velocity
makes it an appealing parameter.

Precisely how the two methods correlate with each
other is not completely understood, but both approaches
suggest that it is more difficult to entrain particles that
are either smaller or larger than medium sand. This
helps explain the commonly observed phenomenon of
sand-sized debris being transported across stationary
material of a smaller size. There is, however, some evidence to suggest that the importance of shear stress and
flow velocity to the entrainment process may vary with
particle size. Rubey (1938), for example, suggested that
critical bed velocity becomes more important in the entrainment process as particle size increases from fine
sand to pebbles (fig. 6.10). Smaller sizes move more as a
function of the *dS* product and seem to be relatively independent of velocity. Thus, the shear stress approach
may be completely valid only for smaller sizes or low-
velocity flows, and very fine-grained sediment requires
higher velocities for its entrainment than the sixth-power
law would predict.

In a related, but different type of approach, Bagnold
(1973, 1977) proposed that entrainment and transportation of bedload can be analyzed in terms of stream
power. **Stream power** is defined as

$$\omega = \gamma Q S$$

where ω is stream power and γ, Q, and S are specific
weight, discharge, and slope, respectively. If power is
considered per unit area of the streambed, it essentially
becomes a combination of shear stress and velocity because

$$\omega = \gamma Q S / \text{width} = \gamma d S V = \tau V$$

where V is mean velocity. Where available stream power
is greater than that needed to transport load, scour of bed
alluvium (entrainment) will occur. As a result, stream
power has become an important parameter in characterizing the erosional capability of rivers.

Most of the studies of entrainment whether utilizing
the shear stress, critical bed velocity, or stream power,
have been based on flume studies. Flumes are not useful
in the study of competence when particles are larger than
pebble size. Most competence investigations of coarser
sediment have, therefore, been made in natural rivers.
These investigations demonstrate that in natural channels
particles of a given size may be entrained by widely
varying flow conditions. Much of this variation comes
from the fact that the river bed is not composed of clasts
of a uniform size, shape, and composition, but is a mixture of particles whose characteristics may vary over a
considerable range. It is now known, for example, that in
channels with poorly sorted bed material, the finer particles are shielded from the flow by the larger particles.
Exposure of fine clasts may be particularly reduced
by microtopographic features, such as pebble clusters
(fig. 6.4) (Brayshaw 1985). The result of these "hiding
effects" is that the larger clasts tend to be more mobile
and the finer clasts less mobile than would be predicted
for material of uniform size (Parker et al. 1982; Andrews
1983; Paola and Seal 1995). Entrainment may also be
complicated in coarse-grained channels by the burial of a
fine-particle layer by a coarser layer, a process that accentuates hiding effects and allows the larger particles to
be more readily available for entrainment (Paola and Seal
1995). These observations suggest that the relative size
of a particle in the mixture may be as important to entrainment as its absolute size. In fact, some investigators
argue that the effects of particle-hiding and sediment layering may be so significant that all clasts in the mixture
become mobile at about the same shear stress, a concept
referred to as the *equal mobility hypothesis* (Parker et al.
1982; Andrews 1983; Andrews and Erman 1986).

A common phenomenon along many river systems
is for particle size to decrease quasi-systematically
downstream, although the reduction in size may be altered by the local influx of tributary sediment (Knighton
1980; Pizzuto 1995). Ashworth and Ferguson (1989)

point out that if precise equal mobility held for the entire length of a channel, downstream fining could occur only through abrasion during transport, or by weathering during periods of rest. While these processes may play an important role in the downstream reduction of a particle size, selective transport (in which progressively larger particles are entrained as shear stress increases) appears to be the predominant control on downstream fining along many rivers (Ashworth and Ferguson 1989; Werritty 1992; Ferguson et al. 1998). In addition, precise equal mobility would suggest that the size distribution of the material being transported as bedload should be the same as that of the bed material load. However, at low shear stress transport is commonly characterized by marginal transport conditions in which easily dislodged clasts are occasionally entrained (Andrews and Smith 1992) and coarse materials are significantly underrepresented in the bedload. At higher shear stresses, the bedload tends to become progressively coarser (Komar and Shih 1992). Thus, true equal mobility may occur only during extreme flood conditions (Ashworth and Ferguson 1989; Komar and Shih 1992).

Bank Erosion

The processes of entrainment determine the type and magnitude of erosion that occurs on the channel floor. It is incorrect, though, to assume that the only significant erosion is vertically directed. Bank erosion, which pro-

ceeds laterally, not only contributes to the sedimentary load but, through its control on channel width, exerts a direct influence on other channel processes. A large number of studies have identified the major processes involved in bank erosion (Thorne 1982; ASCE 1998; Simon et al. 1999). Invariably researchers of this phenomenon arrive at the striking conclusion that bank erosion is rarely, if ever, accomplished by a single process but instead involves some combination of processes unique to the individual setting. In general terms, bank erosion is related to two major types of processes: fluvial entrainment and the weakening and weathering of bank materials which enhance the potential for mass wasting (Thorne 1982).

Fluvial entrainment promotes bank erosion in two ways. First, sediment may be entrained directly from the bank surface by the forces generated in river flow, a process usually referred to as *corrasion.* In this case, erosion is primarily related to the flow velocities in the near-bank environment as well as the type, density, and root system of vegetation along the channel margins. Second, corrasion often produces an overhanging ledge of cohesive sediment because noncohesive layers in the bank are eroded more rapidly than the cohesive materials (Thorne and Tovey 1981). The overhanging bank sections, called *cantilevers,* are mobilized when the undercut cohesive material finally fails and drops to the surface below (fig 6.11). A similar process operates where vertical fractures, called *tension cracks,* exist in

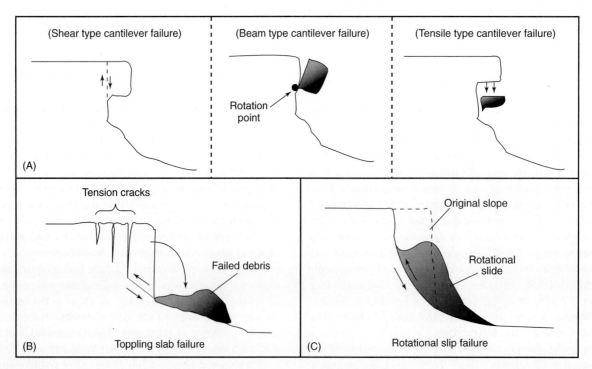

Figure 6.11
Types of bank failure mechanisms commonly observed along rivers with cohesive banks.
(Adapted from Thorne 1982)

the floodplain sediment. In these cases, lateral trimming of the bank near the surface of the water intersects a tension crack and produces downward failure along the fracture plane. This type of movement has been called *soil fall* (Brunsden and Kesel 1973), *earth fall* (Twidale 1964), *slab failure* (Hagerty 1980), and *shallow slip* (Thorne 1982). Normally, slab failure does not involve as much bank material as other mass movements discussed below, but the process is very significant because it occurs frequently (Thorne and Tovey 1981).

Weakening and weathering processes tend to reduce the strength of bank materials and thereby promote instability and mass movements. The mechanics of failure depend on many variables such as the geometry and stratigraphy of the bank and the physical-chemical properties of the bank material. The most important control on weakening of bank material, the soil moisture condition, depends on both climate and bank properties. Soil moisture is transformed into bank erosion by processes that (1) reduce strength within the bank and (2) act on the bank surface to loosen and detach particles and their aggregates. For example, where saturated banks are found in poorly drained, cohesive sediment, positive pore pressure can decrease the strength of the bank material (see chapter 4). This is especially true in high, steep banks after prolonged precipitation or in the case of rapid drawdown of the river level. Under these conditions bank failure may occur by rotational sliding (fig. 6.11c). In some cases, the stratigraphy of the floodplain sediment plays an important role in bank failures, especially where cohesive layers rest above and below a noncohesive layer. Usually the noncohesive layers consist of permeable gravel, sand, or silt in contrast to the cohesive material which is normally richer in clay. The coarse-grained, cohesionless zone often serves as an avenue of pronounced seepage of underground water. Sapping (or piping) may occur, with the seepage actually transporting material away from the noncohesive unit (Deere and Peck 1959; Hagerty 1980; Hagerty and Hamel 1989; Ullrich et al. 1986; Odgaard et al. 1989). The removal of the noncohesive material by sapping creates overhanging blocks of upper bank sediment and leads to bank instability and failure. Ullrich et al. (1986) note that the overhanging bank sediments usually fail along tension cracks that form immediately following a mass wasting event. Thus, failure, by creating tension cracks, may prepare the bank materials for future erosion by sapping and mass wasting processes.

In addition to producing overhanging blocks of bank material, seepage may create a lubricated surface immediately below the noncohesive unit that serves as a sliding plane for overlying material. This type of failure, referred to as a *planar slide*, usually functions on sloping surfaces, but where shear resistance on the plane of sliding is very low, the movement can occur on a horizontal surface. Planar slides (fig. 6.12) have been recog-

nized as important factors in the erosion of bluffs along the Mississippi River (Brunsden and Kesel 1973) and, in combination with other types of movement, are probably quite common (Varnes 1958). They require, however, that vertical fractures (tension cracks) exist in the bank sediment.

The effects of water on the mechanics of bank erosion are most pronounced when the bank materials are saturated. Moisture may, however, have a significant effect on the shear strength of the bank materials even in the absence of saturated conditions. For example, along rivers with cohesive banks it is not uncommon for part of the bank material to reside above the water table. During dry conditions, the pores within these sediments are filled with both water and air, and the shear strength of the material is partly dependent on the *matrix suction,* defined as the difference between the air pressure and the water pressure in the unsaturated pores. Increases in matrix suction enhance the apparent cohesion of the material and, thus, its shear strength. During precipitation events, decreases in matrix suction caused by infiltrating rainwater may be sufficient in many cases to initiate bank failure (Simon and Curini 1998; Simon et al. 1999).

The maximum rate of bank failure does not generally correspond to periods of peak discharge, but occurs during the waning stages of the event. The seemingly delayed response of the banks has been attributed to (1) the movement of water into the bank sediments during the rising stage of the flood, a process that increases both pore pressure and the weight of the alluvial deposits, priming the banks for failure, and (2) the release of pressure from the channel banks as water levels in the river recede. Many researchers have also noted a distinct seasonality associated with bank erosion rates, winter and spring rates being considerably greater than those in the summer (Wolman 1959; Thorne and Lewin 1979; Simon et al. 1999). Presumably this results from higher moisture contents during the winter and spring months, and its influence on pore pressure and matrix suction. Simon et al. (1999), for example, argue that during the dry summer months, rainstorm events may be unable to destabilize the banks because of high antecedent values of matrix suction. However, during prolonged wet periods when matrix suction is low, as is commonly the case during the winter and spring months, even small precipitation events may lead to bank failure. The effects of frost action may also result in seasonal differences in bank erosion.

If river processes did not remove the debris from the base of the banks, the accumulated sediment would inhibit further failure (Pizzuto 1984; Nanson and Hickin 1986). From this perspective, the rate of bank erosion is controlled by the rates at which the failed debris can be transported from the base of the banks by the prevailing hydrologic regime. Nonetheless, the above discussion illustrates that in many cases the processes of river bank

Figure 6.12
Bank erosion along the Osage River in central Missouri by lateral spreading and planar sliding.
(Dale Ritter)

erosion have little to do with rivers. Often it is a mass movement phenomenon controlled by the texture and stratigraphy of floodplain sediment and triggered by the movement of groundwater. In light of this, the rate of bank erosion in alluvial channels can be enormous or minuscule, depending mainly on the character of bank materials. In general, banks that are composed of fine-grained sediment or densely vegetated (Hadley 1961; D.G. Smith 1976) have more resistance to corrasion than channels with sandy or gravelly banks. The actual process of erosion, however, may differ; clay-rich banks usually retreat by undercutting and subsequent block failure (Stanley et al. 1966; Laury 1971), while more coarse-grained banks erode by dislodgement and slough-ing of individual particles. Even highly cohesive banks may therefore erode rapidly if the dominant process is undercutting and/or mass failure.

Erosion of Bedrock Channels

Our attention thus far has been on the entrainment and transport of sediment in rivers formed in alluvial de-posits. Far less is known about the erosional processes operating along bedrock rivers that are most commonly found in headwater areas or tectonically active oro-graphic belts (Montgomery et al. 1996). In fact, a uni-versal definition of what constitutes a bedrock channel has yet to emerge. Nonetheless, it is clear that bedrock rivers exhibit a variety of forms that are not necessarily devoid of loose debris. Many bedrock channels contain thin, discontinuous accumulations of unconsolidated sediment along the channel bed, but they differ from al-luvial rivers in that all of the loose bed sediments can be mobilized during flood events. Bedrock rivers, then, are capable of moving significantly more sediment than is available for transport, and the nature of bed and bank erosion is controlled by the erosionally resistant sub-strate that forms part or all of the channel perimeter. In most cases, the erosional resistance of the bounding rock substantially limits changes in channel form during indi-vidual floods, and notable alterations in channel geome-try can only be observed over periods of decades or cen-turies (Tinkler and Wohl 1998). Thus, channel form represents the integration of processes acting over long periods of time.

Because of the difficulties of directly measuring the changes in bedrock channels over a period of a few years, erosion rates are typically estimated using numer-ically based approaches (Howard and Kerby 1983; Kooi and Beaumont 1994). One of the most widely used ex-

Figure 6.13

Large potholes in the lower Susquehanna River gorge near Holtwood, PA, formed by macroturbulent flow during catastrophic paleofloods. The large pothole contains erratics eroded from rock formations more than 50 km upstream. Smaller potholes can also be seen with their axes tilted slightly upstream (to the left).

(R. Craig Kochel)

pressions describes the total rate of erosion (ε) in terms of basin area and channel gradient:

$$\varepsilon = KA^m S^n$$

where m and n are positive constants, A is drainage basin area (a surrogate for discharge), S is channel slope, and K is a dimensional coefficient of erosion, which is a function of lithology, climate, channel width, flow hydraulics, and sediment load (Whipple et al. 2000). The expression is based on the postulate that the rate of erosion is a function of the shear stress applied to the channel bed (Howard and Kerby 1983). It is now known, however, that the exponent for channel slope (n) is directly dependent on the erosional processes that operate along the channel (Whipple and Tucker 1999). Thus, its universal application to rivers characterized by differing downstream processes of erosion is still in question (Wohl and Ikeda 1998; Whipple et al. 2000).

The primary processes involved in the erosion of bedrock channels include abrasion, plucking, corrosion and other forms of chemical weathering, and cavitation (Wohl 1998, 1999; Whipple et al. 2000). *Abrasion,* or the slow, incremental wearing away of the rock surface, may produce a variety of sculptured forms in the substrate such as potholes, flute marks, or longitudinal grooves (fig. 6.13). Rates of abrasion are increased substantially where sediment is available as a grinding tool. There is some debate, however, regarding the relative importance of bedload versus suspended load in the abrasion process (Sklar and Dietrich 1998; Whipple et al. 2000), but it is likely that the significance of each varies with the local conditions.

The entrainment and transport of bedrock blocks from the channel floor is referred to as *plucking.* The

size of the blocks is typically defined by fractures, joints, or bedding planes (fig. 6.14). Thus, plucking is particularly important in eroding channel bottoms formed in thinly bedded and densely jointed rocks that result in small clast sizes (Wohl 1998). In most cases, the blocks must be loosened in place prior to their removal by high-magnitude flows. Where carbonate rocks are present, the joints and bedding planes may be enlarged by dissolution (corrosion), a process that is concentrated on the edges and corners of the blocks where surface area is relatively high (Miller 1991; fig. 6.14). In other rock types, the propagation and enlargement of fractures and joints can occur by the hydraulic wedging of sand and gravel into cracks, by high instantaneous stresses associated with the impacts of large saltating clasts, and by fluctuations in pressure resulting from intense turbulence (Whipple et al. 2000).

Cavitation results from the formation and implosion of vapor bubbles in water. The implosion of the bubbles near the water-rock interface generates shockwaves that can pit or weaken the bedrock (Wohl 1998). Cavitation is a well-known and documented process of erosion in dam spillways, tunnels, and turbines, but its importance along natural bedrock channels is less certain. Recent investigations suggest, however, that it may be more important than originally envisioned (Whipple et al. 2000).

The dominant erosional process operating within any given reach of a bedrock channel appears to be closely linked to the lithology and structure of the underlying substrate (Miller 1991; Wohl and Ikeda 1998; Wohl 1998). Whipple et al. (2000), for example, argue that plucking is the dominant process along reaches that are prominently fractured. In more massive rocks devoid of closely spaced fractures, abrasion appears to be dominant. Moreover, they found that the lack of plucking allows for the development of topographic irregularities in the channel bed. These features tend to generate intense vortices in the flow that bring suspended load into contact with the channel floor, creating incipient potholes and flutes. Once these topographic features have started to form, a positive feedback mechanism may be established as a vortex stabilizes over the depression, accelerating the abrasion process and enhancing the topographic irregularities even further.

The influence of rock structure on erosional processes is not as obvious as suggested above along all bedrock rivers (Tinkler and Wohl 1998; Wohl 1998). In some cases, similar erosional patterns have been observed across a diversity of substrate types, and morphologic changes in channel configurations can appear to be independent of the underlying strata (Wohl 1993). This led Wohl (1998) to suggest that bedrock channels exhibit a continuum of forms in which features associated with flow hydraulics are most fully expressed in relatively weak, lithologically homogenous rocks, but are poorly expressed in heterogenous and resistant substrates.

(A)

(B)

Figure 6.14

(A) Bedrock blocks exposed in channel bed of Potato Run, southern Indiana. Fractures which define the limestone blocks have been enlarged by the dissolution and by the wedging of sand and gravel particles into the fracture by hydraulic forces. (B) Erosion of the channel bed by the plucking of blocks at a knickpoint in Potato Run.

(A, B: Jerry Miller)

Deposition

Transportation of entrained sediment cannot continue forever and, thus, particles must come to rest during the depositional phase. Suspended rock and mineral fragments tend to settle to the bottom at a rate that depends on the density of the water, the fluid viscosity, and the size, shape and density of the sediment. The distance any suspended particle will move in one event depends on its fall velocity and on whether its downward settling is off-set by turbulent forces in the water column. Coarse parti-

cles tend to be deposited first as flow velocities decrease, and may be deposited during minor fluctuations in velocity (see fig. 6.7). The net effect of these differences in deposition is to create downstream and vertical variations in grain size of the bed material. It should be recognized, however, that the constant fluctuations in flow make the channel floor a dynamic interface where some particles are being entrained while others are simultaneously being deposited. The net balance of this activity, referred to as *scour* or *fill*, depends on local conditions rather than on the average cross-sectional hydraulics; in fact, local effects may be different from those that occur over a long reach of the channel (Colby 1964).

long-term

Fluvial deposition is important in geomorphology in several ways. On a long-term basis, continued deposition, called *aggradation,* results in landforms that reflect distinct periods of geomorphic history. The sedimentology and stratigraphy of the associated deposits indicate the types of rivers involved in the aggradational phase (Schumm 1977) and provide clues to the environmental conditions present at the time of the aggradational event.

short-term

On a short-term basis, deposition creates bed forms and other microtopographic features such as dunes, bars, and riffle-pool sequences that are closely related to channel pattern and the character of flow within the channel (for example, see Schumm et al. 1982). Finally, you should recognize that the short- and long-term mechanics of deposition have implications beyond the boundaries of geomorphology. They are clearly basic to sedimentology and stratigraphy and, interestingly, may be key factors in subdisciplines of economic geology such as the exploration for valuable placer deposits (Schumm 1977).

The Frequency and Magnitude of River Work

At this juncture we can logically ask when and how fluvial work is done. Is it the super event of very high discharge that happens once in a millennium that causes rivers to do what they do, or is it the normal flow that is repeated time and time again? The answer to this question rests firmly on the concept of geomorphic work.

Geomorphic work is usually estimated in one of two ways. Wolman and Miller (1960) suggest that the work done by a river can be estimated by the amount of sediment it transports during any given flow. They concluded that in most basins 90 percent of the total sediment load (i.e., 90 percent of the work) is removed from the watershed by the sum of rather ordinary discharges that recur at least once every five or ten years. While megafloods transport an abnormally high sediment load, they occur so infrequently that their contribution to the total amount of sediment that is transported out of the basin over a period of years is minimal. In contrast, flows of limited magnitude are incapable of transporting significant loads. Thus, the discharges that transport the most sediment are those that are able to move debris at a moderate rate and that occur relatively frequently.

The Wolman and Miller hypothesis has been extensively examined for a wide range of river systems. These studies demonstrate that the most effective transporting discharges vary significantly from one region to another; considerable variability may even exist within any given region (Ashmore and Day 1988; Nash 1994). Andrews and Nankervis (1995), for example, examined 17 gravel bed rivers in the western United States and found that the most effective flows for transporting bed

material load over a period of years ranged from 0.8 to 1.6 times the bankfull discharge. Nonetheless, the basic tenet of the Wolman-Miller hypothesis—that most of the sediment transported by rivers is performed by moderate, relatively frequent discharges—appears to hold true for the majority of the rivers investigated.

The second way to estimate geomorphic work, with perhaps greater implications, is to assess the conditions under which rivers make adjustments to or maintain their channel morphologies. Wolman and Miller (1960) suggest that river channels form and reform within a narrow range of flows. The lower flow limit is set by the demands of competence. Clearly, the shape of the channel cannot be modified by erosional processes if the flows are incapable of transporting the bed and bank material. The upper limit is defined by the flow that exceeds bankfull and is no longer confined to the channel. From this perspective, channel configuration is presumed to be a direct indication of river work, and its precise form is perceived to be the product of high-frequency events. This hypothesis has also received considerable support and, indeed, was reinforced by studies that suggested that channel morphologies are adjusted during flows having a recurrence interval of 1.1 to 2 years, and that approximate bankfull discharge (Kilpatrick and Barnes 1964; Dury 1973). Therefore, the discharge that determines the characteristics and dimensions of a channel, known as the **dominant discharge,** has been implicitly accepted to have a frequency and magnitude equivalent to the bankfull condition.

It seems justified to say that river channel morphology is maintained in all environmental settings by geomorphic work done during a dominant discharge or within a distinct range of flows. However, it should be recognized that the recurrence interval of the bankfull discharge can vary significantly, potentially exceeding 1–2 years by an order of magnitude (Williams 1978). Moreover, it is now questionable as to whether bankfull discharge is the dominant discharge for all rivers. For example, Harvey and his colleagues (1979) found that river flows in northwest England redistributed bed material between 14 and 30 times a year and changed overall channel form from 0.5 to 4 times a year. In coarse-grained rivers, low to moderate flows may be incapable of entraining the bed and bank material. Thus, only rare, high-magnitude events may be able to effect a change in channel form (Baker 1977).

The concept of a dominant discharge is further complicated by the realization that the effect of major floods on channel configuration, referred to as *geomorphic effectiveness,* varies with the environmental setting (Costa 1974b; Gupta and Fox 1974; Baker 1977; Moss and Kochel 1978). This prompted the suggestion that the Wolman-Miller principle should be modified to include factors that control the work of floods in different environments (Wolman and Gerson 1978).

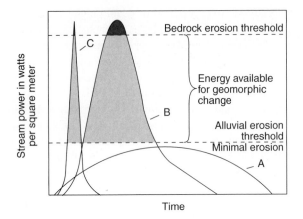

Figure 6.15
Hypothetical stream-power graphs associated with different kinds of floods. The most geomorphically effective floods are those characterized by curve B that exceed the threshold of erosion for significant periods of time.

(From Costa and O'Connor 1995)

Historically, the impact of floods on channel morphology has been related to flood magnitude, a parameter that varies greatly with basin morphometry and climate (see chapter 5). Kochel (1988) points out, however, that the most significant effects of flooding have been associated with peak discharges that are several times greater than the mean annual discharge. Thus, the *difference* between peak flood discharge and the discharge that is normally experienced by the channel may be more important in controlling the extent to which the channel is modified than the *absolute* magnitude of the event. In addition, Costa and O'Connor (1995) found that some dam-burst floods generated exceptionally high instantaneous stream powers, but produced few geomorphic effects. They argue that the limited effects of these events are related to the fact that flow duration was relatively short and the total energy expended was minimal. Long-duration flows may be necessary to wet and disaggregate the soils, thereby reducing the shear strength of the bank materials. Thus, the maximum effects of flooding may be associated with some optimal combination of flow magnitude (stream power), duration, and total energy expenditure above the initiation of particle transport (Costa and O'Connor 1995)(fig. 6.15). The geomorphic response will also be influenced by the erosional resistance of the materials that comprise the channel perimeter.

In light of the above, it should be clear that the magnitude of channel modification during an event is dependent upon the complex interplay between a large number of parameters. Kochel (1988) has subdivided these controlling parameters into two categories, which he refers to as drainage basin factors and channel factors. Figure 6.16 shows that these factors interact in such a way that the most significant effects are generally concentrated along high gradient, coarse-grained channels in headwater areas, particularly those characterized by abundant bedload.

There is a growing realization that an individual basin having constant physical/biological properties can experience different geomorphic responses in successive floods of similar magnitude (Newson 1980; Beven 1981; Kochel et al. 1987). This indicates that effectiveness is partly controlled by factors other than the nature of the flow and the channel characteristics. The most important factor seems to be recovery time (Wolman and Gerson 1978). *Recovery time* is essentially the time needed for a river to recover its equilibrium form after a major flow event has disrupted the channel configuration (for alternative definitions, see Pitlick 1993). Implicit in this perception is that major hydrologic events may be able to affect the form of a channel, and that changes produced may be long-lived or may be quickly erased as the system reverts to its pre-event condition. Thus, the effectiveness must be related to the time needed to obscure the impacts of the event on the river. Moreover, the effects of any given event may be dependent on whether the channel has fully recovered from the impact of the previous flood. Kochel (1988), for instance, documented the responses of the Pecos River of west Texas to catastrophic floods in 1954 and 1974. He found that the 1954 flood resulted in the massive redistribution of channel bed gravels and the severe erosion of the channel margins. In contrast, the 1974 event resulted in few channel changes. Presumably, the recovery times in this area were sufficiently long that the channel was still largely adjusted to the high discharges of the 1954 flood. Kochel's conclusions indicate that the effectiveness of any event is dependent upon both the actual time between successive floods and the time required for the system to recover. The healing interval is generally thought to be climatically controlled. In humid areas, recovery times appear to be short, whereas arid and semi-arid regions usually have much longer recovery times (Wolman and Gerson 1978).

THE QUASI-EQUILIBRIUM CONDITION

Every river strives to establish an equilibrium relationship between the dominant discharge and load by adjusting its hydraulic variables (e.g., channel width and depth, velocity, roughness, and water slope). This normal fluvial condition has been aptly referred to as a "quasi-equilibrium" state (Leopold and Maddock 1953; Wolman 1955) because the flow variables are mutually interdependent, meaning that a change in any single parameter requires a response in one or more of the others. The difficulty involved in understanding rivers becomes evident when you consider that discharge and load are in continuous flux, and so all the hydraulic variables must always be adjusting. Obviously a river cannot attain equilibrium as a steady-state condition; thus the term quasi-equilibrium.

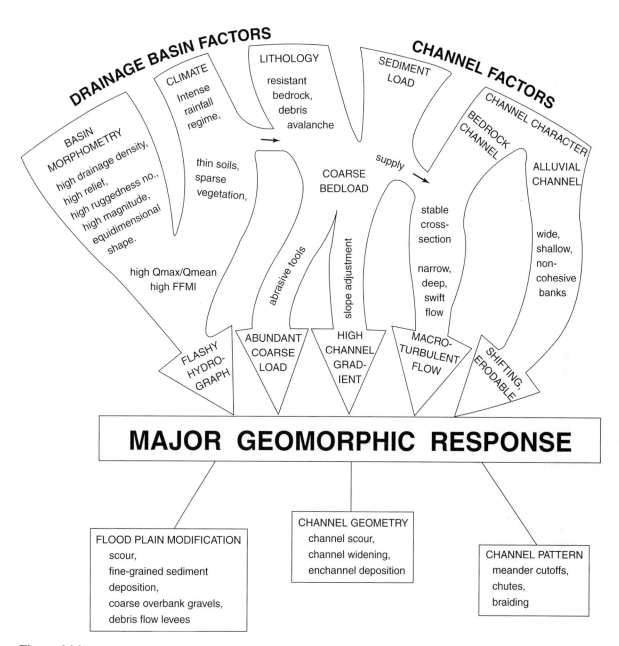

Figure 6.16
Summary of the factors controlling channel and floodplain response to large-magnitude floods.
(From Kochel 1988)

Hydraulic Geometry

The quasi-equilibrium condition was first demonstrated in a landmark study by Leopold and Maddock (1953). Using abundant flow records compiled at gaging stations throughout the western United States, they set out to determine the statistical relationships between discharge and other variables of open channel flow; these relationships are known as *hydraulic geometry* of river channels. Because every river has wide fluctuations in discharge, any given channel cross-section must transport the range of flows that comes to it from the adjacent upstream reach. Discharge, therefore, serves as an independent variable at any station, and the changes in width, depth, velocity, or other variables can be observed over a wide spectrum of discharge conditions (fig. 6.17). At a station each of the factors (w, d, v) increases as a power function such that

$$w = aQ^b$$

$$d = cQ^f$$

$$v = kQ^m$$

where a, c, k, b, f, and m are constants. The exponents b, f, and m indicate the rate of increase in the

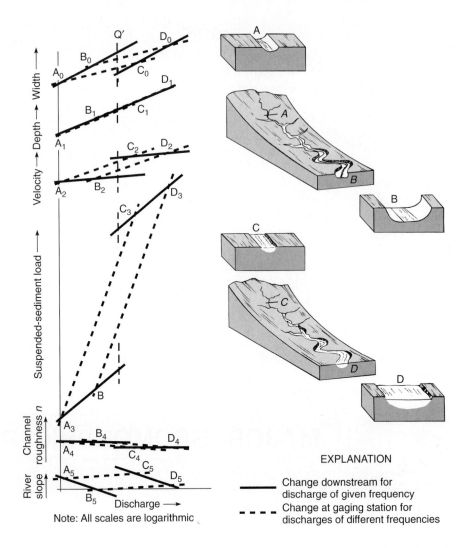

Figure 6.17
Hydraulic geometry relationships of river channels comparing variations of width, depth, velocity, suspended load, roughness, and slope to discharge at a station and downstream.
(Leopold and Maddock 1953)

hydraulic variable (w, d, v) with increasing discharge. Because discharge (Q) equals the product of width, depth, and velocity, the relationship can be expressed as

$$Q = aQ^b \times cQ^f \times kQ^m$$

or

$$Q = ackQ^{b+f+m}$$

and it follows that $(a \cdot c \cdot k)$ and $(b + f + m)$ must equal 1. Leopold and Maddock found that the average at-a-station values of b, f, and m for a large number of midwestern and western streams were 0.26, 0.40, and 0.34, respectively. Essentially the at-a-station exponents tell us what portion of the increase in discharge will be caused by an increase in each of the component variables. It should be recognized, however, that the exponent values represent average values and, thus, will not fit any particular stream. In fact, Phillips (1990) suggested that the proportion of the increase in discharge accounted for by each of the component variables may not even be consistent from one flow to the next.

Discharge also increases with the expansion of drainage area, and so on most rivers it must increase

downstream. The question is how much of the downstream increase in discharge results from width, depth, and velocity. To make this analysis, care must be taken to ensure that the variables are measured during the same flow conditions. On a given day, for example, a disastrous flood with high w, d, and v values may be occurring in an upstream reach whereas flow conditions far downstream are normal. A comparison of the hydraulic variables in these two widely divergent frequencies of flow would be misleading. Obviously the frequency of the discharge must be considered for any observations of downstream hydraulic geometry to be valid.

In sum, then at-a-station and downstream hydraulic geometry differ in that one (at-a-station) compares flows of vastly different frequencies whereas the other (downstream) analyzes variables at the same frequency of Q even though the absolute values of Q differ between downstream stations.

Leopold and Maddock (1953) found that width, depth, and velocity increase downstream with increasing mean annual discharge (fig. 6.17). The average values of b, f, and m for western streams are 0.5, 0.4, 0.1, respectively. In general the rate of change in depth (f) is relatively consistent in both downstream or at-a-station geometry, whereas width usually increases much more rapidly and with more consistent values downstream than at a station (Knighton 1974; Williams 1978). Velocity increases more rapidly at a station than it does downstream.

The suggestion that mean velocity increases downstream came as a shock to most geologists, who intuitively "knew" that water in small tributaries flowing on steep slopes must be traveling faster than water in the low-gradient trunk rivers. Their surprise at this new interpretation of velocity probably resulted from geologists' inclination to consider slope as the major, if not the overriding, control of velocity. Nonetheless, the possibility of a downstream increase in velocity should have been suspected because Manning's equation tells us that depth plays a greater role than slope in determining velocity. In a stream with a constant roughness, increased depth can overcompensate for the loss of velocity resulting from a decrease in slope.

While it is clear that velocity increases downstream along some rivers, the exponents derived for downstream analysis vary from region to region and even for any particular stream within a region.

Carlston (1969), for example demonstrated that on large rivers downstream velocity is probably constant, but on smaller streams it may increase or decrease according to local controls. Thus, the downstream analysis of hydraulic geometry represents the exposition of a general trend which may not be applicable to all rivers. The utility of hydraulic geometry in geomorphic studies has yet to be satisfactorily documented. In fact, Park (1977) found that variations in sets of b-f-m values do not even distinguish between rivers in diverse climates. Nonetheless, Rhodes (1977) argues that hydraulic geometry may provide a relatively simple means of describing local variations in geomorphic process. He believes that all rivers can be categorized on the basis of various ratios of exponent values or other river properties (e.g., Froude number, roughness) that are controlled by the exponential values. The groups proposed by Rhodes reflect basic fluvial mechanics, and this approach suggests that hydraulic geometry should be useful in predicting how any particular river will work. For example, one group includes rivers in which the rate of increase in velocity (m) exceeds the combined changes in width and depth ($b + f$). Such rivers should experience a rapid increase in competence with rising discharge, a condition that is probably needed to entrain coarse bedload (Wilcock 1971).

In a slightly different approach, Williams (1987) was able to define temporal changes in the unit hydraulic geometry (i.e., geometry based on discharge per unit channel width) of selected stream reaches in the western United States. He argued that these changes were the direct result of channel adjustments

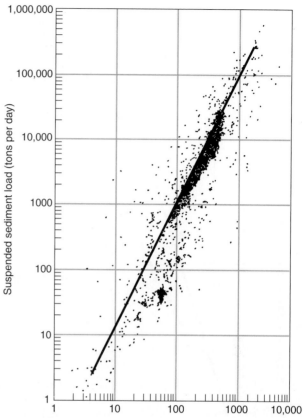

Figure 6.18
Relation of suspended load to discharge in Powder River at Arvada, Wyo.
(Leopold and Maddock 1953)

to natural or anthropogenic disturbances. Thus, the analysis and comparison of hydraulic geometry for different time periods may represent a valuable means of assessing the impact of environmental change on river systems.

In addition to examining the variations in width, depth, and velocity with increasing discharge, considerable attention has been given to the changes in suspended sediment loads as discharges fluctuate at a site. Within most rivers, the amount of suspended sediment at a station increases directly with discharge (fig. 6.18) and can be expressed as the simple power function in which

$$L = pQ^j$$

where L is suspended load and p and j are constants. The at-a-station value of j is often > 1, indicating that the influx of sediment to the river is greater than the addition of water. Interestingly, the dramatic increase in sediment content is not necessarily caused by scouring of the channel floor. Several studies have shown that scouring can occur at peak flow, during rising flow, or in the waning part of the flood (Leopold and Maddock 1953; Foley 1978; Andrews 1979).

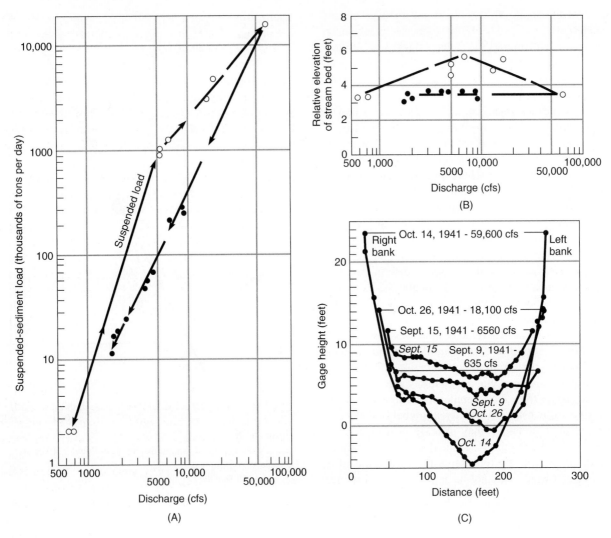

Figure 6.19
Changes in (A) suspended load, (B) streambed elevation, and (C) water-surface elevation with discharge during the
September–December 1941 flood of the San Juan River near Bluff, Utah.
(Leopold and Maddock 1953)

In those situations where scouring occurs at peak
flow or during the recession phase, deposition may
take place in the rising stages of the flood, precisely
when the suspended load is increasing rapidly
(fig. 6.19). Given the channel bed deposition is occurring, the suspended sediment cannot be derived from
the erosion of the channel floor. This observation
means that the bulk of sediment added to a river during
a flood is derived from the valley-side slopes of the
watershed, channel bank erosion, or tributary input.
Moreover, figure 6.19A shows that for a given discharge, the suspended sediment load is greater during
the rising limb of the flood than when the flood waters
are receding. Similar observations have been made for
the relations between sediment concentration (suspended load/unit volume H_2O) and discharge. This
phenomenon, referred to as *hysteresis,* helps explain
the notable variation in suspended load at any given
discharge that is evident in figure 6.18.

Not all rivers exhibit higher suspended sediment loads
during rising flood stages. Williams (1989) has identified
five relationships that may exist between suspended sediment concentration and discharge for any given river
(fig. 6.20). The differences in these relations between
rivers, or even between reaches of the same river, have
been attributed to the interaction of a large number of factors. These include the intensity and areal distribution of
precipitation within the basin, the amount and rate of
runoff, distance of the gaging station from sources of
water and sediment production, differences in the transport
rates between water and sediment, the amount of sediment
stored within and along the channel, and the depletion of
easily eroded debris within the channel or the surrounding
uplands (Williams 1989). Because many of these variables
change seasonally, it is possible that the relationships between sediment load and discharge for an event will also
change during the year (although such trends have not
been adequately documented).

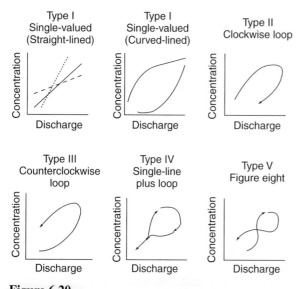

Figure 6.20

Types of suspended sediment concentration-discharge relationships recognized in natural channels.

(Adapted from Williams 1989)

The Influence of Slope

Channel slope has always been recognized as a prime adjustable property of rivers, and there is abundant evidence to substantiate the importance of slope in a river striving to maintain balance. Nevertheless, it is clear that gradient represents only one of the many variables that may be altered to maintain the quasi-equilibrium condition as changes in sediment load and discharge occur. In fact, observations at gaging stations show that the slope of the water surface remains relatively constant during flows of different magnitudes. Therefore, we cannot call on a dramatic increase in slope to produce the relatively high rate of increase in velocity (m). The increasing velocity must be generated by an increase in depth, a decrease in roughness, or both. Downstream the channel gradient does exert an influence, because in most rivers there is a notable decrease in slope. Roughness, however, usually remains fairly constant (Leopold and Maddock 1953) because of the offsetting effects of a decrease in particle size (decreased n) and a decrease in sediment concentration (increased n). As a result, any increase in velocity downstream can best be justified by the increase in depth, explaining the low exponent values of m in that direction.

The relationships between slope and other hydraulic parameters reveal the complexities of quasi-equilibrium, but they do not explain why slope usually decreases downstream or what external factors may control the form of the longitudinal profile. As early as 1877, G. K. Gilbert concluded that slope was inversely related to discharge, and because Q increases with basin area and stream length, it is axiomatic that slope should decrease downstream. However, in most rivers particle size gen-

erally diminishes downstream, prompting many observers to suggest that channel gradient adjusts to the size of the bed material. Actually both factors are probably involved. Rubey (1952) demonstrated that if channel shape is constant the slope will decrease with (1) a decrease in particle size, (2) a decrease in total load, and (3) an increase in discharge. Rubey concludes that the channel gradient at any point along the river is a function of both sediment and discharge. If Rubey is correct, then slope is dependent, or partially so, on all hydraulic variables because they are also related to discharge.

Many studies subsequently have shown the correctness of Rubey's analysis. In one of these studies, comparing stream profiles in areas of differing geology, Hack (1957) found no consistent correlation between slope and bed-material size when all sample localities from a geologically divergent region were considered together. Only after Hack added a third variable, drainage area, to the analysis did a significant relationship become apparent (fig. 6.21), and slope could then be defined by the equation

$$S = 18(M/A)^{0.6}$$

where M is the median size of the bed material in millimeters, A is area in mi^2, and S is slope in ft/mi. Because basin area can normally be used as an index of discharge (Leopold et al. 1964), Hack's study reinforces Rubey's contention that both Q and sediment are determinants of slope. It does not indicate which factor is the principal determinant; indeed, one would expect the relationship to be defined by different mathematical equations in different physical settings. In addition, there is some evidence to suggest that the factors controlling channel gradients may be scale dependent. Prestegaard (1983b) found that within some gravel bed streams, particle size and bed configuration (topography) were the major determinants of water-surface slope. Particle size was the most important determinant over a range of distances from local (one to three times the channel width) to an entire reach (100–300 m). Bed configuration exerted an influence only on a scale relating to the entire reach length.

The interaction of the factors controlling channel gradient ultimately results in the river's longitudinal profile (change in elevation with increasing length). Within alluvial channels it is generally accepted that a concave-up longitudinal profile is associated with rivers in equilibrium (fig. 6.22). However, the concave-up form is not a necessary requirement of the equilibrium condition (Sinha and Parker 1996), and recent studies have demonstrated that along rivers formed in bedrock, diverse longitudinal profiles can be maintained over graded time (Pazzaglia et al. 1998). In contrast to rivers developed in alluvium, bedrock channels are capable of transporting more sediment than is available. This suggests that for bedrock rivers, sediment may not be the most critical factor controlling the shape of the

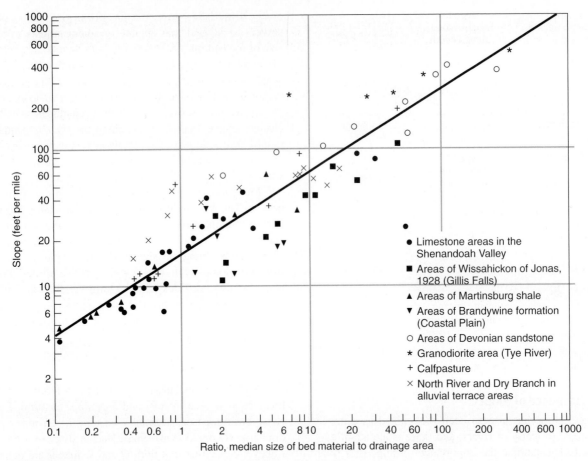

Figure 6.21
Relation between slope and the ratio of median size of bed material to drainage area in selected streams, Maryland and Virginia.
(Hack 1957)

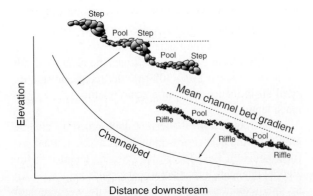

Figure 6.22
Concave-up longitudinal profiles are characteristic of alluvial rivers in an equilibrium state. At the reach scale, channel bottoms commonly exhibit step-pool sequences in high-gradient headwater areas, and pool-riffle sequences along lower-gradient downstream reaches.

longitudinal profile. The profile may instead be adjusted to maintain rates of vertical incision that are equal to the rates of tectonic uplift, while transporting the material delivered to the channel from the adjacent hillslopes. Pazzaglia et al. (1998), for example, studied a number of bedrock channels located in a diversity of tectonic, geologic, and climatic settings (table 6.4). They concluded that in tectonically active environments, uplift led to steep slopes and, because of orographic effects, increased precipitation. The net result of these products was the generation of high peak discharges and stream power that was capable of developing concave-up longitudinal profiles while maintaining rates of incision that were equal and opposite to the rates of rock uplift. In contrast, rivers exhibiting convexities in their longitudinal profile are associated with drainage basins that generate low discharges and stream power, and, therefore, are unable to maintain rates of incision that are equal to

TABLE 6.4 | **Summary of Basin Characteristics and Incision Rates for Selected Bedrock Channels.**

River	Location	Rock-type	Precipitation (cm/yr)[a]	Drainage Area (km²)	Q/Q_p at Mouth (m³/s)[b]	Q/Q_p per Unit Area (m³/s/km²)	Rate of Rock Uplift[c] (m/ka)	Profile Shape	Rate of Channel Incision (m/ka)
Susquehanna	Northeastern United States			62,000	1025/9900	0.016/0.16		convex	
upstream		carb./schist	100				$\sim$1*10^{-2}		6*10^{-3}
middle		schist/gneiss	100				$\sim$5*10^{-3}		1.2*10^{-2}
downstream		schist/gneiss	100				<5*10^{-3}		7*10^{-3}
Jemez	North central New Mexico		20–30	1,200	2.25/51[d]	0.002/0.04		straight to concave	
middle		sandstone					0.15–0.2		0.17
Clearwater	Olypmic Mts., Washington State		150	390	—/600	—/1.54		concave	
upstream		sandstone					0.9		0.8
middle		siltstone					0.6		0.4
downstream		sandstone					0.1		0.1
Mattole	Coastal, Northern California		100	655	36/1100[e]	0.05/1.68		straight to concave	
upstream		mixed sed.					$\sim$2–4		0.7–1.8
downstream		mixed sed.					2–4		$\sim$2
Rio Aranjuez	West coast, Costa Rica	volcanics	$\sim$200	210	—/—	—/—		strongly concave	
upstream							$\sim$0.5		0.1
downstream							$\sim$0.5		0.7
Rio Barranca	West coast, Costa Rica	volcanics	$\sim$250	470	—/—	—/—		strongly concave	
upstream							$\sim$1.5		1.1
downstream							$\sim$1.5		0.5
Rio Narajno	West coast, Costa Rica	forearc clastics	$\sim$350	270	—/—	—/—		strongly concave	
upstream							$\sim$2.5		0.5
downstream							$\sim$2.5		0.2
Texas streams	Southern Texas								
Colorado		carb./coastal Plain seds	variable ~ 75	109,885	84/673[f]	0.0008/0.006	—	concave w/ knickpoint	—
Guadalupe		carb./coastal Plain seds	variable ~ 75	13,522	59/696[f]	0.004/0.05[g]	—	concave	—
Nueces		carb./coastal Plain seds	variable ~ 75	43,340	22/605[h]	0.0005/0.014	—	concave	—

From Pazzaglia et at. (1998).

Notes:

a = Representative basinwide average.

b = Mean annual discharge (mean of daily discharge records over entire period of record) and peak annual discharge (Q_p, annual peaks only) compiled from 1980–1997 gage data.

c = Rates of rock uplift are inferred from geologic and modeling data other than fluvial incision including regional rates of rock exhumation (fission track thermochronology), regional stratigraphic relationships, and geodynamic models.

d = More representative measure at Jemez Pueblo gage, not at mouth.

e = Gage located at Petrolia, CA, 4 km upstream from mouth.

f = Q_p is storm-cell size-limited.

g = The Rio Guadalupe has a huge subsurface drainage that far exceeds the surface divides. This subsurface drainage likely contributes to the large peak discharge per unit area.

h = Represents 1950–1980 data as 1980–1997 data are unavailable or incomplete.

the rates of uplift (table 6.4). These rivers most commonly occur in tectonically stable environments with resistant rock types, or where there is little annual variation or seasonality in precipitation.

Channel Shape

Logic tells us that, unless velocity is completely unrestrained, rivers with a large mean annual discharge have greater cross-sectional areas than streams with smaller average flows. This fact has been verified repeatedly by casual observation and documented by the relationships exposed in hydraulic geometry. However, many rivers with the same mean annual discharge have different cross-sectional areas, and even when the total area is the same, the width-depth ratio may vary considerably. Obviously, factors other than discharge alone influence the shape of river channels.

Schumm (1960) presented cogent arguments to suggest that channel shape, as defined by W/D ratio is determined primarily by the nature of the sediment in the channel perimeter. Where perimeters have a high percentage of silt and clay (particles <0.074 mm), channels tend to be narrow and deep. In contrast, wide, shallow channels seem to be characteristic of rivers having coarse-grained perimeters. Schumm's data, collected from semiarid and arid climate streams, show that W/D ratio is related to the percent silt-clay by the equation

$$F = 255M^{-1.08}$$

where F is the width-depth ratio and M is the percent silt and clay in the channel perimeter. The magnitudes of the mean annual discharge or the mean annual flood do not seem to affect this relationship (Schumm 1971).

The relationship developed by Schumm is partly related to sediment transport mechanics. Bank materials are generally a reflection of the type of sediments transported through a given stream reach. In streams that are not aggrading or degrading, coarse-grained bedload is more efficiently transported in a wide, shallow channel because higher velocities (and steeper velocity gradients) are located near the channel floor (Pickup 1976; Morisawa 1985). In contrast, suspended loads are carried best in channels having lower width-depth ratios (narrow, deep channels). This relationship between the nature of the load in transport and channel shape is so striking that it prompted Schumm (1963a) to classify rivers on the basis of load types (fig. 6.23). More important, it demonstrates another viable method by which a river can be adjusted to alterations in sediment and discharge. A channel reach suddenly burdened with a different type of load than the one it previously carried may as easily alter its channel shape to accommodate the new load as change its gradient by deposition or erosion.

It is also clear that the W/D ratio is influenced by the erosional resistance of the channel banks. Although bank strength cannot be defined by a single parameter, it does depend on the cohesion of the bank material, which can be expressed by the silt and clay content of the channel sediments (Knighton 1987). Recall, however, that alluvial bank deposits often are vertically stratified with fine-grained, cohesive sediments overlying a coarse-grain basal layer. In such cases, bank erosion may be related more to the ability of the stream to remove the basal, noncohesive layer than the corrasion of the fine-grained alluvium (ASCE 1998). In addition, the erosional resistance of the bank may be related to mass wasting processes and have little to do with fluvial mechanics.

Factors other than the character of sediment also can affect the cross-sectional shape of channels (Miller 1990). For example, root systems of riparian vegetation drastically decrease bank erosion and therefore control the channel width (Smith 1976). In contrast, large trees

that fall into a channel may increase bank erosion by diverting flow around a tree jam into an unprotected bank (Keller and Swanson 1979; Keller and Tally 1979).

CHANNEL PATTERNS

The discussion of adjusting mechanics in rivers has thus far centered on the balancing factors that function in a channel cross-section or a series of cross-sections considered in a downstream direction. Rivers also have characteristic plan-view forms that display a distinct geometric pattern over long stretches of their total length. The plan-view pattern that a river adopts is now recognized as another manifestation of channel adjustment to the prevailing conditions of discharge and load.

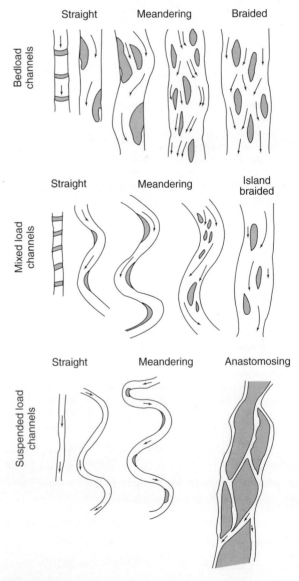

Figure 6.23
Classification of channel patterns presented by Schumm (1981, 1985).

In that sense, channel patterns are another important aspect of river mechanics.

Historically, channel patterns were typically classified as straight, meandering, or braided (Leopold and Wolman 1957). Many geomorphologists now recognize a wider range of river patterns and believe that this traditional classification scheme is insufficient in describing the types of patterns that exist in nature (Knighton 1998). Schumm (1981), for example, combined the original nomenclature (straight, meandering, and braided) with the type of load transported by the river, ultimately recognizing 14 different categories (fig. 6.23). Others have divided stream patterns into two broad groupings including single-channeled (straight and meandering) and multi-channeled (braided and anabranching) systems (Knighton 1998). This latter classification scheme emphasizes the existence of a number of patterns that are distinct from the classical braided configuration in which the flow moves through an interconnected network of channels.

Regardless of the classification system that is utilized, it is clear that the boundaries between types are arbitrary and indistinct. The distinction between straight and meandering, for example, is based on a property called **sinuosity,** which is the ratio of stream length (measured along the center of the channel) to valley length (measured along the axis of the valley) (Schumm 1963b; see Leopold and Wolman 1957 for a slightly different definition). The transition between straight and meandering streams is usually placed at a sinuosity of 1.5, but this value has no particular mechanical significance.

[margin note: value for if a stream is meandering or straight]

The braided pattern, characterized by the division of the river into more than one channel, is more easily discerned. The designation becomes vague, however, when only part of the river's total length is multi-channeled. How much of the river must consist of divided reaches to constitute a braided system is an individual decision. Another complication is that some single channeled rivers become distinctly braided in times of high flow, requiring that stage be considered when deciding on the pattern of classification. In addition, there is no reason to expect that a river will display the same pattern for its entire length. Because patterns are a function of discharge and load, minor variations in those factors downstream may easily generate different patterns, especially in very large watersheds.

Straight Channels

In spite of the fact that a broad range of river patterns is now recognized, the traditional subdivision into straight, meandering, and braided provides a useful framework for examining the processes that are responsible for creating the types of river patterns that exist. We will, therefore, utilize this traditional scheme, starting our discussion with straight channels. We will then conclude our discussion by examining the anabranching pattern, which is now considered a distinct type of plan obtained by some rivers. Most streams do not have straight banks for any significant distance, making the straight pattern a rather uncommon one. It may seem strange, then, that straight streams display many of the same features as the more common meandering rivers. As figure 6.24 illustrates, straight reaches often contain accumulations of bed material, known as *alternate bars,* that are positioned successively downriver on opposite sides of the channel. A line connecting the deepest parts of the channel, called the *thalweg,* migrates back and forth across the bottom. In rivers with a poorly sorted load, the channel floor undulates into alternating shallow zones called *riffles* and deeps called *pools* (figs. 6.22, 6.24). The pools are directly opposite the alternate bars, and the riffles are about midway between two successive pools. Clearly, a straight channel implies neither a uniform streambed nor a straight thalweg, and the spacing of bars, riffles, and pools is closely analogous to that in a meandering channel (Leopold et al. 1964).

Sequences of pools and riffles are very important manifestations of how bedforms, flow, and sediment transport are interrelated in rivers to maintain quasi-equilibrium. This is true regardless of channel pattern, indicating that the tendency to develop bars and/or pool and riffle sequences must be due to some fundamental

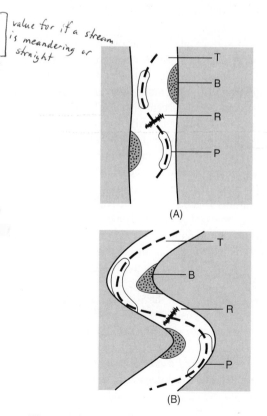

Figure 6.24
Features associated with (A) straight and (B) meandering rivers. T = thalweg, B = bar, R = riffle, P = pool.

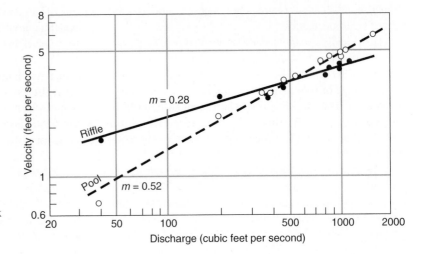

Figure 6.25

Rate of increase in mean velocity over a pool and riffle with increasing discharge. East Fork River, Wyo.

(Andrews 1979)

property of moving water. Pool-riffle sequences have distinct spatial characteristics that are largely independent of the material type forming the channel perimeter, even if the channel is cut in bedrock (Keller and Melhorn 1978). Successive riffles in straight channels are usually spaced about five to seven channel widths from one another and, in alluvial channels, the surface materials are composed of more coarse-grained sediment than the intervening pools. Riffles also tend to be wider and shallower than pools (Richards 1976a).

At low discharge, riffles are characterized by rapid flow and steep water surfaces; in contrast, pool velocities are low and surface gradients are gentle. If we assume that coarse particles cannot be moved under the low flow conditions existing in the pool, the pools should be destroyed by aggradation when sediment moves downstream from adjacent riffles. Consequently, the question arises as to how the riffle-pool sequence is continuously maintained. Keller (1971) argued that as discharge increases, the rate of increase in bottom velocity and mean velocity becomes greater in the pools than in the riffles, a trend that was subsequently observed in other river systems (Richards 1976b; Andrews 1979). Thus, a velocity reversal occurs at some particular discharge above which the pool velocity is greater than the riffle velocity (fig. 6.25). As a result, coarse particles entrained from the riffles during rising discharge can be moved through the pool when discharge exceeds the point of velocity reversal. In the recessional phase, the largest particles are deposited on the riffles while discharge is still above the velocity reversal. Ultimately, fine sediment is eroded from the riffles and trapped in the adjacent pools, leaving a coarse-grained lag on the riffle and a thin, fine-grained deposit in the pool. Thus, pools and riffles seem to be formed and maintained by scouring (pools) and deposition (riffles) under conditions of relatively high discharge. Presumably this discharge occurs with moderate frequency and therefore is probably not associated with overbank flooding.

More recent studies have demonstrated that the changes in velocity with rising discharge over pools and riffles are more complex than originally described by the velocity reversal hypothesis (Carling 1991; Thompson et al. 1996, 1998), and some investigators have questioned its universal application to all river systems (Carling and Wood 1994). One concern, for instance, is the realization that for a constant channel roughness, cross-sectional area must increase faster at the riffles than at the pools as discharge increases. This follows because for any given stage of flow, the discharge moving through a pool-riffle sequence should remain constant and, thus, changes in velocity observed between the pools and riffles should be accommodated by alterations in width and depth of the flow. Field observations indicate, however, that pools along some rivers can exhibit larger cross-section areas than riffles, and should therefore have lower mean velocities (Carling 1991; Carling and Wood 1994). Various hypotheses have been put forth to explain these apparent contradictions to the velocity reversal hypothesis (Carling and Wood 1994; Thompson et al. 1996), but the simple fact is that the competence of the flows within the pools at high discharges must exceed that observed over the riffles in order for the pool-riffle sequence to remain intact.

The pool-riffle distributions observed along a river are somehow related to the fundamental mechanics of open channel flow that produce secondary water movements. The secondary motion occurs as cells circulating in planes that are transverse to the normal downstream component of flow (Leliavsky 1966; Leopold 1982). Although the patterns of flow vary, the direction of cell circulation over the pools is generally opposite of that operating over the riffles (Keller and Melhorn 1973; Richards 1982), and the surface portion of the cells can be noted as zones of converging or diverging flow. In *convergent flow,* secondary surface water flow is toward the axis of the channel, and the maximum velocity occurs near the bottom because isovels in the water are de-

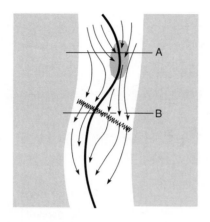
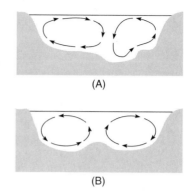

Figure 6.26
Schematic diagram of convergent flow and secondary circulation over a pool (A) and divergent flow and secondary circulation over a riffle (B) in a straight channel.

pressed. This facilitates scour (fig. 6.26). In *divergent flow*, surface water tends to spread outward toward the banks. In this case bottom velocities are retarded and deposition is common (fig. 6.26). Significantly, convergent flow is common over pools and divergent flow is normally located over riffles (fig. 6.26).

In gravel-bed streams with gradients greater than about 2–5 percent, pools and riffles are commonly absent. They may be replaced, however, by a repetitive sequence of steps that are separated by relatively fine-grained pools that form a staircaselike structure (fig. 6.22). Although the steps may be composed of woody debris or bedrock (Grant et al. 1990; Wohl and Grodek 1994; Wohl et al. 1997), they most often consist of tightly interlocking accumulations of cobbles and boulders that are oriented across the channel (Chin 1998; 1999). Flow over the step-pool sequence is characterized by a quasi-systematic pattern of accelerations and decelerations as water falls over a step and moves through the low-gradient plunge pool below. The net result is the considerable loss of energy through turbulent mixing (Whittaker and Jaeggi 1982).

Until very recently, step-pool sequences had received little attention in comparison to their pool-riffle counterparts found along lower-gradient reaches, and their formation remains a topic of considerable debate. The available data indicates, however, that the spacing between successive pools (or step wavelength) is related to channel slope, becoming shorter as channel gradient increases (Grant et al. 1990). Moreover, the spacing between pools appears to consistently occur at distances of one to four channel widths, although there is considerable variability in the data (Grant et al. 1990; Chin 1999). The significance of these relationships is that step-pool sequences appear to form through the orderly scour and erosion of sediment and, thus, represent another component that may be altered to accommodate changes in sediment load and discharge (Chin 1999).

Meandering Channels

The most common river form by far is the meandering pattern (fig. 6.27). Meandering reaches contain the same physical components observed in straight channels (pools, riffles, bars), distributed in a similar way (see fig. 6.24). The thalweg also migrates back and forth across the channel, impinging against the outer bank of the meander bends and crossing to the opposite side near the riffles.

Secondary circulation is a prime property of flow in a meandering system. Presumably, centrifugal force acting on the water as it moves around a meander bend causes a slight elevation in the water surface along the outside of the meander (fig. 6.28). The difference in water elevation across the channel sets up a pressure gradient that gives the flow a circulating motion. The water moves along the surface toward the undercut bank and along the bottom toward the point bar. This corkscrew motion, called *helical flow* or *helicoidal flow,* has traditionally been thought of as a single rotating cell that reaches its greatest velocity slightly downstream from the axes of the meander bends at the position of the pools. More recent studies have suggested, however, that three regions of secondary flow may exist near the meander apex (Markham and Thorne 1992). The midsection of the channel, which may contain as much as 90 percent of the discharge (Markham and Thorne 1992), is characterized by the classical form of helicoidal flow (fig. 6.28). Near the outer bank, the interaction of the water with the bank sediments generates differences in pressure that cause the water to move upward along the bank and then outward away from it (Markham and Thorne 1992). The flow direction of this cell is opposite to that observed in the midchannel areas. Thus, the two rotating cells meet at the surface in a zone of convergence that promotes scouring close to the undercut bank (fig. 6.28). Along the inside of the bend, there is generally a net outward component of flow over the top of the point bar (Dietrich and Smith 1983; Dietrich 1987).

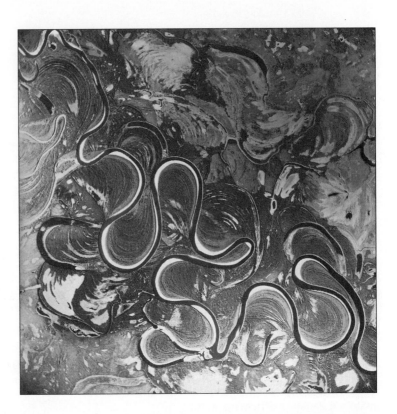

Figure 6.27
Meandering channel in the Arctic plains, Alaska.
(U.S. Geological Survey)

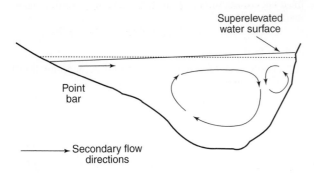

Figure 6.28
Directions of secondary flow at a meander bend.
(From Billi, in DYNAMICS OF GRAVEL-BED RIVERS by A.J. Markham and
C.R. Thorne, Figure 22.2, p. 436. Copyright © 1992. Reproduced by permission
of John Wiley & Sons Limited.)

An examination of the secondary flow cells between successive meanders shows that the circulation patterns are opposite to one another because the orientation of the meanders themselves has been reversed (fig. 6.29). This suggests that the polarity of the rotating cells must change between the meander axis and the next downstream inflection point. The nature of the secondary flow patterns in these transition zones appears, however, to be less consistent than that observed at the meander apex. Indeed, flow patterns observed in different rivers have exhibited different configurations (compare A and B in fig. 6.29)(Hey and Thorne 1975; Thompson 1986; Dietrich 1987; Markham and Thorne 1992). The variability in the nature of the lateral flow between meanders is apparently related to differences in channel planform, W/D ratio, and flow stage (Dietrich 1987; Markham and Thorne 1992).

Meandering rivers shift their positions across the valley bottom by eroding on the outer banks of meander bends and simultaneously depositing point bars on the inside of the bends. Even though the location of the river varies with time, there is no compelling reason to suggest that the shape or hydraulic properties of the river stray far from average values as long as the prevailing controls of climate and tectonics remain unchanged. In fact, meanders in rivers of all sizes are dimensionally similar and have consistent geometric and hydraulic relationships, as shown in table 6.5 and figure 6.30.

The most revealing geometric property is the meander wavelength (λ), which relates to many other variables including discharge, width, and the radius of curvature (r_m). The relationships are in many cases almost linear and undoubtedly reflect basic mechanical principles. For example, using the equations in table 6.5 and considering them as linear, we find that where all parameters are measured in feet,

$$r_m/w = \frac{\dfrac{\lambda}{4.7}}{\dfrac{\lambda}{10.9}} = 10.9/4.7 = 2.3$$

Actual measurement of this ratio shows most rivers to have values between 2 and 3 (Leopold and Wolman 1960), suggesting that the relationship probably is a function of channel curvature exerting an influence on

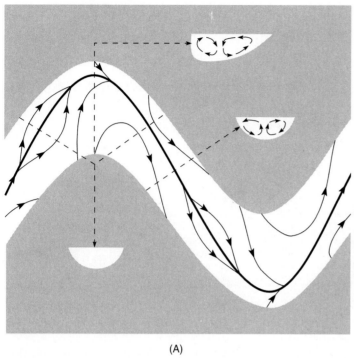

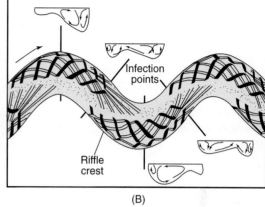

(A)

(B)

Figure 6.29

Models showing secondary flow cells and surface flow lines in a meandering stream.

(A): (From "Secondary Flows in River Channels" by D. Hey and C. Thorne in AREA, Vol. 7, p. 193. Copyright © 1975 Royal Geographical Society (with the Institute of British Geographers). Reprinted by permission of Blackwell Publishers Journals.)

(B): (Figure 4, p. 636 by A. Thompson in "Secondary Flows and the Pool-Riffle" from EARTH SURFACE PROCESSES AND LANDFORMS, Vol. 11, 1986, pp. 631–641. Reproduced by permission of John Wiley & Sons Limited.)

TABLE 6.5	**Empirical Relationships Between Parameters that Define Meander Geometry.**

Dependent Relationship	**Source**
Wavelength	
$\lambda = 6.6\, w^{0.99}$	Inglis (1949)
$\lambda = 10.9\, w^{1.01}$	Leopold and Wolman (1957)
$\lambda = 4.7\, r_m^{0.98}$	Leopold and Wolman (1957)
$\lambda = 30\, Q_{bf}^{0.5}$	Dury (1965)
Amplitude	
$A = 18.6\, w^{0.99}$	Inglis (1949)
$A = 10.9\, w^{1.04}$	Inglis (1949)
$A = 2.7\, w^{1.1}$	Leopold and Wolman (1957)

λ = wavelength (ft)
w = width (ft) at bankfull stage
r_m = radius of curvature (ft)

A = amplitude (ft)
Q_{bf} = bankfull discharge (cfs)

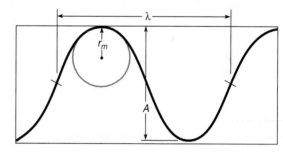

Figure 6.30

Geometric parameters of a meander: λ = wavelength, A = amplitude, r_m = radius of curvature.

flow. Bagnold (1960) showed that as the radius of curvature decreases (decreasing r_m/w), the main filament of flow tends to shift toward the outer bank, causing a concomitant decrease in resistance on the inside of the bed. Greater curvature will continue to decrease resistance until a critical value of r_m/w is attained, when flow along the inner bed becomes unstable and breaks away from the boundary (fig 6.31). These conditions create eddy currents along the inside boundary, increasing the energy dissipation and so effectively establishing a minimum resistance for the flow. In most fluid systems, eddying begins when the curvature ratio is between 2 and 3, suggesting that the large number of real meanders having these values probably represents a quasi-equilibrium between flow and geometry. This suggestion is supported by the flume experiments of Hooke (1975). He argued that when r_m becomes too large, shear stress along the outer bank will cause erosion to increase along the upstream limb of the meander, thereby leading to an increase in the curvature. When r_m is too small, changes

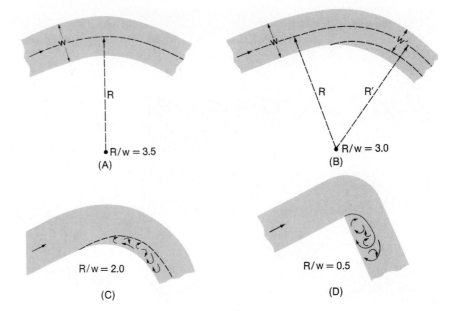

Figure 6.31

Relation between radius of curvature, width, and flow properties. Decreasing ratio r_m/w in (B) causes flow to break away from inside of bend and create eddying zone (C and D). $R = r_m$.

(After Bagnold 1960)

in shear stress cause the downstream limb to migrate more rapidly, ultimately producing a less arcuate bend.

In addition to controlling the erosional and depositional patterns observed along meanders, there is some evidence to suggest that bend geometry may affect that rate of meander migration. The maximum rates of migration correspond to r_m/w ration of between approximately 2 and 3, although there is a significant amount of scatter in the data (Hickin 1974; Nanson and Hickin 1983, 1986; Hooke 1987).

Before the recent developments in fluvial geomorphology, most geologists accepted the premise that random diversions of flow by slumped boulders or fallen trees were the ultimate causes of meanders. Such aberrations undoubtedly can start meanders, and once an initial bend develops, the sinuous nature is transmitted downstream and causes more bends to form (Friedkin 1945). Callander (1969) showed that any deflection of downstream flowlines on the river surface creates hydraulic instability because secondary flow is initiated. Sediment on the channel floor begins to move toward the inner bank. This results in the formation of an incipient bar, thalweg migration, and scouring of the bottom which produces an incipient pool.

Although it may be tempting to write off meandering as the result of random perturbations of flow direction, the change from straight to meandering is dependent on other factors operating within the system. First, the meandering pattern will not develop without bank erosion. Moreover, bank erosion must be a local phenomenon rather than the widespread retreat of both banks, a process that would lead to the creation of a wide, shallow, and straight channel. Second, the development of a meandering pattern requires certain energy

adjustments within the flow. Bank erosion and flow around curves dissipate energy. Therefore, the transition from a straight to a meandering pattern implies that energy utilization within the river must change. Most analyses of river energy indicate that meandering streams are probably closer to an equilibrium condition than straight streams because (1) meandering tends to dissipate energy in equal amounts along the length of the channel, and (2) under constraints of (1), meandering tends to minimize the total energy expenditure (to do the least work) or the rate of energy consumption. The efficient utilization of energy, as suggested above, is accomplished by adjusting the bend geometry or gradient of the channel. Third, bedload transport is required in alluvial channels to redistribute the material eroded from the channel banks and create the sequence of bars, pools, and riffles associated with the meandering pattern. This last requirement is supported by field and laboratory observations that show that the development of incipient meanders is closely associated with the formation of a sinuous thalweg as well as the construction of pools and bars (Rhoads and Welford 1991).

A large number of hypotheses have been put forth to explain the development of the meandering pattern. Rhoads and Welford (1991) note that these hypotheses are generally based on the assumption that meandering results from periodic oscillations in the flow, and that they can be classified into two groups: (1) those that consider flow oscillations as an inherent property of turbulent flow in open channels, and (2) those that view the oscillations as the product of water and sediment interactions along the channel perimeter. Currently, however, there is no single, satisfactory explanation of why meanders form, or how the meandering pattern develops.

Figure 6.32
Braided channel of the Robertson River, Alaska.
(Photo courtesy of D. Germanoski)

Braided Channels

A basic part of the braided pattern is the division of a single trunk channel into a network of branches (fig. 6.32) and the growth and stabilization of intervening islands, often called *braid bars*. Detailed studies of braided systems in both flumes and rivers (Leopold and Wolman 1957; Fahnestock 1963; Church 1972; Rust 1972; N. D. Smith 1970, 1974) show that divided reaches have different channel properties than adjacent undivided segments. Braided zones are usually steeper and shallower, total width is greater although each channel may be narrower than the undivided trunk, and changes in channel positions and the total number of channels are likely to be extremely rapid (Fahnestock 1963; Church 1972; N.D. Smith 1974). Fahnestock, for example, documented lateral shifting of channels up to 122 meters in eight days in the braided segment of the White River in Washington.

Any explanation of the origin of braids is necessarily oversimplified because it involves the simultaneous interaction of a number of factors. According to Fahnestock (1963), the most important of these are the following:

1. *Erodible banks.* Most investigators of channel patterns believe that bank erosion is perhaps the most necessary factor in creating a braided system. If bank erosion is prohibited by material cohesiveness or vegetation, it is unlikely that a braided pattern will develop.
2. *Sediment transport and abundant load.* Almost every braided river transports large volumes of bedload, and much of the channel shifting is prompted by temporary deposition of sediment in bars. It is incorrect, however, to assume that braided rivers are overloaded, because in many cases the braided pattern can be maintained when the channel is actively being eroded (Leopold and Wolman 1957; Germanoski and Schumm 1993).
3. *Rapid and frequent variations in* Q. Fluctuations in discharge tend to increase the magnitude of the alternating patterns of erosion and deposition that seem to be a necessary part of braiding mechanics. However, laboratory studies (e.g., Ashmore 1991; Germanoski 1989) have produced braids under constant discharge, indicating that discharge may sustain the pattern but not cause it.

Although slope and other channel properties of braided streams are different from those of meanders, they probably are not factors in the origin of braids. More likely, they represent the geometric modifications brought about by particular sediment and discharge requirements. For example, Parker (1976) emphasized that slope and the width-depth ratio are important manifestations of the controls leading to braiding versus meandering. If slope and *W/D* are high under conditions of dominant discharge, the pattern will probably be braided. In addition, braids result from a river's tendency to form bars (Parker 1976), but the type of bars developed may be a significant factor in maintaining a braided pattern (Church and Jones 1982). Therefore, braids do not necessarily connote instability. The pattern simply represents another condition a river may establish in response to external controls. It may be maintained for a long period of time and possibly is as close to true equilibrium as the meandering pattern.

Historically, the development of braided patterns *[Central Bar Theory]* was thought to follow a distinct sequence of events (fig. 6.33) first observed during laboratory studies by Leopold and Wolman (1957). During high flow a portion of the coarse load being transported is deposited near the center of the channel because of some local change in transport conditions. This initial accumulation becomes the locus of an incipient longitudinal bar because reentrainment of the particles requires a greater velocity than did their transportation and deposition (see fig. 6.7). Continued deposition allows the bar to grow both upward and in the downstream direction. As particles move across the reach, they are deposited on the lower end of the bar where depth suddenly increases and velocity decreases. Most smaller particles move easily over the growing bar, but some may be trapped in the interstices between the larger grains.

As the expanding bar begins to occupy a significant portion of the channel area, the channel is no longer wide enough to contain the total flow. Flow is then deflected around the bar and the banks are eroded. Simultaneously, the bar itself may be trimmed and the channel somewhat deepened. These processes combine effectively to enlarge the channel on both sides of the bar, allowing the water level to be lower at any equivalent discharge. Eventually the bar emerges as an island flanked

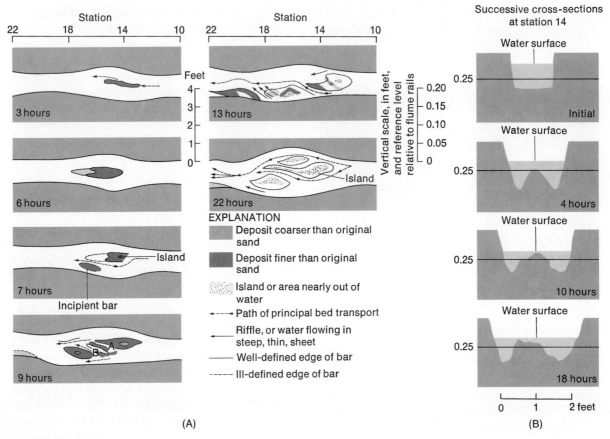

Figure 6.33
Stages in the development of a braid in a flume channel. (A) Sequential development of the pattern at various times. (B) Cross-sections at one station along the flume.
(Leopold and Wolman 1957)

by two distinct channel branches. Bars and islands do not necessarily remain fixed in position or shape, because they are also susceptible to later erosion (N. D. Smith 1974).

There is no doubt that the braided pattern may develop through the processes described above, which are typically referred to as *central bar theory*. Nonetheless, it is now clear that there is more than one mechanism of braided pattern formation (Rundle 1985; Ashmore 1991; Germanoski 1989, 1990; Ferguson 1993). One of the more common processes is through the formation and subsequent dissection of migrating bedforms that stall along the channel floor. This process has been described in a number of ways, including transverse bar conversion (Ashmore 1991; Ferguson 1993), the dissection and accretion of stalled linguoid bars (Germanoski 1990), and lobate bar conversion (Germanoski 2000). Regardless of the terminology used, the mechanism is envisioned as a dynamic process resulting from the interaction of water, sediment, and bedforms, which exert a continuous influence on each other. While the exact morphology of the bedforms may vary, they commonly assume a tongue-shaped geometry described as a lobate (or transverse unit) bar (fig. 6.34). These bedforms are characterized by an axial depression that is created by flow convergence and bed scour. Further downstream, the flow diverges over the bar, and bedload is either deposited near the bar crest as flow depths and shear stress are reduced, or rolls down the avalanche face where the sediment is carried away by flow in the surrounding channels (fig. 6.35). Thus, the interaction of the bar and fluid produces a pattern of flow convergence and scour followed downstream by divergence, deposition, and sediment avalanching that allows the bar to retain its overall shape while migrating along the channel as a coherent entity (Germanoski 2000). During bar migration it is not uncommon for one or both flanks of the lobate bar to build in elevation as sediments are deposited near the crest of the avalanche face. Continued deposition shifts flow away from the elevated region until the entire lobate bar stalls, after which one or both flanks of the bar are dissected by the deflected flow (fig. 6.35). The bedform then acts as a nucleus of deposition as other bar forms stall adjacent to the original feature and are partly an-

Figure 6.34
Lobate Bar, Sand Creek, Colo.
(Photo courtesy of D. Germanoski)

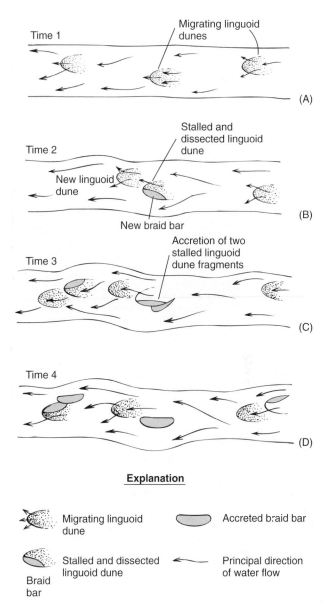

Explanation

Figure 6.35
Stages in the development of a braid associated with the migration of transverse unit bars (linguoid dunes). Migrating linguoid dunes (A) stall and are dissected by flow, creating a braid bar (B). Subsequently, other bar forms stall adjacent to the original feature and are annealed to it (C). The braid bars are streamlined by erosion and by the deposition of grains along the bar margins (D).
(Germanoski 1989)

nealed to it. Eventually, the accumulated bar forms are streamlined by erosion and by the grain-by-grain deposition of bedload along the bar margins (fig. 6.35).

Laboratory experiments clearly show that the formation of braid bars by lobate bar conversion can occur under constant discharge conditions (Ashmore 1991; Germanoski 1990). There is little question, however, that bar dissection is facilitated by decreasing discharge. This may explain the common occurrence of braiding by lobate bar conversion in many glacial outwash channels, which are typically characterized by diurnal fluctuations in flow magnitude.

Other mechanisms of braid bar formation have also been described, including the erosion, through chute-cutoff processes, of point or alternate bars (Ashmore 1991), and the reoccupation of abandoned channels (Eynon and Walker 1974; Germanoski 2000). While one mechanism of braiding may dominate a given river, it is important to recognize that several of these processes may simultaneously operate along a braided reach at any given time.

Anastomosing and Anabranching Channels

Historically, the terms *braided* and *anastomosing* were used interchangeably. It is now well accepted that braided and anastomosing rivers are characterized by distinctly different patterns. As described above, braided rivers exhibit the separation and convergence of flow around rapidly evolving bars. In contrast, anastomosing rivers are characterized by flow through an interconnected network of channels that are separated by islands of relatively stable materials. Moreover, the classical anastomosing rivers described in the earlier literature exhibited channels with low gradients, low W/D ratios, and fine-grained bed and bank sediments, all of which differ from the characteristics of the braided stream (Smith and Smith 1976; Smith 1986).

Anastomosing rivers are rather uncommon, and until recently, their formation and evolution received little attention. It has become clear, however, that not all of the rivers exhibiting flow through an interconnected network of channels occur in low-gradient environments, or are composed of fine-grained sediments (Miller 1991; Knighton and Nanson 1993). These rivers are now referred to as anabranching systems to distinguish them

Figure 6.36
Anastomosing channel of the Carson River, Nevada. Stable floodplain sediments are located between
the individual channel branches.
(Jerry Miller)

from the classical anastomosing channels previously described in the literature. The anabranching pattern has been defined by Nanson and Knighton (1996) as an interconnected network of channels separated by relatively stable alluvial islands that allow flows to be divided and combined at discharges up to bankfull (fig. 6.36). Nanson and Knighton (1996) delineated six types of anabranching systems, one of which represents the classical anastomosing pattern (table 6.6). They are found in nearly every climatic regime and range from low-gradient suspended load to high-gradient gravel and boulder dominated rivers.

The mechanisms through which the interconnected network of channels is produced have yet to be fully understood, and are likely to vary as a function of the tectonic, climatic, and geologic setting. Nonetheless, the formation of individual channels appears to be related to at least two fundamentally different processes (Nanson and Knighton 1996). First, channels may be formed by avulsion during which the local occurrence of overbank flow cuts a new trench into the existing floodplain deposits or scours out and reoccupies an abandoned channel. Second, a network of channels may be produced when deposition forms an enchannel ridge that diverts flow into two directions, or during channel extension associated with delta progradation. The development of the anabranching network by both processes appears to be promoted by (1) stable, cohesive banks that limit channel widening, (2) a flood-prone hydrologic regime, and (3) one or more

mechanisms (e.g., channel sedimentation, or the occurrence of vegetation- or ice-jams) that promote localized overbank flooding (Knighton and Nanson 1993; Nanson 1993; Nanson and Knighton 1996).

The association of these multichanneled rivers with aggradation and avulsion suggests that they represent systems in a state of disequilibrium. There is little question, however, that anabranching rivers can exist for long periods of time (Knighton and Nanson 1993). Moreover, Nanson and Knighton (1996) argue that the anabranching pattern may possess a fundamental advantage over other patterns in environments where there is little or no opportunity to increase gradient. In these environments, the development of a semipermanent network of channels allows for the concentration of stream flow and, therefore, the maximization of bedload transport. Thus, anabranching may represent a form that provides for the most efficient movement of water and sediment through the reach.

The Continuity of Channel Patterns

It is generally agreed that channel patterns form a continuum of plan view morphologies rather than a distinct set of pattern types. Nevertheless, distinct differences in pattern exist and channel reaches can be categorized according to the straight, meandering, braided, and anabranching classification. It follows, then, that if channel patterns reflect specific adjustments to a set of

TABLE 6.6	Types of Anabranching Rivers.

Type of Channel	Description
Type I *Cohesive-sediment anabranching rivers (classical anastomosing pattern)*	Characterized by channels with low gradients, fine-grained, cohesive banks, and low w/d ratios; stream power is very low and channels commonly exhibit aggadational regime that leads to avulsion.
Type II *Sand-dominated, island-forming anabranching rivers*	Similar to type I channels, but possess less cohesive bank sediments; bank stability is related to low stream powers and stabilizing vegetation.
Type III *Mixed-load, laterally active anabranching rivers*	Transport a mixed load of sand, mud, and occasionally, fine gravel; individual channels commonly migrate across floodplain.
Type IV *Sand-dominated, ridge forming anabranching rivers*	Characterized by subparallel channels separated by narrow, steep-sided sand ridges that are stabilized by vegetation; strongly dominated by sand deposition; moderate stream powers.
Type V *Gravel-dominated, laterally active anabranching rivers*	Gravel bed channels characterized by high stream powers that promote lateral migration; anabranching largely produced by avulsion channels that incise into floodplain deposits, or the vertical growth of bars stablized by vegetation.
Type VI *Gravel-dominated, stable anabranching rivers*	Gravel bed channels with steep gradients, and high stream powers; banks stabilized by vegetation; mechanisms of anabranch formation similar to type V rivers.

Source: Nanson and Knighton (1996).

controlling variables, the boundaries between the patterns should be definable in terms of those variables. Furthermore, each pattern must be stable within a certain range of conditions set by the controlling factors. When this range of stability is exceeded, a viable fluvial response would be a pattern change. Patterns, then, may be ephemeral fluvial properties, especially in segments where the values of the controlling factors are critically close to the threshold condition.

Perhaps the most precisely, although empirically, defined transition is between the meandering and braided patterns. Several studies have suggested a threshold boundary between these two forms based on the relationship between slope and discharge (Lane 1957; Leopold and Wolman 1957; Ackers and Charlton 1971). Although the limiting values are not consistently the same (fig. 6.37), at the threshold an increase in slope at any given discharge (or an increase in discharge at any given slope) will change a meandering pattern to a braided pattern. The components that define this threshold (slope and discharge) are a reflection of stream power ($\omega = \gamma QS$) and, therefore, flow strength. Nanson and Knighton (1996) demonstrated that slope-discharge relations can also be used to differentiate different types of anabranching rivers.

Other factors also influence the change in channel patterns. For example, pattern is strongly influenced by resistivity of the banks to erosion (Ferguson 1987; Knighton and Nanson 1993). If the banks are highly resistant and erosion is limited, channels are likely to re-

main straight. Meandering streams require localized erosion, and braiding occurs only in areas of extensive erosion. It has also been argued (Ferguson 1987; Knighton and Nanson 1993) that the transition between patterns is related to sediment size as well as the relative sediment supply; that is, the load supplied to the reach relative to the river's ability to transport the sediment downstream. This suggestion is consistent with the conclusions of Schumm and Khan (1972), who defined a threshold between patterns by a slope-sediment load relationship.

In sum, then, the combined effects of flow strength, bank erodibility, and sediment type and supply delimit the stability range for any channel pattern. Slope is probably not the inducing agent in pattern change but more likely adjusts as a dependent variable, along with the pattern, to changes in the controlling factors.

RIVERS, EQUILIBRIUM, AND TIME

The physical operations within rivers are driven by their tendency to establish and maintain the most efficient conditions for transporting water and sediment. Because every river has a unique combination of these two factors, the parameters that define the equilibrium state must differ from river to river and even in various segments of the same river. In addition, sediment load and discharge are not constant, and thus the element of time

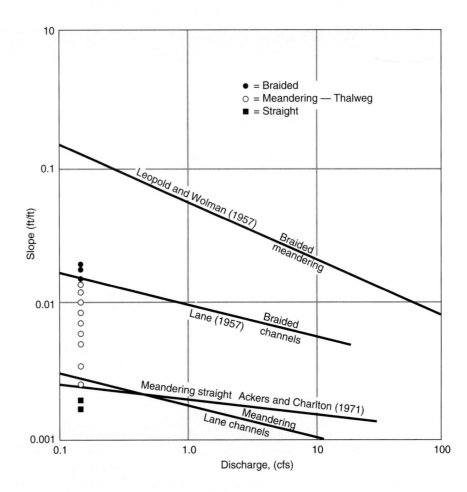

Figure 6.37
Relation of slope and discharge.
Lines represent threshold slopes at
various discharges as determined in
different studies.
(Schumm and Kahn 1972)

becomes a significant factor in any consideration of fluvial mechanics. Remember that discharge and load are
not, in themselves, independent variables because they
are ultimately a function of climate, geology, and tectonics (see chapter 5). Furthermore, discharge and load
commonly change concurrently because both factors are
related to the same, more basic, controls. A change in
climate, for example, will surely alter discharge, but it
may also prompt a simultaneous change in the character
of the load because vegetation and weathering will likewise adjust to the new climatic regime.

The type of fluvial variable most likely to act to
maintain equilibrium depends on the time span being
considered. Figure 6.38 shows a gradual increase in
mean annual discharge provided to a river over a period
of several hundred years. Presumably the load characteristics are also changing for the reasons stated above.
Within the period, normal variations in discharge and
load are accommodated by instantaneous adjustments of
hydraulic geometry. Major floods may occasionally alter
the valley topography or divert the river to a new position, but the channel itself will reorganize (recover) according to the average flow and load conditions. At
some point, however, the gradually changing mean values of load and discharge can no longer be balanced by
hydraulic variables under the prevailing channel config-

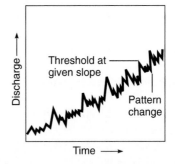

Figure 6.38
Diagram showing gradually increasing discharge with time.
Discharge variations are accommodated by variables of
hydraulic geometry until threshold is reached. At threshold, a
major change in the river character, such as a pattern change,
is required to carry the increased average discharge.

uration or pattern. One flow event, perhaps not even a
major flood, will eventually exceed the stability limits of
the original channel morphology; a fluvial threshold is
passed and a major rearrangement of the channel pattern
or its configuration must take place.

During this long time interval the river was in a perpetual quasi-equilibrium condition, if one considers
only the instantaneous responses of hydraulic geometry

(*w, d, v,* sediment concentration, roughness) to short-lived events such as floods. During the same interval, however, the river was *approaching* a different equilibrium condition that required a particular channel morphology or pattern to balance the new mean values of discharge and load.

Recognizing the importance of different geomorphic time spans, Schumm and Lichty (1965) argued that some variables of fluvial geomorphology are dependent or independent according to the time span being considered. Channel morphology, for example, is an independent variable during steady time and exerts a direct control on the hydraulics of flow. During graded time channel morphology is a dependent variable, adjusted to the conditions of discharge and load.

Our interest here is in the adjustments a river might make to counterbalance changes in discharge and load that occur over a period of years to hundreds of years, the time interval known as graded time (Schumm and Lichty 1965). This is consistent with our suggestion in chapter 1 that thresholds should be defined as graded-time phenomena. Channel morphology is the main dependent variable on this temporal scale and is largely determined by mean values of the controlling factors.

The idea of channel adjustment over a period of years has its roots in the concept of a graded river. Mackin (1948) provided the first clear definition of a graded river as

> one in which, over a period of years, slope is delicately adjusted to provide, with available discharge and with prevailing channel characteristics, just the velocity required for the transportation of the load supplied from the drainage basin. The graded stream is a system in equilibrium; its diagnostic characteristic is that any change in any of the controlling factors will cause a displacement of the equilibrium in a direction that will tend to absorb the effect of the change.

The concept of grade as an equilibrium condition is valuable in understanding fluvial mechanics even though it probably overstates the role of slope. Every altered load condition does not have to be countered with a modification of declivity. This has tempted geomorphologists to consider the possibility that all rivers flowing in alluvial channels are graded, in that they adjust their slope or other characteristics to transport their loads. In this sense, graded becomes analogous to quasi-equilibrium, and the graded condition represents the most probable state for the channel configuration and flow properties (Langbein and Leopold 1964).

In spite of the advantages of equating the graded condition with quasi-equilibrium, we must remember that slope in a graded river adjusts "over a period of years"; this may be the true distinction of a graded river (Knox 1976). For example, actively downcutting streams are still in quasi-equilibrium as defined by their hydraulic geometry; that is, the hydraulic variables are perfectly adjusted to flow, and their measurement would not indicate that any fluvial response is occurring or, for that matter, that any change requiring a response has occurred.

A significant ongoing objective for many fluvial geomorphologists is to *predict* the nature of the adjustments in rivers to changes in sediment load and discharge induced by either natural processes (e.g., climate change and tectonism) or human activities (e.g., urbanization, dam construction, or channelization). Predictive models of river adjustment are primarily based on observed alterations in channel form to past changes in the environment. In this sense, the uniformitarian principle ingrained in the geological literature is reversed so that the past becomes the key to the present and future (Knighton 1998). Given the number of variables that can be altered, predicting channel adjustments to changes in the controlling factors has proven to be extremely difficult. In fact, some rivers may absorb minor alterations in the sediment load and discharge and remain in a state of equilibrium, while others go through a complete metamorphosis in form. It has even been shown that different segments of the same river have responded differently to similar alterations in the controlling factors (Schumm and Brakenridge 1987). These contrasting responses are at least partly attributed to variations in landscape (or landform) sensitivity.

Sensitivity can be defined as the tendency of a river to respond to an environmental disturbance by attaining a new equilibrium state (Schumm and Brakenridge 1987). It is related to the interaction of the driving and resisting force. Thus, it is controlled by such factors as the erosional resistance of the channel-forming materials, basin relief and morphometry (e.g., drainage density, stream frequency, and basin shape), and watershed hydrology (Gerrard 1993). While sensitivity varies between basins, it also varies within a basin. As a result, asynchronous geomorphic responses may occur because thresholds are crossed under different magnitudes of disturbance. Perhaps the most consistent variations have been observed between headwater and downstream reaches of a given river (Knighton 1998). Headwater areas, with their closer proximity to the sources of water and sediment production, their lower sediment and water storage capacities, and their greater potential energy, tend to be more responsive than downstream reaches (see, for example, Balling and Wells 1990).

When a threshold is crossed, it is typically assumed that the alteration in channel form will initially be rapid and decrease with time until a new equilibrium state is attained. The changes in river morphology will not precisely coincide with the onset of the disturbance because the system requires time to react to the disrupting event, a period referred to as the *reaction time*. It is now clear that the reaction time will vary between system components (Bull 1991), in part because the responses are

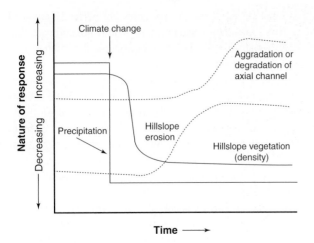

Figure 6.39
Potential responses of selected variables to climate change. The initial response to a decrease in precipitation is a decline in vegetation density. The alteration in vegetation cover leads to increased erosion of hillslope materials, which ultimately results in channel aggradation. Each component in the watershed is characterized by a different response time, and is related to the other variables by means of process linkages.
(Adapted from Bull 1991)

linked through a sequence of cause and effect relationships. For example, upland vegetation may respond rapidly to a change in climate (although it may take a considerable amount of time for it to reach a new equilibrium state)(fig. 6.39). Changes in hillslope erosion may not occur until vegetation has been significantly altered, and adjustments along the river may lag even further behind, responding to the changes in hillslope processes (fig. 6.39). Moreover, alterations in channel form (e.g., width, depth, and roughness) that require minor expenditures of energy and limited transfers in sediment may occur relatively rapidly. In contrast, changes requiring large expenditures of energy (e.g., the transformation of the longitudinal profile) may require long periods of time (fig. 6.40). The significance of these relationships is that the initial adjustments in the river to a change in the environment may not be the final response. Moreover, it illustrates the concept of process linkage in that the responses of biotic, hillslope, and river systems are intimately interconnected to one another.

Adjustment of Gradient

A common response of channel morphology to changes extending over a graded time span is the alteration of slope. In alluvial rivers the normal downstream decreases in gradient promote a concave-up longitudinal

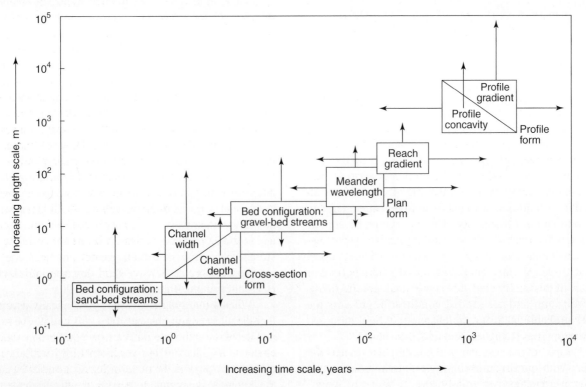

Figure 6.40
Time and spatial scale of adjustment for various channel form parameters in a hypothetical basin of intermediate size.
(From Knigton 1998)

Figure 6.41
Knickpoint developed in bedrock along
Texas Creek, south-central Indiana.
(Jerry Miller)

profile (see fig. 6.22). The concavity, however, is usually not perfectly smooth in detail but is commonly interrupted by perturbations. These can be caused by reaches where the channel is floored by bedrock or by local zones of erosion or deposition. Local filling may be initiated by an influx of bed material load that is too coarse or too great in volume to be transported on the preexisting gradient. For example, incision by upstream tributaries might provide so much additional load that deposition would occur in the downstream reaches. A change in particle size of the load impressed on a trunk river may also occur if tributary basins are experiencing a change in fundamental controls. This is especially common where tributaries drain mountains that are subject to spasms of glaciation or uplift.

If a coarse load is produced, deposition should occur at the confluence of the two rivers until the local gradient is increased sufficiently to allow the bedload to be transported. As the channel floor is raised by deposition, the slope of the river upstream is effectively lowered, and a wave of filling may spread through the channel network. In contrast, a change that produces finer or less load may induce a river to entrench its channel. The most apparent cause of this response is the construction of large dams (Williams and Wolman 1984). These effectively starve the river of bedload in the reaches immediately downstream from the construction, and the river uses its excess energy to lower its gradient by downcutting. The entrenchment, however, may be stopped if the channel floor becomes armored by particles in the alluvium that are too large to be entrained on a lower slope.

Presumably, then, gradients will adjust to counteract changes in the load-discharge relationship. Even if a change in slope occurs, however, there is no requirement that it will be propagated throughout the length of the system. As pointed out earlier, other responses involving the channel cross-section are equally able to move the system toward a new equilibrium condition.

A clear indication that a river gradient is undergoing active readjustment is a short, oversteepened segment of the longitudinal profile known as a knickpoint (fig. 6.41). *Knickpoints* are created by any process that lowers the base level. Such an event causes pronounced channel incision immediately upstream from the site of base-level decline because the river is attempting to establish a new equilibrium condition. It is generally accepted that knickpoints will migrate upstream with time and, in doing so, may even initiate a wave of erosion throughout a river basin. In fact, rates and distances of the headward erosion are amenable to modeling (Pickup 1977) and have been expressed in mathematical terms (Begin et al. 1981). Knickpoint behavior can also be studied experimentally (Brush and Wolman 1960; Begin et al. 1980; Gardner 1983).

Both flume and field data suggest that in channels formed of cohesionless material, pronounced knickpoints will be smoothed out after only a short distance of upstream migration (Ritter et al. 1999). Where the channel is composed of bedrock or cohesive sediments, the knickpoint may retreat for a considerable distance and still preserve a vertical headcut, although Gardner (1983) showed that this does not always occur. Thus, the details of the headward erosion depend on the character

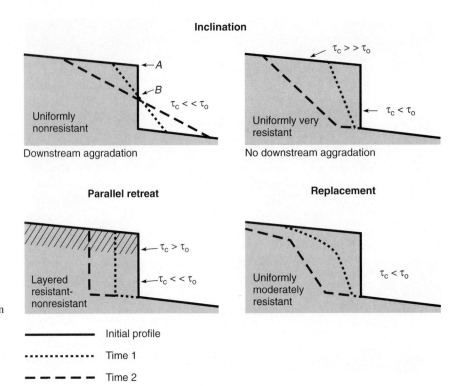

Figure 6.42
Models of knickpoint evolution for various types of bed material. τ_c is critical bottom shear stress needed to initiate erosion. τ_o is actual bottom shear stress. The knickpoint lip, shown at A, is the break in slope where the channel becomes oversteepened. The knickpoint face at B extends from the lip to the base of the knickpoint.
(Gardner 1983)

of the geologic and hydrologic setting. Miller (1991), for example, found that the distribution, shape, and headward migration of bedrock knickpoints along streams in southern Indiana are highly dependent on the stratigraphic and structural characteristics of the rock into which they are cut.

Gardner (1983) suggests that knickpoint evolution can occur in any of three different modes, depending on the balance between the shear resistance of the bed material and the shear stress produced in the river flow (fig. 6.42). In *knickpoint inclination* a uniform change in slope of the knickpoint face occurs, the details of which depend on the material resistance and whether the river can transport the sediment away from the front of the knickpoint. Where the shear stress needed to erode this debris (τ_c) is less than the actual shear stress in the river (τ_o), aggradation will occur and the fill becomes part of the readjusted gradient. *Parallel retreat* is characterized by retreat of the near-vertical knickpoint face without a change in its inclination. This mode of migration is produced best if layering exists in the parent material such that a more resistant zone overlies a less resistant zone. The third type of migration, *knickpoint replacement*, occurs when erosion of the bottom takes place upstream from the knickpoint lip as well as along the face. It results in a knickpoint profile consisting of two distinct zones in which the original slope has been modified.

Adjustment of Shape and Pattern

In the ideal case, any prediction of the adjustments in channel shape would include an understanding of the possible direction of change in the hydraulic variables (width, depth, sinuosity, etc.), the magnitude of the changes, and the rates of alterations in response to a threshold crossing event. Given our current understanding of fluvial processes, it appears that the best we can do is to decipher the potential directions in which the form variables will adjust. The most extensively utilized model for this purpose has been presented by Schumm in his concept of *river metamorphosis*. In a series of papers, Schumm (1965, 1968, 1969) pieced together many of the empirical equations we have examined into a comprehensive, though qualitative, model of possible river adjustments to altered hydrology and load. Four equations are produced when changes in both sediment load and discharge are considered. They are expressed as follows (Schumm 1969):

$$Q_w^+ Q_t^+ = w^+ \, L^+ \, F^+ \, P^- \, S^\pm \, d^\pm$$
$$Q_w^- Q_t^- = w^- \, L^- \, F^- \, P^+ \, S^\pm \, d^\pm$$
$$Q_w^+ Q_t^- = d^+ \, P^+ \, S^- \, F^- \, w^\pm \, L^\pm$$
$$Q_w^+ Q_t^- = S^+ \, F^+ \, d^- \, P^- \, w^\pm \, L^\pm$$

In these equations Q_t is the percentage of the total load transported as bed material load (sand-sized or larger), and Q_w can be either the mean annual discharge or the mean annual flood. The other variables are width (w), depth (d), slope (S), meander wavelength (L), width-depth ratio (F), and sinuosity (P). The plus or minus exponents indicate whether the dimensions of the variables are increasing or decreasing.

To exemplify the use of these equations, assume that a large area is clear-cut of its natural forest cover. We can expect an increase in Q_w, because infiltration rates will be lowered and direct runoff will increase, as well as an increase in Q_t, because coarse sediment normally stabilized on the slopes by roots now makes its way to the channel. The coarse sediment also will be moved more frequently because of the increased peak discharge. With both Q_w and Q_t increasing, we can expect increases in width, wavelength, and W/D and a decrease in sinuosity. Depth and slope may vary in either direction. Slope will probably increase, however, because the channel becomes straighter, and depth will probably be constant or decrease since both w and W/D increase.

It is important to note that the equations presented by Schumm do not specify the magnitude of the changes in any given variable, or even which variables will respond along any given river. Rather, they simply indicate the most likely direction of change if a particular variable were to respond to a disturbance. These types of predictive models may seem archaic given our current access to high-speed computers and the progress that has been made toward modeling climate change, rainfall-runoff processes, and other phenomena. To be sure, a number of more quantitative, numerically intensive models have been put forth and appear promising (see, for example, Chang 1986), but their application to adjustments over long periods of time has largely gone untested. Perhaps the best we can do at this time is to assess how a given river, or group of rivers, has responded in the past, and use this information to predict adjustments in the future.

SUMMARY

River action, like all geomorphic processes, responds according to the driving and resisting forces built into the system. For example, a river will entrain, transport, or deposit sediment depending on the driving energy given to the water by velocity, depth, and slope, and on the amount of that energy consumed by the resistance to flow offered by elements such as channel configuration, particle size, and sediment concentration. The work demanded of a river, the amount of load it must handle under prevailing discharges, is determined by the geological and climatic character of the drainage basin. Each river develops a particular combination of shape, gradient, and hydraulic variables (called the hydraulic geometry) that allows it to accomplish its work most efficiently. The river will attempt to maintain its high efficiency by adjusting the above properties whenever discharge or load varies. Because discharge and load fluctuate continuously, equilibrium as a steady-state condition can never be attained, and the river variables must perpetually be adjusting. Nonetheless, the normal variations of discharge and load are accommodated by hydraulic geometry.

The type of channel pattern (straight, meandering, braided, or anabranching) a river displays and the longitudinal profile are other fluvial characteristics controlled by the basin environment. Each pattern originates in a specific manner, and its geometric form is designed to facilitate the work of a river, measured as the prevailing values of discharge and load. Once established, the pattern will be maintained as long as the normal variations in load and discharge can be absorbed by the mechanics of hydraulic geometry.

Major long-term changes in climate or basin tectonics may alter the average discharge and/or load to a point where adjustments of hydraulic geometry can no longer maintain the most efficient system. When those threshold values of discharge or load are reached, major fluvial responses in the form of pattern changes, degradation or aggradation, or dramatic revisions of the width-depth ratio will occur to reestablish the greatest fluvial efficiency. Our present knowledge, however, does not allow us to predict which of the possible adjustments will occur in response to major changes in the fundamental controls.

SUGGESTED READINGS

The following references provide greater detail concerning the concepts discussed in this chapter.

Ashmore, P. E. 1991. How do gravel-bed rivers braid? *Canadian Jour. Earth Sci.* 28:326–41.

Billi, P., Hey, R. D., Thorne, C. R., and Tacconi, P. 1992. *Dynamics of gravel-bed rivers.* London: John Wiley and Sons.

Hey, R. D., Bathurst, J. C., and Thorne, C. R. 1982. *Gravel-bed rivers.* New York: Wiley-Interscience.

Knighton, D. 1998 *Fluvial forms and processes: A new perspective.* London: Arnold.

Nanson, G. C., and Knighton, A. D. 1996. Anabranching rivers: Their cause, character, and classification. *Earth Surface Processes and Landforms* 21:217–39.

Schumm, S. A. 1969. River metamorphosis. *American Soc. Engrs. Proc., Jour. Hydrology Div.* HY 1:255–73.

Schumm, S. A., Mosley, M. P., and Weaver, W. E. 1987. *Experimental fluvial geomorphology.* New York: John Wiley and Sons.

Tinkler, K. J., and Wohl. E. E., eds. 1998. *Rivers over rock: Fluvial processes in bedrock channels.* Washington, D.C., American Geophysical Monograph 107.

FLUVIAL LANDFORMS

Courtesy of
Nick Lancaster

Introduction
Floodplains
    Deposits and Topography
    The Origin of Floodplains
        The Humid-Temperate Model
        Spatial and Temporal Effects
Fluvial Terraces
    Types and Classification
    The Origin of Terraces
        Depositional Terraces
        Erosional Terraces
    Terrace Origin and the Field Problem
Piedmont Environment: Fans and Pediments
    Alluvial Fans
        Fan Morphology
        Fan Deposits and Origins
    Pediments
        Morphology and Topography
        Processes
        Formative Models
Deltas
    Delta Classification, Morphology, and Deposits
    Delta Evolution
Summary
Suggested Readings

INTRODUCTION

In chapter 6 we examined the basic mechanics of fluvial processes and found that the activity within a stream channel is generally related to the energy possessed by the river and to the ways that energy is utilized to carry water and sediment most efficiently. Rivers, however, are more than natural sluices; they also mold the geologic setting into descernible topographic forms. They accomplish this primarily through the erosional capability inherent in the movement of sediment-laden water, and through the deposition of debris that occurs when the transporting energy is less than the demands being made on it. Some fluvial features are purely erosional; the topographic form is clearly one of sculptured rock, and little, if any, sediment is associated with the feature. Others may be entirely depositional, and the exposed topography is formed by the burial of an underlying surface that existed before the covering sediment was introduced. In these cases, the bedrock framework may have no influence on the surface configuration. Many features spring from some combination of both erosion and deposition; the pure cases are probably end members of a continuum of possible forms.

FLOODPLAINS

Floodplains are perhaps the most ubiquitous of fluvial features. They are found in the valley of every major river and in most tributary valleys. However, a precise definition of a floodplain is more difficult than one might expect. Topographically and geologically, a floodplain is the relatively flat surface occupying much of a valley bottom and is normally underlain by unconsolidated sediment. The sediments of most valley bottoms are not necessarily a function of the river occupying the valley but may be deposited there by a variety of geomorphic processes. Nonetheless, to be considered as part of the floodplain, the surface and the sediments must somehow relate to the activity of the present river. The definition often has a hydrologic connotation, because the floodplain is a surface subject to periodic flooding. The floodplain can easily be defined in terms of hydrology as the water level attained in some particular stage of the river. Detailed analyses (Wolman and Leopold 1957) demonstrate that most *topographic floodplains* formed by perennial rivers in humid climates are subjected to flooding nearly every year or every other year. The recurrence interval of bankfull stage, for example, averages about 1.5, indicating that most rivers leave their channels two out of every three years. If the surface flanking the river has any relief or is produced under different conditions, part or all of the topographic floodplain will not be inundated by the annual or biannual flood that marks the *hydrologic floodplain*.

For the purposes of process geomorphology, it may be beneficial to link the definition of a floodplain, and the classification of floodplain types to parameters that reflect flow mechanics. In that sense, Nanson and Croke (1992, p. 460) have defined a *genetic floodplain* as "the largely horizontally-bedded alluvial landform adjacent to a channel, separated from the channel by banks, and built of sediment transported by the present flow regime." They suggest further that floodplain development essentially reflects some accommodation between stream power (see chapter 6) and sediment character, thereby linking floodplain genesis to factors of basin geology and river mechanics.

Nanson and Croke (1992) also present a floodplain classification based on the relationship between a river's ability to entrain and transport sediment and the erosional resistance of surfaces that form the channel boundary, including the channel banks. Three primary types of floodplains are designated according to the fluvial energy and bank cohesion in the system, and numerous suborders are suggested within the major categories. Figure 7.1 shows examples of floodplain types found in each major group.

Regardless of how it is defined, a floodplain plays a very necessary role in the overall adjustment of a river system. It not only exerts an influence on the hydrology of a basin (e.g., lag) but also serves as a temporary storage bin for sediment eroded from the watershed. Floodplains, therefore, are both the products of the river environment and important functional parts of that system.

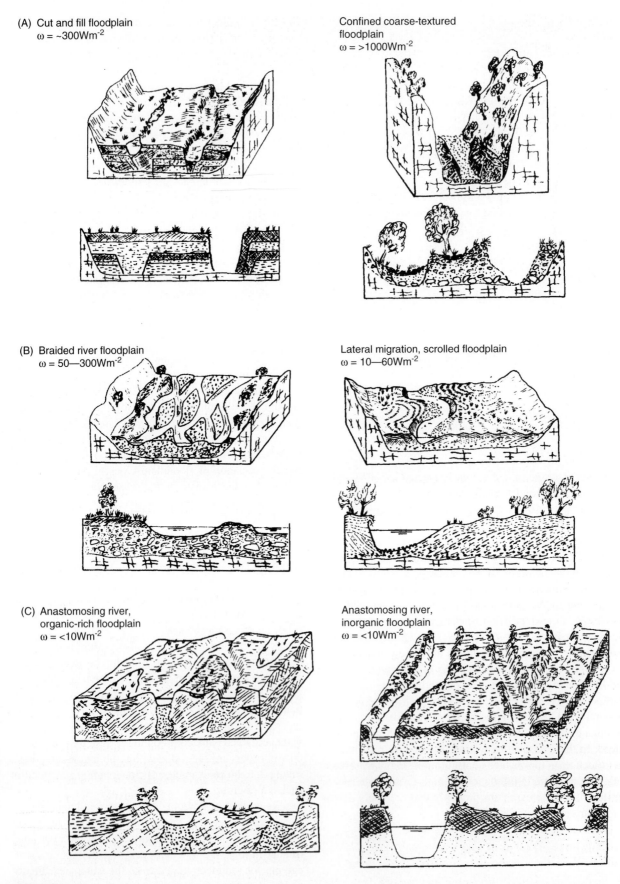

(A) Cut and fill floodplain
$\omega = \sim 300 \text{Wm}^{-2}$

Confined coarse-textured floodplain
$\omega = >1000 \text{Wm}^{-2}$

(B) Braided river floodplain
$\omega = 50\text{---}300 \text{Wm}^{-2}$

Lateral migration, scrolled floodplain
$\omega = 10\text{---}60 \text{Wm}^{-2}$

(C) Anastomosing river, organic-rich floodplain
$\omega = <10 \text{Wm}^{-2}$

Anastomosing river, inorganic floodplain
$\omega = <10 \text{Wm}^{-2}$

Figure 7.1

Examples of floodplain types in the classification of genetic floodplains. (A) High energy, noncohesive; (B) Medium energy, noncohesive; (C) High energy, noncohesive. (After Nanson and Croke 1992.)

(Figures 1-3 from pp. 471, 474, 478 by Nanson and Croke in GEOMORPHOLOGY, Vol. 4, pp. 459–486. Copyright © 1992. Reprinted by permission of Elsevier Science.)

Figure 7.2
Normal sequence of floodplain stratigraphy. Lower half of deposit is coarse-grained, laterally accreted point bar gravel. Gravel in lowest unit dips gently to the right. Upper half of deposit is vertically accreted silt deposited during overbank flow. Sexton Creek, Shawnee National Forest, southern Illinois. The man is pointing to overbank gravel layer deposited during extreme flood event.

Deposits and Topography

Floodplains are composed of a variety of sediments that are created by diverse processes and accumulate in distinct subenvironments within the valley bottom. Most floodplain sediment can be differentiated into deposits of channel fill, channel lag, splays, colluvium, lateral accretion, and vertical accretion. Near the valley sides, *colluvium,* resulting from unconfined wash and mass wasting, may be prominent in the floodplain sequence. Toward the axis of the valley, these deposits grade into alluvial-type deposits. Coarse debris from which the fines have been winnowed are interpreted as *channel lag* deposits, in contrast to *channel fill,* which consists of a poorly sorted admixture of silt, sand, and gravel. *Splay* deposits are composed of material spread onto the floodplain surface through breaks in natural levees and usually are more coarse-grained than the overbank sediments they cover. The most important deposits in the floodplain framework are those of **lateral accretion** and **vertical accretion,** which in some cases can be separated on the basis of particle size. Laterally accreted sediment usually consists of sands and gravels and is normally more coarse-grained than the vertically accreted silts and clays (fig. 7.2). Point bars, however, the most common deposit of lateral accretion, sometimes have the same texture as the overbank sediments (Wolman and Leopold 1957; Nanson 1980). Therefore, particle size is not an infallible criterion for distinguishing between vertical and lateral accumulation.

Most valleys of large rivers are occupied by well-defined floodplains that consist primarily of laterally and vertically accreted deposits. These are usually associated with specific depositional environments. In the Mississippi River valley, for example, floodplain deposits have been broadly categorized as channel types and overbank types, consisting of splays, natural levees, and back-swamps (fig. 7.3A). The surface of the floodplain may have considerable microrelief that reflects the fluvial mechanics in the various depositional environments.

The floodplain surface is most irregular in a zone close to the river where point bars are molded into alternating ridges and swales which Leopold and his colleagues (1964) refer to as *meander scrolls* (see fig. 7.1; 7.3B). The characteristic scroll topography may start as a longitudinal bar with a narrow trough to its rear or simply as a low ridge of sediment that accumulates on the inside of a meander bend during bankfull flow. When the high discharge subsides, the ridge is exposed and rapidly vegetated, becoming, for all practical purposes, the new channel bank. The next high flow repeats the process. As the river shifts across the valley by undercutting on the outer bank, the successive ridges and intervening swales that characterize meander scroll topography develop simultaneously.

It is certain that scroll topography can develop in other ways (i.e., Nanson 1980). Nonetheless, regardless of the precise origin, flow occasionally will break across the point bar surface, often occupying a particular swale and scouring the surface into more pronounced low channels called **chutes.** Wolman and Leopold (1957) measured velocities of up to 1 mps in chute channels, a flow that is capable of eroding the surface and transporting coarse sand. Chute erosion tends to accentuate the scroll topography, and even though the slough areas may be silted in by overbank deposition, the scalloped profile of scroll topography (fig. 7.4) can be preserved for hundreds of years (Hickin and Nanson 1975). Furthermore, the internal depositional framework seems to be similar to that shown in figure 7.1B (Leclerc and Hickin 1997).

In addition to the irregularity caused by scrolls on the inside of meander bends, some topographic relief near the channel results from the formation of **natural**

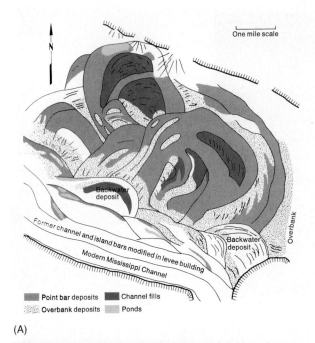

(A)

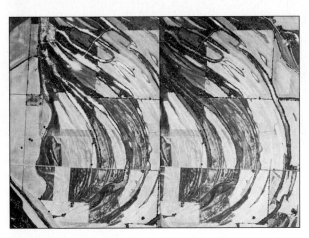

(B)

Figure 7.3
(A) Map showing complex distribution of various types of deposits on a portion of the Mississippi River floodplain near Grand Tower, Ill. (B) Meander scroll topography on abandoned point bar of the Mississippi River in Pike County, Ill. Local relief between ridges and swales is about 3 meters.

levees. Natural levees stand along most major rivers as low ridges that commonly are broader than one might expect, often extending away from the bank for hundreds of meters. Levees are usually highest near the active channel and slope gradually toward the valley sides. They owe their character to the retardation of flow velocity when rivers leave the channel, which causes the largest suspended particles to be deposited adjacent to the bank (Middelkoop and Asselman 1998). Levee deposits are, therefore, more coarse-grained than most other overbank sediment and are accreted more rapidly (Mertes 1994). Kesel and his colleagues (1974) found a net addition of

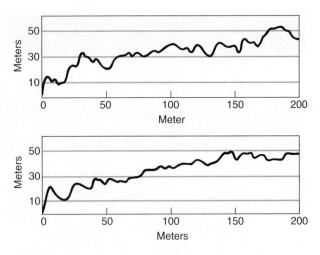

Figure 7.4
Diagram of two profiles of meander scroll topography that has been preserved for considerable periods of time. Beatton River, Canada.
(Hickin and Nanson 1975)

53 cm on natural levees during the two-month overflow of the Mississippi River in 1973, while only 1.1 cm accumulated in the backswamp area during the same period. This phenomenom is most pronounced where the flow velocity in the channel is markedly greater than it is in waters crossing the floodplain. Where floodplain velocities are high and closer to values in the channel flow, lateral decrease in size and thickness of overbank deposits is less dramatic (Wyzga 1999).

As one proceeds away from the river, floodplain topography becomes much more regular, and its flatness is interrupted only by oxbow channels or by splay deposits that have been carried onto the backswamp surface. The zone away from the active river is characterized by gradual accumulation of overbank sediment that tends to subdue any existing relief. Abandoned channels represented by oxbows or oxbow lakes gradually fill with silts and clays, leaving *clay plugs* as the only evidence of the former channel. Old sloughs, chutes, or cutoffs can also fill with overbank sediment, adding to the general reduction of relief away from the channel. Not all overbank sediment is fine-grained, however. Where banks are cohesive, coarse gravel can be transported onto the surface of the floodplain (e.g., McPherson and Rannie 1967; Costa 1974a; Ritter 1975, 1988; Teisseyre 1978; Stene 1980; see fig. 7.5).

In general, the relationship between floodplain deposits and river processes would be straightforward if rivers would stay in the same place for extended periods of time. Actually, as the rates in table 7.1 indicate, most rivers migrate laterally across the valley bottom quite rapidly, forcing the depositional environments also to shift their location with time. A backswamp region, for

Figure 7.5
Large lobe of overbank sand and gravel deposited on the floodplain surface in December 1982 flood of Gasconade River near Mt. Sterling, Mo. Lobe sediment is 1.0 to 2.0 m thick and 100 m wide.

example, may include remnant deposits of a channel. The displacement of one environment by another adds to the complex maze of floodplain deposits, emphasizing the point that floodplains are dynamic rather than static fluvial features (Brakenridge 1984, 1988).

The Origin of Floodplains

The Humid-Temperate Model   It is now generally accepted that two dominant fluvial processes act simultaneously to develop most floodplains, leading to a general model for floodplain construction by meandering rivers in humid-temperate environments. In this model, maximum erosion takes place on the outer bank just downstream from the axis of curvature. At the same time, sediment accumulates in point bars that build up along the inside of the meander bend. Detailed study of these processes has documented that bank erosion and point bar accumulation are volumetrically equal during any given period of lateral and downvalley migration of the meander beds (Wolman and Leopold 1957). In addition, data from the same study show that the point bars tend to increase in height until they reach the level of the older part of the floodplain. A meandering river can shift

its position laterally without changing the channel shape or dimensions.

In channels where coarse sediment is an important part of the load, the point bars tend to collect sediment that is easily distinguished from that of overbank origin. During low flow, sediment of all sizes may be temporarily trapped on the channel floor, but at the peak of bankfull discharges this material will be removed along with any debris eroded from the undercut banks. The coarse sediment is deposited on the point bars, which are now submerged. As figure 7.2 shows, these deposits commonly display cross-beds dipping into the channel. Over a period of years, the point bars expand laterally, spreading progressively across the valley bottom as a thin sheet of sand or gravel (Mackin 1937; Leopold et al. 1964). If the load is not characterized by coarse-grained particles, the spreading of lateral accretion deposits proceeds in exactly the same way (fig. 7.6), but the point bar sediment may be more difficult to distinguish from the overbank materials.

The maximum thickness of laterally accreted deposits is determined by the depth to which a river can scour during recurring floods. Natural channels are

TABLE 7.1	**Rates of Lateral Migration of Rivers in Valleys.**

River and location	Approximate Size of Drainage Area (square miles)	Amount of Movement (feet)	Period of Measurement	Rate of Movement (feet per year)
Tidal creeks in Massachusetts		0	60–75 yr	0
Normal Brook near Terre Haute, Ind.	±1	30	1897–1910	2.3
Watts Branch near Rockville, Md.	4	0–10	1915–55	0–0.25
	4	6	1953–56	2
Rock Creek near Washington, D.C.	7–60	0–20	1915–55	0–0.50
Middle River near Bethlehem Church, near Staunton, Va.	18	25	10–15 yr	2.5
Tributary to Minnesota River near New Ulm, Minn.	10–15	250	1910–38	9
North River, Parnassus quadrangle, Va.	50	410	1834–84	8
Seneca Creek at Dawsonville, Md.	101	0–10	50–100 yr	0–0.20
Laramie River near Ft. Laramie, Wyo.	4,600	100	1851–1954	1
Minnesota River near New Ulm, Minn.	10,000	0	1910–38	0
Ramganga River near Shahabad, India	100,000	2,900	1795–1806	264
	100,000	1,050	1806–1883	14
	100,000	790	1883–1945	13
Colorado River near Needles, Calif.	170,600	20,000	1858–83	800
	170,600	3,000	1883–1903	150
	170,600	4,000	1903–1952	82
	170,600	100	1942–52	10
	170,600	3,800	1903–42	98
Yukon River at Kayukuk River, Alaska	320,000	5,500	170 yr	32
Yukon River at Holy Cross, Alaska	320,000	2,400	1896–1916	120
Kosi River, North Bihar, India		369,000	150 yr	2,460
Missouri River near Peru, Nebr.	350,000	5,000	1883–1903	250
Mississippi River near Rosedale, Miss.	1,100,000	2,380	1930–45	158
	1,100,000	9,500	1881–1913	630

From Wolman and Leopold (1975), USGS Professional Paper 282-C.

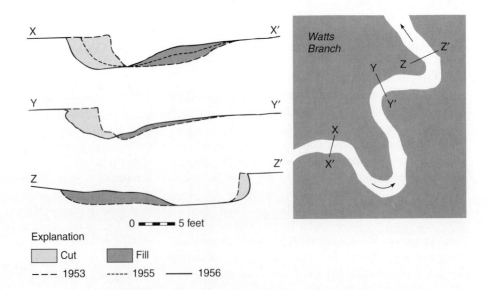

Figure 7.6

Progressive lateral erosion and point bar deposition (cross-sections) in Watts Branch near Rockville, Md.

(Wolman and Leopold 1957)

Explanation

☐ Cut   ☐ Fill

- - - 1953   ----- 1955   ——— 1956

0 ▬▬ 5 feet

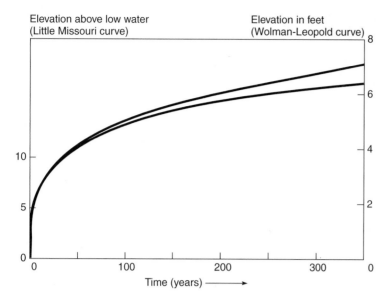

Figure 7.7
Increase in elevation of floodplain with time. Lower curve from empirical data collected on floodplain of the Little Missouri River; upper curve was derived theoretically for Brandywine Creek (Pa.) by Wolman and Leopold (1957). Note different vertical scales.
(Everitt 1968)

probably scoured to a depth 1.75 to 2 times the depth of flow attained during a flood. This rule of thumb generally fits the scouring observed in a variety of perennial rivers (Leopold et al. 1964, p. 229), but where the channel width is prevented from expanding by bridge supports or extremely resistant banks, the scour depth may reach 3 or 4 times the water depth. The thickness of lateral accretion deposits should, therefore, increase gradually downvalley along with the normal increase in discharge and depth noted on most rivers.

The second dominant process in the origin of floodplains is overbank flow. As rivers wander across the valley bottoms, they normally leave the channel confines during periodic flooding and deposit fine-grained sediment on top of the floodplain surface. Because of this, the floodplain is vertically accreted, and in most floodplain stratigraphy a thin layer of overbank silt and clay rests on the laterally accreted point bar deposits described above (fig. 7.2). Assuming the same vertical increment with each flood, the level of a floodplain built entirely by overbank deposition should increase at a progressively decreasing rate, as shown by the curves in figure 7.7. The initial growth would be rapid because flooding would occur frequently, and perhaps 80 to 90 percent of the floodplain construction would take place in the first 50 years (Wolman and Leopold 1957; Everitt 1968; Nanson 1980). However, as the surface grows higher relative to the channel floor, the stage needed to overtop the banks also increases (Moody et al. 1999). The surface is inundated less frequently, and the rate of growth is drastically retarded. Ideally, then, the entire floodplain sequence consists of a relatively thin accumulation of laterally and vertically accreted sediments that have been spread more or less evenly across the valley bottom. The two types of deposits can differ in textural properties (although the process mechanics does not require it), but they will be essentially the same age.

The model of floodplain origin just described raises the question as to which process—lateral migration or overbank flow—plays the dominant role. There is probably no universal answer to this question, because each system obeys its own unique combination of controlling factors. Nonetheless, evidence suggests that most humid-temperate floodplains result primarily from the processes associated with lateral migration. Perhaps the most persuasive support for this conclusion comes from a comparison of the rates involved in the two competing processes. Table 7.2 is a random sampling of sediment increments measured during individual floods and over various intervals of time. Although incomplete, the data show that deposition during a single flood tends to be higher than long-term average values. Nonetheless, some caution should be used when making conclusions based on single-flood accumulation rates because they can vary considerably from one flood to the next (Gomez et al. 1998). For example, the 100-year flood of the Mississippi River occurring during the summer of 1993 deposited less than 4 mm of overbank sediment on the floodplain surface. This accumulation, significantly less than deposition during earlier floods of equal magnitude, indicates that accretion values are subject to many different controls and may be influenced by event sequencing (Gomez et al. 1995).

In contrast to single events and assuming that long-term vertical accretion rates are valid, a 3 mm thick floodplain sequence would probably take thousands of years to accumulate. Therefore, it is instructive to note again the rates of lateral migration given in table 7.1, which indicate that most rivers migrate across their valley bottoms quite rapidly. This suggests that the magnitude of vertical accretion depends primarily on the rate at which the river migrates laterally. The total thickness will approximate the vertical accretion that can be accomplished in the time the river takes to migrate the

TABLE 7.2	Rates of Overbank Deposition in Floodplain Development.

Increment Rates of Overbank Deposition in Major Floods

River Basin	Flood	Average Thickness of Deposition (cm)
Ohio River	Jan.–Feb., 1937	0.24
Connecticut River	March, 1936	3.42
Connecticut River	Sept., 1938	2.16
Kansas River	July, 1951	2.94

Selected Long-term Vertical Accretion Rates

Locality	Period (yr B.P.)	Rates (cm/yr)
Galena Cr., WI[a]	3610–150	0.040
Ti Valley, OK[a]	2350–1850	0.020
	720–320	0.200
Lower Ohio River[a]	7000–6000	0.027
	3000–2000	0.060
Delaware Canyon, OK[a]	2750–1900	0.650
Delaware River, PA[b]	3468–2718	0.115
	5590–5198	0.077

From Wolman and Leopold (1975), USGS Professional Paper 282-C. See this work for references to individual data sources on major flood deposition.

See Orbock Miller et al. (1993 [a]) and Ritter et al. (1973 [b]) for references to data sources on long-term accretion.

entire width of the valley. For example, if a floodplain is a kilometer wide and the river shifts laterally at a rate of 2 m a year, it will take 500 years for the river to complete one swing across the valley. At any given locality, perhaps several meters of overbank sediment will accumulate in that time, but the entire deposit will be reworked by lateral erosion when the river reoccupies that position. The apparent preeminence of lateral processes in floodplain construction does not mean that overbank deposition is unimportant. Indeed, vertical accretion may be the dominant process involved during the initial stage of floodplain development (Schumm and Lichty 1963; Everitt 1968; Moody et al. 1999), even though lateral erosion may subsequently rework the sequence. Furthermore, in some cases rivers may lack the widespread lateral movement needed to rework the entire floodplain, and portions of the surface could continue their unimpeded growth by overbank deposition (Ritter et al. 1973; Kesel et al. 1974; Smith and Smith 1976; Nanson and Young 1981).

Although lateral migration controls the total thickness of overbank sediment, the *rate* of vertical accumulation depends on how much sediment is produced and yielded from the watershed, a factor controlled by geologic and climatic characteristics. In addition, human activities directly influence the rate of vertical accretion (Lajczak 1995; Brown and Quine 1999). For example, land use can be an important factor in accelerated overbank accumulation (Trimble and Lund 1982; Knox 1987; Orbock Miller et al. 1993). Knox (1987) demonstrated

that the vertical accretion rate increased by an order of magnitude during the interval of time when prairies and forests in southwest Wisconsin were converted into agricultural lands. The rate decreased significantly when better land management practices were subsequently employed.

It is important to stress again that the model of floodplain development discussed primarily applies to meandering rivers in humid climates. Where the fluvial setting is different, the origins of floodplains may vary considerably. For example, it is now recognized that anastomosing river systems occur in a wide variety of climatic environments. Floodplains developed in these systems (fig 7.1C) result from the presence of cohesive banks and low stream power rather than exceptional vertical accretion caused by stabilizing bank vegetation (Nanson and Croke 1992). However, hydrologic and sedimentologic differences between anastomosing systems found in various climatic settings may lead to some variations in the development of the associated floodplains (Smith et al. 1989; Knighton and Nanson 1993), Knighton and Nanson 1993).

The systemic relationship between flow and floodplain is more dynamic and less predictable in arid climates (Graf 1988) or in rivers with transitional braided-meandering patterns (Gottsfeld and Gottsfeld 1990). Arid-climate rivers often do not flow except in response to local thunderstorms or annual snow melt. This tends to create flashy discharge which will be unevenly distributed in time and magnitude over a drainage basin. Because discharge fluctuates drastically during short time intervals, trends towards a process-form equilibrium may never by completed (Graf 1988). In fact, as Graf (1988) points out, flood discharge is commonly contained within the channel because rapid erosion of noncohesive banks increases the channel size. Thus, no overbank flow occurs and flood damage is caused primarily by erosion rather than inundation of a floodplain surface. Floodplains may be absent in these fluvial systems.

In other braided river systems, bars and bank erosion are not restricted to one particular side of the channel, and the river can shift its position without laterally eroding the intervening material. Abandoned channels and islands gradually coalesce into a continuous floodplain surface. Normally, floodplain sediments in a braided stream system can be expected to be less thick and more irregular.

Spatial and Temporal Effects   Geomorphologists tend to conceptualize the origins of floodplains as being controlled by the regional climate and geologic framework. The difficulty with this perception is that a floodplain formed by any given river may result from a variety of process mechanics in different longitudinal segments, a spatial control exerted by minor changes in geomorphic and/or geologic characteristics. As discussed earlier, it is also clear that human activities may produce a distinct spatial overprint on floodplain process and characteristics. The various spatial effects have been documented in numerous studies set in a variety of climates, which indicate that the floodplain type, dominant genetic processes, and magnitude of overbank sedimentation varies directly from upstream to downstream reaches or where valley width is significantly altered (Grant and Swanson 1995; Mertes et al. 1996; Lecce 1997; Miller et al. 1999; Wyzga 1999).

In general, storage of overbank sediment increases with greater valley width and decreases with higher cross-sectional stream power. This relationship has a direct effect on the dominant floodplain process (figs. 7.8, 7.9). For example, Lecce (1997) found that in the Blue River watershed of Wisconsin, meandering belts associated with high stream power and rapid lateral migration are located in upstream reaches. In downstream reaches, however, the low gradient river lacks the stream power to produce meander belts and lateral migration. Thus, the downstream reach is dominated by vertical accretion and storage of alluvium on the floodplain (fig. 7.9).

In addition to spatial controls, we emphasize that floodplains are also subjected to temporal variations. Even though Nanson and Croke (1992) suggest that genetic floodplains develop under the present flow regime, they fully recognize that floodplains usually form over periods of time and commonly reflect processes that are time transgressive. Thus, although they are dominantly formed or reformed by contemporary processes, floodplains may contain elements developed during prior

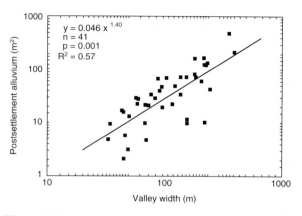

Figure 7.8

Relationship between cross-sectional area of alluvial storage and valley width, Blue River, Wis.

(Figure 5, p. 269 by S.A. Lecce from GEOMORPHOLOGY, Vol. 18, pp. 265–277. Copyright © 1997. Reprinted by permission of Elsevier Science.)

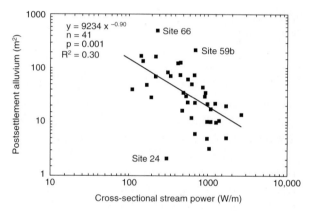

Figure 7.9

Relationship between cross-sectional area of alluvial storage and cross-sectional stream power, Blue River, Wis.

(Figure 6, p. 269 by S. A. Lecce from GEOMORPHOLOGY, Vol. 18, pp. 265–277. Copyright © 1997. Reprinted of Elsevier Science.)

flow regimes. Often the earlier flow conditions are terminated by significant climate change or tectonic activity and result in fluvial terraces that are distinguishable from the modern floodplain. However, unless a clear elevational difference is developed, it may be an onerous task to separate the alluvium deposited under the various flow conditions. In fact, there is no compelling reason to expect that a river channel must exist at different elevations during the variable flow regimes brought on by a change in climate. This phenomenon was clearly documented in the floodplain history of the Brazos River in Texas (Waters and Nordt 1995) where distinct changes in processes occurred at the same valley-bottom level during late Quaternary climatic fluctuations.

In summary, most floodplains in humid regions appear to be formed by balanced fluvial systems and also serve as integral parts of the system. They are constructed by simultaneous processes of lateral migration and overbank flooding. The deposits of lateral accretion are spread in a rather even sheet across the valley bottom, whereas overbank deposits accumulate over the entire floodplain surface away from the channel. The floodplain acts as a storage area for sediment that cannot be transported directly from the basin when it is eroded. In arid climates, floodplain development is more complex and may not manifest a process-form equilibrium between river flow and the feature. In many cases overbank flows are rare or absent. Braided rivers tend to shift laterally without the undercut banks and point bar phenomena so common in humid-climate meandering patterns. Abandoned channels and islands occupy much of the valley bottom, and these are also subject to periodic shifts of position.

Floodplains are usually considered to be associated with stable rivers, but there is no overriding reason why they cannot be present when a channel is undergoing long-term aggradation or degradation. In fact, the observed frequency of overbank flooding can continue during valley filling if the channel floor and the floodplain surface are raised at the same rate. Once the thickness of the valley deposits exceeds the limits of a reasonable scouring depth, however, the sediment below that depth can no longer be considered part of the active floodplain. In a degrading channel, the floodplain becomes a terrace when channel incision prevents the river from inundating the surface on a regular and frequent basis.

FLUVIAL TERRACES

Terraces are abandoned floodplains that were formed when the river flowed at a higher level than at present. Topographically, a terrace consists of two parts: a **tread,** which is the flat surface representing the level of the former floodplain, and the **scarp,** which is the steep slope connecting the tread to any surface standing lower in the valley (fig. 7.10). The surface of a terrace is no longer inundated as frequently as a normal, active floodplain. It is important to remember, however, that floodplain surfaces elevated by vertical accretion also have a low frequency of inundation. Thus, it may be problematic to define a floodplain (and by contrast, a terrace) on the basis of the recurrence interval of overbank flow. Most geomorphologists would recognize a surface constructed by continuous vertical accretion as a floodplain; that is, the surface and the river are still hydrologically related even though overbank flow rarely occurs. For our purposes then, the distinguishing factor in terrace formation is that the river channel floor associated with the abandoned floodplain surface (tread) must have previously been at a higher level. In fact, the presence of a terrace demands an episode of downcutting (channel entrenchment), and indicates that some significant change must have occurred between the conditions that prevailed during development of the tread and those that produced the scarp. Usually the downcutting phase begins as a response to climatic or tectonic changes, but these are not always necessary. The tread surface normally is under-

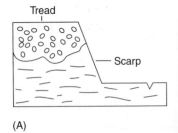

(A)

Figure 7.10
(A) Parts of a fluvial terrace.
(B) Terraces along the Madison
River upstream from Ennis, Mont.

(B)

lain by alluvium of variable thickness, but these deposits are not a true part of the terrace. To avoid confusion, it is better to limit the term to the topographic form and refer to the deposits as fill, alluvium, gravel, and so on.

Types and Classification

In general, terraces are broadly categorized as erosional or depositional. **Erosional terraces** are those in which the tread has been formed primarily by lateral erosion. If the lateral erosion truncates bedrock, the terms *bench, strath,* or *rock-cut terrace* are commonly used. If the erosion crosses unconsolidated debris, the terms *fill-cut* or *fillstrath* may be used. **Depositional terraces,** the second major grouping, are terraces where the tread represents the surface of a valley fill. Figure 7.11 illustrates both types.

Erosional terraces, especially rock-cut types, are identifiable by the following, rather distinct, properties (Mackin 1937): (1) they are capped by a uniformly thin layer of alluvium in which the total thickness is controlled by the scouring depth of the river involved, and (2) the surface cut on the bedrock or older alluvium is a flat mirror image of the surface on top of the capping alluvium. In contrast, the alluvium beneath the tread of depositional terraces varies in thickness and commonly exceeds any reasonable scouring depth of the associated river. Although the tread surface may be flat, the surface beneath the fill can be very irregular.

The classification of terraces as erosional or depositional is clearly a genetic distinction, and must be supported by evidence of the formative processes. Bull (1990) refines this approach by suggesting that major erosional terraces (straths) are the fundamental tectonic stream-terrace landform; that is, major straths are tectonically controlled. In contrast, terrace treads resulting from significant depositional events are the fundamental climatic stream-terrace landform; they are related to climatically controlled aggradation.

Another classification scheme is based on the topographic relationship between terrace levels within a given valley, as illustrated in figure 7.12. In this method, terrace treads that stand at the same elevation on both sides of the valley are called **paired** (matched) **terraces** and presumably are the same age. If the levels are staggered across the valley, they are said to be **unpaired** (unmatched) **terraces.** Most investigators interpret unpaired terraces as erosional types formed by a stream simultaneously cutting laterally and downcutting very slowly. Levels across the valley, therefore, are not exactly equivalent in age but differ by the amount of time needed for the river to traverse the valley bottom.

The Origin of Terraces

Depositional Terraces   The development of a depositional terrace always requires a period of valley filling and subsequent entrenchment into or adjacent to the fill. This cyclic pattern is necessary because the alluvium at the tread surface takes its form from purely depositional processes. The tread, in fact, represents the highest level attained by the valley floor as it rose during aggradation. The initial entrenchment that forms the terrace scarp is primarily vertical, and so the tread surface is virtually unaffected by subsequent lateral erosion at a lower level (see fig. 7.11).

Valley filling occurs when, over an extended period, the amount of sediment produced in a basin exceeds the amount that the river system can carry away. Prolonged aggradation is usually triggered by (1) glacial outwash, (2) climate change, or (3) changes in base level, slope, or load caused by rising sea level, rising local or regional base level, or an influx of coarse load because of uplift in source areas. Where tectonics are ruled out, the balance between load and discharge is determined primarily by climatic processes, although it may be driven by glaciation and may be complexly interrelated with sea level changes. Although entrenchment has been considered

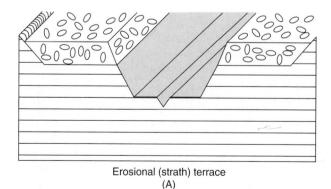

Erosional (strath) terrace
(A)

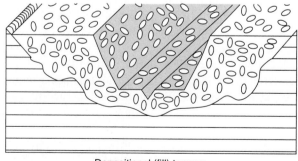

Depositional (fill) terrace
(B)

Figure 7.11

(A) Erosional (strath) terrace. Thin alluvial cover with truncation of underlying bedrock along smooth, even surface.
(B) Depositional (fill) terrace. Terrace scarp underlain by alluvium that is highest level of fill deposited in valley. Note thickness of alluvium and irregular bedrock surface beneath the fill.

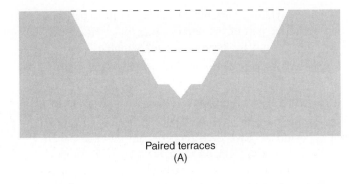

Figure 7.12
Terraces classified on basis of topographic relationships.
(A) Paired terraces have treads at same level on both sides of
the valley. (B) Unpaired terraces stand at different elevations
on either side of the valley.

Paired terraces
(A)

Unpaired terraces
(B)

theoretically (Foley 1980a) and studied experimentally (Shepherd and Schumm 1974), details of the mechanical processes involved are still not understood. Nonetheless, incision, like filling, can be triggered by tectonic events or climate change.

Spontaneous filling and cutting can also result from physical processes that have no relationship to tectonics or climate (see Foley 1980b). For example, small streams that rise in the plains surrounding high mountains often have gentler valley gradients than do the larger rivers that head in the mountains. This trait develops best where the piedmont area is underlain by easily eroded siltstones and shales. Streams originating there adjust their gradients to the fine-grained sediment released from the weakly resistant rocks. The mountain rivers, however, must transport coarse bedload derived from the resistant rocks in the mountain core, and they do so most efficiently by developing a steeper channel gradient. Because of this unique physical control, the main river stands at a higher elevation than its tributaries at an equal distance upstream from their confluence (fig. 7.13). It is well established that such a lithologic and drainage distribution leads to repeated stream captures when headwardly eroding tributaries intersect the position of the master stream (see Ritter 1972 for references to early studies documenting this phenomenon). The mountain stream is diverted into a lower tributary valley and is contained there until the process begins again.

The sudden influx of coarse load into the valley of the capturing tributary produces an untenable fluvial condition because the master stream cannot transport its oversized debris on the low valley gradient established by the tributary. The obvious result is filling of the valley until the gradient increases to an incline capable of transporting the mountain load under the prevailing discharge. Subsequent downcutting, often along the valley side, produces a depositional terrace (Ritter 1972). The enigma of depositional terraces formed in this manner is that the eroded surface beneath the gravel was formed by one river (the tributary) while the filling was caused by another (the mountain stream).

Erosional Terraces One sometimes wonders if any aspect of fluvial processes escaped the genius of G.K. Gilbert. The following statement is contained in his remarkable discussion of the origin of floodplains:

> The deposit is of nearly uniform depth, descending no lower than the bottom of the water-channel, and it rests on a tolerably even surface of the rock or other material which is corraded by the stream. The process of carving away the rock so as to produce an even surface, and at the same time covering it with an alluvial deposit, is the process of planation. (Gilbert 1877, pp. 126–27)

Clearly Gilbert presupposed the process of lateral erosion long before any detailed understanding of its mechanics existed. Certainly he provided a theoretical base for the early analyses of fluvial terraces, and his thinking probably represents a cornerstone in the classic model of rock-cut terraces developed later by Mackin (1937) and illustrated in figure 7.14.

Erosional terraces are those in which lateral erosion is the dominant process in constructing the tread.

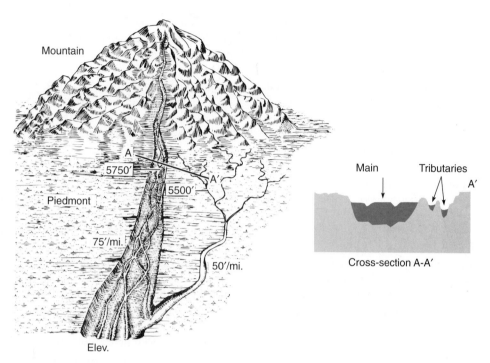

Figure 7.13

Physiographic and geologic controls of stream piracies in piedmont regions. Main river coming from mountain carries coarse-grained load on a high gradient. Tributaries that head in piedmont region carry fine-grained load on a gentle gradient. At an equal distance upstream from juncture of the two rivers, the tributaries stand at lower elevation and are in position to capture the poised master stream.

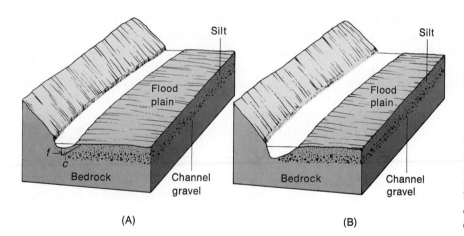

Figure 7.14

Stages in the development of a rock-cut terrace. (A) During low water stages, fine sediment (f) is deposited during normal flow and coarse sediment (c) is deposited at the end of a high-water event. (B) High-water stage entrains all the channel sediment and scours the underlying bedrock before coarse detritus is deposited again on the channel floor. (Mackin 1937)

Mackin (1937) presented an excellent description of terrace origin (see that work for details of the process). Briefly, as rivers migrate across the valley bottom, they erode one bank while simultaneously depositing point bar debris near the other (see discussion above concerning floodplain deposits formed by lateral accretion). The bar sediment later becomes the capping terrace alluvium. It is usually thin and of constant thickness, and it sits on a flat surface eroded across the underlying bedrock or sediment. The buried surface is carved during floods when scouring penetrates the debris lying on the channel floor. For this to occur, the scouring depth of the river must be great enough to remove the entire pile of channel alluvium and expose the suballuvial material to short-lived erosion (fig. 7.14). Continual shifting of the channel position back and forth across the valley, combined with the occasional scouring, creates beneath the alluvium the bevelled surface that is a mirror image of the plane surface on top of the deposit.

The sheet of alluvium is almost always present in an erosional terrace, but it is not a prerequisite and is certainly not the paramount characteristic of the feature. That role falls to the laterally eroded surface. Thus, it is probably acceptable to ignore the alluvium and consider the cut surface to be the terrace tread. This approach to terraces, like any other, may or may not lead to difficulties in the field depending on the particular situation.

Erosional terraces are normally thought of as the "equilibrium" model of the terrace line. It would seem that the development of a tread surface by lateral planation should require not only time but also a long period of stability during which base level and channel functions are constant, and no vertical disruptions by filling or cutting occur. Nonetheless, even this logical rule of thumb has exceptions (Born and Ritter 1970; Womack and Schumm 1977).

Terrace Origin and the Field Problem

Understanding local terraces and establishing a regional pattern of terrace development are not only basic in historical geomorphology, they are also useful in providing information for regional planning, land management, allocating water supplies, and locating sand and gravel for building materials. Acquiring such knowledge, however, is a painfully slow procedure requiring field study and correlation of surfaces within a valley or between valleys. Determining terrace origins is not easy. Although we postulate guidelines for recognizing terraces of different origins, terraces in the real world develop in such a variety of ways that they defy generalization. Each terrace sequence must be examined according to its own geologic, climatic, and tectonic setting without preconceived ideas about its origin.

Using terraces to interpret geomorphic history is a monumental task for several fundamental reasons. First, terraces are rarely preserved intact along the length of a valley but instead are segmented into isolated and physically separated remnants, often kilometers apart. Reconstruction of the original longitudinal profile of the terrace surface requires correct correlation of the remnants, and every method used in that procedure is burdened with fundamental assumptions that may be invalid in certain situations (for details see D. W. Johnson 1944; Frye and Leonard 1954). In bedrock streams, longitudinal profiles often have shapes that reflect the vigor of tectonic activity. Tectonically active regions tend to produce profiles that are concave up, contrasting to convex forms in tectonically inactive areas with slow uplift rates (fig. 7.15). Pazzaglia et al. (1998) suggest that strongly concave profiles result when the ability to generate stream power is in excess of that needed for incision to keep pace with rapid uplift and also transport the introduced bedload. In that setting, strath terraces will be produced because a portion of the stream power will be utilized in lateral erosion, a phenomenon similar to that found in drainage basins that generate high peak and mean annual discharges. In contrast, convex or straight profiles presumably reflect the inability of the system to produce excess stream power, and characterize basins that have more narrow valleys and less frequent terrace

formation (see fig. 7.15). Other studies have also suggested that the development of strath terraces may be controlled by the spatial distribution of stream power (Merritts et al. 1994). Note, however, that the linkage of stream power with terrace analyses should be done with some caution (Sklar and Dietrich 1998).

Terrace development and the character of longitudinal profiles may also be directly affected by eustatic sea-level change (Merritts et al. 1994; Pazzaglia and Gardner 1993, 1994; Pazzaglia et al. 1998). For example, in the tectonically active region along the Pacific coast of California, detailed terrace analyses (Merritts et al. 1994) revealed that the lower reaches of river valleys are characterized by fill terraces that form in response to oscillations of sea level. Valley aggradation occurs during high stands of the ocean followed by incision of the fills during low stands. This eustatically controlled process overrides the effect of tectonically produced uplift and strath-terrace development that prevails in the middle and upper reaches of the river valleys. The sea-level effect has also been noted along passive continental margins (i.e., Pazzaglia and Gardner 1994).

Second, more than one terrace can result during a period of downcutting. This indicates that entrenchment, representing the response to a threshold-exceeding change, is not a continuous, unidirectional erosional event that produces a lowered river gradient. Instead, as discussed in chapter 1, the response is complex (Schumm 1977). It often involves pauses in a downcutting phase during which the river may form erosional terraces by lateral planation. Downcutting reduces local river slope which causes particles larger than the competent size to accumulate on the channel floor while smaller sediment is transported downstream. The accumulated layer of unentrainable clasts is known as a *streambed armor,* which tends to protect underlying material from erosion. Entrenchment and armoring are common phenomena downstream from dams (Williams and Wolman 1984). Armoring also seems to facilitate lateral erosion (Rains and Welch 1988; Bull 1990) and therefore may be an important ingredient in the development of erosional terraces associated with complex response. Complex response may also be involved in the development of depositional terraces. The most commonly cited examples result from responses of drainage basin systems to base-level lowering (Schumm and Parker 1973; Womack and Schumm 1977). The point here is that a complex response results in multiple terraces which are formed during the adjustment to a single, equilibrium-disrupting external event.

The interpretation of depositional terraces is further complicated by the fact that alternating cut and fill events can occur without any external forcing. Responses are controlled by factors existing within the system, and channel incision or aggradation is triggered by crossing some threshold value of these factors (Patton

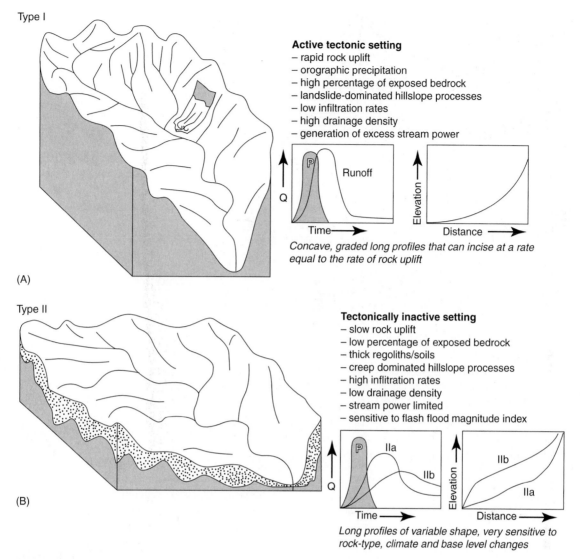

Figure 7.15
Summary of geomorphic, hydrologic, and long-profile characteristics. (A) Tectonically active areas; (B) Tectonically inactive settings. P indicates a unit precipitation event.

(From Pazzaglia et al. In Rivers over rocks, fluvial processes in bedrock channels. Edited by K. Tinkler and E. Wohl. Fig. 113, p. 230. Copyright © 1998 by the American Geophysical Union.)

and Schumm 1975, 1981; Schumm 1979). Thus, terraces may form repeatedly under intrinsic threshold controls during periods when a relatively constant base level persists and no significant climatic or tectonic changes occur (fig. 7.16). It should be clear from the above that it is not safe to assume that terrace sequences in adjacent basins, or even in tributary valleys of the same basin, correlate in time and mode of origin (i.e., Patton and Boison 1986; Rains and Welch 1988; Taylor and Lewin 1997).

There is a third difficulty in using terraces to interpret geomorphic history. Erosional terraces may not seem to be that different from depositional terraces. After all, both are usually covered with alluvium, and if one walked across that alluvial surface there would be nothing to indicate what type of terrace lay beneath. Nonetheless, a very important difference does exist. When an erosional ter-

race forms, the capping alluvium is deposited *at the same time* that the underlying surface is eroded. In contrast, the surface beneath a depositional terrace was present before the influx of the alluvial fill; a finite time gap separates the deposition from the cutting of the underlying surface (fig. 7.17). Failure to recognize this subtle distinction between erosional and depositional terraces can lead to very different reconstructions of geomorphic history. A classic example of this interpretive problem can be found in the studies of Mackin (1937) and Moss and Bonini (1961). Both studies examined the same terrace sequence near Cody, Wyoming, but each came to different conclusions concerning terrace ages and origins because they had different interpretations about the thickness of alluvium under the terrace tread and the configuration of the bedrock surface beneath the alluvial cover.

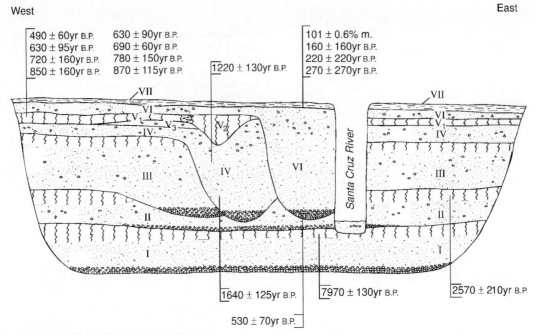

West   East

490 ± 60yr B.P.   630 ± 90yr B.P.
630 ± 95yr B.P.   690 ± 60yr B.P.
720 ± 160yr B.P.   780 ± 150yr B.P.
850 ± 160yr B.P.   870 ± 115yr B.P.

1220 ± 130yr B.P.

101 ± 0.6% m.
160 ± 160yr B.P.
220 ± 220yr B.P.
270 ± 270yr B.P.

1640 ± 125yr B.P.   7970 ± 130yr B.P.   2570 ± 210yr B.P.

530 ± 70yr B.P.

Figure 7.16
Generalized composite geologic cross-section of the northern portion of the San Xavier reach of the Santa
Cruz River. Not to scale. Note that the level of entrenchment prior to fill IV and VI is the same as previous
channel bottoms.
(Waters 1988)

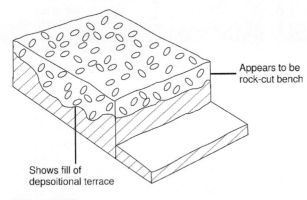

Appears to be
rock-cut bench

Shows fill of
depsoitional terrace

Figure 7.17
Difficulty in interpreting terrace origin from field data. If
downcutting exposes only part of fill, terrace may appear to be
rock-cut. Data across the terrace are needed to determine true
thickness of the fill.

PIEDMONT ENVIRONMENT: FANS AND PEDIMENTS

The topography of almost every region reflects an adjustment between dominant surficial processes and
lithology. When the rocks have diverse resistances, geomorphic processes tend to maximize the relief between
regions of greatest and least resistance. Nowhere is this
more apparent than in areas where mountains and plains
adjoin, especially where the climate is arid or the region
has undergone recent tectonism. Aridity buffers the

smoothing effects of vegetation; vertical tectonic activity accentuates relief by bringing more resistant base-
ment rocks toward the surface, where they become the
cores of topographic mountains.

The sloping surface that connects the mountain to
the level of adjacent plains is the **piedmont.** It extends
from the mountain front to a floodplain or playa, either
of which can mark the base level for geomorphic
processes that function on the piedmont surface
(fig. 7.18 A & B). Piedmonts consist of a number of
geomorphic landforms, but most commonly they are
composed of eroded bedrock plains called **pediments**
and depositional features called **alluvial fans.** The relative percentage of the total piedmont area occupied by
either of these features probably depends on a unique
combination of local geomorphic variables.

Alluvial Fans

Alluvial fans have been investigated most extensively and
in great detail in regions of arid or semiarid climate. This
does not mean that fans are absent in other climatic zones.
On the contrary, humid-climate fans and deposits have
been examined in such diverse settings as humid-glacial
(Boothroyd and Ashley 1975), humid-periglacial (Ryder
1971a, 1971b; Wasson 1977), humid-tropical (Mukerji
1976; Wescott and Ethridge 1980), and humid-temperate

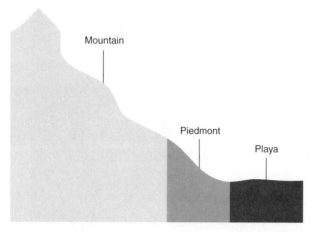

Figure 7.18A

Physiographic components of a mountain-basin geomorphic system showing the position of the piedmont zone, which contains pediments and alluvial fans.

Figure 7.18B

Photo of mountain front environment in an arid basin of central Nevada: Mountain is in rear and playa is white area on basin floor. Piedmont zone in center.

(Photo by D. Germanoski)

TABLE 7.3	Generalized Characteristics of Alluvial Fans Formed in Different Environments.			
Parameter	**Arid fans**	**Humid-glacial Fans**	**Humid-tropical Fans**	**Virginia Humid-temperate Fans**
Fan morphology				
Plan view	Broad, fanlike, symmetrical	Broad, fanlike, symmetrical	Broad, fanlike, symmetrical	Broad, fanlike to elongated
Axial profile	Segmented (20–100m/km)	Smooth (1–20m/km)	Smooth	Segmented (40–100m/km)
Thickness	Up to 100s m	Up to 100s m	Up to 100s m	5 to 20 m
Area	Small	Very large	Large	Small
Depositional processes				
Major processes	Debris flow Braided stream Sheet flood Sieve flood	Braided stream	Braided stream Debris flow	Debris flow (avalanche)
Return interval	1–50 yrs, discrete events	0–few days, seasonally constant	Seasonally constant to discrete	3000–6000 yrs, discrete events
Fan area activated	10–50%	80–100%	30–70%	10–70%
Triggering processes	Heavy rain Snow melt	Meltwater Outwash	Heavy rain Monsoon	Heavy rain Hurricane
Discharge	Flashy	Seasonal	Seasonal	Flashy

After Kochel and Johnson (1984). Used with permission of the Canadian Society of Petroleum Geologists.

(Hack and Goodlett 1960; Williams and Guy 1973; Kochel and Johnson 1984; Wells and Harvey 1987). Fans developed in every climatic setting are linked together by a similar plan-view geometry, but other aspects of morphology and depositional processes may vary considerably (table 7.3). More recent and excellent reviews of fan processes and deposits can be found in Blair and McPherson (1994a, b) and Harvey (1997).

Alluvial fans are one end of an erosional-depositional system, linked by a river, in which rock debris is transferred from one portion of a watershed to another. Fans are largest and most well developed where erosion takes place in a mountain and the river builds the fan into an adjacent basin. Deposits tend to be fan-shaped in plan view and are best described morphologically as a segment of a cone radiating away from a single point source (fig. 7.19). Adjacent fans often merge at their lateral extremities; the individual cone shape is lost, and a rather nondescript deposit is formed covering the entire piedmont. The coalesced

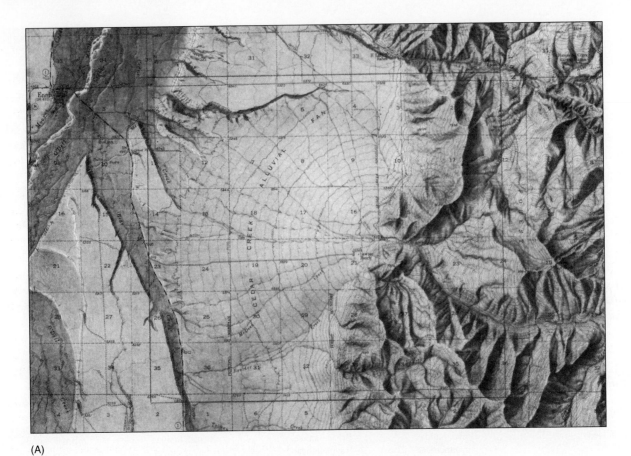

(A)

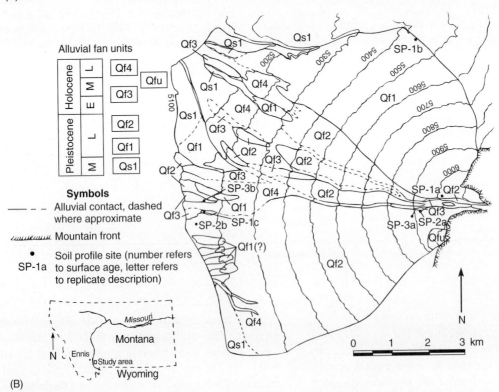

(B)

Figure 7.19

(A) Shaded relief edition of the Ennis, Mont., quadrangle (U.S.G.S. 15′) illustrating the classic fan morphology of Cedar Creek alluvial fan. (B) Geologic map of Cedar Creek alluvial fan illustrating spatial distribution of Quaternary alluvial fan deposits. Contours, in feet, were redrawn from the Ennis, Mont., quadrangle.

(B): (J. Ritter et al. 1993)

fans are commonly referred to as *bajadas, alluvial aprons,* or *alluvial slopes.* Alluvial fan processes and deposits are readily distinguished from those associated with other sedimentary depositional environments (Blair and McPherson 1994a).

The components of an alluvial fan system are shown on figure 7.20. The upland part of the system consists of the drainage basin that provides sediment that accumulates in the lower part of the system represented by the fan. The fan and drainage basin are connected by the main channel of the system known as the *feeder channel.* The highest and most proximal part of the fan is called the *apex,* and represents the point where the feeder channel emerged from the highlands during deposition of the initial fan sediment. Commonly, the feeder channel has been entrenched to a level below the original fan apex. This *fanhead trench* normally develops when a threshold value of slope is exceeded (Schumm 1977). The new and lower equilibrium slope probably evolves in stages by integration of disrupted and isolated channel segments that formed during the instability caused by the threshold crossing (Scott and Erskine 1994). The fanhead trench is often referred to as the *incised channel* where it extends into and partially through the older fan alluvium. Thus, flow leaving the upland area is prevented from spreading onto the original fan surface in the apical zone because it is confined within the fanhead trench and incised channel. Clearly, where entrenchment has taken place, parts of the total fan surface are no longer burdened by active sediment accumulation.

At some point downfan, the incised channel intersects the original fan surface because the slope on the fan is greater than the gradient of the incised channel. At this juncture, called the *intersection point,* flow leaves the incised channel and spreads onto the surface of the original fan. Thus, the segment downfan from the intersection point may be the only part of the fan system that experiences aggradation at any given time, and is aptly referred to as the *active depositional lobe.* It is possible for the intersection point to shift longitudinally when deposition and backfilling occur within the incised channel. This will result in a position shift of the active depositional lobe.

Many abandoned channels and discontinuous gullies exist on the older portion of the fan. These depressions are temporarily disconnected from the feeder channel, and therefore, characterize parts of the fan system that are currently inactive. However, discontinuous gullies are known to erode headwardly, eventually encountering the incised channel and shifting its position by stream capture. This process represents another way that the location of active deposition can be changed.

In a genetic sense, fans have been categorized by Blair and McPherson (1994a,b) as Type I (debris flow/mudflow fans) and Type II (fluvial fans) on the basis of flow type and the character of sediment being deposited. Characteristics of Type I and Type II fans are shown in table 7.4 and detailed field descriptions of the two major types are available (Blair 1999 a,b). The key factors that determine which type of fan will develop are (1) the water-to-sediment ratio (w/s) of the mix introduced into the depositional zone and (2) the availability of mud-sized sediment (Harvey 1997). High w/s values, especially when accompanied by low concentrations of fine-grained sediment, tend to promote fluid transport of coarse clasts and deposits having fluvial characteristics. As w/s values decrease and fine sediment concentrations increase, clasts become matrix supported and debris-flow transport can occur. Geomorphically, Type I fans dominated by debris flow deposits tend to form where watershed areas are small and steep, and where the geologic framework produces abundant fine-grained sediment (Harvey 1997). Furthermore, debris flow fans tend to be more common in arid regions where spasmodic rainfall is the norm. In contrast, Type II fans are more common in humid climates and in large basins that produce minor amounts of fine sediment.

Regardless of the perceived environmental links to different fan types, it is important to stress that all combinations of controlling factors can develop under any prevailing climate. It is also true that water-to-sediment ratios can change during the tenure of any given storm, and they may vary significantly from proximal to distal portions of a fan. Thus, a variety of transitional flows and deposits are possible in any setting, creating enormous complexity in the interpretation of the evolution of a given fan.

Fan Morphology The longitudinal slope of an alluvial fan generally decreases downfan even though its precise value at any point depends on the load-discharge characteristics of the fluvial system. Near the mountain front, slopes are commonly very steep, especially where freefall accumulations are part of the fan complex (Blair and McPherson 1994a). Fans gradually flatten to their lower extremity, called the *toe,* where gradients may be as low as 2 m per kilometer. The steepest gradients are usually associated with coarse-grained loads, low discharges, high sediment production in the source area, and transport processes other than normal streamflow. These factors often conflict in the same region. In Fresno County, California, for example, fans derived from basins underlain by mudstones or shales are 33 to 75 percent steeper than fans of the same size related to sandstone basins (Bull 1964a, 1964b). The low gradient expected because of the small particle size is offset by a high rate of sediment production. Fan gradients may also be related to parameters of system morphometry such as drainage basin area. However, considerable variation of fan slope exists at the same value of drainage basin

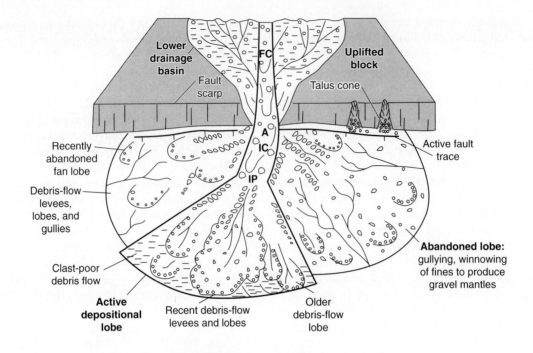

(A)

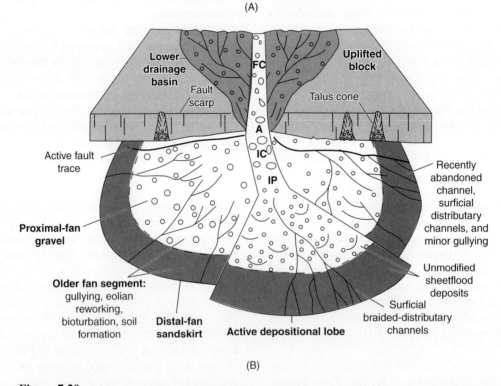

(B)

Figure 7.20

Schematic diagrams showing the lower drainage basin, and primary and secondary depositional features of alluvial fans, including those dominated by (A) debris-flow processes and (B) sheetflood processes. Abbreviations: A = fan apex; FC = drainage-basin feeder channel; IC = incised channel on fan; IP = fan intersection point.

(Figure 1. p. 455 from "Alluvial fans and their natural distinction from rivers" by Blair & McPherson, JOURNAL OF SEDIMENTARY RESEARCH, Vol. 64, pp. 450–489. Copyright © 1994. Reprinted by permission of SEPM (Society for Sedimentary Geology) and American Association of Petroleum Geologists.)

TABLE 7.4	Characteristics of Type I versus Type II Alluvial Fans.	

Feature	Type 1 Alluvial Fan	Type II Alluvial Fan
Dominant primary process and facies	Debris flows, especially lobe facies	Sheetfloods, especially couplet facies
Minor primary processes	Rockfall, rock slide, rock avalanche, colluvial slide, incised channel	Rockfall, rock slide, rock avalanche, colluvial slide, incised channel, noncohesive debris flow
Dominant secondary processes	Winnowing by overland flows and wind to produce desert pavements, boulder mantles, gullies, and shallow channels	Winnowing by overland flows and wind to produce desert pavements, gullies, and shallow distributary channels
Typical grain sorting and size	Very poorly sorted clayey boulder, pebble, and cobble gravel	Poorly sorted sandy and bouldery cobble to pebble gravel
Downfan trend in maximum clast size	Relatively constant	Typically decreases from boulders to pebbles or sand
Typical grain shape	Angular	Angular to subangular
Typical stratification style	Poorly or subtly stratified except for secondary winnowed surfaces	Well-stratified coarse gravel and sandy fine gravel couplets
Presence of granular or sandy interbeds	Rare	Common
Presence of a distal sand-skirt facies	Rare	Common
Presence of depositional matrix clay	Common	Rare
Drainage-basin size	Small to moderate	Small to large
Feeder channel length	Short to moderate	Moderate to long
Typical bedrock lithology underlying the drainage basin	Pelitic metamorphic rocks, shale, aphanitic volcanic rocks, or mafic plutonic rocks; also granitic or gneissic rocks weathering under humid climate	Quartzite, quartz-rich conglometric or sandstone; also granitic or gneissic rocks weathering under arid climate.
Clay abundance in the drainage-basin colluvial slopes	Moderate to abundant	Rare
Common average slope value	5 to 15°	2 to 8°
Downfan slope style	Constant or straight	Distally decreasing or planoconcave
Premeability	Low	High
Porosity	Low	High
Connectivity of permeable units	Low	High

Table 4 from "Alluvial fans and their natural distinction from rivers" by Blair & McPherson, JOURNAL OF SEDIMENTARY RESEARCH, Vol. 64, p. 478. Copyright © 1994. Reprinted by permission of SEPM (Society for Sedimentary Geology) and American Association of Petroleum Geologists.

area (fig. 7.21). Thus, slope values at any given point probably reflect a myriad of controls in the fan system (Hooke and Rohrer 1979). Indeed, Milana and Ruzycki (1999) suggest that fan slope is best related to transport efficiency, a factor derived from the combined effect of drainage basin area and mean annual rainfall.

Two slope characteristics deserve special attention. First, the gradient of most fans near the mountain front is approximately the same as that of the mountain river where it merges with the fan apex. Deposition on the upfan surface, therefore, is not initiated by a dramatic decrease in gradient as the master river passes from the mountain onto the fan. Second, although fans are concave-up from the apex to the toe, their longitudinal profiles are usually not a smooth exponential curve. Instead, on

many fans the concavity stems from a junction of several relatively straight segments, each successive downfan link having a lower gradient.

The changes in fan slope, represented by individual segments, may be genetically related to changes in the channel of the trunk river upstream from the fan apex. For example, in many fans intermittent uplifts have increased the stream gradients, and in response to each event, a new fan segment has formed, gradually adjusting its slope until it approximates the newly formed steeper slope of the trunk river (Bull 1964a, 1964b; Hooke 1972). Under this particular control, each segment farther up the fan is steeper and younger, and its deposits are graded to the level of the next lower segment. It is also possible that segmentation may result

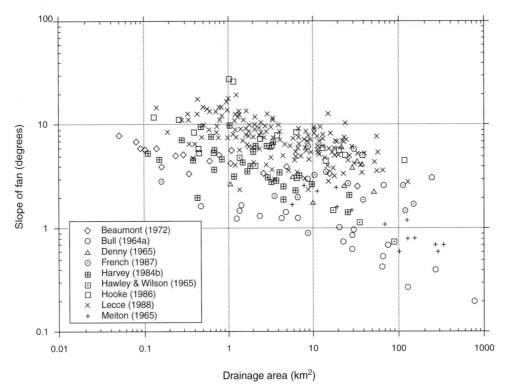

Figure 7.21
Log–log plot of average fan slope versus drainage basin area.

(From GEOMORPHOLOGY OF DESERT ENVIRONMENTS (Blair and McPherson 1994) by Athol D. Abrahams and Anthony J. Parsons. Figure 14.23 p. 393. Copyright © 1994. (References to studies shown on plot can be found there.) Reprinted by permission of the authors.)

from climatically induced changes in the load-discharge balance, or even from a single fan-forming event when variations in the depositional setting lead to discontinuous segments of the fan surface (Blair 1987). Thus, the overall longitudinal profile may be sensitive to historical changes in controlling factors (Bull 1964, 1968) or to irregularities in the physical environment on a short-term basis.

The area of a fan is statistically related by a simple power function to the area of the basin supplying the sediment such that

$$A_f = cA_d^n$$

where A_f is the area of the fan and A_d is the area of the drainage basin. The exponent n is the slope of the regression line in a full logarithmic plot of the two variables; it measures the rate of change in fan area with increasing drainage basin area. The coefficient c indicates how much the fan spreads out.

The mean value of n varies from 0.7 to 1.1 when A_f and A_d are measured in square kilometers (fig. 7.22). The coefficient c, however, seems to vary widely, reflecting the effect on fan dimensions of geomorphic factors other than drainage basin size. Chief among these are climate, source rock lithology, tectonics, and the original space available for fan growth in the collecting

basin. For example, Hooke and Rohrer (1977) suggest that c probably relates to the competition for space in the depositional zone. A particular fan that receives a large volume of sediment from its drainage basin would tend to thicken faster than its neighbor and spread outward at the expense of the neighboring fan area. Thus, even where drainage basins in the same region have equal areas, the areas of their fans may differ by as much as an order of magnitude if these other characteristics differ greatly.

Several specific examples demonstrate the effect of these factors and the significance of fan morphometry. Bull (1964) showed that fans derived from basins underlain by fine-grained sedimentary rocks are almost twice as large as those derived from sandstone basins of equal size. The regression lines have approximately the same slope ($n = 0.91$ and 0.98), but the effect of particle size shows up in the value of the coefficient c, which varies from 0.96 for sandstone to 2.1 in the mudstone drainage basins. The effect of tectonics may be evident in the fans of Death Valley, where eastward tilting of the valley permitted fans on the west side of the valley to grow larger while those on the east side were stunted by burial beneath the playa (Denny 1965). The c values are 1.05 for fans on the west side of the valley and 0.15 for those on the east side.

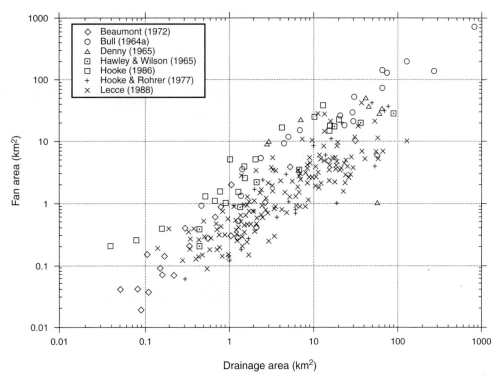

Figure 7.22
Log-log plot of drainage basin area versus fan area.

(From GEOMORPHOLOGY OF DESERT ENVIRONMENTS (Blair and McPherson) by Athol D. Abrahams and Anthony J. Parsons. Figure 14.22 p. 392. Copyright © 1994. (References to studies shown on plot can be found there.) Reprinted by permission of the authors.)

Fan Deposits and Origins

Deposits and Depositional Processes　As discussed earlier, the movement of sediment from source areas to depositional sites involves a variety of flow types, ranging from highly viscous debris or mudflows to normal water flow. The type of flow during any given event depends primarily on the lithology of the basin and its degree of weathering, and secondly on the magnitude of the precipitation causing the flow.

On many fans (especially in arid regions), as flow leaves the confines of the trunk channel, deposition is initiated by changes in hydraulic geometry, not by a sudden decrease in gradient (Bull 1964). Generally, when the flow becomes unconfined on the fan surface, the width increases so dramatically that both depth and velocity decrease to a level where the flow can no longer transport the load. The effect of changing hydraulic geometry is reinforced by a loss in discharge if the fan surface is permeable and water seeps into the underlying deposits. It is not unusual, for example, for the entire flow to disappear underground before it can traverse the length of the fan.

In events involving highly fluid streamflow, lateral shifting of the loci of deposition allows the braided stream system to deposit a sheet of poorly bedded sand and gravel in which individual beds can be traced laterally for only short distances. This sheetlike configura-

tion may be interrupted by thicker deposits that represent an occasional channel entrenchment into the fan surface and subsequent backfilling. Deposits within these larger channels are generally more coarse-grained. However, even within incised, active washes a microtopography may exist that is directly related to variations in textural properties of fan deposits.

Debris flows or mudflows usually follow more well-defined channels because the confining limits of the channel ensure the depth of flow needed to offset the high viscosity of the fluid. Debris flows are so dense and viscous that only the very largest particles can settle from the mass during flow. Nonetheless, they are capable of transporting extremely large boulders for considerable distances on lower gradients than normal streamflow would require. The high viscosity, however, effectively restricts the distance of transport (in comparison with water flow), and the forward movement may simply stop even though the flow is still confined in a channel. Therefore, deposits from debris flow are poorly sorted with boulders embedded in a fine-grained matrix. In contrast to water-transported sediment, they are usually lobate and have well-defined margins often marked by distinct ridges (fig. 7.23).

In a detailed study of modern, storm-generated fans in England, Wells and Harvey (1987) recognized a

Figure 7.23
Debris-flow ridges (levees) produced by large flow event on fans of the Santa Rosa Mountains in the Anza Borrego Desert, California. Older levees developed on small varnished fan in background.

variety of fan deposits that can be placed in distinct facies defined by morphology and sediment texture (table 7.5; fig. 7.24). End members of their facies types are viscous debris flows and fluvial sheet deposits that have the textural and morphological properties of deposits generated in debris flows and fluid streamflow (described above). Wells and Harvey recognized a variety of facies between these end members, and each facies could be related to the relative fluidity (water/sediment ratio) of the transporting medium. What is extremely significant is that all of these facies were formed during the same storm and many within the same drainage basin. Detailed documentation of the variety of facies resulting from the same fan-forming event under the same geomorphic controls (climate and geology) is instructive as we consider the origins of fans and what they mean in the interpretation of geomorphic history.

Fan Origins and Geomorphic History   Although our purpose in this book is to examine process, we would be remiss not to point out that the historical aspect of alluvial fans is one of the more perplexing problems facing Quaternary scientists. This is because most fans experience episodes of erosion or nondeposition, and long-term, continuous growth on all parts of a fan is a rare, if not nonexistent, phenomenon. The interpretive problem with this spasmodic development stems from different perceptions of what is driving the erosional episodes. Many earlier studies viewed alternations of deposition and erosion as part of a continuing equilibrium state that is maintained by internal fan mechanics. In contrast, others believe that variations in fan dynamics are impressed on the system by external forces related to climate change. Although tectonic activity can cause en-

trenchment, climate change seems to be the most important external driver (Ritter et al. 1995).

In the first model, morphometric relationships between basin and fan are maintained in a long-term, dynamic equilibrium state (Denny 1965, 1967; Hooke 1968). In this concept, episodes of entrenchment, creation of abandoned zones, and relocation of the active depositional lobe are simultaneous events driven by shifts in the location of the feeder channel brought on during stream captures. Thus, both erosional and depositional processes operate continuously within the fan and are not dependent on the timing of external changes. In addition, intrinsic threshold crossings may alter the longitudinal position of the feeder channel and the active depositional lobe. For example, Schumm (1977) demonstrated that fanhead trenching and backfilling on some fans may be part of a continuous erosional/depositional cycle controlled by an intrinsic threshold of slope.

The above suggestions received some support with the growing recognition that all types of deposits and flows can form in any given climate or single-flow event. This fact effectively removes a change in deposit type as being an absolute indicator of climate change (see Blair and McPherson 1994b). Nonetheless, many studies suggest that the alternation of fan deposition and erosion are related to episodic climate change, with aggradation usually occurring during late Quaternary glacial climates and dissection during the Holocene (Harvey 1984a, 1984b, 1987, 1988; Dorn et al. 1987; Dorn 1988). For example, Harvey and Wells (1994) note that in a dated sequence of southern California fans, debris flow deposition dominated under a colder/wetter Late Pleistocene climate, which produced enhanced mass movement and high sediment supply in the catchment. This changed with the Holocene climate transition

TABLE 7.5	Morphologic and Sedimentologic Field Criteria for Distinguishing Facies Types on Alluvial Fans.			
Facies types	**Morphology**	**Depositional Relief (m)**	**Texture and Stratification**	**Characteristics of Clast Fabric**
Debris flow (D1)	Lobate to digitate Narrow Steep front and flanks Flat tops with low relief pressure ridges	High (0.8–1.5)	Matrix-rich (muddy) Matrix supported clasts Poorly sorted bmax range 80–210 mm Stratification absent	Elongate clasts oriented parallel to flow boundary, forming a push fabric (terminology after Lawson 1981)
Dilute debris flows (D2)	Thin, lobate Broad, flat top Gentle lobe fronts and flanks	Moderate (0.3–0.5)	Matrix-rich (muddy) Matrix-supported clasts Poorly sorted bmax range 60–230 mm Stratification absent	None observed
Traditional-flow deposits (T1)	Stacked lobes Broad, small superimposed mounds Small collapse depressions	High (0.5–1.5)	Clast support with no matrix in upper few centimeters Matrix (sandy) increases with depth bmax typically < 180 m Moderately sorted Stratification present	Collapse packing
Fluvial boulder bar and lobes (S1)	Linear bars to transverse lobes	Moderate to high (0.5–0.8)	No matrix Clast support Front-to-tail Sorting bmax typically > 200 mm	Imbrication
Fluvial longitudinal bar (S2)	Linear bars	Moderate (0.2–0.5)	Class support Matrix (sandy) increases with depth Marked front-to-tail sorting More poorly sorted than type S1 bmax typically < 120 mm	Strong imbrication
Fluvial sheet deposits (S3)	Broad and flat Some fan-shaped Subdued bar and swale forms	Low (~0.1)	Clast support Little matrix (sandy) Well stratified Normal grading in some strata Moderate sorting in each stratum bmax typically <100 mm	Weak imbrication

From Wells and Harvey 1987, with permission of the author.

that enabled hillslope stabilization and concomitant reduction of sediment supply. The result was an episode of fan dissection. It is notable, however, that fan margins expanded during the dissection because sediment released by trenching was transported and deposited farther downfan.

In some instances the climate control is more direct. For example, the Cedar Creek fan (see fig 7.19) expanded by the coalescence of lobes that were constructed from outwash emitted during several glacial episodes. Interglacial periods were characterized by en-

trenchment and soil development on the older deposits. The link between fan growth and glacial episodes clearly demonstrates that most of the Cedar Creek fan is relict and unrelated to modern conditions.

Part of the problem is understanding the historical meaning of fans rests in the fact that until recently we did not know enough about the criteria being used as the basis of our interpretations. For example, we now know that all fans do not necessarily develop gradually over long time intervals (Blair 1987), nor do their characteristics necessarily reflect average climatic or

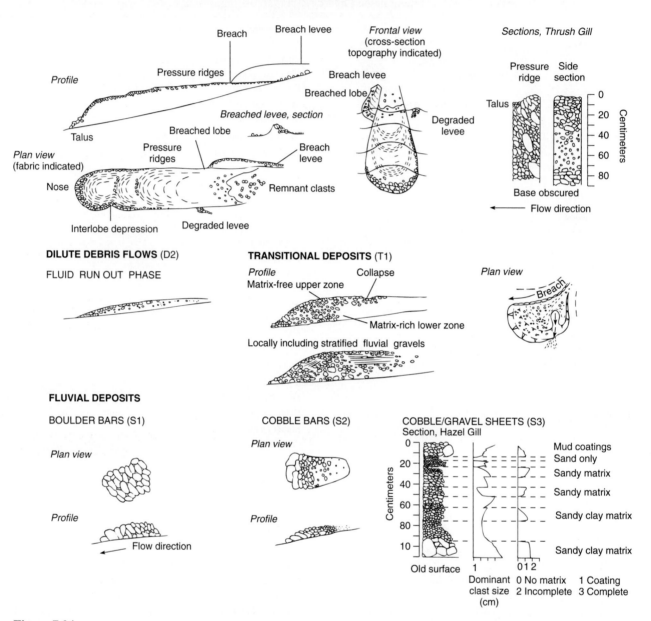

Figure 7.24
Geomorphic and sedimentologic features of six major facies types: D1 (viscous debris flow), D2 (dilute debris flow), T1 (transitional flow), S1 and S2 (fluvial bars and lobes), and S3 (fluvial gravel sheets). Vertical sequences of facies types D1 and S3 illustrate typical end-member facies stratigraphy. See table 7.5 for characteristics.
(Wells and Harvey 1987)

tectonic controls. This is especially true when we consider fan deposits as surrogates of climate, and fan entrenchment as a process that occurs only in response to climate change or tectonic activity. For example, we noted earlier that any type of flow and the characteristics of the associated deposit can be generated in any given climate during a single depositional event (Beaty 1963; Blair 1987; Wells and Harvey 1987). Wells and Harvey (1987) further demonstrated that the important ratio of water to sediment content (relative

fluidity) can change dramatically during any storm-generated flow event (fig. 7.25). This reinforces the earlier suggestion (Nilson 1982) that debris-flow, transitional-flow (hyperconcentrated flow), and streamflow deposits are probably characteristic of almost all fans. Thus, interbedded debris-flow and streamflow deposits do not always indicate responses to climatic fluctuations, but instead may represent normal variations of relative fluidity within a system during a single flow event.

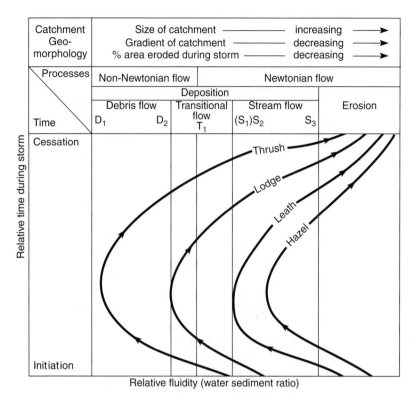

Figure 7.25
Conceptual model showing the changes in the water sediment ratio, the sequence of depositional and erosional events, and associated flow conditions during the 1982 storm for the four study fans in the Howgill Fells. Bold lines and associated arrows illustrate the relative timing of facies deposition, resulting stratigraphic sequence, and relative timing and amount of erosion; for example, processes on Thrush Gill fan started with streamflow and transitional-flow deposition, which was followed and dominated by debris-flow deposition during most of the storm and ended with streamflow deposition and erosion.
(Wells and Harvey 1987)

In addition to the need for care in interpreting fan deposits, we now recognize that fan entrenchment can be either temporary or permanent, and distinguishing between the two seems to be critical in the analysis of fan origin. We saw earlier that many fan-head trenches appear to be rather ephemeral; they show evidence of alternating episodes of trenching and filling that probably result from processes controlled by local conditions (Schumm 1977).

However, entrenchment can also be a permanent phenomenon. Sometimes fan-head trenches are incised to depths that cannot be easily backfilled, often being cut to levels greater than 30 m below the fan surface. The fan surface standing above the trench is no longer involved with active fan processes. Soils may develop on the alluvium, and incipient drainage networks may be established. In other cases where the basin of deposition is open and the base level for the fan is the floodplain of a through-flowing river, down-cutting of that river may initiate a wave of fan incision that is propagated upslope from the toe of the fan. Eventually the entire fan is dissected when the incision reaches the apex and captures the trunk river as it emerges from the mountain. This process is especially effective on small valley-side fans of glaciated valleys (fig. 7.26) and commonly relates to climatic fluctuations and the glacial cycle (Ryder 1971a, 1971b; Ritter and Ten Brink 1986). In contrast to temporary fan-head trenching that is driven by intrinsic thresholds, permanent en-

trenchment is controlled by factors that are external to the fan system. Therefore, this phenomenon probably represents a valid criterion for using climate change or tectonic activity in interpreting the origin or meaning of a fan.

Pediments

Since Gilbert first described "hills of planation" in the Henry Mountains of Utah, geomorphologists have been intrigued with their origin, and these features called pediments have been discussed endlessly during the last century. Our discussion of these interesting features will necessarily be brief, but excellent reviews of the topic are available in Tator (1952, 1953), Tuan (1959), Hadley (1967), Cooke and Warren (1973), Twidale (1978), Dohrenwend (1994), and Oberlander (1997). Definitions of the term *pediment* are as numerous as the workers who have studied this feature, and descriptions range from general to rather graphic and precise. For example, Denny (1967, p. 97) employs the term in reference "to the part of the piedmont that is more or less bare rock surface." On the other hand, R.U. Cooke (1970, p. 28) suggests that "pediments are composed of surfaces eroded across bedrock or alluvium, are usually discordant to structures, have longitudinal profiles usually concave upward or rectilinear, slope at less than 11°, and are thinly and discontinuously veneered with rock debris."

To understand what pediments are, it may be better to summarize the characteristics that are universally recognized as salient properties of pediments rather than adopt a formal definition:

1. Pediments are erosional surfaces that abut and slope away from a mountain front or escarpment.
2. They are entirely erosional in origin and commonly form in a direction that diverges from the trend of the regional structures.
3. The surfaces are usually, but not necessarily, cut on the same rocks that make up the mountain. They may truncate both bedrock and alluvium, but they are best developed and preserved on bedrock, especially resistant types such as granite or related crystalline rocks.
4. Pediments are generally larger and more continuous in areas having marked vertical tectonic stability (Bull 1977; Dohrenwend 1994).
5. Pediments may or may not have a thin covering of sediment, which presumably represents load that is in transit. This characteristic has traditionally created problems because the question arises of how much alluvium can be tolerated before a pediment must be recognized as a fan, younger in age than the pediment surface and therefore divorced from the process of pedimentation. Cooke (1970) restricts the

pediment to only that part of the eroded surface not continuously covered by alluvium (fig. 7.27). The erosional surface beneath the continuous debris cover is called the *suballuvial bench,* a term first used by Lawson (1915), and the cover itself is referred to as the *alluvial plain.* The pediment, then, is bounded upslope by the mountain front and downslope by the alluvial plain (fig. 7.27).

6. Pediments are usually found in arid regions, although most workers would not restrict the processes of pedimentation to that climate. Note that we said they are *found,* not formed, in arid climates. Field data suggest that some pediments in the Mojave Desert may be relict features that formed under a more humid, Tertiary climate. In fact, many geomorphologists believe that pediment expansion is presently occurring in regions having greater precipitation than most deserts and often in seasonally wet environments (see Oberlander 1989).

Morphology and Topography   A multitude of hypotheses has been offered concerning the processes of pedimentation, but little reliable data are available to support the various ideas. After a century of study, confusion and disagreement still exist about every aspect of pedimentation. However, we can make some definitive remarks about pediment morphology and topography.

Figure 7.26
Dissected fan in Rock Creek valley, Beartooth Mountains, Mont. Grassy areas represent original fan surface. Tree-covered zones are in portions of the fan that have been entrenched when the master river (Rock Creek) cut to lower level.

Size and Shape Pediments vary in size from less than 1 square kilometer to hundreds of square kilometers, probably depending on fundamental geomorphic controls. Shape is also variable; pronounced irregularities occur when the rocks cut by the pediment surface have wide differences in resistance to erosion. Generally they tend to be fan-shaped in plan view, narrowing toward the mountain front and widening downslope. Across the pediment, the shape can be either convex or concave.

Surface Topography Contrary to lay opinion, pediments are not monotonous, smooth, flat surfaces but are dissected by incised stream channels and dotted with residual bedrock knobs, called **inselbergs,** that stand above the general level of the pediment itself. Inselbergs have been investigated repeatedly with regard to pedimentation (e.g., Twidale 1962, 1978; Kesel 1973, 1977; Twidale and Bourne 1975). In some cases, these residual hills might be the last unconsumed vestiges of a landscape that has been totally pedimented. More likely, however, most inselbergs represent areas of rock that are more resistant to weathering and erosion (Kesel 1977; Twidale 1978).

The frequency and size of both incised valleys and inselbergs seem to increase toward the mountain front, sometimes giving the topography the aspect of gently rolling hills and valleys. Some of the channels and other depressions may be filled with alluvium up to 3 m thick (Cooke and Warren 1973), giving the false impression that the bedrock surface is smooth and perfectly planed.

The Piedmont Angle The upper boundary of the pediment is usually marked by an abrupt change from the steep slopes of the mountain front to the low declivity of the pediment surface. In plan view the boundary is usually linear, but embayments into major valleys of the mountain front can give the trace a rounded or crenulate appearance. The angle formed by the junction of the two surfaces is the **piedmont angle** (fig. 7.28). Its development and maintenance have traditionally been cited as evidence in theoretical models of pediment origin.

In detail, the piedmont angle can take the form of a narrow zone of intense curvature rather than a distinct angle. Twidale (1967) reports mountain front slopes of 22° changing to pediment gradients of 3° over a transition zone 100 m wide. Both the magnitude of the piedmont angle and the sharpness of the angular relationship are probably related to structural or lithologic control (Denny 1967; Twidale 1967; Cooke and Reeves 1972). For example, the piedmont angle is most distinct when formed in granitic and gneissic rocks, perhaps because these rocks tend to break down into bimodal debris consisting of sand and boulders. The assumption is that the piedmont angle represents an adjustment of the two slopes to the size of debris they are required to transport; the boulder size of the mountain front is presumably controlled by joint spacing. Piedmont angles seem to be less well-defined in other rock types, but exceptions do occur (Selby 1982). Other factors such as joint density and subsoil weathering may affect the character of the piedmont angle (Mabbutt 1966; Twidale 1967, 1978).

Slope The longitudinal profile of almost all pediments is slightly concave-up, although local convexities do occur. Overall longitudinal convexities have been suggested as a theoretical possibility if the suballuvial bench is also considered (Lawson 1915), but available observation and geophysical data (Langford-Smith and

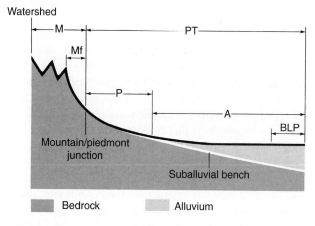

Figure 7.27

Landforms in the mountain-basin geomorphic system: M = mountain area, Mf = mountain front, P = pediment, PT = piedmont plain, A = alluvial plain, BLP = base-level plain.

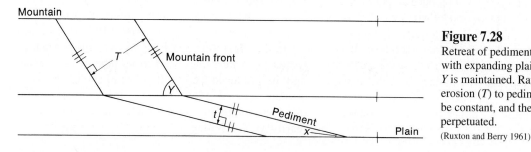

Figure 7.28

Retreat of pediment surface parallel with expanding plain. Piedmont angle at *Y* is maintained. Ratio of hillslope erosion (*T*) to pediment erosion (*t*) must be constant, and the rectilinear profile is perpetuated.

(Ruxton and Berry 1961)

Dury 1964) have not demonstrated the actual presence of such a form. Slope angles on pediments range from 0.5° to 11° but seem to average about 2.5°.

It is widely accepted that pediment slopes are controlled entirely by the size of the material they are required to transport, and thus are "slopes of transportation" (Bryan 1922). Some studies have demonstrated a strong correlation between pediment slope and particle size in debris mantling the eroded surface (Akagi 1980). However, detailed measurements show that the relationship is not as straightforward as previously supposed (Melton 1956b; Dury 1966b; Cooke and Reeves 1972). Where particle size decreases in an orderly way downslope, the rate of decline in pediment gradients is often greater than the rate at which particle size is reduced. Part of the lack of correlation between slope angle and particle size may be explained by the complication introduced by changing environmental conditions. Parsons and Abrahams (1987) suggest slope and size are related only in that portion of the load that is small enough to be mobile under present surficial processes. Therefore, lack of correlation between the two parameters may indicate that the load characteristics were generated under conditions that no longer exist.

It might also be logically assumed that slope should be related in a significant way to the area of the mountain drainage basin or to the length of the pediment. It has been noted, for example, that pediments have lower gradients where they are associated with large rivers or canyons. Although data are limited, neither of these assumptions was substantiated in other morphometric studies of pediments in California and Nevada (Mammerickx 1964; Cooke 1970). In addition, Twidale (1978) showed that in sequences of pediments having the same source, pediment slopes may decrease with time. The oldest and highest pediments have steeper gradients than the younger pediments. Twidale stressed that in order for this to occur the mountain front scarp must be notably stable. He also demonstrated that slopes are related to structural and lithologic controls.

Processes Any viable model of pediment genesis must explain the morphologic and topographic elements of the pediment association. The *pediment association* includes the pediment, the mountain area adjacent to it, and the area of the related alluvial plain. Because these show considerable variation, it seems unlikely that any one combination of processes will produce all pediments or that any single evolutionary model will suffice. Over the years, a number of models utilizing a few basic processes have emerged as the prime hypotheses for pediment origin. However, pediment processes are not easy to study over large areas or during short time periods, and so most of the proposed mechanics of formation are based on intuition rather than solid observational data.

There is no doubt that water flows across pediment surfaces in several different forms. The presence of drainage patterns on undissected surfaces provides unmistakable proof that true streamflow occurs on pediments. Many authors have suggested it as a dominant process of pedimentation (Gilbert 1877; Paige 1912; D.W. Johnson 1932; Rahn 1967; Warnke 1969), especially when the streams migrate laterally across the surface while simultaneously planating the rocks. Although most experts recognize the efficacy of lateral planation in developing part of the eroded surface, many believe this process is incapable of producing and maintaining the piedmont angle in areas remote from the main drainage lines. Streamflow erosion of these interfluve regions would require that rivers emerging from the mountains occasionally flow perpendicular to the pediment slope—a maneuver that defies the law of gravity.

Besides normal river flow, unconcentrated flows in the form of sheet and rill wash or floods also traverse pediment surfaces. The phenomenon of sheet-flooding was first observed by McGee (1897), who saw the flow as the erosive mechanism in the formation of pediments. Although this concept was adopted by some later workers (Lawson 1915; Rich 1935), it was refuted by others who believed that sheetflow acting alone could not create large pediments, especially across resistant lithology, because the smooth surface itself is a necessary prerequisite for the development of unconcentrated flow. Observations of storm runoffs in Arizona (Rahn 1967) indicate that large discharges on pediments occur as streamflows rather than sheetflows. It may not, in fact, be particularly important to know whether the flow at the head of the pediment is an unconfined type such as sheetwash or whether it moves in small rills or channels. What is important is that flow at the base of the mountain front appears to be a capable transporting agent. Therefore, flow plays a significant role in pedimentation by preventing the accumulation of debris and, by doing so, perpetuating the character of the piedmont angle. This fact has been documented beyond question in the development of miniature pediments where the underlying rocks are relatively nonresistant (Schumm 1962).

The processes of weathering have also been noted, with different degrees of emphasis, as important factors in pedimentation. The objection that sheetwash may be ineffective as an erosive agent is partly overcome if the initial strength of the pedimented rock is lowered by weathering. Weathering profiles of considerable thickness have been recognized on pedimented rocks (Mabbutt 1966; Twidale 1967; Oberlander 1972, 1974), leading to the suggestion that the rock surfaces beneath the weathered mantle, and even some suballuvial benches, may be produced by weathering rather than by water erosion. Mabbutt (1966) suggested that pediments developed on granite and related rocks may be formed primarily by continuing subsurface weathering that, be-

Figure 7.29
Inselberg standing above general level of pediment surface in Namibia.

cause of slight differences in resistance, produces an uneven bedrock surface beneath the mantle. In a series of papers, Twidale (see 1990 for references) has reinforced this conclusion and has suggested that a genetic relationship exists between weathering and a wide variety of erosional features found in granitic terrains. The contact between the unconsolidated mass produced by weathering (the regolith) and the fresh rock beneath the regolith is called the **weathering front.** Erosional stripping of the regolith exposes the weathering front, and the unaltered rock surface assumes a myriad of shapes known as **etch forms** (Twidale 1990). Etch forms, therefore, require two separate events in their development; they are considered polygenetic because the creation of the weathering front and its subsequent exposure result from different geomorphic processes. Regarding pediments, Twidale views the etching process as being responsible for the vast Australian plains. High areas that stand above those surfaces are not inselbergs (fig. 7.29) left unconsumed by continuously expanding pediments. In-

stead, bedrock hills are born that way during stripping of regolith from an uneven surface of fresh granite. The higher zones of the undulating granitic surface are the result of a structural control in the rocks, primarily the wide spacing of joints.

Dohrenwend (1994) points out that part of the reason pediments are so common in granitic terrains is the fact that subsurface weathering of this rock type produces grus (see chapter 4), a residual material that is particularly nonresistant to erosion. Thus, when discharge increases in piedmont streams, lateral erosion of banks composed of grus quickly increases sediment load to the point where the excess energy needed for incision is not available. The result is that entrenchment of granitic surfaces is more unusual than in piedmont zones underlain by other rock types.

In other rock types, weathering may be more rapid at the surface, and the products are removed by sheetwash to maintain the relatively flat pediment surface. Some evidence supports the conclusion that subsurface weathering is most pronounced at the junction of the pediment and mountain front (Mabbutt 1966; Twidale 1967). Thus, headward extension of the pediment and the maintenance of the piedmont angle may be intimately related to the type of rocks involved and the efficiency of the weathering processes.

Formative Models Cooke and Warren (1973) point out correctly that the critical boundary in piedmont areas lies between zones that are primarily depositional and those that are predominantly erosional. The position of this alluvial boundary and the presence or absence of pediments are determined by the amount of sediment produced in the mountain relative to the ability of processes to transport the material across the piedmont zone. If supply exceeds transportation, the alluvial boundary may abut the mountain front and no pediment will be found. In the opposite case, the boundary may be well down the piedmont slope and pediments will be present. If equilibrium exists between rates of supply and removal, the alluvial boundary will be stable and not necessarily parallel to the mountain front. Once established, the boundary can change its position in response to climatic or tectonic alterations. Pediments may be modified after their formation by regrading, or they may be isolated from the processes that are adjusted to the new piedmont setting, as when the surface is entrenched or is buried beneath an alluvial cover. With these possibilities in mind, Cooke and Warren (1973) and Cooke et al. (1993) have categorized situations under which pediments form. Discussions of the most commonly cited formative models are presented in these references.

There is compelling evidence to believe that climate and time cannot be ignored in interpreting pediments (Cooke et al. 1993). Oberlander (1972, 1974) presented a thought-provoking argument that the granite pediments of the Mojave Desert were formed in their entirety before the region became as intensely arid as it is today. He concluded that boulders included in the mantles covering the pediment were originally isolated as corestones in very deep chemical weathering and have reached their present position by subsequent stripping of the weathered zone. The evidence for this conclusion is quite strong. The bouldery mantle can be traced into a well-developed weathering profile, including corestones in a grus matrix, which is preserved beneath basalts dated at > 8 m.y. The pediment, therefore, is not an exhumed, fluvially eroded surface but, instead, most likely represents the weathering front formed under a semiarid Tertiary climate. The mantle cover is not material in transit but is the remaining part of a weathered profile that was progressively stripped after the region became more arid in the late Pliocene and Quaternary.

The recognition of inherited desert pediments merits special attention as geomorphologists search for genetic models. Oberlander and Twidale recognize the polygenetic nature of many pediments, although they differ on the magnitude of headward extension of the pediment surface and on how the positive elements of the landscape (e.g., mountain remnants, inselbergs) are developed. Nonetheless, their work demonstrates that some landforms do not yield significant relationships based on analyses of process and form. The failure of pediments to reveal any morphometric consistency may be attributed to the fact that some pediments developed under different morphogenetic conditions than those of the present. They are not equilibrium forms but may be relicts from the distant past that are disequilibrium anomalies in their modern surroundings.

The relationship between weathering and pediment development probably applies only to granitic terrains. Nonetheless, it demonstrates once again the irrefutable importance of climate and geology in geomorphic systems and resurrects time in considering equilibrium. Granites in an arid climate may require imponderable time spans before their external form reflects an adjustment between processes and geology.

DELTAS

Because sediment being transported by rivers must ultimately come to rest, deposition at or near a river mouth represents a bona fide component of the fluvial system. The most important geomorphic feature produced in that environment is called a delta. Deltas, also important sedimentary entities, are studied by geologists for the depositional sequence and complex facies relationships included in the deltaic mass. Our interest in deltas is only in the geomorphic processes that develop their form. The details of delta sedimentation will be left to the sedimentologists.

The term *delta* is usually applied to a depositional plain formed by a river at its mouth, where the sediment accumulation results in an irregular progradation of a shoreline (Coleman 1968; Scott and Fisher 1969). The feature was first named 2500 years ago by the historian Herodotus, who noted that the land created at the mouth of the Nile River resembled the Greek letter Δ (delta). Modern deltas, however, display a great variety of sizes and shapes. At the apex of a delta, the trunk river divides into a number of radiating branches, called *distributaries,* that traverse the delta surface and deliver sediment to the delta extremities. In plan view some deltas look like alluvial fans and, in fact, a *fan-delta* generally means an alluvial fan prograding into a body of standing water. Considerable attention has been given to fan-deltas because they are composed of coarse-grained deposits, and sedimentologists find utility in distinguishing coarse deltaic materials according to depositional characteristics. For example, the term *braid-delta* has been suggested to indicate deltas having a coarse subaerial component consisting entirely of braided-river or braid-plain facies (McPherson et al. 1987). These deposits have different properties than normal fan-deltas and may be more important as reservoir rocks in the geological record. This terminology, however, is not universally accepted (Dunne 1988).

Historically, sedimentologists have been most interested in portions of the delta that are continuously submerged and, therefore, have studied in detail the marine deposits and processes affecting the feature. Geomorphologists are more interested in the feature as produced by fluvial processes and usually concern themselves more with the subaerial portion (e.g., see Morton and Donaldson 1978). Thus, even though fans and deltas are related in form and process, they differ in several important respects: (1) deposition on deltas results from a reduction of river velocity as the flow enters a body of standing water, which can be the ocean or a lake of any size or origin; (2) delta expansion in a vertical sense is finite, the base-level water body being the approximate limit of upward growth; (3) the gradient on the delta surface is notably flatter than that on most fans. (4) Perhaps the greatest difference is that all major marine deltas are Holocene in age and seem to have their beginnings between 8000 and 6500 B.P. The key factor causing this worldwide phenomenon was deceleration of the sealevel rise that began when the last major glacial episode ended (Stanley and Warne 1994).

Delta Classification, Morphology, and Deposits

Most major rivers of the world develop deltas, and each has its own unique properties reflecting some balance between the fluvial system, the climate, tectonic stability, and shoreline dynamics. For those interested in detailed analyses of the major deltas of the world, we refer

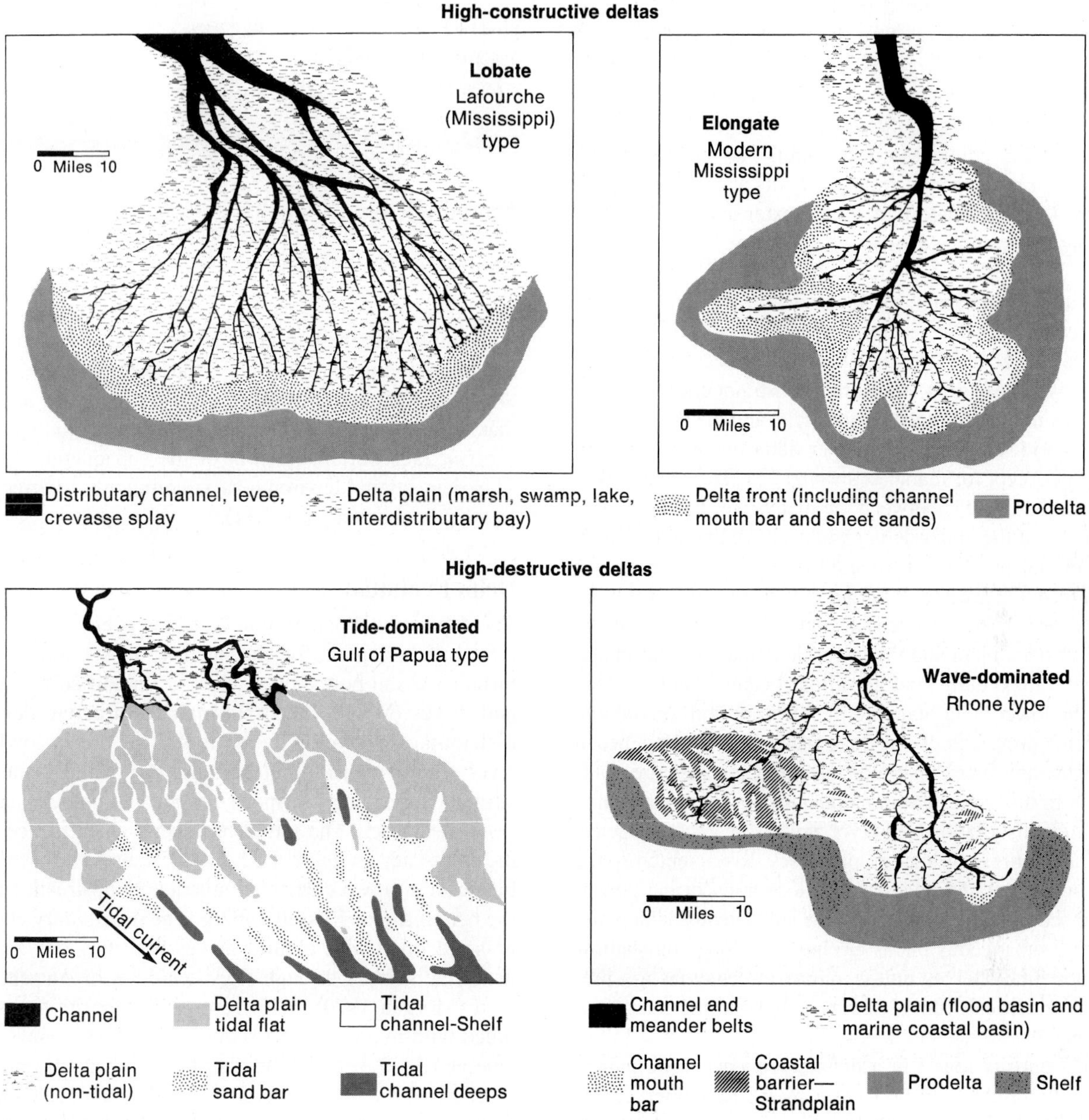

High-constructive deltas

Lobate
Lafourche
(Mississippi)
type

0 Miles 10

Elongate
Modern
Mississippi
type

0 Miles 10

■ Distributary channel, levee, crevasse splay
Delta plain (marsh, swamp, lake, interdistributary bay)
Delta front (including channel mouth bar and sheet sands)
Prodelta

High-destructive deltas

Tide-dominated
Gulf of Papua type

Tidal current

0 Miles 10

Wave-dominated
Rhone type

0 Miles 10

■ Channel
Delta plain tidal flat
Tidal channel-Shelf

Delta plain (non-tidal)
Tidal sand bar
Tidal channel deeps

■ Channel and meander belts
Delta plain (flood basin and marine coastal basin)

Channel mouth bar
Coastal barrier—Strandplain
Prodelta
Shelf

Figure 7.30
Classification and geomorphic characteristics of basic delta types.
(Scott and Fisher 1969)

you to the excellent collection of reviews presented as a tribute to J. P. Morgan, one of the pioneers in deltaic research (Stone and Donley 1998). As these reviews show, deltas come in a multitude of plan-view shapes, dependent primarily on the balance between sediment supply and marine processes. In general, however, we believe that several types serve as model forms and can be used as a broad basis for classification. **High-constructive deltas** develop when fluvial action is the prevalent influence on the system and therefore are commonly called fluvially dominated or river-dominated deltas. As figure 7.30 shows, these deltas usually occur in one of two forms: an *elongate* type exemplified by the modern

birdfoot delta of the Mississippi River or a *lobate* type exemplified by the now-abandoned Holocene deltas of the Mississippi River system. Both types have high sediment input relative to the marine dynamics. Elongate deltas have a higher mud content and tend to subside rapidly when they become inactive, thereby preserving the upper sand facies. Lobate deltas sink slowly upon abandonment, and much of the sand that was prograded in the upper zones is reworked by marine processes (Scott and Fisher 1969).

High-destructive deltas originate where ocean or lake energy is high and much of the fluvial sediment is reworked by waves or currents before its final deposition, and thus are also called marine dominated. Figure 7.30 shows two types of these deltas. In *wave-dominated* types, such as those of the Nile and Rhone rivers, sediment is accumulated as arcuate sand barriers near the mouth of the river. In *tide-dominated* types tidal currents arrange the sediment into sand units that radiate linearly from the river mouth. Muds and silts accumulate inland from the segmented bars where extensive tidal flats or mangrove swamps evolve.

Deltas consist of three physiographic parts called the upper delta plain, the lower delta plain, and the subaqueous delta. The *upper delta plain* begins as the river leaves the zone where its alluvial plain is confined laterally by valley walls. When this constraint ends, the river breaks into a multitude of channels, and the depositional plain widens. The entire upper delta plain is fluvial in origin except for marshes, swamps, and freshwater lakes that exist in areas between the many river channels. The surface of the upper delta plain is above the highest tidal level and thus is not affected by marine processes. In contrast, the *lower delta plain* is occasionally covered by tidal water. For this reason, the boundary between the upper delta plain and the lower delta plain is determined by the maximum tidal elevation. Features and deposits in the lower delta plain are the result of both fluvial and marine processes. Tidal flats, mangrove swamps, beach ridges, and brackish-water bays and marshes are common in this zone. The *subaqueous delta plain* is located continuously below sea level, and its character is determined entirely by marine processes. Rivers entering the ocean in well-defined channels lose transporting power, and sediment is deposited as the subaqueous delta plain. Large subaqueous plains are best developed on shallow continental shelves unless submarine canyons are present which can funnel sediment beyond the margin of the shelf.

Deposits found in a deltaic sequence were named topset, foreset, and bottomset by G. K. Gilbert in his 1890 report on Lake Bonneville (fig. 7.31). Although Gilbert examined small deltas along the margins of the ancient lake, the stratigraphic sequence he observed is similar to that found in large marine deltas. *Topset* beds are a complex of lithologic units deposited in various subenvironments of the upper and lower delta plains. Layers in the topset unit are almost horizontal. *Foreset* deposits accumulate in the subaqueous delta plain. These deposits are usually coarser at the river mouth and become finer as they radiate seaward into deeper water. Strata in the foreset unit are inclined seaward at an angle reflecting that of the delta slope. In large marine deltas, the beds rarely dip more than 1°, but where load is coarse, as in fan- or braid-deltas, foreset beds may be inclined at angles greater than 20°. Foreset layers are bevelled at the landward positions by topset beds, which expand horizontally as the entire delta advances into the ocean. At the seaward extremity, foreset beds grade im-

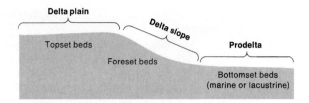

Figure 7.31
Primary depositional environments and their associated layering in a classic delta.

perceptibly into the bottomset strata. *Bottomset* deposits are composed primarily of clays that were swept beyond the delta front. These beds usually dip at very low angles that are consistent with the topography of the continental shelf or lake bottom in front of the subaqueous delta. This depositional environment is commonly referred to as the prodelta zone (fig. 7.31).

Delta Evolution

The depositional pattern that develops at any individual river mouth depends on the intensity of spreading and turbulence and how those factors are modified by tides and waves (Wright 1977). In high-constructive deltas, distributaries develop at the mouth of the inflowing river where longitudinal bars are deposited because bedload cannot be transported when the velocity suddenly decreases. The initial bar exerts an influence on the flow and, as figure 7.32 shows, causes the river to bifurcate into two channels immediately upstream from the bar crest (see Russell 1967b). The distributary channels are lined with natural levees that may begin beneath the surface through slow accretion of suspended load (Morgan 1970). With continued deposition, combined with minor channel scouring, the levees and bars emerge (fig. 7.32), and the distributary channels extend farther into the basin. The process may be repeated frequently, giving the basin a veinlike appearance and prograding the delta front.

As the delta progrades, shorter routes to the ocean become available. These pathways often begin far inland from the delta front, usually developing when the river is diverted through a breach in the levee called a *crevasse*. These diversions of sediment and water are associated with dramatic shifts in the course of the river, resulting in periodic relocations in the position of deltaic deposition. This phenomenon, known as *delta switching*, has produced five distinct delta complexes by the Mississippi River during the Holocene (fig. 7.33). Delta switching within the Mississippi system seems to occur with a frequency of 1000 to 2000 years. A sixth complex, related to diversion of the Atchafalaya River, is presently in an early stage of development (fig. 7.33; Roberts 1997; Coleman et al. 1998). The Atchafalaya, a distributary breaking off the Mississippi upstream from

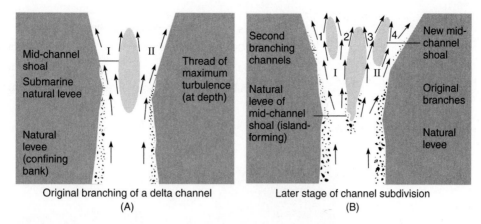

Figure 7.32

Stages in bifurcation of trunk river and creation of distributary channels of a delta.

(Russell 1967)

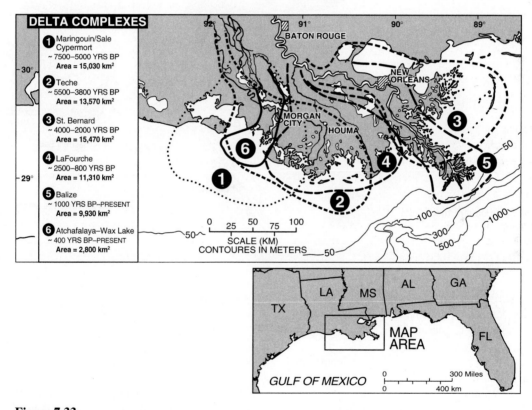

Figure 7.33

Location map of the Mississippi River delta and coastal plain built from previous Holocene deltas.

(Figure 1. p. 701 by Coleman et al from JOURNAL COASTAL RESEARCH. Vol. 14 No. 3. pp. 698–716. © 1998. Reprinted by permission of Coastal Education & Research Foundation.)

Baton Rouge, Louisiana, carries about 30 percent of the Mississippi River flow.

During flood events considerable load is transported down the Atchafalaya, which has progressively filled in shallow lakes in the lower Atchafalaya basin. Since the early 1950s most of the sediment has been reaching Atchafalaya Bay (approximately 160 km west of New Orleans), where it is actively building the new deltaic lobe (Shlemon 1975; Rouse et al. 1978). Many scientists believe that total diverson of the Mississippi into the Atchafalaya is inevitable because that route to the ocean is about 300 km shorter than the present course. This gives the Atchafalaya a distinct advantage, and capture will occur unless humans use heroic measures to maintain the status quo (for an excellent discussion see McPhee 1989). When lobes are abandoned during

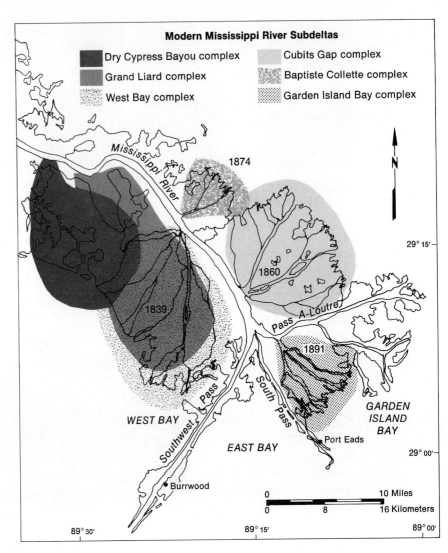

Figure 7.34
Subdeltas of the modern birdfoot delta
of the Mississippi River.
(Morgan 1970)

shifting of the river course, they are no longer fed by in-coming river debris and immediately become vulnerable to erosional attack by the ocean. Thus, as a new lobe de-velops, older lobes are being destroyed.

It appears from recent analyses that the depositional architecture of a delta may consist of offset and/or stacked lobes. Within any lobe are smaller depositional units called subdeltas (fig. 7.34), which, in turn, are built when sediment is diverted through a breached levee and accumulates in even smaller units known as crevasse splays. Thus, a distinct depositional hierarchy exists in most deltas of the Mississippi type. Four such subdeltas have formed in historic time (fig. 7.34). As channels in the subdeltas bifurcate and prograde, their gradients de-crease and they lose the ability to transport load. When their gradients approach that of the main trunk channel, it is no longer advantageous for the river to utilize the crevasse system. Sedimentation in the subdeltas ends, and the ocean begins to inundate the area as subsidence and compaction lower the subdelta surface. At the same

time, however, a new subdelta may be developing some-where else within the birdfoot system.

Delta formation can be viewed on several time scales. On a short-term basis only a limited area (sub-delta) receives any sediment, but the position of the ac-cumulation shifts repeatedly. On a longer time scale the entire active delta (lobe) has a periodic migration. What we observe as the Mississippi River delta is in fact a monstrous area created by the coalescence of a number of major lobes over a long period of time. Active deposi-tion occurs on only one lobe at any given time, and on only a minor portion of that lobe. Even so, the evolu-tionary processes occurring on any time scale are proba-bly similar in that they involve channel bifurcation, levee development, and crevassing (see Coleman 1988). In fact, delta complexes and lobes, as well as their smaller components, seem to follow a cyclic pattern consisting of three phases of growth and abandonment. Roberts (1997) suggests that the stages are: (1) rapid growth with increasing-to-stable discharge, producing

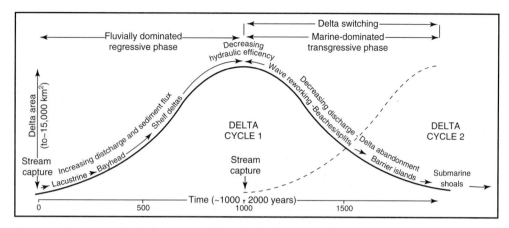

Figure 7.35   A graphic representation of the systematic changes associated with delta growth and abandonment: the delta cycle.

(Figure 5. p. 703 by Coleman et al from JOURNAL COASTAL RESEARCH. Vol. 14. No. 3. pp. 698–716. © 1998. Reprinted by permission of Coastal Education & Research Foundation.)

coastal aggradation and delta plain accretion, (2) relative stability during the beginning of waning discharge when deposition balances the effect of subsidence, and (3) abandonment during delta switching, followed by dominance of marine processes that rework the delta perimeter (fig. 7.35).

Even though this evolutionary model is probably correct, it is valid only for deltas of the high-constructive type. This model is known in such detail only because geologists have studied the Mississippi delta for many years. Other delta types have not been so closely examined, and their mode of origin is not nearly so well known. Furthermore, geomorphologists are far from understanding precisely how changes in climate, tectonics, or any of the other fundamental controls affect the mechanics of delta evolution.

SUMMARY

The character of each major fluvial landform results from the manner in which a particular set of geomorphic controls influences the river mechanics. The features can be predominantly erosional, be predominantly depositional, or derive from the combination of both types of processes.

Floodplains in humid, temperate regions originate by lateral migration of meanders and by periodic overbank flooding. The sediment composing a floodplain sequence is mainly laterally accreted point bar deposits accumulated as the river shifts its position across the valley bottom. The point bar material is usually capped by a thin layer of silt and clay deposited as vertically accreted sediment during overbank flooding of the river. The total amount of vertical accretion is probably controlled by the rate of lateral migration of the river. Because many rivers move across their valley floors rapidly, most floodplains are developed by lateral accretion, but overbank deposition may be important in the initial phase of construction. In arid climates floodplains form in a different manner and may not be in equilibrium with the usual hydrologic characteristics that control humid-climate floodplains.

Terraces are merely abandoned floodplains. They form when entrenchment places the river at a lower level, thereby removing the former floodplain surface from the hydrologic activity of the river. The origin of a terrace usually refers to how its flat tread (the former floodplain level) was formed. The tread of an erosional terrace is produced by the lateral migration of a river and is capped by the thin point bar and overbank deposits associated with floodplains formed in that manner. In contrast, the surface of a depositional terrace represents the upper level of the sediment deposited in an episode of valley filling. The use of terraces to reconstruct geomorphic history demands some knowledge of their origins, because the salient properties of the various terrace types develop through different sequences of events.

Piedmont regions are characterized by depositional features (alluvial fans) and plains of erosion (pediments). Both features manifest some accommodation between the amount of sediment derived from a source area and the ability of the river to transport the sediment across the piedmont zone. The processes that develop fans and work on their surfaces are so complex that little agreement exists regarding the meaning of fans in geomorphic history. Pediments also seem to defy genetic generalization. They are usually interpreted to be the result of lateral planation, weathering, and rill wash, or some combination of the various processes. As with fans, however, little agreement can be found on the origin of pediments. Some may be relict features that are completely unrelated to modern geomorphic controls. It is clear that piedmont landforms cannot be placed into all-inclusive genetic models. Their properties vary too much to be explained by one mode of origin. Because of this, every piedmont region must be examined and interpreted according to local tectonics, climate, geology, and geomorphic history.

Deltas represent the accumulation of sediment as a transporting river enters a body of standing water. At its mouth the river bifurcates into distributaries and constructs levees. This allows debris to be transported farther into the ocean or lake basin and permits the delta to expand by prograding into the basin area. The form and size of the delta, however, depend on the balance reached between river flow and the counteracting energy of currents and waves in the ocean or lake. Detailed studies of deltas reveal a very complex growth history in which the site of active sedimentation shifts periodically through crevasses in the natural levees. Active sedimentation occurs on only a small part of the feature at any given time.

SUGGESTED READINGS

The following references provide greater detail concerning the concepts discussed in this chapter and also contain more extensive bibliographies on the various topics.

Blair, T., and McPherson, J. 1994a. Alluvial fans and their natural distinction from rivers based on morphology, hydraulic processes, sedimentary processes, and facies assemblages. *Jour. Sed Res.* 64:450–89.

Bull, W.B. 1977. The alluvial fan environment. *Prog. Phys. Geog.* 1, 222–70.

Bull, W. B. 1990. Stream-terrace genesis: Implications for soil development. *Geomorphology* 3:351–67.

Coleman, J., Roberts, H., and Stone, G. 1998. Mississippi River Delta: an overview. J. Coastal Res. 14, 698–716

Dohrenwend, J. 1994. Pediments in arid environments. In Abrahams, A. and Parsons, A., eds. *Geomorphology of desert environments,* pp. 321–53. London: Chapman and Hall.

Harvey, A. 1997. The role of alluvial fans in arid zone fluvial systems. In Thomas, D., ed., *Arid zone geomorphology: Process, form and change in drylands,* pp. 231–59. 2nd ed. Chichester: Wiley.

Knox, J. 1987. Historical valley floor sedimentation in the Upper Mississippi valley. *Ann. American Assoc. Geog.* 77:224–44.

Nanson, G., and Croke, J. 1992. A genetic classification of floodplains. *Geomorphology* 4:459–86.

Oberlander, T. 1997. Slope and pediment systems. In Thomas, D., ed., *Arid Zone Geomorphology: Process, Form and change in Drylands.* pp. 135–64. 2nd ed. Chichester: Wiley.

Pazzaglia, F., Gardner, T., and Merritts, D. 1998. Bedrock fluvial incision and longitudinal profile development over geologic timescales determined by fluvial terraces. In Tinkler, K., and Wohl, E., eds., *Rivers over rock: Fluvial processes in bedrock channels,* pp. 207–35. Wash., D.C. Am. Geophys. Un.

Schumm, S. 1977. *The fluvial system.* New York: Wiley-Interscience.

Stone, G., and Donley, J., eds. 1998. Tributes to James Plummer Morgan. *Jour. Coastal Res.* 14 (3): Thematic Section I, 695–858; Thematic Section II, 859–915.

Twidale, C. 1990. The origin and implications of some erosional landforms. *Jour. Geology* 98:343–64.

Waters, M. and Nordt, L. 1995. Late Quaternary floodplain history of the Brazos River in east-central Texas. *Quat. Res.* 43, 311–19

Wells, S., and Harvey, A. 1987. Sedimentologic and geomorphic variations in storm-generated alluvial fans, Howgill Fells, Northwest England. *Geol. Soc. Amer. Bull.* 98:182–98.

WIND PROCESSES
AND LANDFORMS

Courtesy of
Nick Lancaster

Introduction
The Resisting Environment
The Driving Force
Entrainment and Transportation
    Processes
    The Effect of Saltation
    Erosional Features
Deposits and Features
    Ripples
    Dunes
    Fine-Grained Deposits
Summary
Suggested Readings

INTRODUCTION

Scientists have not always agreed on the role of wind in geomorphology. Nonetheless, proof that wind is capable of significant geomorphic work is becoming irrefutable, because definitive, quantitative data concerning eolian processes and features are being collected at an increasing rate, and scientific syntheses of knowledge concerning eolian phenomena are now appearing on a regular basis. In addition, geological studies indicate that the results of eolian processes are recognized throughout the stratigraphic record (e.g., Glennie, 1970; Kocurek 1981; Loope 1984; Clemmensen et al. 1989). Thus, it is safe to say that wind can be an effective geomorphic agent given the proper physical conditions. Regions having sparse vegetation and unconsolidated sediment not tightly bound by rooting systems are most susceptible to wind attack. Extensive evidence of eolian processes, therefore, is found in those areas, such as modern deserts, where vegetation growth is stunted by lack of water or immature soil development. As is true of any process, however, the proficiency of wind to do geomorphic work depends on whether the driving force can exceed resistance of the surficial material. Thus, assigning wind processes to one particular environment alone is a gross perversion of the principles governing geomorphic processes. For example, dune development adjacent to beaches or stream channels is a common phenomenon that is driven by the same underlying process mechanics that generate many major desert landforms. The emphasis placed on deserts here is simply a pedagogical tool, not a geomorphic necessity.

There are some compelling reasons for the study of eolian processes in desert zones. More than one-third of the land on our planet is characterized as arid or semi-arid, and even a generation ago over 600 million people lived in these areas. That number is certainly larger today. Understanding the processes that function within these vast regions is critical for their future utilization. In fact, conversion of productive land into deserts (desertification) is a major global problem having enormous ramifications. We know that eolian processes play a significant role in desertification because dried and unvegetated topsoil is easily eroded by wind action. Therefore, in periods of extended drought, arable land is changed into infertile desert areas.

In addition to human concerns, our knowledge of eolian processes has been utilized extensively in the interpretation of features and environments on other planets. Much of the analysis of Mars, for example, has been based on our understanding of eolian work on Earth. At the same time, space technology now allows scientists to interpret the processes and landforms in large tracts of deserts that would be difficult, if not impossible, to synthesize through land-based studies (Greeley et al. 1989; Blount and Lancaster 1990; Forman et al. 1992; Greeley et al. 1997; Ramsey et al. 1999). It is clear that these efforts will increase in the future. Nonetheless, the viability of such analyses will still depend on the continuous expansion of knowledge concerning the details and complexities of eolian systems. Detailed reviews of the physical basis of wind action and its geomorphic results are available (Bagnold 1941; Greeley and Iverson 1985; Pye 1987; Pye and Tsoar 1990; Cooke et al. 1993; Pye and Lancaster 1993; Lancaster 1995; Tchakerian 1995; Livingstone and Warren 1996; Thomas 1997; Goudie et al. 1999) and much of the following has been liberally excerpted from those excellent treatments.

THE RESISTING ENVIRONMENT

Contrary to popular thought, deserts are not the barren tracts of shifting sand depicted in Hollywood epics. Only one-fourth to one-third of most desert surfaces is occupied by sand, which usually occurs in large, sandy plains. Normally, deserts display a variety of erosional and depositional landforms in a diverse topography that ranges from flat plains to rugged mountains (table 8.1, fig. 8.1). Deserts also exist in different temperature zones. Polar deserts are common, although relatively unstudied. Specific processes differ in relative importance in polar deserts and hot deserts, and so the similarity of landforms that sometimes exists between the two regions is not infallible proof of an identical genetic history. The same external form may be a function of a myriad of basic processes combined in slightly different ways. Unless stated otherwise, the discussion that follows refers only to hot desert conditions.

Deserts are by definition arid. They receive less than 25 cm of annual precipitation and have enormous evaporation rates, commonly 15 to 20 times greater than the precipitation (Stone 1968). Although deserts have a meager plant cover, the diversity of vegetal types is surprising, ranging from shrubs and grasses to true woodlands. This diversity is created by minor variations in soil moisture that relate to elevation. It is

TABLE 8.1	Area and Percentage of Total Desert Region of Major Components in the Arid Zone of Australia (Compare with Map in Fig. 8.1.).	
	Km³	**Percentage of Arid Zone**
Mountain and piedmont deserts	930,000	17.5
Riverine desert	210,000	4.0
Stony desert	640,000	12.0
Desert clay plains	690,000	13.0
Sand desert	1,680,000	31.0
Shield desert	1,200,000	22.5
Total	5,350,000	

Source: J. A. Mabutt (1971).

important because humus content and soil binding differ with particular species and because some plants are annual and others are perennial. The type of flora influences the resisting setting. Woodlands, for example, are significantly denser than other vegetation types and protect more surface area. In general, mountainous zones with stabilizing vegetation and thicker weathering mantles are less susceptible to wind attack than is the lower level of the desert environment.

Weathering and soil-forming processes are intimately involved in the character of a land surface and how it resists or succumbs to wind erosion. The angular, unaltered debris so prevalent on desert surfaces is presumed to be the residuum of rock disintegration, leading to a general concurrence that mechanical weathering predominates in the desert environment. This does not mean that mechanical weathering functions to the total

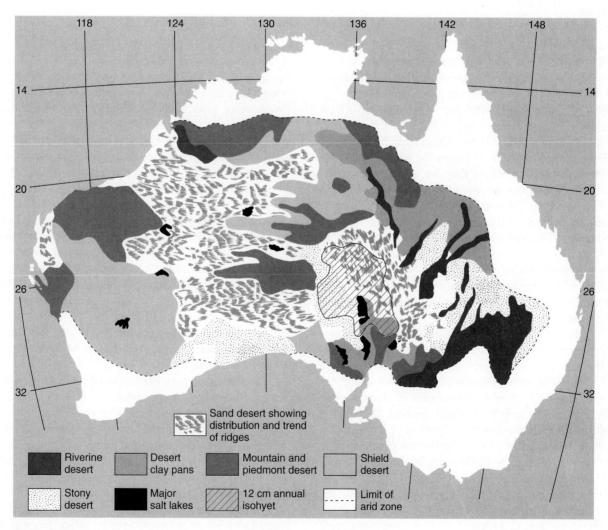

Figure 8.1
Map of Australian arid regions showing diverse physiography of deserts.
(Mabbutt 1971)

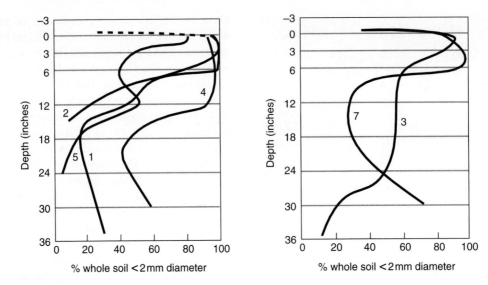

Figure 8.2
Percentage of soil composed of particles less than 2 mm in diameter. Analyses of soils in the
Lahontan Basin, Nev.
(Springer 1958)

exclusion of decomposition. Chemical processes, in fact, do function in the desert, as evidenced by the ubiquitous desert varnish and the common red coloration of desert soils, both of which probably require some degree of iron mobility. In many desert areas, surfaces are armored with a thin layer of stones that protects an underlying zone of sand, silt, and clay. These covers are particularly common in hot deserts where they are given a variety of names such as *stone mantles, gibber, hammada, reg,* and *desert pavement.* As mentioned in chapter 7, desert pavement is commonly associated with alluvial fans and other deposits of unsorted alluvium. The surface armor is only one or two stones thick and consists of whole clasts, or disintegrated parts of clasts, found in the coarse fraction of the underlying alluvium. In most cases, a decimeter thick silt and clay layer exists between the pavement and the underlying sediment (fig. 8.2).

Over the years, researchers have suggested numerous processes for the concentration of stones at the surface. The most commonly cited in earlier studies were (1) removal of surface fines by wind deflation and/or surface runoff and (2) upward migration of clasts through the fine-grained zone by repeated episodes of shrinking and swelling (for reviews see Cooke et al. 1993; Dixon 1994). More recent work suggests that pavements are born at the surface by colluviation of clasts from topographic highs into depressions filled with eolian silts, and/or detachment and uplift of bedrock clasts as dust accumulates in fractures (McFadden et al. 1987). These suggestions were given added credence from cosmogenic dates indicating that clasts are not concentrated randomly over geologic time as required by earlier interpretations of pavement origins (Wells et al. 1995).

It now seems clear that stones remain at the surface because of vertical inflation caused by deposition of wind-blown dust and subsequent soil-forming processes (Wells et al. 1995; McFadden et al. 1998). In addition, most researchers now agree that stone pavements, regardless of origin, enhance the entrapment of eolian dust. This leads to formation of soil horizons that are not normally created by shallow chemical weathering. Most important of these is a horizon characterized by well-developed vesicles and a columnar structure. This zone, called the *vesicular horizon* (designated A_v), seems to be formed only where desert pavement is present. Thus, formation of the A_v horizon, as well as other desert soil horizons (see Reheis 1987a), is uniquely related to the creation of pavement and the influx of eolian dust. Furthermore, the magnitude of dust accumulation and the downward translocation of clays and mobile ions (Ca, Na, etc.) in soil profiles are directly related to Quaternary climate changes (Reheis 1987a, b; McFadden et al. 1998; Reheis et al. 1995).

Details concerning the origin of arid-climate soils were presented earlier in chapter 3. For our purposes here, we note that desert soils (even in an immature state) are important determinants of the facility of wind action. In many cases soils form during a period of landscape stability that is broken by episodes of eolian erosion and/or deposition. The soils are interbedded with eolian deposits (Holliday 1989b; Forman and Matt 1990), and upper portions of the soil profile are often truncated prior to sediment accumulation (Blair et al. 1990). These truncations are called *bounding surfaces* and may be used as correlation tools between stratigraphic sections that contain eolian depositional units. In general, regardless of the presence or absence of pavement, A horizons in desert soils are quite thin, mechanically weak, and more permeable than lower horizons (Cooke et al. 1993). In fact, even though illuviation proceeds slowly, it may cement B horizons to the extent that they become highly indurated zones (Gile et al. 1965, 1966; Gile 1966). Thus, the long-term result of desert weathering is to promote a surface that is susceptible to wind erosion and a subsurface that is eminently resistant. These resistant zones, referred to as *duricrusts,* are usually composed of silica, calcium carbonate, or gypsum (see Cooke et al. 1993). Stripping of the surface horizon will expose the lower, resistant zones, which are preserved as new stable surfaces for extended periods of time. Stripping at one place is commonly balanced by deposition of eolian debris somewhere else in the area.

In contrast to regions of armored surfaces and/or resistant soil horizons, deserts also have areas of abundant fine-grained sediment that are unprotected by vegetal or stone cover. Many of these areas have been affected by human activity such as consumptive use of surface and ground waters, and removal of vegetation during construction and development. These activities produce effects similar to those brought on by climate change, and in many basins have resulted in dessication of lakes, formation of playas, and reactivation of eolian processes (Gill 1996). It is well documented that playas are especially susceptible to significant erosion by deflation (Reheis and Kihl 1995; Cahill et al. 1996), even though it is difficult to ascertain the magnitude of such erosion. Several factors contribute to the problem of predicting the extent of the deflation process. First, erosion may be retarded because fine sediment at the surface is commonly fabricated into a cohesive crust that can protect unconsolidated fines immediately beneath the surface. This occurs because crust production creates aggregates of particles that collectively become larger than the entrainable size expected under any given wind velocity (see Chepil and Woodruff 1963; Cooke et al. 1993). Second, the relationship between factors producing entrainment (e.g., saltation, threshold velocities) and the flux in particle mass during transport are still not completely understood, even though considerable progress has been made toward that end (e.g., Shao et al. 1993; Gillette et al. 1997a, 1997b; Marticorena et al. 1997).

In summary, the geomorphic effect of wind action is regulated to a significant degree by the properties of the resisting framework. For wind processes to have any geomorphic consequence, the surface must be unvegetated and littered with noncohesive sediment smaller than gravel size. Any region producing sediment of this type, and lacking the climatic, topographic, and geomorphic conditions needed to protect it, may be open to pronounced wind attack. Because hot deserts are most susceptible to wind action, they are taken as the model for the discussion of eolian processes. However, coastal regions, unvegetated semiarid or subpolar zones, and areas in front of active glaciers are also often modified by wind processes. In any situation, the amount of geomorphic work actually accomplished by the wind depends on the second prime variable—the character of the wind itself.

THE DRIVING FORCE

Global wind systems are related to the large pressure differentials associated with worldwide circulation patterns. In contrast, local winds are generated by anomalies in the physical characteristics within a specific area. For example, in many cases winds reflect a difference in the thermal properties of surface materials, and so we can expect the greatest wind activity in those environments where large temperature variations exist in the air layer immediately above ground level. Such conditions are common in deserts, along seacoasts, and in areas with pronounced diversity in elevation, such as the juncture of mountain ranges and the low plains fringing them (Tyson and Seely 1980).

Certain attributes of the wind, mainly its direction, velocity, and degree of turbulence, are responsible for most geomorphic effects. In areas with large thermal contrasts, *wind direction* is more or less predetermined by the temperature gradient of the near-surface air. However, temperature gradients often change from daytime to nighttime because of variations in rates of cooling and heating. In deserts, heating of the bare desert surfaces incites winds to move toward them from surrounding areas that have a vegetal cover. At night, however, the desert surface cools more rapidly than the other regions, and the winds reverse their direction. A similar diurnal direction change occurs along coasts because the ocean warms less slowly than the adjacent land, prompting onshore winds during the day. The ocean, however, gives up stored heat less readily at night, and winds blow offshore. The importance of wind direction is most evident in the development and preservation of eolian bed forms constructed from loose, transportable sand.

Wind velocity is important because it is the prime determinant of what material will move under wind attack and what will remain stationary. Wind velocity increases with height above the ground because it is slowed at the surface by friction. The change in velocity with height can be expressed in a number of ways exemplified by the equation (Bagnold 1941)

$$V_z = 5.75 \, V_* \log \frac{z}{k}$$

where V_z is the velocity at any given height z, V_* is a parameter called **drag velocity** (or friction velocity), and k is a constant relating to surface roughness. V_* originally designated as a drag phenomenon, is actually a shear velocity. It therefore relates to shear stress and commonly is called the *surface friction speed* (Greeley and Iversen

(V_*)

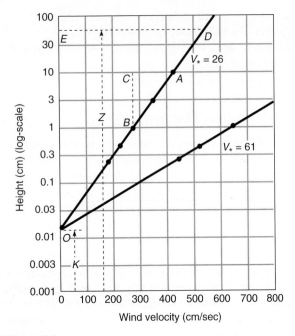

Figure 8.3
The relationship of wind velocity, height above surface (z)
and shear or drag velocity ($V*$).
(Bagnold 1941)

1985). The parameter k is the height of a thin zone immediately above the surface in which velocity is zero. The thickness of this zone depends on the surface roughness, but on a flat, granular bed it tends to be about 1/30 of the diameter of the surface grains. The general relationship between height and velocity plots as a straight line on semilogarithmic paper (fig. 8.3).

As the wind blows harder, $V*$ increases and exerts a greater stress on particles exposed at the surface. This follows because the magnitude of shear (drag) across any unit area of surface is related to $V*$ such that

$$V_* = \sqrt{\tau/\rho}$$

or

$$\tau = \rho \, V_*^2$$

where τ is shear/unit area and ρ is the density of the air. Thus, for a surface with constant particle size, a stronger wind will increase the value of $V*$, and therefore the height-velocity relationship is defined by a number of straight lines, each having a different $V*$ value (fig. 8.3). Friction velocity in this sense represents the *rate* at which velocity increases with height and can be thought of as a velocity gradient. The value of $V*$ can be determined as the tangent of the straight line, shown on figure 8.3, divided by the constant of proportionality, 5.75. For example, the tangent of line OD is $\dfrac{AC}{CB}$ or $\dfrac{150}{\log 10}$.

Therefore,

$$V_* = \frac{150}{1}/5.75 = 26 \text{ cm/sec}$$

Interestingly, all lines of different $V*$ values merge at the ordinate where velocity is zero. This occurs simply because k is a function of surface conditions, such as particle size, rather than wind velocity.

Once the value of $V*$ is known, it is possible to calculate the wind velocity at any height above the surface. For example, if $V* = 26$ cm/sec, the velocity at 4 cm above the surface is

$$V_z = 5.75 \, V_* \log \frac{z}{k}$$

$$= 5.75 \, (26) \times \log z - \log k$$

$$= 150 \, [0.6 - (-1.82)]$$

$$= 363 \text{ cm/sec}$$

A similar equation

$$\frac{u}{u_*} = \frac{1}{0.4} \ln \frac{Z}{Z_0}$$

is commonly used to demonstrate the relationship between surface friction speed and wind velocity at any given height (see Greeley and Iversen 1985). In this equation u equals the velocity at any height Z. u_* is the surface friction speed ($= V_*$) and Z_0 is the roughness height ($= k$).

Surface conditions may vary significantly along the path of wind flow. The roughness height (k, Z_0) will therefore change drastically. For example, k values range from 1 mm or less over the ocean to 5 m over the center of a large city. Variations in roughness such as these cause significant changes in the surface friction speed and influence eolian processes (Greeley and Iversen 1985; Blumberg and Greeley 1993; Wolfe and Nickling 1993).

Most winds are characterized by the irregular motions associated with turbulent flow. As in rivers, the transition from laminar to turbulent flow occurs at some critical Reynolds number above which winds are able to accomplish geomorphic work. Turbulence that occurs in the form of eddy currents has a direct influence on the entrainment process. Although it is most important in lifting particles smaller than sand (Bagnold 1941; Chepil and Woodruff 1963), grains up to 0.125 mm in diameter may be affected by wind turbulence. Most likely, turbulence helps in the molding of desert landforms, but the types of eddy motion are numerous and mechanically complex.

ENTRAINMENT AND TRANSPORTATION

Processes

As wind blows across a surface composed of loose sediment, a critical drag (friction) velocity exists, called the

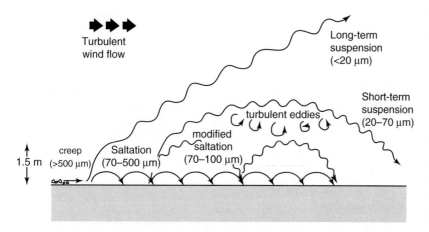

Figure 8.4
Modes of transport for particle size ranges under moderate windstorm.

threshold friction velocity ($U_{*\tau}$; or $V_{*\tau}$), at which particle motion begins. In general, the entrainment process is controlled by the relationship between particle size and wind velocity, but in detail other factors may be more important. Marticorena et al. (1997) present convincing evidence that surface roughness (indicated by roughness length, z_0) is the primary physical factor influencing the value of $U_{*\tau}$. In addition, other factors such as soil moisture, salt precipitation, and development of resistant crusts cause significant variations in the threshold velocity under any given roughness length (McKenna Neuman and Nickling 1989; Pye and Tsoar 1990, pp. 97–99; Marticorena et al. 1997).

The critical friction velocity can be estimated by the equation

$$V_{*_\tau} = A \sqrt{\frac{\rho_s - \rho_a}{\rho_a} gD}$$

where $V_{*\tau}$ is the threshold value of friction velocity, ρ_a is the density of the air, ρ_s the density of the sediment, D is the particle diameter, g is gravity, and A is a constant of proportionality, which for air is 0.1. Most desert sands have a threshold velocity of about 16 km/hr, but the precise value for any size varies with other factors.

The vertical velocity of grains leaving the surface and the height they attain is determined by the ratio between friction velocity and the downward terminal velocity produced by the gravitational force acting on the grains. The relative strength of the oppositely directed velocities is extremely important in controlling how particles are transported. The general relationship between particle size and the mode of transport is shown graphically in figure 8.4. When the upward velocity exceeds the terminal velocity, turbulent eddies are able to lift particles to high elevations, perhaps kilometers. This type of transport, called *suspension,* usually involves only silt and clay-sized particles (figs. 8.4 and 8.5). In contrast, when the upward component of drag velocity is much lower than the terminal velocity, grains will rise only centimeters above the surface and travel downwind

in a spasmodic bouncing motion known as *saltation.* Bouncing particles commonly dislodge other grains when they strike the surface (fig. 8.6). This type of transport usually involves only fine to medium sand, although in extremely strong winds saltation grains may dislodge larger particles into the airstream (Sakamoto-Arnold 1981).

Some particles are too large to be lifted from the surface by the wind itself. However, coarse sand and fine gravel may be transported by rolling, sliding, or small, nearly imperceptible hops as the surface is impacted by saltating grains (see fig. 8.4). This motion is referred to as *surface creep* (Sharp 1964) or, more recently, *reptation* (Ungar and Haff 1986; Mitha et al. 1986; Anderson 1987; Werner and Haff 1988).

The Effect of Saltation

The threshold equation clearly indicates that entrainment velocity varies directly with particle diameter; nevertheless, as figure 8.7 shows, the relationship becomes invalid when size is less than 0.1 mm because of the greater interparticle cohesion and low roughness associated with finer particles (Bagnold 1941; Smalley 1970). Furthermore, grains greater than .84 mm are moved with great difficulty, and that size may represent a logical upper limit for unaided wind entrainment. Once motion begins, however, the surface is subjected to a continuous rain of moving particles that, on impact with stationary grains, produce entrainment at a velocity lower than the fluid threshold. This reduced threshold velocity, called the *impact threshold* (fig. 8.7), becomes progressively more significant as size increases, and particles greater than .84 mm may be entrained even when the velocity is well below the fluid threshold. Indeed, an exceptionally strong 1977 windstorm in southern California carried particles as large as 7 mm at a height 240 cm above the surface (Sakamoto-Arnold 1981). In addition, it is important to recognize that fine-grained sediment can also be entrained even when threshold velocities or direct

Figure 8.5
Fine-grained sediment being lifted into suspension by winds blowing across a playa near Owens Lake, California.
(Photo by Nick Lancaster)

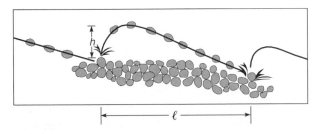

Figure 8.6
Diagram of the saltation process. Moving grain strikes
the surface and dislodges a particle to elevation h. The particle
moves downwind and repeats the process. l represents the
distance of travel of the dislodged particle before striking the
surface.

lifting by turbulence are not able to produce its erosion.
In fact, impact of saltating grains is probably the most
important factor in the entrainment of silts and clays and
in determining the total amount of dust transported by
the wind (Shao et al. 1993; Cahill et al. 1996; Gillette et
al. 1997a, 1997b). It appears, therefore, that when grains
begin to move at the friction threshold, the entrainment
process is self-generating because the wind speed is already above the impact threshold. Although saltation
will begin at the friction threshold velocity, the total entrainment of grains at the initiating velocity is not great.
Wind tunnel studies, however, show that only a small
velocity increase above the friction threshold value can
change limited saltation movement into large-scale cas-
cading transport (Nickling 1988). Each moving grain is
not only a product of the system but also an integral
component of the entrainment mechanics through the
process of saltation, as depicted in figure 8.6.

When entrainment occurs, the moving sand directly
influences the characteristics of the wind itself because
the saltating grains exert extra shear on the system
(fig. 8.8). The friction velocity (V_*) is a measure of the
velocity gradient of the wind, or the rate at which velocity increases with height. The stronger the wind blows,
the greater the shear on the surface and the greater the
divergence of the friction velocity curves away from the
vertical. The slope of each line on the figure, therefore,
reflects its rate of increase in velocity with height (given
as friction velocity, V_*). The higher V_* values indicate a
more rapid increase in velocity with ascension above the
surface.

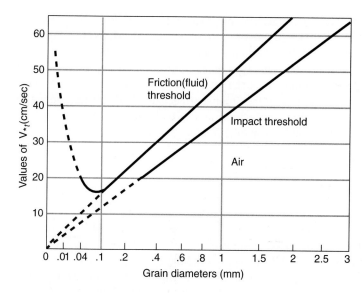

Figure 8.7
Relation of particle size to threshold velocity and impact threshold.
(Bagnold 1941)

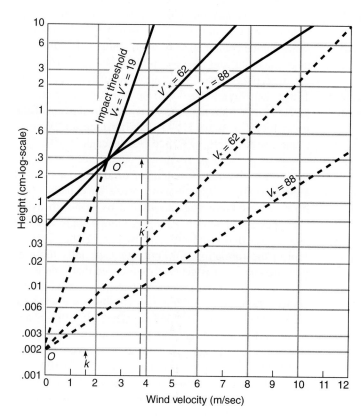

Figure 8.8
Change in friction velocity caused by saltation. Dashed lines show friction velocity before saltation begins. Solid lines represent friction velocity after saltation starts.
(Bagnold 1941)

The lines on figure 8.8 represent the changes impressed on the system after saltation begins. Note that under saltating conditions k rises dramatically, and it no longer represents a zero velocity but assumes a constant velocity value at the height k' where all the gradient lines merge. The wind velocity at any height (z) is then defined by the equation

$$V_z = 5.75 \; V_*' \log \frac{z}{k'} + V_t$$

where V_*' is the drag velocity under the conditions of saltation, k' is the new height of the focus, and V_t is the threshold velocity (impact threshold) measured at the height k'.

The effect of saltation is obvious from the graph. For example, a wind with a velocity gradient $V_* = 62$ blowing across a completely resistant surface has a velocity of 970 cm/sec at a height of 2 cm. Once motion begins, however, the velocity at that level, read along the gradient line $V_*' = 62$, has decreased to approximately 530 cm/sec. It is important to stress again that the value of k or k' is dependent on the roughness components of the surface, and so figure 8.8 defines curves that are valid for only one particle size. Nonetheless, significant advances have been made in understanding the effect of saltation on wind velocities in boundary layers and its relationship with roughness height. We refer you to Gillette (1999) for a historical perspective in the development of our current thinking.

Saltating grains travel in a characteristic path in which particles dislodged or bounced from a surface move at first in a predominantly vertical direction (see fig. 8.6). The subsequent travel path is probably controlled by the particle's initial upward velocity and the velocity gradient above the surface. Bagnold (1941) suggested that most of the forward momentum is gained from the wind when a grain is near the top of its path, and from then on the actual trace of the particle movement depends on the magnitude of the wind thrust and the settling velocity of the grain (for discussion see McEwan 1993).

Exactly how fast particles leave a surface (liftoff speed) and how high they rise is a function of many variables, including particle size, packing, and wind speed. Bagnold's early work suggested that initial vertical speed would be comparable to the friction speed. Thus, theoretical maximum heights of a particle trajectory for any given friction speed can be estimated if the effect of gravity is considered (Owen 1964). Although many estimates of maximum trajectory height are reasonable, we now know that some grains attain greater elevations than predicted. This occurs because liftoff speeds are increased when grains impact a hard surface; grains bouncing off such surfaces take on a spinning motion that imparts a greater lift force to the particle. The increased lift caused by spinning is known as the *Magnus effect,* and its consideration in motion equations provides better agreement between theoretical and observed trajectories (White and Schulz 1977). Regardless of the specific mechanics, it is clear that most grains saltate within a thin near-surface zone. Observations suggest that an overwhelming percentage of the total saltation load is carried within 2 m of the surface (Bagnold 1941; Chepil 1945; Sharp 1964; see discussion in Pye and Tsoar 1990 for details of the saltating phenomena).

Theoretically, it should be possible to estimate the total amount of saltation load being transported by integrating particle concentrations and mass flux at all heights above the surface (Anderson and Hallet 1986). This follows because the trajectories of all grains within a saltating screen are basically a function of liftoff factors and wind velocity. Actually, the derivation of equations that provide precise values of saltation load is extremely difficult and may be complicated by the fact that the response of different grain sizes to fluctuating wind velocities must be analyzed as a stochastic rather than deterministic process (Anderson 1987). In fact, even though most sand moves by saltation, some of the load never enters the realm of airstream mechanics but moves

as surface creep. Creep (reptation) results when the impact of saltating grains spasmodically shoves or rolls surface particles forward without displacing them upward. This type of motion becomes progressively more important as the grain size increases (Sharp 1964) and is probably the dominant mechanism of forward motion in very coarse sand and gravel. Nonetheless, in average windblown sand, ranging from 0.15 to 0.25 mm, creep seems to account for 7 to 25 percent of the total load.

Erosional Features

Geomorphic features manufactured by wind action are due to abrasion and deflation, the two main erosive processes. Abrasion results when sand particles carried by the wind act as grinding tools to physically wear away exposed surfaces of solid rocks or rock fragments. Maximum abrasion usually occurs within 2 meters of the surface where the saltating curtain is dominant. The magnitude of erosion, however, seems to be determined by some unique combination of wind velocity, resistance of target material, and the concentration of hard particles in the wind (i.e., Sharp 1964, 1980; Suzuki and Takahashi 1981). Our review here will be necessarily brief. For greater depth we refer you to excellent treatments of the topic found in Laity (1994), Livingstone and Warren (1996), Breed et al. (1997, and Goudie et al. (1999).

Perhaps the most frequently cited evidence of wind abrasion is the development of eroded stones called **ventifacts** (fig. 8.9), which in a broad sense includes all objects that have been modified by wind abrasion (Whitney and Dietrich 1973). Although there is some suggestion that wind-blown dust can be the abrasive tool in the formation of ventifacts, especially those of delicate texture, it is generally conceded that sand is the dominant agent used in the production of these features (Laity 1994). The most common ventifacts are wind-cut faces called facets, polished surfaces, pits, flutes, and grooves (fig. 8.9). *Facets* are produced on rock surfaces that are oriented nearly perpendicular to the prevailing wind. The incessant bombardment of wind-driven sand gradually cuts a smooth face into the windward portion of the original surface and forms a sharp edge between it and the leeward side of the particle. The slope on the faceted surface is normally between 30° and 60°, but enough variability exists in these angles to suggest that some evolutionary sequence is involved in their development. It is equally feasible, however, that various heights of pebbles above the surface relative to various levels of maximum abrasion potential in the wind can produce a number of facet inclinations and shapes (Sharp 1964). Most facets are cut by sand moving in the saltation zone. Maximum abrasion, however, seems to occur at about 10 to 15 cm above the ground surface (Sharp 1964, 1980; Greeley and Iversen 1985; Anderson 1986).

Figure 8.9
Ventifact features on mafic boulders on the Tolman Ranch Terrace surface. (A) Boulder standing above the surface with indistinct ventifaction. Coin is silver dollar. (B) Faceted boulder. Quarter for scale. (C) Pitted surface. Nickel for scale. (D) Pitted and fluted surfaces. Penny for scale.
(Ritter and Dutcher 1990)

Facets do not form on surfaces that parallel the wind direction, and therefore the ubiquitous presence of boulders with more than one face and edge has created some interpretive furor. It is tempting to regard multifaceted ventifacts as evidence of a shift in the prevailing wind direction, but it is equally possible that the stones themselves are occasionally turned over or rotated. Overturning, of course, would result in a new portion of the original pebble being subjected to wind erosion, and thus any number of facets could form in a unidirectional wind.

Small ventifact features—pits, grooves, flutes—can occur on any rock surface exposed to wind action, but they are most commonly associated with or superimposed on faceted boulders (fig. 8.9). *Pits* are small depressions eroded into the surface by grain impact. They are usually circular or concentric but may assume any shape. Pits are best developed on steeply inclined surfaces that face into the prevailing wind. They often

begin at and grow from points of weakness (e.g., soft minerals, microfractures) in the target rock.

Grooving of bedrock surfaces is also a common eolian phenomenon. The abrasive processes that result in parallel furrows seem to operate at different orders of magnitude, ranging from tiny elongate flutes on facets (Sharp 1964), to linear, U-shaped depressions, called *grooves,* which may be centimeters wide and extend in length for tens of centimeters. These large varieties commonly occupy linear weaknesses in the target rock such as veins or fractures. *Flutes* seem to be transitional between pits and grooves in that they often originate in a pit and extend from there across the rock surface as shallow, linear furrows. They usually occur no higher than 2 m above the surface and are aligned parallel to the strongest winds (Sweeting and Lancaster 1982), but not necessarily the prevailing wind. When measured correctly, however, flutes seem to be the most reliable ventifact for determining the prevailing wind direction at the time of their formation (Sharp 1949; Laity 1987).

On a larger scale, abrasion may fabricate entire rock outcrops. The most common large feature is known as a **yardang,** a wind-eroded ridge that is elongated parallel to the prevailing wind. Yardangs (fig. 8.10) normally stand less than 10 m high but under ideal conditions may attain heights greater than 100 m and extend longitudinally for kilometers (McCauley et al. 1977). The windward end is normally blunt, high, and steep with general lowering to a pointed leeward end (see fig. 8.10B). Yardangs are found in deserts on all the continents except Australia and are well developed in the equatorial plains of Mars (Ward 1979; Greeley and Iverson 1985). They are most prominent in regions underlain by relatively soft rocks and tend to occur in groups in which individual ridges are separated by round-bottom, wind-eroded troughs (Haynes 1982). Blackwelder (1934) emphasized the erosion of the intervening "Yardang trough" rather than the ridge itself. He believed that most erosion occurs at the lowest level, decreasing in magnitude and efficiency toward the ridge crest. As a result the troughs are eroded more rapidly, so that sometimes the ridge sides are slightly undercut and the ridge crests are left rather ragged.

Detailed studies and reviews (McCauley et al. 1977; Ward and Greeley 1984; Greeley and Iversen 1985) suggest that the geometry of yardangs represents an equilibrium shape established by gradual erosion. The ideal form is streamlined with a width to length ratio of about 1:3 or slightly larger which offers minimum resistance to the prevailing wind (Ward and Greeley 1984; Laity 1994; Goudie 1999). The ideal shape may be affected by other factors such as rock lithology and structure, topography, and supply of abrasive tools, but in general the form is similar to that of other streamlined features. For greater detail concerning yardangs, we refer you to Laity (1994), Livingstone and Warren (1996), Breed et al. (1997), and Goudie (1999).

The process of deflation has been suggested as the prime mechanism in the genesis of many enclosed desert basins. These basins range in size from small hollows to vast expanses measured in hundreds of square kilometers. They are often elongated in a downwind direction and may contain subcircular extinct or ephemeral lakes called *pans.* In the Kalahari Desert, the depressions are normally 5 to 20 m deep and have centrally located pans that are grass-covered or bare. The pans are underlain by a sequence of deposits up to several meters thick including clays, sandy clays, and sands. The depositional sequence probably reflects changing hydrologic conditions within the depression, and, in fact, the extent of pan deposits indicates that the depressions once held water bodies 2 to 3 times larger than present (see Cooke et al. 1993). General morphometric relationships of pans can be found in Goudie and Wells (1995) and Sabin and Holliday (1995).

There seems to be general agreement that wind-erosion by deflation is involved in the development of pans. This follows because many but not all pans are closely linked to the formation of lunette dunes on the downwind margin of the pan (Lancaster 1978; Livingstone and Warren 1996; Goudie et al. 1999). However, initial creation of the depression that becomes a pan may involve other processes (i.e., Albritton et al. 1990), and some pans and their related dunes may be controlled by fluctuating climates during the later Quaternary (Lancaster 1978).

DEPOSITS AND FEATURES

The most striking features associated with eolian processes occur in vast, sandy deserts called **sand seas,** or **ergs** in the Sahara of North Africa. They have been recognized throughout the geologic record (Kocurek 1981; Loope 1984; Clemmensen et al. 1989). These areas have an enormous volume of sand and are marked by varied assemblages of depositional features (fig. 8.11). Complex dune fields are common but are not the only component of sandy deserts. Large areas are often occupied by tabular bodies of sand, called **sand sheets,** or by relatively flat interdune areas of various types (see Ahlbrandt and Fryberger 1982). Sand sheets are characterized by the absence of dunes and therefore have little surface topography. They are often marginal to dune fields, however, suggesting that some factor in the region of their development may interfere with the processes of dune formation (Kocurek and Nielson 1986). The largest sand sea in the Western Hemisphere is the Nebraska Sand Hills, an area of eolian features that was formed in the Holocene and covers almost 57,000 km^2 of northwest Nebraska (Ahlbrandt and Fryberger 1980; Stokes and Swinehart 1997).

Sand seas initially form in depositional sinks where wind energy that is used to transport sand from a source area is drastically lowered. The system processes are controlled by sediment supply, transport capacity of the wind, and availability of sediment for transport. These variables are ultimately determined by tectonics, climate, and sea-level fluctuations (Lancaster 1999; Kocurek and Lancaster 1999). The enormous volume of sand contained in sand seas (table 8.2) accumulated over thousands, and perhaps, million of years, during which climatic cycles have influenced the distribution of sand and the character of features (Holliday 1989a, 1989b; Forman et al. 1992; Blount and Lancaster 1990; Lancaster 1999). In fact, dune patterns within sand seas are the product of changes in wind systems and sand supply that result in the formation of different types of dunes. Furthermore, these alternating conditions are clearly associated with distinct episodes of dune formation. Thus, the complex assemblages of features found in the modern sand seas are reflecting multiple generations of

(A)

(B)

Figure 8.10
Photos showing development of yardangs in the Karga region of central Egypt. (A) Wind erosion of limestone in initial stage of yardang formation. (B) Yardang in final stage. Marginal playa deposits deflated by winds moving from left to right.

(Photos by Ted Maxwell)

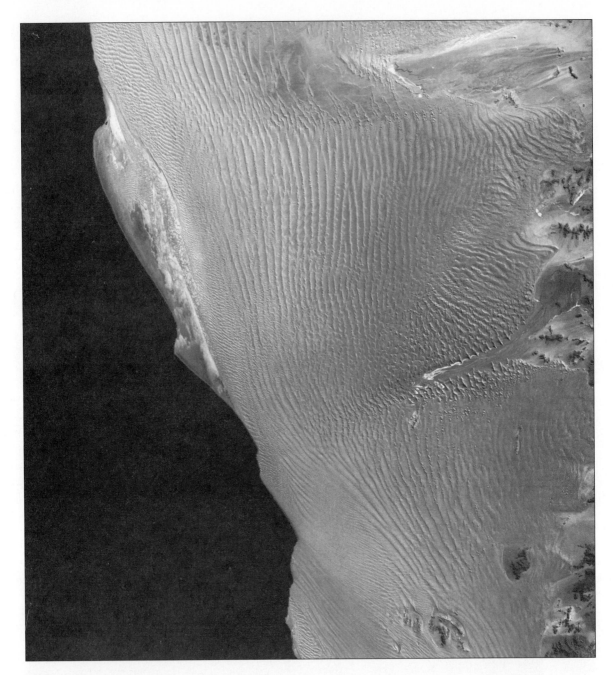

Figure 8.11
Landsat photo of Namib sand sea. Crescentic dunes in upper left. Linear dunes in central area, and star dunes in center right.

development (Lancaster 1992, 1999), and some historical perspective is needed to understand their meaning.

Within the sand seas, surface features commonly referred to as bed forms are spaced with pronounced regularity and come in a variety of sizes ranging from tiny **ripples** to giant forms called **draa.** Draa are actually special varieties of **dunes** that are the most common depositional bed form and are intermediate in size between the other types. No clear dimensional boundaries separate the geometric types, which are described in table 8.3, and it is not unusual for the smaller features to be superimposed on parts of the larger ones. For example, draa are built from superimposed dunes in which dune types may be the same (called compound dunes) or different (called complex dunes). Furthermore, after draa growth reaches a critical size, individual (simple) dunes tend to develop readily on the flanks of the draa (Lancaster 1988a). The factors that control the spacing of bed forms are not completely understood and will be discussed in the section treating dunes. Suffice it to say here that no simple relationship explains the spacing or dimensions of bed forms.

Ripples

Wind ripples range in amplitude from .01 to 100 cm and may be spaced up to 20 m apart. The dimensions depend primarily on wind velocity, particle size (Sharp 1963), and the type of ripple. Some ripples are formed purely by the shear stress of the wind acting on the surface (aerodynamic ripples), but the overwhelming majority of ripples result from the surface bombardment of saltating grains and its associated creep (impact ripples).

Bagnold (1941) suggested that the process of forming impact ripples is strongly influenced by the angle at which saltating grains strike the surface. Since saltating grains descend in a nearly uniform manner, most variations in the angle of incidence are caused by minor topographic irregularities of the surface (fig. 8.12). The intensity of creep in any surface area is proportional to the number of impacts produced by saltating grains. Thus, in the natural surface hollow depicted in figure 8.12, the number of grain collisions is much greater on the windward face of the depression (*BC*) than the lee side (*AB*). Thus, more grains are transported up the slope *BC* than are replenished by movement down the slope *AB* into the depression. Not only is the hollow maintained, but grains also begin to accumulate at point *C* because they are delivered there faster than they can be removed on the adjacent level surface. Eventually the accumulation at *C* produces a second lee slope *CD,* which reinitiates the mechanics that operated in the original hollow. Repetition of this sequence inexorably propagates the ripple form downwind. Because of this distinct formative process, Bagnold (1941) felt that surfaces covered by unprotected sand and having no bed forms are probably unstable.

The process just described should create an ever-rising ripple crest and a gradually deepening hollow. Actually, the height of the ripple is effectively limited because it eventually rises to a level where wind velocity is much lower. In this way the ripples assume a consistent geometry that, for well-sorted sands between 0.19 and 0.27 mm, has a height/wavelength ratio that is normally only 1:70 and never lower than 1:30 (Bagnold 1941).

The exception is where coarse grains are concentrated at the ripple crest (Bagnold 1941; Sharp 1963). This allows the ripple to grow higher than expected, because the large grains continue to be deposited while the smaller particles flow over the crest into the hollow. This commonly results in a height/wavelength ratio as low as 1:10 and an asymmetric ripple shape in which windward slopes are notably more gentle than lee slopes. The asymmetry, however, may be lost if a high percentage of the ripple is composed of very large grains. These "granule ripples" (Sharp 1963) have very large wavelengths and may be burdened by a high proportion of grains larger than sand size.

The origin and spacing of ripples is complicated by particles that travel only short distances when they are dislodged by saltating grains. These particles, moving very close to the surface in a mass known as the **reptation population** (Ungar and Haff 1987; Mitha et al. 1986), are probably very important in determining ripple spacing (see fig. 8.4). In fact, Anderson (1987) suggests that the fastest growing wavelength in spacing is roughly six times the mean reptation length. This strongly indicates that the process of reptation is the primary factor leading to the formation and translation of ripples, and the reptation length is a much more important parameter in the development of ripple wavelength than the much longer trajectory associated with normal saltation.

TABLE 8.2	**Estimates of Sediment Volume Contained in Some Sand Seas.**		
Sand Sea	**Area (km²)**	**Average Thickness (m)**	**Sediment Volume (km³)**
Erg Oriental	192,000	26	4992
Issaouane-n-Irarraren[a]	38,500	43	1655
Erg Occidental[a]	103,000	21	2163
Simpson Desert[a]	300,000	1	300
Namib Sand Sea[b]	34,000	20	680

Source: Lancaster (1999).

[a]Data in Wilson (1973)

[b]Data from Lancaster (1989)

From N. Lancaster, in AEOLIAN ENVIRONMENTS, SEDIMENTS AND LANDFORMS by Goudie, Livingston and Stokes, p. 50, table 3.1. Copyright © 1990. Reproduced by permission of John Wiley & Sons Limited.

TABLE 8.3	**Eolian Bed Forms and Their Geometry and Possible Origin.**			
Wavelength	**Height**	**Orientation**	**Possible Origin**	**Suggested Name**
300–5500 m	20–450 m	Longitudinal or transverse	Primary aerodynamic instability	Draa
3–600 m	0.1–100 m	Longitudinal or transverse	Primary aerodynamic instability	Dunes
15–250 cm	0.2–5 cm	Longitudinal or transverse	Primary aerodynamic instability	Aerodynamic ripples
0.5–2000 cm	0.05–100 cm	Transverse	Impact mechanism	Impact ripples
1–3000 cm	0.05–100 cm	Longitudinal	Secondary vortices	Secondary ripple sinuosity

From I. G. Wilson (1972). Used with permission of *Sedimentology*, Blackwell Scientific Publications.

Dunes

Of all desert and wind phenomena, sand dunes have received the greatest scientific attention (for reviews see Lancaster 1994, 1995). Many dunes attain a characteristic equilibrium profile that can be logically divided into three components: the backslope or windward surface, the crest, and the slip face or lee slope (fig. 8.13). Measurements show that the backslope declivity, normally between 10° and 15°, is in stark contrast to the slip face, which always stands near the angle of repose for sand, between 30° and 34°. The crest, separating zones of erosion and deposition on the dune, is usually convex-up.

The occurrence of different dune types is generally controlled by vagaries in wind direction combined with wind speed, sand supply, vegetative cover and particle size (Fryberger 1979; Lancaster 1983; Wasson and Hyde 1983a). Most dunes range in height from less than 3 m to 100 m. Dune height and width seem to correlate well with crest-to-crest spacing (Breed and Grow 1979; Lancaster 1983, 1988b; Wasson and Hyde 1983a, 1983b). The relation between dune height and spacing (fig. 8.14) is regular enough to be represented by the power function.

$$D_H = cD_s^n$$

where D_H is dune height and D_S is spacing. The value of the exponent is subject to considerable variation but apparently is sensitive to the availability of sand and the wind regime. These factors determine whether dunes will accrete vertically, migrate, or extend, and thus are closely related to the type of dune that develops (Lancaster 1988b). In fact, much of the scatter shown in figure 8.14 is a function of dune type and becomes less pronounced when the height/spacing relationship is considered for separated dune species (fig. 8.15). It is generally assumed that the height and spacing will increase until a dune stabilizes in some steady-state or quasi-equilibrium condition, but this perception of dune evolution has been seriously challenged (Werner 1995; Werner and Kocurek 1999).

The pronounced regularity in dune spacing hints at some prevailing atmospheric motion that is capable of maintaining dune forms and their spacing in an equilibrium state. The processes involved, however, are complex because dune growth itself interferes with the air flow and creates eddy currents or, alternatively, fixes a regularly spaced pattern of turbulence downwind from the intruding dune. Eddy currents unquestionably exist in the lee of dunes, but considerable questions remain as

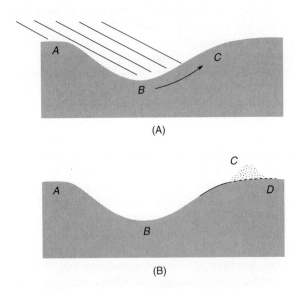

Figure 8.12
Process of creep in the development of ripples in a natural surface hollow. (A)Minor topographic irregularity increases the incidence of saltating grains on the windward face *BC*. (B) Creep builds ripple crest at *C* and a second lee slope *CD*.
(Bagnold 1941)

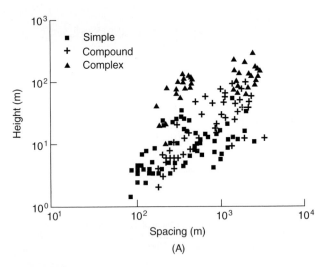

Figure 8.14
Relation between height and spacing of eolian dunes. For all dune types ($r = 0.80$). Note overlap in height and spacing between simple, compound, and complex varieties.
(Lancaster 1988)

Figure 8.13
Cross-profile of normal dune showing common geomorphic components.

to whether they are modified by other flow mechanics and, more important, whether they possess the erosive power needed to regulate the spacing or shape of dunes. Recent studies suggest that other complex airflow patterns exist beyond the eddy zone on the lee side of dunes (Frank and Kocurek 1996). It also seems certain that different scales of air flow may exist in the same regional wind regime, and the morphology of different dune types is adjusted accordingly. This explains why the heights of draa in the Namib sand sea are inversely proportional to wind-related, potential sand-transport rates, while individual dunes superimposed on their flanks vary directly with those same potential rates (Lancaster 1988b).

The growth and form of any dune is fundamentally controlled by changes in sediment transport rates that are directly related to erosion and deposition on the dune surface. Early studies attempted to estimate potential rates of sand flux from wind velocity profiles (e.g., Howard et al. 1978; Mulligan 1988). This procedure has been questioned because wind profiles that are derived from conventional anemometry do not include that part of the boundary layer that pertains directly to sand transport (Frank and Kocurek 1996; Lancaster et al. 1996). Direct measurements of sand flux and near-surface flow show that wind speeds increase significantly toward the crest on the stoss side of dunes, and that this velocity increase is accompanied by an increase in sediment transport by

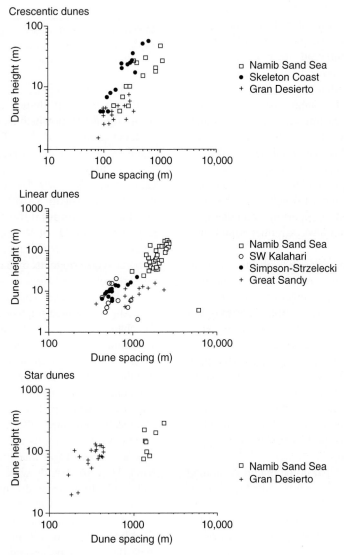

Figure 8.15

Relations between height and spacing of different dune types (Lancaster, 1994).

(From GEOMORPHOLOGY OF DESERT ENVIRONMENTS by Athold D. Abrahams and Anthony J. Parsons, Figure 18.6, p. 478. Copyright © 1994. Reprinted by permission of the authors.) (Lancaster 1994)

one to two orders of magnitude (Lancaster et al. 1996; McKenna Neuman et al. 1997). The crestal zone is, therefore, the most active dune area and should theoretically exist at a sharp angle between the stoss slope and slip face. However, this geometry is most prevalent only where dunes are subjected to bidirectional wind regimes (McKenna Neuman et al. 1997), and, in other cases, the crestal shape is partly conditioned by flow patterns on the lee side of the dune.

It is axiomatic that maintenance of the equilibrium shape requires forward movement of the entire feature, because erosion from the rear slope must be volumetrically balanced by deposition on the lee slope. There seems to be general agreement that dunes retain their original form as they advance and that the shape components, especially the height, influence the rate of forward migration. The actual rate of migration, however, varies greatly between individual dune types and larger features such as draa. In fact, Lancaster (1988b) points out that the time needed for the different varieties to migrate one wavelength in the Namib sand sea may vary by several orders of magnitude. Simple or superimposed dunes may travel that distance within tens of years and, therefore, are probably adjusted to annual or seasonal wind conditions. In contrast, draa migration is so slow that draa are relatively insensitive to short-term wind characteristics and sand availability. The implications of these differences are significant because the size and spacing of draa may have developed under regional wind patterns and sand transportation rates which no longer exist. Therefore, unlike the simple or superimposed dune varieties, draa properties do not always reflect equilibrium with modern conditions.

The cross-sectional characteristics of dunes tend to be somewhat changeable because complex wind motions and variable surface conditions interfere with their ideal development. Even more complex are the plan views of dunes, or **dune patterns.** Pattern classifications are almost as diverse and bewildering as the forms they claim to classify. Like other features, dunes have been grouped on the basis of shape, genesis, wind types involved, surface conditions, and the like, but each attempt somehow fails to account adequately for nature's incredible diversity. For our purposes, we will use a modified form of the classification devised by the U.S. Geological Survey (table 8.4, fig. 8.16).

Hack (1941) suggested that in the region he studied three basic dune forms exist, which he called transverse, parabolic, and longitudinal. Transverse dunes are usually free of vegetation. They may be the most probable dune pattern if winds are unidirectional over a limited supply of sand and the sand is free to migrate. Actually, the mechanics involved in maintaining this dune form may be controlled by the wind velocity (and its associated turbulence) needed to transport the largest particles in the sand (Lancaster 1982). The normal transverse

	TABLE 8.4	**Terminology for Basic Types of Dune Forms.**

Form	Number of Slip Faces	Name Used in Ground Study of Form Slip Faces and Internal Structure
Circular or elliptical mound	None[a]	Dome
Crescent in plan view	1	Barchan
Row of connected crescents in plan view	1	Barchanoid ridge
Asymmetrical ridge	1	Transverse ridge
Circular rim of depression	1 or more	Blowout[b]
U shape in plan view	1 or more	Parabolic[b]
Symmetrical ridge	2	Linear (seif)
Asymmetrical ridge	2	Reversing
Central peak with 3 or more arms	3 or more	Star

Modified from McKee (1979) USGS Professional Paper 1052.

[a]Internal structures may show embryo barchan type with one slip face.

[b]Dunes controlled by vegetation.

dune is a crescent-shaped feature, called a **barchan dune** (fig. 8.17), in which tapering edges or horns of the crescent point downwind. Transverse dunes may also stand as simple ridges oriented perpendicular to the wind, or they may form a sinuous asymmetrical ridge, called a *barchanoid ridge,* which is composed of connected crescents (fig. 8.16).

Parabolic dunes differ from transverse dunes only in that they are U- or V-shaped, and their horns or tapered extremities point upwind. When the lateral edges of a transverse ridge become anchored by vegetation, the wind removes sand from the central zone and deposits it on the leeward slope. This, of course, allows the middle segment to advance relative to the edges and develops the characteristic parabolic shape. In addition, the dune surrounds a scoop-shaped hollow, called a *blowout,* from which the sand in the dune was derived, indicating that the pattern is partly erosional in origin. In fact, parabolic dunes may exist in a variety of forms that probably reflect a complex pattern created during reactivation of older stabilized dunes (Wolfe and David 1997).

Longitudinal forms are called **linear dunes.** They exist as narrow ridges that extend parallel to the forming wind. They are usually wider and steeper at the upwind end, gradually tapering downwind until they merge with the desert surface. In the Navajo country, in northeastern Arizona, linear dunes are separated from one another by sand-free flats up to 100 m wide. Both the flats and the dune flanks may be vegetated, leaving only the sand of the ridge tops bare of vegetal cover and susceptible to wind transport. In fact, it is often difficult to change the shape of a linear dune (Werner 1995). A special variety of linear dune is called a **seif dune.** Bagnold (1941)

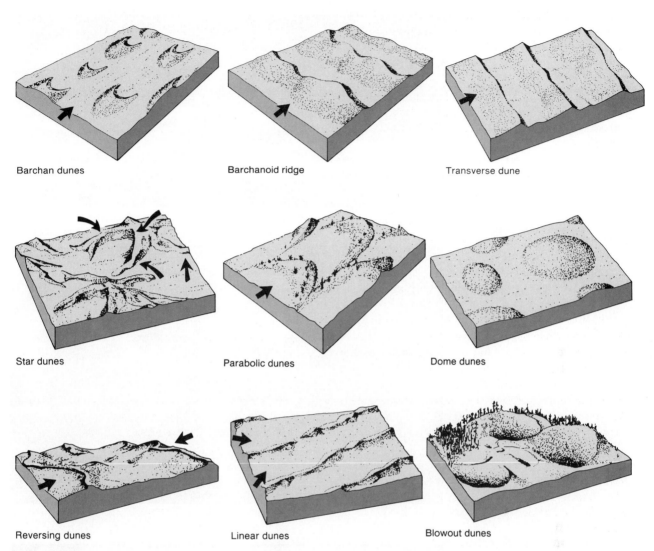

Figure 8.16
Common dune types. Arrows indicate direction of formative winds.
(McKee 1979)

considered it to be one of only two true dune forms, the other being the barchan. Seifs are elongate, sharp-crested ridges that often consist of a succession of oppositely oriented, curved slip faces that give a sinuous or chainlike appearance to the dune crest. In the Sinai Desert, the pattern is accentuated because the angle between wind direction and the dune crest line controls whether erosion or deposition will occur. Because the crest line is sinuous, this angle of incidence varies along the length of the dune. Thus, both erosion and deposition can occur simultaneously in different crestal zones during the same unidirectional wind storm (Tsoar 1983). In many cases seifs attain spectacular lengths up to 300 km.

In addition to the basic patterns described above, other types have been reported including *dome-shaped, reversing,* and *star* (McKee 1966, 1979), as well as coppice dunes (Melton 1940) which are fixed by clumps of vegetation. It is important to understand that all dune

types may be present side by side in a single region with the same prevailing wind, indicating that other controls are significant in pattern development. For example, the development of star dunes in Mexico seems to involve seasonally changing wind directions that interact with other simple dune forms (Lancaster 1989). Thus, variables of multiple wind directions, topography, size and abundance of sand, and vegetation may be so changeable that complex patterns are the rule, and the simple patterns of our classification the exception. It is also possible that the complexity may reflect many episodes of dune formation under a variety of environmental conditions (e.g., Warren and Allison 1998). It seems reasonable, therefore, to use the simplest ideal forms—barchans and linear types—as models to demonstrate how controlling factors may influence dune patterns. This choice is predicated on the generally accepted idea that dunes frequently form perpendicular to the wind or

(A)

(B)

Figure 8.17
(A) Barchan dune in Sherman County, Ore., September 1899. (B) Asymmetric wind ripples on side of compound barchan dune. Dawson County, Mont., September 1928. (C) Dune forms in the Tularosa Basin, Otero County, N.M. Wind direction is from lower right. Dunes are crenulated transverse in lower right of photo. Dune forms progressively change to individual barchans and finally become U-shaped in the downward direction.

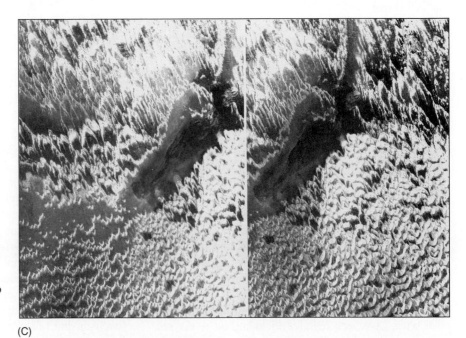

(C)

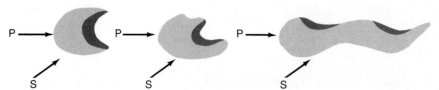

Figure 8.18
Stages of seif development from barchan under influence of secondary wind direction: P = primary wind; S = secondary wind.
(Bagnold 1941)

parallel to it. However, dunes oriented at oblique angles to prevailing winds also are found in every region where dunes are reported. Thus, even though linear and transverse dunes are most easily related to wind action, they may be less important in understanding processes than the explanations of the oblique forms.

Bagnold (1941) described barchans as forms that develop in an area where sporadic sand patches exist within a desert pavement. Certain conditions must be met to mold the barchanoid form: (1) a constant, unidirectional wind, (2) a rate of sand supply that is symmetrically distributed on either side of the longitudinal axis, and (3) a slip face that is completely sheltered so that all sand crossing the crestal zone is trapped on the lee slope (Bagnold 1941, p. 222). Under these constraints, sand movement will be fastest near the lateral edges where the sand patch thins to meet the surface of the adjoining desert pavement. The greatest vertical growth of the sand body occurs in the middle segment, decreasing gradually to the edges. Thus, the lateral extremities will advance more rapidly than the center, and the crescentic form will develop as the equilibrium state is attained. As Cooke and Warren (1973) point out, however, Bagnold's hypothesis does not explain why barchans seem to have a regularly repeated width, nor does it account for the tremendous diversity in the length and direction of the barchan horns. Those authors suggest that divergence from the ideal form is a function of complex secondary flow patterns in the wind and of differential sand supplies. In fact, as suggested above, ideal barchans probably develop only where there is a limited supply of sand and unidirectional winds. Abundant sands lead to linear dunes rather than individual barchans.

Migration rates of barchans suggest a quasi-equilibrium condition. Haynes (1989) showed that one of the dunes originally studied by Bagnold has maintained a relatively constant 7.5 m/yr forward movement since 1930. This advance occurred even though the height of the dune decreased nearly 5 m during the same interval, suggesting that some complex interaction of factors controls the equilibrium migration rate. Indeed erosion of the underlying sand sheet, Holocene climatic variations, and local storms may be involved.

The origin of linear dunes and draa has traditionally revolved around their relationship to the prevailing wind direction. Draa seem to develop by vertical growth of existing simple dunes (Lancaster 1985). In fact, the suggestion has been made that draa should be defined as

any eolian bed form consisting of superimposed dunes (Havholm and Kocurek 1988). This removes the size connotation normally associated with draa, even though in actuality most draa are extremely large. In contrast to the vertical growth of draa, Bagnold (1941) suggested that seifs develop when barchans are modified by crosswinds so that one of the horns is dramatically extended, as shown in figure 8.18. According to this idea, these linear dunes are not aligned parallel to the prevailing wind, but their trend represents the resultant direction of more than one wind. The secondary wind influence may arise from storm winds, diurnal reversals, or seasonal changes in wind direction. The multiwind hypothesis has received support from workers investigating dune formation in a variety of physical environments (Cooper 1958; McKee and Tibbitts 1964; McKee 1966; Lancaster 1980).

The possibility of a resultant wind being generated by the influence of crosscurrents on the prevailing wind has never been questioned, but considerable argument remains as to how much and what type of geomorphic work it can accomplish. Substantial evidence has been reported to support the conclusion that seif chains are aligned with the resultant of several prominent wind directions (Cooper 1958; McKee and Tibbitts 1964; Brookfield 1970; Warren 1970). However, other data argue against a major genetic role for resultant winds. For example, seifs are often oriented in more than one direction within the same region, and some are distinctly oblique to the resultant wind direction (Brookfield 1970; Warren 1971).

Although the relationship between wind direction and dune pattern is complex (Sweet 1992), eolian features are generally so wind-dominated that landform characteristics can be utilized to make interpretations concerning wind direction and velocity. Where features are large, Landsat imagery can be employed to make such interpretations (Greeley et al. 1989), and normal aerial photography is more suitable for documenting wind properties from smaller features (for general discussions see McKee 1979; El-Baz and Maxwell 1982; Marrs and Kolm 1982).

In summary, depositional processes and features reflect a multitude of interacting geomorphic systems. Small features usually occur as ripples and simple dunes, but mega-scaled linear forms (seifs and draa) are normally found only in sand seas. Each feature of the hierarchy occurs in patterns that are mainly a function of

Figure 8.19
Loess deposit on east side of Mississippi River valley near Chester, Ill.

the wind properties. The major dune patterns are probably oriented parallel and perpendicular to the prevailing wind direction, but enough forms trend obliquely to suggest the possibility of a more complicated relationship involving form adjustment to multidirectional winds. In addition, vagaries of grain size and sorting, vegetation, topography, and sand supply introduce additional complications, and geomorphologists are only beginning to understand the influence of environmental changes on the genesis of dunes (Twidale 1972; Haynes 1989). Some dunes undoubtedly formed under wind and sand conditions that no longer exist, and these relict forms confuse the issue even more because their geomorphic characteristics may not be in equilibrium with the modern wind and sediment conditions (see Lancaster 1988b; Kocurek et al. 1991).

Fine-Grained Deposits

Our examination of wind action in geomorphology logically concludes with a look at sediment that is not normally moved near to or in contact with the surface. Most silt and clay entrained from the surface is transported by the wind as suspended dust. Eventually the dust comes to rest as a blanketlike deposit called *loess* (see fig. 8.19). Actually, reworking of the original dust by other surface processes is common and minor cementation is required to produce true loess. Excellent discussions of dust and loess are found in Pye (1987) and Livingstone and Warren (1996). Our treatment of these materials here is liberally extracted from those sources. Loess is usually characterized as homogeneous unstratified silt, up to 100 m thick, which is highly porous and has the capacity to maintain vertical or nearly vertical slopes (Lohnes and Handy 1968). It occupies up to 10 percent of all land (Pesci 1968) and covers all surfaces regardless of their topographic position, capping drainage divides as well as valley bottoms. Most loess is moderately well sorted, with nearly 50 percent of the deposit consisting of silt grains between 0.01 and 0.05 mm in diameter (Pesci 1968). Loess usually contains significant amounts of clay (5 to 30 percent) and 5 to 10 percent sand. The mineral composition is fairly consistent, with quartz being dominant and feldspars, carbonates, heavy minerals, and clay minerals present in smaller amounts. Certain constituents vary in percentage according to local controls. For example, the amount of calcium carbonate and soluable salts in windblown dust tends to be high where playas are the source areas (Reheis and Kihl 1995).

Silt and clay particles in windblown dust are entrained by deflation or, as discussed earlier, impact of saltating grains. These processes are needed to lift the fine material into higher velocity zones. When conditions of lift and turbulence are properly combined, dust particles can rise to elevations of several kilometers, and the total amount of material placed in the atmosphere can be enormous. Peterson and Junge (1971) estimated that more than 500 million tons of dust are transported each year on Earth, and dust storms (defined as involving a visibility of less than 1000 m) may account for much of that total. For example, in a 1935 dust storm over the interior plains of the United States, about 5 millions tons of dust were estimated to be in suspension over a 78 sq km area near Wichita, Kansas, and at least 300 tons/sq km of dust were deposited in one day of that storm near Lincoln, Nebraska (Lugn 1962). Although measurement of dust in transport and after deposition can be difficult, techniques have now been developed to accomplish those tasks (for discussion see Livingstone and Warren 1996).

It is certain that increasing the frequency of dust storms has serious environmental consequences (Goudie 1983; Goudie and Middleton 1992). Fortunately, there seems to be no global trend toward an increase in dust-storm frequency (Goudie and Middleton 1992). Instead, dust-storm frequency and sediment flux are probably dependent on (1) regional climate trends such as seasonality and extended droughts and (2) influences of source area characteristics (Reheis and Kihl 1995; Bach et al. 1996). In addition, human activities have clearly altered the frequency of dust storms by creating surfaces that are more susceptible to entrainment of fine sediment by the prevailing winds. Examples are irrigation of the Great Plains of the United States, and abstraction of water from the Owens and Mono basins of California (Goudie and Middleton 1992).

The hypothesis that loess originates as wind-blown dust stems from at least a century of observations. The conditions that facilitate its wind derivation are an abun-dant supply of loose, fine-grained sediment, moderate to strong prevailing winds, and a surface free from a continuous vegetal cover. Deserts obviously meet these requisites (Péwé 1981), but curiously some continents with vast deserts, such as (Africa and Australia), have almost no loess deposits. Also, some regions notably lacking in desert conditions have experienced major loess deposition. Chief among these is the periglacial environment (Péwé and Journaux 1983), especially where glacial meltwater has spread outwash debris in the path of the prevailing winds (Péwé 1955; Westgate et al. 1990). In that setting, silt is winnowed from the outwash before the surface can be fixed by vegetation. In the midcontinental United States, for example, outwash is a logical source for many loess sheets, whose distribution shows a marked affinity to the outwash bodies occupying large valleys of the region. Presumably the loess can grow to its considerable thickness because the silt supply is continuously replenished during annual or even diurnal flooding by the proglacial rivers.

The distribution of loess may not be totally dependent on a continuous supply of silt and wind conditions in the source area. Tsoar and Pye (1987) argue cogently that the absence of loess in deserts is primarily a function of conditions in zones of accumulation, that is, significant thicknesses of loess can form only where topography, vegetation, or soil moisture provide suitable environments for trapping dust. Loess accumulation is inhibited in deserts because vegetation and areas of moist ground are rare, and therefore dust traps are not available.

The topographic effect of loess deposition is different from that of most constructional geomorphic processes; it molds no important landforms but tends to form plains by filling in depressions and thereby smoothing out preexisting relief. Dust and other constituents of loess are carried as particulate matter suspended in the air, and its transport distance depends mainly on the constancy of the wind velocity, both vertical and horizontal, and the settling velocity of the grains. Loess constituents can be transported for hundreds of kilometers and deposited over vast areas (Van Heukon 1977).

Although the shape of grains included in loess deposits and the markings on their surfaces suggest wind transport (Millette and Higbee 1958; Krinsley and Donahue 1968), windblown loess is difficult to distinguish from silts deposited in other ways. In fact, much controversy has arisen in the past because sediments bearing the characteristics of loess have been deposited by processes such as mass wasting, wash, weathering, and fluvial action. This has produced what some authors refer to as the "loess problem." Obviously some confusion results when loess is reworked by other transporting processes and some of the clay and $CaCO_3$ comes from weathering of the original mass. In addition, movement

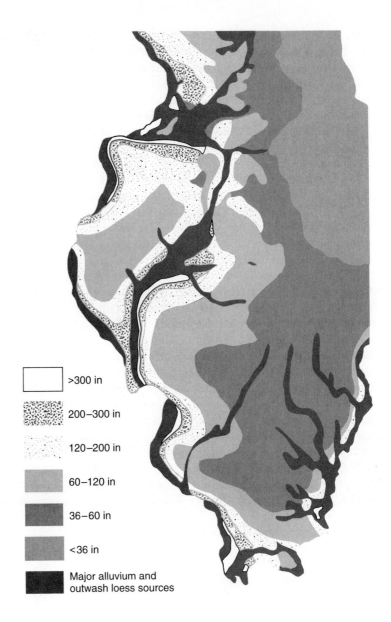

Figure 8.20
Approximate thickness of loess on level, uneroded topography in Illinois.

Legend:
>300 in
200–300 in
120–200 in
60–120 in
36–60 in
<36 in
Major alluvium and outwash loess sources

has been prevalent after the original deposition of the material. Nonetheless, the relationship between widespread blanket deposits and the presumed prevailing winds is often so clear that an eolian origin for the original concentration of the material cannot be denied.

Some of the problem can be resolved if we restrict the term "loess" to deposits that have the characteristics described and are definitely of windblown origin. Proof of an eolian genesis may be circumstantial in some workers' minds, but the evidence of field relationships and deposit properties is often more than convincing. In

the western Great Plains, for example, the thickest loess (up to 70 m) occurs immediately downwind from the Sand Hills region of Nebraska, which contains the largest accumulation of sand dunes in North America. In Illinois, as the map in figure 8.20 shows, the loess is thickest on the bluffs overlooking the Mississippi and Illinois valleys and thins in a gradual and regular manner eastward into Indiana. As might be expected, particle size also decreases downwind, again with a demonstrable regularity (Ruhe 1969).

SUMMARY

Wind is an effective geomorphic agent in regions with sparse vegetation and an abundant supply of unconsolidated sediment. Although the most discernible evidence of wind action is found in deserts, eolian processes function in any locality having strong winds and the proper conditions of vegetation and sediment. The amount of geomorphic work actually accomplished depends also on the properties of the wind, especially its velocity and turbulence.

Particle entrainment occurs when the wind reaches a critical velocity called the fluid or friction threshold. For any given sediment size, the value of the threshold velocity depends on a number of variables such as particle shape, sorting, soil moisture, and surface roughness. Once in motion, however, particles may strike stationary grains and cause their entrainment at wind velocities well below the friction threshold (impact threshold).

Wind-transported sediment moves in suspension, by saltation, or by surface creep. The largest portion of the load is carried within 2 m of the surface.

Geomorphic features associated with wind action are depositional and erosional, ranging in size from microscopic to those measured in kilometers. Erosional features develop primarily from the abrasive action of windblown sand. The most prominent depositional features occur in a hierarchy of bed forms, including ripples, dunes, and draa, that are produced from sand-sized debris. The geometric shape and spacing of simple bed forms probably reflect an equilibrium condition between the wind and the characteristics of the sand, but draa may have developed under environmental conditions that no longer exist. Most windblown silt and clay are deposited in sheets of loess that tend to smooth out topography rather than collect into pronounced constructional features.

SUGGESTED READINGS

The following references provide greater detail concerning the concepts discussed in this chapter.

Bagnold, R. A. 1941. *The physics of blown sand and desert dunes.* London: Methuen.

Cooke, R., Warren, A., and Goudie, A. 1993. *Desert geomorphology.* London: UCL Press.

Goudie, A., Livingstone, I., and Stokes, S. 1999. Aedian environments, sediments and landforms. Chichester: Wiley.

Greeley, R., and Iversen, J. D. 1985. *Wind as a geological process: Earth, Mars, Venus and Titan.* Cambridge: Cambridge Univ. Press.

Lancaster, N. 1995. *Geomorphology of desert dunes.* London: Routledge.

Livingstone, I., and Warren, A. 1996. Aeolian geomorphology: An introduction. Addison Wesley Longman, Harlow.

Pye, K. 1987. Aeolian dust and dust deposits. London: Academic Press.

Pye, K., and Lancaster, N., eds. 1993. *Aeolian sediments.* Spec. Pub. 16, Intl. Assoc. Sedimentologists. Oxford: Blackwell Scientific.

Pye, K., and Tsoar, H. 1990. *Aeolian sand and sand dunes.* London: Unwin Hyman.

Tchakerian, V., ed. 1995. Desert Aeolian processes. New York: Chapman and Hall.

Thomas, D., ed. 1997. Arid zone geomorphology. London: Bellhaven/Halsted Press.

9

GLACIERS AND
GLACIAL MECHANICS

© John Shelton

Introduction
Glacial Origins and Types
The Mass Balance
The Movement of Glaciers
    Internal Motion
        Basic Mechanics
        A Simple Model of Internal Flow
        Extending and Compressive Flow
    Sliding
    Motion with Deforming
        Subglacial Sediments
    Velocity Variations with Time
Ice Structures
    Stratification
    Secondary Features
        Foliation
        Crevasses
Summary
Suggested Readings

INTRODUCTION

Some of the most spectacular landscapes in the world are the result of the erosional and depositional action of glaciers, and every textbook of physical geology and geomorphology includes photographs and descriptions of these remarkable features. Nonetheless, to be true to the theme of this book, we will consider only the processes that form glaciers. It is also tempting to stress the stratigraphic relationship between different glacial deposits and the effects exerted by glaciations on climate, life, and the physical setting. Obviously these are subjects of paramount interest to geomorphologists, because much of the modern landscape is closely associated with physical events that occurred during the alternating glacial and interglacial stages of the Quaternary period. While recognizing the importance of these topics, we realize pragmatically that an attempt to treat them in a general way is destined to fail. Therefore, our goal in this chapter is to understand how glaciers form and how they move. Glacial erosion and deposition, and the landforms resulting from these processes, will be examined in chapter 10.

The study of glacial mechanics is an intimate part of the science of glaciology. Although this field is in its infancy, many of the techniques developed in recent years to study glaciers are extremely sophisticated and involve geophysics, remote sensing, and computer analyses, topics well beyond the scope of an introductory discussion. Excellent reviews of the basic concepts of the discipline can be found in Sugden and John (1976), Raymond (1987), Sharp (1988), Hambrey and Alean (1992), Paterson (1994), Hooke (1998), and Benn and Evans (1998). Discussions of more recent advances in the field are found in current periodicals, especially the *Journal of Glaciology,* which contains articles pertinent to all aspects of our discussion.

GLACIAL ORIGINS AND TYPES

A **glacier** is a body of moving ice that has been formed on land by compaction and recrystallization of snow. Two critical requirements must be met before a mass of ice can be called a glacier. First, the mass must be formed from the accumulation and metamorphism of snow, and second, the ice must be moving, either internally or as a sliding block. Areas in which the winter snow is entirely lost to summer melting and other forms of dissipation cannot bring forth glaciers, even if the amount of snowfall is enormous. In areas where a portion of the snowfall does survive the summer melt, it may be buried by the next winter's accumulation. With continued annual increments, the snow pack grows in size, changes its constituent properties, and finally, with enough mass, begins to move. In any region a specific elevation exists called the *snowline* or the *firn line,* above which some snow remains on the ground perennially and may permit the formation of a glacier. In polar climates, snowlines are usually near or at sea level. In climatic zones characterized by higher mean annual temperatures, snowlines occur at higher elevations. However, temperature is not the only determinant of snowlines because glaciers can form at lower elevations at the equator than they do in desert zones near 30° north. This somewhat incongruous situation arises because of higher annual precipitation in equatorial regions.

When snow accretes over a period of years, changes in the properties of the particles mark the transition of a snow pack to true glacier ice. Newly fallen snow having a very low density (usually 0.05 to 0.07 gm/cc) and a delicate hexagonal crystal structure is transformed into glacier ice through a series of complex but recognizable stages (fig. 9.1). In the initial phase, points of the crystalline flakes are preferentially melted, resulting in a more spherical particle shape and tighter packing due to settling of the grains. At the same time, water produced by melting percolates to the base of the snowpack where it refreezes in the pore spaces (Woo and Heron 1981). As table 9.1 shows, this freezing tends to decrease porosity and dramatically increase density. The time needed for this initial change varies depending on the climate and the pressure added by continuous accumulation of snow. Where the temperature remains near freezing or where partial melting occurs, the transition from fluffy snow to coarse granular snow may take hours or days (Embleton and King 1968). On the other hand, extremely frigid conditions retard the process such that years may be required.

In temperate regions, snow lying on the surface for a complete year becomes granular and usually increases in density to about 0.55 gm/cc through the rounding and settling processes just mentioned. This material, now known as *firn,* is much denser than the original snow but is still permeable to percolating water and is not yet true

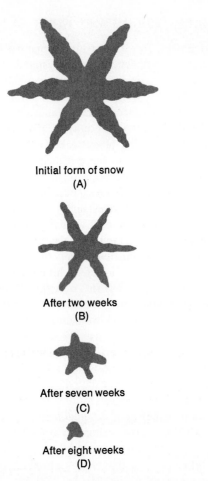

Initial form of snow
(A)

After two weeks
(B)

After seven weeks
(C)

After eight weeks
(D)

Figure 9.1
Changes in the shape of snow crystal in the transition to firn.

TABLE 9.1	Increasing Density of Snow During Transition to Ice.

Materials	Density (gm/cc)
New snow	0.05–0.07
Firn	0.4–0.8
Glacier ice	0.85–0.9

glacier ice (table 9.1). The rate of densification beyond this point is much slower. In addition, the mechanics involved in transforming firn to ice differs from earlier processes of densification. This transformation produces enlarged ice grains (in some cases up to 10 cm long) by recrystallization (fig. 9.2). Gradually, pore space is eliminated by crystal growth or by freezing of the downward-permeating meltwater until the density is approximately 0.8 to 0.85 gm/cc. The only air remaining is trapped as bubbles within the crystal. For all practical purposes, the creation of ice is complete at this stage, even though the bubbles continue to be slowly expelled by compaction, and the density may increase to about 0.9 gm/cc.

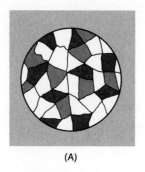

(A)   (B)

Figure 9.2
Increase in grain size caused by recrystallization process. (A) Original texture of water-soaked snow. (B) Texture after application of stress.

The rate of density increase and crystal growth is closely related to the temperature of firn and the rate of increased load on the individual grains. Firn in temperate regions will transform into ice faster and at a shallower depth than it will in polar regions. This fact is probably best demonstrated by comparing the depth-density relationship exhibited by glaciers in temperate and polar regions (Behrendt 1965). In the Antarctic, a 0.85 gm/cc density is not achieved until at least 80 m of firn is accumulated. In contrast, only 13 m is needed to change snow to ice on the temperate Upper Seward Glacier in Alaska. Because accumulation rates are greater in glaciers of temperate regions, it follows that the greater depth to true ice in the Antarctic ice sheet represents the monumentally greater time needed to create glaciers in dry, cold climates.

Glacier ice is not preserved intact after its creation but is susceptible to further changes with continued increase in stress. For example, experimental work shows that the random orientation in polycrystalline ice cannot be maintained at higher stresses. When deformation of the polycrystalline mass exceeds several percent, the process known as recrystallization begins. Theoretically, this process should align the c-axis of the ice crystal within the vertical plane of a glacier, altering what is known as the *ice fabric*. Changes in fabric (c-axis orientation) with increasing depth and shear stress have been noted in a number of glaciers (Gow and Williamson 1976; Russell-Head and Budd 1979; Hooke and Hudlestone 1980, 1981). In many cases, the fabric changes from a weakly oriented c-axis in fine-grained, near-surface ice to coarser ice with a broad, single maximum fabric at some lower level. At still greater depths, the fabric usually develops multiple orientations. Finally, in some, but not all, glaciers the fabric returns to a single maximum orientation at the glacier base (Hooke and Hudleston 1980).

The depth at which fabric transition occurs is highly variable and depends on a complex interaction of density, impurities, temperature, and so on. In fact, a gen-

TABLE 9.2	Morphological, Dynamic, and Thermal Classification of Glaciers.

Morphological classification

Major types

Cirque glaciers	Flowing ice streams restricted to amphitheater-shaped depressions in valley headlands
Valley glaciers	Streams of ice that flow downvalley well beyond the cirque
Ice sheets	Broad, flowing ice masses that are not confined to valleys; ice accumulates to massive thickness in high-latitude continental areas or broad uplands

Intermediate Types

Mountain ice sheets	Valley glaciers that enlarge to form ice sheets that bury all but the highest Alpine peaks
Piedmont glaciers	Glaciers that discharge ice onto broad lowlands located along the base of mountains

Dynamic classification

Active glaciers	Glaciers characterized by high rates of ice movement from accumulation zones to their terminus
Passive glaciers	Glacier exhibiting low rates of ice movement from accumulation zones to ablation zones
Dead glaciers	Glacier possesses no discernible internal ice flow

Thermal classification

Temperate glaciers	Glacier in which the ice is at or near its pressure-melting point throughout the ice mass
Polar glaciers	
Subpolar glaciers	Glacier in which the ice mass is generally below the pressure-melting point except for summer melting of the upper few meters
High-polar glaciers	Glacier characterized by an ice mass that is below the pressure-melting point at all times

eral model describing the formation of ice fabric has yet to be developed (Alley 1992). Nevertheless, the *c*-axis orientation appears to be related to the overall scheme of glacier movement because the orientation of basal crystal planes is probably parallel to the direction of ice flow (Rigsby 1960).

Glaciers have been classified on the basis of many salient properties, but as with other physical phenomena, the appropriateness of the classification depends on the prime purpose of the groupings. Ahlmann (1948) suggested morphological, dynamic, and thermal criteria for classifying glaciers (table 9.2). The morphological classification, which is based on glacier size and the environment of its growth, is most commonly employed, and the different categories, such as valley glaciers and ice sheets, are undoubtedly familiar. Although Ahlmann (1948) recognized many morphological subdivisions, Flint (1971) suggested that glaciers can be placed in three broad categories—*cirque glaciers, valley glaciers, and ice sheets*—and two intermediate categories—*piedmont glaciers* and *mountain ice sheets*. In this abbreviated form the classification is quite useful, especially in a descriptive sense.

For our process orientation, however, two other classifications are probably more useful. The dynamic classification is based on the observed activities of glaciers and consists of three main groups: *active, passive,* and *dead* glaciers. Each type reflects a different balance between losses and gains of ice and probably somewhat

different thermal properties. Active glaciers, like the one in figure 9.3, are characterized by the continuous movement of ice from their accumulation zones to their terminus. The movement may occur in response to normal snow accumulation, or it can be generated by avalanches or ice falls that provide the impulse for the forward motion. As you might expect, a glacier is passive when its movement is minimal. Dead ice has no discernible internal movement.

The thermal classification is based primarily on the temperature of the ice. Glacial behavior is directly influenced by ice temperature, and therefore the thermal classification should lend itself to process analysis. Two types of glaciers exist in this classification: *temperate glaciers* and *polar* or *cold glaciers*. In *temperate glaciers* the ice throughout the entire mass is at its pressure-melting point (the temperature and pressure conditions at which ice begins to melt), although the upper 10 m may freeze in the winter. Meltwater seems to be present in abundant amounts within or beneath the ice mass and, in contact with the underlying rock, often facilitates slippage of the ice over the bed. This causes velocity and erosive action to be generally greater in temperate glaciers than in other types. Near the snout meltwater may emerge as basal streams, or it may be temporarily dammed within the ice; both situations promote extensive fluvial removal of debris from the terminus of the glacier.

Polar glaciers were subdivided by Ahlmann into *subpolar* and *high-polar* types. In *subpolar* glaciers the

Figure 9.3
The La Perouse Glacier in Alaska. A very active glacier that spreads onto lowland when it emerges from a narrow valley. Crevasses are distinct, and annual dark bands (ogives) are clearly visible.

accumulation zone is characterized by a thin layer of firn, perhaps 20 m thick, which contains some water in the summer if temperatures are warm enough to melt the surface ice. The surface of a *high-polar* glacier remains below freezing at all times, resulting in a completely water-free ice mass and a thick firn zone extending to at least 75 m before true glacial ice is encountered. The absence of meltwater within polar glaciers is an extremely significant difference between polar and temperate glaciers, and its geomorphic importance cannot be overemphasized. This condition requires that ice at the base of polar glaciers be below its pressure-melting point and, for all practical purposes, be solidly frozen to the underlying bedrock. Since slippage over the bed is minimal, ice movement is greatly diminished.

The assumption that temperate ice is at its pressure-melting point is perhaps unfortunate because pressure is only one of several factors that determine the temperature within a glacier. Temperature variations can be caused by downward transfer of surficial heat or upward transfer of geothermal heat that is released at the glacier base (Sugden and John 1976; Hooke 1977). Heat is also produced at the base and inside the ice by friction when

the glacier moves. Thus, a temperate glacier may have patches of frozen ice at its base and portions of polar ice sheets may have free water. In addition, the effect of climatic change may be gradually superimposed on the other types of heat variation and may not permeate through an entire glacier for centuries. Thus, most "temperate glaciers" have a temperature distribution that disagrees with the pressure distribution predicted by the thickness of the ice (Harrison 1975), even though the temperature generally increases with depth (Hooke et al. 1980). In some cases the temperature increase may be linear (fig. 9.4); however, this is not always true, and the gradient of temperature increase may vary considerably from glacier to glacier or within a single ice mass (Hooke et al. 1980). In sum, the simple model of pressure-melting offers only an approximation of the temperature conditions within glaciers.

Regardless of the difficulties of assessing thermal characteristics, the condition of basal ice with regard to melting temperature is perhaps the primary determinant of a glacier's ability to do geomorphic work. Glaciers with basal ice at the melting temperature tend to move faster, erode more, and carry greater loads than polar

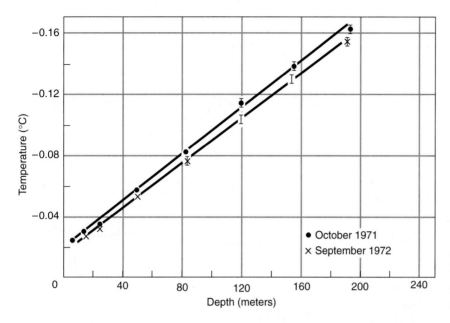

Figure 9.4
Temperature-depth relationship on a temperate glacier.
(Harrison 1975)

glaciers. Therefore, we can expect erosion and deposition to be more pronounced when they arise from temperate ice. In addition, many differences between modern and Pleistocene glacial activity may result from variations in the thermal characteristics of the ice at the different times.

THE MASS BALANCE

The mechanical response of glaciers and the geomorphic work they accomplish are intimately related to their **mass balance,** sometimes called the **glacial budget.** The mass balance is essentially an accounting or budgeting of the gains and losses of snow that occur on a glacier during a specific time interval. The water equivalent of ice and snow added to a glacier during the period in question is called *accumulation* and may result from a variety of processes including snowfall, rain and other water that freezes on the surface, avalanches from the valley walls, and the "freeze-on" of meltwater at the base of the glacier. Processes that remove snow or ice, collectively known as *ablation,* commonly include melting, evaporation, wind erosion, sublimation, and the breaking off of large blocks of ice into bodies of standing water, a process called *calving.* Losses by melting within or beneath the glacier are usually minor compared with surface volumes and are, therefore, neglected in most budget studies.

The time interval used in most balance analyses is the budget year (or balance year), which is ordinarily taken as the time between two successive stages when ablation has attained its maximum yearly value. Usually these values are achieved at the end of the summer season, but two ablation maxima may not occur exactly a year apart, and so the budget year is not necessarily 365 days long.

On a glacier surface two values of accumulation and ablation can be determined: a gross annual accumulation or ablation, representing the total volume of water equivalent added to or lost from the glacier during the budget year, and a net annual accumulation or ablation, representing the difference between the gross values and indicating whether there was an actual gain or loss of mass during the year. The latter value is simply the algebraic sum of accumulation and ablation, and for the budget year at any single measurement locality this value is called the *net specific budget.* If net specific budgets are determined for a network of points distributed over the entire glacier surface, their values can be integrated into the mass balance. Any glacier may have (1) a positive mass balance, meaning that more accumulation has occurred during the year than ablation, (2) a negative mass balance, indicating an excess of ablation, or (3) a mass balance of zero if the volumes of accumulation and of ablation have been precisely the same.

Although accumulation and ablation occur on all parts of the glacier surface, the higher elevations of the glacier usually receive more accumulation and experience less ablation than the lower reaches of the glacier, where the opposite is true. Thus, large areas with either positive or negative net specific budget can be identified on most glaciers, as shown in figure 9.5. The two areas are separated by an *equilibrium line* along which the annual volumes of accumulation and of ablation are equal. The equilibrium line should not be confused with the firn line or snowline (mentioned earlier) because the two may not occur at the same place. On many glaciers, the firn line is clearly marked as a contact between snow above the line and dense blue ice below the line. On others, some of the dense ice downglacier from the snowline may have been formed by refreezing of meltwater

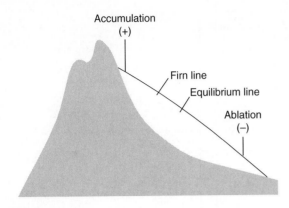

Figure 9.5
Zones of accumulation and ablation on a glacier as determined by budget analyses.

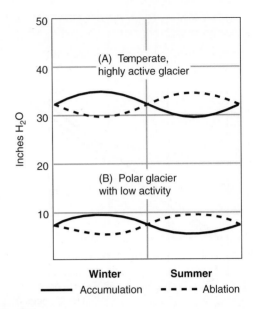

Figure 9.6
Annual budgets on two glaciers in which the net mass budget of each equals zero. Active, temperate glacier (A) has greater total amounts of ablation and accumulation than the low-activity polar glacier (B). Even though both are in equilibrium and their fronts are stationary, (A) does more work because it must transfer more ice during the year from the accumulation zone to the snout.

and, as such, represents a net accumulation of mass. This *superimposed ice* is found in enough cases to warrant the distinction between the firn line and the equilibrium line (fig. 9.5), and it sometimes creates a rather complex transitional zone between the accumulation area and the ablation area. Nevertheless, equilibrium lines and firn lines are usually close enough to approximate one another.

What does all this have to do with glacial mechanics and the resulting processes of erosion and deposition? If glaciers are viewed as open systems, both the mass balance and the absolute total amounts of accumulation and ablation are important factors in the character of glacial movement. When the net budgets are perfectly balanced (total mass balance equals zero), no expansion or shrinkage of the glacier occurs and the glacial extremities remain stationary. This equilibrium condition, however, is seldom maintained for a long period of time, and so the glacial front and sides usually fluctuate constantly. Glaciers with a positive mass balance actively advance and characteristically maintain a relatively steep or vertical front. In contrast, an overall negative budget induces a recession of the glacial front, and the snout will be gently sloping and often partially buried by debris released from the ice. The mass balance, therefore, is closely related to the position and type of morainal system constructed by the ice.

The absolute amount of snow and ice relative to the area of a glacier (gross accumulation and ablation) directly affects its internal activity. Large gross accumulation values on small glaciers promote very rapid flow from the accumulation area to the ablation area; in most cases temperate glaciers are likely to have large accumulation and ablation values and, consequently, high flow velocities. In contrast, polar glaciers, which tend to be rather passive, usually have small values of accumulation and ablation and low internal flow velocities. It is very important to recognize that glaciers with different gross values can have the same mass balance, and if they

are in an equilibrium condition, the glacier front does not advance or recede even though the internal transfer of ice may be enormous. Figure 9.6 is a hypothetical diagram of annual budgets for a polar glacier with low activity and a highly active temperate glacier, both of which have a mass balance equal to zero. The snouts and lateral edges are stationary in both cases, but the temperate glacier, having a large value for accumulation and ablation, is moving at a higher rate internally. This dynamic activity causes pronounced erosion and rapid transportation of debris through the system. With the proper budgets, large moraines may result at the terminal and lateral boundaries. The polar glacier depicted in figure 9.6 will have little if any internal motion of its ice and, as a result, will probably not form any significant depositional features. The depositional and erosional character of every glacier, therefore, is fundamentally determined by the characteristics of the snow and ice added to or lost from its surface.

THE MOVEMENT OF GLACIERS

Several hundred years ago, through direct observations, Alpine residents realized that glaciers move. Measurements of the rate of glacier flow were made as far back as the early eighteenth century. It is now generally accepted that glaciers move by two mutually independent processes: internal deformation of the ice, called *creep,* and *sliding* of the glacier along its base and sides. In addition, where glaciers override a layer of unconsolidated debris, the ice mass may move along with the underlying sediment as it deforms or flows beneath the glacier.

Internal Motion

Basic Mechanics   The first real attempts to explain the physical dynamics of englacial ice motion are found in the work of J.D. Forbes, who in 1843 suggested that glaciers respond to stress much like a substance undergoing plastic deformation. By the beginning of the twentieth century, numerous models based on a plastic flow mechanism had been developed, although other ideas were also being proposed. They included flow caused by differential shearing along numerous, closely spaced planes (Phillip 1920; Chamberlain 1928) and the widely accepted notion that ice behaves as a viscous liquid that deforms in linear proportion to stress (Lagally 1934). Each of these concepts suggests that glacier flow manifests a relationship between ice deformation (strain) and the stress (force/unit area) that produces it. Stress generated at any point can be separated into two parts, hydrostatic pressure and shear stress. *Hydrostatic pressure,* which is related to the weight of the overlying ice, is exerted equally in all directions. *Shear stress,* which causes glacier motion, is a function of the overlying weight and the slope of the glacier surface. For the purposes of demonstration, consider the two simplest possibilities: (1) that ice behaves as a viscous fluid and (2) that ice deforms as a perfectly plastic substance.

If ice behaves as a Newtonian viscous material, and we assume a constant viscosity, the application of stress should result in a linear relationship between the stress value and the strain rate (fig. 9.7). In addition, deformation will begin in the ice as soon as stress is applied and will maintain the linear proportionality regardless of the changes occurring in the stress. In contrast, a plastic substance shows no immediate response to stress but is capable of supporting a certain amount of stress without sustaining any deformation (fig. 9.7). Thus, at low stresses the strain of a plastic will be zero. As stress increases, however, it eventually attains a value, called the *yield stress,* where the ice will experience limitless and permanent deformation. An entire glacier would behave plastically only if the shear stress along its base were equal to the yield stress.

Neither of these simple cases prevails in the mechanics of ice motion. However, in many instances glacial properties predicted according to pure plasticity closely approximate the observed characteristics (Nye 1952b), and thus glacial movement is often referred to as pseudoplastic (Meier 1960; Johnson 1970). Laboratory studies since the late 1940s have shown that single ice crystals under stress deform as soon as stress is applied. Under any given stress the strain will increase

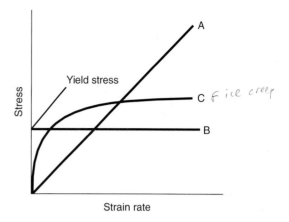

Figure 9.7
The relationship between stress and deformation.
(A) Newtonian viscous fluid. Relationship is normally plotted as a function of stress and the rate of strain. (B) Perfect plastic deformation. The material experiences no deformation until stress is increased to the value of the yield stress. The material will then continue to deform as long as the stress is applied. Normally plotted as a function of stress and strain. (C) Ice creep according to Glen's flow law.

rapidly at first (Glen 1952, 1955), but within a short time (tens of hours) it approaches an almost steady value that plots as a nearly straight line on a graph (fig. 9.7). This continuous deformation with no increased stress is the creep process that allows glacial ice to flow steadily under its own weight.

The rate of strain during the creep process is related to varying stress values by the following equation derived by Glen (1952, 1955) and now commonly referred to as the *power flow law:*

$$\dot{\epsilon} = k\tau^n$$

where $\dot{\epsilon}$ is the strain rate ($d\epsilon/dt$), τ is stress, and k and n are constants. The values of $n,$ determined by a number of investigators, seem to vary from approximately 2 to 4 for individual crystals. In polycrystalline ice they range from 1.9 to 4.5, with the mean value being close to 3. In any case, the n values associated with the power flow law are significantly greater than 1, the value required for linear viscous flow. The power flow law indicates that minor changes in stress can produce a major response in the strain rate. For example, if $n = 4$, doubling of τ increases the strain rate 16 times.

Glacier ice is always polycrystalline. The flow law, although based on the deformation of a single crystal, still seems to predict the response of glaciers (Thomas et al. 1980), although minor modifications of the equation may be needed (Meier 1960; Colbeck and Evans 1973). The value of k in glacier flow might be reduced by interference of adjacent grains and recrystallization. In addition, as shown in figure 9.8, cold ice deforms less readily than temperate ice, mainly because the constant k is

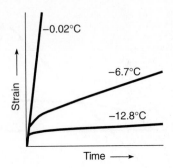

Figure 9.8

The effect of temperature on the creep deformation curve. All curves represent deformation under a stress of 6 bars.
(After Glen 1955)

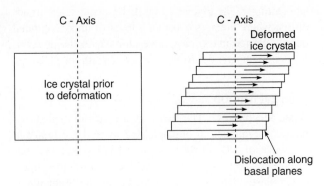

Figure 9.9

Ice deformation involving the dislocation of atoms along the basal planes of an ice crystal. As crystals deform, recrystallization occurs, reestablishing the equidimensional shape of the grains.

also dependent on temperature. In Glen's experiments the value of k decreased by two orders of magnitude (from 0.17 to 0.0017) when the temperature was lowered from 0° C to –13° C. In fact, Paterson (1969) suggests that the strain produced in ice at –22° C by any stress is only 10 percent of its value when the ice is at 0° C. Impurities contained within the ice, including gas bubbles, dissolved constituents (ions), and rock debris, are also known to influence strain rates (Benn and Evans 1998). The effects of rock debris on ice deformation have received the most attention because ice at the base of many glacier contains considerable amounts of sediment. The influence of solid particles on deformation is still unclear, however, as the observed effects on strain rates differ between individual studies (see, for example, Nickling and Bennett 1984; Hubbard and Sharp 1989; Echelmeyer and Wang 1987). In some cases, solid particles appear to harden the ice, in others they soften it. Benn and Evans (1998) point out that these variations are probably due to the numerous factors that are involved in controlling the deformation process including the grain size of the sediment, the orientation between the debris and the ice crystals, and the temperature-related melting of ice around individual particles.

Given the above, it should be clear that ice does not behave like a viscous fluid, although it may approximate a viscous response under low stress when the strain rate is still in its transient phase. At stress <1 bar in temperate ice, for example, n values as low as 1.3 have been measured (Colbeck and Evans 1973). However, when nearly steady strain is attained or the ice is under high stress, the deformation becomes more plastic. Although this condition approximates plasticity, no distinct yield stress is associated with the creep process, indicating that ice is not a perfectly plastic material.

Much research has been done to determine why ice under high stress takes on a nearly plastic behavior. The answer probably rests in the mechanics of strain. Ice deforms by a process involving the dislocation of atoms along planes of weakness inside individual crystals (see

Glen 1958, 1987; Weertman and Weertman 1964). The dislocation process works most efficiently through gliding on the basal planes of the ice crystal (fig. 9.9), the movement somewhat resembling the slip observed in a stack of playing cards (Sharp 1988). If deformation were allowed to continue, it would produce sheetlike ice grains. As the crystals deform, however, recrystallization occurs, reestablishing the equidimensional shape of the grains (fig. 9.9). It was suggested years ago that plastic flow was closely associated with ice crystals lying in a preferred orientation (Perutz 1940). It is now clear that a consistent orientation of grains in a polycrystalline ice mass (ice fabric) produces strain exerted nearly parallel to the basal planes (Russell-Head and Budd 1979; Duval 1981).

In sum, the power flow law derived in Glen's laboratory seems to fit the actual observed motions of glaciers. We cannot expect that polycrystalline ice will respond in precise agreement with the flow law, however, and modifications of the equation are generally required based on the individual situation. In addition, where stress systems are complex because of irregular valley floors, cross-sectional shapes, or nonisothermal ice, the flow law will have to be generalized or modified (Nye 1957, 1965). Attempts to go beyond the general approach and replace the simple power law with more sophisticated mathematical models become extremely difficult (Hutter 1982). Thus, in spite of inherent difficulties, Glen's power flow law provides an excellent approximation of glacial motion, between theoretical predictions and observed flow data, and with proper modification probably best describes the internal mechanics of glaciers.

A Simple Model of Internal Flow   We can now develop a simple model of internal ice flow and determine whether glacial velocities agree at all with the basic mechanics of ice motion. In developing this theoretical model, we will assume that an active glacier remains un-

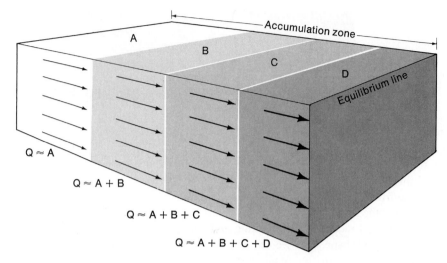

Figure 9.10
Hypothetical diagram showing why the velocity of glacier flow reaches a maximum at the equilibrium line. If no change in cross-section area (width × thickness) is allowed, increasing discharge down-ice requires the increase in velocity shown by arrows.

changed in its total and local dimensions over an entire budget year; that is, the total accumulation equals the total ablation, and all cross-sections of the glacier are equal in area and remain constant in area during the year. To maintain its area, each successive downglacier cross-section must transport from its lower boundary exactly the amount of ice and snow delivered to its upstream boundary. As figure 9.10 shows, in the accumulation zone the cross-section at the highest elevation has only a small area of accumulation above it and so must discharge only a small volume of ice, equivalent to the snow accumulated in that restricted surface area. Each section farther down-ice, however, must transfer a progressively larger volume of ice, since it moves not only the ice-equivalent of accumulation on its surface but also the cumulative volumes of all the higher sections. Without a change in cross-sectional area, the velocity of flow must increase to a maximum at the equilibrium line. This follows because discharge is increasing to that level and because glacier discharge is equal to the cross-sectional area times the velocity. In the ablation area we can similarly expect a gradual decrease in velocity from the equilibrium line to the terminus.

The model assumes that glaciers strive for and maintain some type of equilibrium and that the ice movement obeys the power flow law in some form. Mathematical treatment of even this simple glacial model is quite complicated. Therefore, we will make some further assumptions, as Nye did (1952a), that simplify the quantification of ice flow; specifically, we assume that ice motion is two-dimensional, plastic, and laminar such that the lines of flow are parallel to the bed and surface at all places (fig. 9.11). Under these conditions the shear stress at the base of a glacier, measured along the central longitudinal axis and perpendicular to the surface, is given as

$$\tau_b = \rho g h \sin \alpha$$

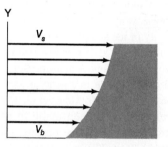

Figure 9.11
Velocity distribution in longitudinal section of glacier. Surface velocity (V_s) is sum total of flow rates at every level within the ice. Strain rate due to shear stress is highest at base of ice and is shown as the basal velocity (V_b).

where τ_b is the shear stress at the glacier base, ρ the ice density, g the acceleration of gravity, h the ice thickness, and α the slope of the surface.

By assimilating Glen's flow law into the equation, the velocity at any depth along the central axis (xy-axis) can be estimated by assuming that shear stress is proportional to depth and that the strain rate directly relates to that stress. Theoretically, then, the internal velocity profile of a glacier in a longitudinal section should show a decreasing rate of flow from the surface to the bedrock floor. Even though the strain rate is greatest where shear stress is the highest (at maximum thickness), the velocity increases from the base to the surface because each internal layer not only moves in response to the shear stress generated at that level but is also being carried on top of, and at the speed of, the adjacent lower layer. Thus, the surface velocity (V_s) is the sum total of strain rates for all the layers within the ice mass (fig. 9.11). The shear stress equation also shows us that internal velocity is proportional to the product of surface slope and ice thickness. The product seems to be fairly consistent, meaning that where the ice is thin the surface gradient will be steep, and vice versa.

Figure 9.12

Velocity distribution in glacier flowing in a channel with a parabolic shape. Diagram represents one-half of the channel cross-section. Numbers represent velocity expressed in any common velocity units. (Nye 1965)

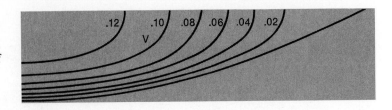

just like velocity does in a channel of H₂O

Using the same basic approach, Nye (1957, 1965) predicted the velocity distribution for an entire cross-section, using different cross-sectional shapes as models, and showed that velocity should decrease from the center of the ice to the lateral boundaries both at the surface and at depth (fig. 9.12). Thus, the simplest ideal model of a glacier, assuming continuity of discharge and shape, and laminar flow, shows velocity decreasing with depth and distance from the central axis. In the downstream direction velocity should increase toward the equilibrium line and decrease away.

The velocity of ice at a glacier surface is obtained by marking the position of stakes driven into the ice relative to some nearby fixed point. Actual measurements of surface velocities substantiate in a general way the theoretical predictions (Meier 1960; Meier et al. 1974), although few measurements have been made in accumulation zones. In most temperate valley glaciers, surface velocities range from 10 to 200 m a year, but vary locally above or below these values. For example, Meserve Glacier, Antarctica, exhibits a maximum velocity of only 2 m per year. In contrast, Jacobshavns Isbrae in Greenland is one of the world's fastest glaciers and had reached flow rates of 8360 m per year at midsummer.

As predicted, velocities tend to increase from the glacier head to a maximum near the equilibrium line (Meier 1960; Meier et al. 1974). The transverse velocities are usually greatest along the central axis, decreasing to the lateral margins (Raymond 1971). In both patterns, however, enough variations occur to indicate that the flow model is infinitely more complex than our original assumptions allowed. Even though flow may show a relatively simple relationship to ice thickness and surface slope as predicted, it may not do so on a local scale. Other factors such as variations in ice temperature, basal sliding, the deformation of subglacial materials and subglacial water pressure may be involved.

Measurements of englacial velocity require that a borehole be drilled through the ice and cased to prevent its closure. The differential movement with depth can then be calculated from an inclinometer, which measures the angle between the axis of the borehole and the vertical. Almost every borehole that has penetrated to a glacier floor or to great depth shows a velocity profile similar to that predicted by theory (fig. 9.13). Velocity does not change significantly with depth in the upper zones of most glaciers, although Meier (1960) has

demonstrated that some differential movement can occur even in the upper portion of a glacier. In the lower half of most glaciers, the velocity decreases more rapidly with depth.

Extending and Compressive Flow In our ideal model we see that allowing no change in cross-sectional area requires that the elevation of each point along the surface of a balanced glacier be maintained. If this is to happen, then the flow cannot consist of laminar horizons moving parallel to the surface because the ice must move slightly downward in the accumulation zone and slightly upward in the ablation zone. (Accumulation tends to elevate the surface and ablation to lower it unless these responses are counterbalanced by motion of the ice.) Thus, the pattern of flow should be as shown in figure 9.14 and not in laminar sheets.

Actually, ice will tend to thicken in some places and thin in others, as compressive and extending flow. Nye (1952a) first investigated these flow types and suggested that rates of accumulation and ablation, as well as changes in the slope of the underlying bedrock, will determine which type will prevail. *Compressive flow* results in a decrease of velocity and occurs where the underlying rock surface is concave-up or where there is a consistent loss of surficial ice. In contrast, *extending flow* has increased downglacier velocity and exists where ice is added at the surface or the glacier bed is convex. At a constant bedrock slope, therefore, accumulation zones should be characterized by extending flow and ablation zones by compressive flow. Within each of these zones, however, topographic irregularities on the underlying surface may produce local changes in the flow type.

This analysis generally fits the velocity distribution that we surmised in our balanced glacier. In ablation zones, with pervasive compressive flow, the velocity should decrease downglacier because the ice is being compressed (Nye 1952a), and in accumulation zones, with extending flow, the velocity should increase downglacier. Nye showed that extending flow and compressive flow should each generate a pattern of potential slip planes that follow the orientation of maximum shear stresses (fig. 9.15). The family of planes is such that their resolution at any point will give the two directions of maximum shear. Thus, they will be perpendicular and parallel to the bed of the glacier and form 45° angles

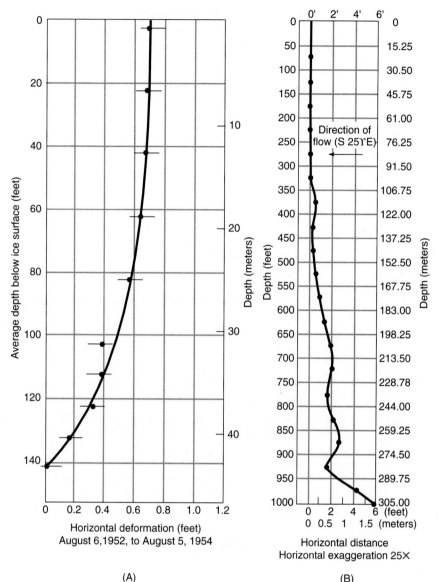

Figure 9.13
Internal velocity of two glaciers.
Horizontal displacement of boreholes
indicates the velocity.
(A) Saskatchewan Glacier
(B) Malispina Glacier.
(A): (Meier 1960)
(B): (Sharp 1953)

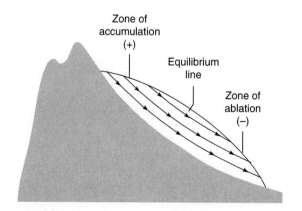

Figure 9.14
Longitudinal flow lines in hypothetical glacier.

with the surface of the ice. The potential slip planes show that in zones of extending flow (accumulation zones), the predominant downglacier slip direction will be downward at the surface, while in compressive zones, such as ablation areas, the low-angle downglacier slip will be upward.

This behavior seems to coincide with the predicted mode of flow in our ideal glacier (see fig. 9.14), but in real glaciers the pattern is much more complicated. An irregular bedrock profile will cause reversals of flow type in the ice (fig. 9.16), and where measurements have been made for an entire glacier, the flow pattern is not the simple one of our ideal case (fig. 9.17). Furthermore, more than one velocity maximum may occur, and the equilibrium line does not necessarily mark a zone of maximum flow velocity. In fact, on the South Cascade Glacier (Washington), the equilibrium line is usually near a zone of lower surface velocity, but the thickness is so great that the total discharge through the section is still very high.

Sliding

Real glaciers seem to possess the general traits that were predicted in idealized internal flow, at least in the distri-

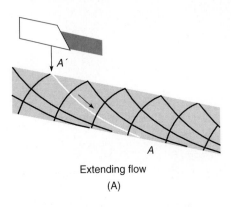

Extending flow
(A)

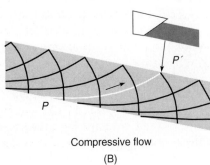

Compressive flow
(B)

Figure 9.15
Potential slip planes under (A) extending and (B) compressive flow. The preferred downglacier slip paths will follow A-A' and P-P'.
(Nye 1952a)

bution of velocity. Nonetheless, we cannot expect glacier characteristics to be predictable in detail. The assumptions made for the balanced glacier are invalid on a local scale, and velocity itself is controlled by several factors that are external to the system defined by the flow laws.

Part of the difficulty in applying the power flow law directly to glaciers is due to the fact that glaciers also move by sliding over the subglacial bed, this motion being in addition to creep within the ice mass. Few direct measurements have been made, but those available show that sliding may account for less than 10 percent of the surface velocity in some glaciers and more than 90 percent in others. In addition, sliding velocities may be extremely variable within different portions of the same glacier. For example, abnormally high velocities are sometimes encountered at the glacier sides (Meier et al. 1974).

Sliding processes operate either at the contact between the underlying glacial bed and the base of the glacier or within the lower ice layers. The processes themselves are poorly understood because direct observation of their action requires tunneling through the ice mass to the bedrock floor, an endeavor that is both costly and extremely difficult. Nonetheless, it is generally accepted that ice temperature, the character of the bedrock-ice interface, and the nature of the subglacial drainage system are the prime factors that determine how sliding actually works.

Perhaps the most intuitive and acceptable explanation of the process of sliding is based on the phenomenon of *glacial slippage,* in which the glacier slips over a water layer that rests between the underlying rock surface and the base of the glacier. This water is derived from precipitation and meltwater on the ice surface that is able to penetrate through openings and fractures to the

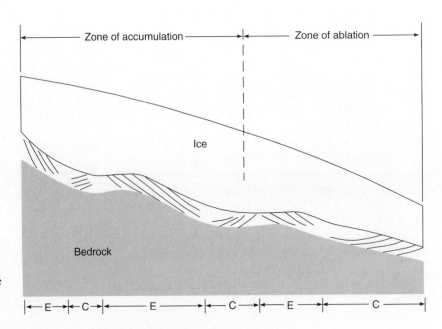

Figure 9.16
Longitudinal section of hypothetical glacier showing irregular bedrock profile and preferred slip planes within the ice. Zones of extending flow and compressive flow indicated by E and C.
(Nye 1952a)

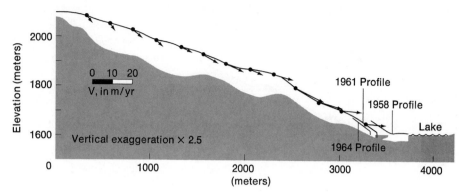

Figure 9.17
Longitudinal section showing average calculated bedrock profile and surface velocity vectors on the South Cascade Glacier, Wash.
(Meier and Tanghorn 1965)

glacier bed or from the melting of basal ice that is near the pressure-melting point. It not only lubricates the underlying surface, but if the subglacial water is under enough hydrostatic pressure, it may partly offset the weight of the overlying ice. This, of course, reduces the resistance to slipping of the ice over the bedrock and increases the movement of the glacier (Weertman 1964). It follows, then, that one of the most important controls on basal sliding is the distribution and pressure of water at the base of the glacier, both of which are dependent on the nature of the subglacial drainage system.

It is now recognized that several forms of subglacial drainage may exist (fig. 9.18) and that drainage type may vary with position across the glacier or with time at any given location. The observed variations in the drainage network appear to depend on the amount of water that is present, the temperature of the ice, and the permeability of the underlying geological materials (Benn and Evans 1998).

In areas where glaciers flow over bedrock, there is abundant evidence to suggest that water may exist as a semicontinuous film along the ice-rock interface (Weertman 1972; Hallet 1979b). Water films were once thought to represent the primary drainage system beneath glaciers (Weertman 1964). However, we now know that their thickness is limited to a few millimeters making them incapable of transmitting large volumes of flow. The films are therefore thought to primarily transport water derived from localized subglacial melting (Hubbard and Sharp 1993). When the availability of water at the glacier bed increases the thickness of the films beyond a few millimeters, they will break up and drain though small channels or fill cavities along the glacier's bed.

The development of cavities beneath the ice was first modeled by Lliboutry (1964, 1968). He hypothesized that ice will separate from its bed downstream from a bedrock obstruction, and the cavities thus formed

will fill with subglacial water. Weertman (1986) argued that these water-filled cavities can exist along the basal contact as isolated pockets. However, with increasing quantities of basal water small channels may form between the cavities, producing an elaborate interconnected network of openings beneath the ice (fig. 9.18) called a *linked cavity system* (Kamb 1987; Hooke 1989). The water-filled cavities may nearly submerge the obstacles at the base of the glacier, thereby reducing the area of contact between ice and rock. In addition, high subglacial water pressure should develop within cavity systems, and this pressure will partially support the weight of the overlying ice. The reduction of overall friction thus produced will allow a drastic increase in sliding velocity.

The degree to which cavities form is dependent on the water pressure at the glacier bed and, therefore, reflects that availability of meltwater from the surface and the extent of subglacial meltwater generation. The water pressures at which cavities begin to develop, called the *separation pressure,* is related to bed roughness, and tends to be greater for beds with numerous high-amplitude bumbs (Schweizer and Iken 1992; Willis 1995). Hooke (1989) points out that linked cavities are likely to be present beneath most glaciers that move by sliding because ice will separate from the bedrock downflow of at least some obstacles, and striations and joints in the bedrock will form connecting channels between the resulting cavities.

With further increases in water pressure, it is likely that the linked-cavity system will switch to some form of channelized flow (fig. 9.19). The character of these channels is a topic of debate (e.g., Hock and Hooke 1993), but regardless of their configuration the downslope drainage of water via a channelized system will lower the hydrostatic pressure at the base of a glacier and reduce the sliding (Chadbourne et al. 1975; Engelhardt et al. 1978; Kamb 1987).

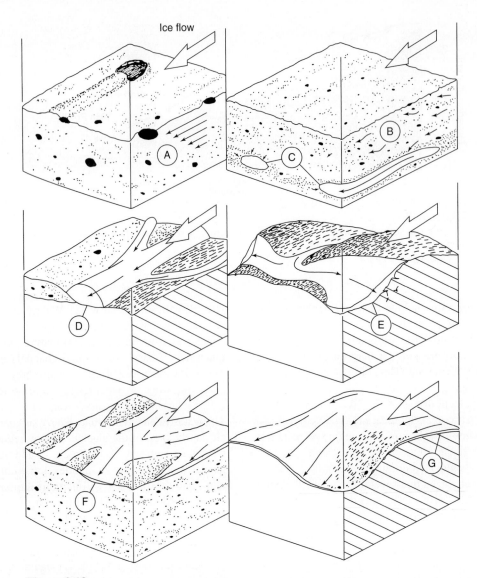

Figure 9.18

Types of subglacial drainage recognized beneath glaciers overlying bedrock or unconsolidated sediments. (A) The movement of water within deforming debris; (B) flow through pores in unconsolidated sediment; (C) flow through pipes developed within unconsolidated materials; (D) flow through discrete channels formed along the ice-bed interface; (E) flow through a linked-cavity system; (F) flow through a braided canal network; (G) the migration of water as a thin film along the glacier bed.

(By D. I. Benn and D. J. A. Evans in QUATERNARY SCIENCE REVIEW, Vol. 15, pp. 23–52. Copyright © 1996. Reprinted by permission of Elsevier Science.)

The presence of pressurized water at the base of a glacier and its fundamental role in the mechanics of sliding has been clearly demonstrated (Hodge 1976, Iken and Bindschadler 1983, 1986; Kamb et al. 1985). In fact, seasonal and diurnal melting has been shown to cause uplift of a glacier surface (Iken et al. 1983; Jansson and Hooke 1989). It cannot be stated, however, that such water is always part of those mechanics because the generation of effective hydrostatic pressure requires that the underlying material be impermeable. Where underlying bedrock is pervasively fractured or the subsurface composed of unconsolidated sediment, free water may drain into the underground system (fig. 9.18). Such downward release of water should lessen the possibility of creating enough hydrostatic pressure to facilitate slippage (see Hobbs 1999 for a discussion of the possible effects of this process during the Pleistocene on the continental ice sheets of southwestern Wisconsin).

Factors other than the distribution and availability of water may also influence basal sliding. For example, the adhesive strength of ice frozen to the underlying bedrock is very high (Benn and Evans 1998) and, thus, freezing can undoubtedly reduce sliding velocities. In fact, it was originally hypothesized that polar glaciers

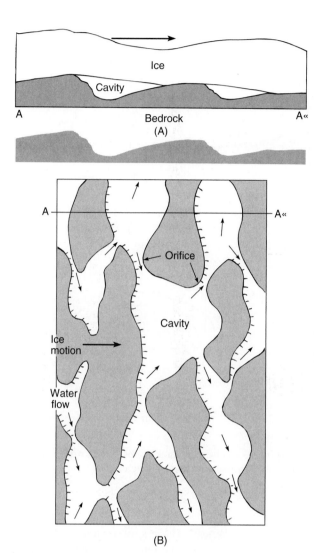

Figure 9.19
Schematic diagram of a linked-cavity system in cross-section (A) and plan view (B).
(Modified from Kamb 1987)

ice can move over an irregular surface by this mechanism alone. The idea of a glacier slipping *up* a slope on a layer of water stretches credibility and suggests that other processes must be involved in the sliding phenomenon. Two such processes that have been recognized are regelation and enhanced creep. The process of *regelation* involves the melting and refreezing of ice because of fluctuating pressure conditions and usually results in a texturally distinct layer of ice, only a few centimeters thick, that rests in contact with the bedrock floor (Kamb and LaChapelle 1964).

The mechanics of *regelation* were first revealed by Bottomley (1872), who demonstrated that a thin wire under tension could be passed through a block of ice without splitting it apart. The pressure exerted by the wire melts the ice beneath it because the pressure depresses the melting point. The water thus released flows in a thin layer to the upper surface of the wire where it refreezes. The heat of fusion released by refreezing above the wire is transferred through the wire to provide the heat needed for the pressure-melting along the leading edge, and so the speed at which the wire moves through the ice block is partly dependent on the rate at which the wire conducts heat. Other objects such as cubes, spheres, and disks have been forced through ice (Kamb and LaChapelle 1964; Barnes and Tabor 1966; Townsend and Vickery 1967; Morris 1976), and wires of various size and composition have been utilized to study the regelation process (Nunn and Rowell 1967; Drake and Shreve 1973). Each experiment has reinforced the correctness of Bottomley's original interpretation of the process, although simple regelation theory seems to predict much greater velocities than those actually measured (Morris 1976). Thus, it seems that any object can be passed through ice without severing the mass and without changing the properties of the ice except along the path of transport. Along that path, the ice develops a new texture, similar in all aspects to that observed in the thin basal layers of glaciers.

In glaciers the process of regelation, shown in figure 9.20, allows ice to circumvent minor irregularities of the bedrock surface over which the glacier is moving. The basal ice melts at the upstream edge of the bedrock knobs where pressure is the greatest. The meltwater thus released flows, probably as a thin film, around the obstacle and refreezes in the region of least pressure, which is along the downstream edge of the barrier. The mechanism seems to be quite effective when the obstacles are small but becomes less important when protuberances are larger (Weertman 1957, 1964). This, of course, begs the question as to what magnitude of bedrock irregularity will render the regelation process inoperative and how thick the layer of regelating ice can be.

The problem is complicated by the fact that ice near the glacier base also may deform according to the flow laws. In that case, ice at the upglacier edge of the obstacle,

characterized by ice below the pressure-melting point could not move by basal sliding. Recent theoretical, laboratory, and field studies have demonstrated, however, that sliding can occur, but at extremely slow rates (Shreve 1984; Fowler 1986; Echelmeyer and Zhongxiang 1987; Cuffey et al. 2000). In addition to adhesion, ice at the base of a glacier may contain considerable amounts of rock debris. When these particles come into contact with the underlying bed it creates a certain amount of frictional drag that can lead to a reduction in slip rates. In general, the amount of drag that is generated is proportional to the strength of the force pressing the particles against the bed—the larger the frictional drag generated, the slower the sliding velocities.

The premise that lubrication of the bedrock surface, aided by hydrostatic pressure, will initiate slippage may seem easy to accept, but the question arises as to how

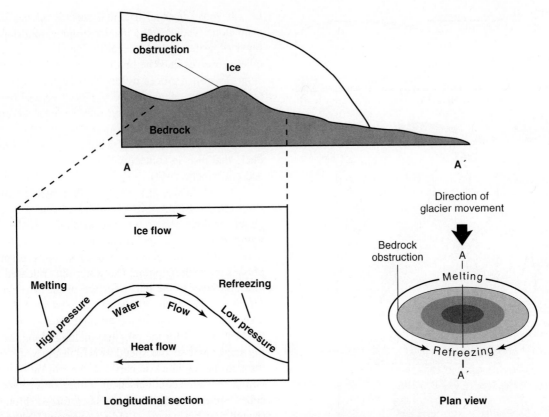

Figure 9.20
The process of regelation as it functions beneath a glacier. Melting occurs on upstream flank of obstruction where pressure is greatest. Refreezing occurs downstream of obstruction where pressure is least, although some cavitation may form.

where pressure is greatest, will have a higher strain rate than that at the downstream edge, and will exhibit *enhanced creep.* Larger obstacles will augment the stress and cause higher flow velocities; that is, the velocity will be directly proportional to the size of the bedrock knob. Because of these complications, Weertman (1957, 1964) suggested that sliding over an irregular bed is produced by components of the two processes: regelation is the predominant control on sliding when the barriers are small, and enhanced creep when the obstacles are large. It follows that some intermediate size of obstacle, called the *controlling obstacle size,* determines which process will function. In the subglacial bedrock topography, ice moving over or around protuberances smaller than the controlling obstacle size will do so mainly by regelation; where surface bulges are larger, creep will probably dominate. Field observations and theoretical studies suggest that the controlling obstacle size is typically on the order of 0.05 to 0.5 m.

Motion with Deforming Subglacial Sediments

The theoretical models of internal flow and basal sliding presented earlier generally assume that the glacier bed is a rigid, impervious bedrock surface. Although such surfaces do exist beneath glaciers, observations of regions recently exposed by glacial retreat illustrate that at least the margins of many glaciers are underlain by unconsolidated debris. In addition, Boulton and Hindmarsh (1987) point out that during the last glacial period, 70 to 80 percent of the continental ice sheets in Europe and North America were underlain by unconsolidated or poorly consolidated sediments of Quaternary or Tertiary age. It is now clear that these unconsolidated, subglacial sediments may deform, allowing the overlying ice to actually move along with the mobilized material. In fact, investigators have recently found that as much as 90 to 95 percent of the forward motion of some temperate glaciers results from the deformation of subglacial sediments (Boulton 1979; Alley et al. 1986; Blankenship et al. 1986; Boulton and Hindmarsh 1987).

The deformation of subglacial sediments is governed in large part by the shear stress applied to the material by the overlying ice and the shear strength of those sediments. Boulton (1979) used a form of the Coulomb equation to express the shear strength of subglacial material in an unconsolidated state:

$$S = C + (P_i - P_w) \tan \phi$$

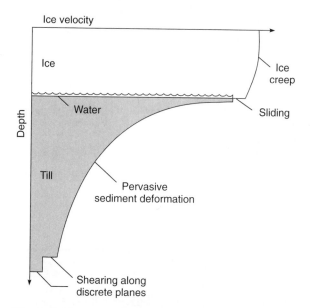

Figure 9.21
Surface ice velocities of glaciers overriding unconsolidated debris are a function of internal creep, basal sliding, and sediment deformation.
(Alley 1991)

where S is shear strength, C is the cohesion of the unconsolidated sediments, P_i is pressure exerted by the overlying ice, P_w is pore water pressure, and ϕ is the angle of internal friction. Conceptually, the expression suggests that deformation of the bed sediments can occur by decreasing shear strength of the unconsolidated material. This is accomplished by increasing the pore water pressure (P_w) to the point where it nearly equals the overburden pressure exerted by the ice (P_i) (Alley 1991). Under these conditions, the expression ($P_i - P_w$) in the equation on page 312, called the effective normal stress, approaches zero and the contact forces exerted between the particles are dramatically reduced. Deformation can also be initiated by an increase in the shear stress applied to the unconsolidated bed sediments.

Some observations suggest that there may be a limit to which increases in pore water pressure will lead to the deformation of the subglacial sediments (Iverson et al. 1995; Benn and Evans 1998). At high basal water pressures, a linked-cavity system may develop beneath the ice (as described in the previous section). The formation of this drainage network will dramatically reduce the shear stress applied to the underlying debris, causing deformation rates and the velocity of glacier motion due to deformation of the subglacial materials to fall. Importantly, however, the absolute rates of glacier motion may not decrease because the development of the linked-cavity system may lead to a dramatic increase in basal sliding at the same time that the rates of bed deformation decline.

The few observations that have been made in tunnels and cavities at the base of glaciers show that pervasive sediment deformation (i.e., deformation spread throughout the sediment mass) may extend for as much as 10 m below the ice-sediment interface. In most cases, both the strain rate and the total amount of deformation decrease with depth (Boulton and Hindmarsh 1987; Alley 1991; fig. 9.21). In addition, movement may occur beneath the pervasively deformed material by shearing along discrete planes. Alley (1991) concluded, however, that the movement along shear planes is not likely to account for a significant portion of the total glacial advance.

The deformation of unconsolidated debris has received considerable attention in recent years in large part because it can account for the rapid movement of some outlet glaciers and ice streams that could not be explained by other mechanisms (Clarke et al. 1984; Alley et al. 1987). Kamb (1991) has argued, however, that the deformation of unconsolidated subglacial sediments should resemble a perfect plastic substance. Such behavior would result in the highly unstable flow, a phenomenon that is not observed on fast moving ice streams. This led Kamb to suggest that maximum ice velocities are not controlled by the deforming sediments but by localized bedrock knobs that support most of the basal shear stress. Observations on other glaciers have indicated that the shear strength of subglacial sediments is highly variable from place to place, consisting of low-strength areas that are separated by high strength zones that support a disproportionately large amount of the overlying ice (Benn and Evans 1996, 1998). These observations have led to the concept of *sticky spots*, in which ice velocities are controlled by small patches of more resistant substrate rather than the deformation of the entire bed.

In addition to influencing the motion of both temperate and subpolar glaciers, the mobilization of subglacial debris is an important factor in the development of glacial features that originate beneath the ice. We will return to this phenomenon in the next chapter when the formation of subglacial landforms is discussed.

Velocity Variations with Time

Flow rate measurements compiled from temperate and polar glaciers show that velocities vary significantly with time. The interval over which velocity variations occur is different from one glacier to another and even in specific locations on the same glacier. For example, many investigators have noted sudden, jerky motions that increase surface velocity for hours, days, or weeks (Meier 1960; Glen and Lewis 1961; Goldthwait 1973; Jacobel 1982; Iverson et al. 1995). These spasmodic movements are normally explained by local controlling factors such as weather conditions, fault slips, or a sudden release of ice that has been retarded in its flow by some obstruction.

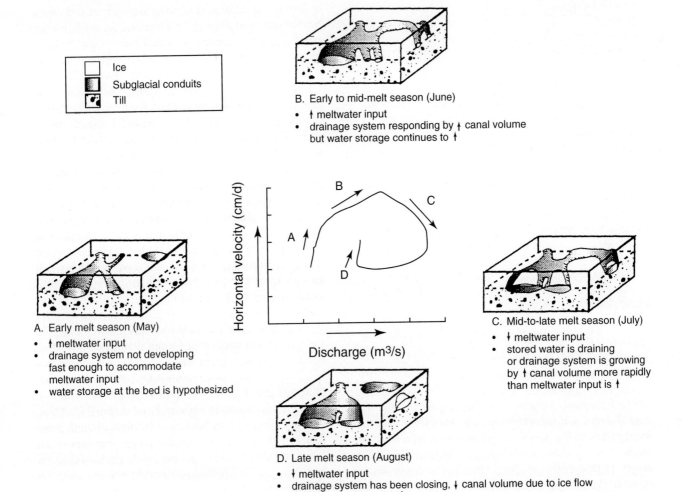

Ice

Subglacial conduits

Till

B. Early to mid-melt season (June)
- ↑ meltwater input
- drainage system responding by ↑ canal volume but water storage continues to ↑

A. Early melt season (May)
- ↑ meltwater input
- drainage system not developing fast enough to accommodate meltwater input
- water storage at the bed is hypothesized

C. Mid-to-late melt season (July)
- ↓ meltwater input
- stored water is draining or drainage system is growing by ↑ canal volume more rapidly than meltwater input is ↑

D. Late melt season (August)
- ↓ meltwater input
- drainage system has been closing, ↓ canal volume due to ice flow more rapidly than input ↓
- water storage ↑
- discharge may cease in some areas long before freeze-up

Figure 9.22
Summary of the changes in ice velocities of the Matanuska Glacier, Alaska, as a function of water discharge from the glacier's terminus. Block diagrams illustrate possible changes in drainage network configuration that accompany the alterations in velocity and discharge.

(From GSA SPECIAL PAPER 337, by Ensminger et al., Figure 12, p. 19. Reproduced with permission of the publisher, the Geological Society of America, Boulder, Colorado, USA. Copyright © 1999 Geological Society of America.)

Seasonal or yearly variations in flow are also common. Velocities in the ablation zone usually increase during the summer. These fluctuations most logically reflect accelerated basal slip or sediment deformation, which is facilitated by the abundance of free water at that time (Meier 1960; Elliston 1963; Paterson 1964; Hodge 1974; Vivian 1980; Anderson et al. 1982). Ensminger et al. (1999), for example, found that velocities of the Matanuska Glacier, Alaska, increased significantly in early June and attained a seasonal high in late June to early July, before decreasing through mid-August. The increase in velocities at the beginning of the summer was due to the storage of water and the buildup of water pressure at the base of the glacier, which facilitated basal sliding (fig. 9.22). The reduction in velocities later in the season was attributed to the formation of a more efficient drainage system that led to

the release of water from the glacier's bed. Interestingly, velocities also increased in late August when the volume of meltwater was decreasing. Ensminger et al. (1999) suggest that the increase in velocity at the end of the summer is due to the partial closure of the canal system by ice creep and the renewed buildup of subglacial water pressures (fig. 9.22).

Short-term variations in velocity are usually recognized because the span of most investigations is long enough to demonstrate their presence. However, this is not true of velocity variations exceeding a few years. Imagine the sampling program needed to observe the long-term fluctuations of velocity that are produced by changes in climate and the associated glacial regime. Nevertheless, any positive change in the glacial budget should increase the discharge of flow, and according to theory, the mechanical response to the increased accumulation will probably involve the propagation of a *kinematic wave* moving down the glacier two to five times faster than the actual particles of ice. The wave will reach the glacier snout long before the new ice formed from the flux in accumulation could possibly be transported that far. It appears that the wave motion has little effect on ice that is extending, but in the ablation zone, where compressive flow dominates and velocity decreases down-ice, the oncoming wave accentuates the compression and the glacier becomes very unstable. As the wave approaches any point in the ablation zone, the ice will thicken and the surface will rise dramatically. As the wave passes, it will quickly thin and subside to its former level. On some glaciers the ice surface may rise and fall more than 100 m in the sequence of events.

Velocity measurements on some glaciers do not fit the predictions based on kinematic wave theory (Lliboutry and Reynaud 1981). Nonetheless, if kinematic waves generally represent the mechanism by which glaciers respond to changes in mass balance (Nye 1960), we must recognize that the time needed for the wave transmission varies from glacier to glacier. For example, the time needed to reestablish a steady state, or the **response time,** in single, temperate glaciers varies from 3 to 30 years. The significance of this fact can be seen in a hypothetical case. Suppose that during the years 1954 to 1958 abnormally high amounts of snow accumulated on two adjacent valley glaciers. In one with a short response time, a kinematic wave rapidly traveled the length of the glacier and the snout advanced dramatically in 1960. The second glacier, having a longer response time, showed no visible effects of the accumulation by 1960; the kinematic wave did not reach the terminal zone until 1975, when the terminus suddenly advanced. The two glaciers were completely out of phase, although both advanced in response to the same event. In addition to problems of mechanics, measurements made on a portion of the ice experiencing the wave action are not documenting the flow of ice but rather the movement of waves.

Kinematic wave velocities may attain catastrophic values of up to hundreds of meters a day (Meier and Johnson 1962; A.E. Harrison 1964), and they have often been cited in the mechanics governing surging glaciers. However, in many cases glaciers surge even though there is no discernible net accumulation of mass to the glacier or any other external stimuli for the movement (Post 1960, 1969; Meier and Post 1969). In addition, the mechanisms inducing kinematic waves may differ from those initiating surges (Sharp 1988). Thus, it is probably best to separate surging from kinematic wave transfer even though the two movements have many of the same characteristics (Palmer 1972).

A *surging glacier* then, is one in which sudden, brief, large-scale ice displacements periodically occur. The ice moves 10 to 1000 times faster than its flow rate

in the quiescent periods between surges (Clark et al. 1986), and thus surging glaciers entrain significantly more debris than normal glaciers (Clapperton 1975). The surge periodicity, which tends to be consistent for any given glacier, seems to range from 10 to 100 years (Clarke et al. 1986) and probably results from unique conditions that create a cyclic instability within the glacier.

Glacial surges are now recognized as fairly common phenomena. In fact, some investigators (e.g., Budd 1975) have suggested that surging glaciers may represent a completely separate type of glacier. However, they have no distinct size, shape, or activity, and they can contain temperate or subpolar ice. Their only unifying characteristics seem to be that surges are initiated in the ablation zone slightly down-ice from the equilibrium line, and a long and pronounced stagnation of ice occurs in the terminal zone in the interval between surges (Post 1960, 1969). During the surge the glacier surface is broken chaotically, and medial moraines and ice bands are intensely contorted.

Direct observations of surges are rare. A notable exception is the 1982–1983 surge of Variegated Glacier in Alaska. During the surge, measurements of ice deformation from a borehole drilled to the base of the glacier demonstrated that 95 percent of the forward motion of the ice was by slip along its base and sides (Kamb et al. 1985). Kamb and his colleagues (1985, 1987) found that the surge was triggered by the buildup of water and water pressure beneath the glacier when a linked-cavity system replaced one or more subglacial channels observed prior to the surge. The linked-cavity system may have formed in response to the accumulation of ice in the reservoir area, and once it had developed, it facilitated the rapid movement of the glacier by sliding, as described earlier in this chapter. The surge presumably terminated when a subglacial channel system was reestablished, releasing the water stored in the linked-cavity system and increasing the resistance to basal slip.

A number of other models have been presented to explain the surging phenomenon (Budd 1975; Robin and Weertman 1973; Fowler 1987; Clarke et al. 1984). Although these models generally agree that the buildup of water at the base of a glacier is involved, they differ in the mechanisms used to explain the sudden advance of the ice. What seems to emerge from these studies is that there is probably more than one mechanism responsible for surging, and that the key element is basal meltwater and its effects on basal sliding or the deformation of subglacial sediments.

ICE STRUCTURES

Glaciers usually display a variety of structures that develop during growth of the glacial mass. These features are called primary structures and are indicative of the nature of accumulation and ablation during the glacier's

Figure 9.23
Saskatchewan Glacier showing outcrop pattern of stratification in Castleguard sector. View upglacier from cliff on south margin below Castleguard Pass. Splaying and en echelon crevasses are also visible. Province of Alberta, Canada.

history. Secondary structures that are closely related to the glacier's mode of flow and stress field can also develop in response to glacial movement. A selected number of the most common primary and secondary structures are described below.

Stratification

Primary structures in glaciers appear as discernible layers or bands within the ice. The layering results from processes that reflect an annual cycle of snow accumulation and ablation above the firn line. During the winter a thick pile of new snow is added to the glacier and, with time, proceeds through the phases of metamorphism that culminate in a layer of clear white ice. In the ablation season, however, the upper portion of the winter accumulation is subjected to melting and refreezing and develops a texture different from that forming lower in the snow pack. In addition, sediment and organic debris may collect in the partially ablated upper zone, making it slightly darker in color. The alternating white and "dirty" layers give the ice a stratified appearance and allow glaciologists to estimate the annual growth in the glacier thickness. The layers tend to be tilted and deformed as the glacier moves. At the surface of the abla-

tion zone, the bands look like truncated beds of plunging folds (fig. 9.23) with the layers dipping into the glacier in an up-ice direction.

Secondary Features

Foliation   A secondary type of layering, called **foliation,** is produced by shear during ice motion. It is sometimes difficult to distinguish from primary stratification because both types may display similar grain size or textures. Foliation usually appears as alternating bands of clear blue ice and white bubble-rich ice. The layers, which dip at all angles, are most prevalent near the ice margins and commonly offset or wrinkle the primary stratification (fig. 9.24). The origin of foliation is poorly understood. Some evidence suggests that it originates where shear stress is greatest, but Meier (1960) argues that it is neither formed nor preserved at great depth, where shear stress would presumably be at a maximum.

Crevasses   Crevasses are cracks in the ice surface that range in size from miniature fractures to gaps several meters wide. The fractures are important in that they provide avenues for surface meltwater to penetrate the interior of the glacier, although the openings are rarely

Figure 9.24
Saskatchewan Glacier showing gently dipping stratification wrinkled and intersected by nearly vertical foliation. Exposed on east wall of a crevasse, 4.5 km below firn limit in midglacier. Province of Alberta, Canada.

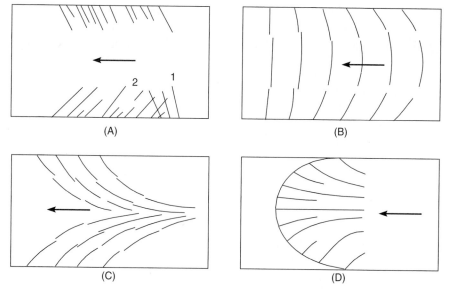

Figure 9.25
Types of crevasses in valley glaciers: (A) marginal or chevron—(1) old rotated crevasses; (2) newly formed crevasses; (B) transverse; (C) splaying; (D) radial splaying. Arrows show flow direction.
(Sharp 1960)

deeper than 30 m. It is generally assumed that crevasses develop perpendicular to the direction of maximum elongation of the ice. On some glaciers, however, the crack directions do not precisely coincide with the measured surface strain rates (Meier 1960). Nonetheless, crevasse types (fig. 9.25) do reflect a local tensional stress environment. *Splay crevasses* or *radial crevasses* form near the flow centerline under compressive flow where spreading exerts a component of lateral extension. In contrast, near the ice margins the shear stress parallels

Figure 9.26
Transverse crevasses developed at icefall on Little Yanert Glacier, Alaska Range.

the valley walls, and crevasses develop diagonally to those sides. Here the crevasses are either *chevron* or *en echelon* types.

Transverse crevasses (fig. 9.26) develop under extending flow where the ice extends in a longitudinal direction. In temperate glaciers these cracks occur most commonly in ice falls where the glacier cascades over convex irregularities in the underlying bedrock slope. In the summer large ice crystals may develop in the transverse cracks when water trapped there repeatedly melts and refreezes. Sediment may also accumu-

late in these openings. At the base of the icefall the crevasses are closed by compression, and a band of dirt-stained ice forms. During the winter, however, ice descending the fall reconstitutes into clear, bubbly ice (King and Lewis 1961). The annual downvalley flow, therefore, tends to produce a distinct series of alternating white and dark bands, called **ogives,** that are prominent at the foot of icefalls (fig. 9.27). These bands are different from primary layering in that they do not dip as strata into the glacier but are purely a surface phenomenon.

Figure 9.27
Surficial ogives forming on the Gilkey Glacier at the base of the Vaughan Lewis Icefall, Juneau Icefield, Southeast Alaska.

SUMMARY

In this chapter we examined the origin and movement of glaciers as well as the factors that might explain the properties associated with the various glacier types. Glaciers can be classified on the basis of their morphology or geographic position, their activity, or their thermal characteristics. Glaciers develop when snow accumulated over a period of years is transformed into ice by compaction, recrystallization, and melting and refreezing. These processes progressively increase the density of snow and firn as the space within the original mass is removed by pressure, crystal growth, and orientation of the grains. When the ice reaches a critical thickness, it is capable of movement.

The amount and type of geomorphic work a glacier can accomplish depends on its mass balance and on the total volume of ice added and lost during a period of time. The mass balance, or budget, is the net difference between accumulation and ablation during the time period in question. A positive mass balance indicates that more ice has been added to the glacier than has been lost, and the transfer of ice from the accumulation zone to the ablation zone causes the glacier front to advance. A negative budget results in a retreat of the front because the amount of ice transferred from the accumulation zone to the ablation zone is not sufficient to replenish the volume of ice lost by ablation. The total volumes of accumulation and of ablation determine the level of glacial activity. Two glaciers may have identical mass balances, but the one having the higher total volumes of addition and loss will be more active and do more geomorphic work.

Glaciers move by internal deformation of the ice and by sliding along the bedrock floor at the base of the glacier. In addition, where glaciers override a layer of unconsolidated debris, the glacier may move with sediments deforming beneath the ice. The internal movement occurs as a type of creep that is mechanically different from viscous flow or pure plastic flow. In most glaciers the deformation caused by flow can be estimated by an equation called the power flow law or some modification of it. The mechanics of sliding is poorly understood but probably consists of the combined effects of regelation, lubrication by water under pressure, and enhanced creep.

Velocity of flow relates well with modified versions of the power flow law, but many characteristics of glaciers complicate a direct relationship. For example, ice does not flow in parallel laminar sheets but thickens and thins along the length of the glacier. This variation develops the compressive and extending flow that causes ice to move downward in the accumulation zone and toward the surface in the ablation zone. In addition, flow velocity varies with time. The phenomenon of surging is difficult to fit into an all-inclusive flow model.

Primary and secondary structures in glaciers are related to the processes of accumulation and ablation and to the stress field generated during ice movement. The major structural features are stratification, foliation, crevasses, and ogives.

SUGGESTED READINGS

The following references provide greater detail concerning the concepts discussed in this chapter.

Alley, R. B. 1991. Deforming-bed origin for southern Laurentide till sheets. *Jour. Glaciology* 37:67–76.

Benn, D. I., and Evans, D. J. A. 1998. *Glaciers and glaciation.* London: Arnold.

Boulton, G. S., and Hindmarsh, R. C. A. 1987. Sediment deformation beneath glaciers: Rheology and geological consequences. *Jour. Geophysical Res.* 92:9059–82.

Ensminger, S. L., Evenson, E. B., Alley, R. B., Larson, G. J., Lawson, D. E., and Stasser, J. C. 1999. Example of the dependence of ice motion on subglacial drainage system evolution: Matanuska Glacier, Alaska, United States. In Mickelson D. M. and Attig J. W., eds., *Glacial processes past and present,* pp. 11–22. Geological Society of America Special Paper 337.

Glen, J. W. 1955. The creep of polycrystalline ice. *Proc. Royal Soc. London,* ser. A 228:519–38.

Hooke, R. L. 1989. Englacial and subglacial hydrology: A qualitative review. *Arctic and Alpine Res.* 21:221–33.

Kamb, B. 1987. Glacier surge mechanism based on linked cavity configuration of the basal water conduit system. *Jour. Geophysical Res.* 92:9083–9100.

GLACIAL EROSION, DEPOSITION, AND LANDFORMS

Courtesy of
P. Jay Fleisher

Introduction
Erosional Processes and Features
    Minor Subglacial Features
    Cirques
        Cirque Morphology
        Cirque Formation
    Glacial Troughs
Deposits and Depositional Features
    Drift Types
        Nonstratified Drift
        Stratified Drift
    The Depositional Framework
    Marginal Ice-Contact Features
        Moraines
        Stratified Marginal Features
    Interior Ice-Contact Features
        Interior Moraine
        Fluted Surfaces and Drumlins
    Proglacial Features
Summary
Suggested Readings

INTRODUCTION

We are now ready to consider the geomorphic significance of glacier mechanics. It would be ideal if we could *directly* correlate geomorphic features with specific ice processes, but we seem to be far from such sophistication. The reasons for this are several: (1) the conceptual models of glacier mechanics are greatly oversimplified and based more on theory rather than extensive observation; (2) much erosion and deposition takes place at the base of glaciers where it is rarely feasible to document the actual process; and (3) we are just beginning to understand the complex feedback mechanics that functions between the basal ice and the geologic framework. For example, we know that glaciers have the ability to erode because we can see the results of that erosion after the ice disappears. What we do not understand is precisely how the erosive process itself is modified by remolding of the subglacial framework. Is it possible, for instance, that the removal of certain obstructions by erosion decrease the subsurface roughness enough to facilitate rapid sliding velocities and might this in turn accelerate further erosion?

Having noted that our understanding of erosion and deposition in the subglacial environment is far from complete, it should be recognized that our knowledge of these processes has improved greatly in recent years. These advances have resulted from the integration of theoretical analyses, laboratory experiments, and direct observations within boreholes, natural cavities, and human-made tunnels in the ice (Benn and Evans 1998). Nevertheless, it seems fair to say that our current understanding of glacial processes is primarily based on the interpretation of the deposits and features that are created. In essence, we must interpret the processes that are operating at the bed of the glacier from the character of its results. This type of deductive reasoning is not completely satisfying, but it has been a part of geology for a long time and is not necessarily incorrect. Moreover, the investigation of glacial features and deposits can provide tangible clues about their origins that become invaluable in deciding how and where to study glaciers in the future. These studies, conducted using new techniques for investigating glacier mechanics, will undoubtedly demand a modification of our present hypotheses concerning the creation of glacial landforms.

EROSIONAL PROCESSES AND FEATURES

Minor Subglacial Features

Glacial erosion is accomplished primarily by two processes, a scraping action called **abrasion** and a dislodgement or lifting action called **quarrying** or **plucking.** Because ice is not a hard mineral (1.5 on Mohs' scale at 0° C), it cannot abrade most solid rock material unless it utilizes fragments of rock carried as load in the ice as grinding tools. Therefore, the efficiency of abrasion and the features it produces depend on the character and concentration of the debris being dragged along the base of the ice and, of course, on the properties of the bedrock being overridden. It also depends on the magnitude of the force pressing the particles in the ice against the bedrock, and the velocities at which the particles are dragged along the bedrock surface. These factors can be combined to estimate the rates of glacial abrasion. For example, a commonly utilized wear law is expressed as

$$A_v = kF_nV_p$$

where A_v is the volume of rock removed from a single fragment per unit time, F_n is the effective contact force exerted normal to the bed surface by the fragment, V_p is the velocity of the fragment parallel to the bed, and k is a coefficient dependent on shape of the fragment and the relative hardness between the fragment and bed material. The parameters assumed to govern the effective contact force (the force pressing the particle against the bed) represent a primary difference between the various models of abrasion that have been constructed in recent years (Boulton 1974; Shoemaker 1988). For example, Boulton (1974) suggests that the contact force is primarily dependent on the weight of the overlying ice. However, glacial ice greater than 22 m thick should theoretically flow around and under the abrading rock fragments. The contact force exerted by active glaciers should therefore be independent of ice overburden pressure. This assumption led Hallet (1981) to argue that the force pressing the particles against the bedrock is dependent on the buoyant weight of the particles in the ice, and the rate at which the fragments are flowing toward the bedrock surface. High contact forces are achieved in areas of relatively rapid downward flow, such as in zones of basal melting on the stoss side of obstructions. It is important to remember, however, that the conditions observed in the subglacial environment are highly variable. Thus, it is likely that no single model can be applied to all glaciers, or, perhaps, even to all locations beneath a given glacier. Nonetheless, the Hallet model seems to work reasonably well in describing the contact force at the bed of a glacier when less than 50 percent of the basal ice layers is composed of sediment (Benn and Evans 1998).

Estimated rates of abrasion usually range from 0.06 to 5 mm a year, but they may be considerably higher under thick, high-velocity glaciers. Boulton (1974) reports that the abrasion of a marble plate inserted beneath the Glacière d'Argentiere in France proceeded at a rate up to 36 mm a year where the ice was 100 m thick and moving at 250 m a year.

Small erosional features produced by abrasion are usually in the form of linear scratches or crescentic marks that show the relationship between the size and composition of the abrading particles and the resistance of the underlying bed. Very fine particles in sufficient abundance produce a smoothly polished surface composed of microscopic scratches (fig. 10.1). As the grain size of the load increases, the scratches become larger in a transitional sequence from polish to striations.

Striations like those shown in figure 10.1 have been noted in every glaciated region of the world. They are best preserved on homogeneous, fine-grained rocks that have not been deeply weathered and on smooth bedding planes that dip gently away from the direction of ice movement (Glasser et al. 1998). Striations and larger linear features can also form in unconsolidated material, such as till or loess, if it is highly compacted (Westgate 1968). Striations are only millimeters deep and are most likely eroded by sand grains or by jutting edges of larger particles carried in the basal ice. They tend to be continuous for only relatively short distances, probably because the sliding clast itself produces a carpet of plowed debris at its forward edge. When sufficient debris accumulates, the scratching particle will ride over the material, thereby interrupting its contact with the solid bedrock until it reaches a fresh surface downstream from the debris cover (Boulton 1974). This effect, however, can be eliminated if circulating subglacial water removes the fines (Vivian 1970). The discontinuous nature of striae may also be related to the propensity of abrading fragments to rotate; the more rapid the rotation, the less likely the fragments will create deep and continuous striations (Iverson 1991a). It is also known that the entire scratching phenomenon becomes ineffective unless particles within the ice mass continuously move downward to replace the original abrasive grains. Those

grains become smoothed during the abrasive action, and failure to replenish the basal ice with new particles having sharp edges and corners will cause abrasion to end (Boulton 1979).

In addition to scratches of all types, a group of small features, generally referred to as *friction cracks* or *chattermarks,* are formed by chipping or grinding of the underlying rock surface (fig. 10.2). Most workers believe these features result when ice flow is temporarily retarded in its forward motion and then is suddenly released. This produces a jerky flow component commonly referred to as slip-stick movement. The various cracks and marks are usually lunate in form, 10 to 12 cm long and 10 to 25 mm deep, and perpendicular to the direction of ice flow as determined by other criteria. Chattermarks are sometimes present on surfaces of minerals within glacial deposits (Folk 1975; Gravenor 1982). These are indicative of the grinding action within a glacier, although chemical etching may produce similar features.

The features of abrasion are thought to reflect the direction of ice movement. Remember, however, that other processes such as mudflows or snowslides can form striations, and even floating ice blocks can cause them in nonresistant materials (Dionne 1974). Moreover, ice will diverge and converge over an irregular bedrock topography, and basal ice may be moving in different directions than surficial ice (Engelhardt et al. 1978). Thus, the orientation of minor abrasion features can be highly variable, and may deviate from the general direction of ice flow. Nonetheless, it is not uncommon for bedrock outcrops to exhibit multiple sets of striae, each with a different average orientation. Many geomorphologists have used these distinct sets of striae to reconstruct changes in the direction of glacial flow (e.g., Veillette 1986; Sharp et al. 1989, Anundsen 1990), although the practice generally requires the analysis of a large number of measurements and a treatment for statistical significance.

In addition to the features described above, bedrock outcrops in glaciated terrains commonly exhibit somewhat larger, smoothed depressions, which Dahl (1965) referred to as *P-forms* or *plastically moulded forms.* The name is related to Dahl's belief that they were produced by plastic deformation of the ice as it moved over the bedrock surface. More recent evidence suggests, however, that their formation may be related to other phenomena, including erosion by subglacial meltwaters (Sharpe and Shaw 1989; Kor et al. 1991; Gray 1992). Thus, Kor et al. (1991) suggest the use of *S-forms* (or *sculpted forms*) for these features to elevate confusion over their possible origin. Whatever they are called, they exhibit a wide variety of sizes and shapes and their orientation with respect to the predominant direction of ice flow depends on the particular feature that is created (fig. 10.3).

Descriptions

(A)

(B)

Figure 10.1
(A) Glacial polish on columnar basalts, Mammoth Lakes, Calif.
(B) Striations on outcrop of limestone in south central New York state.

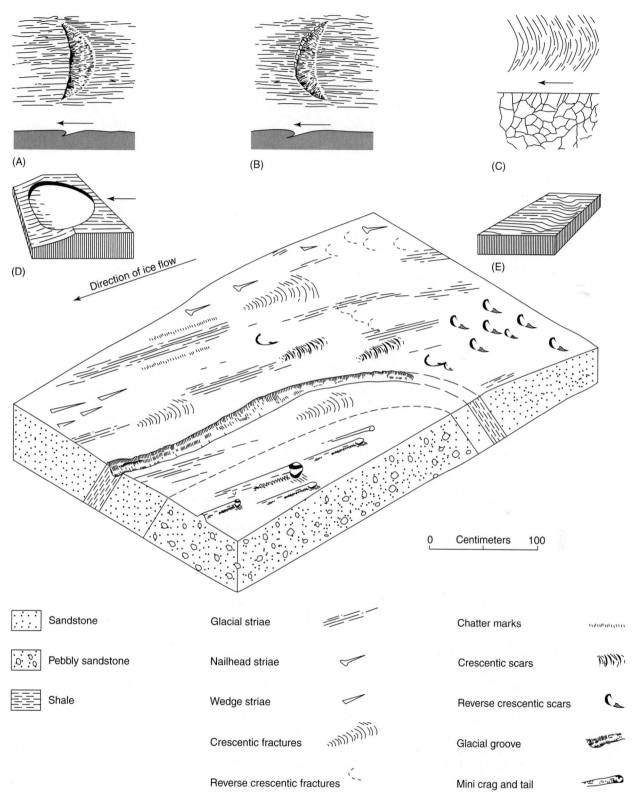

Figure 10.2

Illustrations of small-scale, glacial erosion features. (A) Lunate fracture; (B) crescentic gouge; (C) crescentic fractures; (D) conchoidal fracture; (E) sichelwanne. Block diagram in center of figure shows striae, fractures, grooves, and other erosional features with respect to the direction of ice flow.

(Figure 9.1, p. 314 from GLACIERS AND GLACIATION by Douglas I. Benn and David J. A. Evans. Reprinted by permission of Arnold Publishers, London, England.)

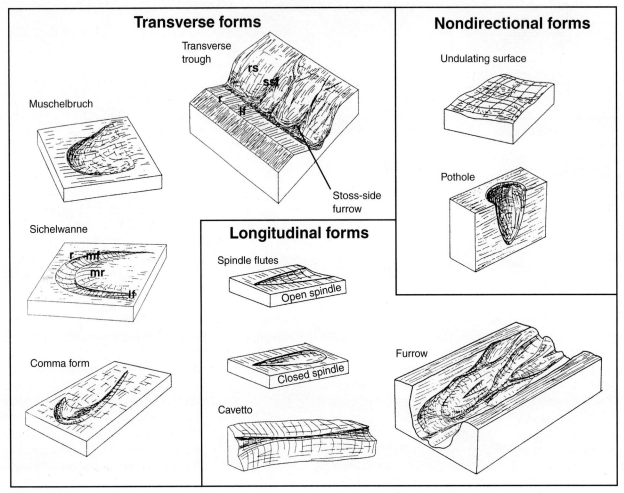

Figure 10.3
Illustrations of smoothed, glacial erosion features referred to as P-forms. The orientation of the features with respect to the direction of ice flow varies between the individual forms.

(Figure 2, p. 627 by Kor, et al. from CANADIAN JOURNAL OF EARTH SCIENCE, 28:623–642, © 1991. Reprinted by permission of National Resource Council of Canada, NRC Research Press.)

Quarrying differs from abrasion in that the functional success of the process depends less on the type of load being transported than on the properties of the underlying rock. In fact, fractures must exist in the bedrock if plucking is to operate at all. A large number of mechanisms have been put forth to account for the formation and growth of fractures in the rock. Historically, significant attention has been given to the processes of pressure release (Lewis 1954; Glen and Lewis 1961), crushing (Boulton 1974), and freeze-thaw, all of which weaken the internal cohesion of the underlying geologic materials. These processes may occur subglacially (Sugden and John 1976; Anderson et al. 1982) or in association with periglacial conditions prior to the arrival of the glacier. In addition to these processes, recent theoretical modeling suggests that fracturing may be produced by concentrating stress in specific areas of the glacier's bed (creating *stress concentrations*), or by developing large stress gradients during periods of short but intense fluctuations in the pressure exerted on the underlying bedrock (Iverson 1991b; Hallet 1996). These latter processes are most commonly associated with the dragging of large particles held in the basal ice layers over the bedrock surface, and with the fluctuation of water pressure in cavities positioned on the leeward side of bedrock obstructions (Benn and Evans 1998). Iverson (1991b), for example, utilized theoretical and numerical data to illustrate how fracturing in bedrock obstructions is promoted by intense changes in water pressure. In this case, rapid deceases in water pressure within cavities located downstream of the obstructions transfers some of the weight of the ice from the subglacial water to the bedrock (fig. 10.4). In addition, the reduction in water pressure results in a loss of lateral support by the water in the cavity on the leeward side of the obstruction (Iverson 1991b). The net result is a change in stress within the bedrock and the growth of preexisting cracks, producing a shattered bedrock zone.

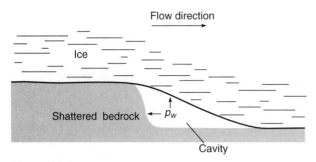

Figure 10.4

Schematic diagram of a subglacial cavity formed downstream of a bedrock obstruction. Localized bedrock shattering may be produced by a rapid decrease in water pressure (p_w) within the cavity, which transfers some of the weight of the ice from the subglacial water to the bedrock and removes the lateral support of the water on the lee side of the obstruction. Subsequently, plucking may occur as subglacial water pressure increases.

Fracturing of the bedrock is clearly a prerequisite for quarrying as it creates particles that can be moved by the overriding ice. It should be recognized, however, that quarrying also involves the entrainment of the loosened particles, a process that requires the ice to exert a shear force on the clast that exceeds the resistance to motion caused by friction (Boulton 1974). In the model put forth by Iverson (1991b), entrainment is promoted by increases in water pressure within the cavities because high water pressures may (1) result in pressure-release freezing between the ice and the glacier bed, enhancing drag on the subglacial materials, (2) decrease the normal stress on the glacier bed, reducing frictional resistance, and (3) accelerate basal sliding, thereby increasing the shear stress on the shattered bedrock surfaces (Iverson 1991b). Thus, Iverson's model provides a unified mechanism for both the shattering of the underlying rock and the plucking of the loosened particles from the glacier's bed.

Utilizing the results of a different theoretically based model, Hallet (1996) argued that the quarrying process is relatively inefficient when the size and number of cavities along the ice-bedrock interface are limited. However, when rising water pressures enlarge the cavities, the basal stresses on the bedrock are applied to a smaller area. This creates stress concentrations and promotes crack growth. Sliding velocities may also increase because high subglacial water pressures are likely to decrease friction at the base of the glacier. This led Hallet to conclude that rapid rates of quarrying results from the combination of high flow velocities, elevated basal water pressures, and the extensive formation of cavities that create stress concentrations. There are very few data sets available to check Hallet's theoretical model. Nonetheless, he notes that it is consistent with at least some data that suggest that rapid rates of quarrying are associated with fast moving gla-

ciers characterized by extensive cavity systems, such as Variegated Glacier, Alaska.

The evidence of plucking action in glaciated landscapes is usually found in erosional features somewhat larger than those produced by abrasion, but commonly the two erosive processes are closely associated in the same landforms. Where abrasion is dominant, the landscape may be indented with smoothly curved elongate surfaces whose long axes are subparallel to the direction of ice flow. Some of these surfaces are distinctly higher at one end, and they taper laterally and longitudinally until they blend into the surrounding ground level, producing a unique teardrop or raindrop shape which Flint (1971) describes as a *whaleback form*. Some whaleback forms may be related to streamlined depositional features, such as drumlins (discussed later in this chapter), in that their shape represents the minimum resistance to flowing ice. The composition of streamlined forms seems to be of little consequence, however, as they can be entirely bedrock, entirely sediment, or a combination of the two. Whalebacks, therefore, may be merely a transitional form in a range of streamlined features from pure bedrock to pure drift (Flint 1971).

When plucking is a significant factor, whalebacks develop a pronounced asymmetry, having a gently sloping upstream surface and a steep rock face on the down-ice side of the feature. Such a form, commonly called *roche moutonnée,* is the result of abrasion on the upstream slope and intense quarrying at the position of the steep, downstream face. The development of this form is due to irregular spacings of fractures within the bedrock. Where joints are widely spaced, abrasion is the dominant process; closely spaced jointing, on the other hand, facilitates plucking and more rapid erosion (fig. 10.5). The distribution of joints may, in some cases, be related to the preglacial structure of the region. It is quite possible, however, that the irregular distribution of fractures in the rock is due to variations in glacial mechanics in the vicinity of bedrock bumps. The flow of ice on the stoss side of an obstruction is toward the bed. This allows debris trapped within the basal ice layers to be pressed against the feature, abrading its surface. In contrast, cavities may form on the leeward side of bedrock bumps where normal stresses tend to be lower (fig. 10.4). Earlier in the chapter it was suggested that water pressure fluctuations within such cavities should lead to localized rock fracture and quarrying (Iverson 1991b), exactly where it is located on roche moutonnée. If this model of landform development is correct, then whalebacks are most likely to form beneath glaciers devoid of subglacial cavities, whereas roche moutonnée will occur beneath glaciers with extensive cavity systems. Benn and Evans (1998) suggest that cavities and, thus, roche moutonnée, are most likely to develop below thin, temperate ice of valley glaciers or the margins of ice sheets.

Figure 10.5
Relationship between joint spacing and
roche moutonnée development in
Yosemite Valley.
(After Matthes 1930)

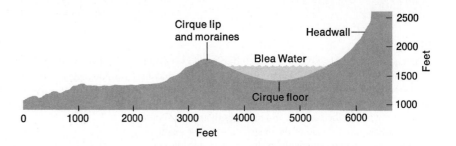

Figure 10.6
Long profile through Blea Water corrie,
a cirque.
(Lewis 1960)

Subglacial erosion is much more complex than our
simple analysis of abrasion and plucking suggests. Many
subglacial processes are involved in fracturing of the underlying bedrock (Gray 1982; Drewry 1986), and these
directly affect the intensity of plucking and the character
of ice-scoured topography (Gordon 1981). Flow of basal
ice around obstacles often reaches extreme velocities.
This fast flow, known as *ice streaming,* produces large,
elongate erosional forms (Boulton 1979; Goldthwait
1979). Water may also be involved in subglacial erosion. For example, in carbonate regions subglacial meltwater may differentially dissolve the bedrock surface
(Hallet 1976a, 1976b). In addition, the fluid mobilization of subglacial till can erode the bedrock into a myriad of forms (Gjessing 1967; Gray 1982), and the features created by the catastrophic release of subglacial
meltwater is just beginning to be understood (Dahl
1965; Sharpe and Shaw 1989; Rains et al. 1993; Beaney
and Shaw 2000; Rampton 2000).

Cirques

Cirque Morphology   The striking landscapes found
in the uplands of glaciated mountains have been
sculpted primarily by the erosive action of ice contained in cirques. The term "cirque" was first used in
the early 1800s to describe the collecting basins for
valley glaciers in the Pyrenees, and locally the feature
has been given a variety of names including *cwm, corrie, kar,* and *botn.* A **cirque** by any name is still a deep
erosional recess with steep and shattered walls that is
usually located at the head of a mountain valley. It is
normally semicircular in plan view, often being described as an amphitheater, and it is floored by a distinct rock basin where the surface has been smoothed
by abrasion. As figure 10.6 shows, the bowl-shaped
rock basins commonly contain lakes, called *tarns,* that

are dammed by a convex-up rock lip that stands as a
threshold boundary between the cirque floor and the
downstream part of the valley. The cirque lip is often
capped by small moraines that contribute to the
damming effect.

Cirques range in size from shallow depressions to
monstrous cavities that are kilometers wide and several
thousand meters high along the rear wall. Their dimensions and geomorphic form depend not only on the type
of rocks into which they are cut, being larger and more
perfectly developed in igneous or high-rank metamorphic rocks, but also on the rock structures (Olyphant
1981), the preglacial relief, and the time span of the formative glaciation. Most maturely developed cirques
seem to possess a reasonably consistent geometry when
length to height ratios are compared, indicating that
cirques probably attain some equilibrium form related to
the processes of formation.

Cirques are often preferentially oriented according
to the direction of solar radiation and the prevailing
winds (Graf 1976), and their elevation is probably (but
not necessarily) related to the snowline at the time of
their formation (Porter 1977; Trenhaile 1977). Thus, although most cirques originate in the headward reaches
of stream valleys, any hollow, regardless of its origin,
that stands at the proper elevation and has the ideal orientation may progressively accumulate snow and finally become a maturely developed cirque like those in
figure 10.7.

The significance of cirque processes in the development of Alpine scenery is that cirque expansion by continued erosion gradually eliminates the preglacial upland
surface. As a number of cirques grow headwardly and
laterally, they progressively consume much of the intervening upland region and leave as its only vestiges spectacular *horns* and *arêtes,* the features so indicative of
mountain glaciation (fig.10.8). With prolonged head-

Figure 10.7
Cirques developed in the Uinta Mountains, Utah.

ward erosion, adjacent cirques may merge, forming *col* depressions in the knife-edged arêtes.

Cirque Formation   Cirques result from two separate groups of processes: (1) mechanical weathering and mass wasting, and (2) erosion by cirque glaciers. The development of a cirque (fig. 10.9) begins with a patch of firn that fills a small depression and stands near the regional snowline. In the ablation season, meltwater released during the day percolates into fractures of the bedrock beneath the firn bank and refreezes there at night. The repeated pressure associated with freezing and thawing presumably wedges out particles of rock that are moved slowly downslope by creep and by water flowing at the base of the firn. The combined processes are commonly referred to as nivation (Matthes 1900). Actually, *nivation* refers to a set of geomorphic processes, including chemical weathering (Thorn 1976), each of which may function more effectively under different controlling factors (Thorn and Hall 1980). Nonetheless, the shape of the original depression is gradually deepened and widened, and eventually it approaches a semicircular form that can logically be called a nivation cirque. The nivation hy-

pothesis also contends that with continued accumulation, the firn changes into true glacier ice, and thereafter erosion by cirque glaciers rather than nivation becomes the dominant process in the development of the cirque. Exactly when this transition occurs is not clear, but it is virtually impossible for rock basins to be formed by physical nivation processes alone because they cannot carry particles upslope to the cirque lip.

Observations made in tunnels excavated into cirque glaciers indicate that such glaciers move by a process known as rotational sliding, in which the ice slides over the arcuate bedrock floor, rotating at the same time around a horizontal axis. The ice exposed in the tunnels displays recognizable yearly accumulation layers that are separated by marked ablation surfaces, giving the entire glacier a banded stratigraphy (Grove 1960). Some of the ablation zones are laden with debris that fell onto the ice surface near the cirque headwall. As figure 10.10 shows, these layers originally dip downglacier at the angle of the ice surface, and with time each layer is incorporated into the ice mass as more snow accumulates in the headwall area. As the ice moves, however, the layers are reoriented so that near the equilibrium line

Figure 10.8
Schwan Glacier in the Chugach Mountains of Alaska. Note dark lobes on glacier surface where rockslide avalanches have moved onto the ice. Medial moraines are displayed as long linear dark bands. Horns and arêtes are shown in mountain uplands.

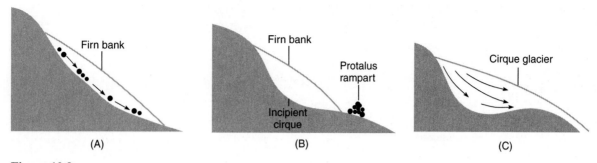

Figure 10.9

Stages of cirque development. (A) Nivation beneath firn bank. (B) Nivation cirque. (C) Cirque with fully developed cirque glacier.

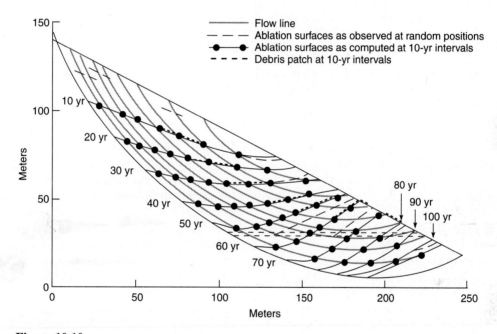

Figure 10.10

Long section through a cirque glacier in Norway showing ablation surfaces and debris patches at 10-year intervals. Note rotation of ablation surfaces in the down-ice direction.

they are almost horizontal, while close to the terminus they dip steeply upglacier. The deformation of the layers, combined with changes in the position of stakes in the tunnels and on the surface (McCall 1952, 1960), makes it clear that flowlines are moving downward near the headwall, parallel to the surface at the firn line, and upward near the terminus. The mechanism of rotational sliding offers an appealing explanation for the scouring of the bowl-shaped depression in the cirque floor, for this process should be capable of carrying the products of abrasion or frost wedging upslope and over the cirque lip.

The creation of Alpine topography requires removal of large amounts of bedrock from the head and side walls of the cirque floor. The surfaces of the cirque walls are assumed to be prepared for erosion by severe frost shattering of the wall rock. In addition, joints often develop parallel to the wall faces when pressure is re-

leased by the removal of outer rock layers (Glen and Lewis 1961). These *dilatation joints* aid in the fracturing process by providing avenues for percolating water and by isolating rock material into discrete units.

Johnson (1904) suggested that surface meltwater gains access to the base of the headwall by percolating down the bergschrund. A **bergschrund** is a crevasselike opening near the headwall that separates actively moving ice of the glacier from nonactive ice frozen to the headwall. Once the water had permeated fractures in the wall rock, it was assumed to produce extensive frost shattering as the water was alternately frozen and melted. The significance of frost shattering has been questioned, however, as actual measurements of temperature fluctuations in bergschrunds (Battle 1960) suggest that the variations are not severe enough to produce fracturing by the expansion of freezing water in the ice-sealed cracks. More

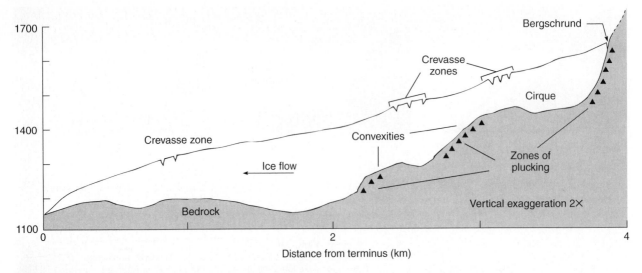

Figure 10.11
Longitudinal profile of Storglaciaren, Sweden, showing the relationship between zones of plucking and points of water input along crevasse zones and the bergschrund. The downward flow of water through these openings may lead to rapid subglacial water-pressure variations and localized plucking.
(Hooke 1991)

detailed analysis of freeze-thaw mechanics, indicates, however, that frost shattering involves more than the simple volumetric expansion of freezing water. Walder and Hallet (1985) argue that at temperatures below 0° C, unfrozen water will migrate to ice bodies in small, preexisting cracks in the bedrock and subsequently freeze. As the water is added to the ice bodies, pressures increase enough to allow the cracks to propagate. Unlike frost shattering by the volumetric expansion of water, shattering by this mechanism is greatest during sustained temperatures ranging from –4° C to –15° C and is enhanced by slow rates of cooling (Walder and Hallet 1985).

In light of the above analysis, frost shattering associated with the movement of water down the bergschrund is plausible (Hooke 1991). Others have argued, however, that the steep headwall morphology of cirques is indicative of erosion below the depths reached by the bergschrund, and therefore the primary mechanism of headwall retreat is by plucking. Hooke (1991) suggests that the flow of rain and meltwater down the bergschrund creates rapid water-pressure variations beneath the glacier in the exact location where erosion is needed to maintain a steep headwall (fig. 10.11). Utilizing a model similar to that proposed by Iverson (1991b) and discussed earlier in the chapter, he argues that rock shattering is initiated along the headwall by abrupt decreases in subglacial water pressure, while subsequent water pressure increases lead to plucking of the shattered rock.

Glacial Troughs

Glacial erosion is not limited to the cirque environment, for ice passing over the cirque lip can also remold the

preglacial valley topography into a characteristic glaciated form. The ability of ice to remove rock protuberances tends to produce valleys with steep, nearly vertical, sides and relatively wide, flat bottoms (fig. 10.12). The transformation of a V-shaped river valley into a U-shaped glacial valley has been the topic of considerable analysis and speculation (Johnson 1970; Boulton 1974). Most recently Harbor (1992) addressed the mechanical development of a glaciated valley utilizing a numerical modeling approach. Harbor's analysis assumes that the rate at which any point on the valley wall is eroded is a function of the velocity at which the basal ice moves over the bedrock. Given this constraint, erosion rates tend to increase from the margins of the glacier toward its center (fig. 10.13). A zone of lower flow should exist, however, near the valley axis. Thus, initially, erosion rates reach a maximum part way up the valley sides. As differential erosion causes the sides to bow outward, the central region of reduced erosion is progressively removed because the velocity distribution changes as a more parabolic cross-valley profile is produced. Eventually, erosion systematically increases toward the axis of the valley, and erosion rates tend to maintain the classic U-shaped morphology (fig. 10.13).

Although the assumption has been questioned (Harbor 1990), a parabolic cross-profile probably aids glacial movement because it exerts the minimum resistance to glacier flow (Flint 1971). It may also allow for the maximum efficiency of glacial erosion (Hirano and Aniya 1988, 1989). The formation of the U-shaped cross-section occurs by both lateral and vertical erosion of the preexisting valley. Whether the parabolic form is derived predominantly by widening or predominantly by

(A)

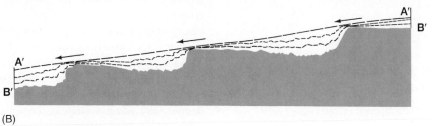

(B)

Figure 10.12

(A) U-shaped cross-profile of glaciated valley and a hanging valley. View up Yosemite Valley from vicinity of Artist Point with El Capitan at left, the Cathedral Rocks and the Bridalveil Falls at right. Yosemite National Park, Mariposa County, Calif.

(B) Longitudinal section of glacial valley in Yosemite National Park illustrating staircase profile. $A'A'$ is preglacial valley floor; $B'B'$ is present valley floor. Broken lines represent intermediate stages of development.

(Photo courtesy of M. Lord)

(After Matthes 1930)

deepening depends on the properties of the rocks and the ice (Augustinus 1992) as well as the extent of preglacial weathering. Harbor (1995), for instance, noted that fluvial processes tend to create valleys that follow relatively weak strata. Thus, he argued that the erosional resistance of the bedrock should be lower near the center of the valley than on the adjacent hillslopes. When these differences in rock strength are taken into consideration, his numerical models indicate that valley glaciers will produce cross-sections that are narrower and deeper than would be the case for uniformly resistant materials. Harbor also found that differences in the underlying geological framework could result in a variety of cross-

sectional shapes, some of which do not exhibit the classical U-shaped morphology. Thus, he suggests that the absence of a U-shaped cross-section does not necessarily mean that the valleys were never glaciated.

Glaciated troughs are also characterized by uniquely irregular longitudinal profiles that essentially represent a series of interconnected basins and steps; the diagram in figure 10.12 shows this typical profile in Yosemite National Park. These features, commonly referred to as *overdeepenings,* may contain lakes, a series of which are sometimes called *paternoster lakes.* Immediately down-valley from the basins or lakes is a gently upward sloping surface called the adverse slope. The adverse slope

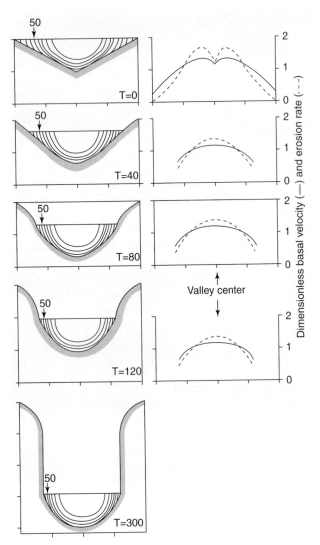

Figure 10.13

Possible sequence of events leading from a V-shaped mountain canyon to a U-shaped glacial valley. Model assumes that erosion rates are dependent on basal ice velocities. Initially, maximum velocities occur along the hillslopes, causing outward expansion of the valley walls. As the U-shaped geometry is developed, the central low-velocity zone is eliminated and the U-shaped cross-section is maintained.

(From GSA BULLETIN, Vol. 104, by J. M. Harbor, Figure 5, p. 1368. Reproduced with permission of the publisher, the Geological Society of America, Boulder, Colorado, USA. Copyright © 1992 Geological Society of America.)

is usually formed in bedrock that shows the effects of intense abrasional smoothing on its relatively flat surface. At the distal end of the step, a sharp break in slope, referred to as the headwall, occurs where plucking has produced a steep scarp that connects the adverse slope to the next downvalley basin.

The staircase profile has intrigued geologists for years and has been the subject of considerable speculation. Hypotheses for its origin include (1) variation of rock structures, especially joint spacing, causing differential erosion, (2) preparation of weak zones in the rock by preglacial weathering, (3) irregularities in the preexisting valley topography, and (4) increased erosive power at the confluence of tributary valleys with the main valley (see Bakker 1965).

Hooke (1991) provided a hypothesis that suggested rock shattering and the subsequent development of overdeepenings were the direct result of glacier mechanics and did not necessarily require a unique geological framework. Rather, he argued, overdeepenings were related to the accentuation of preglacial irregularities by localized rock shattering and plucking processes. Hooke's hypothesis is based on the realization that crevasses often develop over convexities in the glacier bed (see figure 10.11), and that openings at the base of the crevasses allow rain and meltwater to enter the subglacial drainage system. In some cases, the water may penetrate to the glacier's bed where it causes rapid subglacial water pressure variations, which you should now know can lead to rock fracturing. Since the variations in water pressure are dampened out away from the crevassed area, the rock shattering tends to be more pronounced immediately down-ice of the original convexity than in other locations. Quarrying of the shattered bedrock accentuates the preexisting irregularity and leads to the development of an overdeepening characterized by a headwall and adverse slope.

It is important to note that the process envisioned by Hooke creates a positive feedback mechanism where increased depth of the overdeepening enhances the convexity in the glacier's longitudinal profile, which in turn promotes ice crevassing, the flow of water to the glacier's bed, localized rock fracturing, and erosion. Clearly, however, this process cannot continue indefinitely. Rather, it appears that the overdeepenings become deeper, and adverse slopes steeper, until a balance is achieved between the erosion of material from the basin and its removal up the adverse slope (Alley et al. 1999). Once a balanced condition is achieved, bed erosion will be dominated by the migration of the headwall upglacier over time.

Hooke (1991) argued that a primary limitation on the depth to which an overdeepening can be eroded is related to the gradient of the adverse slope. When the slope becomes too great, the water in the subglacial drainage system becomes supercooled, producing ice that clogs the channels (Alley et al. 1999). This reduces the rate at which sediment can be carried out of the basin and armors the adverse slope against erosion. Alley et al. (1999) demonstrate that the supercooling of waters flowing up the adverse slope may not only fill the drainage channels, but also lead to the freeze-on of ice to the base of the glacier in link-cavity systems. Because considerable amounts of debris may be trapped in the ice, freeze-on may allow for the removal of substantially more sediment from the basins than would occur by the subglacial drainage system alone and, thus, create slightly steeper adverse slopes than originally envisioned by Hooke (1991).

A special type of glacial trough exists mainly in high-latitude coastal regions that are underlain by resistant rocks, so that the general land surface stands at considerable elevation above the nearby ocean. These troughs are called *fiords,* and they differ from other types only because they are partially submerged by the ocean. The inundated bottoms of fiords have the same variety of topographic elements, both erosional and depositional, that exist in a normal continental glaciated valley (Holtedahl 1967). Their history may include components of both glacial and fluvial processes, and they may be partly controlled by tectonic and lithologic factors. For these reasons, it is unwise to make sweeping generalizations about their origin.

Perhaps the most salient property of fiords is that part of their development took place when the ice was physically beneath the ocean (Crary 1966). Flint (1971) reminds us that a glacier 1000 m thick with a density of 0.9 will remain in contact with its bed and be fully capable of erosion at water depths up to 900 m. Even at greater depths, when the snout begins to float, high topographic irregularities of the valley might still be eroded (Crary 1966). The water depths in fiords, several hundred meters in many and greater than 1000 m in some, are well beyond that which can be attributed to a postglacial rise in sea level (see Flint 1971), adding credence to the suggestion that much fiord erosion was accomplished in a submarine environment.

DEPOSITS AND DEPOSITIONAL FEATURES

Before glaciers were recognized as viable geomorphic agents, deposits containing boulders that obviously came from a distant source were called *drift.* This term arose because elimination of known processes led to the belief that the anomalous boulders reached their site of deposition by riding on top of floating ice. After glaciers were recognized as the transporting vehicles, the term "drift," or *glacial drift,* was retained and expanded to include all deposits associated with glaciation. Drift covers approximately 8 percent of Earth's surface above sea level and almost 25 percent of the North American continent. The thickness of this cover varies greatly. In New England, for example, only a thin layer of drift (< 20 m) covers most upland areas, but drift may be several hundred meters thick in buried valleys. The drift in the central United States generally varies from 10 to 60 m thick, but once again these are average values. In some places the drift is merely a thin mantle on top of bedrock, and in other regions, such as parts of Illinois and Indiana, it is found in buried valleys and exceeds 150 m in thickness. The exact volume of drift deposited depends on the time span of glacier activity and the thermal regime of the ice, but with high velocities and loads, as much as 30 m of drift can be accumulated in less than 10 years (Flint 1971).

Through the years geologists have been intrigued with the amazing variety of glacial drift and the complex interrelationships that exist between the different types. This complexity arises for several reasons: (1) drift may be deposited from mediums that contain vastly different amounts of water; (2) deposition occurs beneath, within, or on top of the ice, at the glacier margins, in bodies of standing water, or in fluvial settings far from the glacier, the debris being transported there by streams rising in the ice mass itself; (3) the depositional environments and the drift composition all change with time because glaciers themselves are not constant in their properties or fixed in their position; and (4) the glacier may be active or stagnant. Because of these complicating realities, any discussion of glacial drift and the associated depositional features is difficult to organize in a way that is entirely satisfactory. Our approach here is to briefly examine the varieties of glacial drift and then look at the depositional features according to the environment in which they originate. We will also attempt to relate the morphology of features and their sedimentary properties to the dynamics of the system. Throughout this discussion it is important to recognize the distinction between the sedimentological character of drift and the morphology of the features that result from its deposition. Morphological terms such as moraines and kames do not imply a particular drift type. Many features with similar morphology are composed of a number of kinds of drift, especially where the environment of deposition is subject to repeated change.

Drift Types

Over the years glacial deposits have normally been divided into two categories based on their sedimentary characteristics, primarily the presence or absence of layers and the degree of sorting in the deposit. In our discussion, we will follow this typical division and separate drift into stratified and nonstratified types.

Nonstratified Drift The term "till" usually connotes material that has been transported and deposited by the ice itself, a process often indicated by striations or microscopic fractures on the grains (Krinsley and Donahue 1968). Till typically has no discernible stratification and is characterized by a mass of unsorted debris that contains angular particles composed of a wide variety of rock types. Many examples justify this description of till. Absence of layering and poor sorting (often characterized by an almost universal bimodal size distribution) seem to be the most reliably consistent properties. The bimodality observed in most tills is probably due to the differences in grains produced by abrasion and those derived by plucking. In an excellent review, Goldthwait (1971) points out that till is probably more variable than any other sediment that is described by a single name.

Figure 10.14
Late Wisconsinan (Pinedale) till in Rock Creek valley, Beartooth Mountains, Mont. Note rounded boulders in till.

Any of the identifying criteria in our definition may be missing at a particular till locality as a result of varying transporting and depositing mechanics and the heterogeneity in the rocks over which the ice has passed. For example, many clasts in till have a subangular pentagonal or triangular shape, but these forms may be significantly altered by rounding during transportation. This is especially true of particles that have been transported subglacially. In that environment, parent debris is rounded by attrition of edges and corners (Boulton 1978), although the degree of rounding is partly dependent on rock type (Holmes 1960; Vagners 1966), the original clast shape (Drake 1968, 1974), and the distance of transport. In addition, glaciers that override older stream deposits may incorporate boulders that have already been rounded, resulting in a till that is notably less angular than one would expect. Figure 10.14 shows till that contains rounded boulders.

Overall particle size tends to be reduced by attrition in subglacially transported till (Mills 1977; Boulton 1978). Each particular source rock tends to produce a characteristic texture depending on how easily its clasts can be crushed and how far they have been transported (Mills 1977; Dreimanis and Vagners 1971; Humlum

1985). Thus, the overall angularity and texture of till depend on the lithologic heterogeneity of the source rocks as well as the distance of transport from the location of their outcrop.

The unpredictability of till increases still more because the material is carried in different parts of the glacier, and each transport subenvironment produces till with different characteristics. For example, in valley glaciers considerable debris may be shed onto the surface from the valley sides and be transported as supraglacial load. The resulting supraglacial tills have a texture dominated by coarse, angular clasts because the particles are not crushed during transport and fines tend to be washed away as the surface ablates. In contrast, englacial load, generally carried in the basal ice layers, can be deposited directly beneath the glacier under considerable pressure exerted by the overlying ice. In this case, the debris tends to be compact and contains a higher percentage of fine-grained sediment.

The deposition of englacial load at the base of a glacier may lead to three primary types of till. *Lodgement till* results when clasts carried in the basal ice layers are plastered along the bed (Dreimanis 1988). Where the ice overlies bedrock, the lodgement of individual

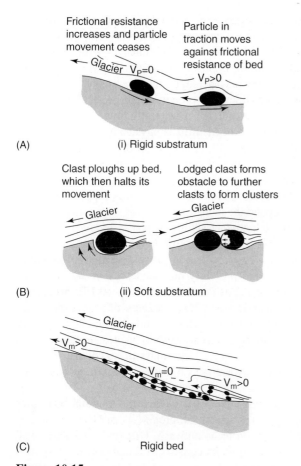

Figure 10.15

Processes involved in the deposition of lodgement till.
(A) Grain-by-grain accumulation of clasts on a rigid bed;
(B) grain-by-grain accumulation of clasts on a bed of
unconsolidated sediment; (C) the lodgement of debris-rich ice
masses along a rigid bed.

(Figures 2b & 2c of Chapter 1 on p. 6 of GLACIOFLUVIAL AND
GLACIOLACTSTRINE PROCESSES by Geoffrey S. Boulton. Copyright
© 1982. Reprinted by permission of the author.)

particles is commonly associated with small irregularities
that increase the friction between the bed and the particles in the ice (fig. 10.15). The particles are deposited
when friction becomes greater than the shear stress applied to the clasts by the overriding glacier (Boulton
1982). In contrast, the lodgement of particles beneath
glaciers resting on unconsolidated sediments usually results from the dragging of particles in the ice through the
bed materials (fig. 10.15). As this ploughing process proceeds, the bed sediments become progressively consolidated until they can no longer be moved (Benn and
Evans 1998). In some instances, deposition may occur
when the particles protrude through the entire thickness
of the deformable subglacial sediments and they become
lodged in the underlying rigid or semirigid material.

Subglacial melt-out till is deposited by the slow release of debris from ice that is not sliding or moving by
processes of creep (Dreimanis 1988). Thus, melt-out tills
are clearly associated with stagnant glaciers. It is also

possible, however, for large masses of debris-rich sediment to detach from the base of an active glacier and to
become lodged in the underlying bed materials
(fig. 10.15). Once deposited, the ice in the detached
mass will slowly melt away and become part of the underlying substrate. While detachment could be considered as part of the lodgement process (Boulton 1982),
Dreimanis (1988) argues that the sediments are actually
released from stagnant ice and, therefore, the resulting
deposits should be considered as melt-out till.

The third primary type of subglacial till is called *deformation till.* It represents unconsolidated sediments or
weak rocks that have been detached from their source
area and in which the primary sedimentary structures
have been distorted or destroyed (Elson 1988). Some
foreign sediment may be added to the till as the debris is
deformed. Hart (1998) argues that significant quantities
of melt-out till can be deposited at the base of glaciers
with deformable bed sediments. However, she notes that
these deposits are likely to be reworked as the bed sediments are overridden by the ice. She suggests, then, that
deformation tills are more likely to be preserved in the
postglacial environment than are melt-out tills.

In addition to being deposited at the base of the
glacier, englacial load can rise within the ice along shear
planes or along the normal upward flow lines in the
terminal zone. This material finally emerges as
supraglacial debris when ablation of the surface ice releases the contained particles. This debris may ultimately form supraglacial melt-out tills, which are usually less consolidated and dense than the tills formed at
the base of a glacier.

Supraglacial melt-out tills are produced by the
downward movement and accumulation of debris during
ablation. Ice covered by a thin veneer of debris ablates
more rapidly than clean ice because it has a lower
albedo and is able to transfer more solar radiation to
heat. However, when the thickness of the surface cover
reaches about 3 cm, the ablation process is severely retarded, and if the surface layer exceeds 1 to 2 m, ablation nearly ceases. Thus, supraglacial melt-out till represents the in situ accumulation of sediment (Boulton
1971, 1972b) and rarely exceeds 3 m in thickness. The
total supraglacial layer can be much greater, however, if
flow-till continues to accumulate on the surface.

Flow-tills result when englacial debris is released at
the surface, and ablation of the underlying ice destabi-
lizes the material, allowing it to move downslope under
the influence of gravity. Generally, the highly mobile
debris moves as a sediment flow, but it may move by
other mass wasting processes such as creep or semiplas-
tic sliding. It is important to recognize that flow-tills are
diamictons (poorly sorted, terrigenous sediments) that
have been remobilized, transported, and redeposited at
sites removed from the original position at which the debris is released from the ice. Therefore, flow-tills are

often difficult to distinguish from sediments deposited by geomorphic agents in nonglacial environments. This has led some geomorphologists to refer to flow-tills as secondary glacial deposits (Coates 1991).

Stratified Drift   The second major category of glacial drift is characterized by sediment that was transported by moving water before its final deposition, thereby acquiring a degree of stratification not normally seen in tills. One common form of such drift is often referred to as **fluvioglacial** because running water is involved in its origin, even though the water may not always be confined in discrete channels. Fluvioglacial deposits can also be distinguished from till because they are usually sorted and the clasts contained in the mass are more rounded. However, the demarcation between some types of fluvioglacial deposits and thoroughly washed melt-out or flow-tills is a matter of degree rather than substance. Exactly where the line between the two is drawn is somewhat arbitrary. Highly saturated flow-till, for example, might move in a nearly fluvial manner.

The layering and sorting in a fluvioglacial deposit depends on precisely where it is formed with respect to the ice that provides the transporting meltwater. Sorting is also partly a function of the energy possessed by the meltwater, the distance of transport, and the continuity of the sorting process. Because the discharge of meltwater is notoriously inconstant, varying drastically with time of day, local climate, and the characteristics of the ice, significant differences in the sedimentology of fluvioglacial deposits can be noted over short distances. These are especially evident where deposits are formed in contact with the ice and the free circulation of meltwater is restricted (see Shaw 1972 and Smith 1985 for sedimentary characteristics). If debris is transported away from the glacier terminus, the sedimentary characteristics tend to vary more regularly (Smith 1985).

We should stress again that the prime requisite of stratified drift is transport by water, much of which is released from melting ice. However, this puts no constraints on the environment in which the sediment is deposited. Fluvioglacial debris can come to rest in stream channels, floodplains, lakes, ocean floors (Rust 1977), deltaic plains, or in any other place where sediment-laden running water loses its transporting energy. In some cases sedimentologic properties in the deposits, such as the degree of roundness and mean grain size, can help identify the depositional environment (King and Buckley 1968), but normally there are too many variables in the system to rely on these criteria alone (Price 1973). Detailed field study is almost always necessary to reach a firm conclusion about the environment of deposition.

Stratified deposits that originate in contact with the ice often contain interbedded bodies of melt-out and flow-till, and the particles tend to be less well rounded and sorted because of the limited distance of transport.

The characteristics of such a deposit often vary from the bottom of the sequence to the top because the environment of deposition repeatedly changes with time, especially if the ice is stagnating. In any deposit, then, a particular layer may be superseded vertically by one with different sedimentary properties that reflect a new depositional environment. The stratigraphy is complicated because much of the drift is physically supported by ice during its deposition, and when the ice dissipates, the support is removed and the sediment collapses. Such a process, as figure 10.16 shows, leads to flexures in some of the layers, minor faults, and beds dipping at angles well beyond the angle of repose for such material. Overall, the ice-contact setting at places produces an interconnected maze of stratified and nonstratified drift in which every conceivable process and environment is possible and probably has been present at some time during the depositional history (fig. 10.17). Moreover, the manner in which the ice ablates influences the resulting deposit. Drift produced while the ice is still active may be quite different, especially in distribution, from that derived from a large mass of stagnant ice that is simply downwasting as it melts and is not influenced by internal glacial movement.

Sediment deposited beyond the terminal margin of the ice is formed in the proglacial environment and is often referred to as outwash. **Outwash** is usually well sorted and normally consists of rounded sand and gravel representing bedload carried and deposited in stream channels. Silt and clay are usually carried as suspended load and are commonly removed from the system unless, as in the lower Mississippi River valley, the transport distance is so great that some of the outwash is silty in texture. Streams transporting outwash do not usually head at the glacier terminus but begin on top of, beneath, or within the ice, well upglacier from the margin. Proglacial features and deposits often can be traced into and through the maze of ice-contact deposits, increasing the complexity of the depositional sequence developed near the ice margin.

The Depositional Framework

Before considering what geomorphic features might be produced during deposition, we must first establish a realistic framework within which we can give some order to the subtle variations of depositional features associated with glaciation. This is done in table 10.1. From our brief look at the nature of drift, it is obvious that both stratified and nonstratified deposits can be formed in contact with the ice. They differ only in whether the ice alone was the primary transporting and depositing agent or whether a stage of water transport intervened between the release of particles from the ice and their final deposition. Thus, the **ice-contact environment** must be considered as one of the major settings of our depositional

(A)

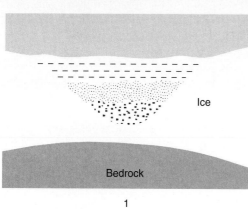

Ice

Bedrock

1

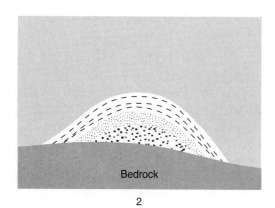

Bedrock

2

(B)

Figure 10.16

(A) Gravel pit cut in kame, south central New York, shows stratification in kame deposits. (B) Structures and deformation of strata in ice-contact stratified drift develop as ice melts and debris collapses and is lowered onto bedrock floor.

Figure 10.17
Marginal ice-contact zone at the terminus of Woodworth Glacier, Alaska.

TABLE 10.1	The Depositional Framework and Associated Features.	
Setting	**Features**	**Type of drift**
Ice contact		
Marginal	End moraines	Till and fluvioglacial
	Kames and kame terraces	Fluvioglacial
	Kettle holes	Fluvioglacial
	Eskers	Fluvioglacial
Interior	Medial and interlobate moraines	Till
	Ground moraines	Lodgement till
	Fluted surfaces	Lodgement till
	Drumlins	Lodgement till
Proglacial environment	Sandar	Fluvioglacial (outwash)
	Kettled sandar[a]	Fluvioglacial (outwash)

[a]May merge with marginal environment.

framework. The ice-contact environment can be subdivided geographically into **marginal** and **interior zones** depending on where the drift was originally deposited (table 10.1).

Features formed in the ice-contact environment can be composed of either stratified or nonstratified drift (till or fluvioglacial sediment) or a combination of both types. The region beyond the terminal edge of the glacier is classified as the **proglacial environment.** In contrast to the ice-contact setting, features formed there are composed almost entirely of stratified drift. These sediments may be deposited directly by proglacial streams, or they may be deposited by other geomorphic agents operating in front of the glacier, such as those associated with proglacial lakes (e.g., see Teller 1987).

The distinction between marginal and interior zones in the ice-contact environment is problematical for several reasons. First, glacial margins migrate forward and backward with time according to the glacier's mass budget. An active glacier with a negative mass balance should have a terminal margin that is progressively receding toward its source. Marginal features will be formed in regions that were interior when the ice was at its greatest extent, and deposits of the two zones may be complexly intertwined. Second, the processes operating near the contact between the marginal and interior zones are dependent on the characteristics of the subglacial environment. For example, the outer 2 to 3 km of ice sheets are commonly frozen to the underlying surface even though farther upglacier the basal ice may be temperate. Where marginal ice is frozen to the bed and high pore water pressure exists, thrusting along preexisting planes of weakness in the substrate can inject large blocks of subglacial material into the ice (Moran et al. 1980). These thrusted blocks tend to concentrate where ice margins rest on aquifers and on the upslope edges of upland areas. Proceeding upglacier toward the interior zone, the ice is free to slide because the ice-bed contact is unfrozen. Thrust blocks in the transitional area between the marginal and interior zone are smaller and smoothed over by the sliding ice. In the true interior zone, no thrusting will occur and the subglacial terrain is characterized by streamlined forms (Moran et al. 1980). Clearly, the position of the frozen ice-thawed zone contact may migrate with time, changing the subglacial environment. Thus, in any given glaciated area, the inner portion of the marginal zone may be transitional into the interior zone rather than marked by a well-defined contact between the two.

Regardless of the problems associated with precise boundary locations, the depositional framework fits our purposes because the suite of features found in each zone is a direct reflection of the genetic processes involved. The utility of the classification is shown in figure 10.18. In this region of Wisconsin the terminal moraine, marking the ice-contact marginal zone, is char-

acterized by a maze of small hills and depressions (kames and kettles) that distinguish deposits formed near the boundary between active and stagnating ice. North of the moraine the relatively low, flat area is underlain by material that was deposited beneath the active glacier (ground moraine); it represents the ice-contact interior zone. South of the moraine the proglacial zone is marked as a plain composed of debris (outwash) transported and deposited by meltwater streams heading within or on top of the ice.

Marginal Ice-Contact Features

Moraines   The term "moraine" originated several hundred years ago as a local name for ridges of debris found at the edges of glaciers in the French Alps. Since then, many definitions have appeared, but for our purposes we can think of a **moraine** as a depositional feature whose form is independent of subjacent topography and is constructed by the accumulation of drift, most of which is ice-deposited (Flint 1971). A precise morphological definition is not possible because, as table 10.2 shows, moraines take many different forms and have a variety of dimensions. The term "moraine" is not synonymous with "till," as many have suggested, but refers to a suite of topographic forms on which the only restriction is that they must be composed of drift.

The most spectacular moraines develop at or near the edges of active glaciers and so are designated as **end moraines.** The end moraine constructed at the downstream edge of the ice at the farthest point of advance is called a **terminal moraine.** In valley glacier systems, it merges imperceptibly into **lateral moraines** on both sides of the valley because ablating ice deposits debris along the glacier's lateral extremities as well as at its terminus (fig. 10.19). Ice sheets form terminal moraines that tend to be long (often hundreds of kilometers), linear, topographic highs marking the forward boundary of the ice, but lateral moraines are absent because there is no side to the ice sheet. Where ice sheet movement was distinctly lobate, however, end moraines called *interlobate moraines* (Flint 1971) may develop along the junction of two lobes (of ice).

Ideally, end moraines assume a rather narrow, ridgelike shape, but actually the form and size depend directly on the amount of glacial load, the mass budget, and the volume of meltwater circulating in the system. Temperate valley glaciers tend to build higher and more massive terminal moraines (some reaching 300 m in height) because these glaciers have higher flow velocities, loads, and total budgets. Ice sheets seem to generate moraines that are less dramatic in size, seldom exceeding 50 m in height. Moraine ridges formed in the marginal zone are developed by three processes known as *dumping, squeezing,* and *pushing* (Price 1973). Creation of an end moraine by dumping requires that englacial or

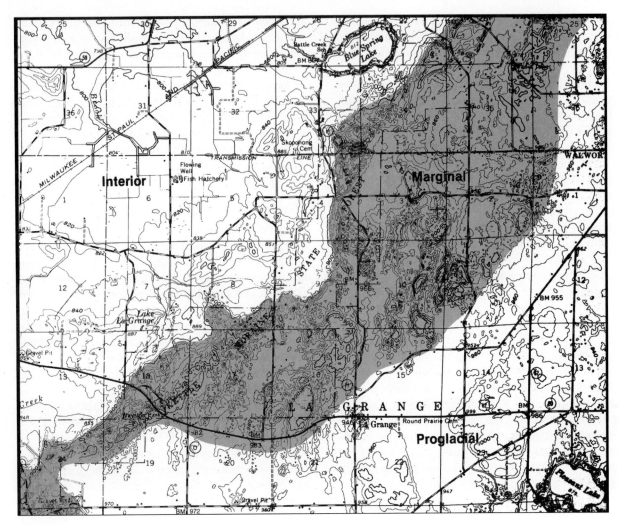

Figure 10.18

Map view of the depositional framework after ice has disappeared. Part of the Whitewater, Wis., quadrangle (U.S.G.S. 15′). Marginal zone contains terminal moraine and ice-contact stratified features, and interior zone is characterized by ground moraine. Proglacial zone is a large outwash plain or sandur.

TABLE 10.2	Moraine Types.
End Moraines	Moraines produced at front or sides of an actively flowing glacier.[a]
Terminal moraines	Mark the farthest advance of an important glacial episode.
Lateral moraines	Deposited at or near the side margin of a mountain glacier.
Recessional moraines	Formed at glacier front during temporary halt or readvance of ice in a period of general recession.
Ground Moraine	Gently rolling surface formed of debris released from beneath the ice.
Interior and Minor Varieties	
Washboard moraines	Small, parallel ridges oriented transverse to direction of ice movement. Also called moraine ridges or cross-valley moraines.
Interlobate moraines	Formed at junction of two ice lobes.
Medial moraines	Elongate ridge developed at junction of two coalescing valley glaciers.
Rogen moraines	Large sequence of ridges transverse to ice flow. Formed in the interior zone.

[a]Lateral moraines may be excluded by some geomorphologists because end moraines are commonly considered only as topographic features developed at the front of a glacier.

(A)

(B)

Figure 10.19
(A) Terminal moraine of Pinedale glaciation, East Rosebud valley near Roscoe, Mont. Low ridge above road in left center of photo is lateral moraine merging with terminal zone.
(B) Recently formed lateral moraine in Rocky Mountains of Canada.

subglacial debris be transported to and released on the ice surface. The sediment cover induces differential ablation and is often associated with subsequent flow movements of the material. Debris can be moved surfaceward along shear planes (Goldthwait 1951, fig. 10.20) or along the normal flow lines found in active glaciers (Boulton 1967). Regardless, the evolution of morainal topography occurs in a zone where debris dumped at the surface triggers a complex sequence of ablation and flow. Because active glacier fronts shift downvalley or upvalley according to the glacial budget, the position of dumped sediment moves, and the morainal topography may exist in a belt rather than a single ridge.

Moraines formed by stagnant ice in the marginal zone are not always characterized by distinct ridges that have developed transverse to the direction of ice flow. These moraines, commonly called *disintegration moraines* (Gravenor and Kupsch 1959), have local relief up to 70 m and develop from the release of supraglacial drift in the lower part of the ablation zone. When the ice in glacier margins stagnates, it often breaks into isolated blocks of wasting ice covered by debris (fig. 10.21). Although the depositional environment is ice-contact marginal, the morainic topography gradually develops as the glacier surface down-wastes over dissipating ice cores. This ice-cored moraine is different from ridges or linear belts formed from active ice that oscillates back and forth. In the stagnant marginal zones, flow till, supraglacial melt-out till, and fluvioglacial deposits can all coexist on the wasting surfaces, and because the ice cores melt at different rates, they may become mixed

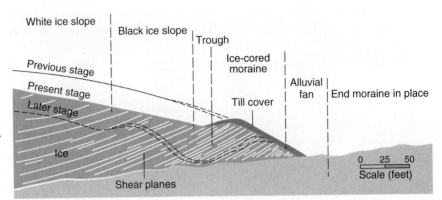

Figure 10.20
Diagram showing retreating margin of Barnes ice cap. Material moves up inclined shear planes and accumulates as till cover on core of ice.
(Goldthwait 1951)

within the chaotic surficial expression (fig. 10.22). Although disintegration moraines evolve in a way somewhat similar to dump moraines, they differ in that debris is not necessarily transported to the surface before the ablation process begins.

In contrast to moraines formed by the dumping process, moraine ridges are also developed by the squeezing of drift originally deposited beneath the ice. Ridges generated by squeezing are usually smaller than dumped moraines, normally standing less than 10 m high, although they can be higher where the subglacial till is highly saturated. Theoretically, the process involves the response of water-soaked subglacial till to pressure exerted by the weight of the overlying glacier. The till will move from under the ice and emerge along the ice front, or it will be squeezed into zones of low pressure within the ice, represented by crevasse openings (Price 1970; Mickelson and Berkson 1974). In either case the ridges seem to be typified by (1) steeper distal slopes than proximal slopes, (2) a plan view outline consisting of linked arcuate segments that parallel the ice front, and (3) till with pebbles oriented perpendicular (or nearly so) to the ridge crests, though occasionally oriented parallel to the trend of the ridge (Mickelson and Berkson 1974).

A third mechanism capable of forming a moraine ridge is the collision of advancing ice with older deposits, which deform them into a ridgelike feature. Such end moraines, called *push moraines,* are only partly composed of sediment carried by the ice and may include blocks of older rocks of a nonglacial origin (Kaye 1964b; Andrews 1980; Humlum 1985). Observations of the internal structure of push moraines have demonstrated that the deformation of proglacial sediments during ice advance occurs by two distinct mechanisms that may work in isolation or in conjunction with each other. In some cases, the advancing ice simply "bulldozes" sediment accumulating at the front of the glacier by dumping, squeezing, or fluvioglacial processes into morainal ridges that usually measure less than 10 m in height. This bulldozing effect is often most pronounced during the winter when additions to the glacier's mass

lead to the advance of the ice front. This yearly advance of the ice (and its retreat during the summer) can result in a series of push moraines at the front of the glacier. In other cases, the advancing ice can lead to the ductile or brittle deformation of large blocks of proglacial sediment, a process referred to as glacitectonics (Aber et al. 1989). Brittle deformation generally involves the thrusting of individual layers of sediment along well-defined failure planes. These thrust sheets may become stacked in front of the ice and comprise a significant portion of the moraine (fig. 10.23). In contrast, ductile deformation involves the shearing away of sediment from the ground lying in front of the ice, and its intense deformation into large (sometimes overturned) folds and faults of all kinds (Mills and Wells 1974). Individual thrust blocks can also be affected by ductile deformation. These types of glacitectonic disturbances can impact sediments located as much as 200 m below the ground surface (Aber et al. 1989), but deformation is more commonly limited to a few meters in depth. It is not uncommon, however, for glacitectonic processes to create large moraines that are tens of meters in height.

It is now clear that the morphology of moraines associated with bulldozing differ from those formed by glacitectonic activity. This led Benn and Evans (1998) to utilize the term "push moraine" for ridges formed by bulldozing of the proglacial sediment and *thrustblock moraine* or *composite ridges* for deposits created by glacitectonic activity, the latter terms depending on the composition of the deformed material. Whatever terminology is used, it should be remembered that bulldozing and glacitectonic activity may occur simultaneously as the ice advances and, thus, the created ridges may contain features related to both mechanisms.

Stratified Marginal Features As pointed out earlier, many glaciers are characterized by marginal zones of thin, stagnating ice. Within these zones a suite of genetically related ice-contact features is developed that is composed predominantly of stratified drift and is morphologically distinct from the moraines just described.

Figure 10.21
Ice-cored moraine, Yanert Glacier, Alaska.

Figure 10.22
Hummocky topography in terminal moraine. East Rosebud valley at front of Beartooth Mountains near Roscoe, Mont.

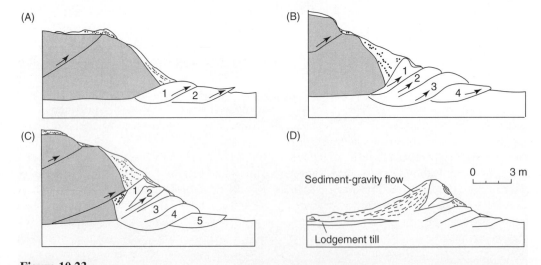

Figure 10.23
Push moraine of Hofdabrekkujökull Glacier, Iceland, formed by glacitectonic activity. Layers of sediment are thrusted upward and stacked at the front of the glacier.

(Figure 11.42a, p. 466 from GLACIERS AND GLACIATION by Douglas I. Benn and David J. A. Evans. Reprinted by permission of Arnold Publishers, London, England.)

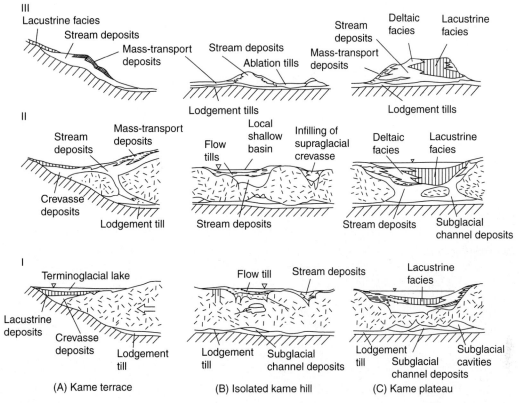

Figure 10.24
Possible mechanisms involved with the formation of (A) kame terraces, (B) kame hills, and (C) kame plateaus. Stage I–III illustrate evolutionary steps in kame development.
(Figure 126, p. 195 by Brodzikowski and Van Loon from GLACIGENIC SEDIMENTS. Copyright © 1991. Reprinted by permission of Elsevier Science.)

These features form by deposition of drift (1) where water flows through openings in and beneath the ice, or in ice-surface channels, (2) in spaces between the ice and the bedrock of the valley sides, and (3) where disseminated sediment is passively concentrated by melting of the encasing ice.

Many of the channelways for the flowing water originate when stagnating ice breaks into individual segments along planes of structural weakness in the ice, and so the features are genetically related to many of the ice-disintegration forms described by Gravenor and Kupsch (1959), Stalker (1960), Clayton (1967), Parizek (1969), and many others. Our discussion, however, is restricted to those features that consist primarily of fluvioglacial drift, even though till may be a minor ingredient. Their morphology is entirely constructional, although they can be affected by slumping as the ice walls that support the drift during its deposition are melted away.

Kames and Kettles   *Kames* are moundlike hills of layered sand and gravel that vary in size from minor swells to conical protuberances standing up to 50 m high and extending 400 m along their base. Kame material can accumulate at the ice-substratum interface and also in cavities located within stagnating ice or on its surface

(fig. 10.24). The englacial or supraglacial debris can be lowered onto the ground surface as the ice dissipates (Cook 1946; Holmes 1947). In addition, some kames form when debris collects as fans or small deltas built against the ice or outward from the ice, with the apex resting at the stagnant margin. In either case, melting of the supporting ice allows the drift to collapse into a kame, a process evidenced by slumped strata within the deposits.

Kames are only one of many forms with essentially the same origin. They are transitional into eskers or minor ridges and circular forms that Parizek (1969) calls ice-contact rings and ridges, or into other features whose names utilize the term "kame" as a descriptive adjective, such as kame delta, kame moraine, and so on. Perhaps the most common of the latter type of feature is a *kame terrace*. Kame terraces originate from drift deposited in narrow lakes or stream channels between the valley side and the lateral edge of the stagnating ice. When the supportive ice disappears, the inner edge of the deposit collapses into the terrace scarp. Kame terraces differ from normal river terraces in that they are restricted in their longitudinal extent and are usually narrow. The tread may slope gently into the valley, and the surface may be dimpled by kettle holes (McKenzie 1969).

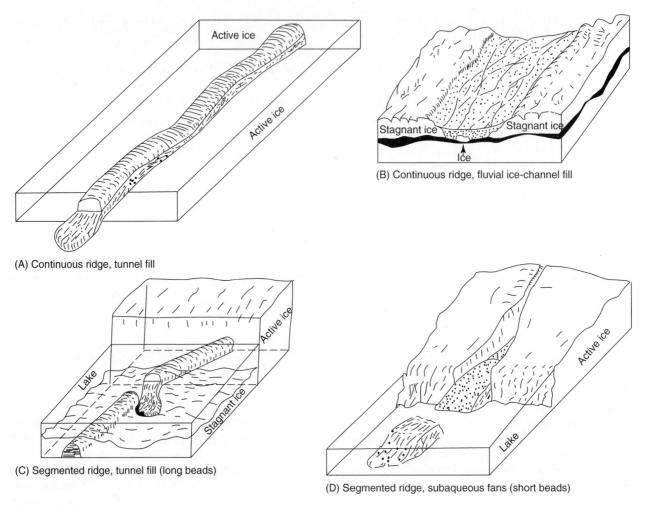

Figure 10.25
Selected mechanisms of esker formation. (A) Esker formed by the filling of subglacial or englacial conduits in the ice and their exposure following ablation; (B) esker created by the deposition of sediment in subaerial channels on the glacier bed; (C) segmented esker formed during the episodic retreat of the glacier front; (D) discontinuous esker formed by the subaqueous deposition of sediment during glacier retreat.

(Figure A64 from JOURNAL OF SEDIMENTARY RESEARCH, pp. 433–449 by W. P. Warren and G. M. Ashley. Copyright © 1994. Reprinted by permission of SEPM [Society for Sedimentary Geology] and American Association of Petroleum Geologists.)

Kettle holes are circular depressions that are formed in a variety of ways (Fuller 1914) but most commonly by the burial and subsequent melting of isolated blocks of ice in stratified drift. The gradual ablation of the ice leads to a gentle downward flexing of the sediment layers as they settle over the dissipating mass. Some kettles are almost 50 m deep and up to 13 km in diameter (Flint 1971), but these giants are exceptions to the normal kettle size of less than 8 m deep and 2 km wide. Kettles usually form in association with kames and other related features, producing an irregular surface described as kame-and-kettle topography, but similar surfaces can develop in the absence of either or both of these features.

Eskers The term *esker,* evidently stemming from the Gaelic word for "crooked" or "winding," has been applied to a wide variety of ridged ice-contact features (fig. 10.25). *Eskers* range in shape from the single, nar-row, sinuous ridge that is the classic form to a com-plexly intertwined maze of branching and joining ridges, some of which terminate in fans or deltas (Huddart and Lister 1981; Shreve 1985; Gorrell and Shaw 1991; War-ren and Ashley 1994). Eskers seem to form most com-monly in stagnating margins of large ice sheets where the underlying surface is broad and relatively flat. In ad-dition, they are generally, but not always, associated with temperate ice and an abundant supply of meltwater (Fitzsimons 1991). It seems certain that eskers result from sediment accumulation in a variety of openings such as in (1) ice cavities including (e.g., crevasses or spaces between stagnant blocks), (2) tunnels beneath or within the ice, (3) supraglacial channels, and even, in rare cases, (4) narrow longitudinal embayments of the ice front (Cheel 1982). The subglacial origin is made possible by meltwater that descends from the glacier's surface to the base through fractures and holes in the ice.

Along the subglacial bed a network of interconnected tunnels passes water and sediment toward the terminus, and normal fluvial deposition fills or partially fills the openings. Filling probably takes place during the last stage of deglaciation when internal ice flow is minimal or stagnant (Shulmeister 1989; Shreve 1985). In addition to forming in subglacial channels, field studies and photographic analyses of esker development by the Casement Glacier, Alaska (Price 1966; Petrie and Price 1966) and the Breidamerkurjokull Glacier in Iceland (Price 1969) provide rather conclusive evidence that englacial and supraglacial deposits can be transformed into eskers. In those areas, some of the eskers are definitely ice-cored and were measurably lowered in elevation during the period 1948 to 1963 by wastage of the buried ice. Presumably the final melting of the ice will deposit sediment on the original subglacial floor.

Eskers are not necessarily continuous but may consist of crudely connected linear segments. In broad valleys eskers usually parallel the slope of the valley floor, but this is not always the case. The flow direction of *pressurized* subglacial channels depends on the ice surface topography and, to a lesser extent, on the slope of the glacier's bed. Thus, these subglacial channels can locally have an upward sloping longitudinal profile and their infilling can create eskers that ascend the valley sides, transect the divide, and descend the flanks of the adjacent valley (Shreve 1985). Such eskers can also be created as stagnant ice ablates, draping sediment in supraglacial or englacial channels across the glaciated terrain.

Eskers show dramatic inconsistencies in dimension. They range from 2 m to more than 200 m in height, from several meters to as much as 3 km in width, and from tens of meters up to 500 km in length (if gaps are considered in the total distance). In cross-profile, they usually have rather sharp crests and steeply inclined sides (up to 30°), but broad eskers can maintain rather flat upper surfaces or may be pitted by kettle development. Price (1973) suggests that the height and width of eskers may be directly related to their overall length, longer ridges being proportionately higher and wider than shorter ones.

Interior Ice-Contact Features

Behind the marginal zone in the interior portion of a glacier, the predominant features are deposited at the base of the ice. There pressure from the overlying ice either spreads till rather evenly across the ground surface or molds water-soaked till already in that position into distinct morphologic shapes. Supraglacial drift is somewhat rare in the interior zone, but where two valley glaciers join, the debris dragged along their lateral edges may coalesce into *medial moraines.* Although these moraines appear as striking linear belts on the surface of the ice (see fig. 10.8), they are superficial in that the de-

posits are shallow. Not all of these linear features represent former ice margins. Some may have an interior origin (Small et al. 1979) and sediment sources that are both surficial and englacial (Eyles and Rogerson 1978). Medial moraines are rarely preserved on the ground surface because they are let down in the middle portions of valleys where meltwater streams are likely to destroy them. The dominant interior ice-contact features are those deposited beneath the ice such as ground moraines, drumlins, and fluted surfaces.

Interior Moraine   In contrast to end moraines, **ground moraine** is distinguished by its apparent lack of topographic expression. It is accepted as a moraine despite its low relief and a complete absence of transverse ridges (Flint 1971) because its surface expression is independent of the topography it covers. Ground moraine occupies much of the surface covered by major Pleistocene ice sheets in North America and Europe. The moraines usually exist as smoothly undulating plains, like that in figure 10.26, seldom exceeding 10 m in total relief. They range in size from small areas interspersed among younger marginal features to regions covering thousands of square kilometers behind the terminal moraine.

In most cases the primary building material of ground moraine is lodgement till, deformation till, and subglacial melt-out till. However, the deposits from which ground moraine is constructed also include other forms of till (e.g., supraglacial melt-out till and flow till) and interbeds of fluvioglacial deposits. These sediments may originate in the same glacial advance (Kirkby 1969; Boulton 1972a, 1972b) or in more than one episode of glaciation.

In addition to producing the rather featureless ground moraine, subglacial processes can create moraine ridges of various sizes and shapes (Sugden and John 1976). For example, large transverse ridges, called *Rogen moraines* or *ribbed moraines,* are 10 to 30 m high, are more than 1 km long, and tend to develop as a series of separate hills spaced 100 to 300 m apart. Significantly, the ridges often occur in direct association with flutes or drumlins; therefore, it is reasonable to assume that they all have a common origin. While a number of theories explaining their formation have been proposed, most attention has focused on two possible mechanisms: (1) the molding of saturated subglacial sediment into linear ridges, and (2) the thrusting and localized accumulation of debris-rich ice at the base of a glacier (fig. 10.27). The former theory suggests that the ridges reflect an interaction of basal debris, pore water pressure, and ice temperature and that the Rogen system develops in response to variations in the shear strength of the subglacial sediment and transverse differences in stress at the glacier bed. In contrast, the latter theory suggests that Rogen moraines are produced as

Figure 10.26
Flat till and loess plain in southern Illinois. Gently undulating topography on ground moraine has been smoothed by influx of younger loess.

debris-rich blocks of ice are stacked up along thrust planes in zones of compressive flow (Shaw 1979). When the glacier ablates, the sediment in the ice blocks forms a ridge that may be overlain by other types of drift. This mechanism is supported by the fact that Rogen moraines are commonly found in depressions within the glacier bed, exactly where zones of compressive flow are expected to occur.

The pushing and squeezing phenomena that occur at the front of a glacier may also relate to smaller varieties of subglacial moraines. Within the marginal zone the locus of deposition shifts periodically, giving rise to ridges, mounds, and depressions of varying size. The ridges in the sequence, usually small and composed of till, commonly parallel the orientation of the ice front. They have been given many names, such as *cross-valley moraines* (Andrews 1963; Andrews and Smithson 1966) and *washboard moraines* or *moraine ridges* (Price 1970), and several theories have been suggested for their origin (Elson 1968). The ridges, instead of exhibiting a regular undulation of ridge crests and intervening sags, are usually segmented or interwoven with many deposits of stratified drift, giving the entire topography a chaotic expression.

Fluted Surfaces and Drumlins   As indicated, the monotony of the topography associated with ground moraine is sometimes broken by ridges or elongate hills that manifest the mechanics functioning at the base of glaciers, especially in temperate ice sheets. The most common features developed in the subglacial environment are fluted surfaces and drumlins. Flint (1971) refers to them as streamlined molded forms but others (Sugden and John 1976) consider them as special types of moraine ridges that form parallel to the direction of ice flow.

Fluted surfaces consist of narrow, regularly spaced, ridges. The ridges are normally less than 5 m high and several hundred meters long, although individual ridges may be considerably larger. Small ridges are usually composed of till, and their origin may be related to pressure-squeezing of saturated debris into longitudinal cavities that open on the lee side of large obstructions on the bed (Hoppe and Schytt 1953; Sharp 1985; Boulton 1976; Benn 1994; Eklund and Hart 1996). The flow of till into the cavity occurs because unloading along the lee side of an obstruction sets up a pressure gradient in the till that causes flow toward the opening

Stage 1: Formation of initial folds in compressive zone

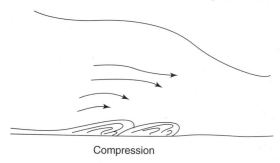

Compression

Stage 2: Development of major folds and thrusts

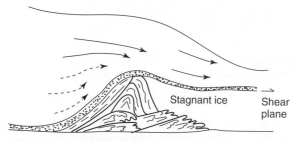

Stagnant ice

Shear plane

Stage 3: Stagnation and undermelt

Debris bands from upglacier folds

Supraglacial complex >active layer

Stagnant ice

Stagnant ice

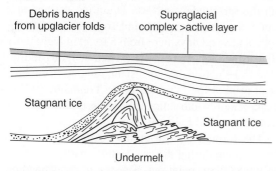

Undermelt

Stage 4 (with change in scale): Development of large thickness of fluvial deposits and slumping from the exposed ridge

Supraglacial complex deposits

Foliated or stratified melt-out or sublimation till

Flow till

Fluvial deposits

Supraglacial complex deposits

} Melt-out tills
  Flow tills
  Stratified deposits

Slumping

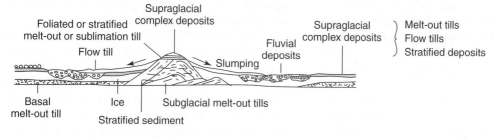

Basal melt-out till

Ice

Stratified sediment

Subglacial melt-out tills

Stage 5: Final landform and sediment complex

Transverse moraine ridge

Dead ice topography

Marginal kettle

Fluvial deposits

Escarpment

Supraglacial complex deposits

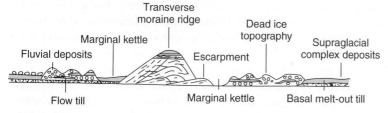

Flow till

Marginal kettle

Basal melt-out till

Figure 10.27

Possible stages of Rogen moraine development at the bed of a glacier.

(Reprinted from "Genesis of the Sveg Tills and Rogen Moraines of Central Sweden: a Model of Basal Melt-out" by J. Shaw from BOREAS, 1979, Vol. 8, pp. 409–426 figure 13, p. 422, www.tandf.no/boreas, by permission of Taylor & Francis AS.)

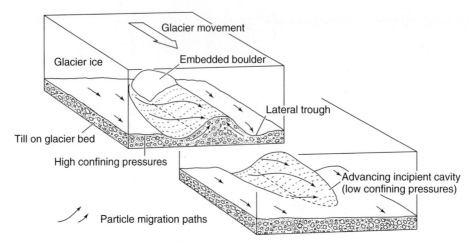

Figure 10.28
Formation of a flute by the squeezing of sediment into a cavity downstream of a bed obstruction.
(From D. I. Benn, figure 9, p. 290, SEDIMENTOLOGY, Vol. 41. Copyright © 1994. Reprinted by permission of Blackwell Science Ltd.)

and because the till is more readily capable of flowing than the ice. Once in the cavity, the unfrozen debris may be carried downglacier while additional sediment is added to the opening. The net result is the creation of a flute (fig. 10.28). Alternatively, flutes may form when new sediment is added to the cavity after previously injected sediment freezes to the basal ice and is carried away. This model of formation is supported by the fact that a large number of flutes are observed to head immediately downglacier from large boulders or other obstructions and the till comprising flutes commonly exhibits folds and faults suggestive of sediment deformation.

Some larger ridges are composed of material other than till (Lemkke 1958; Gravenor and Meneley 1958), and the surfaces between adjacent ridges are often noticeably grooved. These characteristics have led many workers to believe that the origin of fluted surfaces may be related not only to the deposition of ridges but to erosion of grooves into the subglacial materials, or a combination of the two processes. Gravenor and Meneley (1958), for example, suggest that alternating high- and low-pressure zones in the basal ice produce the unique fluted surface. In their model, grooving occurs in the material beneath the high-pressure zones, and debris eroded from there is moved not only downglacier but also upward into regions of low pressure.

Drumlins also are elongated parallel to the direction of ice flow, their long axes deviating only slightly from the average trend of the glacier movement (fig. 10.29A). They have been described as having a plan view shape that is similar to leminscate loop (Chorley 1959) or an ellipsoid (Reed et al. 1962) and in long profile a form that Flint (1971) likens to an inverted bowl of a spoon (fig. 10.29B). The exact shape, however, is probably variable enough that no particular

model will fit all drumlins. In any case, drumlins are higher and wider near their rear edges and they narrow and thin downstream until they merge imperceptibly with the surrounding surface. Drumlins average in size from 1 to 2 km in length and from 400 to 600 m in width, and stand anywhere from 5 to 50 m high; individuals can be smaller or larger. Their length/width ratio seems to be reasonably consistent, ranging from 2 to 3.5 (Reed et al. 1962; Vernon 1966; Trenhaile 1971), although length/width ratios exceeding 6 have been measured (Boyce and Eyles 1991).

The formation of drumlins has received considerable attention during the last century, partly, as Menzies and Rose (1989) point out, because an understanding of their development provides insights into the processes operating at the base of the glacier. Nevertheless, their origin is not fully understood.

Any hypothesis concerning the origin of drumlins must consider both their sedimentologic and spatial character. In terms of sedimentology, drumlins display a variable internal composition. Many are fabricated entirely from clay-rich till, but others have obvious cores of solid rock or preexisting drift that may or may not be stratified. Geographically, drumlins are located in only a small number of glaciated areas, but where they exist, they rarely occur as individuals but instead cluster together in fields that are commonly wider than most morainal belts (Gravenor 1953). The density of drumlins within a field seems to be inversely related to size; that is, very dense clusters are composed of relatively small drumlins (Doornkamp and King 1971). Some fields may contain as many as 10,000 individuals (Sharp 1985). In addition, many studies show a strong probability that drumlins within any field are spaced in a nonrandom manner; that is, spacing between neighboring individuals is somewhat regular (Reed et al. 1962; Vernon 1966;

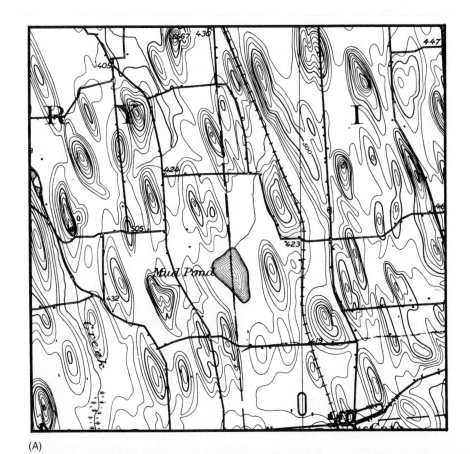

(A)

(B)

Figure 10.29
(A) A portion of the drumlin field located near Weedsport, N.Y. From the northeast quarter of the Weedsport, N.Y., quadrangle
(U.S.G.S. 15′). Contour interval 20 feet. (B) Drumlin shown in longitudinal profile near Rochester, N.Y.

Smalley and Unwin 1968; Trenhaile 1971). Drumlin fields also seem to be located in zones close to but behind the terminal moraines that mark the limit of a particular glaciation (Patterson and Hooke 1996).

The models used to explain drumlin genesis fall into three main groups: (1) drumlins are erosional features developed when moving ice streamlines preexisting drift or rock (Gravenor 1953; Kupsch 1955; Lemke 1958; Whittecar and Michelson 1979; Bouchard 1989; Aylsworth and Shilts 1989; Habbe 1992); (2) drumlins are depositional features formed when a moving glacier deposits till and, as a result of spatial differences in debris rheology caused by dilatancy, pore water dissipation, localized freezing, or grain size, locally molds the material (and perhaps preexisting sediment) into streamlined forms (Smalley and Unwin 1968; Boulton 1979, 1987; Aylsworth and Shilts 1989; Menzies 1989; Boyce and Nicholas 1991; Hanvey 1992); and (3) drumlins are depositional features derived from the infilling of cavities carved into the basal ice by meltwater (Shaw and Sharpe 1987; Shaw et al. 1989; Shaw 1983, 1985, 1989; Shaw and Kvill 1984; Sharpe 1987; Hanvey 1992).

Perhaps the most widely accepted hypothesis is that drumlins result from the erosion and redeposition of deformable subglacial drift (Boulton 1987). This argument is based on the realization that subglacial sediments possess varying shear strengths and that those with a high shear-strength should deform at a slow rate (or perhaps, in the case of bedrock, not at all). In contrast, weak materials will deform rapidly. Thus, it is possible that as the ice overrides an area, the rapidly deforming sediments are wrapped around a core of slowly deforming or nondeforming material creating a streamlined drumlin.

The above model seems to be consistent with the formation of drumlins near the ice margin. Patterson and Hooke (1996) note, for example, that the margins of many ice sheets are frozen to the bed. As a result, water located along the base of the glacier may not be able to drain from the beneath the ice, creating high pore pressures within the subglacial sediments, thereby lowering the shear strength of the bed materials. These conditions are ideal for the deformation of subglacial drift. In addition, the model can explain why drumlins are commonly composed of an out carapace of till and an inner core of more resistant material such as stratified fluvioglacial sediment or bedrock.

Shaw and his colleagues have proposed a radically different model to explain the origin of drumlins as well as flutes, Rogen moraines, and other types of subglacial erosional features. They suggest that these features are the product of erosional and depositional processes associated with the catastrophic flow of subglacial meltwaters (Shaw and Sharpe 1987; Shaw et al. 1989; Fisher and Shaw 1992; Shaw 1993; Beaney and Shaw 2000). For example, they hypothesize that drumlins and Rogen moraines are created by the infilling of large scour holes cut into the base of the ice by major subglacial floods. Clearly, this model requires the accumulation and release of large volumes of water, and the identification of such subglacial sources using independent data has been slow in coming (Benn and Evans 1998). Nonetheless, there is reasonable evidence to support their conclusions and the creation of subglacial bedforms by large discharge events clearly warrants additional study.

Proglacial Features

The large volume of water released from glaciers carries with it a tremendous quantity of sediment that is deposited in a number of environments beyond the margin of the ice. This debris usually accumulates in stream channels and associated floodplains that, because of their continuous lateral shifting, spread the sediment into a large plain called a **sandur** (from Icelandic; plural *sandar* or *sandurs*). Downstream from the sandar, the meltwater streams may empty into bodies of standing water and construct deltas, beaches, and other geomorphic features from the fluvioglacial sediment. These forms do not differ appreciably from their counterparts developed in normal fluvial, lacustrine, or marine environments and so will not be discussed here. Sandar are somewhat analogous to alluvial fans, but they are unique in that the hydrology of the streams that form them is controlled by intense seasonal variations in the melting of ice. In addition, because each glacial advance develops its own related sandur, the surface of which is generally located at a different elevation, the distribution and age of the alluvial surfaces are extremely helpful in unraveling the Pleistocene history of a glacier. Thus, the origin of sandar and the geomorphic criteria for recognizing their form in ancient settings deserve our attention.

The study of sandar began in the nineteenth century in Europe and North America. They are now recognized as consisting of two primary types. A **valley sandur,** which originates within well-defined valleys, is created by one main river and its anabranches; the entire system rarely occupies the total valley bottom at any given time. In the United States, valley sandar have been called **valley trains** and are usually associated with individual mountain glaciers (fig. 10.30). The second type of sandur, called a **plain sandur,** differs in that it develops with no lateral constraints but represents the form of a massive plain. In North America, these are often referred to as **outwash plains** and are usually associated with large ice sheets.

Most sandar are composed predominantly of gravel, although the deposits may include lenses of sand. A general decrease in particle size is sometimes, although not always, apparent in the downstream direction (Fahnestock 1963; Church 1972). The deflation of very fine material on active sandar that are unvegetated helps to

Figure 10.30
Valley sandur in front of the Scott Glacier, Alaska.

produce a coarse-grained surface and simultaneously produces loess (see chapter 8).

In general, the long profiles of sandar are similar to river profiles, being concave-up in form and expressible as a simple exponential function (Church 1972). The concavity, however, may not be perfect because of the presence of linear segments similar to those characterizing alluvial fans. Cross-profiles tend to be convex-up, but the exact form may be irregular or may slope continuously in one direction. In addition, the cross-profile shape seems to depend on where it is measured in relation to the ice margin.

Krigstrom (1962) has recognized on sandar three distinct zones relative to the ice front, called proximal, intermediate, and distal zones, each of which has different surficial characteristics. The *proximal zone,* closest to the ice, is usually traversed by only a few main rivers that flow in well-defined entrenched channels. These rivers may originate beneath or within the ice (Gustavson and Boothroyd 1987) or pass continuously onto the ice mass itself. In some localities, the stratified drift extends onto and buries extensive areas of stagnating ice. As the ice subsequently melts, kettle holes form and the proximal sandur takes on a rough, pitted configura-

tion. These **kettled sandar** (or pitted outwash plains) are difficult to place in our framework classification because they form in the marginal ice-contact environment, but they are continuous with the surface of the proglacial sandur and develop with the same original mechanics (described in Price 1973). In addition, on many sandar the proximal surface stands well above the elevation of rivers emerging from the ice. Several hypotheses have been suggested to explain why rivers in the proximal zone are so entrenched in their own deposits: (1) the rivers may be regrading in response to hydrologic and load characteristics (Fahnestock 1969), parameters that vary significantly in surging glaciers (Sharp 1988); (2) the sediment is supraglacially deposited and left elevated as the ice front recedes and rivers emerge at a continuously lowered level; (3) modest incision may be normal in the proximal zone, with the sandur surface simply representing the high flow level; or (4) the channels may be downcutting as the ice margin is uplifted by glacitectonic processes. It is conceivable that each of these interpretations is correct.

In the *intermediate zone,* the channels become wide and shallow and distinctly braided, and the entire depositional network shifts its position rapidly from side to side. This active lateral migration leaves a maze of abandoned channels with a relief of one or two meters impressed on the surface topography of the plain. Commonly the main channel is aggraded to a higher level than the smaller channels, facilitating rapid changes in the position of the river. Downstream the system changes gradually into the *distal zone,* where channels become so shallow that the rivers may merge into a single sheet of water during high flow. The flow here commonly feeds deltaic growth when the river enters a body of standing water. However, the sandur may extend itself downstream by growing over the rear portion of the delta, while the deltaic front simultaneously progrades (Church 1972).

Sandar originate from the combined effects of a large sediment supply and the high floods associated with melting ice. Most of the abundant load is derived from older drift, morainal deposits, and the continual delivery of new debris to the ablation zone and its release from the ablating ice. In some cases, however, the debris may be transported to the proglacial environment by meltwater flowing in subglacial tunnels (Gustavson and Boothroyd 1987). The greatest fluvioglacial work occurs near the ice margin where floods are produced by summer melting or as sudden releases of lake water dammed within the ice called *jokulhlaups* (from an Icelandic word pronounced "yokel-lawp"). These floods are characterized by a rapid and drastic increase in discharge (Church 1972; Waitt 1980; Booth and Hallet 1993).

The bulk of **aggradation** on a sandur takes place during high flow events as channel fills, sandur levee deposits, and overbank sedimentation. Overbank deposits are more prevalent in the intermediate zone where channels are shallower and interchannel reaches are covered more frequently by floods. Although high flow does initiate pronounced channel scouring, the amount of aggradation during the peak and waning stages of a flood simply obliterates the scour channels. Thus, aggradation may be rather rapid. For example, Fahnestock (1963) measured a net elevation gain of 0.36 m in a two-year period on the sandur produced by the Emmons Glacier (Washington).

In general, then, sandar can be considered as transport surfaces that aggrade during high flows but are probably eroded and changed in form when discharge and load are at normal volumes. The seasonal variations in load and discharge may also be accompanied by changes in the river pattern (Fahnestock 1963). Therefore, the ultimate size and properties of a sandur are probably related to a quasi-equilibrium condition established by the balance between meltwater volumes and the quantity and size of the sediment made available for transportation. The surface will always be a montage of flood sediments, but the exact topography will change incessantly.

SUMMARY

In this chapter we examined the landforms developed by the process of glacial erosion and deposition. Erosional features range in size from minor embellishments of exposed bedrock to major forms that dominate the landscape. Minor features such as striations, roche moutonnées, and friction cracks are a function of the subglacial mechanics that control abrasion and plucking. Major erosional landforms develop in two environments. In mountain uplands, the expansion of cirques produces features such as arêtes and horns that give glaciated mountains their characteristically rugged appearance. Cirques are created by nivation and rotational sliding of cirque glaciers. Cirques increase in size by erosional retreat of their walls which is facilitated by frost shattering or possibly by quarrying at the foot of the headwall. Glaciated valleys are created by large-scale abrasion and plucking that cause the dominant staircase longitudinal profile and the U-shaped cross-profile.

Deposits associated with glaciation, called drift, consist of either stratified or nonstratified material. Stratification requires that some sediment be transported by meltwater after the debris is released from the ice. Nonstratified drift is deposited directly by the ice. Certain types of depositional landforms tend to accumulate in particular geographic positions with respect to the ice front. In the marginal zone, moraines and stagnant ice features are most common. In the interior zone behind the ice margin, ground moraine, fluted surfaces, and drumlins are most conspicuous. Downstream from the glacial margin, in the proglacial zone, all drift is fluvioglacial and usually accumulates in the form of large plains called sandar. Processes operating in each of these depositional environments have been discussed with regard to how they might generate the landforms developed in the specific regions.

SUGGESTED READINGS

The following references provide greater detail concerning the concepts discussed in this chapter.

Alley, R. B., Cuffey, K. M., Evenson, E. B., Strasser, J. C., Lawson, D. E., and Larson, G. J. 1998. How glaciers entrain and transport basal sediment: Physical constraints. **Quaternary Science Reviews** 16:1017–1038.

Brodzikowski, K., and van Loon, A. J. 1991. *Glacigenic sediments.* Amsterdam: Elsevier.

Coates, D.R. 1991. Glacial deposits. In Kiersch, G. A., ed., *The heritage of engineering geology: The first hundred years,* pp. 299–322. Centennial Special, vol. 3. Boulder: Geological Society of America.

Hallet, B. 1996. Glacial quarrying: A simple theoretical model. *Annals of Glaciology* 22:1–7.

Hooke, R. R. 1991. Positive feedbacks associated with erosion of glacial cirques and overdeepenings. *Geol. Soc. Amer. Bull.* 103:1104–8.

Rampton, V. N. 2000. Large-scale effects of subglacial meltwater flow in the southern Slave Province, Northwest Territories, Canada. *Canadian Journal of Earth Science* 37:81–93.

Shreve, R. L. 1985. Esker characteristics in terms of glacier physics, Katahdin esker, Maine. *Geol. Soc. Amer. Bull.* 96:639–46.

Walder, J., and Hallet, B. 1985. A theoretical model of the fracture of rock during freezing. *Geol. Soc. Amer. Bull.* 96:336–46.

11

PERIGLACIAL PROCESSES AND LANDFORMS

Aerial view of rockfall slopes and Galera Creek Rock Glacier, Absaroka Mountains, Wyoming.
(Photo by R. Craig Kochel)

Introduction
Permafrost and Ground Ice
    Definition and Thermal Characteristics
    Distribution, Thickness, and Origin
    Hydrology
Periglacial Processes
    Frost Action
        Frost Wedging
        Frost Heaving and Thrusting
        Frost Sorting
        Frost Cracking
    Nivation
Pedogenesis in Permafrost Terrain
    Mass Movements
        Frost Creep
        Solifluction (Gelifluction)
Periglacial Landforms
    Landforms Associated with Permafrost
        Ice Wedges and Ice-Wedge Polygons
        Pingos
        Thermokarst
    Patterned Ground
        Classification
        Origin
    Landforms Associated with Mass Movement
        Stratified Slope Deposits
        Gelifluction Features
        Rock Glaciers
        Blockfields
        Cryoplanation Terraces
    Relict Periglacial Features and Their Significance
Environmental and Engineering Considerations
    Building Foundations
    Roads and Airfields
    Utilities: Water and Sewage
    Pipelines
    Implications for Global Warming
    Applications to Planetary Geology
Summary
Suggested Readings

INTRODUCTION

A group of processes and features called **periglacial** characterize regions having extremely cold climates. The term "periglacial" was first used to describe the processes operating and the features developed in zones adjacent to ancient or modern ice sheets (Lozinski 1912). Since then, the original connotation of "near-glacial" has been expanded, and most workers now accept the term as encompassing all nonglacial phenomena that function in cold climates, even if glaciers are not present (Dylik 1964; Butzer 1964; Washburn 1973). Washburn (1980) adopts the term "geocryology" to indicate the study of frozen ground processes and phenom-

ena. However, he suggests continued use of periglacial as a descriptive adjective. The periglacial environment is difficult to define in terms of precise temperature and precipitation values, although several attempts have been made to do so. For example, Peltier (1950) suggested that average annual temperatures range from −15° to 2° C (10°–35° F) and precipitation values vary from 50 mm to 1250 mm for the precipitation values.

Some scientists believe that permanently frozen ground, or permafrost, is an essential ingredient (if not a prerequisite) for periglacial conditions (Tricart 1967; Péwé 1969). Certainly some periglacial features occur in close association with frozen ground, and a complete explanation of periglacial systems therefore requires a fundamental understanding of permafrost. However, many significant periglacial processes can also function in the absence of permafrost. Most students of periglacial phenomena do not view permafrost as a requirement of a periglacial system.

The two points that all investigators agree on are that (1) nearby glaciers are not necessary for the processes to function, and (2) the fundamental controlling factors of geocryology are intense frost action and a ground surface that is free of snow cover for part of the year. Many regions meet these requirements, ranging from polar to subpolar lowlands to high-elevation mountains that may rise in any regional climate including temperate and tropical zones.

Interest and research on periglacial geomorphology has surged in recent decades because of the effects of global environmental changes on sensitive permafrost ecosystems and the increased pressures of resource exploration and development in these frontier regions (Dixon and Abrahams 1992).

Human occupation of high-latitude regions of the world is not new; Inuit and other nomadic people have wandered these areas for millennia. These inhabitants, however, lived in rather simple social organizations, and their life-styles brought them into little conflict with their natural surroundings. In more recent times, the emplacement of defense installations in periglacial regions, the discovery of oil above the Arctic Circle, and increased exploration for mineral wealth brought the realization that developing a highly technical infrastructure in these regions is an engineering and scientific nightmare. For inhabitants of temperate zones, shattered highways and frozen or broken water pipes are inconveniences of a rigorous winter and the subsequent spring thaw. Multiply these problems by a thousand; add disruption of services, construction difficulties, and complications from all the trappings associated with a modern civilization; and you will begin to understand the difficulty of living in a periglacial environment.

Governmental agencies in several countries, including the United States and Russia, have conducted research related to the problems of cold climates, and symposia and reports in scientific journals increasingly treat such topics (for example, Wolfe 1998). Nonetheless, we are far from a detailed understanding of geomorphic processes in cold climates. Faced with an inevitable human migration into cold regions, geomorphologists must devote additional efforts to this discipline. Only within the last decade have long-term process studies, spanning 10 years or more, been completed. Initial results from long-term studies, such as Price (1991) on solifluction, have added considerably to our understanding of processes in periglacial regions.

Recent investigations have largely focused on (1) the identification and/or recognition of permafrost, (2) understanding the chemistry and physical mechanics important in prominent periglacial processes, (3) identification of landforms indicative of a periglacial climatic regime, and (4) reconstruction of paleoenvironments through the study of relict periglacial phenomena (Thorn 1992). We have the basic concepts to understand geomorphology in these regions because the processes responsible for periglacial effects are not greatly different from those in temperate climates. Frost action and mass movements form the core of most periglacial processes.

The main difference between periglacial and temperate phenomena lies in the magnitude of the processes and the manner in which they are combined in systemic operations on the landscape. Periglacial geomorphic systems are distinctive because of overwhelming driving processes and landform development that emphasizes accelerated freeze-thaw and frost weathering processes and the growth and decay of subsurface ground ice in the regolith. Thorn (1992) notes that in periglacial environments, with and without permafrost, frost-related and mass movement processes are intensified so that a distinctive suite of landforms and processes is created. Thus, as L. W. Price (1972) stresses, if we are to understand the periglacial environment completely and provide viable information for future planning, we must cast aside our provincial, mid-latitude approaches to the study of this system.

Periglacial processes are not only important in shaping the cold climate regions of remote high latitudes and high altitudes today, they appear to have had an important, and in some areas dominant, role in the Quaternary geomorphic evolution of hillslopes and piedmont areas of mid-latitude regions that now experience temperate climates. Numerous investigators have suggested that Appalachian Mountain hillslope morphology is largely a product of intense periglacial activity during the Pleistocene and that geomorphic processes during the Holocene were relatively ineffective (Clark and Ciolkosz 1988; Ciolkosz et al. 1990; Braun 1989). Similar notions on the efficacy of paleoperiglacial processes compared with those of modern climates have been noted in Europe (Harris 1987) and other areas where major sedimentation episodes in montane areas have been attributed to

periglacial episodes (Boardman 1992). Boardman (1992) rightly called for more process-rate studies and for geomorphologists working in periglacial regimes to devote more attention to magnitude and frequency concepts. We will return to the importance of paleoperiglacial climates in mid-latitude areas toward the end of this chapter because their impact is often neglected in the assessment of Holocene processes in many of the well-populated and intensely studied areas that account for nearly half of the subaerial portions of the planet. As in all geomorphic systems, understanding the rate of processes is central to defining the operations and historical evolution of landforms in cold climate regions.

The terminology within the subdiscipline of geocryology is often confusing, is contentious, and contains numerous unique terms. Evidence of terminology problems can be found in the recent publication "Glossary of Permafrost and Related Ground-Ice Terms" (National Research Council of Canada 1988). The terminology adopted here is relatively simple and streamlined for the purpose of providing an overview of major processes and landforms associated with periglacial environments. Be advised that the classification of landforms varies considerably between authors for many of the features discussed in this chapter. For greater detail, several excellent texts and reviews are available: Embleton and King 1968, 1975b; Price 1972; Washburn 1980; Tricart 1969; Péwé 1969; National Research Council of Canada 1978; National Academy of Sciences 1983; French 1976; Williams and Smith 1989; and Evans 1994.

PERMAFROST AND GROUND ICE

Definition and Thermal Characteristics

Although permafrost is not a requirement for the functioning of periglacial processes, the most troublesome engineering problems occur in regions underlain by zones of perennially frozen ground. Many periglacial features are related to abundant ground ice. Permafrost was originally defined in terms of temperature only (Muller 1947) as being soil or rock that remained below $0°$ C continuously for more than two years. This definition has been accepted in many subsequent discussions of the topic (Washburn 1980). By this definition water is not a necessary component of permafrost, and "dry" permafrost has been recognized (Bockheim and Tarnocai 1998a). In most cases, however, ice is so important in the mechanics of periglacial processes that some workers (e.g., Stearns 1966) believe that moisture must be present before any material can be considered as true permafrost. Other investigators have placed the temperature criterion for permafrost below $0°$ C (Ferrians 1965; Brown 1970). For geomorphic effect the presence of ice, regardless of the temperature value, seems to be critical. The ice may exist as a cement between soil particles or as larger masses of pure ice. With dry permafrost, water contents are typically less than 5 percent and insufficient to develop a cement. Bockheim and Tarnocai (1998b) suggested that dry permafrost may have developed from sublimation of moisture from ice-cemented permafrost over thousands of years.

Where distinct ice masses are prevalent, they usually occur as horizontal lenses, but they may also fill cracks in the parent material and stand as vertical veins or wedges (fig. 11.1). Permafrost, usually a subsurface phenomenon, is a zone of permanently frozen ground that can extend downward to incredible depths. The upper surface of the permafrost, the **permafrost table** (fig. 11.2), is overlain by a thin layer of material (15 cm to 5 m in depth) that freezes and thaws on a seasonal basis. This uppermost layer, the **active layer,** is thickest in the subarctic region (Price 1972), becoming shallower toward both the pole and the mid-latitudes. In any area, however, significant variations in thickness of the active layer can occur over short distances depending on environmental conditions (Owens and Harper 1977). Recent studies have focused on the influence of water and vegetative cover on the geometry of the active layer (e.g., Repelewska-Pekabwa and Gluza 1988). The active layer is usually thicker in sands and gravels than in more fine-grained soil materials. Insulative covers such as standing bodies of water and peat generally limit active layer thickness.

The mechanics in the active layer are similar in most respects to the normal seasonal freezing and thawing in temperate climate zones. In the permafrost regions, however, water released by thawing cannot percolate into the solidly frozen substrate, which accentuates the effects of frost action and mass movements. In this way, following the summer months, unique phenomena related to the development of increased hydrostatic pressures result from water trapped between the downward-advancing freezing plane and the permafrost. For example, many studies have shown dramatic increases in the occurrence of shallow landslides with slip surfaces at the base of the active layer, where shear strength is reduced during summer melt (e.g., Mackay 1973). Lewkowitz (1992), working on Ellesmere Island in northern Canada, documented an average of 3.5 such failures per year over a 40-year period ending in 1990 within a small area of only 5.3 km^2.

Even within the permafrost itself, the temperature fluctuates with the seasons. Temperature variations, however, become less radical with depth until some level is reached, generally at 10 to 30 m deep, where the temperature never changes (fig. 11.3). This level is known as the **zero annual amplitude.**

Thermal operations in permafrost are affected by so many factors that they are understood only in general terms. For example, thermal conductivity varies with material composition and may cause freezing or thawing to progress unevenly from the surface downward. Thus,

as freezing proceeds after a thaw season, unfrozen lenses may develop between the permafrost table and the solidly frozen ground surface. This phenomenon causes unusual pressures in the soil water and depresses its freezing point. In addition, heat of fusion is generated when the zone around the ice-free lenses freezes. The result is that pockets of free water may exist in the ground for considerable time after the surface is frozen, possibly for several months.

Further complications in the distribution of frozen ground occur because the permafrost table tends to mirror the surface topography, rising beneath hills and lowering under valleys. In addition, the permafrost table is influenced in various ways by the position of surface water (fig. 11.4), depending on whether the rivers or lakes freeze over in the winter (Ferrians et al. 1969).

Distribution, Thickness, and Origin

Where annual temperatures average 0° C or below, ground freezing during the winter will penetrate deeper than the depth of summer thawing. Each passing year will produce another increment of perpetually frozen ground, and the permafrost zone will gradually thicken. The process, however, cannot go on indefinitely because the thermal regime within Earth will exert controls on the depth to which freezing can penetrate. Heat affecting permafrost issues from the sun and from the interior of Earth as geothermal heat flow. Therefore, cold filtering downward from the surface is counteracted by heat escaping upward from inside Earth. The interaction results in the **geothermal gradient,** the rate of temperature increase with the depth below the surface. Thus, under a constant climate, at some depth the temperature will be kept above 0° C by the geothermal heat flow (fig. 11.5), and the total thickness of permafrost will relate to the position of this temperature level.

The geothermal gradient, and therefore thickness of permafrost, can be quite variable under areas having identical surficial climates. This occurs because the geothermal gradient is affected by the thermal conductivity of the parent material. Where mean surficial temperatures are the same, permafrost will extend to greater depths in materials of higher conductivity. The variations in geothermal gradients along northern Alaska, shown in figure 11.5, are explained by this phenomenon.

Permafrost today extends to formidable depths (table 11.1). Average depths occur between 245 and 356 m in North America and tend to be slightly thicker in Eurasia. The thickest known permafrost in Siberia reaches a depth of approximately 1500 m. The great thicknesses of permafrost reflect complex interactions between modern climate conditions and inheritance from former conditions.

Several lines of evidence indicate that some permafrost must have originated during the Pleistocene

Figure 11.1
Ice wedge (ground ice) in permafrost exposed by placer mining near Livengood, Alaska, about 50 miles northwest of Fairbanks.
Tolovana district, Yukon region, Sept. 1949.

(Photo by O. J. Ferrians, U.S. Geological Survey)

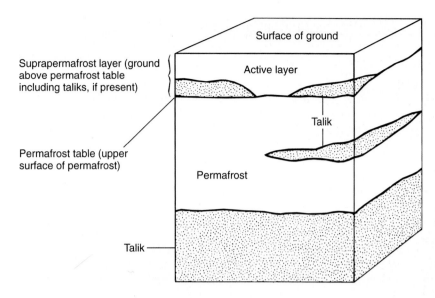

Figure 11.2
Profile in a region underlain by permafrost. Taliks are zones of unfrozen ground within or beneath the permafrost or between the permafrost table and the base of the active layer.
(After Ferrians et al. 1969)

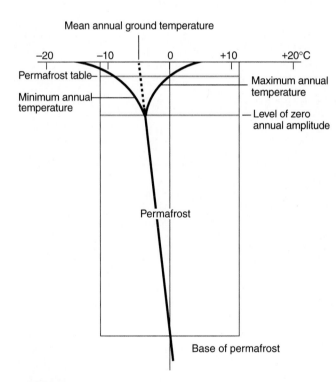

Figure 11.3
Ground temperature change with depth in permafrost regions.
(After Williams 1970)

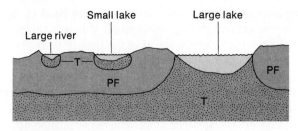

Figure 11.4
Schematic cross-section showing the effect of surface water on the distribution of permafrost. PF = permafrost, T = taliks.

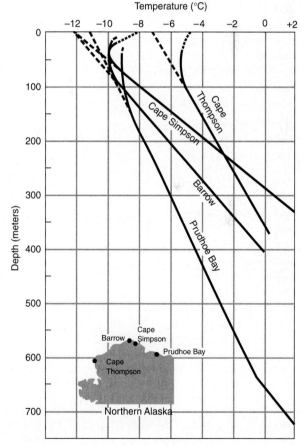

Figure 11.5
Generalized profiles of measured temperature on the Alaskan arctic coast (solid lines). Dashed lines represent extrapolations.
(After Gold and Lachenbruch 1973)

TABLE 11.1	Selected Permafrost Thicknesses in the Northern Hemisphere.

Location	Mean Annual Air Temperature	Thickness of Permafrost
Alaska		
Prudhoe Bay (70°N, 148°W)	–7 to 0° C	609 m
Barrow (71°N, 157°W)	–12 to 7° C	405 m[a] 16 km inland
Umiat (69°N, 152°W)	–12 to 7° C	322 m 235 m under Colville River
Cape Thompson (68°N, 166°W)	–12 to 7° C	306 m[a]
Bethel (60°N, 161°W)	–7 to 0° C	184 m 13 m under Kuskokwim River
Ft. Yukon (66°N, 145°W)	–7 to 0° C	119 m 5.5 m under Yukon River
Fairbanks (64°N, 147°W)	–7 to 0° C	81 m
Kotzebue (67°N, 162°W)	–7 to 0° C	73 m
Nome (64°N, 165°W)	–12 to –7° C	37 m
McKinley Natl. Park, East Side (64°N, 149°W)	extremely variable	30 m
Canada		
Melville Island, N.W.T. (75°N, 111°W)	—	548 m near coast, probably thicker interiorward
Resolute, N.W.T. (75°N, 95°W)	–16.2° C	396 m
Port Radium, N.W.T. (66°N, 118°W)	–7.1° C	106 m
Ft. Simpson, N.W.T. (61°N, 121°W)	–3.9° C	91 m
Yellowknife, N.W.T. (62°N, 114°W)	–5.4° C	61–91 m
Schefferville, P.Q. (54°N, 67°W)	–4.5° C	76 m
Dawson, Y.T. (64°N, 139°W)	–4.6° C	61 m
Norman Wells, N.W.T. (65°N, 127°W)	–6.2° C	46–61 m
Churchill, Man. (58°N, 94°W)	–7.1° C	30–61 m
Russia		
Upper Reaches of Markha River (66°N, 111°E)	—	1500 m
Udokan (57°N, 120°E)	–12° C	900 m[a]
Bakhynay (66°N, 124°E)	–12° C	650 m
Isksi (71°N, 129°E)	–14° C	630 m
Mirnyy (63°N, 114°E)	–9° C	550 m
Ust'-Port (69°N, 84°E)	–11° C	425 m
Salekhard (67°N, 67°E)	–7° C	350 m
Noril'sk (69°N, 88°E)	–8° C	325 m
Yakutsk (62°N, 129°E)	—	195–250 m
Vorkuta (67°N, 64°E)	—	130 m

From Price (1972). Various sources, but chiefly the following: Brown (1967), map; Ferrians (1965), map; Yefimov and Dukhin (1968).[a] A calculated depth not actually measured.

(see Washburn 1980, pp. 60–61), especially in areas that were not glaciated. In areas covered by Pleistocene glaciers, permafrost probably formed after the ice dissipated. In polar regions permafrost is forming today where glacial retreat has exposed unfrozen ground (Washburn 1980). This indicates that some present-day climates are cold enough to accumulate modern permafrost and certainly maintain ancient permafrost. Permafrost does, however, respond to climatic change (Mackay 1975b). For example, in the upper parts of the temperature profiles from northern Alaska (fig. 11.5), the distinct curvatures are caused by climatic warming

after the permafrost was formed. The equilibrium geothermal gradient is shown by the straight parts of the curves, and their upward projection indicates the prevailing surficial temperatures at the time of permafrost formation (Lachenbruch and Marshall 1969; Gold and Lachenbruch 1973).

The destruction or protection of permafrost varies with factors other than climate, such as the type of surface cover. The presence of ice or dense vegetation has a decided influence on permafrost thickness. In Antarctica, for example, some areas beneath the ice sheet have no permafrost (Ueta and Garfield 1968), whereas zones

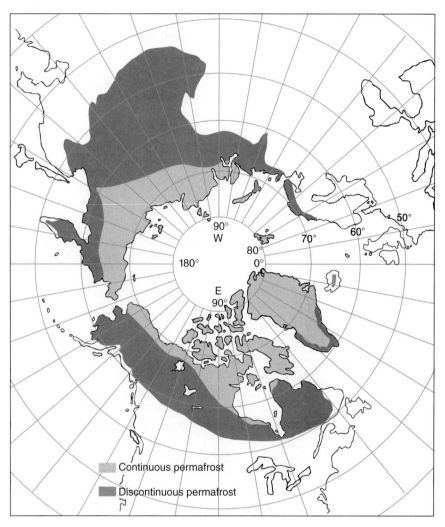

Figure 11.6
Extent of continuous and discontinuous permafrost in the Northern Hemisphere.
(After Ferrians et al. 1969)

that are glacier free have permafrost up to 150 m thick (J. R. Williams 1970).

Permafrost currently underlies 26 percent of the land surface on Earth (Black 1954) and so is not, as we may tend to think, an unusual phenomenon. During cold intervals in the Pleistocene, permafrost was much more extensive. Permafrost is also known to exist beneath the ocean in many nearshore polar areas, although it probably formed on land and was submerged during a subsequent rise in sea level. As the map in figure 11.6 shows, most of the known permafrost exists in the polar regions of the Northern Hemisphere, extending to a southernmost limit at a latitude of approximately 55° N in North America and Eurasia. In China, however, latitudinally controlled permafrost is known to exist as far south as 46° N. Permafrost in these regions is divided into continuous and discontinuous types (Ray 1951). **Continuous permafrost** usually consists of thick layers of perennially frozen ground that are spread rather evenly under a wide areal surface. The continuous nature of the permafrost alters only where it thins under deep, wide lakes or rivers (see fig. 11.4). In contrast, **discontinuous permafrost** is shallower and contains unfrozen zones

within the frozen ground or wide gaps that remain unfrozen (fig. 11.7). These unfrozen areas, called **taliks,** increase in size and number southward until true permafrost exists only as isolated patches. Taliks appear as islands exposed at the surface, lenses or layers within the permafrost, or unfrozen ground beneath the permafrost (see figs. 11.2, 11.7). The southern limit of continuous permafrost coincides in a general way with the –6° C annual isotherm and the discontinuous boundary with the –1° C annual isotherm (Brown 1970).

We should not assume that permafrost exists only in high latitudes. Isolated zones of frozen ground, called **sporadic permafrost** (fig. 11.7), can be found far south of latitude 55° N and are probably relics of a once colder climate. Pockets of permafrost persist also where elevations are sufficiently high. In the United States, for example, permafrost is found near the summit of Mount Washington in New Hampshire (Goldthwait 1969) and in a couple of other summit areas in the northern Appalachians. Ives and Fahey (1971) mapped sporadic permafrost on numerous sub-summit plateaus throughout the Rocky Mountains. Schmidlin (1988) found that the lower limit of permafrost in eastern North America

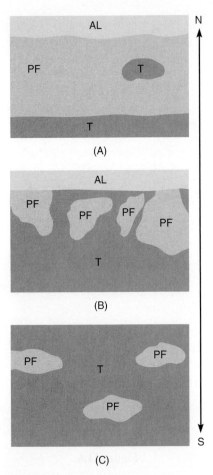

Figure 11.7
Types of permafrost zones: (A) continuous, (B) discontinuous, (C) sporadic. AL = active layer, PF = permafrost, T = talik.

roughly corresponds to the 0° C mean annual isotherm and generally extends below tree line. Permafrost has even been reported near the summit of Mauna Kea, Hawaii (Woodcock et al. 1970; Woodcock 1974), at an elevation of 4170 m. In some areas (for example, the high-altitude plateaus of China), permafrost is so widespread that it is appropriately considered to be continuous or discontinuous, rather than sporadic.

The general but imprecise relationship between average air temperature and permafrost boundaries suggests that modern climate determines the distribution of frozen ground (Harris 1983), even though most of the included ice formed at an earlier time. This conclusion fits the interpretation that the recent retreat of the discontinuous permafrost boundary in Manitoba (Canada) is related to a climatic shift that began only about 120 years ago (Thie 1974). As Washburn (1980) suggests, discontinuous permafrost may be in such a delicate equilibrium with the present climate that only minor changes in climate or surface condition will produce drastic effects.

All other things being equal, the climate most amenable to preservation of permafrost is one with cold, long winters followed by cool, short summers and low precipitation in all seasons (Muller 1947). Other variables, such as vegetation type and density, composition of surface materials, topography, and surface water, influence the spatial character of permafrost. For example, most permafrost lies beneath the northern boreal forests or, north of the tree line, under a tundra vegetation dominated by low sedges, grasses, and mosses. The boundary between continuous and discontinuous permafrost often parallels the southern limit of the tundra. Because the thermal properties of vegetation control how efficiently temperature can penetrate the ground (Corte 1969; Price 1972), it is not clear whether the climate or the vegetation produced by the climate is the determining factor in permafrost distribution. Probably all of the above factors exert some control, and therefore we cannot expect a precise correlation between air temperature and permafrost boundaries.

Hydrology

Because flowing water operates only for a few brief months during the summer in periglacial regions, its role as a geomorphic agent has traditionally been neglected. Infiltration is generally low in permafrost regions because of the decreased hydraulic conductivity of frozen regolith. Groundwater occurs in suprapermafrost (active-layer), intrapermafrost, and sub-permafrost environments. Through permafrost zones groundwater responds in a manner not unlike the flow through lithified bedrock, tending to follow avenues of secondary permeability. Water exploits fractures and passages in permafrost enlarged by the thermal erosion of relatively warm water (Van Everdingen 1990). We will see later that groundwater plays a significant role in the development of several landforms unique to permafrost terrains, because water originating outside the permafrost zone creates differential hydraulic pressure in the permafrost zone.

Rivers in periglacial climates typically experience peak flows during late summer and have a markedly seasonal hydrology dominated by snowmelt. Sources of water for stream flow are highly variable depending on snow distribution from year to year (Woo et al. 1994). Church (1974, 1988) developed a comprehensive classification of streams in cold climates based on flood hydrology. Apart from the dominance of snowmelt, periglacial rivers are characterized by other processes unique or prominent in cold climates. Lateral migration and bank erosion are not simply a function of the interaction of hydraulic forces and abrasion of the bank stratigraphy but are supplemented by the process of **thermal erosion** from relatively warm water (even if only a few degrees above freezing) in contact with permafrost bank sediment. High lateral migration rates such as 4 to 6 m/yr reported on the Yukon River (Riddle et al. 1988) are the rule rather than the exception. Major slumps along river banks typically occur in response to

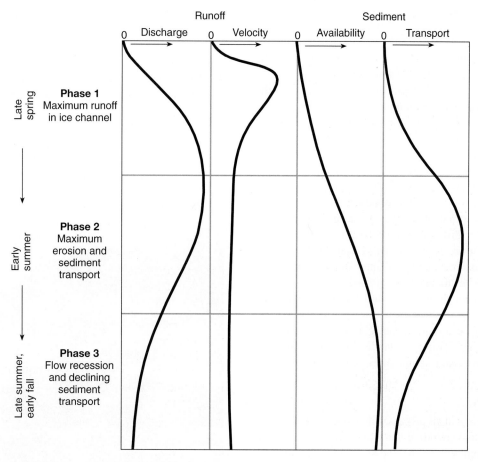

Figure 11.8
Dynamics of runoff and sediment during snowmelt in the Northwest Territories, Canada.
(Priesnitz 1990)

thermal erosion in rivers on permafrost regions. Burn (2000) detailed one of these retrogressive slumps and demonstrated the importance of subsurface conditions on near-surface ground temperatures. Depending on conditions, either aggradation or degradation of permafrost can occur in the slump blocks. Significant erosion is also accomplished by ice flows during late spring breakup floods (Smith 1980). Although ice-related erosion is important along many mid-latitude areas outside the periglacial realm, thermal erosion is significant only in areas hosting ground ice.

Recent studies have compared the load transported by rivers in periglacial regions with those in temperate climates. Priesnitz (1990) distinguished evolutionary phases of snowmelt floods related to variations in discharge, sediment transport, and geomorphic effectiveness (fig. 11.8). In contrast to temperate climates, the transport of sediments in periglacial rivers appears to be supply-limited and less important than the exports of dissolved load. Clark et al. (1988) observed that large rivers experience sediment depletion because there is commonly a long lag between ice breakup and the flood

peak. Small basins, experiencing rapid rise to the peak flood, were more likely to have a lag in sediment transport because of the ice freezing and armoring on the channel bed and banks until after the flood peak. In a long-term study of water quality in the Colorado Front Range, Caine (1992) discovered that solute loads were generally low (5–20 g/m^3 per year). Nonetheless, solute loads exceeded suspended sediment loads by an order of magnitude. Caine's (1992) study also showed that solute loads were highest in basins experiencing the greatest winter snow accumulation, suggesting that chemical weathering is more efficient under heavier snow cover. This conclusion helps explain why hollows develop on the landscape beneath permanent snowfields, facilitating an erosional process known generally as nivation.

Extensive discussion of periglacial hydrology is beyond the scope of this text, but studies in this field are on the increase (e.g., see Dixon 1986, Prowse and Ommaney 1990, Caine and Swanson 1990, Caine 1992) and will likely bring significant new understanding to long-understudied phenomena.

PERIGLACIAL PROCESSES

Among students of periglacial phenomena, the consensus is that the periglacial regime is dominated by the processes of enhanced frost action, the growth and decay of ground ice, and mass movement amplified by the combination of sediment supply and hydrology unique to these environments. These processes have special importance in the periglacial setting. Frost action encompasses a group of processes—wedging, heaving, thrusting, and cracking—that all serve to prepare bedrock or soil for erosion. Mass movements transport the loosened debris. These two groups of processes function in periglacial environments as they do in temperate zones; however, major differences arise because frost action is considerably more severe in periglacial zones, and because mass movements may be intensified during thawing because the material is saturated with excessive water that cannot drain downward through the system. The combination of these factors results in geomorphic features that are unique to periglacial regions and in rates of hillslope erosion and landform modification that can greatly surpass those in temperate regions.

Frost Action

The driving force in all processes included in the realm of frost action is the growth of ice within a soil or rock. Intuitively we might expect the explanation of freezing in a porous substance to be relatively simple, but the thermodynamics of the process are very complex (see Miller 1966; Anderson 1968, 1970). In addition, other factors within the system influence the final geomorphic effect of frost action because they respond to freezing in different ways. For example, segregation of ice into discrete lenses generates a stress field that differs from that produced by the freezing of disseminated pore water, and the processes triggered by this also are dissimilar. Whether ice lenses actually form depends mainly on the rate of freezing, the ability of water to be drawn to a central freezing plane or point, and the size of the pore spaces. Figure 11.9A illustrates well the complexities involved in developing a semi-empirical model from field measurements to predict the rate of bedrock frost shattering (Matsuoka 1991). The model indicates that frost-shattering rates are a function of the freeze-thaw frequency, moisture content, and the tensile strength of the bedrock (fig. 11.9B), which when exceeded, of course, results in disintegration. Moisture content was the most significant parameter noted in this study (Matsuoka 1991). The role of moisture, however, is not unequivocal. A study comparing dry interior regions of Antarctica with moist maritime environments there found comparable low weathering rates despite the presence of additional moisture where more dynamic weathering was anticipated (Hall 1992).

Ground ice occurs in a wide variety of shapes and thicknesses. The volume of ice varies depending upon

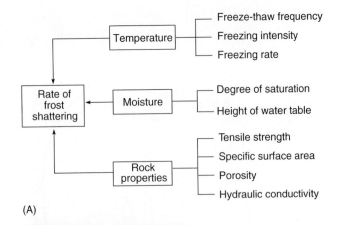

(A)

(B)

Figure 11.9
(A) Schematic of factors affecting frost shattering. (B) Frost-shattered outcrop along a glacial valley in Iceland. Note the abundance of joints for frost-riving to take advantage of.
(A): (Matsuoka 1991)
(B): (Photo by R. Craig Kochel)

sediment texture and the morphology of the ground ice, but it is typically significant. For example, Couture and Pollard (1998) found an average of 31 percent ground ice in the upper 6 m of permafrost in the Canadian Northwest Territories, with ranges between 2 and 69 percent.

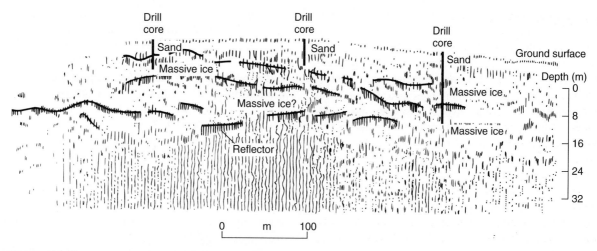

Figure 11.10
Ground-penetrating radar transect and drill locations at Ya Ya Lake, Richards Island, Northwest Territories, Canada. GPR reflectors graphically show the variations in ground-ice structure common in permafrost sediments.
(Dallimore and Wolfe 1988)

Ground ice in discrete bodies up to 20 m thick have been observed in glaciofluvial sediments (fig. 11.10). The origin of ground ice can range from buried glacial ice (Dallimore and Wolfe 1988) to ice formed in situ by the various ice-segregation processes mentioned above. The in-situ ice can be from a meteoric source or from groundwater. Harris (1990) provides a useful review of how surface landforms can be used as a guide to the type and origin of ground ice present in the regolith.

Generally, lenses of ice are smaller and less common as depth increases because the greater pressure lowers the freezing point. However, exceptions to this generalization occur when the soil properties are conducive to ice lens development. The water in fine-grained sediment (having small pore spaces) freezes at lower temperatures and also has a greater propensity to suck water to a central freezing plane. Because of these factors, and the tendency of fines to hold more water, segregated ice masses are most common in fine-grained sediment, and coarse sediment usually contains interstitial ice. On the other hand, extremely fine-grained sediment may be impermeable. Therefore, ice lenses form preferentially where suction potential and permeability are most advantageously combined, usually in deposits of uncemented silt (discussed in Washburn 1980).

The movement of water to a freezing plane is more complicated than the foregoing discussion might suggest and depends on many factors other than grain size (Konrad and Morgenstern 1983; Rieke et al. 1983). Konrad and Morgenstern have identified a parameter called the segregation potential to indicate whether a soil will readily form ice lenses (see Konrad and Morgenstern 1983 for discussion and references). The **segregation potential** is the ratio of the rate of water migration to the temperature gradient near the frost front. It assimilates all of the controlling factors and provides an index that can be

applied in predicting the susceptibility of soils to frost-related problems.

The expansion associated with freezing exerts pressures that can produce results as variable as shattering of solid rock or physical lifting of the ground surface. Precisely what event will transpire during freezing and subsequent thawing is determined by a multitude of complex interdependent variables that operate within the system.

Frost Wedging Frost wedging, the prying of solid material by ice (Washburn 1973), is synonymous with splitting, riving, and shattering. The precise mechanisms of **frost shattering** are extremely complex (for a discussion, see McGreevey 1981) and well beyond our introductory treatment. Exactly what conditions (rate of freezing and absolute temperatures) will result in shattering are quite variable because the process is also controlled by the internal properties of the rocks (Douglas et al. 1983). Most evidence for the wedging process has been derived from laboratory experiments or is based on theoretical models. It is doubtful that these are directly applicable to the field situation; in fact, Thorn (1979) found little correlation between the experimental work and measurements of shattering in the natural setting. Furthermore, he suggests the possibility that porosity, saturation of the rocks, and freezing intensity may combine to generate a number of disruptive mechanisms. This idea tends to reinforce the idea that hydration shattering (White 1976a; see chapter 10) may be an important partner in frost-wedging mechanics.

In porous substances that are saturated, frost wedging is facilitated by rapid freezing of water in the near-surface pores. This rapid freezing tends to close the system, allowing the buildup of some extra pressure that is needed to cause shattering. Slow freezing seems to be an effective wedging process only if the water is free to migrate to a freezing plane where excess pressure is generated by growth of large ice crystals. As discussed in chapter 10, wedging in cracks takes place only if freezing proceeds from the surface downward (Battle 1960).

The effect of frost wedging is commonly linked to the number of freeze-thaw cycles. However, numerous factors influence frost wedging besides the rate of freezing and the closure of the system. A high water content increases the strain on rocks (Mellor 1970) and most rocks shatter more readily if they are immersed in the fluid (Potts 1970). This probably explains why wedging is often more pronounced at the base of sheet cliffs where groundwater is available than at the top. Support for this argument was given by Coutard and Francou, (1989) who found that north-facing slopes experienced more frost shattering than other cliffs in the French Alps. Water supply on the north-facing slopes was more or less constant, freezing was of longer duration, and freezing was more intense. Shorter-duration freezes, although perhaps more frequent, were less effective on south-facing exposures partly because many of the freezes did not coincide with wet conditions in the bedrock fractures. Caution is required, in correlating freeze-thaw frequency and wedging however, because air temperature is not always a good predictor of rock surface temperature or the temperature within cracks in those rocks (Douglas et al. 1983). Furthermore, a knowledge of rock properties is important because water may freeze at a temperature other than 0° C depending on those characteristics. Thus, prolonged freezing with extremely low temperature may have more geomorphic significance than frequent freeze-thaw cycles (Rapp 1960; Washburn 1973). Given an abundant supply of water, the characteristics of parent material will control the extent of wedging. For example, larger pore size and porosity in a rock will increase its susceptibility to wedging. Sedimentary rocks with fissility due to micaceous minerals generally allow easier migration of water, thereby promoting greater wedging.

Recent work by Hall (1998) questions the ubiquitousness of freeze-thaw processes. Instead, he suggests greater importance should be given to the potential role of thermal shock and thermal stress fatigue of rocks. He believes that higher frequency measurements, on the order of one-minute intervals, may be required to demonstrate the importance of such processes due to wind fluctuations and other high-frequency variables.

In exceptionally arid regions, like the Dry Valleys of Antarctica, the effectiveness of freeze-thaw processes may also be limited (French and Guglielmin 1999). Because of the lack of moisture and shallow active layer, debris production may result more from wind erosion, salt weathering, tafoni production, and (in granitic rocks) grusification.

The products of frost wedging are notably angular and range in size from blocks as large as buildings to fine-grained debris. It has been traditionally accepted that the terminal size of frost shattering is silt (Hopkins and Sigafoos 1951; Taber 1953). Although this is probably true in most situations, some workers now believe that clay-sized particles may be formed under certain conditions (McDowall 1960) or that any terminal size is only rarely attained (Potts 1970). In any case, coarse angular debris is common in periglacial regions, accounting for the prevalence of talus rubble at the base of Alpine slopes. Wedging also breaks particles loose from bedrock covered by soil, and these progressively make their way to the surface, creating problems for farmers working the land.

Although most frost weathering can be attributed to the increase in volume upon freezing, many studies have shown that accelerated weathering is possible where crystallization pressures are generated by salt growth in rocks (Williams and Robinson 1991). Coastal environments typically provide enhanced salt loads through sea spray. Increased frost shattering attributed to the influence of salts is facilitated by (1) surface sealing on rock exteriors that leads to internal stresses from trapped water as freeze-up occurs, (2) the combined growth of salt and ice crystals if the system is below the eutectic point, and (3) osmotic pressures developed in fine-grained rocks where salts can block pore spaces (Williams and Robinson 1991).

Frost Heaving and Thrusting   **Frost heaving** refers to vertical displacement of matter in response to freezing, whereas **frost thrusting** connotes horizontal movement (Eakin 1916). In a natural setting the two processes are virtually indistinguishable. Heaving is directly responsible for several phenomena that are commonplace and accentuated in periglacial environments. First, heaving causes the ground surface to move vertically as ice formation expands the ground material. The extent of this displacement depends on the physical variables within the system (Rieke et al. 1982), and the surfaces of adjacent local areas may be lifted at disparate velocities and in different amounts. Such differential heaving causes building foundations and other types of constructions to crack. Second, heaving has the ability to sort heterogeneous debris by forcing larger particles to migrate surfaceward relative to finer ones. The particles may move upward as much as 5 cm a year (Price 1972). Any objects inserted into the ground or dispersed within the near-surface mass are subject to the mechanics of heaving. These include not only stones of various sizes that emerge in farmlands or on slopes but also fence posts, telephone poles, pilings, coffins, or anything else that people might conceivably shove into the ground. The heaving process also functions in indurated solids and is capable of displacing joint-bounded blocks of bedrock, even though these are held rather tightly by the surrounding mass. These upheaved blocks may project up to 1.5 m above the ground and, where soil is absent, give the surface a jagged appearance.

In several classic studies of heaving mechanics, Taber (1929, 1930) showed that heaving in a closed system is limited to that produced by the 9 percent volume increase caused when water freezes. In open systems, heaving is much greater because additional force is gained from crystal growth along the freezing plane. The amount of extra heaving depends on how much water can be drawn to the point of freezing from the surrounding material, and is a function of the segregation potential. The heave rate will be directly related to the temperature gradient at the freezing front (Nixon 1991). A mass balance approach presented by Nixon (1989, 1991) offers a method of predicting the thickness and location of ice lenses based on cooling rates, the physical character of the regolith, and projected suction potential. Forland et al. (1988) presented a theoretical treatment of frost heave, representing it as a process of thermal osmosis. In this model heat from the enthalpy of melting ice is transported by water to colder zones, which leads to increased pressure. The building pressure ultimately serves as a check to counter the migration of water and thus regulates the rate of frost heave. Taber (1930) also showed that heaving stress from crystal growth is exerted normal to the freezing isotherm and not in the direction of least resistance. Larger particles of varying compositions will have different thermal conductivities and thus may locally influence the orientation of the cooling surface. This process certainly introduces lateral movements in the expansion that are not perpendicular to the freezing isotherm. Some workers disagree with Taber's conclusion that differences in resistance to expansion do not influence the direction of the heave (Beskow 1947).

Smith and Patterson (1989) conducted an instrumented field study that showed the evolution of frost heaving as a freezing front advanced through the soil. Their study, using ring magnets as floating sensors of ground motion, showed how soil strain varies with both depth and time as temperature conditions change in the soil. These experiments illustrated that frost heave is a complex process that is heavily dependent on the thermodynamics of water, ice, temperature, and pore pressure, modified by the physical and hydraulic properties of the regolith such as texture, porosity, and permeability.

Frost Sorting One of the peculiar characteristics of regolith in permafrost regions is its propensity to sort sediment vertically and/or laterally (fig. 11.11). Any theory attempting to explain why large particles are moved surfaceward in a nonsorted soil mass must consider not only the forces that lift the clast but also the reasons it does not return to its original position during contrac-

tion. Details of two often-suggested theories concerning heaving mechanics, called frost-pull and frost-push, are discussed by Washburn (1973). Briefly, the frost-pull hypothesis suggests that stones are lifted vertically along with the fines when the ground expands during freezing. The fine sediment, being more cohesive, is brought downward quickly upon thawing and collapses around the larger clasts while the bases of the stones are still frozen. Cavities formed when a large particle is heaved (fig. 11.12) may also be partly closed by thrusting into this zone of lowered resistance.

The frost-push hypothesis is based on the fact that individual stones are better conductors of heat than porous soil. As a result, stones will cool more quickly, and the first ice to form along the freezing plane will be adjacent to and at the base of the stones, thereby pushing them upward. The phenomenon of ice forming preferentially near stones embedded in a silty soil has been observed and is not conjecture, nor is it conjecture that thawing occurs first around the stones. Presumably, however, material thawed adjacent to the top of the stone will collapse against the upper part of the clast while the base remains frozen and thus prevents its return to the original position. The shape of the stone may become important in this procedure; for instance, wedge-shaped particles with the narrow edge projecting downward will have greater difficulty returning to the original level.

Evidence exists to support the frost-pull and frost-push theories, and probably both processes function simultaneously in a heaving environment. Washburn (1973) believes, however, that rapid heaving or movement that breaks a vegetal cover probably requires that frost-push be dominant. Slow ejection of clasts can probably be accomplished by frost-pull with little necessity for rapid ice buildup beneath the stones. Mackay (1984) advanced our understanding of frost-sorting processes by suggesting that sorting was facilitated as a freezing front moved upward from the permafrost table.

A significant amount of detailed research has been conducted during the past decade that generated support for an earlier idea to explain frost sorting by a convectionlike circulation in the active layer driven by buoyancy differences developed during freezing (Hallet and Waddington 1992). This theory suggests that vertical soil motions operate in a diapiric fashion, driven by buoyancy forces that are developed seasonally in the freezing and thawing of ice-rich regolith in the active layer. Because thaw proceeds downward from the ground surface, there is progressively less time for regolith compaction in deeper portions of the active layer. This leads to higher soil density near the ground surface. This density stratification causes gravitational instability within the active layer during most of the thaw season and provides the conditions required for diapiric motion from below. Unlike free convection in fluids, soil

Figure 11.11
Sorting created in sporadic Alpine permafrost in conjunction with ice-wedge activity, Beartooth Plateau, Mont. The largest rocks have been selectively concentrated in distinctive rings and moved toward the ground surface.
(Photo by R. Craig Kochel)

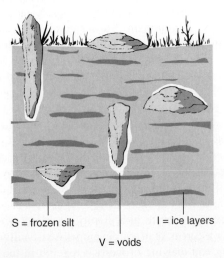

S = frozen silt   I = ice layers

V = voids

Figure 11.12
Displacement of stones during the freezing process.
(After Taber 1943)

convection would only operate intermittently during the period in which the soil is thawed (Hallet and Prestrud 1986). Field observations by Hallet and Prestrud in Spitzbergen showed that it is often the case that fine-grained regolith remains less dense and is forced upward throughout the summer thaw season, while buoyancy differences can be expected to be less in coarse-grained materials. Additional field support for the buoyancy-driven convection model was provided by Anderson (1988) in a study of frost-sorted landforms in Spitzbergen that showed that strain produced by frost heave decreases with depth. Figure 11.13 shows an example of the horizontal displacements measured across sorted circles by Hallet and Prestrud (1986) and illustrates convergence of material toward the coarse circle borders.

Frost Cracking   Frost cracking is the development of fractures at very low temperatures. The process functions best in permafrost regions, although it has been reported in other environments. Frost cracking is usually considered a frost action phenomenon, but Price (1972) points out that it is different because it results from thermal contraction rather than expansion associated with freezing.

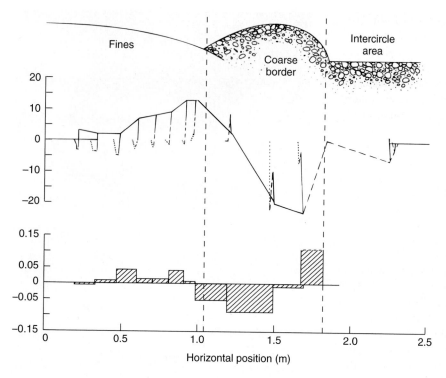

Figure 11.13
Radial displacements and strain measured during 1984 across active sorted circles in Spitzbergen. Inward displacements represent movement toward the centers of circles, and outward motion is toward circle rims.
(Hallet and Prestrud 1986)

At very low temperatures frozen ground often evolves into a polygonal network of contraction fractures, but cracking may depend more on the rate of cooling than on the absolute temperature value at the moment of fracture (Lachenbruch 1966; Black 1963, 1969). Cracking at the surface may extend to a depth of 3 m and can become progressively wider and deeper. The results of frost cracking will be discussed more in a later section when we examine polygonal ground features.

Nivation

Workers in alpine periglacial regions have observed the existence of concave depressions below semi-permanent or permanent snow fields and have attributed their existence to enhanced weathering associated with the presence of the snow; referring to the process as nivation (See review in Guilcher 1994). Considerable debate has arisen as to whether the hollows preceded the snowfield or developed as the result of the snow. Guilcher (1994) suggested that both mechanical and chemical weathering may be enhanced beneath snowfields. It has long been recognized that additions of meltwater from the snow would be expected to enhance frost weathering of bedrock if it is sufficiently jointed. Guilcher noted that because CO_2 solubility increases as temperature declines, the aggressiveness of carbonic acid should be enhanced under snowfields to assist nivation weathering processes. The CO_2 solubility at $0°$ C is three times its value at $30°$ C.

Pedogenesis in Permafrost Terrain

Frost has long been known to be a major pedogenic agent in periglacial environments (Washburn 1980). Cryoturbation is the dominant process involved in the formation of the soil order Gelisol. Chief in the recognition of cryoturbation are the irregular, discontinuous subhorizontal layers, chaotic folds or involutions, deformation of layers from ice-wedge expansion, diapers, silt cappings, ice and sand wedges, oriented stones, and preferential accumulation of organic matter at the permafrost table (Bockheim and Tarnocai 1998). Involutions develop downward in imperfectly drained soils in response to the rigid effect of the freezing surface layer during fall freeze-up (Van Vilet-Lanoe 1998a,b).

Care must be taken in paleoclimatic interpretations to differentiate cryoturbation morphology from that produced by other processes that commonly disturb soil profiles, such as seismicity, tree throw, and mass wasting. Bockheim and Tarnocai (1998a,b) provide an extensive discussion of criteria useful in recognizing cryogenic effects in soils. Mass wasting typically results in the development of slope-parallel fabric (i.e., creep) and stonelines. Tree throw by wind normally produces pits and mounds with fragmented but well-developed soil horizons. Seismic disturbances typically display offset along fault planes or expulsion features associated with liquifaction and sand boils.

Van Vilet-Lanoe (1998a) provides a detailed treatment of frost as a pedogenic agent in cryogenic soils. Based on microscopic and macroscopic features, cryogenic soils can be readily distinguished from those formed in temperate environments. Van Vilet-Lanoe (1998b) discusses the peculiar hydrology and physics of soils in permafrost areas related to the evolution of frost boils and hummocks. Swanson (1996) showed that cryogenic soil catenas can be recognized, showing systematic variation in soil morphology with a gradient from warm-dry mineral soils to cold-wet organic soils, depending on slope position and soil texture. In general, soils on concave slopes tend to retain moisture longer, while those on convex slopes tend to be drier. Likewise, the influence of texture is strong, with fine-grained soils tending to hold moisture better (Swanson 1996).

Mass Movements

Any variety of mass movement (discussed in chapter 4) can occur in a periglacial environment, but frost creep and solifluction dominate the cold regime. The movement of debris occurs simultaneously with the frost action processes, and it is questionable how effective the erosion would be if it operated alone. Mass movements, however, are responsible for a variety of landforms that are unique to periglacial regions and often result from the interaction of ground ice with these phenomena.

Frost Creep **Frost creep,** like any form of near-surface creep, is the downslope movement of particles in response to expansion and contraction and under the influence of gravity. It is unique only because freezing and thawing generate cycles of expanding and contracting. The freezing front in the soil usually parallels the ground surface. Individual soil particles are heaved perpendicular to the surface by needle-ice (fig. 11.14A), but during thawing they are affected only by gravity and should contract in a vertical direction (see fig. 4.30). Ballentyne (1996) showed that the limiting conditions for development of needle-ice include: (1) temperatures below −2°C at the surface; (2) relatively high soil moisture; and (3) soil texture conducive to rapid migration and freezing of water, typically achieved in soils rich in silt and clay. Actually, a component of upslope migration has been observed in the contracting phase (Washburn 1967; Benedict 1970). This motion, called retrograde movement, probably stems from the attraction of fine-grained particles and the cohesive strength developed between them. On the other hand, some downslope flow may occur during the thaw, and so the overall route followed by any soil particle involved in frost creep probably resembles the path shown in figure 11.14B.

Detailed studies in northeast Greenland (Washburn 1967) showed frost creep to be the dominant process where soils are silty and slopes are between 10° and 14°. A recent attempt to investigate the influence of slope angle on rates of frost-creep indicated that surface velocity is proportional to the second power of the slope gradi-

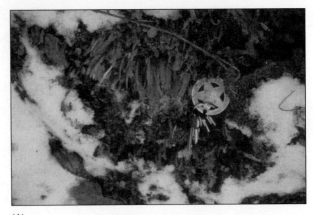

(A)

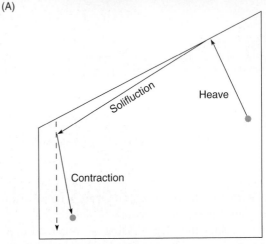

(B)

Figure 11.14

(A) Needle-ice during expansion phase, lifting particles perpendicular to the hillslope. (B) Route of downslope movement followed by a particle under combined frost creep and solifluction.

(Photo by R. Craig Kochel)

ent (Matsuoka 1998). Steeper slopes are also dominated by rolling of debris loosened by downslope-bending ice needles. Washburn (1967) noted, however, that in any given year other processes could dominate the system. Therefore, it may not be safe to generalize about the conditions that promote frost creep because soil texture, moisture content, vegetation, and freeze-thaw frequency all have some bearing on the efficacy of the process.

Solifluction (Gelifluction) The term **solifluction** was first proposed to describe the action of slow flow in saturated soils. The original meaning carried with it no climatic restrictions, but over the years the term has generally been used to connote a process that functions in periglacial regions. We will follow this practice here, although it is technically incorrect because solifluction can occur anywhere. The term **gelifluction** (Baulig 1957) refers to soil flow associated with frozen ground and as such is a specific type of solifluction.

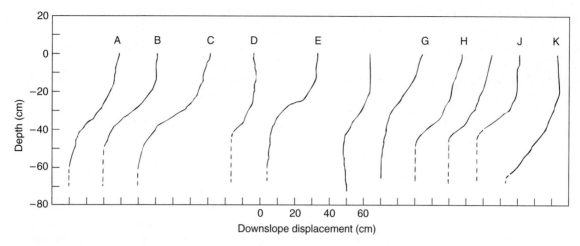

Figure 11.15
Displacement profiles representing differential velocities with depth in active solifluction on Yukon Territory slopes between 1967 and 1988. In contrast to creep, in which peak velocity occurs in the uppermost layers, these profiles show greater differential displacement occurring at depth.
(Price 1991)

Solifluction is most dramatic in permafrost zones because water released in the active layer during thawing cannot penetrate below the permafrost table. Rate studies demonstrate that solifluction is most active in late spring and early summer when soils are usually saturated; because saturated soils with increased pore pressure behave like viscous fluids, there is a loss of friction and cohesion. In some periglacial environments, solifluction is more geomorphically effective than the combined work of rockfall and snow avalanche processes (Smith 1992). Bennett and French (1991) showed that solifluction and permafrost creep may operate on the same regolith mass in a time- and process-independent manner. Permafrost creep rates progressively decrease with depth and experience plastic deformation peaks in the late summer. Another long-term study (Price 1991) summarized 20 years of observations and found distinctive mechanical differences with depth. Whereas the greatest total downslope movement occurred near the ground surface (in the upper 20 cm), the greatest differential movement occurred in a layer 20 to 50 cm below. Depths below 50 cm near the base of the active layer showed very little movement (fig. 11.15). These observations underscore the distinction between two separate mechanisms involved in solifluction (Price 1991): (1) viscous flow during periods when high pore water pressures and saturated conditions develop, and (2) shearing along planes of low strength localized along segregated ice layers that become partially thawed. Solifluction flow can occur on slopes as gentle as 1°, but maximum displacement seems to occur where the gradient is between 5° and 20°. Soils on slopes steeper than 20° tend to drain easily, and water escapes as surface runoff.

Detailed fabric analyses by Matthews et al. (1986) show macrofabric patterns exhibiting a strong orientation of clasts parallel to slope and some radial patterns diverging toward the margins of solifluction lobes in plan view. Some studies report a shift from downslope to cross-slope clast orientations toward the distal parts of solifluction deposits, probably in response to compressive stresses (Wahrhaftig and Cox 1959; Potter and Moss 1968).

Abundant moisture is the overriding factor in the solifluction process (Washburn 1967; Chambers 1970; Price 1972). Many students of periglacial geomorphology believe that significant flow can occur only when the moisture content equals or exceeds the liquid limit, but this interpretation cannot be accepted unequivocally because solifluction has been observed at lower moisture contents (Fitze 1971). Obviously, other factors such as grain size, slope angle, and vegetation play a modifying role on a local basis (Harris, 1973). For example, gravel and coarse sands are so permeable and easily drained that they virtually never flow. Clays, on the other hand, are extremely cohesive. Thus, silty soils, especially those with a bimodal size distribution, are most susceptible to solifluction. Nonetheless, the overwhelming importance of water content was clearly demonstrated in Greenland (Washburn 1967), where the highest rates of movement commonly occurred on the most densely vegetated areas of the slopes and at gradients that are lower than the dry portions of the slopes. Thus, moisture can overcome both the binding effect of vegetation and the apparent deficiency of a transporting gradient.

The rates of combined frost creep and gelifluction in northeastern Greenland and other regions are shown in table 11.2. It is difficult to ascertain how much of the total movement is accomplished by frost creep and how much can be attributed solely to flow. A 10-year study in the Canadian Rockies (Smith 1992) found average solifluction rates of 0.47 cm/yr between 1980 and 1990.

TABLE 11.2	Representative Values of Combined Frost Creep and Gelifluction.	
Rate (cm/yr)	Slope	Source
2.0	15°	Rapp 1960
0.9–3.7	10°–14°	Washburn 1967
1.0–3.0	3°–4°	Jahn 1960
5.0–12.0	7°–15°	Jahn 1960
10.0	20°	P.J. Williams 1966[a]
2.5	6°–13°	Benedict 1970
2.6	2.5°–4°	French 1974

Note: List is a random sampling and is not complete. Great variability exists depending on differences in ground moisture, vegetation, depth, slope face direction, grain size, soil-texture, and so on.
[a]Pebbles on bare slope surface.

The interesting aspect of this study came from a comparison of the first three years of data with the entire period. The 1980–83 record showed a bias of 166 percent over the long-term average, causing Smith (1992) to suggest that there is a 3- to 4-year stabilization period associated with the installation of inclinometer tubes for the system to return to equilibrium.

Recent work investigating the sedimentology and stratigraphy of solifluction sediments provides good evidence of accelerated gelifluction activity within the Holocene (fig. 11.16). These periods of increased activity have sometimes been associated with minor climatic variations and have been calibrated by radiocarbon dates from buried paleosols incorporated in the solifluction deposits. Active Holocene solifluction intervals, excluding the present, include 300 to 600 B.P., 900 to 1700 B.P., and 2300 to 3000 B.P. (Matthews et al. 1986; Morin and Payette 1988). Marron et al. (1995) present a detailed study of the dynamics of a Holocene debris slope in subarctic Canada. They were able to make strong connections between slope activity by solifluction, frost-creep, rockfall, and external forcing factors such as fire and climate change.

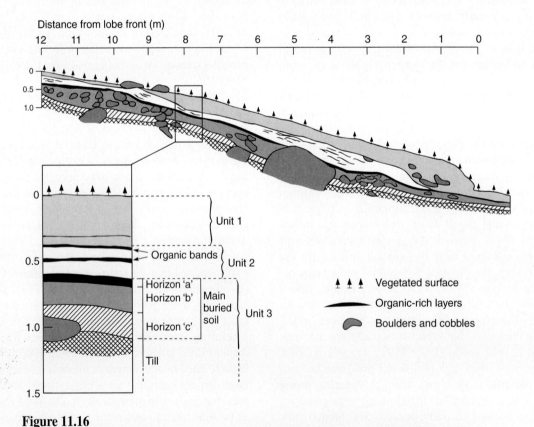

Figure 11.16
Example of multilayer stratigraphy common in gelifluction lobes. This Norwegian example from Matthews et al. (1986) also shows textural variations and buried organic soil horizons that have been radiocarbon-dated. Their work also contains detailed fabric analyses that show how gelifluction processes modified the preexisting underlying colluvium.
(Matthews et al. 1986)

PERIGLACIAL LANDFORMS

Landforms Associated with Permafrost

Some periglacial features are so closely allied to the distribution of permafrost that a genetic relationship can hardly be denied. Observation of these features preserved in temperate regions affords the best evidence known for reconstructing the areal extent of ancient permafrost boundaries. The basic processes involved in developing these features are no different from those that form other periglacial landforms; that is, they rely on frost action and/or mass movement. The permafrost layer, however, places a distinctive imprint on the system and allows us to consider the geomorphic features as a separate group.

Confusion over terminology applied both to active and relict features is especially common for landforms considered to be of periglacial origin. To discuss features without using some form of terminology is difficult. Therefore, we will use names for features simply to facilitate the discussion, but readers should be advised that they are by no means accepted as standards by all students of periglacial processes and landforms. Disagreements over the specifics of individual terms will be left for those wishing to pursue this field in greater detail.

Ice Wedges and Ice-Wedge Polygons Frost cracking begins when thermally derived stresses generated upon cooling exceed the strength of the surface materials (Mackay 1986). These fractures penetrate into the active layer and the upper portion of the permafrost. At first the cracks are only millimeters wide, but they may extend vertically downward to depths of several meters. The actual moment of cracking seems to occur sometime between January and March when temperatures in the ground reach the annual minima (Mackay 1974, 1992) and the rate of temperature drop may be extreme. During spring and early summer the crack may be partially or totally filled with snow that has filtered down from the surface and with water released by the onset of thawing in the active layer. As figure 11.17 shows, this mixture freezes below the permafrost table and occupies the initial crack as an ice veinlet. Because the initial fracture now represents a zone of weakness within the permafrost, subsequent cold winters produce cracking along the trace of the original opening, and new ice is added as before. Over the years, this accumulation of relatively pure and vertically oriented ice takes the form of a downward-tapering wedge, called an **ice wedge** (see also fig. 11.1). In dry permafrost, the material from the contraction fissures may be sediment, especially wind-transported sand or loess associated with extreme aridity (e.g., see Carter 1983). The resulting feature is commonly called a sand wedge (Péwé 1959).

Ice wedges range from 1 cm to 3 m wide and from 1 to 10 m deep, with the depth to top width ratio normally ranging from 3:1 to 6:1. They are best developed in fine-grained soils with a high ice content (Black 1976). The host sediments are usually upturned within 3 m of the wedge, in response to the horizontal compression generated as the wedge grows. Ice in wedges contains oriented air bubbles and a complicated fabric of ice crystals (Black 1974).

The growth of an ice wedge does not necessarily proceed on an annual basis. Mackay (1974, 1975a) observed that only 40 percent of the ice wedges on Garry Island (Northwest Territories, Canada) cracked in any given year, the frequency of cracking apparently being controlled by the depth of snow cover. Even if cracking does occur, the openings often narrow before ice-veinlet growth begins. The size of a veinlet thus is dependent on the timing between closure of the winter cracks and the appearance of meltwater (Mackay 1975a). In most cases the thickness of an annual increment is considerably less than the size of the thermal crack.

Ice wedges are commonly connected in a polygonal pattern that is similar in most respects to the more familiar mud cracks. **Ice-wedge polygons,** however, are much larger, having diameters that range from several

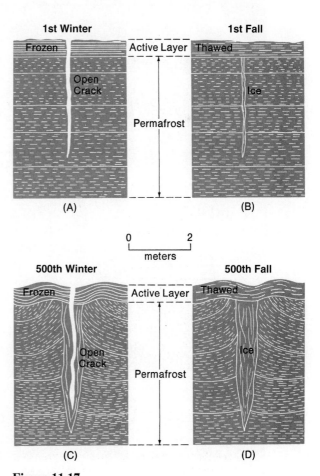

Figure 11.17

Evolution of an ice wedge by thermal contraction.
(Lachenbruch 1966)

meters to more than 100 meters (fig. 11.18). They contract and expand according to the laws that govern temperature effects on ice (Black 1974). The perimeters of some polygons are accentuated as minor ridges, so that the boundary is elevated relative to the center of the feature (these are called low-center ice-wedge polygons). Others (high-center polygons) are depressed along their edges by melting erosion of the ice wedges, leaving the central core higher than the rim (Péwé 1966; Black 1974, 1976).

Actively growing ice wedges are restricted to areas of continuous permafrost; the southern boundary seems to be the –6° to –8° C mean annual isotherms (Péwé 1973, Wayne 1991). Farther south in the discontinuous permafrost zone, they become inactive and disappear. As ice wedges melt, the surrounding sediment may collapse into the space opened by the disappearance of the ice, thereby creating an ice-wedge cast, a feature that is best preserved in gravels (Black 1976). Although ice-wedge casts are probably the best proof of an earlier permafrost condition, other wedgelike features unrelated to permafrost look very similar. For example, features called **soil wedges** are commonly reported in the Russian literature. They apparently originate by frost cracking in the active layer and infilling with local sediment. Jetchick and Allard (1990) provide evidence from the coincidence of fire horizons and periods of soil-wedge initiation in Quebec that suggests the soil wedges originated from the disruption of vegetation by fire, which led to thermal cracking of the ground. It is easy to misinterpret these features as ice-wedge casts, which would incorrectly suggest that a permafrost environment existed in the past and was subsequently destroyed. An old ice-wedge cast can become part of a modern soil wedge if frost cracking operated in the present active layer. To avoid confusion, Washburn (1980) suggests that any wedge or wedge-cast type should be designated as being either fossil or modern.

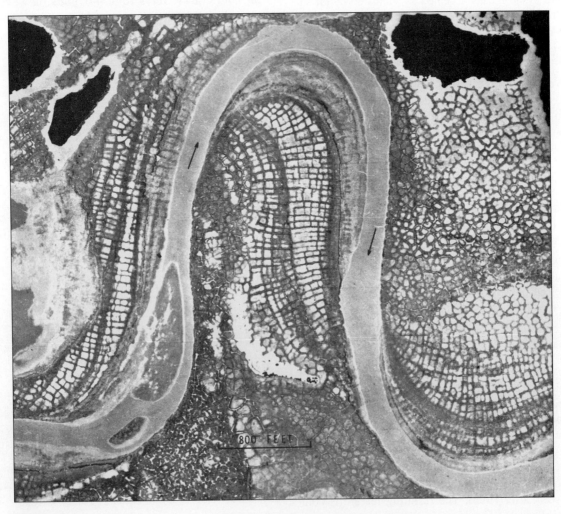

Figure 11.18
Polygonal markings on ground surface caused by ice-wedge polygons in the vicinity of Meade River, about 35 miles southeast of Barrow, Alaska. Barrow district, northern Alaska region, July 1949.

Pingos   A variety of landforms peculiar to the dynamics of near-surface hydrology in permafrost has been studied in periglacial regions, including string bogs or mires, (Koutaniemi 1999) frost blisters, palsas, and pingos. The term **pingo** (Inuit for "mound" or "conical hill") was first used by Porslid in 1938 for large, ice-cored, domelike features that exist only in permafrost regions. They are most perfectly developed in areas of continuous permafrost, although they do form in the discontinuous zone (Holmes et al. 1968). Pingos are common landforms in some periglacial areas. For example, more than 1000 pingos occur along the northern coastal plain of Alaska, about 70 occur in the Brooks Range, and over 300 occur in discontinuous permafrost areas of central Alaska (Ferrians 1988). Active pingos rise from a few meters up to heights of 70 m and have basal diameters from 30 to 1000 m (Washburn 1980; Ferrians 1988). They are oval or circular in plan view, as in the photograph in figure 11.19. The ice core in a pingo is typically massive and is often exposed by tension fractures that develop at the summit of the mound as it rises. Ice up to 7 m thick has been reported from pingos in Sweden (Lagerback and Rodhe 1986). Sometimes the exposed ice melts, and small freshwater lakes occupy the craters bounded by the tensional cracks.

Many varieties of pingos are known (Pissart 1970; Flemal 1976) but most can be placed in two major genetic categories. Closed-system pingos develop in level, poorly drained, shallow lake basins. They are most distinctly preserved in the Mackenzie delta (Canada) where a clutch of 1450 pingos gives evidence that pingos tend to form in fields rather than as single individuals. In general, the draining of a lake allows the permafrost table to rise to the level of the former lake floor. As the table rises, water is trapped in the saturated soil. The cryostatic pressure (pressure from ice formation) displaces the water upward until it also freezes and becomes the core of the pingo (see Flemal 1976). The size and shape of the pingo often reflect that of the original residual pond. The pingo may also grow in distinct stages (Mackay 1973). Mackay (1990) has been able to distinguish seasonal growth bands in pingo ice. Mackay (1998) summarizes some 25 years of research on the closed-system pingos in the Mackenzie delta. He concludes that most involve subpingo water lenses where pore ice grows by downward freezing. Mackay (1998) provides a good discussion of the mechanics of pingo growth and morphologic criteria useful in recognizing closed-system pingos. He noted that distinctive morphology includes (1) peripheral ruptures and normal faults; (2) frost mounds; (3) spring flow; and (4) oscillations in height. All of these are elements associated with hydrofracturing.

Open-system pingos develop most readily on slopes where free water under artesian pressure is injected into the site of the pingo. As the water approaches the surface, it freezes. The continuing introduction of water

Figure 11.19
The Discovery Pingo. A dense stand of large birch trees on the pingo contrasts sharply with the open stand of small black spruce in the surrounding muskeg. Goodpaster district, Yukon region, Alaska, Aug. 1960.

Mechanism and References	Simplified Graphic Representation	Fabric and Texture	Ion Inclusion Pattern	Miscellaneous, Structure, Stratigraphy
Ice segregation Lundqvist, G., 1951 French 1976 Seppala 1982, 1986	Summer 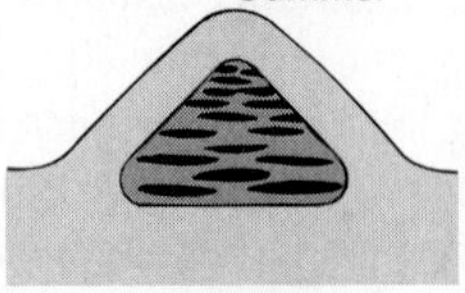	Small crystals; *c*-axes random to preferred normal often with mineral inclusions. Texture diverse, usually granular, subhedral to anhedral crystals.	Very complex, due to concentration of solutes and unidirectional flow of water.	Thin ice lenses parallel to ground surface, within peat or mineral soil; lenses tend to thicken with depth.
Buoyancy Zoltai 1972; Outcalt and Nelson 1984	Summer 	Pore ice and/or segregation ice lenses.	As in ice segregation mechanism.	Low elevation floating peat bodies; water in bore holes.
Hydraulic van Everdingen 1978, 1982 Pollard and French 1983, 1984, 1985	Winter (1st) 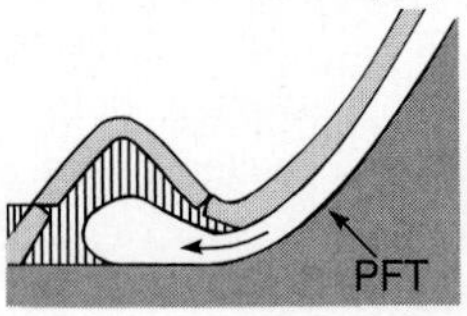 PFT	Upper chill zone; bubbles along crystal boundaries; *c*-axes preferred normal, crystals elongated parallel to freezing direction; subhedral to anhedral.	Relatively high in chill zone; increasing coreward until rupture and flushing introduce new source of water.	Varied ice stratigraphy; probable structural deformation of ice; restricted spatially to areas of high relief, associated with perennial springs.
Hydrostatic Akerman 1982 Brown et al. 1983 Hinkel et al. 1987 Hinkel 1988	Winter (1st) 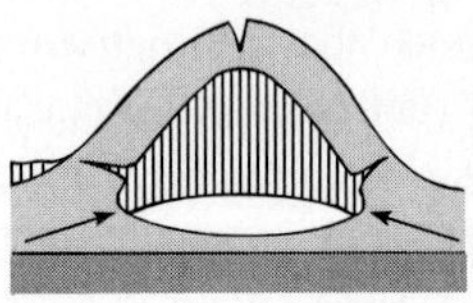	Radially inward toward unfrozen core; possible chill zone and/or unfrozen chambers; bubble density increases inward. Vertical prismatic texture; anhedral crystals; *c*-axes preferred normal.	Initially decreasing, then increasing toward center as melt progressively enriched.	Condordant injection of water, often evidence of overburden deformation; unconformities common.
◯ Water Seasonally frozen ⬤ Permafrost ⬚ Ice				

Figure 11.20
Summary of mechanics, morphology, ice character, and structure of palsas.
(Nelson et al. 1992)

from below under hydrostatic pressure builds up the ice mass and domes the surface material into the pingo shape (see Muller 1963). The initial growth of some pingos may involve ice lenses above the water table, but their continued growth requires a proper combination of hydrology and soil texture (Ryckborst 1975). Limits appear to exist on the artesian pressures conducive to the formation of open-system pingos. For example, Yoshikawa (1998) showed that sites with spring discharges above 3l/s did not have pingos. Open-system pingos are much more common in discontinuous permafrost where taliks and freely circulating water are not unusual.

A variant of the pingo is the palsa. **Palsas** are generally regarded as small-scale versions of pingos, usually less than 10 m in height, with peat-rich interiors. Their evolution remains somewhat sketchy, but research has identified four major processes important in the mechanics of palsa growth: ice segregation, buoyancy, hydraulic pressure (for open systems), and hydrostatic pressure (Nelson et al. 1992). Figure 11.20 illustrates the main aspects and varieties of palsa growth. Harris (1998) provides a genetic classification of palsalike mounds on

western Canada, which includes five major groups: (1) floating palsas developed in peat; (2) minerogenic palsas formed in thin peats, extending into the mineral substrate; (3) lithalsas, developed totally in the substrate; (4) floating peat plateaus; and (5) anchored peat plateaus.

The rate of pingo growth is variable, ranging from a few centimeters to more than a meter a year (Washburn 1980; Mackay 1973, 1990). This indicates that many large pingos are quite old. In fact, carbon-14 dates have shown two closed-system pingos to be 4000 and 7000 to 10,000 years old (Muller 1962). On the other hand, some pingos are growing today (Mackay 1973), suggesting that pingo formation did not occur in a unique or specific time in postglacial history but is probably a continuing process.

Fossil pingos (Wayne 1967; Mullenders and Gullentops 1969; Flemal et al. 1976; Marsh 1987) provide good evidence of an extinct permafrost environment in regions that are now temperate. Proving that features are fossil pingos, however, requires detailed field and laboratory analyses. The DeKalb mounds (northern Illinois), for example, consist of approximately 500 circular or elliptical mounds shaped from late Pleistocene sediments. Good evidence of a pingo origin for these mounds is the large cluster of features with centers composed of lacustrine silts and clays. Topographically, however, most of the mounds are less than 5 m higher than the surrounding ground level, and many are up to 300 m in diameter. Careful field examination was necessary to recognize these features as fossil pingos. Worsley and Gurney (1995) caution on using hollows as evidence of former pingos because morphology alone remains ambiguous

Pingos and palsas may prove to be very useful in paleoecological reconstructions and in efforts to monitor the effects of global environmental changes. The dynamics of the aggradation and degradation appear to be very sensitive to changes in the thermal character of the regolith as regulated by climate or vegetation (Nelson et al. 1992).

Thermokarst   Thermokarst features are a variety of topographic depressions that result from thawing of ground ice. These features are similar in many respects to those found in normal karst regions, but the fundamental process is melting rather than solution of rock material (fig. 11.21). They are more abundant where ice wedges are present and where the thermal equilibrium is somehow disturbed. The destruction of ground ice results from either lateral degradation (back-wearing), where lateral erosion of surface water exposes the ice, or vertical degradation (down-wearing), in which the surface thermal properties are altered (discussed in Czudek and Demek 1970).

A thermokarst lake is a water body that occupies a closed depression, known as an **alas,** created by subsidence following the decay of ground ice. Once initiated, thermokarst lakes grow by positive feedback using the thermal energy of the water to perpetuate the expansion of the lake along its margins (Burn 1992). Extensive bank erosion is common along thermokarst lakes and along rivers attributed to the process of thermal erosion. The combination of thermal energy, together with normal mechanical and hydraulic erosional processes can yield uncommonly high rates of bank and shoreline erosion along streams and shorelines (e.g., Riddle et al. 1988; Walker 1988; Burn 1992). Carter (1987) provides a comprehensive collection of shoreline recession rates for arctic coastal regions influenced by thermal erosion processes.

Using accelerated episodes of thermokarst lake formation as an indicator of global warming is difficult because some climate models indicate that increased CO_2 would result in increased high-latitude snowfall (Hansen et al. 1984). More snow cover could further insulate permafrost and actually impede thermokarst expansion. In an extensive survey of thermokarst lake history in the Yukon Territory, in northwestern Canada, Burn and Smith (1990) discovered distinct periods of accelerated thermokarst during the Holocene, centered on 8500,

3900, and 2300 years B.P., but they could find no correlation with climate. They concluded that although climatic warming can increase the likelihood of thermokarst development (by bringing the regolith close to the threshold of formation to set the stage for the process), the specific location of thermokarst lakes is determined by other factors such as fires and local disturbances in ground conditions. These might include the clearing of forest vegetation for farming or the shifting of stream channels, which could breach an insulating cover of peat or vegetation and upset the thermal balance. Mackay (1970), for example, reports a remarkable case in which trampling of vegetation by a dog led to subsidence of 18 to 23 cm in a small area within two years. Observations such as this demonstrate the distinctly fragile equilibrium of the periglacial system.

Patterned Ground

Periglacial regions are often characterized by a peculiar arrangement of surface materials into distinct geometric shapes. The features, collectively known as **patterned ground,** include such diverse shapes as polygons, circles, and stripes. They are common in permafrost regions, but perennially frozen ground is not a prerequisite for development. They can be found anywhere within the periglacial regime or even in other morphogenetic regimes. Since patterned ground was first described in the late 1800s, almost every worker in periglacial areas has noted its striking appearance, and theories concerning its origin are almost as numerous as the features themselves. Perhaps the most extensive review of patterned ground and its origin is provided by A. L. Washburn (1956, 1980) whose excellent work provides the foundation for our knowledge about these forms.

Classification   According to Washburn (1956), patterned ground features can be placed in a descriptive classification on the basis of two primary criteria: (1) its geometric shape (e.g., circle, polygon), and (2) whether the material composing the feature has been sorted (table 11.3). Sorting separates the larger-sized fractions from the fine particles and usually generates a feature rimmed by stones and centered with fines (for a review see Goldthwait 1976). The most common forms seem to be circles, polygons, and stripes, with steps and nets as minor transitional types. For a more detailed discussion see Washburn (1980).

The areal morphology of patterned ground landforms is affected significantly by slope aspect (which regulates microclimatic parameters such as moisture and temperature) and slope angle (fig. 11.22). Odegard et al. (1988) discussed the effect of these parameters on the morphology of patterned ground. They found that patterned ground landforms were more common on east-facing slopes of less than 10°.

Figure 11.21 Thermokarst.
(A) Experimental thermokarst depressions formed in ice-cemented sand undergoing thaw. (B) Schematic showing cross-section of thermokarst degradation along the boundaries of ice wedges forming a depression known as an alas.

(B): (French 1976, modified from Soloviev 1962, 1973).

(Photo by R. Craig Kochel)

(A)

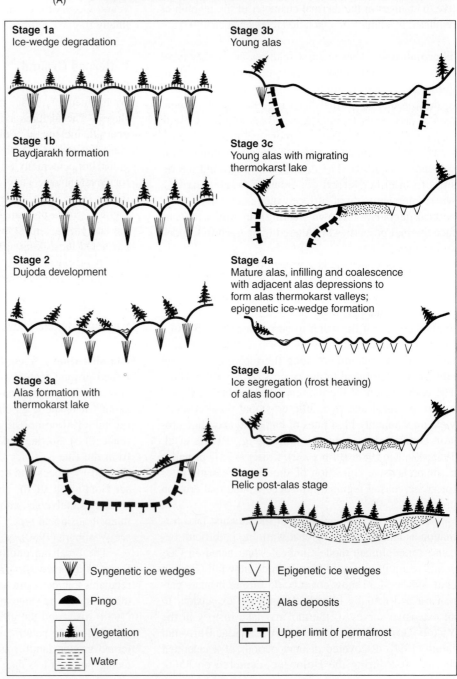

(B)

TABLE 11.3	Simplified Classification of Patterned Ground Features.	

Types	Subtypes	Processes
Circles	Sorted Nonsorted	Features subdivided on basis of necessity of cracking (frost cracking, permafrost cracking, dilation, joint control). Where cracking is not essential, heave and mass displacement is important.
Polygons	Sorted Nonsorted	Includes ice and sand wedges. Cracking of all types. Heave, mass displacement, and thaw processes are important in noncracked types.
Nets	Sorted Nonsorted	Includes earth hummocks. Cracking of all types except joint controlled types. Heave and mass displacement in noncracked types. Thaw also important in sorted varieties.
Steps	Sorted Nonsorted	Cracking unimportant. Heave, mass wasting, and displacement in nonsorted. Nonsorted frost sorting and thaw also important in sorted varieties. Terracette form.
Stripes	Sorted	All types of cracking important. Heaving, mass wasting, and displacement in nonsorted. Frost sorting and thaw also important in sorted types.

From Washburn 1970, in *Acta Geographica Lodziensia.* Used with permission of Institute of Geography, Poland.

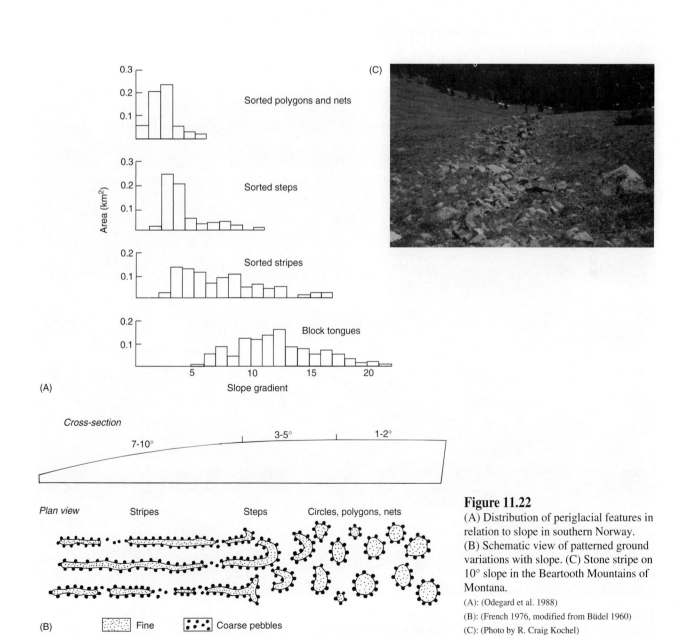

Figure 11.22

(A) Distribution of periglacial features in relation to slope in southern Norway. (B) Schematic view of patterned ground variations with slope. (C) Stone stripe on 10° slope in the Beartooth Mountains of Montana.

(A): (Odegard et al. 1988)

(B): (French 1976, modified from Büdel 1960)

(C): (Photo by R. Craig Kochel)

Polygons Polygons are best developed on flat, nearly horizontal, surfaces. The sorted variety is bounded by straight segments composed of stones that surround a central core of finer material. They range in diameter from a few centimeters to over 10 m and always occur in groups rather than individually. The stones marking the boundary are often oriented parallel to the direction of the rim. The stones increase in size with the dimension of the polygon and decrease in size with depth.

Nonsorted polygons differ in several ways from the sorted type: (1) their polygonal borders are devoid of stones, the geometric form being marked by furrows; (2) they can be considerably larger than the sorted variety, often reaching 100 m in diameter; and (3) they have been found on slopes as steep as 31°. Nonsorted polygons are usually associated with ice-wedge polygons (discussed earlier). As such, they are primarily related to permafrost. However, they can also form by desiccation, even in solid rock (Walters 1978).

Circles Sorted circles are stone-rimmed circular forms that range from a few centimeters to several meters in diameter (fig. 11.23). As with polygons, the stone size correlates with the dimension of the circle and decreases with depth. Circles, unlike polygons, can exist singly or in groups. Nonsorted circles lack the coarse, bouldery rim and are bounded instead by vegetation surrounding a central core of bare soil. In some cases the barren core develops because scouring and winnowing action by the wind preferentially erodes and initiates the circular form (Fahey 1975).

Stripes Stripes are linear alignments of stones, vegetation, or soil on slopes. Sorted stripes are elongated strips of stones separated by intervening zones of fine sediment or vegetation. Stripe fabric often occurs where the long axes of clasts are oriented parallel to the slope (Perez 1992). Stripes are common on steeper slopes and appear to have a lower limit of 3° (Odegard et al. 1988). This does not mean that stripes never form on gentle

Figure 11.23
Sorted stone circles caused by frost action, in Alaska Range, south-central Alaska, June 1968.

slopes; they do (Evans 1976). Such occurrences are rare, however, and as gradients increase, stripes are the most common form of patterned ground. Transitional forms of nets or steps often develop at angles between 2° and 7°. The precise angle at which stripes develop depends on other factors such as soil texture and moisture and the formation of needle ice (Mackay and Matthews 1974; Washburn 1980). Stripes usually range in width from several centimeters to several meters. The length varies but is sometimes greater than 100 m. As in other sorted features, larger stones are associated with bigger stripes, and the largest particles are found at the surface of the accumulation. Nonsorted stripes consist of vegetation or soil with intervening zones of bare ground. They are usually not as long as the sorted variety and are sometimes discontinuous.

Origin Patterned ground probably originates in a variety of ways. Washburn (1956) discusses 19 major hypotheses and concludes that patterned ground is polygenetic and that some forms result from a special combination of processes. Despite the conflicting opinions, the properties of the major features do imply that patterned ground exists in a range of transitional types. The characteristics of the different types reflect variations in a few controlling factors rather than completely different origins. Although details are still missing, it does seem certain that several processes are basic to the formation of patterned ground: (1) cracking or desiccation, (2) heaving and its associated sorting, and (3) gravity as expressed by the slope inclination. Cracking is instrumental in developing polygonal geometry and is primary to sorted and non-sorted types.

Sorting of sediments in patterned ground is achieved by frost heaving, frost-push, frost-pull, or convective activity driven by buoyancy differences. Van Vilet-Lanoe (1991) first suggested differential frost heave as a unifying process to explain the development of patterned ground. Kling (1997) believes that this process will work as long as sorted circles are in a phase

of thermal degradation and suggests that circulation results from the process of loading when higher clay content lower in the active layer creates a negative frost-susceptibility gradient. These frost-sorting processes commonly create a fine-grained nucleus. Eventually, adjacent sorted cells coalesce, and the coarse stones become concentrated as polygonal or circular rims along the lines of interference.

Many of the recent studies are converging on a convectionlike model driven by the buoyancy differences resulting from differential compaction during vertically progressive annual thaw within the active layer (Ray et al. 1983a, 1983b; Hallet and Prestrud 1986; Hallet et al. 1988). Density gradients established during uneven compaction may appear to be great enough to drive convection and transport clasts. Bulk density tends to decrease with depth as moisture gradients develop during thaw and the consolidation of ice-rich soils (Hallet and Prestrud 1986). Figure 11.24 illustrates how this convective motion can produce a concentration of fine-grained sediments in the centers of coarse rims and a geometric pattern of regular spacing. These models do not suggest that clasts are physically moved by convecting water; instead, circulating water influences the shape of the underlying ice front. This, in turn, determines where sorting mechanisms will work most effectively until the surface pattern is a mirror image of the undulating ice front. In a three-year study of horizontal soil displacements in instrumented sites, Spitzbergen monitored tilt and stress fields in the soil of active sorted circles. This study documented lateral divergence of markers from fine-grained centers and convergence across the coarse boundaries (Hallet et al. 1988). Hallet (1998) presents a model of soil circulation in fines with central upwelling. A similar model is presented by Harris (1998a, b) from the results of an instrumented site in Canada. He noted that ice-segregation in fine centers appears to be the primary force driving circulation.

The rate of lateral movement of large clasts is not well documented. Vitek (1983) suggests that particles

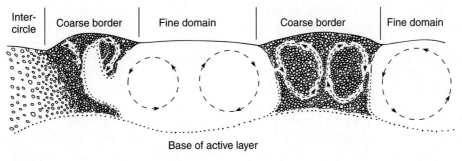

Figure 11.24
Schematic model for buoyancy-driven convection within the active layer shown with respect to sorted circles.
(Hallet et al. 1988)

larger than 1.3 cm in length are transported to the polygonal margins at a rate ranging from 0.45 to 0.89 cm/yr. Hallet et al. (1988) note that convective activity appears to be intermittent (operating on time cycles of hundreds of years) but can result in displacement rates up to 10 mm per year. Larger particles move at a slower rate, probably by frost action and the growth of needle ice and/or lacustrine ice.

Most studies of patterned ground agree that the mean annual temperature needs to be at least as low as –6° to –8° C for it to be well developed. Stripes and nets apparently can develop in the absence of cracking as shown in a study of small-scale patterned ground features in Ireland (Wilson 1992). The presence of such relict patterned ground landforms has generally been taken as an indicator of former periglacial climatic conditions (e.g., Clark 1968; Clark and Ciolkosz 1988; Johnson 1990; Clark and Hedges 1992).

Landforms Associated with Mass Movement

In contrast to patterned ground features, which are shaped primarily by frost action, certain periglacial forms result when the dominant mechanism is transportation downslope. Frost action is a necessary prelude to the mass movement, and it is likely that the two processes function simultaneously to facilitate motion and fabricate the ultimate landform. Nonetheless, the final geometry of these features reflects the motion of rock and soil particles rather than a static, in-situ disintegration of the parent material (for reviews see Benedict 1976 and S. E. White 1976b).

Stratified Slope Deposits　Deposits of stratified platy, subangular shale chips are common and locally extensive on hillslopes throughout temperate climates in the Appalachians. An extensive body of literature on the deposits exists, including field and laboratory experimental studies (French 1976; de Wolf 1988; Bertran et al. 1994). Washburn (1980) and Clark (1988) suggest diverse origins for these deposits, typically involving Pleistocene periglacial conditions. Most studies agree that the formation of stratified slope deposits requires a continuous supply of frost-shattered sediments moving along valley slopes by sheetwash or a variety of mass movement processes. A cogent review of process and sedimentary facies is presented in Van Steijn et al. (1995). Deposition of the sediments appears to be episodic or cyclical, and is perhaps related to climatic episodes when freeze-thaw activity produces an abundant source of gelifracts (Gardner et al. 1991). Fabric and textural studies of a deposit in central Pennsylvania indicate that deposition resulted from high-viscosity debris flows and low-viscosity sheetwash (Gardner et al. 1991). The presence of a periglacial climate is supported by the occurrence of frost cracks, soil wedges, and ice-wedge casts. Other studies of stratified slope deposits (sometimes referred to as gréze litées) point toward solifluction processes as a major influence (Ozouf et al. 1995). This conclusion is supported by observations of the following characteristics: (1) alternation of matrix and open fabric, (2) sharp bed boundaries, (3) vertical sorting, (4) microflow structures within the matrix, (5) orientation of clasts parallel to slope, and (6) lobate forms (Francou 1990). Ozouf et al. (1995) caution that a diverse set of processes can play an important role in the development of gréze litées depending on bedrock source type and topography. These include (1) solifluction, (2) rockfall, (3) grain flow, and (4) debris flow. Significant recent interest in stratified slope deposits is evidenced by an entire issue of *Periglacial and Permafrost Processes* elevated to the topic in 1995 (vol. 6). The difficulties in determining process origin for paleoclimatic interpretations reviewed by Van Steijn et al. (1995).

Gelifluction Features　The process of gelifluction operates on gradients as low as 1° to 2°. Gelifluction is controlled by both slope and water supply, but is accelerated by frost-creep and cryoturbation. Where the process is active a number of deposits and surficial forms result: gelifluction sheets, gelifluction benches, gelifluction lobes, and gelifluction streams. Although these features can be separated on the basis of individual morphology (see Washburn 1980), they share common genetic mechanics. Variations in form result from differences in texture, gradient, and soil moisture. Material in gelifluction deposits is usually poorly sorted, but a crude stratification may be present in some deposits. Angular clasts are normally oriented with long axes parallel to the direction of movement.

In regions covered by arctic or alpine tundra the most common feature seems to be the **gelifluction** (solifluction) **lobe** (fig. 11.25). The large, tonguelike masses of surface debris may occur as a single lobe, usually 30 to 50 m wide, or as one of many individuals that are joined laterally into a much broader field. In long profile an individual lobe is marked by a pronounced scarp or bulge at its leading edge that ranges from 1 to 6 m high. Commonly, however, a succession of lobes develops on a slope in which each upstream lobe overlaps the rear of the next downslope lobe, giving the entire sequence a staircase profile. Internally the lobes consist of angular, unsorted debris, with particle sizes varying from silt to coarse boulders. The long axes of the boulders are usually oriented parallel to the direction of movement. Various types of gelifluction lobes and related forms (e.g., turf-banked lobes and terraces, stone-banked lobes and terraces) are discussed in detail by Benedict (1970, 1976). Gelifluction lobes, as you might expect, move primarily by gelifluction processes, as discussed earlier in this chapter.

Figure 11.25
Gelifluction lobe on the Beartooth Plateau, Mont.
(Photo by R. Craig Kochel)

Rock Glaciers Rock glaciers are large tongue-shaped or lobate features generally composed of angular boulders (fig. 11.26). Their overall appearance resembles that of a glacier in that they occur mainly high in Alpine valleys and have surface microtopography, including ridges and furrows, and a steep snout (Potter 1972).

Although considerable confusion occurs in the terminology applied to rock glacier features (see reviews in White 1976; Martin and Whalley 1987; Giardino and Vitek 1988; Hamilton et al. 1995; Stieg et al. 1998), two main types appear to be commonly recognized in the model of Potter (1972). Rock glaciers containing subsurface ice with a superficial coating of rock fragments (up to several meters) are considered ice-cored. The coarse block layer may act as a thermal filter to protect the permanently-frozen rock glacier core if snow cover is thin or absent (Humlum 1997). Ice-cemented rock glaciers contain a mix of rock materials cemented by interstitial ice. Both styles appear to represent some transitional phenomenon between a nonglacial process and a true glacier with a mantle of rocky debris, or alternately an ice-cored moraine (see Barsch 1971; Ostrem 1964). The controversy surrounding the genesis of rock glaciers revolves around the question of whether or not a true glac-

ier is a required precursor (Outcalt and Benedict 1965). In other words, is the ice contained within the rock glacier necessarily glacier ice, or could it have accreted in a periglacial setting by gradual freezing of pore water? Some relict rock glaciers in the southwestern United States, developed in regions with no evidence of glaciation, support a completely periglacial origin for at least some of the features (Blagborough and Farkas 1968; Barsch and Updike 1971). It seems reasonable, however, that most tongue-shaped rock glaciers are somehow related to true glaciers. Lobate types may be ice-cemented and move by creep that is generated by the weight of the bouldery mass on the interstitial ice (White 1976b; Washburn 1980; Wayne 1981). Nonetheless, enough disagreement exists about the origin and transport mechanics of these deposits for us to refrain from sweeping conclusions at this time (see Whalley 1983).

Research efforts investigating the rheology of rock glacier movements have been hampered by the presence of debris cover. Many of these studies have attempted to interpret rheological processes from observations of surface microrelief, assuming that microrelief provides evidence of formerly active processes. A recent example of this type of study (Johnson 1992) suggests that

Figure 11.26
Rock glacier on Sourdough Mountain near McCarthy, Wrangell Mountains, Alaska.

microrelief develops from a variety of processes, including (1) deformation of the ice core, (2) movement of surficial cover along the debris-ice interface, (3) deformation caused by a period of glacial advance, and (4) changes in the hydrologic balance. Surface morphology must be used with caution because it may misrepresent the actual internal dynamics of the rock glacier. Johnson's studies observed isolated lobe activity within Yukon rock glaciers, providing support for the hypothesis that the rock glacier core was composed of pure glacier ice. Barsch (1994) showed that the crevasses on the rock glacier surface are not formed by melting ice but by internal movement like the well-known ridge and furrow patterns. Like studies of glacial mechanics, borehole data has been instrumental in providing information on process dynamics (e.g., Haeberli 1985).

An excellent overview of rheological studies of rock glaciers is provided by Giardino and Vitek (1988). Flow models range from viscous (Wahrhaftig and Cox 1959) or pseudoplastic flow (Barsch et al. 1979) to mechanisms emphasizing creep and basal sliding (Barsch 1977; Haeberli 1985). Some of these studies have attempted to estimate rates of mass transport of debris related to geomorphic work considerations (for example, Barsch and Jakob 1998). Because rock glaciers cannot be considered homogeneous like glaciers the mechanical analysis becomes more complex. One of the most frequently cited models, discussed in detail by Haeberli (1985), uses deformation of the ice as the major movement mechanism. Therefore, this model considers rock glaciers as long-term creep phenomena. The basal sliding model (Wahrhaftig and Cox 1959; Haeberli 1985) points to failure between the frozen mass and the bedrock by direct shear and notes that sliding is enhanced by hydrostatic pressure (Giardino 1983). Recent studies have begun to focus on the hydrology of the system, and have led to a conceptual model (fig. 11.27; Haeberli 1985, 1990; Gardner and Bajewsky 1987). The inputs for this system include direct precipitation, runoff from slopes, ice and snow avalanche, groundwater, and initial glacier ice. The outputs are surface runoff, groundwater discharge, surface seepage, and sublimation and evaporation (Giardino et al. 1992).

Modern, active rock glaciers are known in polar or subpolar regions and in the high mountains of the midlatitudes. They vary in thickness from 15 to 50 m and

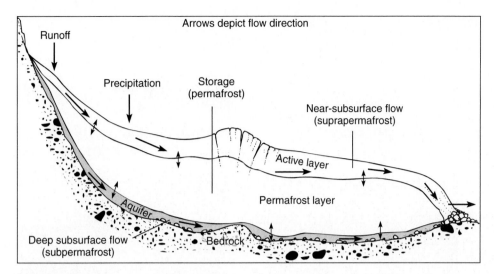

Figure 11.27
Schematic of the hydrology of a rock glacier.

move continuously at rates usually ranging from centimeters to a meter per year, depending on local conditions. Velocity generally decreases downvalley and laterally across the glacier (Benedict et al. 1986). In some cases these rock glaciers are known to be advancing in areas where glaciers are presently retreating, perhaps because the surface rubble insulates the underlying ice and slows its response to climatic change (Osborn 1975). Inactive rock glaciers are widely distributed in regions that now have temperate climates. Whether they have paleoclimatic significance depends on precisely how such features originate. Although a variety of origins have been suggested, only those processes that produce a continuous forward motion are acceptable. Concerned with the paleoclimatic significance of relict rock glaciers, many workers are studying active features to determine the climatic and topographic constraints on rock glacier initiation, (this approach is similar to attempts to define the glacial equilibrium line, as discussed in chapter 9). Humlum (1988) provides evidence for determining the rock glacier appearance level (RAL) and the rock glacier initiation altitude (RILA). Humlum (1998) noted that while glaciers in Greenland and Antarctica occur at sites with local high accumulation of snow relative to the input of talus, the opposite is true for the locations favorable to rock glacier development. Thus, he concludes that differences between the two are primarily controlled by topographic climatic differences rather than due to regional climatic differences. In a review of the climatic zonality of rock glaciers, Harris (1994) used meteorologic data to show differences between the conditions where gelifluction, rock glaciers, and active blockfields dominate the local style of mass wasting (fig. 11.28). Gelifluction appeared to dominate in dry climates with mean annual temperatures below –5° C; rocky glaciers in moist climates with temperatures approximately

–1° C; and blockstreams in colder, dry climates (Harris 1994). In the Italian Alps, for example, rock glaciers do not occur where the mean air isotherm is above –2° to –1° C and where precipitation is above 250 cm. These criteria imply that rock glaciers occur within the climatic constraints for periglacial climates set forth by Peltier (1950). Recent work on O^{18}/O^{16} ratios indicate the possibility of obtaining paleoclimate signals from the geochemistry of ice cores from rock glaciers (Steig et al. 1998).

Blockfields Blockfields (often called felsenmeer) are usually thought of as broad, relatively level areas covered by moderate to large angular blocks of rock (Sharpe 1938). Similar accumulations of blocky debris also are found on slopes and are variously referred to as block slopes, block streams, rubble sheets, and rubble streams (see Washburn 1973). Our focus is on block accumulations that occur on slopes, either concentrated in valleys or disseminated across a wide area. Although we may employ the term "blockfield" in reference to these accumulations (as have many others), some workers would consider this technically incorrect. True blockfields, as originally described, are not the product of mass movement but are produced in situ from frost wedging and heaving of the underlying bedrock. In this discussion we will actually be treating block streams or block slopes, but the term "blockfield" has been so widely applied to these features that we will employ it to mean any accumulation of blocky debris, including those found on slopes and subjected to mass movement.

Blockfields on slopes tend to be elongate with widths averaging between 60 and 120 m and lengths from 350 m to 1.3 km. Maximum and minimum values vary drastically depending on the block size, the slope angle, and the distribution of the bedrock source. The

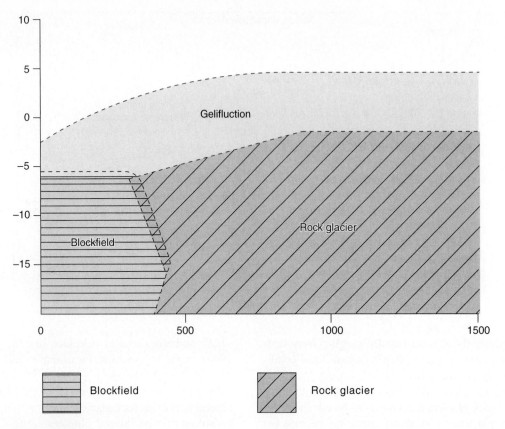

Figure 11.28
Coincidence of periglacial landforms and processes with climate in all pure regions.
(From Harris 1994)

gradient on the accumulation ranges widely, from 1° to 20°, although it is most commonly between 3° and 12°. Some workers feel that an appropriate upper limit for these features is 10° to 15° (Caine 1968; White 1976). Blocky slope accumulations normally display an internal fabric indicative of movement; that is, the long axes of individual blocks are oriented downslope (Harris et al. 1998). The orientation may become transverse to the slope near the distal end of the deposit (Caine 1968; Kochel 1975), or it may diverge locally from the norm. Accumulations seem to be 4 to 20 m thick, but very few observations of the total thickness have been made. One deposit that was totally exposed displayed an internal layering characterized by an open-textured surface zone and a increase in matrix with depth (Caine 1968). This type of fabric distribution, however, cannot be accepted as a general trait until more observations become available. Individual blocks contained in boulder masses are striking in their enormity, most being 1 to 3 m in intermediate diameter and some as large as 6 m (fig. 11.29).

A number of mass movement processes, ranging from avalanches or landslides to catastrophic floods, have been proposed for the development of block slopes or streams. The distinct probability exists that block accumulations have more than one mode of origin. The most commonly cited transporting mechanism is gelifluction. Caine (1968), for example, concluded that the block slope at a site in Tasmania material moved when the entire deposit contained abundant matrix. The open texture in the surface zone (up to 3 m deep) could have been created after movement ceased when the matrix was removed by groundwater or surface water (Caine 1968; Potter and Moss 1968). The gelifluction model will work efficiently only if the lower portions of the accumulation contain a high content of fine-grained matrix. Such a layer was observed by Caine (1968) and suggested by Denny (1956) and Sevon (1969). However, it has not been proven to be a universal property of blockfields, and until more conclusive data are found, the solifluction mechanism can be only a viable hypothesis. Rea et al. (1996) argued that considerable chemical weathering may be required to produce gibbsite found in the fine-fraction of the matrix of some blockfields in Norway. Nonetheless, the gelifluction model would explain the terrace and lobe development noted on some block streams (Denny 1956; Potter and Moss 1968). Furthermore, the orientation of blocks is similar to that characteristically produced by gelifluction transport. Figure 11.30 illustrates the morphology often cited in support of solifluction processes for a major blockfield in east-central Pennsylvania. Additional support for a

Figure 11.29
River of Rocks blockfield at Hawk Mountain Sanctuary, Berks County, Pennsylvania Slope is 5°. Blocks in foreground are 3–4 meters in average length.
(Photo by R. Craig Kochel)

periglacial origin for these features includes (1) a distinct solifluction-like fabric with clasts oriented downslope, (2) contributory block tongues or lobes along the margins, and (3) the occurrence of crudely sorted patterned-ground features on blockfield surfaces (fig. 11.31). It is difficult to be more certain about emplacement processes for most of the Appalachian blockfields because of the lack of fines within block interstices. Because of extreme porosity, water flows under the blocks as they are concentrated in hollows and likely removes any sediments that may have accompanied their emplacement.

Some investigators have suggested that slope blockfields are emplaced by movement as rock glaciers, accumulations containing interstitial ice rather than a soil matrix (Patton 1910; Kesseli 1941; Blagborough and Farkas 1968). Such a mechanism has the advantage of easily accounting for the open texture in the blocky debris, because no matrix sediment must be removed. The ice simply melts away as the rock glacier becomes inactive. Rock glacier transport, however, fails to explain why slope blockfields lack the normal characteristics of rock glacier deposits such as steep fronts and pronounced transverse ridges and furrows. In addition, a greater shear stress is required in moving a rock glacier than in gelifluction of a fine soil, and so, to attain the proper mobility, rock glacier accumulation should be considerably thicker than blockfield deposits (Wahrhaftig and Cox 1959; Potter and Moss 1968).

Recent block movements on these features can be determined by careful inspection of clast weathering and interactions with hillslope vegetation. For example, active blockfields contain an abundance of clasts that either lack lichen cover or have a more or less uniform lichen cover, uniform pitting and weathering, and generally fresh-appearing clast surfaces. In addition, active block movements normally destroy downslope trees or create impact scars on the trunks (e.g. Hupp 1983). Because of the wide distribution of active block features in modern periglacial regions, inactive deposits in temperate zones or at altitudes well below functioning periglaciation have been widely interpreted as evidence of earlier periglacial conditions. For example, many blockfields have been described in the middle Appalachian Mountains where present climates are patently not periglacial (Peltier 1950; H.T.U. Smith 1953; Denny 1956; Potter and Moss 1968).

Cryoplanation Terraces A significant body of literature exists on cryoplanation since the term was introduced by Bryan (1946) as altiplanation surfaces (fig. 11.32). (See review by Guilcher 1994.) There

(A)

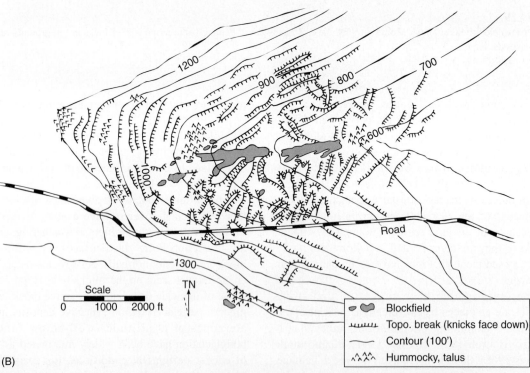

(B)

Figure 11.30

River of Rocks blockfield, Pennsylvania. (A) Oblique aerial photo of blockfield located in the axis at a major plunging anticline. Vertical stripe in the left third of the photo is a road. (B) Sketch map showing morphology of the valley floor in the forest surrounding the blockfield. Features indicate concentrations of blocks thicker than the very thin surficial mantle. Note the asymmetry of form with slope aspect. North-facing slopes are dominated by relict solifluction lobes containing blocks oriented parallel to slope. South-facing slopes are marked by terraces or steps. Similar slope asymmetry is common with exposure direction throughout central Pennsylvania (from Kochel 1975).

(Photo by R. Craig Kochel)

(A)

(B)

Figure 11.31

Devil's Potato Patch blockfield, Pennsylvania. (A) Oblique aerial photo showing microrelief on the blockfield surface. (B) Ground view of these forms showing quasi-sorted blocks resembling relict sorted nets. Scene is approximately 3 meters wide.

(Photos by R. Craig Kochel)

(A)

(B)

Figure 11.32

(A) Cryoplanation surface commonly referred to as the sub-summit surface or plateau in the Beartooth Mountains of Montana.

(B) Cryoplanation surface in foreground of this Icelandic plateau shows active solifluction lobes. Glaciers on the Vatnajökull icecap are visible in the distance.

(Photos by R. Craig Kochel)

exists considerable confusion over their formative processes. Landforms vary from small nivation hollows several meters across to major terraces several kilometers long in mountainous regions. Most researchers agree that first-action and periglacial weathering processes are involved, along with a variety of mass transfer processes such as solifluction and other frost-creep processes. Hall (1997) suggests that freeze-thaw may play a role but may not be the dominant formative process. Instead, he suggests rock fatigue from thermal stress and believes the role of solar radiation on cold regions may be significant. In this model, bench initiation may be related to extensive jointing from dilation following the retreat of glaciers or snow fields (Hall 1997). Regardless of their formative processes, significant low-relief surfaces are quite common in many alpine periglacial regions such as the extensive subsummit surface in the Beartooth Mountains of Montana. The combination of periglacial weathering and mass wasting processes appears to be very effective in producing regional low-relief surfaces.

Relict Periglacial Features and Their Significance

Many of the landforms active in modern periglacial climates have relict counterparts in milder temperate climates today. Common relict features include blockfields, rock glaciers, solifluction lobes and terraces, patterned ground, and pingos. Inactive (relict) features in temperate zones or at altitudes well below current periglacial conditions are widely accepted as evidence of earlier periglacial conditions. Whereas active periglacial processes currently affect only about one-fourth of the subaerial landmass on Earth, largely in sparsely populated regions, Pleistocene periglacial processes have played a significant role in shaping the landscapes of about half the globe, including many heavily populated areas.

The geomorphic significance of Pleistocene periglacial conditions on the present morphology of Appalachian hillslopes may be considerable. In fact, some students of Appalachian geomorphology believe that these episodes of periglacial activity account for almost all of the significant denudation and morphological development of slope and piedmont landforms that occurred during the entire Quaternary Period, leaving interglacial climates like the Holocene largely ineffective. In this view, the modern slope morphology is the legacy of former periglacial conditions, and is essentially armored from significant change under the normal range of active processes. If this perspective is anywhere close to being correct, it has enormous implications for our understanding of the efficacy of "observable" processes on Appalachian hillslope-piedmont landforms today.

Braun (1989) examined the contributions of glacial and periglacial erosion in denudation of the north-central Appalachians over the past 2.3 million years, including some eight glacial-periglacial episodes during the last 850,000 years. He concluded that glacial erosion must have lowered the Appalachian Mountains at an average denudational rate of 0.5 to 0.8 m/1000 years (between 120 and 200 m total to account for the Pleistocene marine sediment record) and at periglacial erosion rates of 0.15 to 0.3 m/1000 years in areas beyond the glacial border. In Braun's model ridge summits would have been lowered by tens of meters during the Pleistocene, an order of magnitude faster than modern rates of denudation estimated from studies of sediment yield in the region's rivers. Others working in the Appalachian region (e.g., Clark and Ciolkosz 1988; Ciolkosz et al. 1990; Clark and Hedges 1992) agree with a concept suggested by Ciolkosz et al. (1986) that the present slopes are so well adjusted to periglacial conditions that they appear to be extraordinarily stable under the present climate conditions. A growing consensus suggests that much of the ubiquitous colluvial debris on Appalachian summits and slopes is the result of periglacial activity. The relative ineffectiveness of temperate fluvial processes in the Appalachians is indicated by the preservation of periglacial landforms including colluvial slope mantles, fans, and blockfields (Clark and Ciolkosz 1988). Waltman (1985) supports this contention by noting the common occurrence of argillic horizons and fragipans in colluvial deposits, which are indicative of slope stability during the Holocene.

Some caution is advised, however, before concluding that Holocene processes are geomorphically ineffective on Appalachian hillslopes. The role of rare high-magnitude catastrophic events has yet to be fully assessed. For example, the Hurricane Camille (1969) rainfall and Rapidan Flood (1995) in Virginia (Williams and Guy 1973) (see chapter 4) triggered hundreds of significant debris flow and debris avalanches, reactivating fans that have subsequently been shown to have been active several times during the Holocene (Kochel and Johnson 1984; Kochel 1987, 1990; Eaton 1999). Detailed investigations of fans in the northern Appalachians to determine the age of those features and the level of Holocene activity have not yet been made. Recent studies of geomorphic effectiveness in Appalachian streams have cited the important role of rare catastrophic floods in development of valley floor landforms (e.g., Kochel et al. 1987; Jacobsen et al. 1989). We may simply not have enough experience with these infrequent events to be able to accurately estimate the potential of Holocene processes to effect change on hillslope and piedmont landforms. Continued studies of the effects of large-magnitude floods and paleoflood investigations may help resolve these uncertainties.

ENVIRONMENTAL AND ENGINEERING CONSIDERATIONS

Human expansion into the arctic and subarctic regions is inevitable. This impending population growth has generated a growing concern as to whether our knowledge of geomorphic systems operating in the areas is sophisticated enough for us to develop the regions and still preserve their fragile environmental balance. The concern is most evident among the engineering community that deals with construction problems in arctic regions on a day-to-day basis (for references see National Academy of Sciences 1983). The technical problems are serious, and our ability to cope with them depends on how well we understand the processes functioning within the system and how those processes are ultimately related to the environmental variables. Excellent reviews of the myriad of environmental problems and hazards associated with human occupation of Alpine periglacial regions is provided by Haeberli (1992) and by Wolfe (1998).

Most engineering problems are associated with permafrost. According to Brown (1970), any one of four basic engineering approaches may be used when dealing with the permafrost condition: (1) disregard of the permafrost, (2) an active approach in which the permafrost in the near surface is eliminated by keeping the soil continuously thawed or by removing the natural soil and replacing it with permafrost-resistant material, (3) a passive approach in which permafrost is preserved by keeping the soil frozen at all times and thereby eliminating the annual thaw, and (4) use of design structures that can withstand frost action.

Which of the above methods is best depends on the local situation. In discontinuous permafrost, it may be possible to ignore the condition if the parent material is bedrock or a sandy or gravelly soil. Active approaches, such as simply removing the vegetation and its insulating effect, are also feasible on a local scale. In zones of continuous permafrost, the potential problems cannot be disregarded. There the most practical method is to keep the area continuously frozen, primarily by ventilation or insulation.

Some heaving and settling will occur in the first several years after construction regardless of the method employed, because the permafrost requires some time to adjust into a new equilibrium condition. Thus, any project should prepare for the initial response in the system with a design that can withstand the stresses generated. Often the periglacial features described earlier give tangible clues about the near-surface ground conditions at any locality (table 11.4). Recognizing these is an impor-

---

TABLE 11.4	Landforms that Indicate Ground Conditions in Arctic and Near-arctic Regions.

Feature and Description	Associated Ground Conditions
Polygonal ground (ice wedges)—Usually indicates the presence of a network of ice wedges—vertical wedge-shaped ice masses form by the accumulation of snow, hoarfrost, and meltwater in ground cracks that form owing to contraction during the winter. Wedge networks are also common in wet tundra where no surface expression occurs. (Subject to extreme differential settlement when surface disturbed.)	Typically indicates relatively fine-grained unconsolidated segments with permafrost table near the ground surface; also known from coarser sediments and gravels where wedge ice is less extensive.
Stone nets, garlands, and stripes—Frost heaving in granular soils produces netlike concentrations of the coarser rocks present. If the area is gently sloped, the net is distorted into garlands by downslope movement. If the slope is steep, the coarse rocks lie in stripes that point downhill.	Indicate strong frost action in moderate, well-drained granule sediments that vary from silty fine gravel to boulders. Surficial material commonly susceptible to flowage.
Solifluction sheets and lobes—Sheets or lobe-shaped masses of unconsolidated sediment that range from less than a foot to hundreds of feet in width and may cover entire valley walls; found on slopes that vary from steep to less than 3°.	Indicate an unstable mantle or poorly drained, often saturated sediment that is moving downslope largely by seasonal frost heaving. On steeper slopes they often indicate bedrock near the surface, and on gentle slopes, a shallow permafrost table.
Thaw lakes and thaw pits—Surface depressions form when local melting of permafrost decreases the volume of ice-rich sediments. Water accumulates in the depressions and may accelerate melting of the permafrost. Often form impassible bogs.	Usually indicate poorly drained, fine-grained unconsolidated sediments (fine sand to clay) with permafrost table near the surface.
Beaded drainage—Short, often straight minor streams that join pools or small lakes. Streams follow the tops of melted ice wedges, and pools develop where melting of permafrost has been more extensive.	Indicates a permafrost area with silt-rich sediments or peat overlying buried ice wedges.
Pingos—Small ice-cored circular or elliptical hills that occur in tundra and forested parts of the continuous and discontinuous permafrost areas. They often lie at the juncture of south and southeast-facing slopes and valley floors and in former lakebeds.	Indicate silty sediments derived from the slope or valley, also groundwater with some hydraulic head that is confined between the seasonal frost and permafrost table or is flowing in a thawed zone within the permafrost. Those in former lakebeds indicate saturated fine-grained sediments.

From Ferrians et al. (1969), USGS Prof. Paper 678.

tant first step in a site analysis (Thomas and Ferrell 1983), and air photos and remote sensing data interpreted by a geomorphologist can be very useful in the planning stage.

A considerable body of literature exists regarding the environmental and engineering problems associated with development in periglacial regions. Although the material is scattered through scientific and engineering publications, excellent encapsulations of the advances made in this area can be found in the periodic proceedings of the International Permafrost Conference. There have been six of these (often multi-volume) publications between 1973 and 1993.

Building Foundations

An excellent up-to-date review of human-induced problems with construction and disturbance of the thermal regime of permafrost is provided by Haeberli (1992) with a focus upon occurrences in sporadic or Alpine permafrost settings. Our basic problem arises from ignorance of the delicate thermal balance characteristic of permafrost regions. The bearing strength of a surface (ability to support load without plastic deformation) varies significantly with the type of material, when it is frozen and when it thaws (Swinzow 1969). The loss of strength during thaw is most dramatic in clay and silt soils that are not permeable and so retain a large portion of the water released by melting. Bedrock or coarse-grained deposits are reasonably stable during the thaw phase because they drain easily. Because water cannot permeate below the permafrost table, the active layer in fine-grained soils becomes a quagmire that cannot support weight and, therefore, allows differential settling of overlying buildings, bridge abutments, and so on. In addition, the lack of lateral drainage and the high segregation potential will accentuate differential heaving during the next freeze. Building foundations, therefore, present enormous problems. In the case of small dwellings it may be adequate simply to live with the frost activity by pursuing an annual maintenance program. Large buildings, however, become unsafe if they are allowed to endure substantial heaving and settling, and preventive measures must be taken. Lobacz and Quinn (1963) were able to show that residual thaw zones persist beneath buildings that are not ventilated at the base, and that when buildings are elevated seasonal frost will extend from the surface to the permafrost table.

Because of these conditions, most builders today in areas of continuous permafrost follow the passive approach and attempt to maintain the permafrost intact. The floor of the building is built on top of pilings driven into the permafrost layer, so that it normally stands 1 to 2 m above the ground surface, and air can circulate freely in the open space between the floor and the ground (e.g., Kutvitskaya and Gokhmen 1988). This keeps the ground frozen in the winter because the heat radiating from the building is removed by cold air moving through the open space. Thawing does occur in the summer but it is minimized because the surface is shaded by the building. Nonetheless, the piles supporting the building are susceptible to some heaving in the active layer; Johnston (1963) suggests they should be driven to a depth at least twice the thickness of the active layer in order to keep the heaving effect to a minimum. Other methods of air cooling or liquid cooling (Cronin 1983) have been utilized to maintain the frozen condition.

Whenever possible, buildings should be located on a sand and gravel substrate, for reasons explained above. In coastal regions some communities, built prior to any understanding of periglacial processes, have been built on silty, deltaic plains. Further development may be impossible in such a setting; in fact, some small towns may have to be relocated if the problems associated with the permafrost cannot be controlled. Relocation was actually accomplished in the case of one community that originally existed on the deltaic plain of the Mackenzie River. The Canadian government physically moved the community 56 km to a new townsite that was chosen because it had abundant sand and gravel for building sites and roads. Although construction of the new town was costly, its success demonstrated that a reasoned approach and proper regulations can allow development to proceed without harmful degradation of the system (see Pritchard 1962; Price 1972; Cooke and Doornkamp 1974).

Roads and Airfields

The outbreak of World War II made the U.S. and Canadian governments aware that defending Alaska and the Northwest Territories from attack would be a monumental task. In an attempt to reduce this vulnerability, many necessary but hastily conceived projects were completed to facilitate the transportation of men and supplies to the area. Railroads, pipelines, and airfields were constructed, and the Alcan Highway, stretching almost 3000 km, was completed in only eight months, a truly remarkable engineering feat. The highway, however, proved to be a case study in permafrost mechanics, as the construction and maintenance problems were often baffling and disruptive (Richardson 1942, 1943). Hayley (1988) presents a review incorporating 60 years of experience in railway design through permafrost in the Hudson Bay region.

We now know that problems associated with roads or airfields vary greatly because they are affected by microenvironmental conditions. In general, the most

Figure 11.33
Gravel road near the Umiat Airstrip showing severe differential subsidence caused by thawing of ice-wedge polygons in permafrost. Anaktuvuk district, Northern Alaska region, Alaska, Aug. 1958.

common ailments are (1) differential heaving and settling that creates a washboard surface (fig. 11.33), (2) sinking of portions of the surface, (3) destruction of bridges by spring floods, (4) burial of the pavement by slumping or other mass movements, and (5) icing. Icing is by far the worst problem. It occurs when water runs onto road surfaces during the winter, quickly freezes over large areas, and often accretes to a considerable thickness. Groundwater under hydrostatic pressure may continue to flow through most of the winter and, with improper planning, may also run onto the road surface if road cuts open springs and seeps. Even something as innocuous as a river freezing can cause icing of a nearby highway surface. This happens when the river freezes from the surface downward and creates a pressure head on the bottom water. The water, thus mobilized, moves to the channel sides where it rises to the surface and overflows the channel banks onto the highway.

The best road protection requires wise planning and careful pavement design to reduce frost action in the active layer and prevent excess water from reaching the road surface. The highway route should avoid seeps and springs by diverting around hills. If cuts must be placed in hillslopes, the pavement can be protected from icing by integrating a system of culverts and drainage ditches to divert free water away from the road.

Pavement design usually begins with a layer of gravel 0.6 to 1.5 m thick spread directly on the tundra to insulate the underlying surface and so minimize thawing. A layer of insulating material such as peat (McHattie and Esch 1983) or polystyrene (Johnston 1983) may also be inserted within or beneath the gravel cover. Some melting must be expected, nonetheless, and the design should allow for reduced strength in the summer and for heaving in the winter (Linell and Johnston 1973; Hennion and Lobacz 1973). Even painting the pavement surface white may help to reduce the thaw of the substrate.

Utilities: Water and Sewage

Some of the most burdensome problems facing communities in permafrost regions involve utilities and services that are taken for granted in the mid-latitudes. Water supply, for example, is complicated by deep winter freezing, and large lakes may be the only sufficient, continuous, and dependable water source. Taliks within the permafrost may provide some water, but the distribution is not regular and they may not contain enough water to support even a small settlement. Although taliks beneath the permafrost can provide large amounts of water, using them is very expensive. Drilling through the permafrost is costly, and pipes must be cased to prevent freezing. In addition, some deep subpermafrost water is brackish or highly mineralized (J. R. Williams 1970).

Locating an adequate, continuous water supply solves only half the problem, because the water must still be delivered to the place where it is to be used. People in tiny settlements have traditionally obtained water by carrying it in tanks from a nearby river or lake. In the winter, when the source freezes over, blocks of ice are melted. Larger communities cannot rely on such primitive methods and most have piped-in water. Either the pipes must be insulated or the water must be continuously circulated to prevent freezing. Many areas now utilize an enclosed heated and insulated conduit, called a **utilidor,** for water delivery. The utilidor complex, placed above the ground, is designed to hold water pipes, electric cables, and sewage disposal pipes (see Zirjacks and Hwang 1983; Kennedy et al. 1988), but installation is very expensive.

Sewage disposal is still another perplexing problem. Some small villages utilize individual buckets as collectors of all kinds of waste, human and otherwise. These are carried to a specified location on a lake or river where they are dumped in expectation of the spring thaw. Essentially, nature cleans the system at winter's end. For sanitation larger communities must be more sophisticated and most now employ the utilidor system. In practice, however, utilidors solve only the immediate problem of removing wastes from individual households. The final disposal of wastes is a most difficult problem. Few communities have efficient treatment plants, and most wastes are ultimately dumped as raw sewage into the nearest waterway. The problem is not nearly as acute in discontinuous permafrost areas, where septic tanks and cesspools are common. In continuous permafrost zones, however, alternative techniques must be developed. Sewage lagoons (large dugout hollows) do allow waste to decompose anaerobically (Brown 1970) but are becoming a common disposal method (Martel 1988). One novel idea, tried at a military base, was to use fuel oil instead of water in the sewage system. When the wastes are collected, they are sufficiently mixed with the oil to be injected directly into a furnace where incineration disposes of the wastes and simultaneously generates heat and electricity for the base (Alter 1966).

Pipelines

The discovery of oil on the north shore of Alaska near Prudhoe Bay was exciting news to an energy-addicted nation. The conflict between energy and environment that ensued created a furor in which opinions ranged from reasoned arguments to pure emotionalism. Nonetheless, it did force us to analyze in a systematic way the problems associated with construction of a pipeline in a permafrost region.

The Trans-Alaska pipeline was not the first pipeline constructed by the United States in a periglacial environment. The Canol pipeline system was built during World War II across 2575 km of discontinuous permafrost. Its purpose was to pump crude oil along the main pipe to a refinery located along the Alcan Highway near Whitehorse, B.C. From there gasoline was pumped to airfields and other military establishments. The Canol project, like the Trans-Alaska pipeline, was steeped in controversy (Richardson 1944), and its use was discontinued in May 1945 after only 13 months of operation, presumably because of the high cost of maintenance and its decreased importance in the war effort.

The Trans-Alaska pipeline is different in most respects from the older Canol project, and little applicable knowledge was derived from the earlier work. The Trans-Alaska line is much farther north, and much of its 1270 km route from Prudhoe Bay to Valdez crosses zones of continuous permafrost. The pipe, 1.2 m in diameter, is larger than the Canol pipe and, fully loaded with oil, weighs over 900 kg per meter (Harwood 1969). The crude oil enters the pipe at a temperature of about 58° C, but friction along the flow path increases this temperature considerably. Cooling devices are placed along the route to keep the temperature at a constant 63° C, but the oil is still hot and this heat presents the most difficult problem in preserving the permafrost environment intact.

Planners of the pipeline had to consider the possibility of major changes in the physical environment as well as a host of engineering problems (Lachenbruch 1970; Kachadoorian and Ferrians 1973). A pipe resting on the surface and carrying hot fluids would certainly cause thawing of the underlying permafrost. Without preventive measures, the thaw would probably expand outward with time, as shown in figure 11.34, perhaps at a decreasing rate but possibly never reaching an equilibrium state (Lachenbruch 1970). The soil could also become liquefied and unable to support the heavy pipe, and any

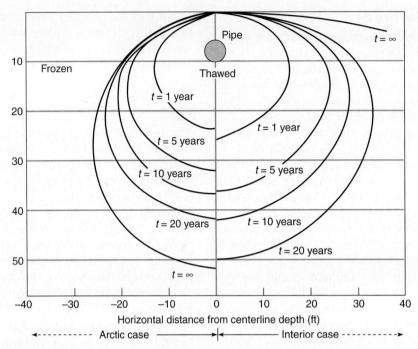

Figure 11.34
Theoretical growth of thawed cylinder around a heated pipe buried in silty soil. The pipe is 1.2 m in diameter and kept at 80° C. The curves at the left apply to conditions near the arctic coast of Alaska; curves on the right apply to the southern limit of permafrost.
(Lachenbruch 1970)

gelifluction induced by the liquefaction might carry the pipe along with the soil, producing ruptures and oil spills. As the pipe crosses materials of diverse texture and composition, differential heaving and settling are inevitable, and stresses generated in this manner could bend and rupture the pipe. This problem is accentuated where the pipeline crosses slopes underlain by ice wedges which melt more quickly and undermine the support rapidly. In addition, the actual construction could have inadvertently altered the delicate environmental balance. For example, destruction of the surface vegetation or diversion of river courses would disrupt the thermal regime and change the thickness of the active layer.

Faced with these potential hazards, the designers of the pipeline anticipated the problems based on the best available knowledge of how the system would respond. In general three approaches were employed in different places: (1) the pipe was buried, (2) the pipe was suspended above the ground, and (3) the pipe was placed along the edge of a road that parallels the route of the pipeline. Burying the pipeline is perfectly safe in coarse-grained soils that are well drained and contain little ground ice. In other zones a refrigerant is run along the pipe to keep the ground frozen. Suspending the pipe above the ground is necessary where the soil is particularly susceptible to thawing and also where the

route crosses the path of migratory animals such as caribou. The above-ground suspension requires a dense network of pilings to provide the needed strength, and these are inserted to great depth. In addition, some of the supporting piles are refrigerated and others are distributed in a zigzag fashion to absorb the effect of differential movement by heaving, gelifluction, or earthquakes (fig. 11.35). Careful planning of the overall project thus far has been able to provide a continuous supply of crude oil with a minimum of hazard to the environment.

Implications for Global Warming

Changes in global climate can be expected to be manifested in periglacial regions because of the extreme sensitivity of these regions to climate. Climate warming may have set the stage for thermokarst expansions, which were acutely triggered by local disturbances such as fire or vegetation disruption. Fedorov (1996) reviewed the effects of recent climate change on permafrost in Yakutsk, Russia, and provides a good discussion of the myriad of physiographic and geocryological conditions involved. He documented four notable episodes of cryolithologic stress in the past century, most resulting in destruction of alas terrains.

Figure 11.35
Trans-Alaska pipeline crossing the Tolovana River valley near Livengood, Alaska.

It is likely that continued research in periglacial regions will reveal indicators sensitive to global climate changes that may be useful in monitoring the effects of human activity. Burn (1992) observed changes in mean winter temperature and snowfall in the Yukon Territory that appear to have caused local ground warming from –2° to –1.3° C. Air temperature seems to be the most important single factor influencing permafrost temperatures, with seasonal snow cover coming in a close second. Efforts to develop physical models of the dynamic interactions between climate and permafrost response such as the work of Zhang and Stamnes (1998) may improve our ability to predict future changes in response to climate change. Although such changes appear slight, they can have major consequences in areas such as these that are already close to the threshold of instability near the 0° C isotherm. An important note from the work of Burn (1992) is that the changes in ground conditions lagged about 20 years behind the changes in air temperature because of the slow exchange in latent heat. This lag needs to be accommodated in models designed to predict reactions to global change.

Applications to Planetary Geology

Periglacial processes can be expected to be effective geomorphic agents on any planetary surface, provided the climate is cold enough to promote active frost weathering and sufficient water is available to permit the development of ground ice. Because most of the rocky planetary bodies in our solar system lack an atmosphere or have very thin ones, periglacial processes may well be one of the most common processes on other planets, provided there is or has been in the past sufficient moisture in the atmosphere or regolith. Viking spacecraft imagery of Mars has revealed a tremendous variety of landforms that resemble terrestrial periglacial features. The presence of periglacial features on Mars was suggested even before the remarkable Viking journeys, based on the interpretation of features from Mariner imagery (Anderson et al. 1967; Wade and DeWys 1968; Sharp 1973; Gatto and Anderson 1975).

Excellent discussions of the significance of periglacial processes in shaping the surface of Mars and the enormous array of landforms believed to be influenced by them can be found in the following texts: Carr (1981), Baker (1982), Carr et al. (1984), and Hamblin and Christiansen (1990). It is beyond the scope of this text to discuss the role of periglacial processes on other planets in detail, but we can provide a brief sampling of the types of features thought to be evidence of periglacial activity on Mars.

Numerous studies have pointed to the enormous array of channels on Mars that are thought to have once

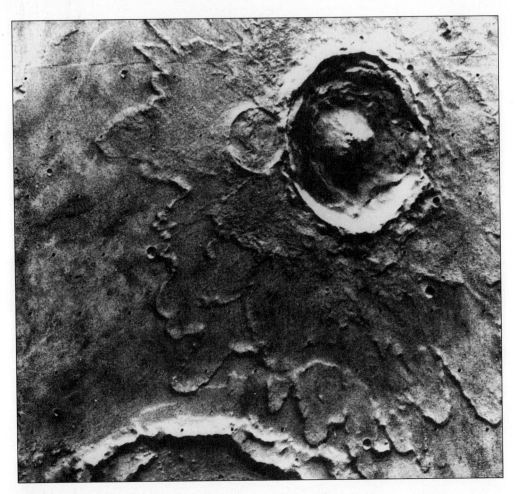

Figure 11.36
Viking image showing evidence of permafrost on Mars. Splosh debris flow ejecta from crater Yuty,
which fluidized upon impact, is likely from melting of ground ice in the regolith.
(Photo by NASA)

carried great volumes of water early in Mars history. Most planetary researchers feel that much of this water infiltrated into the Martian regolith and may be present today as permafrost. Figure 11.36 shows flow ejecta patterns common around impact craters on Mars. In contrast to the ray pattern created during impact of dry planetary surfaces like the Moon and Mercury, these splosh craters are thought to signify impact-caused melting of ice-rich regolith and mobilization of flow from the craters. Additional evidence for permafrost within the Martian regolith includes large-scale polygonal fracture patterns that may represent ice wedges, their terrestrial counterparts, and closed depressions and backwasted escarpments that are thought to have been formed by thermokarst in cases where evidence of fluvial erosion is absent. Thermal erosion has been suggested along Martian outflow channels as a major widening process. (Costard et al. 1996). Thermokarst processes have been widely suggested as a major process not only for downwasting of the surface

but in backwasting of major escarpments, in much the same manner as on Earth (Czudek and Demek 1970). In fact, a major planetwide escarpment on Mars separating the northern lowland plains from mountainous and ancient cratered terrain contains abundant landforms indicative of thermokarst and related processes associated with ground-ice decay (fig. 11.37). These images depict aprons of debris, which appear to have flowed from upland escarpments, displaying numerous flow features such as radial ridges and furrows and compressional ridges formed around obstacles. The Nilosyrtis region of Mars (fig. 11.38), also along the planetary escarpment, shows debris-filled valleys between upland remnants marked by prominent longitudinal ridges. These features may be evidence of glaciation or rock glaciers (Colaprete and Jakosky 1998). Table 11.5 depicts the wide range of possible periglacial features recognized on Mars, demonstrating the similarities with cold-region landscapes here on Earth.

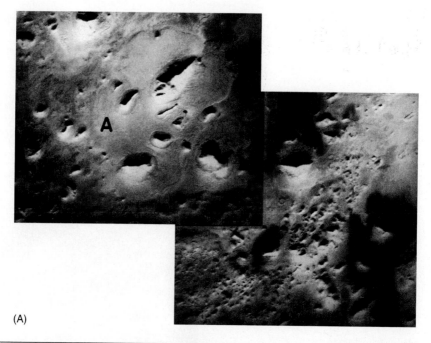

(A)

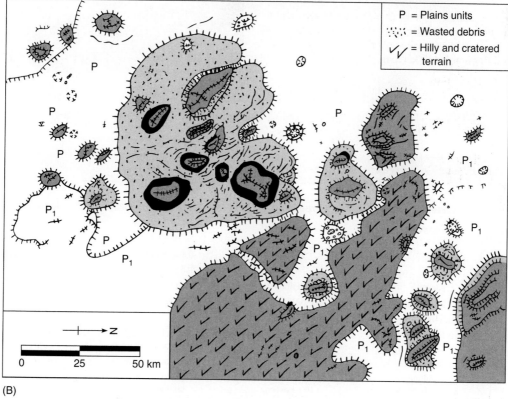

(B)

Figure 11.37

Debris aprons along the major planetary escarpment in the northern mid-latitudes of Mars in the area known as the fretted terrain. (A) *Viking* photos 338 S31–32 in the Protonilus Mensae region showing debris aprons and flow lines. (B) Interpretive geomorphic map of the north-central part of the mosaic (based on *Viking* frame 058B61). Note the paucity of craters on the debris, indicating that the debris is very young compared to the cratered uplands. P = younger plains unit.

(Photo from NASA)

Figure 11.38
Viking photo mosaic 18086 from the Deuteronilus Mensae region of the Martian fretted terrain. This image shows cirquelike valleys filled with debris containing longitudinal ridges, tributary flows, and debris fans. Bar is about 10 km long.
(Photo by NASA)

TABLE 11.5	Probable Periglacial Features on Mars.

Process	Feature	Size on Mars	Size of Terrestrial Analogs	Reference
Patterned ground processes	Polygons	20 km diameter	1–100 m diameter	Carr and Schaber (1977)
	Stripes	1–2 km wide; 1–2 km spacing	0.1–1.5 m wide; 3–4 m spacing	Carr and Schaber (1977)
Thermokarst	Alases	10 km diameter	Several km diameter	Theilig and Greeley (1978)
	Coalescing alases (alas valleys)	10–100 km long	Several 10's km long	Carr and Schaber (1977)
	Fretted terrain	Scarps 1–3 km high	Relief of 10–100 m	Gatto and Anderson (1975)
	Chaotic terrain	100 km diameter	Several km diameter	Sharp (1973b)
	Sapping and ground-ice collapse features	Several km high; several 10's km long	Variable, but smaller than Martian counterpart	Sharp (1973b)
Hillslope processes	Spur-and-gully topography	Several km high	Up to 1 km high	Lucchitta (1978a)
	Hillslope chutes	Several km high	Up to 1 km high	Sharp (1973a)
Mass movement	Debris lobes	10–50 km long	Variable, up to several km long	Squyres (1979)
	Landslides	1–100 km wide; 1–180 km long	1–10 km wide; 1–20 km long	Lucchitta (1978c, 1979)
	Solifluction	10's km long	Variable, up to several km long	Lucchitta (1978a)

SUMMARY

The processes and landforms found in periglacial environments have been briefly examined. In general, periglacial conditions are those existing in any cold, nonglacial setting regardless of latitude, but most regions of intense periglaciation are polar or subpolar. In many cases periglacial zones possess a unique property of permanently frozen ground, called permafrost. Permafrost adds a complicating factor in the activity of periglacial processes and leads to the formation of diagnostic features such as ice-wedge polygons, thermokarst, and pingos, which develop only where permafrost is present.

The driving processes in periglaciation are frost action and mass movements. Frost action includes wedging, heaving, sorting, and cracking. Mass movements usually involve frost creep and a form of soil flow called gelifluction. Both processes act in the development of most periglacial features, although certain forms seem to be more common where a particular process is dominant.

The rigorous climate and the presence of permafrost in cold regions create unusual difficulties in urban development that require engineering techniques not employed in other regions. The expected population expansion in arctic zones demands that we develop a sophisticated understanding of periglacial processes in order to maintain the fragile environmental balance.

SUGGESTED READINGS

The following references provide greater detail concerning the concepts discussed in this chapter.

Barsch, D. 1993. Periglacial geomorphology in the 21st century. *Geomorphology* 7:141–63.

Benedict, J. B. 1976. Frost creep and gelifluction features: A review. *Quat. Res.* 6:55–77.

Black, R. F. 1976. Periglacial features indicative of permafrost: Ice and soil wedges. *Quat. Res.* 6:3–26.

Dixon, J. C., and Abrahams, A. D., eds. *Periglacial geomorphology.* New York: Wiley.

Embleton, C., and King, C. A. M. 1975. *Periglacial geomorphology.* New York: Halsted Press.

Evans, O. J. A., 1994. *Cold climate landforms.* New York: John Wiley.

Ferrians, O. J., Kachadoorian, R., and Greene, G. W. 1969. *Permafrost and related engineering problems in Alaska.* U.S. Geol. Survey Prof. Paper 678.

French, H. M. 1976. *The periglacial environment.* London: Longman.

Giardino, J. R., Shroder, Jr., J. F., and Vitek, J. D., eds. 1987. *Rock glaciers.* Boston: Allen and Unwin.

Goldthwait, R. P. 1976. Frost-sorted patterned ground: A review. *Quat Res.* 6:27–35.

Haeberli, W. 1992. Construction, environmental problems, and natural hazards in periglacial mountain belts. *Permafrost and Periglacial Processes* 3:111–24.

Hallet, B., and Prestrud, S. 1986. Dynamics of periglacial sorted circles in western Spitzbergen. *Quat. Res.* 26:81–99.

National Academy of Sciences. 1983. *Permafrost, Proceedings of the 4th international conference.* Washington, D.C.: National Academy Press.

Péwé, T. L. 1969. *The periglacial environment.* Montreal: McGill-Queen's Univ. Press.

Senneset, E., ed. 1988. *Permafrost: Proceedings of the 5th international conference.* Trondheim, Norway. Trondheim: Tapir Pub.

Steig, E. J., Clark, D. H., Potter, N., Jr, and Gillespie, A. R. eds. 1998. The Geomorphic and climatic significance of rock glaciers: Geografiska Annaler Ser. A., v. 80, n. 3–4.

Tricart, J. 1969. *Geomorphology of cold environments.* Translated by Edward Watson. New York: St. Martin's Press.

Washburn, A. L. 1980. *Geocryology.* New York: John Wiley & Sons.

Williams, P. J., and Smith, M. W. 1989. *The frozen earth.* Cambridge, Cambridge Univ. Press.

12

KARST—PROCESSES AND LANDFORMS

Introduction
    Definitions and Characteristics
The Processes and Their Controls
    Karst Rocks—The Resisting Framework
        Lithology
        Porosity and Permeability
    The Driving Mechanics and Controls
        The Solution Process
        Solution Rates—The Controlling Factors
        Spatial Variations in Limestone Solution
Karst Hydrology and Drainage
Characteristics
    Surface Flow
    Karst Aquifers and Groundwater Flow
    The Relations Between Surface and Groundwater
        Morphology of Karst Drainage
Surficial Landforms
    Closed Depressions
        Dolines
        Doline Morphometry
        Uvalas and Poljes
    Karst Valleys
        Allogenic Valleys
        Blind and Dry Valleys
        Pocket Valleys
    Cockpit and Tower Karst
Limestone Caves
    Cave Physiography
        Entrances and Terminations
        Passages and Passage Morphology
    The Origin of Limestone Caves
Summary
Suggested Readings

INTRODUCTION

In regions of carbonate rocks and evaporites, weathering and erosion produce unique landforms called karst or, when widespread, karst topography. In contrast to most processes discussed in earlier chapters, karstification (the processes that develop karst topography) is not easily observed because much of the geomorphic work is accomplished well below the ground surface. In fact, some modern textbooks of geomorphology completely ignore karst or treat it in a few descriptive paragraphs because it is primarily driven by the solution process and can justifiably be considered under the topic of chemical weathering. However, the magnitude of solution in the development of karst is so great, and the unique topography resulting from the process so widespread, that it deserves special treatment.

Several excellent books deal specifically with the topic of karst (Jennings 1985; White 1988; Bosak et al. 1989; Ford and Williams 1989; Klimchouk et al. 1996; Drew and Hoetzl 1999), and the information contained in those sources provides the framework for this chapter.

Much of the classical literature dealing with karst was published in languages other than English. *interesting*

Definitions and Characteristics

Karst is defined by Jennings (1985) as "terrain with distinctive landforms and drainage arising from greater rock solubility in natural water than is found elsewhere." This definition stresses two main points: (1) a distinctive landscape developed on highly soluble rocks and (2) a unique type of drainage pattern resulting from karst processes. As all rocks are soluble to some extent, karst must develop only on those rocks that are particularly susceptible to solution. In such situations, the solution process can create and enlarge cavities within rocks. This leads to the progressive integration of voids beneath the surface and allows large amounts of water to be funneled into an underground drainage system while simultaneously disrupting the pattern of surface flow. Because hydrology exists in physical "symbiosis" with solution, it is a very important aspect of karst phenomena. Solution integrates spaces, allowing pronounced underground circulation of water that, in turn, promotes further solution. A true karst area, therefore, possesses a predominantly underground drainage with a poorly developed surface network of streams. The same processes may apply in many areas that have surface drainage but have not developed a strong enough topography to be considered as karst. Importantly, however, the area may still be affected by karst-forming processes.

The term "karst" is a German adaptation of the Slavic word *Kras* or *krs* and the Italian word *carso*, which literally mean "a bleak waterless place" (Monroe 1970) and also connote a bare rock surface. The early and classical description of karst was derived from a high plateau area near the Adriatic Sea between northwest Italy and the area formerly known as Yugoslavia (Gams 1993). This region is characterized by irregular topography containing many closed depressions and interrupted stream valleys; thus, early observers of the region considered karst as a geomorphic freak with chaotic and disordered topographic expression. Most geologists and geographers thought of karst as a curiosity rather than a topic for serious scientific investigation. It was not until 1893, when Jovan Cvijíc (rhymes with "screech") published his book *Das Karstphanomen*, that karst geomorphology was given true scientific status. Cvijíc's work clearly defines karst landforms and demonstrates the predominance of solution in their development, although the solution effect had been alluded to in earlier studies (Sawkins 1869; Cox 1874).

Karstlands (or karst landforms) are found in almost every region of the world, including arctic and arid zones, but they are most likely to occur in temperate or tropical climates. In the United States, karst has developed wherever conditions are favorable, but the major concentrations *in the US*

Figure 12.1
Major karst areas of the United States.
Boundaries are generalized, and some regions
shown in solid black contain some nonkarst
areas.
(Palmer 1984)

of features exist in several general areas (fig. 12.1): (1) the Valley and Ridge province of the Appalachian Mountains (Pennsylvania, Maryland), (2) central Florida, (3) the plateau region of east-central Missouri, (4) a belt extending from south-central Indiana into west-central Kentucky, and (5) the Edwards Plateau (Texas).

The field of karst geomorphology has its roots in central and eastern Europe, and contributions by English-speaking geomorphologists were rather meager during the formative years of the discipline. In recent years, however, these geomorphologists have added greatly to our knowledge of the discipline, especially by recognizing that analysis of karst processes and karst landscapes requires the integration of hydrologic, chemical, and mathematical concepts as well as traditional geomorphology (Palmer 1984). In addition, the modern emphasis in karst geomorphology is largely on process and involves the application of conceptual and numerical models to practical problems involving engineering, water supply, and aquifer contamination.

The historical development of karst science explains the proliferation of terminology that has arisen for karst features and processes. Although the terrain in the former Yugoslavia inspired the study of karst, not all the names for individual karst features come from there, and almost every country has developed a particular terminology of karst geomorphology. Table 12.1 presents an abbreviated glossary of karst terminology adopted by the U.S. Geological Survey, which is adopted in this discussion. The complete listing of terms can be found in Monroe (1970).

THE PROCESSES AND THEIR CONTROLS

Karst Rocks—The Resisting Framework

It is tempting to say that karst develops primarily on limestones and leave it there without further elaboration.

Although technically correct, this statement is grossly misleading because some limestones are not potential harbingers of karst topography, whereas karst landforms may develop in other lithologies. Nevertheless, most karst is associated with limestones that exhibit a unique combination of properties that allow them to succumb to karstification. We will briefly examine what rock properties are most conducive to karstification and why.

Lithology   As a general rock group, limestones show great variability, but the accepted definition is that limestone is a rock containing at least 50 percent carbonate minerals, most of which occur in the form of calcite ($CaCO_3$). Although very young limestones may contain some aragonite, the two most common carbonate minerals in limestones are a low-magnesium calcite, containing 1 to 4 percent magnesium, and dolomite (Sweeting 1973). If more than 50 percent of the carbonate minerals are calcite, the rock is called limestone; if more than 50 percent of the carbonate minerals are dolomite, the rock is called dolomite (Leighton and Pendexter 1962). The purer the limestone is with respect to $CaCO_3$, the greater will be its tendency to form karst. Corbel (1957), for example, suggests that 60 percent $CaCO_3$ is needed before any karst will form, and about 90 percent is required to expect a fully developed karst region. However, even pure limestones may not produce a karst terrain because the processes also depend on other factors.

Given enough time, rocks other than limestone can produce karst if ancillary conditions are proper and the material is sufficiently soluble. Dolomites are commonly karstified, but unless they are very pure their porosity and permeability tend to be low. Moreover, dolomites tend to be much less soluble than rocks containing calcite, inhibiting the development of fully karsted terrain. Evaporites such as gypsum and halite are also prone to karstification (Klimchouk et al. 1996). In fact, gypsum can be dissolved about 100 times faster than pure lime-

TABLE 12.1	Abbreviated Glossary of Common Karst Terminology.

Term	Definition
Aggressive water	Water having the ability to dissolve rocks, especially water containing dissolved CO_2.
Blind valley	A valley that ends suddenly where its stream disappears underground.
Cave	A natural underground room or series of rooms and passages large enough to be entered by a person.
Chamber	The largest order of cavity in a cave or cave system.
Closed depression	Any closed topographic basin having no external drainage, regardless of origin or size.
Cockpit	(1) Any closed depression having steep sides. (2) A star-shaped depression having a conical or slightly concave floor.
Cockpit karst	Tropical karst topography containing many closed depressions surrounded by conical hills. Similar to cone karst.
Cone karst	Tropical karst with star-shaped depressions at base of many steep-sided, cone-shaped hills.
Doline	A basin or funnel-shaped hollow in limestone ranging in diameter from a few meters to a kilometer and in depth from a few to several hundred meters. May be distinguished as "solution" or "collapse" if precise origin is known. In United States, most dolines are referred to as sinks or sinkholes.
Exsurgence	Point at which underground stream reaches the surface if the stream has no known surface headwaters.
Karren	Channels or furrows caused by solution on massive bare limestone surfaces. Synonym *lapiés*.
Karst plain	A plain on which closed depressions, subterranean drainage, and other karst features may be developed. Also called *karst plateau*.
Karst topography	Topography dominated by features of solutional origin.
Karst valley	(1) Elongate solution valley. (2) Valley produced by collapse of a cavern roof.
Karst window	Depression revealing a part of a subterranean river flowing across its floor, or an unroofed part of a cave.
Karstic	Adjective form of karst.
Karstification	Action by water, mainly chemical but also mechanical, that produces features of a karst topography.
Mogote	A steep-sided hill of limestone generally surrounded by nearly flat alluviated plains. Generally used for karst residual hills in the tropics. Synonym *pepino*.
Polje	A very large closed depression in areas of karst topography, having flat floors and steep perimeter walls.
Resurgence	Point at which an underground stream reaches the surface. Reemergence of a river that has earlier sunk upstream.
Room	A part of a cave system that is wider than a normal passage. Similar to chamber.
Speleothem	A secondary mineral deposit formed in caves.
Swallet, swallow hole	A place where water disappears underground in a limestone region. A swallow hole generally implies water loss in a closed depression or blind valley. A swallet may refer to water loss in a streambed even though there is no depression. Also called *ponor, sink, sinkhole, stream sink*.
Terra rossa	Reddish-brown soil mantling limestone bedrock; may be residual in some places.
Tower karst	Karst topography characterized by isolated limestone hills separated by areas of alluvium. Towers generally steep-sided and forest-covered hills, often with flat tops.
Uvala	Large closed depression formed by the coalescence of several dolines; compound doline.

After Monroe (1970), USGS Water Supply Paper 1899k.

stone and, thus, the formation of karst landforms in some evaporates can occur within a few decades (Benito et al. 1995). In general, however, the occurrence of karst in dolomites and evaporites is minor compared to the widespread distribution of limestone karst.

Porosity and Permeability In addition to lithology, the creation of karst depends on how much water any rock can hold and how easily the water moves through the rock system. Porosity is a measure of the water-storing capacity of a given rock, and it is usually expressed as the percentage of void spaces in the rock:

$$P = V_v/V \times 100$$

where P is porosity in percent, V_v is the volume of voids, and V is the total volume of the material. Presumably, open textures and higher porosities facilitate the solution process and the development of karst because the rock will hold more water. Some evidence exists to support this contention (Sweeting 1973). More important, however, is the type of porosity; that is, whether it is primary or secondary. *Primary porosity* relates to intergranular void spaces created during formation of the rock. This type of porosity commonly decreases with time by precipitation of cement, recrystallization, and changes in mineralogy. In older limestones, for example, calcite tends to be replaced by dolomite, a process that usually decreases the primary porosity (Powers 1962). In addition, diagenesis or metamorphism may decrease carbonate porosity by inducing an increase in calcite grain size and reducing void size by pressure. Thus, primary porosity may be an ephemeral characteristic of limestones and perhaps is not as important in karstification as secondary porosity and permeability.

Secondary porosity comes from openings in rock that occur along bedding-plane partings or as fractures, such as joints and fault zones. The nature and pattern of these openings may be the single most important factor in karstification. Not only do they allow the rocks to hold more water, but they also promote circulation within the system by increasing permeability. Because permeability (the capacity to transmit water) depends on the continuity of voids, even rocks with high primary porosity may develop little karst if no avenues of flow are present.

Porosity, then, has real importance only if the system is also permeable. Zones of weakness increase porosity but, more important, they have a decided influence on the permeability. This explains why permeability rates vary by as much as five orders of magnitude depending on the size and interconnectedness of fractures and partings.

Joints are important avenues of water transport and enhanced permeability. Usually joints occur in patterns with one dominant direction and a secondary set of joints intersecting the main set at angles between 70° and 90°. The spacing of the joint sets can be very significant in the genesis of karst. If intersecting planes are too close, the rock may be highly permeable but too weak to allow the full development of karst.

Faults also transmit water effectively, but their precise role in karstification varies according to local conditions (Stringfield and LeGrand 1969). For example, fault zones sometimes have low permeability where voids are occupied by secondary mineralization associated with ore deposits. Such a deterrent to karstification, however, may be partly offset if the ore includes sulfide minerals, because their oxidation produces sulfuric acid, which makes permeating fluids more aggressive in the solution process (Pohl and White 1965; Morehouse 1968).

In summary, the full development of karst depends primarily on whether water capable of solution passes through a rock sequence along discrete flow paths with enough discharge to create significant solution openings. Given this requirement, the process is aided by having a thick sequence of pure, crystalline limestone that is not interrupted by major insoluble beds. In addition, some relief must be available to permit free circulation of water in the system.

The Driving Mechanics and Controls

The Solution Process The solution process itself is in reality the critical function in the entire analysis of karst. Regardless of how conducive climate, lithology, fractures, and other variables are to karstification, karst topography would never develop if the solution process were somehow rendered inoperative. Its function or malfunction is fundamental to the topic, and we must attempt to understand its mechanics, at least in its simplest terms.

Laboratory studies reveal that the mineral calcite, like all common minerals, is soluble in pure water. At saturation it is soluble to the extent of about 12 to 15 ppm depending on the temperature of the water. This solubility is rather startling when compared to natural river waters where concentrations of Ca^{+2} and bicarbonate (HCO_3^-) are much greater (Picknett et al. 1976), indicating a substantial increase in the solubility. Since we are considering the same substance (calcite), it is obvious that the solvent in the natural system is not pure water. Rainwater is not pure because it incorporates a variety of chemical constituents as it passes through the atmosphere. The most important of these is carbon dioxide (CO_2), which is soluble in pure water.

$$CO_2 \overset{H_2O}{\rightleftharpoons} CO_{2(dissolved)} \tag{1}$$

Some of the dissolved CO_2 reacts with the water to form a weak acid (H_2CO_3), called carbonic acid:

$$CO_{2(dissolved)} + H_2O \rightleftharpoons H_2CO_3 \tag{2}$$

At pH values typically found in karst terrains, this acid rapidly dissociates into its ionic state (fig. 12.2) and the above reaction can be expressed more realistically as

$$CO_{2(dissolved)} + H_2O \rightleftharpoons H^+ + HCO_3^- \tag{3}$$

The amount of CO_2 actually dissolved in water depends on the partial pressure of carbon dioxide (P_{CO_2}) in the air standing at the air-water interface and on the temperature of the water. The air in contact with the water can be in the atmosphere, in spaces within the soil, or in subterranean cavities such as caves. In any case, the amount of CO_2 dissolved in the water increases as the P_{CO_2} of the air increases and as the temperature of the water decreases. Colder water will dissolve more CO_2 than warm water at any given P_{CO_2} value. In the atmosphere P_{CO_2} is rather small, having values averaging about 0.03 percent of volume (3×10^{-4} bar). Higher values, ranging from 0.05 to 2 percent of volume, are common in caves (Jennings 1985). Anomalously high P_{CO_2} values are found in air contained within soil or the vegetal litter covering it. Values of 1 to 2 percent of volume are common, and some poorly ventilated tropical soils may contain more than 10 percent (Jennings 1971). The abnormal CO_2 values in soil air, and the resulting large amounts of dissolved CO_2 in the soil water, stem largely from the respiration of plant roots and microbial action involved in the decomposition of vegetal matter. This biogenic CO_2 is regarded by most karst experts as being the prime ingredient in the solution process.

Calcite itself is dissociated into an ionic state (fig. 12.2) such that

$$CaCO_{3(calcite)} \rightleftharpoons Ca^{+2} + CO_3^{-2} \tag{4}$$

However, the CO_3^{-2} ion produced quickly reacts with the H^+ formed when CO_2 is dissolved in water and thus

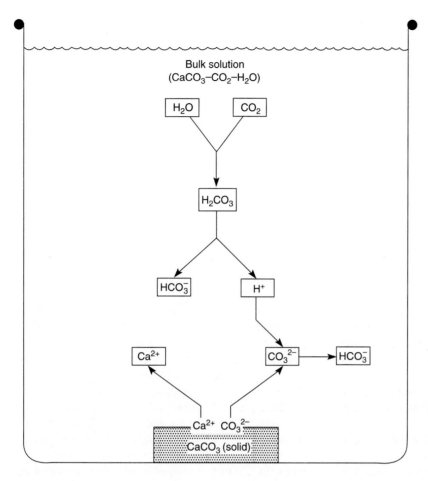

Bulk solution
(CaCO₃–CO₂–H₂O)

Figure 12.2
Steps involved in the dissolution of calcite.

the dissociation of calcite also produces a bicarbonate ion because

$$CO_3^{-2} + H^+ \rightleftharpoons HCO_3^- \qquad (5)$$

These reactions demonstrate that the solution of limestone revolves around the $CaCO_3$-CO_2-H_2O chemical system. This system is extremely complicated, and its mechanics are much more sophisticated than this introductory treatment can show. We are not dealing with a single reaction that produces solution of calcite, but a process that involves a series of reversible and mutually interdependent reactions, all proceeding at different rates and each regulated by different equilibrium constraints. We will therefore explain the process only in the most general terms.

If we combine reactions 3 and 4, a general form of the process can be expressed by the following:

$$CaCO_3 + H_2O + CO_{2(dissolved)} = Ca^{+2} + 2HCO_3^- \quad (6)$$

The bicarbonate ions (HCO_3^-) are derived from two sources shown in equations 3 and 5. In equation 5, the reaction of carbonate ions (from the dissociation of $CaCO_3$) and the H^+ (from the dissolving of CO_2 in water) produces a disequilibrium between the P_{CO_2} in the air and the P_{CO_2} in the water. This causes more CO_2 to diffuse from the air into the water and allows further

solution of the calcite because reaction 6 is driven to the right side of the reversible equilibrium equation.

To a large extent the amount of CO_2 dissolved in water regulates the solubility of limestone. Where the amount of CO_2 dissolved in water is high, the fluid will aggressively attack calcite. Soil water with its high P_{CO_2} is therefore the most effective solution agent (Drake and Wigley 1975).

Another important mechanism in the solution process is mixing corrosion first described by Bögli (1964). Essentially, when two bodies of water that are saturated with calcite, but with different CO_2 contents are mixed, the resulting fluid needs less CO_2 to establish equilibrium than the sum of the CO_2 contained in the two original fluids. Some CO_2 is released and becomes available to promote further solution (fig. 12.3). Similarly, the mixing of a saturated solution and an unsaturated solution, or two unsaturated solutions may result in more aggressive waters as long as the original CO_2 contents differ. However, the effect of mixing corrosion is favored by large differences in CO_2 content, especially if one fluid exhibits low CO_2 concentrations (Palmer 1991). Dissolution is reduced or inhibited if either of the fluids is supersaturated (Thrailkill 1968). In natural waters, mixing corrosion typically produces solutions that are capable of dissolving 1–2 percent more calcite, but

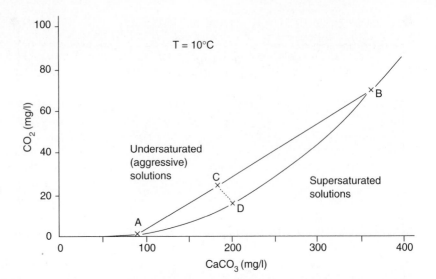

Figure 12.3
The equilibrium curve (ADB) for CaCO$_3$ at 10° C in water charged with CO$_2$. The mixing of two solutions (A and B) produces a solution (C) that is undersaturated, resulting in the additional dissolution of CaCO$_3$.

(Figure 9.4, p. 377 from SOLUTE PROCESSES by S. T. Trudgill. Copyright © 1986. Reproduced by permission of John Wiley & Sons Limited.)

in the extreme case, as much as 20 percent more calcite may be dissolved (Gunn 1986).

Ford and Williams (1989) note that some investigators question the significance of mixing corrosion on the basis that groundwater is rarely saturated with respect to calcite and where undersaturated the effects are reduced. Nevertheless, mixing corrosion provides an appealing explanation for the commonly observed enlargement of joints and bedding planes at or immediately below the water table, in the zone where vadose and phreatic waters mix.

Effects similar to mixing corrosion are produced by the merging of fluids with differing salinities (Back et al. 1984; Wigley and Plummer 1976). This process is very significant where seawater mixes with fresh water in carbonate regions. In that situation, mixing produces brackish water that is undersaturated with respect to calcite and enhances the solution process (Runnells 1969; Plummer 1975; Hanshaw and Back 1980). Thus, coastal mixing zones can represent sites of intensive karstification.

Solution Rates—The Controlling Factors Normally, karst in regions of low rainfall and high temperature is quite subdued and features such as collapsed and deranged surface drainage are not as striking. In extremely cold arctic or subarctic regions, water is generally present. However, it is often frozen and free circulation is inhibited (Smith 1969; Ford 1987); therefore, karst is a rather anomalous phenomenon. Karst landscapes are dominated by depressional features in humid, temperate climates and are abundant in areas of soluble rock because free water is available to circulate through the strata. However, the greatest abundance and variety of karst landforms are found in tropical climates with lush vegetation (Jennings and Bik 1962; Monroe 1976), but the most prevalent forms are residual hills and not depressions characteristic of temperate karst.

Given that the type and intensity of karst landform development varies with climate, precipitation and temperature have historically been seen as the primary factors controlling karstification. Water not only serves as the solvent in the development of karst but also encourages the growth of vegetation and enhances soil microbial activity that adds extra CO$_2$ to the system. Similarly, the production of CO$_2$ increases with temperature (Miotke 1974) allowing more carbonate dissolution.

Solution rates within any given region are usually estimated by direct measurement of the dissolved load in streams leaving a karst area, although other techniques can be employed (for an excellent discussion see Jennings 1983). The rate is equal to the concentration of dissolved load multiplied by the stream discharge such that

$$S = CQ$$

where S is the solution rate in units of mass/time, C is the solute concentration (mass/volume), and Q is stream discharge (volume/time). Many measurements can be integrated over time and converted into rates of surface lowering, or *karst denudation*.

Solution rates have now been measured in a number of drainages located in diverse climates. Examination of the compiled results shows that the most substantial effect of climate is related to runoff; the more water available to circulate through the system, the higher the rates of limestone removal (Smith and Atkinson 1976; Jennings 1985). A somewhat surprising result is that the estimated solution rates *for a given discharge* are similar between warm, tropical environments and humid, temperate regions. This suggests that the role of temperature and CO$_2$ in creating differences in calcite solution rates between climatic regimes may be limited. Such a conclusion is supported by the work of White (1984) who examined the relative importance of temperature, P_{CO_2} and runoff on karst denudation using a geochemical modeling approach. White's model predicts that runoff should be the primary factor controlling karst denudation. The significance of P_{CO_2} is muted in the model by a cubed root function (White 1984; Ford et al. 1988). Thus, a change in P_{CO_2} by a factor of 100, as might occur as water moves from the atmosphere into warm, humid soils, would increase denudation rates by only about a factor of 5 (White 1988). Ford et al. (1988) point out that the changes in soil P_{CO_2} from tropical to humid temperature regions are much less and, therefore, changes in denudation rates as a result of different P_{CO_2} should be limited. Temperature was found by White (1984) to be the least important factor controlling denudation rates. In sum, then, runoff appears to be the most significant parameter controlling calcite dissolution on a regional scale, followed by the P_{CO_2} content of the water, and, finally, water temperature.

Spatial Variations in Limestone Solution   During the past decade significant effort has been placed on understanding the spatial variations in the rates of limestone dissolution that occur in a given area. As water percolates from the surface downward through fractures, the solution rates decrease with flow distance. This occurs because water gets closer to saturation with respect to calcite the farther it travels. Arguments have been made that if the decrease in the solution rate changes linearly with the percentage of saturation, aggressive, percolating groundwaters would obtain saturation only after a short distance of downward travel. Some observations support this contention. For example, in many areas 70 percent of limestone corrosion occurs within the first 10 m of the ground surface (Ford and Williams 1989). Nonetheless, it is clear that if linear rates of saturation existed in nature the transport distance for complete saturation would be so short that long, linear caves would be difficult, if not impossible, to develop (Palmer 1984). More recent studies have demonstrated that in significantly undersaturated conditions solution rates are linear, but as equilibrium is approached the rates decrease dramatically (fig. 12.4). The change in corrosion rates occurs at about 85 percent saturation (Dreybrodt 1990, 1998; Ford et al. 1988). The abrupt change in solution rates allows corrosion to continue for a greater time over a longer distance but at a much reduced rate. Changes in corrosion rates that occur as saturated conditions are approached not only provide an explanation for enhanced solution near the surface but also help to explain the origin of elongate cave systems.

As our knowledge of the solution process grows, we are beginning to recognize that the enlargement of joints and bedding planes (commonly referred to collectively as fractures) is a much more complex phenomenon than originally envisioned. The enlargement process depends not only on the degree of saturation, but also on the hydraulic gradient, the length of travel from the recharge to the discharge point, and the initial width of the fracture through which the water is moving (Dreybrodt 1990). Moreover, the evolution of a fracture into a large conduit system involves two primary phases of development that are controlled by different processes and that function at distinctly different rates (Dreybrodt 1990; Groves and Howard 1994; Dreybrodt et al. 1997).

Initially, fracture widths are quite narrow, measuring in many carbonate terrains from 50 to 500 μm (White 1999). Water flows through the fractures at a very slow rate under laminar conditions from a site of recharge to a point of discharge. During this initial phase of enlargement, saturated conditions are quickly approached as the water moves through the fracture allowing slightly aggressive waters to be carried significant distances through the aquifer. Calcite dissolution occurs through the entire length of the fracture, but at a slow rate. Eventually, however, the width of the fracture increases to the point where waters with less than 85 percent saturation can travel from the site of recharge to the point of discharge. This typically occurs when the fractures reach approximately 1 cm in diameter, and is referred to as *breakthrough*. Following breakthrough, the dissolution rate increases dramatically allowing for rapid enlargement of the cavity and turbulent flow conditions that are capable of carrying clastic debris (White 1999). In addition, corrosion rates become nearly independent of the discharge of water through the conduit, but are primarily controlled by the rates of reaction at the bedrock surface and, thus, the degree of undersaturation (Palmer 1991; Dreybrodt et al. 1997).

Most carbonate aquifers contain only a limited number of conduit systems. This appears to be related to the fact that of all the fractured flow paths that could potentially be enlarged, the one with the largest average initial width will reach breakthrough first. Once breakthrough occurs (and dissolution rates increase), it will grow at the expense of the others as it becomes the dominant flow path. The time required for breakthrough under the hydraulic gradients that are typically found in nature appear to range from about 10,000 years to a few million years (Dreybrodt et al. 1997). Groves and Howard (1994) argue, however, that if the initial fracture widths are too small (<100 μm), breakthrough cannot occur within any reasonable geologic time frame.

KARST HYDROLOGY AND DRAINAGE CHARACTERISTICS

Surface Flow

Karst terrains are commonly classified as either holokarst or fluviokarst. **Holokarst** was initially described by Cvijíc (1893) while studying landforms in the thick limestone sequences of the former Yugoslavia. These limestone terrains are characterized by the full array of closed depressions occurring at a variety of scales. Within holokarst little evidence exists that fluvial processes have previously affected the landscape, and precipitation is drained underground with little or no channelized flow (Jennings 1985; White 1990). Most karst terrains are classified as **fluviokarst** (initially referred to as merokarst by Cvijíc). Regions of fluviokarst are characterized by channelized surface water flow, and it is generally clear that karst landforms have been superimposed on a former fluvial landscape.

Karst drainage networks, including those in fluviokarst, characteristically are disrupted, and few rivers can traverse such areas in a continuous and unsegmented manner. The reason for this strange fluvial behavior is the facility with which surface flow can be diverted into the underground system. Depending on soil types, vegetation, and joint spacing, overland flow on hillslopes may be drastically reduced by infiltration, and in extreme cases no water will enter the nearby channels.

Rivers also lose water when some of the flow descends into **swallow holes** or **swallets.** These are nothing more than open cavities on the channel floor that are capable of pirating a portion of a river's discharge or even the entire river (especially during low flow) into the underground system. Thus, a large part of the total flow of rivers in karst regions may follow a subsurface route that may or may not parallel the path of the river valley on the surface. Rivers that are able to cross a karst terrain as continuous surface entities have distinct hydrologic characteristics. For example, flood records for 114 basins of different size in Pennsylvania show that the mean annual floods (with a recurrence interval of 2.33 years) in carbonate basins were considerably lower than those in basins underlain by different rock types (White and Reich 1970). Commonly, the peak flow is spread over a longer period as the subsurface water is slowly released to the rivers. It appears likely that the precise hydrologic character of surface rivers in karstic areas depends greatly on the state of development of the underground drainage, especially the degree of interconnection between subsurface passageways (Ede 1975).

Karst Aquifers and Groundwater Flow

In our brief look at hydrogeology in chapter 5 we examined the flow of groundwater through porous and permeable materials. The movement of groundwater beneath karst terrains differs from flow through these previously discussed materials in that karsted strata typically exhibit much larger variations in porosity and permeability. This is due to the fact that water in karst aquifers can move through (1) intergranular pores within the unfractured bedrock (matrix permeability), (2) joints and bedding planes imparted to the strata following deposition and lithification (fracture permeability), and (3) conduits (with widths > 1 cm) that have been enlarged by aggressive solutions (conduit permeability) (White 1999). In addition, the dimensions of fractures and conduits and, thus, permeability of the aquifer materials, can change through time, a feature that is generally not observed in other types of aquifers.

Historically, karst geomorphologists have viewed the hydrology in karstic systems in discordant ways. Most conflicting opinions concerned the type of water movement, the depth to which groundwater penetrates, and most important, whether or not a water table as we normally conceive it exists in karstic terrains. Some investigators completely deny the existence of a water table and with it the

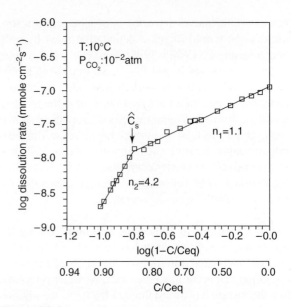

Figure 12.4

Dissolution rates for calcite in closed system at a P_{CO_2} content of 0.01 atmospheres. (From Dreybrodt 1998)

(Figure 2.2, p. 34, Chapter 2 "Dissolution rates for calcite in a closed system at a P_{co_2} content of 0.01 atmospheres" by Wolfgang Dreybrodt from GLOBAL KARST CORRELATION, © 1998 by Yan Daoxian and Liu Zaihau. Reprinted by permission of VSP International Science Publishers, The Netherlands.)

distinct vertical zonation of water below the surface of the Earth. Workers accepting this view believe that the distribution and movement of groundwater are controlled entirely by the spatial characteristics of the interconnected network of passageways within the rocks. Several points support this conclusion: (1) adjacent wells drilled into limestone often have hydrostatic levels that are significantly different in elevation; (2) tunnels reveal dry fissures immediately next to cracks that are filled with water; (3) tracing of water with dyes shows that paths of movement often cross one another (obviously flowing at different levels) and even pass under surface streams from one side of a valley to the other, a physical impossibility in nonkarsted, unconfined aquifers; and (4) poljes, periodically flooded karst depressions, at the same level do not behave in a similar manner; some flood in winter while others remain dry. Thus, the underground system is thought to be a collection of conduits functioning like three-dimensional rivers.

There is little question that extreme changes in porosity, permeability, and flow allow water levels to fluctuate dramatically over short distances beneath karst terrains. However, it is now generally accepted that the water table concept is valid in karsted areas provided it is applied on a regional scale (Palmer 1990; 1991; White 1988). In fact, detailed mapping of water-table levels has been performed in a number of karsted regions (e.g., the Mammoth Cave area by Quinlan 1982). These maps demonstrate that large conduit systems generally form groundwater troughs that collect waters flowing through the matrix and fractures of the surrounding bedrock as well as from openings at the ground surface (fig. 12.5).

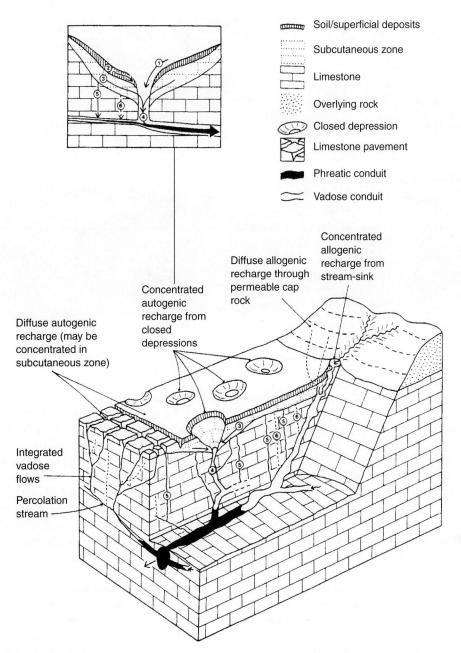

Soil/superficial deposits
Subcutaneous zone
Limestone
Overlying rock
Closed depression
Limestone pavement
Phreatic conduit
Vadose conduit

Diffuse allogenic recharge through permeable cap rock

Concentrated allogenic recharge from stream-sink

Concentrated autogenic recharge from closed depressions

Diffuse autogenic recharge (may be concentrated in subcutaneous zone)

Integrated vadose flows

Percolation stream

Figure 12.5
Conceptual model of karst hydrology.
(Figure 9.8, p. 400. from SOLUTE PROCESSES by S. T. Trudgill. Copyright © 1986. Reprinted by permission of John Wiley & Sons Limited.)

The conduits subsequently discharge their waters at springs. During flood, however, flows can be reversed so that the conduits can recharge the surrounding rocks, and in some cases, flood flows can reoccupy dry conduits that exist at higher elevations, creating highly complex patterns of groundwater flow.

The Relations Between Surface and Groundwater

Assuming that flow through conduits is the unique characteristic of karst hydrology, we should briefly examine how water is exchanged between the surface and the underground system and vice versa, and how the properties of the passageways control the movement. In many karst areas, runoff from catchment areas having rocks with low solubility flows onto the limestone terrain where it is diverted into the groundwater system through vertical openings and swallets. This input of water to karstic aquifers is referred to as *allogenic recharge* (fig. 12.5). In addition, precipitation falling directly onto karsted rocks may percolate slowly through interconnected spaces in the zone of aeration as *vadose seepage* where it is added to the total accumulation of groundwater. Beneath

Figure 12.6
Subcutaneous zone in limestone along Route 55 near Perryville, Mo.

closed depressions of a karst landscape, water input to the underground system is through a highly permeable zone at the base of the depression. The permeable zone usually represents an intersection of vertical joints or cylindrical solution openings called *shafts.* Downward percolation here usually occurs rapidly as *vadose flow.* Precipitated water reaches the input zone by one or more of the possible transmission routes associated with the sloping surface of the enclosed depression. These routes are similar to those discussed in chapter 5 for "normal" fluvial systems, with the added process of flow through the subcutaneous zone. The subcutaneous or epikarstic zone stands above the phreatic zone, but it stores water and is periodically saturated, especially after storms. The subcutaneous zone develops by enhanced solution immediately beneath the soil that enlarges cavities and fractures and creates high permeability and porosity (fig. 12.6). The solutional enlargement decreases with depth, however, and at some level openings are too small to transmit water at the same rate as in the overlying, more permeable rock. As a result, water is stored in the form of a perched water table. The perched water table slopes toward points of rapid vertical percolation (e.g., major joints, faults, shafts). Thus, a lateral component of flow develops in the subcutaneous zone and converges at those points. As we will show during the discussion of solutional doline formation, subcutaneous flow has a significant influence on the hydrologic characteristics of the system and on landforms developed in karst regions. Water entering the groundwater system by moving only through the limestone as either vadose seepage or vadose flow represents *autogenic recharge* (fig. 12.5).

An alternative method of describing groundwater recharge has been presented by Smart and Hobbs (1986). In their model, emphasis is placed on the mode of entry into the karst aquifer rather than the source of the water, and recharge is viewed as a continuum ranging between two end-member states: *concentrated recharge,* characterized by large inputs at discrete points, and *dispersed recharge* defined by smaller inputs at a large number of sites. The type of recharge that predominates depends on the materials overlying the karsted rock, the initial porosity of the bedrock, and the magnitude of aquifer karstification. In general, aquifers with conduits are associated with concentrated recharge.

The various possible combinations of flow types exhibit different water chemistry. For example, allogenic recharge, derived from flow off rocks of low solubility, is often highly aggressive because it has not come into contact with carbonate rocks. In contrast, vadose seepage may be saturated to supersaturated with respect to calcite (White 1990). Variations in water chemistry have prompted many workers to suggest that the type of flow will have a decided effect on the extent of solution and the morphologic evolution of the karst (Gunn 1981, 1983; Ford and Williams 1989; White 1990). Dreybrodt et al. (1997) argue, however, that during the initial phase of fracture enlargement, dissolution rates are primarily dependent on the discharge of water through the fracture. Therefore, karstification will not be substantially enhanced by allogenic waters. However, dissolution rates should be increased by allogenic recharge during the second phase of enlargement (i.e., following breakthrough).

Figure 12.7
Spring emergence along solution-enlarged bedding planes and
joints in limestone along the wall of a sinkhole in Nippenose
Valley, northcentral Pennsylvania.
(Photo by R. Craig Kochel)

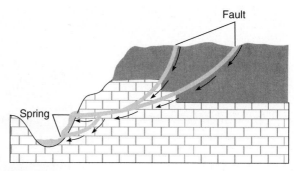

Figure 12.8
Fault control on the position of springs. Rainwater enters
shales exposed at the surface along faults. Water follows fault
zones into and through limestones until it emerges as springs.
(Arrows show movement paths of water.)

The manner in which groundwater leaves the system is also dependent on the aquifer and flow type. In aquifers dominated by diffuse flow, water will typically emerge according to the rules that govern groundwater flow in nonkarsted terrains. In conduit systems, however, water usually reaches the ground surface in the form of springs.

Springs in karst areas are usually larger and more permanent than those in other regions because of the high infiltration associated with karst rocks (fig. 12.7). Those springs or seeps stemming from flow in diffuse aquifers are called **exsurgences** (fed by seepage), in contrast to those fed by groundwater moving through distinct conduits, which are called **resurgences.** This distinction is often blurred since normal diffuse flow may encounter cavities during its movement and ultimately emerge mixed with water derived from sinking streams (Thrailkill 1972). In general, however, resurgences and exsurgences can be differentiated on the basis of their variability in discharge and chemistry, thereby providing important information on the distribution and properties of the aquifer itself. For example, Shuster and White (1971), analyzing 14 springs in the

central Appalachians, suggest that diffuse-flow aquifers maintain a constant hardness and have spring water nearly saturated with respect to $CaCO_3$. Aquifers that utilize conduits are undersaturated and show considerable variability in hardness (Thrailkill 1972; Drake and Harmon 1973). More recent studies in other regions have shown that the variations in water chemistry are more complex than those observed by Shuster and White (1971). Nonetheless, springs characterized by large variations in hardness must be fed to some degree by conduits, the sudden declines in hardness being due to injection of low-hardness storm water from sinking streams (White 1999). Thus, fluctuations in the hardness of spring waters provide a good indication of the proportion of water derived from allogenic and internal conduit flow compared with that contributed from diffuse flow.

The physical controls on spring position are quite variable. In general, however, springs emerge at base level which is determined in many cases by the elevation of major entrenched rivers that are able to traverse the limestone terrain or by sea level when the karst is located in coastal areas. Spring positions are further constrained by large structural features or by stratigraphic relationships. Faults sometimes serve as the locus of artesian springs, especially if the structural relationships produce a large potential difference between the recharge area and the fault (Burdon and Safadi 1963). In other situations, faults may deflect underground flow until it emerges into a valley that stands well below the level of the recharge zone (fig. 12.8). An excellent example of this is in the area of the Kaibab Plateau in Arizona (Huntoon 1974). The plateau north of the Grand Canyon supports no surface streams, and all precipitation not consumed by evaporation infiltrates into porous surface rocks. Huntoon (1974) observed that water collected in those rocks is transmitted to faults that act as drains, funneling the water downward to carbonate rocks that stand almost 1000 m below the plateau surface. The water is then discharged through springs at the base of the canyon.

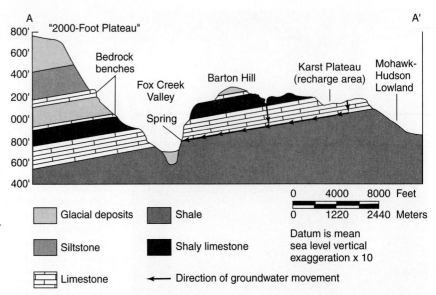

Figure 12.9
Stratigraphic control on the movement of groundwater. Water moves down-dip at contact of limestones and impermeable shale.
(From Baker 1973a)

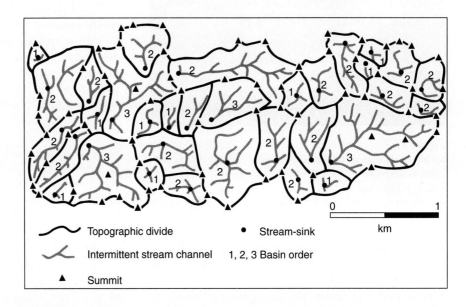

Figure 12.10
Components of karst morphometry and hydrology in New Guinea as ordered by P. W. Williams.
(Williams 1972b)

Regional stratigraphy and the dip of beds may also control the position of springs. Baker (1973a) showed that in New York State spring placement often results from water moving down-dip in well-integrated conduits floored by rocks with low permeability (fig. 12.9).

Morphology of Karst Drainage One trend in karst geomorphology has been the attempt to describe karst drainage and landforms in quantitative terms (Baker 1973a; P.W. Williams 1966, 1971, 1972a, 1972b; La Valle 1967, 1968; Troester et al. 1984; Miller et al. 1990). For example, Williams devised a method of ordering stream segments, following the Strahler method, such that every swallet would accept the drainage of a particular order (fig. 12.10). After each sink is ordered, measurements of morphometric parameters are made and plotted against the ordered hierarchy to demonstrate a statistical relationship. Baker (1973a) extended

the Williams methodology by including in the analysis underground links between the swallets and the springs that were identified by mapping and dye tracing. Stream orders so designated also showed a high correlation to the number of streams and total length of streams (fig. 12.11). In addition, streams flowing in conduit systems seem to maintain longitudinal profiles at regional spatial scales that are similar to those of surface stream channels in the same area (White and White 1983).

Results of these studies show karst drainage to be well organized rather than having the chaotic nature assumed by early investigators. In fact, karst drainage systems in many parts of the world have a similar spatial organization (Ford and Williams 1989). In addition, these morphometric studies illustrate that in areas of fluviokarst, surface and subsurface drainage systems co-evolve as a single entity (Kastning 1990).

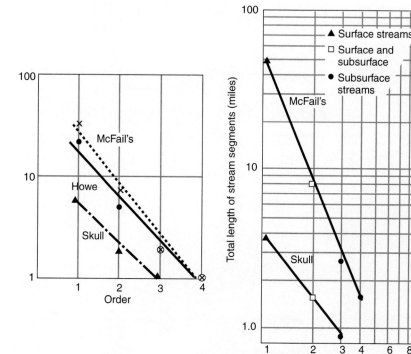

Figure 12.11
Morphometric relationships of the karst hydrology, including subsurface segments, in an area of eastern New York state.
(From Baker 1973a)

SURFICIAL LANDFORMS

The solution process, working on rocks with diverse properties, results in a number of surficial landforms that define a true karst. The features range in size from tiny modifications of exposed limestone outcrops to large depressions and hills that dominate the topography. However, they all manifest the corrosional process.

Closed Depressions

If someone were to ask what kind of landform best typifies a karst terrain, the answer would have to be closed depressions. Although these depressions range in size from tiny holes to those covering wide areas, they all have in common the property of supporting no external surficial drainage. In addition, it seems unlikely that widespread surficial depressions can develop unless they are connected to an underground conduit system in which water is free to flow to a spring outlet at a lower level (Palmer 1984; Ford and Williams 1988).

Dolines By far the most common karst landform is a closed hollow of small or moderate size called a *doline* (in the United States often called a *sink* or *sinkhole*). Dolines are usually wider than they are deep, having diameters ranging from 10 m to 1 km and depths between 2 m and about 500 m. In plan they are circular or elliptical, but their cross-profile shape can vary considerably from the normal funnel-like form to shapes resembling a disk, bowl, or cylinder. Occasionally isolated dolines

occur, but more commonly they are abundant enough to provide a karst terrain with a strongly pitted appearance, such as the one in figure 12.12. Doline densities exceeding 2500 per km^2 have been reported (Ford et al. 1989), but more commonly they range from 1 to 9 per km^2. As figure 12.13 illustrates, dolines can justifiably be considered the fundamental element of karst because when present in large numbers they substitute for the valleys that dominate a fluvial landscape.

The term "doline" has had so many different connotations that some authors have called for its elimination, but its use is so widespread that dropping the term is virtually impossible. In fact, Cvijíc (1893) used the term in his classic book on karst, and classifications of doline types are firmly established (Cramer 1941). A variety of doline types have been described, including *solution dolines, collapse dolines, alluvial dolines,* and *solution subsidences* (Cramer 1941). We will focus on the solution and collapse dolines because they are predominant forms. Although these are treated as separate features, most dolines are a combination of the solutional and collapse types.

Solutional Dolines Simply stated, **solution dolines** form as water infiltrating into joints and fissures enlarges the cracks by corrosion. More intensively corroded zones begin to develop a closed surface depression called a solution doline. Many reports have documented this surface-downward origin, and it seems clear that this form represents the paramount doline type.

Figure 12.12
Ground surface pitted by dolines, Monroe County, Ill.

In many cases, but not all, the development of a solution doline is related to flow in the subcutaneous zone (fig. 12.14). Water infiltrating from the surface is ponded as fissure openings narrow downward, forming a perched water table at relatively shallow depths. Locally, however, fractures have been enlarged, allowing rapid downward flow. A cone of depression similar to that of a pumping well is produced in the perched water table as downward flow rates in the enlarged fractures exceed that of the surrounding area (Ford and Williams 1989). Lateral flow toward the fractures results and the increased movement of water intensifies corrosion. It now seems clear that the dimensions of the developing solution doline are controlled by the radius of the subcutaneous drawdown cone (Ford and Williams 1989).

Like all geomorphic features, solution dolines develop best where controlling factors are combined in a particular way. The factors most conducive to the formation of solutional dolines are these:

1. *Slope*. Because ponding or retardation of lateral surface flow accelerates infiltration, the frequency of solution dolines is inversely proportional to the surface gradient. Steep slopes promote rapid flow across the surface, and so valley floors or gently undulating plains are the best places for solvent action to initiate the process. Dolines formed on steeper slopes tend to be asymmetric; however, that characteristic can also be produced in other ways.

2. *Lithology and structure*. Porous limestones are less susceptible to solutional doline formation than dense limestones that are well jointed. The joints allow selective solution rather than a uniform corrosion over the entire surface. Structures tend to

Figure 12.13
Area of southern Indiana showing well-developed karst topography. Note the predominance of dolines and the absence of surface drainage. Part of the Corydon East quadrangle, Ind. (U.S.G.S. 7 1/29′); contour interval = 10 feet.

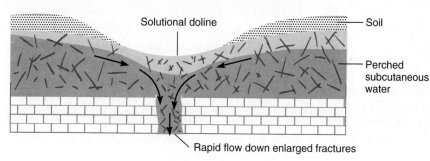

Figure 12.14
Formation of solutional doline aided by subcutaneous flow that is directed to zones having enlarged fractures and high permeability.

align and elongate the dolines parallel to the major trends (LaValle 1967; Matschinski 1968; Kemmerly 1976), but the degree of control depends on many variables.

3. *Vegetation and soil cover.* Soil and vegetal cover usually increase solution activity because of the CO_2 factor. Other factors being equal, solution dolines will develop more rapidly under an organic-rich cover than where surfaces are bare.

Collapse Dolines   Collapse dolines differ from solutional dolines in that the depressions are initiated by solution that occurs beneath the surface. For example, the expansion of caverns, caused by corrosion and by the failure of roof material falling under gravity, may lead to collapse by decreasing the support of the overlying rock material. In a study of Tennessee karst, Kemmerly (1980a) suggested that most collapse dolines occur in the partially weathered residuum overlying the

solid bedrock. In this process vertical fractures beneath the residuum are gradually widened by solution. This creates a bridge of unconsolidated debris, or *regolith arch,* that is supported by pinnacles of the underlying bedrock (fig. 12.15). Further widening of the arch by solution, coupled with the simultaneous spalling of debris from beneath the arch of sediment, makes it impossible to support the overlying mass, and collapse occurs. The size of the opening required to produce roof collapse depends on the thickness of the overburden and its overall strength. In areas where the overlying unconsolidated debris is relatively thick, roof collapse may be facilitated by piping that creates cavities in the overburden above the enlarged solution feature (Benito et al. 1995; Hyatt and Jacobs 1996). The growth of the pipes toward the ground surface, particularly during floods, eventually leads to instability and roof failure.

Morphologically, collapse dolines tend to have greater depth/width ratios than solution dolines. Their

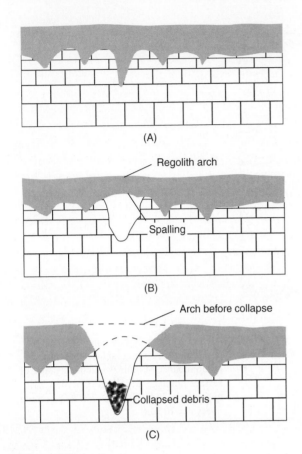

Figure 12.15
Sequential development of collapse doline.

Figure 12.16
Office complex downtown Allentown, Penn., damaged beyond repair by sinkhole collapse in winter 1994. The building was destroyed by demolition crews two weeks after the photograph was taken.

sidewalls are characteristically steep and rocky, and the bottom is filled with fragments of the collapsed debris. Nevertheless, postcollapse modifications in doline morphology by corrosion, mass wasting, and surface drainage often make it impossible to distinguish collapse from solutional dolines.

The collapse process is facilitated in humid environments where underground drainage is well established. Some evidence suggests that collapsing can be initiated by rapid lowering or repeated fluctuation of the water table that alters the buoyant effects of pore water and, thus, stress on the cavity's roof rocks (Kemmerly 1980a; Benito et al. 1995; Hyatt and Jacobs 1996). Although geological processes such as river entrenchment can cause this, the association of drawdown with collapse dolines demonstrates that humans can be a geomorphic agent (Benito et al. 1995) commonly with catastrophic results (fig. 12.16). A classic example of human intervention into the natural balance has been reported by Foose (1967). Near Johannesburg, South Africa, a mining company lowered the water table by an extensive pumping program in order to access deeper ore than existed in the phreatic zone. Several years after the project's completion, large collapse dolines, exceeding 50 m in diameter and 30 m deep, began to form sud-

denly and with disastrous results. In December 1962 an ore-refining plant dropped 30 m into a doline as the surface suddenly collapsed, and 29 men were killed in one terrifying moment.

Doline Morphometry The belief that karst forms are controlled by climate was the prevailing concept in karst geomorphology after World War I. This reliance on climate to explain karst processes and form maintained its dominance until investigators became concerned that too much variability of form existed in the same local region where climatic controls were identical (Jennings and Bik 1962; Verstappen 1964). In the 1960s investigators concerned with the hydrologic and geochemical process of karstification began to study the variations in spatial morphology of karst features. For example, Hack concluded as early as 1960 that doline density varies significantly with rock type in the Shenandoah Valley of Virginia (Hack 1960a).

Although analyses of linear karst features have been made (White and White 1979), most morphometric work has concentrated on the spatial characteristics of dolines, relating those to some geologic or hydrologic variable (LaValle 1967, 1968; Matschinski 1968; Williams 1971, 1972a, 1972b; Kemmerly 1976; Palmquist 1979; White and White 1979; Mills and Starnes 1983; Benito et al. 1995). These studies have been conducted in many different parts of the world and under a variety of geologic and climatic conditions. For example, Williams (1966, 1971, 1972a, 1972b) developed a sophisticated spatial analysis of tropical karst topography. Using aerial photographs or topographic maps, the karst terrain was separated into divides, summits, channels, and stream sinks. Each closed depression was given a number representing the highest stream order of the drainage that disappears into the sink (see fig. 12.10). The topographic divides surrounding the depressions form a polygonal network, indicating that the terrain is completely partitioned into separate and adjoining basins. This topography, which Williams calls *polygonal karst,* is dominated by the hills but dynamically controlled by the position of the sinks. Williams analyzed the pattern of sinks by measuring the average distance from each sink to its closest neighbor and comparing that value to the expected mean distance, which is determined from a density analysis of the sink population. The index ratio $L\bar{a}/L\bar{e}$ (where $L\bar{a}$ is the mean actual distance and $L\bar{e}$ is the mean expected distance) tells whether the sinks have a random or uniform distribution. In New Guinea the stream-sink dispersion is highly uniform, which Williams (1971, 1972a, 1972b) interprets as the best accommodation that can be made as depressions compete for space when the topography evolves under the processes of doline formation.

The example just given reveals the primary assumption of all morphometric studies, that derived relationships will provide insight into the basic factors controlling landform morphometry and, through that insight, an understanding of the formative processes. In fact, Palmquist (1979) identified the independent variables that initiate and enlarge dolines as (1) groundwater recharge, (2) secondary permeability, (3) regolith thickness and shear strength, and (4) hydraulic gradient. From these he proposed a process-response model in which the initiation and enlargement of primary dolines leads to the generation of secondary dolines. The implication is that mixed doline populations can exist within the same karst area.

Kemmerly (1982) demonstrated that such a mixture of primary and secondary dolines is present in the Western Highland Rim area of Kentucky and Tennessee. In that region, one population consists of large, joint-controlled dolines having second-order (or higher) internal drainage and wide spacing. A second population has smaller, first-order swallet depressions that exhibit no joint control. The large dolines are apparently primary in nature and reflect high groundwater recharge, permeability, and hydraulic gradients, while the smaller dolines probably developed where recharge, permeability, and hydraulic gradient were considerably less.

Assuming the preceding models are correct, we are led to several pertinent observations about karst depressions. First, the initiation of secondary dolines in the manner suggested by Palmquist (1979) requires that those depressions must be linked to an unbroken conduit system. It also follows that no depression can form until sediment covering the swallet is flushed into and through the system (Palmer 1984). Second, the distinction of primary and secondary dolines indicates that time is an important factor in the development of a fully Karsted terrain, a fact emphasized by many researchers (Cooke 1973; Palmer and Palmer 1975; Wells 1976; Kemmerly and Towe 1978; Kemmerly 1980).

The emphasis on morphometry does not mean that climate is irrelevant in karst processes. Obviously, it is an important factor, mainly because it produces the water needed for karst processes to work. Thus, what morphometry should demonstrate is that given sufficient time, karst landforms, like most surface features, may approach and possibly attain a dynamic equilibrium, the properties of which are determined by the relationship between process and local controlling factors.

Uvalas and Poljes   Closed depressions of larger size than dolines are called uvalas or poljes. **Uvalas** form as dolines enlarge and coalesce or as smaller dolines develop in older, larger dolines (Jennings 1985). Uvalas are characterized by hollows with undulating floors and more than one topographic low, the irregularity being produced by the differences in size of the integrated dolines. Jennings (1967) reports a single uvala that was constructed from 14 separate dolines of diverse size and shape. Uvalas have no specific size requirements as they range from 5 to 1000 m in diameter and from 1 to 200 m in depth. Their plan shape can be highly irregular as a result of their unique origin.

Poljes are relatively large closed depressions with flat bottoms and steep sides. They are irregular in plan and usually are elongated along the strike of bedding or some zone of structural weakness. Thus, they can be structurally or lithologically controlled, and some expand by pronounced lateral corrosion when they are temporarily filled with water. Gams (1978) places a rather vague minimum size requirement on poljes, suggesting that their flat-floor must equal or exceed 400 m in width.

Poljes often abut against impermeable and nonsoluble rocks, and rivers flowing over those rocks may extend partly across the polje surface before they sink. In times of high flow the sink may not be able to absorb the discharge and shallow lakes occupy the polje basin. In dry seasons evaporation may destroy the lakes. As discussed earlier, polje lakes may also form and disappear with changes in the underground hydrology.

Figure 12.17
Gorge with steep, canyonlike walls, Turkey.

Karst Valleys

The second major topographic group in karst topography is karst valleys, which show a variety of characteristics. In general, they can be divided into several types with clearly discernible properties.

Allogenic Valleys **Allogenic valleys** head in insoluble rocks adjacent to the karstic area. As the chemically aggressive surface flow enters the more soluble terrain, it may incise the karst forming an allogenic valley. Occasionally, surface flows form spectacular gorges with steep, canyonlike walls (fig. 12.17). For instance, the Tarn Gorge in the Grands Causses area of France is 2 km wide and 300 m deep. Such magnificent valleys obviously require considerable discharge to develop, with a combination of solution and fluvial abrasion as the driving mechanics.

Blind and Dry Valleys Most surface rivers traversing a karst surface, no matter where they originate, eventually sink into the underground system. They may disappear into holes within the channel (swallets or swallow holes), or the entire drainage may be inside a doline such

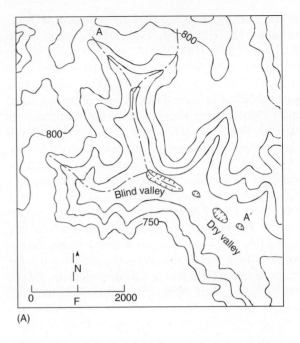

(A)

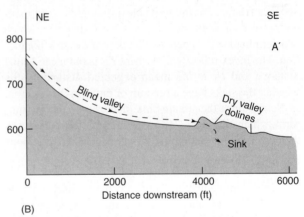

(B)

Figure 12.18
Blind and dry valleys near Moberly, Ind. (A) Plan view.
(B) Longitudinal profile.

that all surface water (confined in a channel or not) sinks into the base of the doline. As figure 12.18B shows, rivers flowing across a karst surface in well-defined valleys tend to lower the valley floor upstream from the sink faster than the reach downstream from the sink because the flow is drastically diminished at the point of infiltration. Eventually a limestone scarp that faces upstream may develop, separating the two channel reaches. Typically, streamflow enters the subsurface at the base of the scarp (fig. 12.19). During floods, however, flow stages may exceed the height of the scarp, allowing floodwaters to continue downvalley on the surface.

Scarp heights can vary from a few meters in small streams to tens of meters in larger rivers (Malott 1939). Usually the size of the cliff increases with age. Rivers terminating at the cliff face are said to be occupying *blind valleys* (fig. 12.18).

Figure 12.19
Sinking stream occupying blind valley in south-central Indiana.

Dry valleys have all the properties of normal fluvial valleys except, as the name implies, they have no well-defined watercourses or carry only ephemeral flow in response to massive floods upstream (fig. 12.18). They probably represent the most common form of karst valley and are the most favorable locale for doline formation because rainwater tends to pond on the valley surface and sink into the rock fractures. Like all karst valleys, their sides tend to be steep, but factors of lithology and age cause some variability in their cross-profile shape. The origin of dry valleys can be complex and varied, but in the simplest and most common case they represent the downstream reach of a blind valley that absorbs all the surface flow at a particular sink (fig. 12.18).

Pocket Valleys **Pocket valleys** are essentially the opposite of blind valleys in that they begin where groundwater resurges rather than where it sinks. They are normally associated with large springs that resurge on top of an impermeable substrate at the foot of a thick exposure of karst limestone.

Pocket valleys, sometimes called steepheads, are usually U-shaped in cross-profile, having steep sidewalls. They characteristically have a steeply inclined headwall that may be recessed by spring sapping that undermines the overlying limestone. Their dimensions

vary depending on the spring discharge and the nature of the karst rocks, but some are 8 km long, 1000 m wide, and 300 to 400 m deep (Sweeting 1973).

Cockpit and Tower Karst

It is safe to say that every landform recognized in karst terrains of temperate regions is present in tropical karst. Nevertheless, depressional karst characteristic of temperate climates contrasts sharply with the spectacular karst in humid-tropical climates that is dominated by steep-sided, cone-shaped residual hills.

The study of the residual hill forms in tropical karst environments has been carried on in diverse regions of the world and, once again, a confusing array of terms has developed to describe the same feature. However, the residual hills, which are numerous and transitional, can be generally subdivided into two topographic components called cockpits and towers, landforms that seem to be diagnostic elements of tropical karst. In fact, the terms "cockpit karst" and "tower karst" are deeply entrenched in the literature.

Cockpits are similar to temperate-climate dolines except that they are usually irregular or star-shaped depressions that surround the residual hills (fig. 12.20). Cockpit karst was first described and named in Jamaica

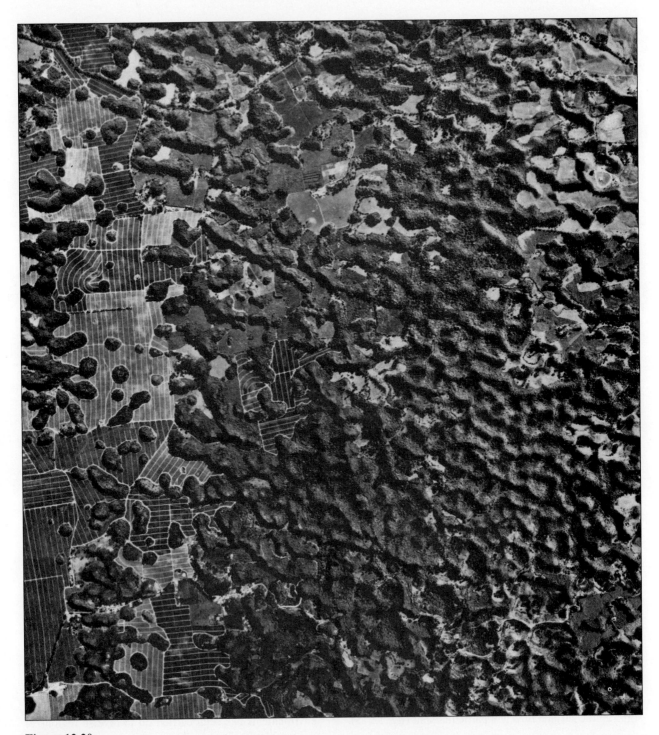

Figure 12.20
Aerial view of cockpit karst.

and was originally ascribed to solution along joints and faults (Lehmann 1936; Sweeting 1958). Later work by Aub (as quoted by Sweeting 1973) and Williams (1971) suggests that the star shape of cockpits arises from gully erosion of centripetal streams flowing into the depressions. However, cockpits, like most normal dolines, develop by solution from the surface downward. Where cockpit karst is prevalent, the residual hills are often joined together to form linear ridges that in turn are cut

by gullies into a sawtooth configuration. The alignment of the ridges and the cockpits is apparently so random that joint control seems to be ruled out, although the gullies may be related to zones of structural weakness.

Residual hills can be steep-sided and many are vertical or even overhanging. However, the steep inclination of these residual hills can change where surface erosion piles talus at the slope base (McDonald 1975). Some classifications distinguish between residual hill

Figure 12.21
Tower karst, China.

forms with vertical as compared to near-vertical sides. Those with vertical sides are referred to as *towers* while hills with near vertical sides are called *cones* (White 1988). For simplicity and because the morphologic differences are highly transitional, we will refer to both forms as towers (fig. 12.21).

According to Jennings (1971), tower karst differs from cockpit karst in the steepness of the residual hills and the presence of swampy plains (often similar to poljes) surrounding the towers rather than the depressed cockpits. Towers can be of dramatic size, sometimes rising several hundred meters above the surrounding plain (Wilford and Wall 1965). Towers have also been called pepinos, haystacks, and commonly *mogotes,* a term used in Cuba, Vietnam (Silar 1965), and Puerto Rico (Monroe 1976). In fact, Monroe (1976) suggests that the term "mogote" is probably more appropriate for the feature and should be universally accepted.

Many of the Puerto Rican towers (mogotes) are asymmetric in cross-profile (fig. 12.22) and considerable attention has been given to the development of this asymmetry. Monroe (1976) attributes it to case-hardening of the limestones on the windward slope by repeated solution and reprecipitation of $CaCO_3$. The downwind slopes, being less resistant, are oversteepened and even

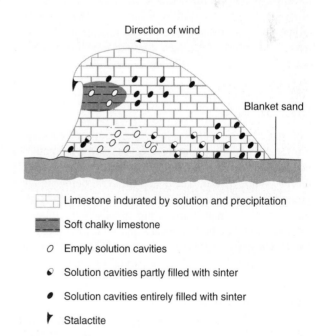

Direction of wind

Blanket sand

☐ Limestone indurated by solution and precipitation

▨ Soft chalky limestone

○ Empty solution cavities

◐ Solution cavities partly filled with sinter

● Solution cavities entirely filled with sinter

▼ Stalactite

Figure 12.22
Diagram showing characteristics of an asymmetric mogote in Puerto Rico.
(Monroe 1976)

overhanging due to slumping of the less-indurated rocks. Day (1978), however, found that the asymmetry occurs in only 35 percent of the mogotes and it does not have a simple pattern reflecting wind direction. Day suggests that the asymmetry is related primarily to erosion at the base, with induration being a secondary factor. The basal erosion is most likely solutional, but mechanical erosion may be possible (McDonald 1979).

Some controversy exists as to what processes cause towers to develop. Williams (1985, 1987, 1990) notes that cockpit karst in southern China occurs where the water table is below the bottom of the enclosed depressions and the vadose zone is thick. In contrast, the plains between the residual hills in tower karst rest near or on the water table. Williams argues that a threshold is reached by vertical lowering of cockpit karst to the water table, where the dominant direction of drainage and corrosion is changed from the vertical to the horizontal. Subsequently, cockpits evolve into towers as the lows between the residual hills are corroded away and undercutting and other processes steepen their sides. From this perspective, towers possess a location, shape, and size that is related to that of the original cockpit karst. However, Yaun (1987) found that, unlike cockpit karst, tower karst near Guilin, China, was most prevalent in the vicinity of allogenic surface runoff. He concluded that towers do not evolve from cockpits but result predominantly from fluvial processes operating at the surface. Other theories of tower formation have also been suggested, which cite intensive corrosion along structural weaknesses in the bedrock, collapse of caves, river erosion, and differential solution of the karst rocks. Given that tower karst is observed in a variety of climatic and geologic settings, they are probably polygenetic (Panos and Stelcl 1968; Williams 1987).

LIMESTONE CAVES

Any survey of karst processes and landforms must include a brief discussion of limestone caves. Caves are natural underground cavities that can be traversed by a human explorer (White 1976). Technically, caves, being underground features, are not part of karst topography. They may, however, create surface topography by facilitating collapse, and they both influence and reflect the mode of karst hydrology that exists in any particular region. So caves can justly be considered as part of the karst system.

Cave Physiography

As defined, caves have entrances, passages, rooms and blockages called terminations. The assemblage of these components in different combinations produces caves with a variety of shapes and patterns.

Entrances and Terminations Cave entrances can be found in places such as doline bottoms, hillsides, spring mouths, roadcuts, and quarries. Perhaps the most spectacular entrances to cave systems are openings into vertical voids called *shafts* or *solution chimneys*. Shafts (Pohl 1955) are cylindrical in shape and evidently form by solution when water moves rapidly down their walls (Brucker et al. 1972). They are most common in flat-lying rock sequences such as those in the Appalachian Plateau, where shafts sometimes are more than 100 m deep and 15 m wide. Shafts characteristically drain at the bottom through a narrow opening that is connected to a larger cave system. "Solution chimney" is a term used to encompass all vertical or nearly vertical openings that do not have the cylindrical shape of a shaft (White 1976, 1988). Commonly solution chimneys follow bedding planes or joints and may exhibit horizontal, sloping, and vertical segments.

Caves terminate when the passages narrow to the point that a person can no longer follow the opening. These terminations result from the collapse of overlying rocks, narrowing of the voids into permeable but impassable units, or filling of caves to the ceiling with silt and clay deposits. Recognize, however, that all caves are part of an integrated network of flow paths even if terminations exist. Water following cave routes still continues through terminations to a spring outlet.

Passages and Passage Morphology Passages are the main physiographic component of caves. They have various shapes, sizes, and patterns, as shown in figure 12.23. In plan view, individual passages are either **linear, angulate,** or **sinuous** depending on hydrologic controls and the presence and orientation of structural features such as joints, bedding planes, and occasionally, faults.

In cross-section, the shape of a passage represents an accommodation between structural properties of the rock that impart the passage with the initial shape of the fissure and hydraulics that tend to modify the passage into smooth, streamlined forms (White 1988). Where flow velocities are high in thick, karsted limestones, the passages are usually controlled by hydraulics and exist mainly as narrow, vertical slits called **canyons** (fig. 12.24) or more circular voids known as **elliptical tubes.** As described below, canyons are usually formed in the vadose zone, while tubes develop in the phreatic zone. Passages that are controlled primarily by rock structure are much more irregular in cross-section (see fig. 12.23). Realistically, many variations exist between the different types, and strict categorization of passage shapes is probably meaningless.

Almost all passages show a variety of small, sculptured markings on their walls, ceilings, and floors (see Bretz 1942 for details). Some are formed by solution and others by abrasion. In addition, many rooms and passages have an abundance of speleothems (chemically precipitated dripstone deposits) that inspire the popular fascination with caves. These features (e.g., stalactites, stalagmites) are composed of a $CaCO_3$ substance called

theories as to how "towers" develop

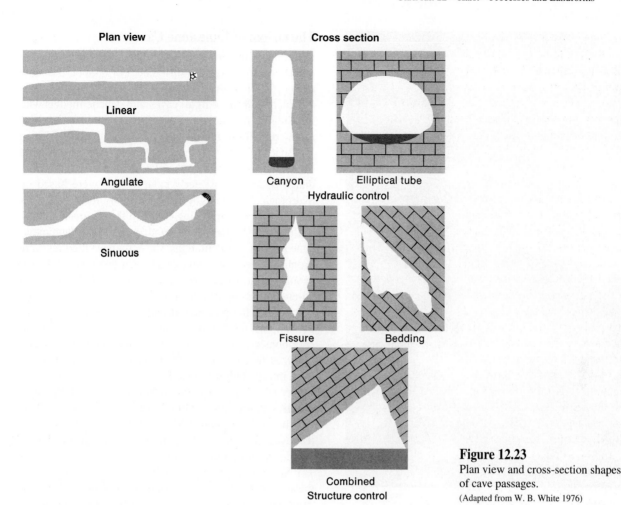

Figure 12.23
Plan view and cross-section shapes
of cave passages.
(Adapted from W. B. White 1976)

travertine. They form when downward-permeating water, saturated with respect to calcite, reaches the cave passage. At this point CO_2 is diffused from the water to the cave atmosphere because the P_{CO_2} in the water is considerably greater (Holland et al. 1964). Some water may also be lost by evaporation. In either case (loss of water or loss of CO_2), $CaCO_3$ must precipitate as shown in the general reversible reaction 6 presented earlier in the chapter.

Passages are usually integrated into three-dimensional conduit systems referred to as cave patterns. In general, the two major types are the branch-work pattern and the maze pattern (fig. 12.25). Although the shape of an individual passage is governed in part by properties of the rock, the type of cave pattern that develops is governed largely by the mode of groundwater recharge (Palmer 1991).

Branchwork patterns, which are by far the most common, are formed by the downward movement of aggressive water into the karsted rock at discrete points. Karst depressions, such as dolines, often serve as water collectors prior to movement into the groundwater system. Branchwork systems consist of tubular or canyon passages that join as tributaries in the downflow direction much like a normal surficial stream network. The pattern develops because, of all the possible flow routes, only a

few are sufficiently open to transport high discharges on low hydraulic gradients. These evolve, via corrosion, into the major passageways. As they do, hydraulic gradients in the major passages are reduced, attracting water in adjacent fissures that are fed in large part by point sources. With time, a branchwork pattern develops.

In contrast to the branchwork pattern, the *maze pattern* consists of many intersecting passages that form closed loops. The pattern has been divided on the basis of geometry into three subtypes: *network mazes, anastomotic mazes,* and *spongework mazes* (Palmer 1975). Maze systems develop as a result of simultaneous, uniform enlargement of openings. Palmer (1991) notes that maze patterns typically form where (1) water infiltrating into soluble, highly porous rock regains its aggressiveness upon mixing with waters of a different chemistry (e.g., in mixing corrosion), (2) diffuse recharge through an insoluble cap rock uniformly transmits aggressive water to the underlying limestones, allowing all fractures to enlarge at approximately the same rate, and (3) water is injected into the fissures surrounding an existing passage under steep hydraulic gradients—a process that is common during flood. Some maze networks may also develop in response to the upward movement of aggressive, hydrothermal waters (Gillieson 1996).

Figure 12.24
Canyon passage, Dixon Cave, Ky.

The Origin of Limestone Caves

Historically, the origin of limestone caves was steeped in intense controversy. The differences of opinion probably stemmed from several factors. First, theorists of cave formation attempted to fit all caves into genetic models with little regard for geologic differences. For example, there is no compelling reason why caves in highly deformed rocks of the Alpine region should develop in precisely the same way as those in the plateau areas of southern Indiana. Second, most early cave models were based on hydrological schemes, yet until recently there was a remarkable lack of hydrologic data in the arguments. Most of the early theories were founded on physiographic evidence. Third, at the time that the early cave theories were proposed, very few caves had been examined in great enough detail to substantiate a local origin, let alone an all-encompassing genetic model.

During the past few decades, studies aimed at understanding the origin of cave systems have increased enormously and the collected information no longer emphasizes cave physiography. Cave development has now been interpreted in terms of the structural and lithologic properties of the rock, groundwater flow, and chemical kinetics (Ford and Ewers 1978; Palmer 1975, 1981; Dreybrodt 1981, 1990; White 1978, 1988). These studies reveal that cave formation includes a highly complex set of processes and sophisticated mechanics.

Several generalized conclusions can be drawn from these investigations. For example, it is now clear that cave development is often guided by rock structure or changes in lithology. In fact, many rock sequences con-

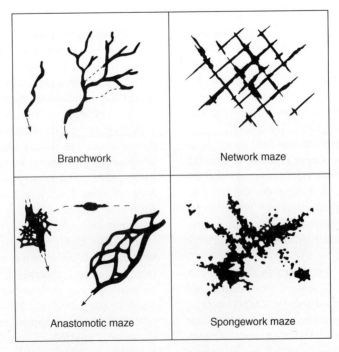

Branchwork

Network maze

Anastomotic maze

Spongework maze

Figure 12.25
Branchwork and maze patterns of cave development.
(Palmer 1991)

tain stratum or horizons, which Lowe (1992) refers to as *inception horizons,* along which cave passages preferentially develop. For instance, caves may preferentially form along the contact of pure and impure limestones or within limestones located immediately above an impermeable unit such as a shale or cemented-sandstone bed (Lowe and Gunn 1997). Perhaps of more importance is the realization that caves tend to develop along groundwater flow routes of least resistance, allowing corrosion to enlarge fissures that transmit relatively high volumes of water. Thus, cave passages may form above, along, or below the water table. In fact, vadose and phreatic caves are interconnected and may develop simultaneously. They are, however, morphologically different, and the transition can often be recognized even when both are currently dry (Palmer 1991). Vadose caves result mainly from gravitational flow that always moves downward

along the steepest available routes as determined by the width of prevailing fissures. Vadose passages, dominated by canyons and shafts, are typically independent of one another, converging only when forced to do so by the controlling properties of the rock (as in the case of flow down the limbs of a syncline). In contrast, phreatic caves, dominated by tubes, develop along hydraulically efficient paths, or routes that can transmit flow with the least expenditure of energy (Dreybrodt 1990; Palmer 1991). Phreatic passages tend to converge and show few consistent trends with rock structure. Changes in passage elevations may be highly irregular, descending well below the water table and subsequently rising in the downflow direction to or near the level of the water table. The frequency and depth of these loops may be related to fissure densities (fig. 12.26); more intense fracturing is associated with numerous, shallow loops or, in

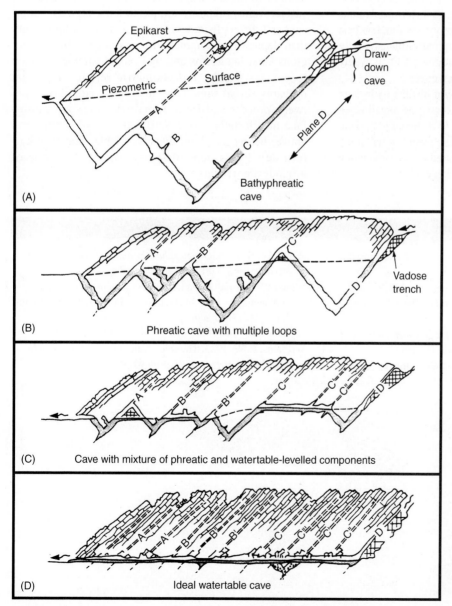

Figure 12.26

Differing types of long cave profiles as a function of fracture density. Fissure density increases from (A) to (D). (From Ford and Williams 1989)

(Figure "Differing types of long cave profiles as a function of fracture density" from KARST GEOMORPHOLOGY AND HYDROLOGY, p. 262 by Derek Ford and Paul Williams, 1989. Reprinted by permission of Derek Ford.)

some cases, ideal water table caves (Ford and Ewers 1978; Ford and Williams 1989).

The development and integration of cave systems is also time-dependent. Gillieson (1996), for example, argues that caves evolve from an initial state in which no conduits are present through a long, slow initiation phase where conduits with dimensions of about 1 cm are created, to a developmental state when conduits are extended and enlarged. During this latter period of evolution, the conduit or passage will respond to external factors of base level change and tectonism, a process that may lead to passage entrenchment or the development of new cave levels. Eventually, the passage(s) will be abandoned and collapse, closing the cave passages of a particular level. It is important to note that the initiation and developmental stages cited in this evolutionary sequence correspond to steps of conduit formation preceding and following breakthrough that were discussed earlier in the chapter (Dreybodt 1990; Groves and Howard 1994; Dreybodt et al. 1997). The evolutionary development of the caves as envisioned by Gillieson (1996) may require in excess of 10^5 to 10^6 years to complete.

SUMMARY

Karst develops by the extensive solution of limestones and other soluble rock types. Limestones that are most susceptible to karstification are crystalline, high in calcite content, and intensely fractured. The limestones should also stand well above base level and have no impermeable units in the stratigraphic sequence. Karst processes function best in humid-temperate and humid-tropical climates where abundant water is available and biogenic CO_2 is freely added to the water from a thick vegetal cover.

Karst areas are characterized by a distinct hydrology in which surface flow is partially or totally disrupted when water is diverted into the underground system. Groundwater movement in karst aquifers differs from normal situations in that the flow is often controlled by the orientation of fractures and interconnected solution cavities.

The most common karst landforms in humid-temperate climates are enclosed depressions called dolines (sinks or sinkholes in the United States). Other distinct forms are larger depressions and a variety of surface valleys that are unique to karst regions. Tropical karst differs from temperate karst in that hills rather than depressions dominate the topography. In most regions karst landforms and hydrologic parameters seem to possess regular morphometric relationships that probably reveal some type of dynamic equilibrium between process and form. Limestone caves are subsurface manifestations of karstification. Caves may form above, below, or along the water table. The cave pattern developed depends primarily on the mode of groundwater recharge.

SUGGESTED READINGS

The following references provide greater detail concerning the concepts discussed in this chapter.

Dreybrodt, W., Gabrovšek, R., and Siemers, J. 1997. Dynamics of the evolution of early karst. *Theoretical and Applied Karstology* 10:9–27.

Ford, D., and Williams, P.W. 1989. *Karst geomorphology and hydrology.* Winchester, Mass.: Unwin Hyman.

Palmer, A. N. 1991. Origin and morphology of limestone caves. *Geol. Soc. Amer. Bull.* 103:1–21.

White, W. B. 1988. *Geomorphology and hydrology of karst terrains.* New York: Oxford Univ. Press.

White, W. B. 1999. Karst hydrology: Recent developments and open questions. In Beck, B. F., Pettit A. J., and Herring J. G. eds., *Hydrogeology and engineering geology of sinkholes and karst, Balkema, Rotterdam.* pp. 3–20.

Williams, P. W. 1983. The role of the subcutaneous zone in karst hydrology. *Jour. Hydrol.* 61:45–67.

COASTAL PROCESSES AND LANDFORMS

R. Craig Kochel

Introduction
Coastal Processes
    Waves
        Wave Generation
        Wave Modification Near the Coast
        Tsunamis and Seiches
    Tides and Currents
        Tides
        Nearshore Currents
Beaches
    The Beach Profile and Equilibrium
        Nearshore Bars
        Beach Morphodynamics
    Shoreline Configurations and Beach Landforms
        Beach Cusps
        Large-Scale Rhythmic Topography and Capes
Coastal Topography
    High-Relief, Erosional Shorelines
    Low-Relief, Depositional Shorelines
Shoreline Change
    Rates of Change
    Causes of Shoreline Erosion
        Rising Sea Level
        Coastal Storms
Barrier Islands
    Distribution and Characteristics
    Geomorphic Processes and Dynamics
Summary
Suggested Readings

INTRODUCTION

In the preceding chapters we have examined a variety of processes that mold the exposed surface of Earth into diagnostic landforms. However, 70 percent of our planet is covered by water that possesses immense energy. The expenditure of this energy on the adjacent land can produce rapid and enormous changes in the nearshore physical environment. Therefore, the erosional and depositional processes that function at the interface between water and the land, in the region known as the *coastal zone,* represent another major realm of geomorphology.

Coastal zones respond to forces like any other geomorphic system. Viewed over the short term, parts of the coast might be considered quasi-equilibrium forms. Repeated movement of sediment and water constructs a beach profile that reflects some balance between the average daily or seasonal wave climate and the resistance of the landmass to this wave energy. Considered over a longer time span (graded time), however, beaches or entire coastlines may be imperceptibly changing toward a larger equilibrium form. On an even longer geologic time scale, marine transgressions and regressions may represent dramatic alterations of the position and character of the coast. Our attention here will focus mainly on the coastal zone during steady and graded time spans.

The study of beaches and coasts is not strictly academic. Most of the world's largest cities are near the ocean. Three-fourths of the people in the United States live in coastal states (fig. 13.1). The concentration of people in coastal regions and the pressures involved in using this land for recreation and development have placed a substantial strain on the system. Coastal zones provide habitat to the majority of migratory waterfowl and shellfish, but they also serve as the nursery grounds for countless marine organisms and some freshwater species. Whether we can utilize the environment completely and still prevent damaging changes in its character depends on how well we understand the geomorphic processes and patterns of the coast. Engineers, geographers, oceanographers, geologists, and other scientists have made significant contributions to our present understanding of the coastal environment. As usual, however, interdisciplinary communication has not been great because each discipline is engrossed in specific problems. Nonetheless, if demands on the coastal system continue to increase, the cooperative efforts of all coastal workers are necessary to avert environmental disaster.

Because the coastal zone occurs at the boundary of the terrestrial and oceanic sectors of the planet, the region is truly dynamic. The coastal zone is affected by variations in tectonic and climatic activity, the two main drivers of geomorphic activity worldwide. Any change in eustatic or isostatic sea level caused by these forces will be reflected in adjustments of shoreline shape, position, and sediment dynamics. The sensitivity of the coastal zone to changes in geoenvironmental parameters can manifest itself on local, regional, and global scales and over a variety of geomorphic time scales. Accordingly, environmental geoscientists are looking at these areas as an important source of information on the Holocene record of past processes such as sea-level change. Additionally, coastal regions will be important in the assessment of the impacts related to global climatic changes in the future.

A coast is a relatively large physiographic zone that extends for hundreds of kilometers along a shoreline and often several kilometers inland from the shore. Like all large physiographic entities, coasts have not escaped the irrepressible urge of scientists to classify, and many attempts have been made to categorize coastal features (see King 1972). Coasts are particularly susceptible to classification because the regional properties are clearly documented on available maps and aerial photographs. In some cases, no extensive field study is needed to systematize coastal properties into a viable classification scheme.

Some of the commonly used classifications of coasts combine, in different ways, a few fundamental criteria. The most important of these are (1) the form of the land-sea contact (i.e., the configuration of the shore-

(A)

(B)

Figure 13.1
Rapid change on the barrier island shoreline near Kitty Hawk, N.C. (A) Aerial view in 1983
showing numerous vacant shorefront lots where homes were destroyed during 1982–1983 winter
storms. (B) House eroded during the near approach of Hurricane Emily in 1993.
(Photos by R. Craig Kochel)

line, in particular, the local relief), (2) the stability or relative movement of sea level, and (3) the influence of marine processes. Some of the classifications are purely descriptive and may have little application in dynamic geomorphology. Others are genetic and so are allied closely with the processes involved in developing the diagnostic coastal properties. For example, Inman and Nordstrom (1971) classify the morphology of coasts on the basis of modern plate tectonic theory. On the other hand, as Russell (1967a) suggests, perhaps we have not yet collected enough precise data concerning coastal properties to warrant any classification at this time. In addition, all coasts have a past as well as a present, and so time and geomorphic history also become important considerations in coastal classification (Bloom 1965). Coasts may reflect evolution and contain relict features

TABLE 13.1	Data Type and Frequency of Measurements at the Coastal Engineering Research Center-Field Research Facility, Duck, N.C.

Data type and collection technique	Frequency of measurement
Meteorological	
Temperature	Continuous
Rainfall	Continuous
Barometric pressure	Continuous
Wind velocity and direction	Continuous
Wave data	
Baylor staff gages on pier	20 min. every 6 hr
Wave rider buoy 2500 m off pier	20 min. every 6 hr
Radar-derived wave height, period, angle	On call; at least 1 per day
Currents	
Longshore surface current direction determined by dye packets at pier end, nearshore, and beach	Daily
Oceanographic	
Surface water temperature	Daily
Water density	
Water visibility	
Water Levels	
Tide levels at pier end gage	Every 6 minutes
Nearshore bathymetry	
Pier lead lines	Daily
CRAB surveys	
Four lines parallel to pier; 500–600 m north and south of pier; up to 1 km offshore	Biweekly and after storms
Complete surveys along 14 lines within ± 600 m of pier	Monthly

Note: On line databases and current real-time conditions at the FRF are available at http://www.frf.usace.army.mil.

that are not in equilibrium with modern, or even Holocene, processes. For instance, marine terraces standing well above the ocean are not related to modern wave attack but do indicate a coast that has undergone movement relative to sea level.

Because of all these considerations, no specific classification will be advocated here, and no attempt will be made to analyze entire coastal regions. Instead, we will examine only parts of the entire coast that are being actively affected by modern processes. In that sense we will talk about "erosional" or "depositional" landforms, but these terms are applied only on a local basis and have no regional application to the associated coastal zone.

Those interested in coasts or beaches will find a wealth of data and many exceptionally good syntheses. The following references treat the topic from a number of different approaches and at many levels of sophistication: D.W. Johnson (1919); Guilcher (1958); Steers (1962); Shepard (1963); Wiegel (1964); Bascom (1964); Ippen (1966); Zenkovich (1967); Manley and Manley (1968); Bird (1969); Muir Wood (1969); Shepard and Wanless (1971); King (1972); Davies (1973); CERC (1973); Komar (1976), (1983b); Davis (1985); Pethick (1984); Carter (1988). Much of the material presented here has been drawn from these excellent sources.

COASTAL PROCESSES

The processes that initiate change in the coastal zone are extremely difficult to study because they are driven by interrelated forces of high energy, each of which may produce a different response in the same coastal environment. These processes are sometimes monitored over short time periods with the installation of temporary arrays of sensors. Long-term observations, however, are not possible without permanent installations designed to provide precise measurements over various time intervals. An example of such a research center is the one at Duck, North Carolina, operated and maintained by the U.S. Army Corps of Engineers. This installation, known as the Coastal Engineering Research Center–Field Research Facility (CERC-FRF) collects data on various oceanographic and meteorological factors, in addition to conducting regular and event-initiated bathymetric surveys of the nearshore environment (table 13.1). Measurements are made from a 561 m steel pier oriented perpendicular to the shoreline (fig. 13.2) and a mobile sampling machine called the CRAB (Coastal Research Amphibious Buggy). The CRAB (fig. 13.3) can be driven as far as a kilometer offshore into water depths up to 9 m.

Figure 13.2
U.S. Army Corps of Engineers Coastal Engineering Research Center–Field Research Facility (CERC-FRF) at Duck, N.C.
(Photo by R. Craig Kochel)

In addition to regularly monitoring process and morphologic data, the CERC-FRF has been host to a series of major experiments focusing on aspects of the nearshore and beach dynamics which bring together large groups of scientists and engineers from various agencies and academic centers. The first of these, known as SUPERDUCK occurred in 1986 (Crowson et al. 1988), and other projects have followed approximately every third or fourth year thereafter. In these cases, coastal researchers can utilize the bountiful instrumentation in place at the FRF, together with their specialized support logistics, to gather detailed information on coastal processes.

The description of CERC-FRF is not meant to suggest that coastal research in the absence of such heroic efforts is inconsequential. On the contrary, many excellent studies have been accomplished without access to such research facilities. Realistically, however, continuous data collection at permanent stations may be needed to integrate information about diverse phenomena operating over a range of time scales into reliable syntheses.

Waves

The paramount driving force in shoreline processes is waves, which expend the energy they obtain in the ocean against the margins of the land and the shallow nearshore seafloor. A thorough analysis of wave theory is steeped in complex physics and mathematics. Here we can introduce only a simplified version of the basic concepts needed to understand the geomorphic role of waves, recognizing that such a discussion is a broad generalization of a very complicated phenomenon and left for specialty texts on coastal processes.

Wave Generation Most waves that do geomorphic work are generated by strong winds blowing across large portions of the open ocean. Precisely how energy is transferred from the wind into ocean waves is not completely understood (see Komar 1976). Nonetheless, empirical studies indicate that wave properties reflecting this energy exchange (fig. 13.4) depend primarily on the wind velocity, the wind duration, and the **fetch,** the distance over which the wind blows. The fetch is particularly important in determining the height of waves and the **period,** or time interval between two successive wave crests passing a fixed point. Although wave height increases with each of the three parameters, extremely high waves with long periods can be generated only when all three factors are at a maximum.

In the area where they are generated, waves are irregular and individual crests are discernible for only

(A)

(B)

Figure 13.3
(A) Coastal Research Amphibious Buggy (CRAB) used at CERC-FRF for offshore sediment sampling. (B) CRAB taking bottom samples near the CERC-FRF pier. Note the prisms on the deck used in laser surveying of the CRAB's position when taking bottom profile surveys.
(Photo by R. Craig Kochel)

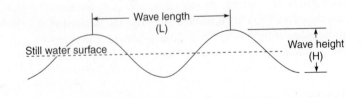

Figure 13.4
Geometry of wind-generated waves.

short distances before they disappear in a maze of interfering waves. This chaotic state occurs mainly because small and large waves generated during the same storm are normally out of phase. Waves passing through the system tend to interfere with one another, adding to the height of some waves and subtracting from others. When waves are enhanced, the heights developed can be truly awesome. Storm waves are commonly more than 20 m high under severe winds and can occur in open water offshore several times annually. The greatest documented wave height, 34 m (112 ft), was measured in a February 1933 storm generated in the open water in the deep part of the South Waves near 30 m high were recorded just off the Mid-Atlantic U.S. coast during the Ash Wednesday Storm of March 1962 and again in the October 1991 storm (of "Perfect Storm" fame).

As waves move away from the generation area, they begin to separate from one another according to various periods. This process, called **wave dispersion,** causes regularly spaced successions of waves with rounded crests to migrate from the source zone. Emerging waves typically have a low ratio of wave height to wave length, a parameter known as the **wave steepness,** and appear as the long, low waves commonly referred to as **swell.** In swell, water particles assume the circular orbital paths that characterize deep-water **waves of oscillation** (fig. 13.5). Although individual water particles in oscillatory waves have little forward motion, the waves themselves advance with a typical trochoid form. That is, the wave form is moving forward, not the ocean water.

The dispersive process, shown diagrammatically in figure 13.6, works because wave length and velocity are a function of the period such that

$$L = \frac{gT^2}{2\pi}$$

and

$$V = \frac{gT}{2\pi}$$

where g and π are the well-known constants, L is the wave length in meters, V is velocity in m/sec, and T is the period in seconds. Because of these relationships, long-period waves with greater length and velocity will separate from short-period waves. For example, a wave with a period of 6 seconds will have a length of 56 m and a velocity of 9 m/sec. Waves generated in the same storm that have a period of 14 seconds have a length of 303 m and a velocity of about 22 m/sec. It is apparent that waves disperse from the generation zone simply because longer waves outrun the shorter ones during any given interval of time.

Waves with identical periods travel away from the source as distinct groups. Storm-generated waves are not the same as waves produced by a point-source impulse such as a pebble tossed into a pond or a tsunami created by rock displacement beneath the ocean floor. In such cases, waves radiate in all directions from the point source. In contrast, ocean swell follows a directional pathway that encompasses a substantial area of the ocean surface but also has finite lateral boundaries. The direction of the well-defined corridor is determined by the direction of the generating wind. Although the wave path does spread somewhat, it is quite possible for swell to strike a few tens of kilometers along a coast while leaving nearby reaches virtually unaffected by the storm.

Studies show that swell can travel great distances across the open ocean without losing much of the original energy (Snodgrass et al. 1966). Most energy loss, especially in the shorter-period waves, occurs near the generation zone. Once the swell condition is attained, the long-period waves experience little additional dissipation of energy and may traverse thousands of kilometers across an entire ocean basin.

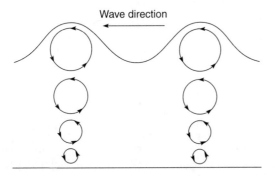

Figure 13.5
Orbital motion of deep water particles in a wave of oscillation.

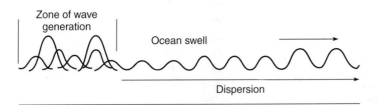

Figure 13.6
Dispersion waves from an area of wave generation. Waves having different periods separate from one another create swell.

Wave Modification Near the Coast   Once established in the generation area, open ocean waves undergo only minor changes during the course of their travels until they reach the coastal zone. There the influence of bottom topography and shoreline irregularities cause major changes to the wave geometry and mechanics.

Changes in Wave Direction   Ocean swell waves entering shallow water will change direction as a result of interference with the bottom topography. Because deflection of the wave crests is a function of the water depth, the waves adjust to the contours of the bottom topography and so gradually bend according to the configuration of the shoreline. This process, known as **refraction,** (fig. 13.7A) works because some sections of a wave crest approaching a coast obliquely are moving in shallow water and at lower velocities than other portions of the same wave, which are traveling in deeper water (Silvester 1966). As shown in figure 13.7B, waves will converge on a landmass that juts into the ocean. When convergence occurs, the energy per unit length of wave, illustrated by the orthogonal vectors, is increased, causing higher waves and increased energy in the headland region. Alternatively, waves may diverge over embayments or submarine canyons, resulting in lower waves and a concomitant spreading of the energy.

Energy can also be transferred along the wave crest during the approach to land. This process, known as **wave diffraction,** is common in the lee of human obstacles (such as breakwaters) or islands and can lead to waves crossing directions, which creates hazardous navigation conditions in the shadow zones of obstacles. Sometimes, these deviations in wave energy become additive to tidal currents moving through narrow passages, resulting in enhanced current energy.

Incident waves that strike cliffs, human barriers such as seawalls, or high-relief, steep beaches may be **reflected** back offshore. The consequences of these reflections, that is, whether they become additive or detractive from subsequent incident waves, is quite variable from place to place and with time but can have significant impact on the arrangement of bedforms and bottom topography within the nearshore zone.

Wave Shoaling   As ocean swell approaches a landmass, the waves begin to interact with the ocean bottom in a phenomenon called **shoaling.** The effective limit of wave influence on the bed, known as the wave base, occurs approximately where the ratio of wave height to wave length equals 0.5. When the water depth is approximately half the wave length, the oscillatory waves begin a transformation into steplike forms called **waves of translation** (fig. 13.8). Wave theory applied to deep-water waves no longer holds once the simple orbital motions of the waves begin to show deformation. Transla-tional waves develop when the orbital path of water particles is intercepted by the ocean floor. The orbits begin to flatten noticeably and the axes of rotation rise to higher levels. Eventually though, the orbital path is destroyed, producing the different wave type. These newly formed waves of translation are different from waves of oscillation in that the water particles have a distinct forward motion without the corresponding backward movement that characterizes oscillation. This asymmetry of orbital pathways is critical in the maintenance of sediment on beaches. Without this shoreward asymmetry, most of the sand entrained would move offshore and shoreline erosion would be greatly accelerated. Once the waves become translational, the velocity and length are determined as

$$V = \sqrt{gh}$$

and

$$L = T\sqrt{gh}$$

where h is the depth of the water.

The change in the wave types during shore approach is accompanied by the phenomenon of breaking waves. Deep-water oscillatory waves remain stable only if the wave steepness (H/L) is lower than 1:7 (.14). As waves approach the shore, water depth decreases progressively and waves begin to "feel bottom." The heights increase as the rounded crests become more peaked, and the velocities and lengths decrease as the wave crests bunch up; only the wave period remains constant (fig. 13.8). The combined effect of an increasing height (H) value oversteepens the wave and it breaks. The distance over which breaking occurs depends primarily on the bottom slope, or how rapidly the water depth decreases toward the shoreline. The width of this wave transformation zone generally decreases with increased bottom slope.

Three common types of breakers, called spilling, plunging, and surging, have been observed and are depicted in figure 13.9 (Wiegel 1964). In spilling breakers, the top of the wave crest becomes unstable first and flows down the wave front as an irregular foam, typically taking the distance of several wave lengths to completely break. In this manner little of the dissipative energy actually impacts the surf-zone floor. A much more violent breaker is the plunging type, where the wave crest curls over the front face and falls with a splashing action into the base of the wave, all within only a couple of wave lengths distance. Plunging breakers thereby concentrate the dissipative energy into a narrow zone and have greater impact on the bottom, normally causing significant scour. In surging, the wave crest remains essentially unbroken, but the base of the wave front advances up the beach. A fourth type of breaker, called collapsing, has been identified by Galvin (1968); its

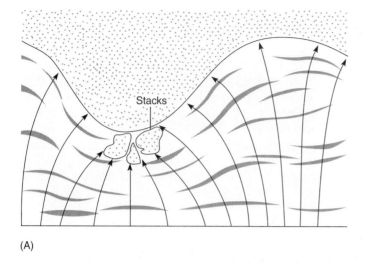

(A)

(B)

Figure 13.7

(A) Refraction of waves along an irregular shoreline. Wave energy converges on headlands that project oceanward while it diverges over recessed embayments. (B) Wave refraction around headlands on the Kohala Coast at the north end of Hawaii. Waves attack the headland, leaving lower energy conditions in the center embayment at Waipio Valley.

(Photo by R. Craig Kochel)

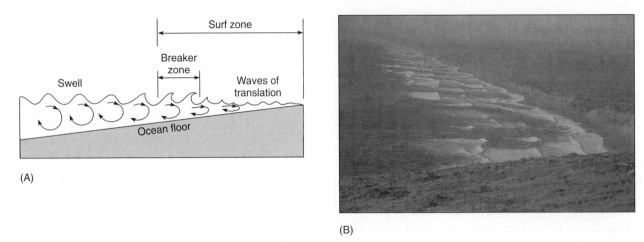

(A)

(B)

Figure 13.8

(A) Wave shoaling in shallow water. As they interfere with the seafloor, they make the transition from oscillating forms to waves of translation. Note that the wave orbitals become progressively flattened before they finally break, giving an onshore asymmetry to the energy. (B) Shoaling waves breaking over multiple nearshore bars, reforming into secondary waves, and finally breaking on the beach at Point Reyes National Seashore in northern California.

(Photo by R. Craig Kochel)

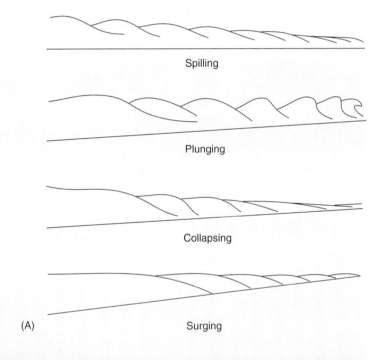

Spilling

Plunging

Collapsing

(A)   Surging

Figure 13.9

Major types of breaking waves. (A) Schematic of types of breaking waves. (B) Spilling waves along the Gulf Coast at Galveston, Texas.

(Photo by R. Craig Kochel)   (B)

characteristics are intermediate between those of the plunging and the surging types. Actually, breaking waves probably occur in a complete spectrum of types that depend on the bottom slope, H/L, and the period. In addition, breaker style at any given locality varies from periods of storms to periods of relative quiet. Thus, the type of breaker may change irrespective of bottom topography.

The significance of breaker style rests in the fact that most types tend to push sediment toward the shore and, therefore, are relatively benign with regard to beach erosion. Plunging breakers, however, are dominantly erosive and will produce large, often destructive, changes in the beach environment as they displace sand toward offshore locations during storms.

Breakers mark the oceanward limit of a zone called the **surf,** in which the original energy given to the waves receives its final transformation. High velocities and substantial impacts occur under breakers, and thus the creation of translation waves provides the water with kinetic energy that is capable of doing geomorphic work. Once formed, the waves of translation (and the water they contain) move forward to inevitable collision with the landmass. The **surf zone** continues onshore until the wave form is lost as it runs up the beach face. Water continues up the beach under its own momentum as **swash** until, at its highest encroachment, the force gathered in the open ocean is finally and totally dissipated. The *swash zone* is alternately covered as the water rushes up the beach face (swash) and exposed as the water moves back down the beach (backwash) under the influence of gravity. Water moving on the beach is generally characterized by shallow, high-velocity, upper flow regime conditions and is associated with significant sediment entrainment over a plane bed or smooth surface (fig. 13.10A). The sedimentology of beaches reflects these conditions with the formation of subhorizontal laminations inclined seaward at approximately the slope of the beach. Alternations in surge energy between swash and backwash normally produce subtle variations in grain size and sorting capability that accentuate the layering in these sediments. Alternations also occur between sediments deposited during the incoming (flood) tide and the outgoing (ebb) tide because of the complex interaction between water depth, wave energy, and groundwater conditions on the sloping beach (fig. 13.10B).

(C)

(D)

Figure 13.9

(C) Plunging breaker along the Maui Coast, Hawaii. (D) Surging breakers on a steep, reflective beach along the northern coast of Hawaii.

(Photos by R. Craig Kochel)

(A)

(B)

Figure 13.10

(A) Wave runup on a North Carolina beach. Swash and backwash can be seen here as exhibiting characteristic shallow, upper-flow regime conditions. (B) Trench showing beach laminations in the upper beach face, Outer Banks, North Carolina. Note the alternations between tidal cycles. The major units, bounded by erosional surfaces, represent major changes in wave climate and show how beach slope has changed in response to the storm history over the past few weeks.

(Photos by R. Craig Kochel)

The position of the surf zone may change frequently because breaking waves reflect the bottom topography but also use their energy to reshape it. A swell approaching a shoreline will break according to the existing shoaling configuration, but at the same time, these breakers may be shifting the bottom sediments to produce a new seafloor topography for the next train of storm-generated waves. In addition, the character of the surf zone changes in response to vertical displacement of the ocean level because of tides, storms, and so forth. Therefore, it may be difficult to characterize the geometry of the surf zone at any location over time. Studies indicate that it may be better to describe the range over which surf zone morphology can be expected to change and address the environmental characteristics that are likely to produce significant changes.

In some cases two sets of swell generated in different areas arrive at a shore simultaneously. This produces a systematic variation in the heights of waves striking the shore, in a phenomenon known as **surf beat** (Munk 1949; Tucker 1950). In surf beat, successive waves gradually increase in height until they reach a maximum, then systematically decrease in height to a minimum value. Thus, a pronounced periodicity develops whereby one large wave will appear with predictable regularity. In many cases the dynamics are such that every sixth to eighth wave will be at the maximum height, but the precise beat depends on the wave periods of the two swell systems and the resulting harmonics. The variation of breaker heights produced by surf beat affects beach processes because it changes the prevailing water level and notably alters the velocity of nearshore currents.

Tsunamis and Seiches Like the waves produced in the open ocean by winds, other waves generated in different ways may have geomorphic significance. **Tsunamis** are waves formed by sudden impulses beneath the ocean that cause trains of waves to radiate in all directions from the point source. Tsunamis are usually initiated by earthquakes of more than 6.5 magnitude on the Richter scale and foci located less than 50 km beneath the ocean floor (Van Dorn 1966). Submarine landslides, volcanic eruptions, and slumping have also been cited as causes. In any case, sudden movement displaces the overlying water column, causing it to oscillate up and down as the water tries to reestablish mean sea level.

The waves leaving the source zone of a tsunami have distinct characteristics. They are extremely long (as much as 240 km), their period may be as much as 1000 sec, and in open water they may be only a meter high. Tsunamis obey the same laws that control shallow-water waves, and therefore their velocity is proportional to the water depth. In 10,000 feet (3000 m) of water, the wave velocity will be

$$V = \sqrt{gh}$$

$$= \sqrt{32 \times 10,000}$$

$$= 566 \text{ ft/s or } 386 \text{ mph (about 618 km/hr)}$$

In the open ocean, such waves pass quickly and with little notice. As waves approach the shore, however, they seem to trigger a harmonic oscillation that does not follow the bottom configuration. In a typical event, a moderate rise or recession of sea level is followed by three to five major wave fronts that are tens of meters high and capable of great destruction. One of the largest tsunamis in the history of Japan, triggered by a magnitude-7.8 quake in the Sea of Japan to the north, caused 185 fatalities in July 1993 with vertical runup values between 15 and 30 m over a 20-km stretch of the Okushiri Island coast (Hokkaido Tsunami Survey Group 1993). N.H. Heck (quoted by Bascom 1964) describes a remarkable case in which a U.S. warship anchored in a Peruvian port in 1968 was picked up by a tsunami wave, carried over the small port city (Iquique), and finally dropped 400 meters inland. After the major waves pass, the nearshore system gradually returns to normal (Van Dorn 1965, 1966). Geomorphic effects of tsunamis are dramatic but probably short-lived. Tsunamis occur rarely, and normal waves rework the coast according to more prevalent controls. Anomalous high-level marine gravels have been ascribed to exceptional tsunamis at some locations along the shoreline of the Hawaiian Islands (Moore et al. 1984).

A **seiche** is another wave type that is not directly related to a prevailing open ocean wind. Seiches are free oscillations of water in enclosed or semienclosed basins. Although originally observed in large lakes, the seiche phenomenon also occurs in harbors, where it is often called surging, and along open coasts with a broad, shallow continental shelf. A seiche is recognized as a repeated rise and fall of the water level. The oscillatory motion begins when some force displaces the water from its equilibrium position. The driving impulse may be heavy rainfall, flood discharge from nearby rivers, long-period waves such as tsunamis or surf beat, or rapid pressure fluctuations associated with storms (B.W. Wilson 1966). In any case, when the initiating force passes, the oscillations gradually decrease until the equilibrium level is once again attained.

Tides and Currents

Although waves are the dominant force influencing the coastal environment, they are not the only significant water motion. Tides and currents each constitute a type of movement that can modify coastal properties. Although our treatment of these forces will be necessarily brief, a complete understanding of nearshore geomorphic mechanics requires detailed consideration of these factors.

Tides As any dedicated beachcomber knows, tides usually occur as a twice-daily rise and fall of sea level. Away from coastal areas, this movement is of little concern to most lay people, but the tidal effect is of consequence to the coastal geomorphologist for several reasons. First, because of the continuous change in vertical range of the

(A)

water level, the lateral position of wave influence during shoaling and the geometry of the surf zone will correspondingly shift. In this manner, the size of the beach and dynamics of wave attack shift throughout the tidal cycle. Second, tides initiate currents that flow into and out of constricted reaches of the shoreline such as bays or lagoons. Many times this ebb and flow can keep drifting beach debris from closing the entrance to the embayment; in fact, some tidal currents are capable of eroding coastal rocks. This is especially true in narrow bays such as the Bay of Fundy in eastern Canada, where the inland constriction produces a maximum range of 15.6 m, the highest value on Earth (fig. 13.11). In some constricted estuaries tides move inland with a pronounced wave front called a *tidal bore* (see Lynch 1982). Often the bores actually break, providing observers with a spectacular show (fig. 13.11). For example, the tidal bore on the Amazon River looks like an 8 m high waterfall advancing upstream for 480 km at a rate of 12 knots.

The tides, of course, are driven by the gravitational effect exerted on Earth by the sun and the Moon (for a nonmathematical discussion, see Bascom 1964; for an understandable mathematical discussion, see Komar 1976). In most coastal regions the lunar influence results in two high tides and two low tides daily. The gravitational attraction of the sun complements or detracts from that of the Moon. As figure 13.12 illustrates, every two weeks the Moon and sun are aligned, causing a higher tide than normal, called the **spring tide.** Midway between spring tides, the moon and sun reach positions that are 90° apart. The solar pull detracts from the lunar effect, and the tide is lower than normal, producing the so-called **neap tide.** Spring tides have tide ranges that

(B)

Figure 13.11
Extreme tidal range (macrotidal) of up to 14 m along the New Brunswick coast of the Bay of Fundy. (A) Low-tide view of the rocky shoreline. Arrow marks the high-tide line. Note the boy standing at the base of the center outcrop for scale.
(B) Dramatic breaking tidal bore rushing upriver at Moncton, several kilometers from the bay.
(Photos by R. Craig Kochel)

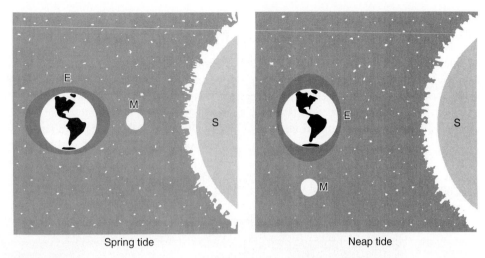

Spring tide   Neap tide

Figure 13.12
Tidal distribution at various stages of the lunar cycle (S = Sun, E = Earth, M = Moon). During spring tide the moon and sun are aligned and cause higher tides. During neap tide the moon and sun are not aligned and tides are lower than average. (Not to scale.)

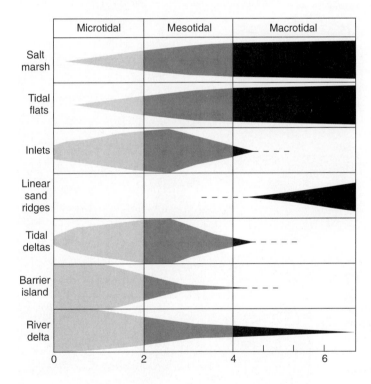

Figure 13.13
Influence of tidal range on the predominance of major coastal landforms. Darker shades and wider funnels indicate increasing dominance of particular landforms. (Hayes 1975)

are about 20 percent greater than the average and neap tides about 20 percent less.

Not all places on Earth experience the tidal motion that pure astronomical and gravitational theory predicts. Some areas, for example, have only one high and low tide, which occur at the "wrong" times. The tide at any locality also varies in magnitude because of other factors, such as perturbation of the lunar orbit, tilting of the axis of Earth, and ocean bottom topography. Tides are generally amplified in areas of broad, low-sloping continental shelves. Davis (1964) has classified tides accord-

ing to their tidal range as *microtidal* (0–6 ft), *mesotidal* (6–12 ft), and *macrotidal* (> 12 ft). Tidal range has been shown to have a considerable impact on the overall geomorphic development of depositional landforms along the coast (fig. 13.13), controlling the dominance and distribution of the major coastal landforms.

Normal tides are not usually destructive. However, some unique tidal events can be devastating, especially if they coincide with storms that produce strong onshore winds and water levels elevated by the storm. For example, because the orbital path of the moon is elliptical, the moon periodically reaches a position where it is closer to Earth than at any other time. This point is known as *perigee*. Occasionally the location of perigee is perfectly aligned with the celestial orientation during spring tide, producing what is known as a **perigean spring tide.** These unique tides not only raise the normal tide levels but, more important, increase the rate at which the tide rises (Wood 1978). In March 1962 the chance combination of a perigean spring tide and a large offshore storm generated enormous flooding and erosion and concomitant record storm damage along the Atlantic coast from the Carolinas to Cape Cod (fig. 13.14). The event resulted in a

loss of 40 lives and $500 million in property damage; in some areas destruction was almost total.

Nearshore Currents   Another type of water motion complicates the nearshore system and our understanding of the mechanics in the surf and swash zones. These littoral or nearshore currents are manifested in flows parallel to and normal to the shoreline and play an important role in shaping topography on the beach and nearshore zones and in moving sediment and water apart from direct wave and tidal forces. We will see that longshore currents provide a linkage between waves, the motion of sediments, and ulti-

Figure 13.14
Overwash flooding of the New Jersey coast during March 1962. Major storm surge combined with the perigean spring tide to cause unprecedented destruction along the entire east coast north of Cape Hatteras, N.C.

mately the evolution of depositional landforms within the coastal zone. The two primary types of wave-induced currents are (1) a cell circulation system consisting of *rip currents* and their associated longshore currents that operate in a cyclic fashion over discrete segments of the shoreline, and (2) longshore currents that are generally unidirectional (during a given wave climate) and are generated by waves striking at an oblique angle to the predominant strike of the shoreline. We should note in passing that related currents can also be caused by tides or storms.

Rip Currents and Cell Circulation   Rip currents are narrow zones of strong flow that move seaward through the surf (fig. 13.15). Shepard and Inman (1950) first recognized the significance of these circulatory currents in returning water to the offshore zone that was moved toward the beach by waves of translation and piled up along the shore. Rips are fed by small currents that move on the beach face parallel to the shoreline. These originate about halfway between two adjacent rips creating elongated circulation cells. The velocity of the longshore feeder current reaches a maximum at the neck or entrance to the rip zone. Field measurements have shown that these velocities exceed the critical values needed for sheetflow transport of sand (Bowman et al. 1992). Velocity in the rip current itself can be significant,

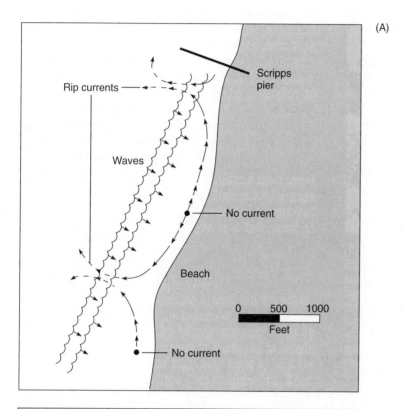

(A)

(B)

Figure 13.15

(A) Nearshore circulation pattern at Scripps Beach, La Jolla, Calif. Note the rip currents and longshore currents that feed the rips, creating small circulation cells. (B) Oblique aerial view of the beach at north Parramore Island, Va. Here at low tide the inner nearshore bar is visible with multiple breaks where rip currents return water to the sea that was stacked onshore during wave runup.

(Photo by R. Craig Kochel)

with measured speeds reaching approximately 2 m/sec (Sonu 1972; Brander and Short 2000).

The development of the cell formed by a longshore current and a rip is related to variations in the rise of mean water level above the level normally attained under still water. This rise is called **wave setup.** Circulation begins because the height of incoming waves varies in a longshore direction, and higher waves create greater wave setup. Longshore currents, therefore, flow from zones of the highest breakers and return oceanward at the position of the lowest breakers. Waves of similar height also initiate cells, but the process probably depends on the type of breakers involved (Sonu 1972). Rip currents are most commonly described in areas of moderate to low tidal range. Brander and Short (2000) showed that the behavior of rips was similar in high-energy, mesotidal areas and indicated that similar sealing relationships characterized them.

Tides and winds blowing in the longshore direction may complicate the system. At low tide, for example, water can be trapped in troughs that parallel the shoreline. Continued spilling of waves into the troughs drives a circulation pattern in which water moves alongshore confined within the troughs until it reaches a rip channel cut through an offshore bar. There the water turns seaward as part of the rip current. In these cases the long-shore current may have a velocity that obeys a mass continuity law rather than a momentum law (see Inman and Bagnold 1963; Bruun 1963; Galvin and Eagleson 1965).

Shepard and his co-workers (1941) realized long ago that the position of rip currents is controlled by the bottom topography in the surf zone, particularly in relation to the pattern of **bars** and **troughs** oriented parallel or subparallel to the shoreline in the nearshore zone (fig. 13.16). We can expect, therefore, that areas of wave convergence, where waves characteristically are higher, are the starting points of the longshore component in the cell. Rips occur away from these zones where breaker heights are smaller.

Although this analysis has been shown to be correct, rip currents also exist on long, straight beaches with smooth bottoms. Bottom topography is not the only cause of cell circulation. For example, ocean swell by itself produces secondary waves in the surf zone (Bowen and Inman 1969; Huntley and Bowen 1973). These waves, called **edge waves,** have crests normal to the shoreline and wave lengths parallel to the shore. They oscillate with an up-and-down motion and appear to resonate between some distant headlands along the coast. These vertical motions are thought to play a role in the differential runup of waves along the shoreline. Some

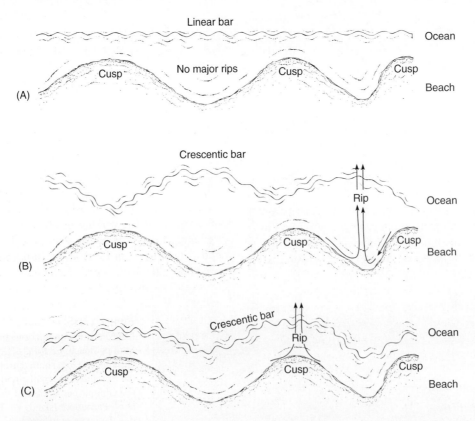

Figure 13.16
Common patterns of nearshore bars, beach cusps, and nearshore currents such as rips.
(A) Linear bar with crescentic beach morphology in the form of cusps. (B) Crescentic beach and bar system, showing out-of-phase alignment. (C) Crescentic beach and bar system, showing in-phase arrangement. Cusp wavelength is generally between 10 m and 100 m.

authors have suggested that edge waves may be the triggering devices for rip currents (see Komar 1976).

The association of edge waves with rip cells is by no means proven. Recent spectral analysis of flow in rip cells along the Mediterranean coast (Bowman et al. 1992) showed that less than half of the values occurred in wavelengths normally associated with edge wave processes. Regardless of the precise origin, rip currents tend to perpetuate themselves because once in place, they can modify the bottom topography so that it provides positive feedback to enhance the circulation pattern. Additionally, edge waves have been called upon to explain the regularity common along the coast in the spacing of zones of maximum shoreline erosion and storm flooding (Dolan et al. 1979, 1980). These processes may become useful in predictive models of shoreline evolution and in providing guidance in predicting erosion patterns along the shoreline. The precise relationship between edge waves and the morphology of the bottom has increasingly been the subject of recent studies. For example, Aagard (1990) found good correlation between bar positions and the theoretical structure of edge waves in a fetch-limited setting along the North Sea coast of Denmark. This study concluded that edge waves were important in the formation and evolution of nearshore bars and even suggested that the number of bars was dependent on the edge wave mechanics.

Longshore Currents Longshore currents generated by waves that strike the beach obliquely are extremely important in beach mechanics because they tend to move sediment parallel to the shoreline for considerable distances and thereby present coastal engineers with innumerable problems. A number of attempts have been made to derive equations that relate wave properties to the velocity of the longshore current (for reviews, see Galvin 1967; Komar 1976; Pethick 1984; Sherman and Bauer 1993). In general, most equations fail in a predictive sense because they either (1) do not distinguish the longshore current associated with cell circulation from that produced by obliquely striking waves, or (2) employ invalid criteria as the fundamental theoretical base.

One approach, based on momentum analysis, has produced good results and probably represents the best hope for a usable predictive model. Momentum, unlike energy, is preserved as waves break and is separated into two components, one directed toward the shoreline and one directed parallel to it. Thus, the flux in momentum directed along the shoreline should be proportionately related to the velocity of the longshore current. This concept has been developed thoroughly by many authors but most completely by Bowen (1969), Longuet-Higgins (1970), Komar and Inman (1970), and Komar (1975). Komar (1976) suggests that the best estimate of longshore currents at the midsurf position follows the equation

$$V = 2.7u_m \sin \alpha \cos \alpha$$

where u_m is the maximum orbital velocity at the breaker zone, V is mean velocity in cm/sec, and α is the breaker angle, the angle of incidence with a line parallel to the shoreline (details of the derivation have been omitted).

The relative contribution of incident waves and longshore currents in the transport of sediment was evaluated in a recent field experiment along the high-energy Oregon coast. Beach and Sternberg (1992) placed velocity meters and suspended sediment measuring devices in cross-shore and alongshore arrays to find that the waves and currents contribute equally to suspended sediment loads. They also observed that the combined wind and current interactions enhanced longshore sediment transport by 50 to 60 percent over transport forced by the waves alone.

BEACHES

The difficulties of associating entire coasts with formative processes tend to direct our attention to smaller components of the coast where processes are more easily understood. The primary coastal landform and the feature most amenable to process analysis is the **beach,** which is most simply defined as the relatively narrow portion of a coast that is directly affected by wave action. It usually terminates inland at a sea cliff, a dune field, or at the boundaries of permanent vegetation. Oceanward, under the constraints of our definition, part of the beach is continuously submerged because it lies beneath low-tide sea level. It is still part of the beach, however, because the bottom is subjected to wave action. In this sense the term "beach" is synonymous with the **littoral zone,** an expression used commonly in geological work.

The location of major beach forms that are significant components of a coastline depends, to a large degree, on the availability of sand. The primary sources of such debris are (1) river sediment delivered to the coast through deltas, (2) the erosion of headlands and sea cliffs, and (3) the reworking of offshore shelf sediment originating from a variety of sources, chiefly by fluvial deposition when sea level stood hundreds of meters lower during the Pleistocene glacial maxima. Small contributions are also made by eolian processes in some coastal areas. The detrital load delivered to the coastal domain by major rivers can be enormous in some cases. For example, the Mississippi River provides more than 3 billion tons of solid material to the Gulf Coast region annually. Sea cliff erosion normally contributes less than 10 percent of the total debris available for accumulation on beaches (Komar 1976).

The Beach Profile and Equilibrium

Beaches represent dynamic systems where loose granular debris is moving steadily under the attack of waves

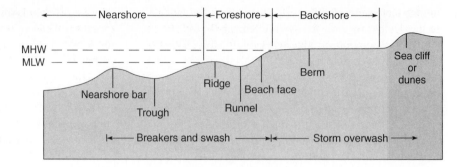

Figure 13.17
Generalized beach profile from the dune to approximate mean wave base offshore.

and currents. If water motion could be held constant, this debris would be molded into a characteristic profile that would reflect an equilibrium between the driving forces (waves and currents) and the properties of the beach sediment, chiefly the grain size and erodibility of the material. We can thus visualize an *equilibrium beach profile* for any set of water and sediment conditions much in the same manner as we viewed the graded river in fluvial systems. Waves and currents, however, do not remain constant but change properties on a daily or seasonal basis, requiring some response in the process of sediment transportation and ultimately in the beach profile. Although we can think of the equilibrium profile as the ideal case, under the dynamics of the natural setting its properties are constantly changing over time along with the driving forces, particularly the wave climate and longshore and cross-shore currents related to the overall energy conditions at the site.

Inman and Bagnold (1963) proposed that there is a position on the beach profile where net sediment transport is zero; that is, the amount of sediment moving onshore is equal to that moving offshore. Therefore, they suggested that the equilibrium concept be looked at as a balance between the force from the angle of internal friction of the sand and the force of gravity working tangentially on the sand over the beach slope. Recently, several experiments and a large number of modeling efforts have sought to improve our understanding and quantify the relationships between the hydrodynamic forces involved in maintaining the equilibrium profile. Field observations of seaward fining of sediment size has been interpreted to mean that particular sediment sizes move across the profile to a position where they come to rest in equilibrium with the wave-generated hydraulic regime. Two major hypotheses have been suggested to explain this shore-normal sorting: (1) the asymmetry of shoaling-breaking waves provides short-duration, high-velocity shoreward motion that are countered by lower-velocity, longer-duration offshore motions; and (2) for every grain size a position on the profile occurs where onshore flow forces balance with the gravitational component driving grains down the beach slope offshore; this is the null point hypothesis

(Horn 1992). Horn's numerical model for testing these hypotheses indicated that mean sediment size can be related simply to the velocity of water induced by wave action, implying that further exploration of this concept will probably not yield much more insight into the cause of cross-shore sediment sorting without a great deal of complexity. Horn (1992) notes that although the model predicts trends in grain size well, it fails to represent the actual sizes. Dean et al. (1993) compared actual profiles with those predicted by a numerical model designed to determine whether sand deficits or excesses occur at 10 New Zealand beaches. This study yielded mixed results, but suggests that the beach profile can be used to make predictions of future shoreline recession or progradation.

The beach profile, as shown in figure 13.17, consists of a number of component parts, each of which develops its own diagnostic characteristics. In its entirety the profile represents a topographic form that induces waves to dissipate energy by breaking. The exact location of breakers is important because it determines where the greatest amount of energy is expended and which part of the beach will be subjected to the greatest change. The position of the area of most rapid geomorphic response changes with time, both on a periodic and predictable basis with fluctuations in water levels driven by tides and aperiodically during elevations of water level and overall energy during coastal storms. The **berm** is a nearly horizontal surface on the backshore portion of the beach. Some beaches have more than one berm; others have none, especially if the mean sediment size is coarser than sand. Landward the highest berm terminates at the base of the sea cliff or foredune, and oceanward it joins the **beach face.** Because the berm is formed by deposition of sediment during backwash primarily during storm or extreme astronomical tidal conditions, its elevation is determined by how high the swash runs up the beach. As swash moves up the beach, it loses velocity because of friction and the loss of water that permeates into the beach debris. Continued upward growth of the berm surface would require ever higher waves and swash runup. In this way berms are analogous to vertically accreted floodplains, and the rate of upward growth must decrease with time for only infrequent storm waves can add sediment to the surface. Lower-energy waves between storms typically erode the berm along its seaward margin and contribute to an overall steepening of the beach face. As Bascom (1964) points out, however, storm waves also erode the front edge of the berm, thereby reducing its horizontal length while simultaneously building the remaining surface to a higher level.

The beach face is the sloping section of the beach profile immediately seaward of the berm. The slope of the beach face is controlled by many factors and so may range from nearly horizontal to gradients that approach the angle of repose for unconsolidated sediment. Sediment moves up the beach face with the swash and down the beach face during the less powerful backwash. The beach face, therefore, represents a surface striving to attain some balance between onshore and offshore sediment transport. Its slope adjusts to provide the balance.

Beach face slopes are directly proportional to the size of the particles being moved in the swash zone. Large particle sizes tend to maintain steep slopes and vice versa (Bascom 1951; Wiegel 1964; McLean and Kirk 1969; DuBois 1972; Wright et al. 1979). We would expect such a relationship because slope provides the backwash velocity needed to transport sediment of any given size, but the phenomenon is much more complicated, and beach slopes probably relate to many factors involved in shoreline dynamics. For example, it is well known that significant adjustments in the slope of the beach face occur during storms. Sector (1995) showed that although significant adjustments of the beach profile took place in response to Hurricane Hugo in South Carolina, beach profiles had completely recovered within five years. Thus, many of these impacts may only be temporary. Typically, storms flatten the beach face as sediment moves offshore. A gradual recovery generally occurs as the beach face steepens again between storms.

Furella and Dolan (1996) noted that most of the erosion of beaches occurred during the first 6 hours of storms in the North Carolina Outer Banks. However, they also observed that 50 percent of the sand lost was recovered within the first 12 hours following the storms. Long-term studies such as those at the CERC-FRF well illustrate the apparent linkages between wave climate and beach profile in the response of the North Carolina barrier beaches to storms. Analyses of 11 years of high-resolution beach profiles surveyed at Duck, N.C., using the FRF's CRAB were analyzed for spatial and temporal patterns by Larson and Kraus (1994). They found that the average profile elevation was symmetrical about mean sea level, but the depth of the active profile was greater during the winter (6 m) compared to summer (4 m). Large charges in the profile always occurred in response to (major) storms

Nearshore Bars   In the zone seaward from the beach face, a submerged **nearshore bar** system is commonly,

but not always, present. An associated trough develops between the bar and the beach face. Nearshore bars may be absent from the profile, especially if the beach is steep. On the other hand, where the beach gradient is low, several bars may be present. The creation of multiple bar systems may be accomplished in a variety of ways: (1) each bar may reflect the breaker position for waves of different sizes; (2) the bars may relate to the shifting of breaker positions associated with high and low tide levels; or (3) oscillation waves that break far offshore on low-gradient beaches may re-form over the trough and break again closer to shore, with each episode of breaking creating an associated bar.

The terminology of nearshore bars can be confusing as numerous classification schemes have been proposed based on bar morphology and attachment versus parallelism to the shoreline. One of the most widely used of these schemes is shown in figure 13.18. In addition, considerable disagreement has arisen concerning the formative processes and processes responsible for the dynamic nature of nearshore bars. Many stocks have observed cycles of bar generation and evolution, generally in response to charging wave and storm conditions over a temporal scale typically ranging from months to years. For example, Ruessink and Terwindt (2000) documented onshore movement of sediment when breakers remained outside the bars. Offshore sand movement coincided with breaking conditions across the entire nearshore bar profile. Hypotheses concerning bar formation are all based on the assumption that breaking waves are intimately involved in the formative mechanics. A number of early laboratory and field studies support this contention (Evans 1940; King and Williams 1949; Shepard 1950), and the relationship is now generally accepted. Usually breakers establish the size, position, and depth of the bars and troughs. Most investigators agree that bars form at locations in the nearshore zone where bottom velocities are relatively low; some of these positions relate to the interaction of incident and reflected waves (e.g., Short 1975) and others relate to the position and interference created by edge waves (e.g., Bowen and Inman 1971).

Once bars gain enough relief to influence wave shoaling, there is much more agreement as to how bars are maintained and how they respond to changes in wave climate. The maximum depth of bar formation depends on the depth to which wave action is able to agitate the bottom sediment. Under average wave conditions, this depth is approximately 10 m below low tide, but large storm waves are known to move bottom sediment in water as deep as 25 m. The deepest bars, molded in the fury of violent storms, maintain their position and shape for long intervals during which normal waves do not influence the bottom. Some bars are continuous parallel ridges that extend unbroken for tens of kilometers. Bars do not usually display such regularity, however,

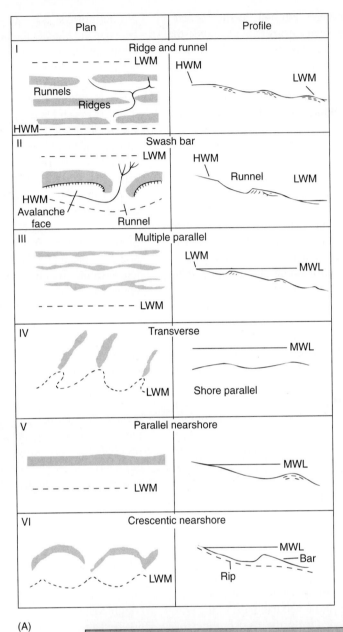

(A)

Figure 13.18

(A) Classification of nearshore bars.
(B) Low-tide view of nearshore bars along southern Assateague Island, Va. Note the exposed inner bar and the outer bar defined by the line of breakers. The onshore area has been steadily growing here at Chincoteague Inlet as the nearshore bars, fed by longshore drift to the south (left) attach onto the island. The linear nature of the topography results from bar attachment, followed by dune growth and vegetation.

(A): (Greenwood and Davidson-Arnott 1979)
(B): (Photo by R. C. Kochel)

(B)

because shifting wave directions realign parts of the bar system and break ridges into segments with a chaotic distribution. Violent storms also produce drastic changes in bar directions and patterns. The mobility of bars has been well documented in studies of the nearshore zone. Bars have been observed to change with respect to their pattern of attachment to the shoreline, plan view (parallel or cuspate), and number. They also migrate shoreward or seaward according to variations in wave climate (wave height, steepness, and breaker type) and shift in a more gradual evolutionary pattern that is probably related to seasonal or quasi-annual changes in wave climate (Davis and Fox 1972; Short 1975; Greenwood and Davidson-Arnott 1979). Long-term observations of bar positions at the FRF in Duck, North Carolina, suggest that the bars rapidly move offshore in response to storm waves followed by slow migration shoreward in periods of relative calm.

In considering bars as products of wave action and as controlling factors in the breaking process, the shallow shoreward bars and troughs are decidedly more important in nearshore wave dynamics. Bars actually serve as natural breakwaters to reduce the energy of waves acting at the shoreline and are major energy dissipaters in the nearshore zone.

Beach Morphodynamics   The ideal beach profile is subject to pronounced changes from deposition and erosion. In addition, the position of the major parts of the profile may be changed according to the dynamics of these processes.

The most pronounced changes in the ideal profile are brought about by storm waves. Where storms are common, the large and vigorous waves tend to destroy or drastically limit the extent of the berm. The eroded material is simultaneously shifted to an offshore position where it augments the nearshore bars. In contrast, during calm periods with a low-energy wave climate, the movement of beach sand is shoreward. As sand moves from the nearshore bars shoreward, it rebuilds the upper part of the beach profile and produces an extensive berm and subaerial beach. Because the frequency of storms is generally a seasonal phenomenon, we can expect the beach profile to change dramatically from one part of the year to another. Many authors, therefore, refer to a *summer profile,* characterized by the absence of bars and a wide berm, and a *winter profile,* with no berm and a series of longshore bars. An example is shown in figure 13.19. Summer and winter profiles may not occur specifically during those seasons. What we are really talking about is a profile formed by storm waves versus a profile generated by swell. These profiles coincide with the seasons only where regular changes in wave climate are associated with seasonal shifts in storm tracks. Many field studies document the offshore movement of sands during storm wave conditions and the landward transport

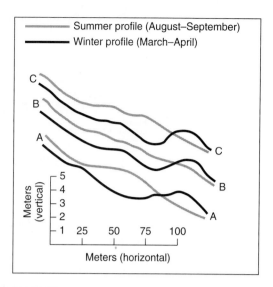

Figure 13.19
Generalized summer and winter beach profiles near Scripps Pier, Calif.
(Shepard 1950)

under swell (Shepard 1950; Bascom 1964; Strahler 1966; Gorsline 1966).

We now know, however, that the summer and winter profiles probably apply best in certain environmental conditions such as those along the California coast. In other areas (eastern U.S. beaches and many Australian beaches), the morphodynamics are much more complex. Evidence of this complexity is revealed by inspection of the extensive data set collected by the CERC-FRF at Duck, North Carolina. Although the surveys often demonstrate a seasonality in the movement of nearshore bars and the overall profile, the pattern varies from year to year depending on the storm history. In years without major winter storms, the bar may fail to retreat significantly offshore. Alternatively, the bar has been observed to move hundreds of meters offshore during a single large storm.

Numerous studies of Australian beaches (Short 1979; Wright et al. 1979; Short and Wright 1981; Wright et al. 1982a; Wright et al. 1982b; Wright and Short 1983) have been able to integrate hydrodynamic and morphologic variables into an organized model describing beach morphology and behavior. These studies have shown how existing and preexisting beach topography affect transporting mechanisms in the surf zone and how fluid motions other than storm waves, such as rip currents and low-frequency standing oscillations (such as edge waves), are extremely important in determining the beach character. The detailed study of Australian beaches has led a classification of beaches that relates morphology to the manner in which energy is dissipated or reflected within the nearshore zone. Six **morphodynamic states** have been recognized (fig. 13.20), each of which has a distinctly different association of

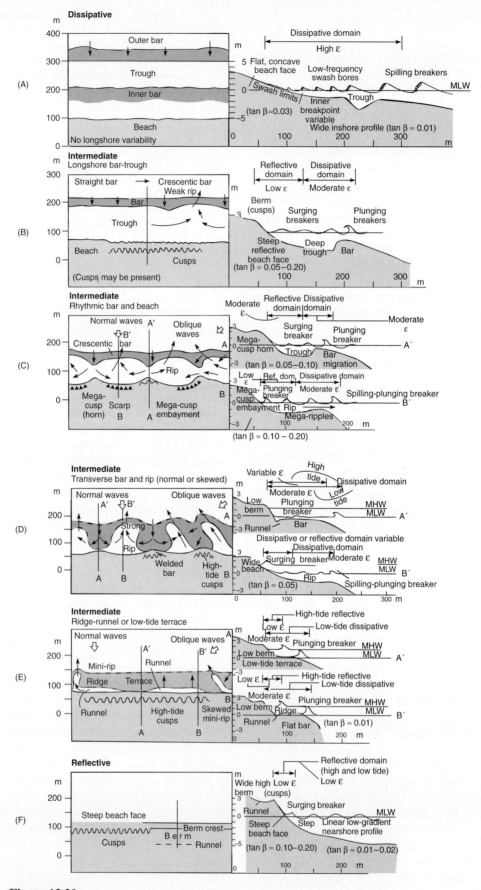

Figure 13.20

Six morphodynamic states of beaches, illustrating the interactive relationships between morphology and process.

(Wright and Short 1983)

morphology, water motions, and sediment characteristics (Short 1979; Wright et al. 1979; Wright and Short 1983). This coherent morphodynamic model recognizes that some beaches persistently occur in a given state, whereas others fluctuate between different states, driven largely by temporal changes in wave climate. In addition, it shows how antecedent conditions of the beach can exert an important influence on the current and future stages of the beach. This process-based scheme developed by the "Australian school" of coastal geomorphologists has since become the standard model used to describe and predict the response of beaches to changes in environmental variables. These models provide a link between beach morphodynamics and environmental conditions in wave-dominated beaches.

The end-member states in this classification are called reflective and dissipative. **Reflective** beaches are characterized by steep (usually greater than 6°), linear beach faces and berms (fig. 13.20F). Much of the sediment in reflective beaches is stored in the subaerial portion of the profile (Wright et al. 1979). Beach cusps (discussed later) are well developed because of the tendency for enhanced edge-wave development as secondary fluid motions become trapped in the narrow breaker zone. The edge waves tend to be associated with surging breakers, high runup, and minimum setup of sea level. **Dissipative** beaches have low-angle, concave-up beach faces attached to a wide, gently sloping, shoaling surf zone (fig. 13.20A). Nearshore topography is dominated by the presence of one or more bar-trough systems, often in multiple-parallel arrangement.

In simplest terms, reflective and dissipative systems differ dynamically because the accumulated wave energy is expended at different places. In the reflective system, most of the incident wave energy is expended at the beach face. In dissipative systems, most wave energy is expended offshore, where energy is lost in turbulence as waves break over the bars. The continuum between dissipative and reflective systems can be distinguished using the **surf-scaling parameter** (Guza and Inman 1975), which is expressed as

$$\varepsilon = \frac{a_b \omega^2}{g \tan^2 \beta}$$

where a_b is amplitude, ω is the incident wave radiant frequency $\left(\frac{2\pi}{T}; T = \text{period}\right)$, b is beach/surf zone gradient, and g is acceleration of gravity.

Strongly reflective beaches occur when $\varepsilon < 2.5$. In these instances, surging waves prevail. Waves reach the beach face without breaking and surge up the beach or collapse over a step at its base. Turbulence is, therefore, confined to the zone of runup on the beach face. Wave energy that is reflected may be trapped and initiate strong, standing-wave motion, commonly in the form of edge waves. This motion promotes the formation of beach

cusps. In extreme dissipative systems, high values ($\varepsilon > 30$) prevail. Spilling breakers occur 75 to 300 m seaward of the beach face, and the wave bores become smaller and lose much energy before they reach the inshore zone.

Four intermediate states have been identified between the two distinct end-member states (fig. 13.20 B–E). The intermediate domains are more difficult to characterize because they incorporate both dissipative and reflective elements. In some cases, the morphodynamic state may change from the offshore to nearshore position or with the high and low tides. This complicates our understanding enormously because different features may develop during the tidal cycle or with distance from the shore. It is also apparent that the tidal range may be significant. In general terms, however, dissipativeness increases with increasing values of ε, increasing swell-wave heights, and decreasing bed slope. Greatest dissipativeness also prevails in areas with the largest amount of inshore sediment and during and immediately following a severe storm. As longshore bars migrate shoreward, transitional states develop, leading to a steepened beach face and reflective conditions.

Because morphodynamic states are commonly transient from one to another, it is probably no longer correct to think of an ideal or equilibrium beach profile. However, when considered over the long term, any beach and surf zone will assume a most frequently occurring condition, which Wright and Short (1983) refer to as the **modal state.** The modal state is dependent on the forcing wave climate (the most frequently occurring wave conditions) and the input of the local environmental setting. Although no all-encompassing equilibrium profile occurs, any given beach will attain its own particular modal state, about which variations will occur as the water motions and bottom configurations continuously change. Finally, we should note that some beaches exhibit a high degree of persistence in a given state. A recent study by Hughes and Cowell (1987) provides a good example of how reflective beaches can respond in a unique way to changing energy conditions such that a reflective profile is maintained over a wide range of wave climate. Their study showed that vertical adjustments could take place through growth of the **beach step,** a narrow zone at the base of the beach face with relatively steep slope. Temporary increases in the height of the step increase the water depth of inshore waters and delays wave breaking, thereby contributing to the maintenance of a reflective profile. Sanderson et al. (2000) extended the approach of Wright and Short's beach morphodynamics to low-wave energy beaches on the soutwest Australian coast. They showed that despite the low energy of sheltered beaches in cores and landward of barrier reefs, there existed a systematic variation in beach processes and morphology. However, they found that as sheltering of the coast increases, there is more impact of tidal and longshore currents upon beaches.

The practical utility of the above morphodynamic classification in shoreline engineering is just starting to be realized. For example, Antia and Nyong (1988) examined the morphodynamic states of five Nigerian beaches to determine the most cost-effective method of sand replenishment. Of concern was where to replenish sand along the beach-nearshore profile to maximize its impact on widening the beach and whether the timing of these applications was critical. Their analysis of beach morphologic evolution showed that minimal sand was lost to offshore zones when the beach was in a dissipative state. Replenishment done in a manner to ensure persistence of the dissipative state of the beach was found to be most effective in maintaining the beach, but it had to be done carefully to ensure that the foreshore and surf-zone slopes were not increased above 7°. Bruun (1988) went on to demonstrate that the nourishment on the beaches was often not very beneficial. He recommended a careful analysis of the equilibrium profile to determine where on the cross-shore profile nourishment using a selected grain size would provide the highest degree of hydrodynamic stability. Unmonitored applications permitting the slope to steepen near the threshold cause the system to shift abruptly toward a reflective mode, resulting in erosion of the shoreline. Consideration of the geomorphic processes appears to be a cost-effective measure for coastal engineers because of the limited duration of many beach nourishment attempts which sometimes last less than a month (Bruun 1988; Clayton 1989).

In addition to the changes discussed so far, large beaches are usually spatially and temporally altered by currents moving in the longshore direction. Although these currents can be generated in a variety of ways, the majority of longshore flow derives from the momentum carried by waves striking the beach in an oblique direction. Because waves are seldom completely refracted, they usually make their final approach at some angle to the shoreline. The velocity of a longshore current varies directly with wave height and the angle between the shoreline and the approaching waves. Velocity also increases with distance from shore, reaching maximum values near the middle of the surf zone. Longshore currents transport sediment that has been entrained by wave action, in a process referred to as **littoral drift.** Because the obliquity of wave incidence tends to retard development of rip currents, longshore currents are generally continuous and have the potential to transport sediment for great distances.

Engineers and scientists have attempted to construct predictive models to estimate rates of littoral drift. Stone and Stapor (1996) developed a sediment transport model for same 400 km of the northeastern gulf coast. In most cases their predictions have relied on empirical correlations between the rate of sand movement and some estimate of the wave power expended in the longshore direction (see Komar and Inman 1970; Komar 1976,

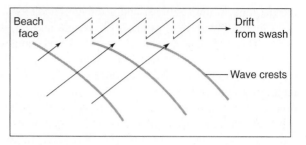

Figure 13.21
Drift of sediment along a beach face caused by the oblique arrival of waves on the beach.

1983). In some cases empirical studies show a good correlation between wave variables and transport rates, and parameters used as indicators of the longshore components of wave power or energy are compatible with wave theory (Komar and Inman 1970; Komar 1971a). A complete understanding of the process, however, is complicated because littoral drift occurs in two different modes of action. When wave steepness is high ($H/L > 0.03$) most transport occurs beneath the breaker zone (Bascom 1964). But when the steepness is low, the drift functions along the beach face in a process known as *swash transport* (beach drift). In this case, waves breaking obliquely to the shoreline push sand grains up the beach face perpendicular to the direction of the wave crests. The backwash, however, affected only by gravity, pulls the water and grains down the beach face in a direction perpendicular to the shoreline. Thus, mobile grains in swash transport move in a spasmodic, zigzag fashion (fig. 13.21). Besides wave steepness, the angle of wave incidence (α) is also important in the drift rate, maximizing the rate at a 30° angle.

The littoral drift process is further complicated because sediment can be moved in suspension or as bed-load. Brenninkmeyer (1975) showed that suspended load is important only in a narrow offshore zone. This analysis seems to contrast with earlier studies, which suggested that suspended load was the dominant mode of transport (Fairchild 1973; Thornton 1973). Perhaps some of the difficulty in identifying load types occurs because fine sediment is commonly winnowed from beach debris and carried away from the nearshore environment by rip currents and normal offshore diffusion. The rate and volume of sediment movement is also dependent on the grain size of the transported load (Duane and James 1980).

Conclusions about littoral drift rates based on empirical or theoretical analyses remain dubious. As Komar notes, "The final uncertainty in the evaluation of the net drift could be much greater than the net drift itself, and quite conceivably the direction of the net drift might be incorrectly evaluated" (1983a, p. 17).

Figure 13.22
Oblique aerial photo of the groin field installed along the Cape Hatteras, N. C., lighthouse to help prevent beach erosion (1982 photo). Groins trap the southerly drifting sand (toward bottom of photo) and widen the beach. Note that the lighthouse was moved almost 1 km landward across the barrier island to safety during the summer of 2000.
(Photo by R. Craig Kochel)

The best evaluation of littoral drift is based on direct measurements of volumetric loss caused by erosion or beach growth caused by deposition of drifting sand. These measurements lend themselves to a budget of littoral sediment in which all sediment contributions to a beach (sources) and all losses from a beach (sinks) are taken into account. The main sediment sources are rivers, sea cliff erosion, artificial beach nourishment, and incoming littoral drift. It is not uncommon to find littoral drift patterns along the coast fragmented into distinct segments or circulation cells. Common sinks are outgoing littoral drift, offshore transport, beach sand mining by humans, and debris moved down submarine canyons. Over a period of years, a particular beach will probably have a positive or negative budget. Obviously, a negative budget indicates more losses than gains, and therefore the beach is eroding. Positive budgets indicate net deposition.

In practice, making reasonable estimates of the components that make up the littoral budget is very difficult. However, the positive or negative character of the budget can be determined by long-term monitoring of erosion and deposition on the beach. This information is very important in predicting the potential impacts of human activity.

The rate of longshore drifting of sand can be significant when considered on a time scale of human life, and the process is often shown in the accumulation of beach debris against human structures such as groins (fig. 13.22). The net littoral drift is the sum total of movements initiated by waves arriving at the shoreline from various directions. Waves may carry debris in one direction for a short period and then, as conditions change, return the material to its original position. Such drift reversals may be random or they may be decidedly seasonal. In California, for example, sand usually drifts southward in the winter and northward in summer. Net littoral drift can be determined only by long-term budget studies that can detail which direction is the prevailing one. Nonetheless, the process of longshore transport causes more problems for coastal engineers than perhaps any other shoreline phenomenon. It places sand where we do not want it, such as across harbors and inlets (fig. 13.23), and it removes sand from locations where

Figure 13.23
Aerial view of Oregon Inlet, N.C., showing spit growth due to southerly longshore drift (toward the right). Note the dredge actively attempting to maintain a navigation channel at the proper position under the Bonner Bridge, which carries the only road linking the developed Cape Hatteras region with other islands and eventually the mainland to the north. A controversy over a plan to build jetties to stabilize this highly mobile inlet has raged for decades (1982 photo). Extensive riprap was deployed along the southern inlet margin (right) in the mid 1990s to help stabilize the inlet. A related beach nourishment program was established to pump sand dredged from the inlet on to beaches of the south (down-drift) island.
(Photo by R. Craig Kochel)

we would like to keep it, such as on resort beaches. Table 13.2 shows net littoral drift along a number of coastlines. The volume of debris involved in the transport demonstrates the tremendous problems engineers face.

Shoreline Configurations and Beach Landforms

In addition to the modal profile that is oriented perpendicular to the shoreline, many coastal geomorphologists now accept Tanner's (1958) premise that beaches develop a shoreline configuration revealing another type of balance between water energy and sediment supply. This plan-view shape is best established where no long-term unidirectional movement of sediment occurs parallel to the shoreline. Like a graded river, the shoreline configuration is developed "over a period of years" and is adjusted to the prevailing wave characteristics.

The most logical environment to preserve a modal configuration would be a protected bay where no dominant longshore current occurs. In such a locale waves might produce minor longshore transport of sediment where the wave crests are not completely refracted. The sediment will continue to drift until the shoreline is reoriented parallel to the attacking wave crests at every segment. Offshore topography thus determines wave refraction, and the waves in turn establish the shoreline configuration. The complications in even this simple model are staggering, however. For example, any additional sources of sand to the beach will prevent complete refraction because some longshore transport away from the source will be necessary. In addition, bottom topography is so variable and so susceptible to change that it seems too much to expect that we can precisely describe the form of an equilibrium shoreline.

In light of the above, it is indeed remarkable that many shorelines, open to all types of waves and currents, contain features that are similar in shape and spaced with a regularity that can hardly be attributed to coincidence.

TABLE 13.2	Representative Rates of Littoral Drift Along Coasts of the United States.

Location	Predominant direction of drift	Rate of drift (yd³ per year)	Method of measure of rate of drift	Years of record
Atlantic Coast				
Suffolk, Co., N.Y.	W	300,000	Accretion	1946–1955
Sandy Hook, N.J.	N	493,000	Accretion	1885–1933
Sandy Hook, N.J.	N	436,000	Accretion	1933–1951
Asbury Park, N.J.	N	200,000	Accretion	1922–1925
Shark River, N.J.	N	300,000	Accretion	1947–1953
Manasquan, N.J.	N	360,000	Accretion	1930–1931
Barneget Inlet, N.J.	S	250,000	Accretion	1939–1941
Absecon Inlet, N.J.	S	400,000	Erosion	1935–1946
Ocean City, N.J.	S	400,000	Erosion	1935–1946
Cold Spring Inlet, N.J.	S	200,000	Accretion	—
Ocean City, Md.	S	150,000	Accretion	1934–1936
Atlantic Beach, N.C.	E	29,500	Accretion	1850–1908
Hillsboro Inlet, Fla.	S	75,000	Accretion	—
Palm Beach, Fla.	S	150,000 to 225,000	Accretion	1925–1939
Gulf of Mexico				
Pinellas Co., Fla.	S	50,000	Accretion	1922–1950
Perdido Pass, Ala.	W	200,000	Accretion	1934–1953
Galveston, Texas	E	437,500	Accretion	1919–1934
Pacific Coast				
Santa Barbara, Calif.	E	280,000	Accretion	1932–1951
Oxnard Plainshore, Calif.	S	1,000,000	Accretion	1938–1948
Port Hueneme, Calif.	S	500,000	Accretion	1938–1948
Santa Monica, Calif.	S	270,000	Accretion	1936–1940
El Segundo, Calif.	S	162,000	Accretion	1936–1940
Redondo Beach, Calif.	S	30,000	Accretion	—
Anaheim Bay, Calif.	E	150,000	Accretion	1937–1948
Camp Pendelton, Calif.	S	100,000	Accretion	1950–1952
Great Lakes				
Milwaukee Co., Wis.	S	8,000	Accretion	1894–1912
Racine Co., Wis.	S	40,000	Accretion	1912–1949
Kenosha, Wis.	S	15,000	Accretion	1872–1909
Ill. State Line to Waukegan	S	90,000	Accretion	—
Waukegan to Evanston, Ill.	S	57,000	Accretion	—
South of Evanston, Ill.	S	40,000	Accretion	—

From J.W. Johnson (1956). Used with permission of the American Association of Petroleum Geologists.

Usually **crescentic,** these features form as periodic seaward projections of the shoreline as curved embayments (fig. 13.24). A complete hierarchy of these features seems to exist, ranging from minor forms with a wavelength of less than a meter to major cuspate-like indentations of the coastline, such as bights and capes, with spacing measured in hundreds of kilometers (Dolan and Ferm 1968; Dolan et al. 1974). Table 13.3 shows this hierarchy.

Beach Cusps The most common crescentic forms are **beach cusps** (fig. 13.25). Cusps develop at the upper part of the beach face and along the outer fringe of the berm. They are usually spaced less than 30 m apart and can form in beach sediment of any size, including boulders and cobbles (Russell and McIntire 1965; Sunamura and Aoki 2000). Some sorting is produced in the formative mechanics because the cusp projections, or **horns,** are usually more coarse-grained than the intervening embayments.

Beach cusps seem to form most readily where wave crests strike parallel to the shoreline and during periods of relatively low wave climate (Antia 1989). This

Figure 13.24
View from Duck, N.C. to the south of the northern Outer Banks showing the long curve of the crescentic shoreline leading to Cape Hatteras. Note the abundance of smaller-scale crescentic landforms, called beach cusps, on the foreground berm.
(Photo by R. Craig Kochel)

TABLE 13.3	Hierarchy of Crescentic Landforms Found on Coasts.				
Form characteristic	Cusplet	Cusp	Sand waves	Secondary capes	Primary capes
Spacing	0 to 3 m	3 to 30 m	100 to 3000 m	1 to 100 km	200 km
Material	Fine sand-gravel	Sand-boulders	Sand	Sand	Sand-gravel
Topographic association	Step	Berm, beach face	Beach berm–offshore bar system	Coastal plains; shores with sufficient sediment	Coastal plain deltas
Rhythmicity	Yes	Yes	Yes	Often	Not always
Motion	Fixed	Normal to beach	Downdrift	Probably downdrift	Slow downdrift
Temporal	Minutes to hours	Hours to days	Weeks to years	Decades	Centuries
Suggested processes	Swash action on beach face, groove erosion	Berm deposition and erosion	Wave action, nearshore circulation cells, back eddies of longshore transport currents	Kinematic nature of sediment transport, circulation cells	Wave action, confluence of coastal currents, backset eddies, and shoals

From Dolan et al. (1974). Used with permission of *Zeitschrift für Geomorphologie,* published by Gebrüder Borntraeger, Stuttgart.

perhaps explains why the features tend to remain in fixed positions, although laboratory and field studies indicate that some longshore migration of the forms is possible if the drift is not excessive (Krumbein 1944). Agreement exists that spacing of cusps is related to the wave height as well as wave direction; the higher the waves, the greater the spacing interval. Even this, however, cannot be pronounced as an inviolate rule; A.T. Williams (1973) found no correlation between wave height and spacing. The spacing of the cusps he studied in a Hong Kong bay were most closely related to the swash distance, that is, the length between the breaker zone and the highest runup of the swash on the beach face.

Because every study of beach cusps seems to reveal some contradiction of earlier studies, it should come as

no surprise that the origin of this feature has been controversial since its earliest description. The simplest explanation of the feature, proposed by D.W. Johnson (1910, 1919) and modified by Kuenen (1948), is based on a process that causes irregular erosion of the beach face. Swash erosion of the beach face initiates cusps, and backwash transports the sediment away. Eroded materials are carried seaward until they are deposited as deltaic projections that stand opposite the excavated hollows. Continued swash action progressively transforms the initial hollows into larger embayments until the water crossing the depressions reaches a critical depth. Swash is then refracted in such a way that coarse sediment is deposited on the cusp horns and finer sediment is transported farther offshore. Bays and horns grow

Figure 13.25
Beach cusps on an intermediate stage beach, Duck, N.C. (April 1988 photo). Cusp spacing here remains fairly constant at around 30 m.
(Photo by R. Craig Kochel)

until the central area attains a limiting depth, swash action is retarded, and the feature assumes an equilibrium condition.

Actually, the process is more complicated, and it may be that cusps can be generated in several ways. Many investigators believe that most cusps are formed in coastal settings that produce a coupling action between incident-surging waves and edge waves (Bowen and Inman 1969, 1971; Bowen 1973; Komar 1973; Guza and Inman 1975; Dolan et al. 1979; Holman 1983). Edge waves augment the breakers systematically in the longshore direction and tend to be well developed in more reflective beach domains. At positions where the breakers are relatively high, recesses are cut into the beach face or berm by the swash-backwash sequence. Thus, cusps are generally associated with reflective beaches, where they play an active role in maintaining the beach state (Antia 1989). If present on dissipative beaches, cusps are thought to be passive and perhaps relict from an antecedent beach condition.

Cusp development can occur within minutes or hours after the generating event begins (i.e., Evans 1938; Komar 1973, and following the event the shoreline will display the rhythmic cusp pattern on the beach surface. At the

CERC-FRF site in North Carolina Miller et al. (1989) documented the destruction of cusps during a major coastal storm in April 1988 and their regrowth within one day following return to normal wave climate. Prior to the storm, cusps were well defined on a slightly reflective beach, spaced at 35 m intervals. Planation by the storm resulted in a lower-sloping dissipative beach devoid of cusps. Within 24 hours following return to average wave conditions, new beach cusps formed sequentially along the beach in the direction of longshore drift. The new cusps adopted a similar prestorm spacing of 35 m with bays coincident with sites of maximum swash runup on the poststorm planar beach, in a manner consistent with edge wave models for this reach of the coast. Figure 13.26 summarizes the sequential destruction and formation of cusps in response to this storm. Holland (1998) examined 9 years of video images of cusp destruction and development at Duck. Similarly, he found that cusps formed within days after major storms as conditions became more reflective and as the spread of incident wave fields became more normal. Mean cusp spacing likewise was measured between 17 and 35 m (Holland 1998).

Cusps perhaps can form in other wave climate settings. For example, DuBois (1978, 1981) and Sallenger

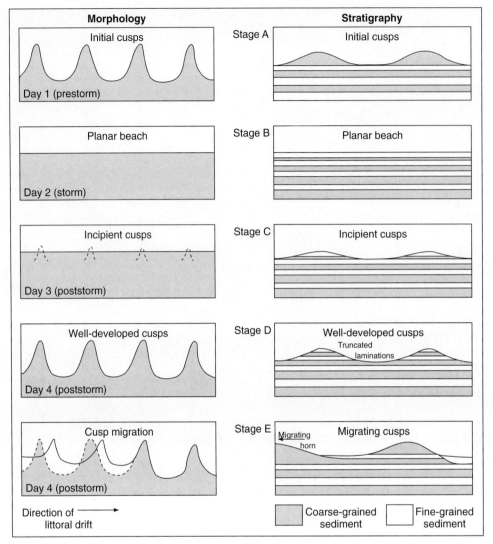

Figure 13.26
Destruction and evolution of new beach cusps during a major northeaster at Duck, N.C., over a four-day period.
(Miller et al. 1989)

(1979) suggest that cusps have developed under plunging wave attack. Cusps in these cases form gradually after swash extending over a berm is ponded in low areas on the berm. As the ponded water is allowed to return seaward, it cuts channels through the berm that are molded into cusps by action during tidal variations. Many studies indicate that cusps develop and evolve with changing wave energy (i.e., Masselink et al. 1997).

Large-Scale Rhythmic Topography and Capes
Larger features in the hierarchy of crescentic forms, called **rhythmic topography** by Komar (1976), consist of two main types: crescentic bars and rhythmic variations that are controlled by rip currents associated with cell circulation. The latter forms have been referred to variously as *sand waves* (see table 13.3), *giant cusps,* and *shoreline rhythms*. These features occur in widely

divergent environments including large lakes (Evans 1938; Krumbein and Oshiek 1950), enclosed seas (King and Williams 1949), and open ocean coastlines in many parts of the world.

Rhythmic topography differs from beach cusps in several important ways. First, rhythmic topographic features commonly migrate parallel to the shoreline. The rate of movement varies but can be a kilometer or more a year. Second, much rhythmic topography is submerged. The emergent forms are usually observed as giant cusps or sand waves that have considerably greater spacing than beach cusps. The wave length of these features ranges from 100 to 3000 m, with horns extending oceanward as much as 25 m. The submerged rhythmic topography is usually in the form of crescent-shaped sand bars or longshore bars that are segmented by rip current channels. The crescentic bars are concave to-

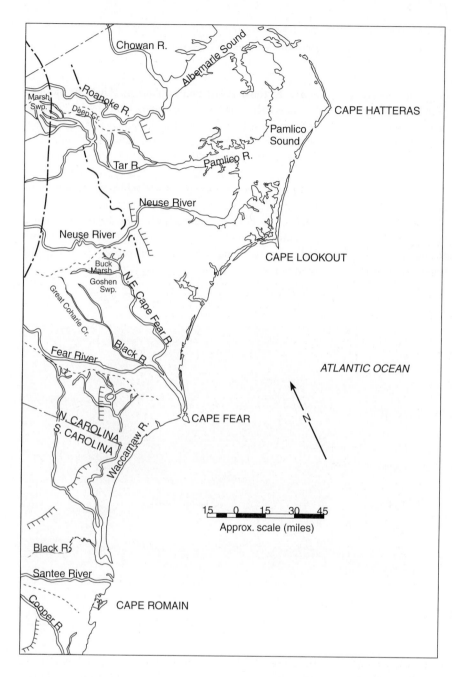

Figure 13.27
Large capes and crescentic recessions (bights) along the south Atlantic coast of the United States.
(White 1966)

ward the shoreline. They commonly stand opposite the horns of large cusps exposed on the beach face itself, but they also exist off straight beaches, especially in protected bays with a small tidal range (Shepard 1952, 1963; King and Williams 1949; Bowen and Inman 1971). A third distinguishing feature of rhythmic topography is that it seems to be more dependent on the bottom configuration in the offshore surf zone than normal beach cusps.

As with beach cusps, there is much uncertainty about the origin of rhythmic topography. Rhythmic topography is a function of wave action, cell circulation with associated rip currents, and longshore transport of sediment driven by obliquely striking waves. Precisely

how these factors combine to construct the shoreline features is simply not clear.

In addition to the cuspate features already examined, many coasts of the world display extremely large shoreline crenulations called **capes.** The map in figure 13.27 shows that capes are prominently developed on the southeast coast of the United States, where spacing is roughly one to two orders of magnitude greater than the spacing of rhythmic topographic features. Capes may be partially relict and therefore not necessarily adjusted to modern conditions. For example, some of the capes shown on the map are fringed by barrier islands that are Holocene in age. Some also coincide with the position of major rivers (W.A. White 1966; Hoyt and Henry

1971), which may indicate that the seaward projections began as ancient deltas.

Many workers believe that these large-scale rhythms are related to rotational cells that are set up as eddy currents along the western edge of the Gulf Stream. Such eddies have not been proven to exist, however, and the direct cause of the capes is unknown.

COASTAL TOPOGRAPHY

As noted earlier, a multitude of classification schemes has been proposed for shorelines. In order to discuss major landforms not treated earlier, we will take a simple approach that groups all shorelines into one of two categories on the basis of local relief and the predominance of erosional or depositional landforms. High-relief shorelines are generally characterized by the presence of sea cliffs and resistant bedrock within the coastal zone (fig. 13.28). These zones commonly occur in areas of active tectonic uplift and volcanism. They are notable for their rocky beaches, small and isolated sand beaches, and landforms created by the erosional and corrasional activity of waves along the cliff. Low-relief shorelines are typically the seaward edge of an extensive coastal plain composed of unconsolidated marine and fluvial sediments sourced from distant bedrock mountainous terrains. The absence of bedrock and the relatively low shelf slope in these settings promotes the predominance of sediment exchange and landforms associated with longshore drift and wave transport processes such as the common barrier island–lagoon system. In general, these

(A)

(B)

Figure 13.28
(A) High-relief coast along Washington's Olympic Peninsula, characterized by steep, bouldery beaches, cliffed shoreline, and offshore sea stacks, separated from the mainland by wave erosion. (B) Low-relief coast along southern Maryland. Assateague Island (foreground barrier island) typifies this coast, characterized by dunes, storm overwash, and abundant sediment in rapid flux. The mainland shoreline is visible in the distance across the lagoon.
(Photo by R. Craig Kochel)

areas occur in regions of inactive tectonics such as along the trailing margin of major plates.

High-Relief, Erosional Shorelines

Emery and Kuhn (1982) have used profiles to classify sea cliffs in southern California as *active, inactive,* or *former.* These forms apparently indicate whether marine or subaerial processes are dominant at any given time. A smooth curve at the base of the cliff (inactive profile) indicates that subaerial erosion is prevailing. In contrast, a sharp-angled basal contact with the beach (active profile) suggests that marine erosion is dominant. A former profile is one that has been removed from the influence of marine processes. Because inactivity is associated with dry climates, the episodic climate shifts in southern California from dry to more humid conditions can initiate a change to active sea cliff erosion. Areas presently near inactive cliffs were extensively developed during one of the driest periods in the past 150 years (Kuhn and Osborne 1989). Erosion problems in these settings became severe in the late 1970s and early 1980s when the climate became considerably more stormy and wet (fig. 13.29). Increased wave activity, combined with ris-

ing groundwater levels in the cliff sediments, was exacerbated by human-induced loading and additions of water to the cliffs. The combination of natural and human-induced changes within the coastal zone have produced one of the most dramatic examples of coastal hazards along the U.S. shoreline.

When considering sea cliff erosion, several facts become immediately apparent. First, retreat of a sea cliff requires wave erosion at the cliff base. Second, erosion at the base leads to increased mass movement of the sea cliff material because of the resulting increase in slope angle and shear stress. Third, the debris of mass movement collects at the cliff base. No base erosion can occur again until this debris is removed and the toe of the cliff is again exposed to wave attack. The erosion is accomplished by a group of geomorphic processes, listed in table 13.4, that function in complex interactions to produce a variety of landforms. The final effect is sometimes controlled by ancillary processes such as the burrowing action of marine organisms (see Ahr and Stanton 1973), frost action, and mass movement. As might be expected, the effectiveness of each process varies with the properties of the shore material and with the particular dynamics of the local ocean or lake system. *Corrosion* affects those rocks that are

Figure 13.29
High-relief coast near San Diego, Calif. (November 1991 photo). Note the rocky beaches, small sandy pocket beaches, and headland cliffs rapidly eroding, exploiting weakened jointed zones in the bedrock.

most susceptible to solution. At normal temperatures seawater is saturated with respect to calcium carbonate and does not dissolve limestone and $CaCO_3$ directly. Solution of these materials, however, may be aided by rainwater or by organisms that create local acidic conditions. As chunks of coastal or lakeshore rocks are released, *attrition* decreases the size of the particles and allows subsequent

TABLE 13.4	Major Erosional Processes Functioning Along Coasts.
Process	**Description**
Corrosion	Solution of coastal rocks by chemical action of seawater.
Attrition	Diminution of rock particles as water rolls, bounces, or slides them on a beach or wave-cut platform.
Corrasion	Physical erosion of bedrock caused by the grinding action of rock fragments that are carried in the ocean waves and currents.
Hydraulic action	Erosion caused by the force of the water itself, includes wave shock pressure and pneumatic quarrying by air trapped in cracks of the headland rocks.

waves to drive the sediment into the cliff face. *Corrasion* then assumes a dominant role in the erosive mechanics. *Hydraulic action* is especially important where the rocks are highly fractured. Not only does the force of the wave exert pressure on the cliff face, but the advancing waves may compress air in the rock cavities, producing a pneumatic effect in the cracks. As the wave recedes, external pressure is instantaneously released, and the compressed air within the rocks exerts an outward stress that may disaggregate the outer zones of the cliff face. Typically, wave undercutting is more rapid than other backwasting processes. For example, in a study of cliff recession along the Chesapeake Bay in Maryland, Wilcock et al. (1998) estimated undercutting ressional rates twice that of cliff decay by freeze–thaw processes. They also were able to show how wind records could be used to estimate cliff recession rates once an index of resistance and wave strength was defined.

As a sea cliff retreats, it leaves behind a beveled surface, called the **wavecut platform,** that stands slightly below water level at high tide (fig. 13.30). Although many processes are involved in its creation (see Wentworth 1938), corrasion at the base of the sea cliff is probably responsible for most of the planating action. The platform is

Figure 13.30
Wave-cut notch eroded into volcanic rocks on the north coast of Oahu, Hawaii (1985 photo). The wave-cut platform is covered by a thin mantle of water-worn cobbles eroded from slump blocks of the bedrock cliff.
(Photo by R. Craig Kochel)

not flat but slopes gently oceanward with a declivity that ranges between 0.02 and 0.01. Therefore, under a stationary sea level the maximum width of the wave-cut platform depends largely on the depth at which wave abrasion is still a viable process. Bradley (1958) suggests that wave-cut platforms can be eroded in water no deeper than 10 m; under the common slope range, platforms wider than 500 m probably form only if sea level is continuously rising. Terraces with a thin veneer of marine sediment over a beveled bedrock surface are common along coasts that have experienced recent uplift. The bedrock surfaces were cut at sea level and represent former wave-cut platforms.

In addition to being influenced by water depth, on low-gradient platforms and with constant sea level the ultimate width is self-controlling because the rate of platform expansion decreases as the sea cliff recedes. Platforms develop rapidly in the early stages of cliff retreat. In time, however, incoming waves lose much of

their energy as they interact with the progressively widening platform. Eventually, cliff retreat and platform expansion cease unless deposition on the original platform surface raises the level of incoming waves and changes the frictional component.

As the shoreline retreats and irregularities appear, a group of landforms develop that are characteristic of coastal erosion along bedrock reaches. **Stacks** (see fig. 13.28A) are formed when narrow oceanward extensions of the coastal rocks are cut into isolated remnants by wave attack. The process is accentuated when waves are refracted around the headland reach. Sometimes less resistant rock zones are exposed to local corrasion or scouring, and notches or **sea caves** are formed. As sea caves grow, they may extend completely through the headland to produce a feature known as a **sea arch** (fig. 13.31). Any or all of these features indicate a local erosional environment. As the headland rocks undergo

Figure 13.31
Sea arch or window along coast of Puerto Rico at low tide. Hole is caused by a combination of wave action and solution of limestone bedrock.

attrition, however, more sediment covers the wave-cut platform. Beach profiles develop, and wave energy dissipates farther offshore as waves break over the longshore bars. This sequence may not occur in every situation. For example, in large lakes such as Lake Erie or Lake Michigan, energy loss over longshore bars is of little or no consequence.

In a series of papers, Sunamura (1975, 1976, 1977, 1978, 1982) has examined cliff erosion theoretically in terms of driving and resisting forces. The assailing wave force (f_w) is initially determined by energy derived in deep water, but it is directly influenced by factors such as water level or tide, bottom and beach topography, and beach sediment. These determine the wave type and height, where the waves break, and at what level they strike the sea cliff. Unfortunately, because f_w cannot be measured directly, some related parameter such as wave height is used as an estimate of the driving force.

Although lithology is the primary determinant of resistance (f_s), it is also influenced by contributing factors such as mechanical strength and geologic structures. Most workers use compressive strength as a surrogate for resistance. Thus, the cliff erosion rate becomes a function of f_w/f_s, which Sunamura (1977, 1983) has expressed as

$$\frac{dX}{dt} = k\left(C + ln\,\frac{\rho gH}{S_c} \right)$$

where H is wave height at the cliff base, S_c is compressive strength, k is a dimensionless constant, and C a constant with dimensions of LT^{-1} This equation suggests that a critical wave height (H_{crit}) exists that is needed to cause sea cliff erosion and can be derived by setting $dX/dt = 0$. This yields

$$H_{crit} = \frac{S_c}{\rho g}\,e^{-c}$$

In areas of relatively homogeneous parent material and distinct wave climate, this approach is quite significant. For example, at Byobugaura, on the Pacific coast of Japan, a homogeneous Pliocene mudstone forms the lower half of a cliff 10 to 60 m high. Sunamura (1982) was able to determine k and C and calculate H_{crit} by using data on cliff strength, wave climate, and erosion rates of two time intervals. The plots shown in figure 13.32 indicate that erosion was caused only under attack of the rarer but larger waves. The smaller, frequent waves produced no sea cliff erosion. This approach led to the identification of a threshold condition that, unlike average rates, has direct application in coastal land management. It also illustrates the problems of using average rates because erosion of the cliff face occurs in spasms of extremely rapid erosion (fig. 13.33) rather than as continual removal of material at a constant rate. Short-term measurements may be far from reality if measured during unusual

times, and long-term analyses may mask the important aspect of prevailing wave climate. Clearly, as Sunamura (1983) suggests, the rate of sea cliff erosion should be determined by the occurrence frequency of waves exceeding the threshold height.

The base of any sea cliff is periodically covered by debris. This occurs during intervals of rapid subaerial erosion when climate change or human activities deliver more debris to the beach surface than can be removed by wave action (Emery and Kuhn 1982; Kuhn and Shepard 1983). This material prevents further undercutting of the cliff until stronger wave action removes it and thereby exposes the cliff face to renewed supercritical wave attack. The implication is that the sea cliff profile will vary with time (Emery and Kuhn 1982; Kuhn and Shepard 1983; Sunamura 1983). These variations tend to offset one another such that, over a long term, the profile may appear to retreat in a parallel manner. However, within that long term the profile will change according to the amount of debris stored on the beach and/or whether the beach elevation causes supercritical waves to strike the cliff face at higher or lower levels. Even though high-relief coasts are dominated by their bedrock features, there can be considerable sediment moving along the shoreline and narrow newshore zone. Figure 13.34 shows a view of a major baymouth bar formed from the accumulating longshore drift in northern California. Figure 13.35 shows a typical pocket beach developed in a recess along a high-relief coast. However, a large percentage of the heartache sediment along high-relief coasts gets lost to the deep ocean in canyons and submarine fans.

Low-Relief, Depositional Shorelines

In coastal areas where the supply of sand is abundant and local ocean forces are capable of transporting sediment, the shapes of some landforms are determined primarily by depositional events, even though erosional processes are very much involved in the entrainment of sand before its final deposition. Large depositional features usually occur in the form of spits, baymouth bars, or barrier islands. **Spits** and **baymouth bars** are definitely related to longshore transport of sediment, or *littoral drift*, with the site of deposition being in tranquil waters of bays, estuaries, or the open ocean, and are common in high-relief and low-relief coastal settings. **Barrier islands** are large elongate features that parallel the shoreline but are not physically connected to it (see fig. 13.28B). Barrier islands are uncommon along the high-relief west coast of the United States, but some 600 of these features occur along the Atlantic and Gulf shorelines (these will be treated in a separate section). Other depositional features may include some of the cuspate and rhythmic forms discussed earlier, especially cuspate forelands associated with sand waves.

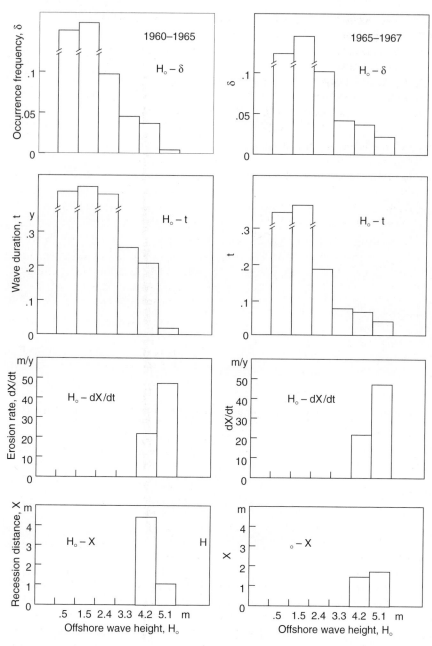

Figure 13.32
Wave occurrence frequency, duration, erosion rate, and recession distance plotted against wave height at Byobugaura, Japan.
(Sunamura 1982)

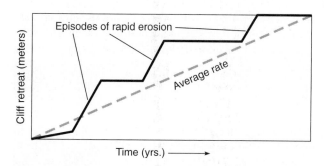

Figure 13.33
Schematic illustration of cliff retreat rates using a long-term average versus the actual continuum of erosion within the long-term interval.

Spits and baymouth bars develop when littoral drift plays a predominant role in the system, provided the drifting sediment enters a zone of slack water where deposition can occur. The erosional and depositional framework in the littoral system might be considered in terms of a littoral power gradient similar to that proposed by May and Tanner (1973). In their study, a wave power model utilizing E, Pl, and dq/dx as the basic parameters (q being the quantity of sand transported and x the distance along the beach) was constructed to predict zones of different littoral transport along a beach. Where dq/dx is greatest, beach erosion is also greatest because a large amount of sediment is being placed in transit. Where dq/dx is negative, deposition is occurring.

Figure 13.34
Northern California coast near the mouth of the Klamath River showing a narrow and fragmented beach. Longshore drift has formed a prominent baymouth bar in the foreground.
(R. Craig Kochel)

Figure 13.35
Sandy pocket beach along the high-relief granitic shoreline in north-central Maine.
(R. Craig Kochel)

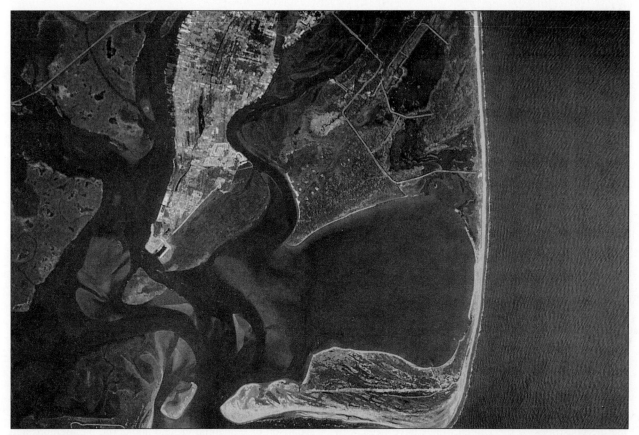

Figure 13.36
Recurved spit at the southern tip of Assateague Island, VA. A very complex history of spit extension can be seen in this infrared image showing at least two formerly active spits to the northwest (upper left) of the active area (the oldest is occupied by the town of Chincoteague). Note the linear strip running east-west on the active spit. This is a stabilized dune line representing the position of the shoreline in the early 1960s.

Spits and baymouth bars are essentially the same feature. They differ only in that spits extend into the open ocean whereas baymouth bars breach the gap between two headland reaches. Baymouth bars are also more likely to initiate lagoonal areas shoreward of the bar deposit. These lagoons may gradually change into tidal marshes or swamps as the embayments fill in with fluvially derived sediment as well as marine sediment washed over the bar by coastal storms.

Although spits and baymouth bars have been attributed to a variety of ocean processes, the origin is clearly and most prominently related to littoral drift. The elongate extensions into open water represent continuations of the beach that rest against the coast. They expand continuously in the direction of littoral drift unless other water motions interfere with the growth process. For example, it is not uncommon for wave refraction around the free end (Evans 1942) or wave trains approaching from different directions (King and McCullagh 1971) to reorient the terminal end of the spit. In these cases the feature may be called a *recurved spit* or a *hook*. Spits or baymouth bars may also widen by progradation associated with the construction of a series of sandy or pebbly

ridges on the oceanward side of the feature (fig. 13.36). These ridges or growth lines, called *beach ridges,* will be discussed in the barrier island section. Finally, spits may link the main coastal region to an offshore island, producing a feature known as *tombolo.* Tombolos, however, also form on normal coastal beaches when wave refraction around the island shapes the beach into a cuspate foreland that eventually extends to the island itself. This process even functions around breakwaters that have been constructed offshore to prevent beach erosion (Inman and Frautschy 1966).

SHORELINE CHANGE

A cover of *Time* magazine in September 1987 asked "Where's the Beach?" and featured an article on the ubiquitous problem of coastal erosion in the United States. Erosion of shorelines constitutes one of the major problems facing scientists, engineers, and land managers throughout the world. In the United States alone, approximately 25 percent of our coastlines have been categorized as seriously eroding (U.S. Army Corps of Engineers

1971). The annual cost of prevention is staggering, and erosion continues in spite of a wide variety of state and federal laws and policies aimed at regulating coastal development.

Erosion not only has an impact on property and human life. Some studies have suggested that erosion rates may control the pattern of vegetation and landform associations along some shorelines. For example, Roman and Nordstrom (1988) statistically evaluated a number of geomorphic process variables and vegetation patterns along Assateague Island (a 40-km-long barrier island on the coast of Maryland and Virginia) and found significant correlations with long-term erosion rates. They found that there was a critical erosion-rate threshold value 4.5 m/yr at which major changes in vegetation and morphologic associations occurred, and that this rate was related to succession and recovery processes following disturbance by recent erosion. When vegetation is erased by erosion, succession occurs over time. Thus, the time frame or time history of erosion can be determined by mapping vegetation communities (Roman and Nordstrom 1988).

Rates of Change

Shoreline changes along the Great Lakes and along the marine coastlines of the United States have been well surveyed (e.g., see May et al. 1983). On a national scale, U.S. shorelines are receding at an average rate of 0.8 m/yr. Erosion rates vary on a regional scale from 0.0 m/yr along the Pacific coast to 0.8 m/yr along the

Atlantic and 1.8 m/yr along the Gulf coast region (table 13.5). Great Lakes shorelines are retreating at 0.7 m/yr. Significantly, rates vary drastically on a local basis where they depend on geology and wave climate. Net accretion occurs commonly on a local scale over the periods of record (fig. 13.37). It is, therefore, unrealistic to use average rates for states or even counties in land management evaluation. More attention needs to be given to local variation if planning is to allow for real erosion rates (fig. 13.38).

Kuhn and Shepard (1983) have shown conclusively that coastal erosion (in this case sea cliff retreat) is episodic, site-specific, strongly related to meteorological conditions, and influenced by human factors. Kuhn and Osborne (1989) provide a nice discussion of the correlation between episodic cliff erosion and climate fluctuation in southern California and offer insights and warnings based on historical records for planners in this region. Clearly, the rates they present are not meant to be utilized in a predictive manner. Much of the erosion included in the rates involves removal of sand from the beach itself. However, because of increased use of the entire coastal zone, engineers and coastal zone managers have given considerable attention to the processes and rates of sea cliff erosion. Cliff erosion rates have been documented by a variety of techniques including comparison of sequential ground and aerial photographs and maps, exposure of pins inserted into the sea cliff (Hodgkin 1964), instruments designed to measure microerosion (Trudgill 1976; Robinson 1977), and detailed air-photogrammatic maps (Norrman 1980).

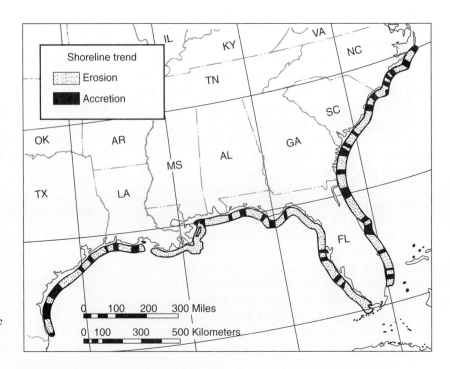

Figure 13.37

Shoreline erosion and accretion trends along the Atlantic and Gulf coasts of the United States.

(May et al. 1983)

TABLE 13.5	Rate of Shoreline Change Along U.S. Marine Coasts and Bays and Lakes (σ indicates within-State Standard Deviation of Rates).

Marine coasts

Region	X (m/yr)[a]	σ	Total Range[a]		N[b]
Atlantic Coast	−0.8	3.2	25.5	−24.6	510
Maine	−0.4	0.6	1.9	−0.5	16
New Hampshire	−0.5	—	−0.5	−0.5	4
Massachusetts	−0.9	1.9	4.5	−4.5	48
Rhode Island	−0.5	0.1	−0.3	−0.7	17
New York	0.1	3.2	18.8	−2.2	42
New Jersey	−1.0	5.4	25.5	−15.0	39
Delaware	0.1	2.4	5.0	−2.3	7
Maryland	−1.5	3.0	1.3	−8.8	9
Virginia	−4.2	5.5	0.9	−24.6	34
North Carolina	−0.6	2.1	9.4	−6.0	101
South Carolina	−2.0	3.8	5.9	−17.7	57
Georgia	0.7	2.8	5.0	−4.0	31
Florida	−0.1	1.2	5.0	−2.9	105
Gulf of Mexico	−1.8	2.7	8.8	−15.3	358
Florida	−0.4	1.6	8.8	−4.5	118
Alabama	−1.1	0.6	−0.8	−3.1	16
Mississippi	−0.6	2.0	0.6	−6.4	12
Louisiana	−4.2	3.3	3.4	−15.3	106
Texas	−1.2	1.4	0.8	−5.0	106
Pacific Coast	−0.0	1.5	10.0	−5.0	305
California	−0.1	1.3	10.0	−4.2	164
Oregon	−0.1	1.4	5.0	−5.0	86
Washington	0.5	2.2	5.0	−3.9	46
Alaska	−2.4	2.0	2.9	−6.0	69

Bays and lakes

Region	X (m/yr)[a]	σ	Total Range[a]		N[b]
Delaware Bay					
New Jersey	−1.9	1.3	0.3	−3.0	13
Delaware	−1.3	2.1	5.0	−3.0	12
Chesapeake Bay	−0.7	0.7	1.5	−4.2	136
Western shore	−0.7	0.5	1.5	−1.9	67
Maryland	−0.7	0.3	−0.1	−1.3	35
Virginia	−0.8	0.7	1.5	−1.9	32
Eastern shore	−0.7	0.8	0.1	−4.2	69
Maryland	−0.8	0.9	−0.3	−4.2	47
Virginia	−0.5	0.4	0.1	−1.2	22
Great Lakes	−0.7	0.5	0.6	−2.7	327
Lake Erie	−0.7	0.6	−0.2	−2.4	98
Ohio	−0.6	0.6	−0.2	−2.2	68
Pennsylvania	−0.3	0.1	−0.2	−0.4	14
New York	−1.4	0.6	−0.5	−2.4	20
Lake Ontario	−0.5	0.2	−0.2	−1.2	58
Lake Huron	−0.4	0.3	−0.3	−1.3	28
Lake Michigan	−0.6	0.8	0.6	−9.9	184
Western shore	−0.6	0.4	0.6	−1.5	62
Eastern shore	−0.7	0.9	0.3	−9.9	122
Wisconsin	−0.7	0.3	−0.3	−1.5	46
Illinois	−0.2	0.4	0.6	−0.9	16
Indiana	−0.4	0.5	−0.3	−0.9	12
Michigan	−0.7	0.9	−0.3	−9.9	110
Lake Superior	−1.3	0.7	−0.3	−2.7	35
Minnesota	−0.8	0.4	−0.3	−1.5	16
Wisconsin	−1.8	0.6	−0.9	−2.7	19

From May et al., *EOS,* issue 35, p. 522 (1983), copyright by the American Geophysical Union.

[a]Negative values indicate erosion; positive values indicate accretion.

[b]Total number of three-minute grid cells over which the statistics are calculated.

(A)

(B)

Figure 13.38
Photos illustrating the magnitude of beach erosion along a portion of the North Carolina coast between Kitty Hawk and Nags Head. (A) Hanging showers no longer reachable along the center of the house (1983 photo). (B) Exhumed foundation and hanging driveways (1993 photo). (C) Exhumed septic tank and drain system to the right of the house (1993 photo).
(Photos by R. Craig Kochel)

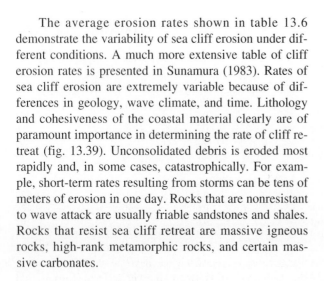

(C)

The average erosion rates shown in table 13.6 demonstrate the variability of sea cliff erosion under different conditions. A much more extensive table of cliff erosion rates is presented in Sunamura (1983). Rates of sea cliff erosion are extremely variable because of differences in geology, wave climate, and time. Lithology and cohesiveness of the coastal material clearly are of paramount importance in determining the rate of cliff retreat (fig. 13.39). Unconsolidated debris is eroded most rapidly and, in some cases, catastrophically. For example, short-term rates resulting from storms can be tens of meters of erosion in one day. Rocks that are nonresistant to wave attack are usually friable sandstones and shales. Rocks that resist sea cliff retreat are massive igneous rocks, high-rank metamorphic rocks, and certain massive carbonates.

Causes of Shoreline Erosion

Apart from human-induced activities, the two major causes of shoreline erosion are rising sea level and coastal storms. Rising sea level provides the impetus for change whereas storms provide the energy or the me-

chanics to accomplish the geomorphic work. Obviously, much of the perception of the problem arises from the construction of "permanent" objects and our attempt to maintain them along coastal environments that can naturally be very dynamic.

Rising Sea Level It is well documented that sea level was more than 100 m lower than present during the peak of the late Wisconsinan glacial maximum, some 18,000 years B.P. Glacial recession was followed by a rapid rise in global sea level to within several meters of modern levels by approximately 4000 to 5000 years B.P. (fig. 13.40). Since that time, sea level has been slowly, but steadily rising. It is beyond the scope of our treatment here to provide a detailed discussion of sea-level change. Excellent reviews can be found in Lisle (1982) and Fletcher (1992). It is clear, however, that sea level has risen some 25 cm since the early part of this century. The magnitude of these changes translate into significant lateral dimensions along low-relief shorelines and also accelerate undercutting of cliffs in high-relief areas.

The rise in sea level has been sufficient to facilitate steady landward migration of barrier island and main-

TABLE 13.6	Representative Rates of Sea Cliff Retreat.		
Location	**Material**	**Rate (m/100 yr)**	**Reference**
New England	Crystalline rock	0	1
Former U.S.S.R.	Volcanics	0	2
England (Cornish coast)	Crystalline rock	0	3
Northern France	Chalk	25	3
England (Yorkshire)	Sedimentary rocks	9	3
England (Yorkshire)	Glacial drift	28	3
Louisiana (coastal islands)	Sands and clays	800–3800	4
Southern California	Alluvium	30	5
Former U.S.S.R.	Clay	1200	2
New Jersey	Sand, clay, and gravel	180	6
Cape Cod, Mass.	Glacial drift	30	7

1. Johnson (1925)
2. In Zenkovich (1967)
3. In King (1972)
4. Peyronnin (1962)
5. Shepard and Grant (1947)
6. Rankin (1952)
7. Zeigler et al. (1959)

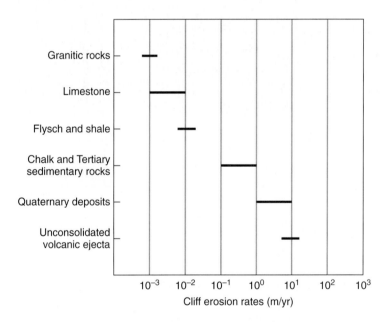

Figure 13.39
Generalized orders of erosion rates for sea cliffs composed of different lithologies and mechanical strength.
(Data from Sunamura 1983)

land coastal environments worldwide. An example of the significance of this change is the common appearance along the barrier and mainland shorelines of the mid-Atlantic coast of dead trees that have succumbed to the invasion of saline groundwater in their root systems. Also, there are numerous examples along the barrier coast within the last century of abandoned farms and agricultural activities where fresh water or subaerial environments have been overrun by salt marshes resulting from the rise in sea level.

Great concern (although tempered by uncertainty over the magnitude of the anticipated effect) exists over the patterns of future sea-level rise that may result from global warming associated with increased introduction of greenhouse gases into the atmosphere. Recent studies have projected that anthropogenic impact of global warming will cause sea level to rise 50 to 150 cm during the next 100 years (Ramanathan 1988). Gornitz and Kancirok (1989) provide a good summary of the types of sea level data currently available in addressing this very complex issue. The impact of sea level rise will vary locally with relief, lithology, landforms, rates of shoreline erosion, beach slope, vertical tectonics, wave climate, and tidal range.

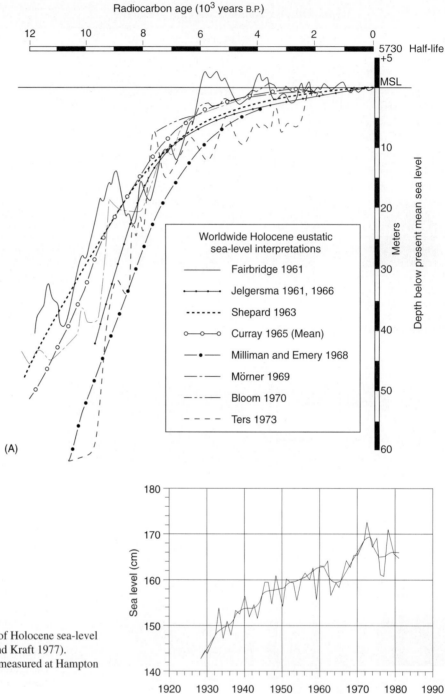

Figure 13.40
Sea-level change. (A) Global record of Holocene sea-level rise (from compilation by Belknap and Kraft 1977).
(B) Twentieth-century sea-level rise measured at Hampton Roads, Va.
(A): (Belknap and Kraft 1977)
(B): (Hicks and Crosby 1974)

Coastal Storms   In addition to the driving factors discussed above, the dynamics of coasts are directly affected by storms that impinge on the shoreline. Coastal storms provide the energy for accomplishing significant erosion over short periods of time. Storms should probably not be viewed as rare or catastrophic processes but as part of the normal continuum of processes affecting shorelines (Bryant 1988).

Coastal storms occur as two types, tropical and extratropical. Both types are cyclones, meaning they in-volve a circular wind pattern, which in the Northern Hemisphere moves in a counterclockwise direction. All cyclones develop because of atmospheric instability and have the effect of removing the instability. *Extratropical storms* are cold-core cyclones that derive energy mechanically in a process associated with the interaction of air motion between zones of high and low pressure. Because they have a cold core, these storms tend not to lose energy with height. They are fueled, instead, by the high-level jet stream which moves mass away from the

(A)

(B)

Figure 13.41

Satellite images of major coastal storms on the mid-Atlantic coast. (A) Hurricane Emily (late August 1993). (B) Significant extratropical storm in mid-April 1988. Note the similar size of these two storms. The extratropical storm typically intensifies off Cape Hatteras and tracks northeastward, lashing the entire eastern coast with high waves for days, while the hurricane usually rushes onshore, impacting only the areas immediately along its right-hand side with exceptional onshore winds and storm surge.

centers, allowing the storms to intensify as barometric pressures are lowered (Davis and Dolan 1993).

In contrast, *tropical storms* are warm-core cyclones fueled by latent heat from the evaporation of water. They evolve as deep, atmospheric low-pressure systems that originate and gain intensity over warm-water marine areas. Tropical cyclones are given a variety of names depending on geographical location and/or wind velocities. In the Atlantic and Gulf coasts of the United States, the term "hurricane" (fig. 13.41) is employed when wind velocity exceeds 74 mph (119 km/hr). In the United

States, hurricanes are the predominant coastal storm south of Cape Hatteras, North Carolina, whereas extratropical storms are much more common north of Cape Hatteras. Some 30 to 40 extratropical storms per year frequent the Atlantic shoreline. Typically 3 to 6 of these significantly affect coastal morphologic processes each year. This frequency distribution is important because the characteristics of the storms are quite different (see table 13.7).

A major difference between the storm types is the magnitude of storm surges they can generate. A **storm**

TABLE 13.7	Characteristics of Tropical and Extratropical Storms.

	Tropical storms	Extratropical storms
Wind speed	Great, > 74 mph	Less, usually < 50 mph
Duration[a]	Short, few hours	Longer, many hours to days
Size	Small, 50–80 km	Large, 100s km
Shape	Circular	Often elongate
Surge	Large, > 15 ft	Small, < 5 ft
Barometric pressure	Low central; greater pressure differential	Higher central; lesser pressure differential
Fetch	Small, 10s km	Large, 100s km
Occurrence	June–Sept.	Oct.–Apr.

[a]Refers to time of effect on coast.

Figure 13.42
Aerial view of overwashed sand from the beach to the back-barrier marsh on Cedar Island, Va. Note the proximity of the two houses to the shoreline, which is retreating at a rate of several meters per year. Storm overwash is a major process facilitating landward migration of barrier islands.
(Photo by R. C. Kochel)

surge is an elevation of normal water level in response to a passing storm. Storm surges result from the combination of the shoreward transport of water from winds blowing onshore and "setting up" water against the coast and the upward bulge of the ocean level caused by the extreme low pressure associated with major cyclones (a 1-inch drop in pressure causes about a 13-inch rise in water level). Extreme surges combined with high surface waves allow water to penetrate inland beyond the beach, in a process known as **overwash** (fig. 13.42) (discussed below), and cover areas normally immune from wave attack (see fig. 13.14). Historically, hurricane-related surges are much more damaging than those associated with extratropical storms, commonly exceeding 6 m, while only the most severe northeasters can produce surges between 3 and 5 m. This does not mean that major flooding and destruction are impossible in extratropical storm surges. On the contrary, the March 1962 storm discussed earlier is excellent proof that extratropi-

cal storms can be significant events. Usually, however, they require some ancillary factor such as coincidence with high astronomical tides to reach full destructive potential. The danger with northeasters, however, rests in duration of impact on the coast (often measured in days) and area of effect along the coast (commonly on the order of 1000 km or more compared to 100 km normally affected by a hurricane). Davis and Dolan (1993) provide an excellent discussion of the importance of northeasters, including a storm-power-damage ranking scheme similar to the one well known for hurricanes.

Concern over the prospect of future global climate changes have generated a number of studies that have examined the record of coastal storms for trends apparent during the past 100 years, when good records have been maintained. Davis and Dolan (1993) clearly show that there has been a significant cyclicity in the pattern of Atlantic coastal storms during this century, with a particularly stormy period occurring between the mid-1940s and mid-1960s. Between the mid-1960s and mid-1970s a notable decline in storminess occurred, followed by a high degree of variability since then. Two of the most powerful storms since records have been kept have occurred within the past five years—one in March 1989 and one, the most powerful storm on record, in late October 1991 (known as the Halloween storm and immortalized in the book and film as "The Perfect Storm"). Although destructive, wholesale damage did not accompany the Halloween storm, as it did the 1962 storm, because it failed to coincide with exceptional astronomical tides as did the latter. One measure of the geomorphic significance of the 1989 storm was provided by Dolan et al. (1988) who calculated the transport energy in the storm. Storm energy was estimated at 11,500 kwH/m of beach, a rate that would account for longshore transport of some 415,000 m³ of sand. This volume compared favorably with an estimate of 360,000 m³ of sand eroded from the north end of Pea Island, North Carolina, adjacent to Oregon Inlet, from a detailed field survey by the CERC-FRF. Large storms such as hurricanes have been known to cause major, instantaneous sand transfers across barrier inlets (i.e., Morton et al. 1985). Likewise, considerable movement of sand can occur between the beach and nearshore, and between the beach or back-barrier during major storms. The sudden and catastrophic impact of storms on coastal sediment budgets can make long-term prediction of shoreline behavior and response to erosion mitigation activities difficult at best. Parsons (1998) discusses how long-term records of major storms may be discerned in freshwater ponds along the coast by changes in the nature of preserved diatoms. This information could be critical in assessing long-term frequency of big storms.

Komar and Good (1989) demonstrated the impact on coastal erosion of an unusually intense El Niño event during 1982–83. The El Niño elevated water levels some 35 cm, on the average, above normal winter levels and narrowed the high-latitude branch of the jet stream such that unusually intense cyclones impacted the Pacific coast that year. The potential impact of El Niño should be considered in long-term planning and prediction of coastal changes.

BARRIER ISLANDS

The most dynamic of our coastal landforms are narrow sandy islands called barrier islands. Their rate of change in response to coastal driving forces can be so rapid that we can observe wholesale changes in a few years or even days.

Distribution and Characteristics

Barrier islands are elongate bodies of sand that are not attached to the mainland but are separated from it by a lagoon or bay (fig. 13.43). The islands normally range in width from a few hundred meters to 5 km and in length from 10 to 100 km, separated from each other by tidal inlets, and are usually less than 6 m in elevation. Sand thickness varies from 2 to 15 m but is often less than 10 m. Barrier islands are commonly large enough to support major cities; Atlantic City, Ocean City, Miami Beach, and Galveston are examples. Barrier islands occur on 13 percent of coasts throughout the world (King 1972) but are concentrated where tidal range and offshore gradients are low and where wave energy is low to moderate. Along the boundaries of the United States, about 300 barrier islands exist along Gulf and Atlantic coasts (Dolan et al. 1980). Oertel and Kraft (1994) provide a nice review of mid-Atlantic barrier island systems.

The origin of barrier islands has always been steeped in controversy, mainly because most of the original interpretations were based on morphology alone. More recent theories rely on subsurface exploration and paleoenvironmental interpretation of facies both shoreward and seaward of the barrier. Although differences of opinion exist, most workers believe that barrier islands

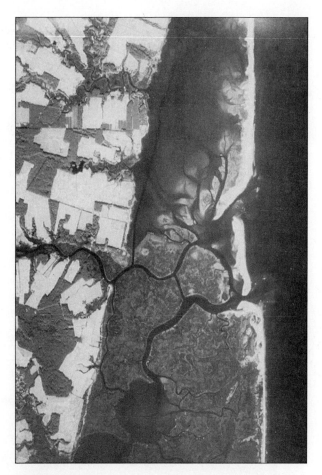

Figure 13.43
High-altitude color infrared aerial photograph of part of the eastern shore of Virginia. Image shows parts of three barrier islands separated from the mainland (left) by a lagoon. Note the smooth ocean-side shoreline compared to the highly irregular back-barrier shoreline. Back-barrier marshes are extensive in this meso-tidal region, as are the flood-tidal delta sediments and channels lagoonward of the inlets.
(Photo courtesy of NASA)

have a long growth history associated with postglacial sea-level rise. During the last full glacial episode, sea level was probably 120 m lower than it is today, and the shorelines possibly 60 to 150 km seaward of their present positions (Curray 1965; Dolan et al. 1980). Beginning as shoreline depositional nuclei during the low sea level of the late Pleistocene, the initial ridges migrated landward as sea level rose. As the rate of sea level rise declined, approaching its present level about 4000 to 5000 years ago, modern barrier islands developed their character. The islands are still evolving because sea level is still slowly rising, and they continue to migrate landward. This movement, however, requires the unique combination of environments, sand supply, and sand transport found in the barrier system.

Barrier islands are composed of distinct geomorphic and vegetative zones (fig. 13.44). On the ocean side, the

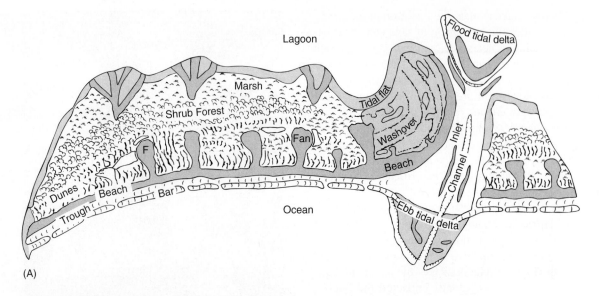

(A)

(B)

(C)

Figure 13.44

Idealized barrier island. (A) Schematic showing barrier vegetation and geomorphic facies. (B) Aerial photo of a typical barrier island on the U.S. east coast (Assateague) showing vegetation and landforms. (C) Idealized cross-profile through an island.

(Photo by R. Craig Kochel)

Figure 13.45
Giant dune at Jockey's Ridge, N.C. (1984 photo).
(Photo by R. Craig Kochel)

islands are characterized by beaches that alternately change configuration during storms and during intervals in which a swell-wave climate prevails. Significantly, the backshore environment of most beaches is characterized by sand dunes rather than a sea cliff. The dune line represents the backbone of the island, typically standing 3 to 6 m in elevation but occasionally reaching much greater heights (fig. 13.45). For example, in a study of southeast African coasts, Orme (1973) reported Holocene dunes climbing to elevations greater than 100 m. Dune elevation is closely related to the prevailing wind and wave direction and sediment supply from the subaerial beach.

Where beach orientation is conducive to littoral drift, dunes may be stunted or even absent. These conditions, especially when they occur in environments where sand supply is limited (due to a lack of headland or river sediment sources), result in low, narrow islands that are extremely vulnerable during storm surges. A good example of this style of barrier island is the chain of undeveloped barriers along the eastern shore of Virginia (fig. 13.46). A similar situation exists along small barrier islands in the Delaware Bay experiencing rapid erosion rates because the volume of longshore drift cannot match the rate of shoreface erosion and landward transport of sand across the islands into the lagoon during storms (Pizzuto 1986). In contrast, where beaches are perpendicular to the prevailing wind and wave direction,

sand is pushed shoreward and eventually is driven by the wind into pronounced dunes.

Behind the dune line occurs a low, flat zone that is covered with grasses and shrub forests (see fig. 13.44). Stands of pine and oak maritime forest are sometimes found here in sheltered areas that have been stable for hundreds of years. This zone grades into salt marshes and tidal flats that abut the lagoon itself. The lagoon, salt marshes, and tidal flats are fed and maintained by tidal inlets that link the open ocean to the sound-side environments.

Geomorphic Processes and Dynamics

In addition to the wind action associated with dune formation and sand transport, two of the most important processes involved in sand transport and morphologic adjustment of barrier island systems are overwash and inlet formation (Dolan et al. 1980). In severe storms, parts of every barrier island are subjected to elevated water levels from storm surge and increased wave energy, permitted in part by the increased water levels along the shoreline. Storm **overwash** can inundate large areas, especially where dune development is minor on low-profile islands, activating large washover flats and terraces (fig. 13.47). Overwash may also occur in narrow washover channels that breach the dune line (see fig. 13.44) and distribute sand to interior and back-barrier locations as distinct geomorphic features known

Figure 13.46
Oblique aerial view of Ship Shoal Island, Va. The beach can be seen overriding former back-barrier marsh sediments. The meandering channel now in the surf zone appears in 1950s photos in the back-barrier marsh. Mean erosion rates here exceed 15 m/yr.

(Photo by R. Craig Kochel)

(A)

(B)

(C)

Figure 13.47
Washover morphology on barrier islands. (A) Washover terrace extending across Assateague Island, Md. (November 1982). (B) Washover fans and islandwide breaches caused by the powerful Halloween 1991 northeaster along the eastern shore of Virginia (view is from northern Smith Island, November 1991). (C) Washover fan that formed by breaching the stabilized foredune (Pea Island, N.C., March 1982).

(Photos by R. Craig Kochel)

as washover fans (Pierce 1970; fig. 13.47). Overwash is usually associated with a major change and disturbance of island landforms and vegetation.

The nature of geomorphic response to overwash depends on several aspects of island morphology and storm characteristics. The most important of these are (1) the orientation of the island to the approach of storm waves, (2) the relative elevation of a given sector of the island (whether it is high or low profile), (3) the morphodynamic state of the beach, especially beach slope and the presence of nearshore bars, (4) the erodibility of the subaerial portion of the island, that is, vegetation cover on the dunes and the occurrence of cohesive muds on the seaward side because of island transgression, (5) the presence of previous island breaches such as healed overwash channels and inlet fills, and (6) the storm surge elevation. Numerous studies of island response to overwash have documented distinctly different responses to the same event along the coast depending on the preexisting morphologic state of island landforms (e.g., Ritchie and Penland 1988; Kochel and Wampfler 1989).

Just as floods are significant drivers in the overall topographic and vegetative evolution of floodplains, overwash serves a similar role on moderate- to low-profile barrier islands. Relationships between the frequency and magnitude of storm overwash play a major role in regulating the topographic and ecological evolution of the islands (Kochel and Dolan 1986; Kochel and Wampfler 1989). High-velocity, shallow surges mobilize large volumes of sand across washover sites, eroding berms and dunes and burying grasses and back-barrier marshes with deposits of significant thickness (fig. 13.48).

Following the planation associated with barrier island overwash, island landforms appear to undergo significant poststorm changes during recovery as they approach a new equilibrium with calm conditions. Washover surfaces are rapidly deflated after a storm, and an armored lag surface is established consisting of shells, exceptionally coarse sand, and heavy minerals (if they are present). Within days to months, incipient dunes form from shell piles, clasts of mud ripped up from the beach face, wrack (dead fragments of Spartina marsh grasses), and other debris that may have been deposited during the overwash event. If there is sufficient time between events, prominent dunes will again form on the site. In addition, minor overwash from smaller-scale storms has been shown to reestablish island topography in less than a year, although the overall position of the island may have been translated landward (Cohn and Kochel 1993). Figure 13.49 shows a schematic model of the idealized response and recovery of barrier islands to storm overwash events. Penland et al. (1989) showed that signature geomorphic responses to hurricane-induced overwash were common on Gulf Coast barrier islands and could be predicted given knowledge of large-scale variations in island morphology prior to the

Figure 13.48

Vibracore through back-barrier washover fan on Wreck Island, Va. Note the transgressive pattern from overwash sediments at the surface, through buried back-barrier marsh, then into back-barrier tidal sediments, and finally into lagoonal mud. Markings are shown in feet below the surface.

(Photo by R. Craig Kochel)

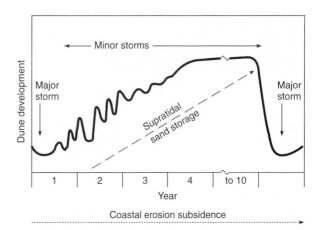

Figure 13.49

Generalized model for subaerial barrier island response and recovery from storms.

(Ritchie and Penland 1988)

Figure 13.50
Ebb tidal delta and flood tidal delta at Brown Inlet, N.C. Ebb tidal delta is marked by offshore shoaling of waves.

storms. They found significant patterns of island response depended heavily on shoreline orientation, storm track and surge elevation, and antecedent beach-nearshore morphology.

The total transport distance of sand during overwash depends on the tidal range, severity of the storm, island elevation, and storm surge penetration. Overwash can traverse the entire dune line and place fans on top of older sound-side deposits. Therefore, overwash represents a major process for the transfer of marine sand from the beach to the lagoon and thus becomes one of the chief agents responsible for the landward rollover (migration) of barrier islands. Once the sand reaches the relatively low-energy environment of the lagoon, it remains until the island itself migrates past that point.

Of equal or perhaps greater importance in the movement of sand and landform evolution within the barrier island system are the major **tidal inlets** between islands (figs. 13.50, 13.43). Inlets are formed rapidly during major storms and commonly have a checkered evolution thereafter. Considerable speculation exists as to the relative role of inlet processes and longshore drift with regard to the volumetric movement of sand in barrier island systems. Using detailed inlet surveys taken from dredges intended to maintain the inlet channel on the Texas Gulf coast, Bales and Holley (1989) estimated that net annual transport though the inlet was nearly an

order of magnitude greater than the volume of sand moving in longshore transport. Because the Gulf of Mexico is a relatively low wave-energy environment, except for the occasional presence of tropical storms, different balances between the two rates may be found elsewhere. For example, the wave-dominated mid-Atlantic shoreline may experience higher rates of storm-driven longshore drift, and thus may have more of a balance between the two. Fenster and Dolan (1996) showed that sand by passing processes at inlets exert significant influence on the adjacent barrier island shorelines in Virginia. The extent of influence is commonly 5–8 km up- and down-drift from the inlet.

Depending on the complexities of such factors as the volume of water exchanged between the ocean and lagoon during each tidal cycle (known as the *tidal prism*) and the rate of longshore drift along the oceanic barrier beach, inlets may rapidly heal up from longshore-deposited sand or they may remain open for many years. Inlets segment the islands and provide a physical connection, to move channel water and sediment between the lagoon and open ocean during tidal exchange. The stability of inlets is highly variable. Dozens of historical inlets have been charted along shorelines where only a few exist today, for example, 37 historical versus 12 active on the New Jersey coast (Nordstrom 1987) and 13 versus 1 on the North Carolina

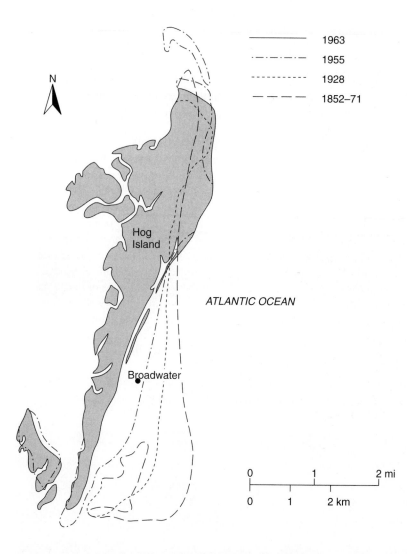

Figure 13.51
Shoreline change history for Hog Island, Va., between 1852 and 1963. Note the rotational behavior and the village of Broadwater now located over 1/2 km offshore. Less than 70 years ago Broadwater was a thriving community of more than 300 tucked away in high dunes complete with a maritime forest.
(Williams et al. 1990)

coast north of Cape Hatteras (Dolan and Lins 1986). Many inlets are ephemeral because they tend to migrate in the direction of littoral drift, extending by spit formation on one side of the inlet and erosion on the other. Spits commonly tend to accrete by the process of attachment or welding of nearshore bars to the island, which eventually become subaerial, vegetated and have dunes established on their tops. This process gives rise to sub-parallel *beach ridges* on the spits whose stratigraphy is based by marine sediments overlain by beach and then eolian sediments (e.g., Carter 1986). Oregon Inlet (shown in fig. 13.23), for example, formed in 1846 and has migrated several kilometers to the south since then.

Other inlets have relatively fixed positions postulated in some cases to be related to the position of buried Pleistocene drainages (Oertel et al. 1992). Islands with fixed inlets, particularly those with significant tidal range (approaching or within the mesotidal range) are often associated with drumstick-shaped barrier islands (Hayes 1975; Fitzgerald 1984), which are characterized by distinctive rotational movements about some central pivot zone. These islands rotate primarily in response to shifting channel positions within the inlets and related

effects on longshore sediment movement across the inlets as it accretes on or erodes the neighboring island. An interesting example of this rotational phenomenon on Hog Island, Virginia, is shown in figure 13.51. A shift in the rotational pattern of the island caused wholesale destruction of a stable maritime forest, dunes, and several kilometers of island retreat, resulting in the destruction of the village of Broadwater, all within the span of about 100 years.

During the flood (incoming) tide, material moves landward through the inlet and deposits a **flood tidal delta** on the inside of the barrier island. During ebb tide, a similar **ebb tidal delta** is formed in the ocean (see figs. 13.44, 13.50). During low-tide events, the lagoon-side shoals are exposed and eventually become the substrate for new salt marshes (Godfrey 1976). The oceanward delta typically remains submerged as a shallow platform because of the higher wave energy, but it has a major impact on the approach of storm waves and on the configuration of island shorelines proximal to the inlet.

Enormous investments have been made in efforts to fix tidal inlets in a single location. Usually jetties are constructed to prevent filling of the channel and to stabilize its position. In many cases these efforts fail without continuous and costly maintenance. A good example is the attempt to maintain a highway bridge across Oregon Inlet (see fig. 13.23) along the Outer Banks of North Carolina. The Bonner Bridge, opened in 1964, has been plagued with problems. Migration of the inlet channel southward has necessitated pier reconstruction and reinforcement several times over the last 30 years. An example of a stabilized inlet is the Ocean City Inlet, Maryland, which breached a barrier island during a 1933 hurricane. Since construction, the jetty system has interrupted the southward net littoral drift (Dean and Perlin 1977). The effect has been a shoreline advance of about 240 m immediately north of the inlet, where drifting sand has been trapped by the northern jetty. The shoreline immediately south of the inlet has eroded landward over 1/2 km because the beach was starved when its supply of littoral sediment was eliminated (fig. 13.52). The inshore portion of the south jetty is so low and permeable that sand being eroded and transported from the beaches south of the inlet is moving over and through the jetty back into the inlet itself. This is producing a shoal area within the inlet and will require major repair of the south jetty (Dean and Perlin 1977). Dean (1997)

provides a review of models for barrier island restoration, arguing cogently for restoration to be based on attention to the dominant processes of sea-level change cross-island sediment movement, and longshore sand transport. Attention to these dynamic processes will improve the chances of engineering projects along barrier shorelines.

Most mid-Atlantic barrier islands are still migrating landward (Fisher and Simpson 1979). Evidence for this transgression is that shells of lagoonal fauna and salt marsh peats are now found on the beach side of the islands, indicating that the present beach is resting on top of older lagoonal sediment (fig. 13.53). Erosion rates at some locations exceed 15 m/yr, but have been measured at over 30 m during a single northeaster (Kochel et al. 1990). So rapid is this retreat rate in some areas that boat docks constructed on the landward sides of barrier island only 50 years ago are now projecting well out into the surf zone on the ocean side. These situations are extremely common along the Virginia barrier shoreline and have been reported along the New York coast as well by Buttner (1990).

Clearly, the barrier island system is dynamic. To paraphrase Orrin Pilkey, who has spent decades studying

Figure 13.52
Inlet and jetties at Ocean City, Md., looking north. Assateague Island (south of the inlet) has migrated landward relative to the fixed position of Fenwick Island to the north. (1981 photo)
(Photo by R. Craig Kochel)

(A)

(B)

(C)

Figure 13.53

Evidence of the rapid rate of landward migration of mid-Atlantic barrier islands. (A) Aerial view of Wreck Island, Va., showing exposed back-barrier peats now in the beach face (November 1991). (B) Ground view of Wreck Island after the Halloween storm of 1991 where some 30 m of shoreline erosion occurred in two days. (C) Benchmarks installed in 1962 on the dune crest of Wreck Island (1990 photo). These markers flapped in the breeze in the surf zone 30 years later as the island marched by. (D) Stabilized dune breaches by overwash on the Outer Banks of North Carolina. Old road surface was eroded by the ocean and buried by recent overwash as the barrier moves landward, forcing displacement of the highway to the back-barrier area.

(Photos by R. Craig Kochel)

(D)

east coast barrier islands, the only constant with barrier islands is their propensity to change. New methods of surveying help keep pace with these rapid changes in barrier environments, such as GPS (Bailey and Kochel 1999), GIS (Hayden et al. 1995, Shao et al. 1998), and airborne laser mapping (Krabill et al. 2000). A recent survey of coastal hazards in North Carolina focused attention on the barrier shorelines and outlined four major sources of problems for study: (1) inlet migration, reorientation, and spit breaching, (2) storm overwash across low-profile islands and through relict inlet fills, (3) mobility of dunes and sand sheets, and (4) the high rate of shoreline erosion (Cleary and Hosier 1987). Cleary and Pilkey (1996) provide a review of coastal processes and applications to environmental geology on barrier islands. These articles provide excellent summaries of the practical issues raised by the dynamics of the barrier island environment. Knowing the ephemeral and migratory nature of barriers, we can see the wisdom of avoiding the use of rigid structures, known as hard stabilization, to defend shoreline development, and instead planning for mobility in the structures themselves or simply avoiding development of the islands entirely (Wright and Pilkey 1989). Islands are eroded on the ocean side while the entrained sediment is transported farther inshore. The lagoonal side of the island system grows by deposition of overwash, and inlet processes along with wind action provide sediment to cover the lagoonal facies. This covering simultaneously creates a new substrate for continued growth of vegetational zones located inshore from the island dunes. This transgression, fueled ultimately by continuing Holocene sea-level rise, is evident in virtually every core (see fig. 13.47) or stratigraphic transect through barrier island-lagoon systems (e.g., see Kraft and Chryzatowski 1975) except where sediment balances may be offset by either tectonics or river sediment supply.

SUMMARY

Coastal zones reflect the balance between the driving forces of water and the resistance offered by the rocks that form the shoreline. Large lakes may experience the same phenomena as ocean coasts. Beaches are the components of coasts that respond most obviously to the dynamics of the system. Energy possessed by waves, currents, and tides is expended on the beach surface. The response to this activity is the creation of a beach profile that is adjusted to the mean values of the wave properties. Waves striking the beach zone obliquely also cause longshore (littoral) drift of sediment.

Coasts and beaches display a wide variety of geomorphic forms that range in size from minor modifications of the beach face to features that encompass kilometers of the shoreline. Commonly the shoreline is indented with a hierarchy of crescentic or rhythmic forms that are related in some way to cellular circulation of the ocean water. Coasts may be typified as erosional or depositional depending on whether the dominant action is causing the shoreline to recede landward or expand oceanward. Each coastal type possesses large-scale features that reflect the prevailing action. Many coastal regions are the sites of rapid and significant geomorphic changes driven by the slow but steady rise in sea level and the energy for erosion and sediment transport provided by powerful coastal storms. Nowhere is this signature of change more evident than in the dynamic nature of barrier islands which fringe many of the low-relief shorelines of the world.

SUGGESTED READINGS

The following references provide greater detail concerning the concepts discussed in this chapter. Each reference cited has an extensive bibliography of topics that may be of particular interest to the reader.

Carter, R.W.G. 1988. *Coastal environments*. New York: Academic Press.

Davis, R.A., Jr. 1985. *Coastal sedimentary environments*. New York: Springer-Verlag.

Davis, R.E., and Dolan, R. 1993. Nor'easters. *Amer. Scientist* 81:428–39.

Dolan, R., Hayden, B., and Lins, H. 1980. Barrier islands. *Amer. Scientist* 68:16–25.

Dolan, R., and Lins, H. 1986. *The outer banks of North Carolina*. U.S. Geol. Survey Prof. Paper 1177-B.

Fletcher, C.H. 1992. Sea-level trends and physical consequences: Application to the U.S. shoreline. *Earth Science Reviews* 33:73–109.

Komar, P.D. 1976. *Beach processes and sedimentation*. Englewood Cliffs, N.J.: Prentice-Hall.

———, ed. 1983. *CRC handbook of coastal processes and erosion*. Boca Raton, Fla.: CRC Press.

May, S., Dolan, R., and Hayden, B. 1983. Erosion of U.S. shorelines. *EOS* 64:521–23.

Pethick, J.S. 1984. *An introduction to coastal geomorphology*. London: Arnold.

Wright, L., Chappell, J., Thom, B., Bradshaw, M., and Cowell, P. 1979. Morphodynamics of reflective and dissipative beach and inshore systems: Southeastern Australia. *Marine Geol.* 32:105–40.

BIBLIOGRAPHY

Aagard, T. 1991. Multiple bar morphodynamics and its relations to low frequency edge waves. *J. Coastal Res.* 7:801–13.

Aber, J. S., Croot, D. G., and Fenton, M. M. 1989. *Glaciotectonic landforms and structures.* Dordrecht: Kluwer.

Abrahams, A. 1980. A multivariate analysis of chain lengths in natural channel networks. *Jour. Geology* 88:681–96.

———. 1984. Channel networks, a geomorphological perspective. *Water Resources Res.* 20:161–88.

Abrahams, A. D. 1972. Environmental constraints on the substitution of space for time in the study of natural channel networks. *Geol. Soc. America Bull.* 83:1523–30.

Abrahams, A. D. ed. 1986. *Hillslope Processes.* Boston: Allen and Unwin.

Abrahams, A. D., and Miller, A. J. 1982. The mixed gamma model for channel link lengths. *Water Resour. Res.* 18:1126–36.

Abrahams, A. D., Parsons, A. J., Cooke, R. U., and Reeves, R. W. 1984. Stone movement on hillslopes in the Mojave Desert, California: A 16-year record. *Earth Surf. Proc. and Landforms* 9:365–70.

Achyuthan, H. 1996. Geomorphic evolution and genesis of laterites around the east coast of Madras, Tamil Nadu, India. *Geomorphology* 16:71–76.

Ackers, P., and Charlton, F. G. 1971. The slope and resistance of small meandering channels. *Inst. Civil Engrs. Proc.,* Supp. 15, 1970, paper 73625.

Ahlbrandt, T. S., and Fryberger, S. G. 1980. Eolian deposits in the Nebraska Sand Hills, *U.S. Geol. Survey Prof. Paper* 1120A.

———. 1982. Introduction to eolian deposits. In *Sandstone Depositional Environments,* pp. 11–48. Ed. by P. Scholle and D. Spearing. Tulsa, OK: Am. Assoc. Petroleum Geologists.

Ahlman, H. W. 1948. *Glaciological research on the North Atlantic.* Royal Geog. Soc. Res. Ser. 1.

Ahnert, F. 1970. Functional relationships between denudation, relief, and uplift in large mid-latitude drainage basins. *Am. Jour. Sci.* 268:243–63.

———. 1987. Process-response models of denudation at different spatial scales. In *Geomorphological modes; theoretical; and empirical aspects. Catena Supplement* 10:31–50.

Ahr, W. M., and Stanton, R. J. 1973. The sedimentologic and paleoecologic significance of Lithotyra, a rock-boring barnacle. *Jour. Sed. Petrology* 43:20–23.

Akagi, Y. 1980. Relations between rock type and the slope form in the Sonora Desert, Arizona. *Zeit. F. Geomorph.* 24:129–40.

Akerman, H. J. 1982. Observations of palsas within the continuous permafrost zone in eastern Siberia and in Svalbard, *Geografisk Tidsskrift* 82:45–51.

Albritton, C., and 3 others. 1990. Origin of the Qattara Depression, Egypt. *Geol. Soc. Amer. Bull.* 102:952–60.

Alexandrowicz, S. W., and Alexandrowicz, A. 1999. Recurrent Holocene landslides; a case study of the Krynica landslide in the Polish Carpathians. *The Holocene* 1:91–99.

Alley, R., and 10 others. 1993. Abrupt increase in Greenland snow accumulation at the end of the Younger Dryas event. *Nature* 362:527–29.

Alley, R. B. 1991. Deforming-bed origin for southern Laurentide till sheets. *Journal of Glaciology* 37:67–76.

Alley, R. B. 1992. How can low-pressure channels and deforming tills coexist subglacially? *Jour. Glaciol.* 38:200–207.

Alley, R. B., Blankenship, D. D., Bentley, C. R., and Rooney, S. T. 1986. *Nature* 322:57–59.

———. 1987. Till beneath Ice Stream B: 4.A coupled ice-till model. *Jour. Geophys. Research* 92:8931–40.

Alley, R. B., Strasser, J. C., Lawson, D. E., Evenson, E. B., and Larson, G. J. 1999. Glaciological and geological implications of basal-ice accretion in overdeepenings. *Geol. Soc. America Bull.,* Special Paper 337, pp. 1–9.

Alter, A. J. 1966. Sanitary engineering in Alaska. In *Proc. Permafrost Internat. Conf.* (Lafayette, Ind., 1963). Natl. Acad. Sci. Natl. Research Council Pub. 1287, pp. 407–8.

Amit, R., and Harrison, J. B. J. 1995. Biogenic calcic horizon development under extreme arid conditions, Nizzana sand dunes, Israel. *Adv. GeoEcol.* 28:65–88.

Amit, R., Harrison, J. B. J., and Enzel, Y. 1995. Use of soils and colluvial deposits in analyzing tectonic events—The southern Arava Rift, Israel. *Geomorphology* 12:91–107.

Anderson, D. M. 1968. Undercooling, freezing, point depression, and ice nucleation of soil water. *Israel J. Chem.* 6:349–55.

Anderson, D. M., Gatto, L. W., and Ugolini, F. C. 1967. An Antarctic analog of Martian permafrost terrain. *Antarctic J.* 7:114–16.

Anderson, R. 1986. Erosion profiles due to particles entrained by wind: Application of an eolian sediment-transport model. *Geol. Soc. America Bull.* 97:1270–78.

———. 1990. Evolution of the northern Santa Cruz Mountains by advection of crust past a San Andreas Fault bend. *Science* 249:397–401.

Anderson, R., and Hallet, B. 1986. Sediment transport by wind: Toward a general model. *Geol. Soc. America Bull.* 97:523–35.

Anderson, R. S. 1987. Eolian sediment transport as a stochastic process: The effects of a fluctuating wind on particle trajectories. *Jour. Geol.* 95:497–512.

———. 1987. A theoretical model for aeolian impact ripples. *Sedimentology* 34:943–56.

———. 1988. The pattern of grainfall deposition in the lee of aeolian dunes. *Sedimentology* 35:175–88.

Anderson, R. S., Hallet, B., Walder, J., and Aubry, B. F. 1982. Observations in a cavity beneath Grinnell Glacier. *Earth Surf. Proc. and Landforms* 7:63–70.

Anderson, S. P. 1988. Upfreezing in sorted circles, western Spitzbergen. In *Permafrost, Fifth International Conference Proceedings,* pp. 666–70. Ed. by K. Senneset. Trondheim: Tapir Publishers.

Andrews, D., and Bucknam, R. 1987. Fitting degradation of shoreline scarps by a nonlinear diffusion model. *Jour. Geophys. Res.* 92:12857–67.

Andrews, D., and Hanks, T. 1985. Scarp degraded by linear diffusion: Inverse solution for age. *Jour. Geophys. Res.* 90:19193–208.

Andrews, D. E. 1980. Glacially thrust bed—An indication of late Wisconsin climate in western New York State. *Geology* 8:97–101.

Andrews, E. D. 1979. Scour and fill in a stream channel, East Fork River, western Wyoming. *U.S. Geol. Survey Prof. Paper* 1117.

———. 1983. Entrainment of gravel from naturally sorted river bed material. *Geological Society of America Bulletin* 94:1225–1231.

Andrews, E. D., and Erman, D. C. 1986. Persistence in the size distribution of surficial bed material during an extreme snowmelt flood. *Water Resources Research* 22:191–97.

Andrews, E. D., and Nankervis, J. M. 1995. Effective discharge and the design of channel maintenance flows for gravel-bed rivers. In *Natural and Anthropogenic Influences in Fluvial Geomorpholgy,* pp. 151–64. Ed. by J. E. Costa et al., Geophysical Monograph 89, American Geophysical Union.

Andrews, E. D., and Smith, J. D. 1992. A theoretical model for calculating marginal bed load transport rates of gravel. In *Dynamics of gravel-bed rivers,* pp. 41–52. Ed. by P. Billi et al. Chichester: Wiley.

Andrews, J. 1991. Late Quaternary glacial isostatic recovery of North America, Greenland, and Iceland; a neotectonics perspective. In Neotectonics of North America, pp. 473–86. Ed. by D. Slemmons, et al. Geol. Soc. Amer. Decade Map, Vol. 1, Boulder, CO.

Andrews, J. T. 1963. Cross-valley moraines of the Rimrock and Isotoq river valleys, Baffin Island. A descriptive analysis. *Geogr. Bull.* 19:49–77.

Andrews, J. T., and Smithson, B. B. 1966. Till fabrics of the cross-valley moraines of north-central Baffin Island. *Geol. Soc. America Bull.* 77:271–90.

Antia, E. E. 1989. Beach cusps and beach dynamics; a quantitative field appraisal. *Coast. Engin.* 13:263–72.

Antia, E. E., and Nyong, E. E. 1988. Beach morphodynamics along the Nigerian coastline; implications for coastal projects. *J. African Earth Sci.* 7:553–59.

Anundsen, K. 1990. Evidence of ice movement over southwest Norway indicating an ice dome over the coastal district of west Norway. *Quaternary Science Reviews* 9:99–116.

Arduino, E., Barberis, E., Marsan, F. A., Zanini, E., and Franchini, M. 1986. Iron oxides and clay minerals within profiles as indicators of soilage in northern Italy. *Geoderma* 37:45–55.

Arkley, R. 1963. Calculation of carbonate and water movement in soil from climatic data. *Soil Sci.* 96:239–48.

Armije, R., Lyon-Caen, H., and Papanastassiou, D. 1992. East-west extension and Holocene normal-fault scarps in the Hellenic arc. *Geology* 20:491–94.

Arvidson, R. E., and Guinness, E. A. 1982. Clues to tectonic styles in the global topography of Earth, Venus, Mars. *Jour. Geol. Educ.* 30:86–92.

ASCE. 1998. River width adjustment, I: Processes and mechanisms by American Society of Civil Engineers Task Committee on Hydraulics, Bank Mechanics, and Modeling of River Width Adjustment. *Jour. Hydraulic Engineering* 124:881–902.

Ashmore, P. E. 1991. How do gravel-bed rivers braid? *Canadian Journal of Earth Science* 28:326–41.

Ashmore, P. E., and Day, T. J. 1988. Effective discharge for suspended sediment transport in streams of the Saskatchewan River basin. *Water Resour. Res.* 24:864–70.

Ashworth, P. J., and Ferguson, R. I. 1989. Size-selective entrainment of bed load in gravel bed streams. *Water Resources Research* 25:627–34.

Atkinson, H., and Wright, J. 1957. Chelation and the vertical movement of soil constituents. *Soil Sci.* 84:1–11.

Atterberg, A. 1911. Die Plastizitat der Tone. *Intern. Mitt. Boden* 1:4–37.

Augustinus, P. C. 1992. The influence of rock mass strength on glacial valley cross-profile morphometry: A case study from the southern Alps, New Zealand. *Earth Surfaces Processes and Landforms* 17:39–51.

Aylsworth, J. M., and Shilts, W. W. 1989. Bedforms of the Keewatin Ice Sheet, Canada. In J. Menzies and J. Rose (eds.), Subglacial bedforms-drumlins, rogen moraine and associated subglacial bedforms. *Sedimentary Geology* 62:407–28.

Bach, A., Brazel, A., and Lancaster, N., 1996. Temporal and spatial aspects of blowing dust in the Mojave and Colorado Deserts of Southern California. *Phys. Geogr.* 17:329–53.

Back, W., Hanshaw, B., and Van Driel, J. N. 1984. Role of groundwater in shaping the eastern coastline of the Yucatan Peninsula, Mexico. In *Groundwater as a Geomorphic Agent,* pp. 281–93. Ed. by R. G. La Fleur. Boston, MA: Allen and Unwin.

Bagnold, R. A. 1941. *The physics of blown sand and desert dunes.* London: Methuen and Co.

———. 1960. Some aspects of river meanders. *U.S. Geol Survey Prof. Paper* 282-E.

———. 1973. The nature of saltation and of bedload transport in water. *Proc. Royal Soc. London,* ser. A:332:473–504.

———. 1977. Bed-load transport by natural rivers. *Water Resour. Res.* 13:303–12.

———. 1980. An empirical correlation of bedload transport rates in flumes and natural rivers. *Proceedings of the Royal Society* 372A:453–473.

————. 1998. An empirical correlation of bedload transport rates in flumes and natural rivers. *Proceedings of the Royal Society* 372A:453–73.

Bailey, N. M., Kochel, R. C., and Carlson, C. R., 1998. Barrier island landform and vegetation response to coastal process variables on the Virginia Coast Reserve: *Geological Society of America, Abs. w. Prog.,* v. 30, n. 4, p. 2.

Baker, V., and Twidale, C. 1991. The reenchantment of geomorphology. *Geomorphology* 4:73–100.

Baker, V. R. 1973a. Geomorphology and hydrology of karst drainage basins and cave channel networks in east-central New York. *Water Resour. Res.* 9:695–706.

————. 1973b. Paleohydrology and sedimentology of Lake Missoula flooding in eastern Washington. *Geol. Soc. America Spec. Paper* 144.

Baker, V. R. 1982. *The Channels of Mars.* Austin: Univ. Texas Press.

————. 1987. Paleoflood hydrology and extraordinary flood events. *J. Hydrology* 6:79–99.

Baker, V. R., and Nummedal, D., eds. 1978. *The channeled scabland.* Washington, DC: N.A.S.A. Field Guidebook.

Baker, V. R., and Pickup, G. 1987. Flood geomorphology of the Katherine Gorge, Northern Territory, Australia. *Geol. Soc. Amer. Bull.* 98:635–46.

Baker, V. R., Benito, G., and Rudoy, A. N. 1993. Paleohydrology of late Pleistocene superflooding, Altay Mountains, Siberia. *Science* 259:348–50.

Baker, V. R., Bowler, J. M., Ehzel, Y., and Lancaster, N. 1995. Late Quaternary Paleohydrology of arid and semi-arid regions. In *Global continental paleohydrology,* pp. 208–31. Ed. by K. J. Gregory, L. Starkil, and V. R. Baker. Chichester: John Wiley.

Baker, V. R., Kochel, R. C., Patton, P. C., and Pickup, G. 1983. Paleohydrologic analysis of Holocene flood slack-water sediments. *Spec. Pubs. Intl. Assoc. Sediment.* 6:229–39.

Baker, V. R., Kochel, R. C., and Patton, P. C. 1988. *Flood Geomorphology.* New York: John Wiley and Sons.

Baker, V. R., Pickup, G., and Webb, R. H. 1987. Paleoflood hydrologic analysis at ungaged sites, central and northern Australia. In *Regional Flood Frequency Analysis,* pp. 325–38. Ed. by V. J. Singh. Dordrecht: Reidel Pub.

Bakker, J. P. 1965. A forgotten factor in the interpretation of glacial stairways. *Zeit. F. Geomorph.* 9:18–34.

Baldwin, M., Kellogg, C., and Thorp, J. 1938. Soil classification. In *Soils and men,* U.S. Dept. Agri. Yearbook, pp. 979–1001.

Bales, J. D., and Holley, E. R. 1989. Sand transport in Texas tidal inlet. *J. Waterways, Port, Coast, and Ocean Engin.* 115:427–43.

Ballantyne, C. E., 1996. Formation and nature of sorted patterns by shallow ground freeze: A field experiment: *Permafrost and Periglacial Processes* 7:409–24.

Balling, R. C., Jr., and Wells, S. G. 1990. Historic rainfall patterns and arroyo activity within the Zuni River drainage basin, New Mexico. *Ann. Assoc. Am. Geog.* 80:603–17.

Barnes, H. A. 1968. Roughness characteristics of natural channels. *U.S. Geol. Survey Water Supply Paper* 1849.

Barnes, P., and Tabor, D. 1966. Plastic flow and pressure melting in the deformation of ice. *Nature* 210:878–82.

Barron, E. J., Hay, W. W., and Thompson, S. 1989. The hydrologic cycle: A major variable during Earth history. *Global and Planetary Change* 1:157–74.

Barsch, D. 1971. Rock glaciers and ice-cored moraines. *Geogr. Annlr.* 53A:203–6.

————. 1977. Nature and importance of mass wasting by rock glaciers in alpine permafrost environments. *Earth Surface Proc.* 2:231–45.

————. 1993. Periglacial geomorphology in the 21st century. *Geomorphology* 7:141–63.

————. 1994. Studies and measurements on rock glaciers at Macun in the Lower Engadine. In *Cold Climate Landforms,* Ed. By D. J. A. Evans. New York: John Wiley, pp. 458–73.

Barsch, D., and Jakob, M. 1998. Mass transport by active rock glaciers in the Khumbu Himalaya. *Geomorphology* 26:215–22.

Barsch, D., and Updike, R. G. 1971. Periglaziale Formung am Kendrick Peak in Nord-Arizona während dere letzten kaltzeit. *Geog. Helvetica.* 26:99–114.

Barsch, D., Fierz, H., and Haeberli, W. 1979. Shallow core drilling and bore-hole measurements in permafrost of an active rock glacier near the Grubengletscher, Wallis, Swiss Alps. *Arctic and Alpine Res.* 11:215–28.

Barshad, I. 1964. Chemistry of soil development. In *Chemistry of the soil,* pp. 1–70. Ed. by F. Bear. New York: Reinhold Book Corp.

Bascom, W. N. 1951. The relationship between sand size and beach face slope. *Am. Geophys. Union Trans.* 32:866–74.

Bathhurst, J. C. 1982. Theoretical aspects of flow resistance. In *Gravel Bed Rivers,* pp. 83–108. Ed. by R. D. Hey, J. C. Bathhurst, and C. R. Thorne. Chichester: John Wiley and Sons.

————. 1988. Velocity profile in high gradient, boulder-bed channels. *Proc. Intl. Ass. Hydraulic Res. International Conference on Fluvial Hydraulics '88,* pp. 29–34. Budapest, Hungary.

————. 1993. Flow resistance through the channel network. Ed. by K. Beven and M. J. Kirkby, pp. 69–98. New York: John Wiley and Sons, Ltd.

Battle, W. R. B. 1960. Temperature observation in bergschrunds and their relationship to frost shattering. In *Norwegian Cirque Glaciers,* pp. 83–96. Ed. by W. V. Lewis. Royal Geog. Soc. Res. Ser. 4.

Bauer, B. 1980. Drainage density—An integrative measure of the dynamics and the quality of watersheds. *Zeit. F. Geomorph.* 24:263–72.

Baulig, H. 1957. Peneplains and pediplains. *Geol. Soc. America Bull.* 68:913–30.

Baumgardner, R. W. 1987. Morphometric studies of subhumid and semiarid drainage basins, Texas panhandle and northeastern New Mexico. Austin: Univ. Texas, *Bur. Econ. Geol., Rept. Invest.* 163, 66p.

Beach, R. A., and Sternberg, R. W. 1992. Susended sediment transport in the surf zone; response to incident wave and longshore current action. *Marine Geol.* 108:275–94.

Beaney, C. L., and Shaw, J. 2000. The subglacial geomorphology of southeast Alberta: Evidence for subglacial meltwater erosion. *Can. J. Earth Sci.* 37:51–61.

Beard, L. R. 1975. Generalized evaluation of flash-flood potential. Austin: *Univ. Texas Center Water Res. Tech. Rpt.* CRWR-124., pp. 1–27.

Begin, Z. B. 1981. Stream curvature and bank erosion: A model based on the momentum equation. *Jour. Geology* 89:497–504.

Begin, Z. B., Meyer, D. F., and Schumm, S. A. 1980. Knickpoint migration due to base-level lowering. Am. Soc. Civil Engineers, *Jour. Water, Port, Coastal and Ocean Div.* 106:369–87.

Behrendt, J. C. 1965. Densification of snow on the ice sheet of Ellsworth Land and South Antarctic Peninsula. *Jour. Glaciol.* 5:451–60.

Belknap, D. F., and Kraft, J. C. 1977. Holocene relative sea level changes and coastal stratigraphic units on the northwest flank of the Baltimore Canyon trough geosyncline. *J. Sedy. Petrol.* 47:610–29.

Benedict, J. B. 1970. Downslope soil movement in a Colorado alpine region. Rates, processes, and climatic significance. *Arc. Alp. Res.* 2:165–226.

Benedict, J. B., Benedict, R. J., and Sanville, D. 1986. Arapaho rock glacier, Front Range, Colorado; a 25-year resurvey. *Arctic and Alpine Res.* 18:349–52.

Benito, G., Perez del Campo, P., Gutiérrez-Elorza, M., and Sancho, C. 1995. Natural and human-induced sinkholes in gypsum terraine and associated environmental problems in NE Spain. *Environmental Geology.* 25:156–164.

Benjamin, M., Johnson, N., and Naeser, C. 1987. Recent rapid uplift in the Bolivian Andes: Evidence from fission-track dating. *Geology* 15:680–83.

Benn, D. I. 1994. Fluted moraine formation and till genesis below a temperate valley glacier: Slettmarkbreen, Jotunheimen, southern Norway. *Sedimentology* 41:279–92.

Benn, D. I., and Evans, D. J. A. 1996. The interpretation and classification of subglacially-deformed materials. *Quarternary Sciences Reviews* 15:23–52.

———. 1998. *Glaciers and glaciation.* Arnold: London.

Bennet, L. P., and French, H. M. 1991. Solifluction and the role of permafrost creep, eastern Melville Island. N.W.T., Canada. *Permafrost and Periglac. Proc.* 2:95–102.

Bennett, P., and Siegel, D. 1987. Increased solubility of quartz in water due to complexing by organic compounds. *Nature* 326:684–86.

Bennett, P., Melcer, M., Siegel, D., and Hassett, J. 1988. The dissolution of quartz in dilute aqueous solutions of organic acids at 25° C. *Geochim. Cosmochim. Acta* 52:1521–30.

Bennett, P., Siegel, D., Hill, B., and Glaser, P. 1991. Fate of silicate minerals in a peat bog. *Geology* 19:328–31.

Berger, G. 1992. Dating volcanic ash by use of thermoluminescence. *Geology* 20:11–14.

Berner, E. K., and Berner, R. A. 1996. *Global environment.* Englewood Cliffs: Prentice Hall.

Berner, R., Holdren, G., and Schott, J. 1985. Surface layers on dissolving silicates: A discussion. *Geochim. Cosmochim. Acta* 49:1657–58.

Berner, R. A., and Holdren, G. R., Jr., 1979. Mechanisms of feldspar weathering-II. Observations of feldspars from soils. *Geochim et cosmochim Acta* 43:1173–86.

Berner, R. A., and Holdren, G. R., Jr. 1977. Mechanism of feldspar weathering: Some observational evidence. *Geology* 5:369–72.

Bernknopf, R. L., Campbell, R. H., Brookshire, D. S., and Shapiro, C. D. 1988. A probabilistic approach to landslide hazard mapping in Cincinnati, Ohio, with applications for economic evaluation. *Bull. Assoc. Eng. Geol.* 25:39–56.

Berry, M. 1987. Morphological and chemical characteristics of soil catenas on Pinedale and Bull Lake moraine slopes in the Salmon River mountains of Idaho. *Quaternary Research* 28:210–25.

Bersier, R., Holdren, G., and Schott, J. 1985. Surface layers on dissolving silicates: *Geochim. Cosmochim. Acta* 49:1657–58.

Berti, M., Genevois, R., Simoni, A., and Tecca, P. R. 1999. Field observations of a debris flow event in the Dolomites. *Geomorphology.* 29: 265–74.

Bertran, P., Coutard, J. P., Francou, B., Ozouf, J. C., and Texier, J. P. 1994. New data on Grézes bedding and their Paleoclimatic implications. In *Cold climate landform,* pp. 437–54. Ed by D. J. A. Evans. New York: John Wiley.

Bertran, P., Hetu, B., Texier, J. P., van Steign, H. 1997. Fabric characteristics of subaerial slope deposits. *Sedimentology* 44:1–16.

Bevan, K. J., and Germann, P. 1982. Macropores and water flow in soils. *Water Resources Res.* 18:1311–25.

Beven, K. 1981. The effect of ordering on the geomorphic effectiveness of hydrologic events. I.A.H.S. Publication 132, Christchurch, pp. 510–23.

Beverage, J. P., and Culbertson, J. K. 1964. Hyperconcentrations of suspended sediment. *J. Hydraul. Div. Am. Soc. Civil. Eng.* 90:117–28.

Bird, E. C. F. 1969. Coasts. Cambridge, MA: M.I.T. Press.

Birkeland, P., and Burke, R. 1988. Soil catena chronosequences on eastern Sierra Nevada moraines, California, U.S.A. *Arctic Alpine Res.* 20:473–84.

Birkeland, P., Berry, M., and Swanson, D. 1991. Use of soil catena field data for estimating relative ages of moraines. *Geology* 19:281–83.

Birkeland, P. W. 1999. *Soils and geomorphology.* 2nd ed. New York: Oxford University Press.

Black, R. F. 1954. Permafrost, a review. *Geol. Soc. America Bull.* 65:839–55.

Blackwelder, E. 1927. Fire as an agent in rock weathering. *Jour. Geology* 35:135–40.

Blagborough, J. W., and Farkas, S. E. 1968. Rock glaciers in the San Mateo Mountains, south-central New Mexico. *Am. Jour. Sci.* 266:812–23.

Blair, R. W. 1986. Development of natural sandstone arches in southeastern Utah. In *International Geomorphology, Part II,* pp. 597–604. Ed. by V. Gardiner. London: John Wiley.

Blair, T. 1987. Sedimentary processes, vertical stratification sequences, and geomorphology at the Roaring River alluvial fan. Rocky Mountain National Park. *Colorado. Jour. Sed. Petrology* 57:1–18.

Blair, T. 1999a. Sedimentary processes and facies of the waterlaid Anvil Spring Canyon alluvial fan, Death Valley, California. *Sedimentology* 46:913–40.

———. 1999b. Sedimentology of the debris-flow-dominated Warm Spring Canyon alluvial fan, Death Valley, California. *Sedimentology* 46:941–65.

Blair, T., and McPherson, J. 1994a. Alluvial fans and their natural distinction from rivers based on morphology, hydraulic processes, sedimentary processes, and facies assemblages. *Jour. Sed. Res.* 64:450–89.

————. 1994b. Alluvial fan processes and forms. In *Geomorphology of desert environments* pp. 354–402. Ed. by A. Abrahams and A. Parsons, London: Chapman & Hall.

Blair, T., Clark, J., and Wells, S. 1990. Quaternary continental stratigraphy, landscape evolution, and application to archaeology: Jarilla piedmont and Tuderosa graben floor, White Sands Missile Range, New Mexico. *Geol. Soc. America Bull.* 102:749–59.

Blankenship, D. D., Bentley, C. R., Rooney, S. T., and Alley, R. B. 1986. Seismic measurements reveal a saturated porous layer beneath an active Antarctic ice stream. *Nature* 322:54–57.

Bloom, A. 1997. *Geomorphology: A systematic analysis of Late Cenozoic landforms.* New Jersey: Prentice-Hall, Inc.

Bloom, A. L. 1965. The explanatory description of coasts. *Ziet. F. Geomorph.* 9(4):422–36.

Blount, G., and Lancaster, N. 1990. Development of the Gran Desierto sand sea, northwestern Mexico. *Geology* 18:724–28.

Bluck, B. J. 1987. Bed forms and clast size changes in gravel-bed rivers. In *River channels: Environment and Process,* pp. 159–78. Ed. by K. S. Richards. Oxford: Blackwell.

Blum, M. D., and Valastro, Jr., S. 1989. Response of the Pedernales River of central Texas to Late Holocene climatic change. *Ann. Assoc. Amer. Geogr.* 79:435–56.

Blumberg, D., and Greeley, R. 1993. Field studies of aerodynamic roughness length. *Jour. Arid Environ.* 25:39–48.

Boardman, J. 1992. Periglacial Geomorphology. *Progress in Phys. Geogr.* 16:339–45.

Bocco, G. 1991. Gully erosion; processes and models. *Progr. in Phys. Geogr.* 15:392–406.

Bockheim, J. G., and Tarnocki, C. 1998a. Nature, occurrence, and origin of dry permafrost. *Collection Nordicana* 7:57–63.

————. 1998b. Recognition of cryoturbation for classifying permafrost affected soils. *Geoderma* 81:281–93.

Bögli, A. 1964. Mischungskor-rosion—ein Beitrag zur Verkarstungsproblem. *Erdkunde* 18:83–92.

Booth, D. B., and Hallet, B. 1993. Channel networks carved by subglacial water: Observations and reconstruction in the eastern Puget Lowland of Washington. *Geological Society of America Bulletin* 105:671–83.

Boothroyd, J. C., and Ashley, G. M. 1975. Processes, bar morphology, and sedimentary structures on braided outwash fans, northeastern Gulf of Alaska. In *Glaciofluvial and Glaciolacustrine Sedimentation,* pp. 193–222. Ed. by B. McDonald and A. Jopling. Tulsa, OK: SEPM Spec. Pub. 23.

Born, S. M., and Ritter, D. F. 1970. Modern terrace development near Pyramid Lake, Nevada, and its geologic implications. *Geol. Soc. America Bull.* 81:1233–42.

Bosak, P., Ford, D., Jerzy, G., and Horacek, I. 1989. Paleokarst, a systematic and regional review. *Developments in Earth Surface Processes.* v. 1.

Bottomley, J. T. 1872. Melting and regelation of ice. *Nature* 5:185.

Bouchard, M. A. 1989. Subglacial landforms and deposits in central and northern Quebec, Canada, with emphasis on Rogen moraines. In J. Menzies and J. Rose (eds.), *Subglacial Bedforms-Drumlins, Rogen Moraine and Associated Subglacial Bedforms, Sedimentary Geology.* 62:293–308.

Boulton, G. S. 1967. The development of a complex supraglacial moraine at the margin of Sorbreen, Ny Friesland, Vestspitsbergen. *Jour. Glaciol.* 6:717–36.

————. 1970a. On the origin and transport of englacial debris in Svalbard glaciers. *Jour. Glaciol.* 9:213–29.

————. 1971. Till genesis and fabric in Svalbard, Spitsbergen. In *Till: A symposium,* pp. 41–72. Ed. by R. P. Goldthwait. Columbus: Ohio State Univ. Press.

————. 1972a. The role of thermal regime in glacial sedimentation—a general theory. In *Polar Geomorphology,* pp. 1–19. Ed. by R. J. Price and D. E. Sugden. *Inst. Brit. Geog. Spec. Paper* 4.

————. 1972b. Modern Arctic glaciers as depositional models for former ice sheets. *Jour. Geol. Soc. Lond.* 128:361–93.

————. 1974. Processes and patterns of glacial erosion. In *Glacial geomorphology,* pp. 41–87. Ed. by D. R. Coates. S.U.N.Y., Binghamton: Pubs. in Geomorphology, 5th Ann. Symposium.

————. 1976. The origin of glacially fluted surfaces—observation and theory. *Jour. Glaciol.* 17:287–309.

————. 1978. Boulder shapes and grain-size distribution of debris as indication of transport paths through a glacier and till genesis. *Sedimentology* 25:773–99.

————. 1979. Processes of glacier erosion on different substrata. *Jour. Glaciol.* 23:15–36.

————. 1982. Subglacial processes and the development of glacial bedforms. In *Glacial Erosion and Deposition,* pp. 1–31. Ed. by B. Fahey, R. Davidson-Arnott, and B. Nickling. Guelph: GeoBooks Norwich.

————. 1986. Push-moraines and glacier-contact fans in marine and terrestrial environments. *Sedimentology* 33:677–98.

————. 1987. A theory of drumlin formation by subglacial sediment deformation. In *Subglacial Bedforms-Drumlins, Rogen Moraine and Associated Subglacial Bedforms,* pp. 25–80. Ed. by J. Menzies and J. Rose. Drumlin Symposium. Balkema, Rotterdam.

Boulton, G. S., and Hindmarsh, R. C. A. 1987. Sediment deformation beneath glaciers: Rheology and geological consequences. *Journal of Geophysical Research* 92:9059–82.

Bovis, M. J. 1993. Hillslope geomorphology and geotechnique. *Prog. in Phys. Geogr.* 17:173–89.

Bovis, M. J., and Jones, P. 1993. Holocene history of earthflow mass movements in south-central British Columbia: The influence of hydroclimatic changes. *Canadian Journal of Earth Sciences* 29:1746–55.

Bowen, A. J. 1969. The generation of longshore currents on a plane beach. *J. Marine Res.* 37:206–15.

Bowen, A. J., and Inman, D. L. 1969. Rip currents, 2. Laboratory and field observations. *Jour. Geophys. Research* 74:5479–90.

————. 1971. Edge waves and crescentic bars. *Jour. Geophys. Research* 76:8662–71.

Bowers, J. E., Webb, R. H., and Pierson, E. A. 1997. Succession of desert plants on debris flow terraces, Grand Canyon, Arizona, USA. *Jour. Arid Environments* 36:67–86.

Bowman, D., Birkenfeld, H., and Rosen, D. S. 1992. The longshore flow component in low energy rip channels; the Mediterranean, Israel. *Marine Geol.* 108:259–74.

Boyce, J. I., and Eyles, N. 1991. Drumlins carved by deforming till streams below the Laurentide ice sheet. *Geology* 19:787–90.

Boyd, K., and Schumm, S. 1995. Geomorphic evidence of deformation in the northern part of the New Madrid seismic zone. U.S. Prof. Paper 1538-R.

Brabb, E. E., and Harrod, B. L. 1989. *Landslides: Extent and economic significance:* Proceedings of the 28th Internat. Geol. Congress, Symposium on Landslides.

Bradley, W., Hutton, C., and Twidale, C. 1978. Role of knihovna salts in development of granitic tafoni, South Australia. *Jour. Geology* 86:647–54.

Bradley, W. C. 1963. Large-scale exfoliation in massive sandstones of the Colorado Plateau. *Geol. Soc. Amer. Bull.* 74:519–27.

Brady, N. C., and Weil, R. R. 2000. *Elements of the nature and properties of soils.* Upper Saddle River: Prentice Hall.

Brakenridge, G. R. 1981. Late Quaternary floodplain sedimentation along the Pomme de Terre River, southern Missouri. *Quat. Res.* 15:62–76.

———. 1984. Alluvial stratigraphy and radiocarbon dating along the Duck River, Tennessee: Implications regarding floodplain origin. *Geol. Soc. Amer. Bull.* 95:9–25.

———. 1988. River flood regime and floodplain stratigraphy. In *Flood Geomorphology,* pp. 139–56. Ed. by V. Baker; R. Kochel; and P. Patton. New York: Wiley—Interscience.

Branden, R. W., and Short, A. D. 2000. Morphodynamics of a large-scale rip current system at Muriwai Beach, New Zealand. *Marine Geology* 165:27–39.

Brandt, C. J., and Thornes, J. B. 1987. Erosional energenetics. In *Energetics of the Physical Environment,* pp. 51–87. Ed. by K. J. Gregory. Chichester: John Wiley.

Brantley, S., and Stillings, L. 1996. Feldspar dissolution at 25° C and low pH. *Amer. J. Sci.* 296:101–27.

Brantley, S., Crane, S., Crerar, D., Hellmann, R., and Stallard, R. 1986. Dissolution at dislocation etch pits in quartz: *Geochim. Cosmochim. Acta.* 50:2349–61.

Braun, D. D. 1989. Glacial and periglacial erosion of the Appalachians. *Geomorph.* 2:233–56.

Brayshaw A. C. 1985. Bed microtopography and entrainment thresholds in gravel-bed rivers. *Geol. Soc. America Bull.* 96:218–23.

Breed, C., and Grow, T. 1979. Morphology and distribution of dunes in sand seas observed by remote sensing. In A Study of Global Sand Seas, edited by E. McKee, pp. 253–303. *U.S. Geol. Surv. Prof. Paper* 1052.

Breed, C., and others. 1997. Wind erosion in drylands. In *Arid zone geomorphology,* pp. 437–64. Ed. by D. Thomas, Chichester: Wiley.

Brenninkmeyer, B. M. 1975. Frequency of sand movement in the surf zone. *Proc. 14th Conf. on Coast. Eng.,* 812–27.

Bretz, J.H. 1942. Vadose and phreatic features of limestone caverns. *Jour. Geology* 50:675–811.

Brizga, S., and Finlaysen, B. 1990. Channel avulsion and river metamorphosis; the case of the Thomson River, Victoria, Australia. *Earth Surf. Proc. Landforms* 15:391–404.

Brookfield, M. 1970. Dune trend and wind regime in central Australia. *Zeit. F. Geomorph.,* Suppl. 10:121–58.

Brown, A., and Quine, T., eds. 1999. *Fluvial processes and environmental change.* Chichester: Wiley.

Brown, G., Everett, S., Rymer, H., McGarvie, D., and Foster, I. 1991. New light on caldera evolution—Askja, Iceland. *Geology* 19:352–55.

Brown, J., Nelson, F., Brockett, B., Outcalt, S. T., and Everett, K. R. 1983. Observations on ice-cored mounds at Sukakpak Mountain, South-Central Brooks Range, Alaska. *Proceedings of the Fourth International Conference on Permafrost* 1:91–96.

Brown, L., Pavich, M. J., Hickman, R. E., Klein, J., and Middleton, R. 1988. Erosion of the eastern United States observed with "SUP 10" Be. *Earth Surface Processes & Landforms* 13:441–57.

Brown, R. J. E. 1967. Permafrost in Canada. Canada Geol. Survey Map 1246A, *Natl. Res. Council Publ.* NRC 9769.

———. 1969. Permafrost in Canada. *Canada Geol. Survey Map* 1246A.

———. 1970. *Permafrost in Canada.* Toronto: Univ. Toronto Press.

Brown, S. L., Bierman, P. R., Lini, A., and Southon, J. 2000. 10,000 yr record of extreme hydrologic events. *Geology* 28:335–38.

Browning, J. M. 1973. Catastrophic rock slides, Mount Huascaran, north-central Peru, May 31, 1970. *Am. Assoc. Petroleum Geologists Bull.* 57:1335–41.

Brucker, R. W., Hess, J. W., and White, W. B. 1972. Role of vertical shafts in the movement of ground water in carbonate aquifers. *Ground Water* 10(6):5.

Brunauer, S., Emmett, P., and Teller, E. 1938. Adsorption of gases in multimolecular layers. *J. Am. Chem. Soc.* 60:309–19.

Brunsden, D. 1984. Mudslides. In *Slope instability,* pp. 363–418. Ed. by D. Brunsden and D. B. Prior. New York: John Wiley.

———. 1993. Mass movement; the research frontier and beyond: A geomorphological approach. *Geomorph.* 7:85–128.

Brunsden, D., and Kesel, R. H. 1973. Slope development on a Mississippi River bluff in historic time. *Jour. Geology* 81:576–97.

Brunsden, D., and Prior, D. B., eds. 1984. *Slope instability.* New York: John Wiley.

Brush, L. M., and Wolman, M. G. 1960. Knickpoint behavior in noncohesive material: A laboratory study. *Geol. Soc. America Bull.* 71:57–76.

Bruun, P. 1963. Longshore currents and longshore troughs. *Jour. Geophys. Research* 68:1065–78.

———. 1988. Profile nourishment; its background and economic advantages. *J. Coastal Res.* 4:219–28.

Bryan, K. 1940. Gully gravure, a method of slope retreat. *J. Geomorph.* 3:89–107.

Bryan, K., 1946. Cryopedology: The study of frozen ground and intensive frost-action with suggestions on nomenclature. *American Journal of Science,* v. 244:622–42.

Bryan, R. B. 1979. The influence of slope angle on soil entertainment by sheetwash and rainsplash. *Earth Surf. Proc. and Landforms* 4:43–58.

Bryant, E. 1988. Storminess and high tide beach change, Stanwell Park, Australia, 1943–1978. *Marine Geol.* 79:171–87.

Buckman, H. O., and Brady, N. C. 1960. *The nature and properties of soils.* New York: Macmillan.

Buckman, R. C., and Anderson, R. E. 1979. Estimation of fault-scarp ages from a scarp-height-slope-angle relationship. *Geology* 7:11–14.

Budd, W. F. 1975. A first simple model for periodically self-surging glaciers. *Jour. Glaciol.* 14:3–21.

Büdel, J. 1960. Die Frosschott-Zone sudorst Spitzbergen. *Colloq. Geographica*. Bonn. 6, 105p.

——. 1968. Geomorphology—principles. In *Encyclopedia of geomorphology*. pp. 416–22. Ed. by R. W. Fairbridge. New York: Reinhold Book Corp.

——. 1982. *Climatic geomorphology*. Princeton, NJ: Princeton Univ. Press.

Budyko, M.I. 1977. *Climatic changes*. Baltimore: Waverly Press.

Bull, W. B. 1964a. Alluvial fans and near-surface subsidence in western Fresno County, California. *U.S. Geol. Survey Prof. Paper 437-A*.

——. 1964b. Geomorphology of segmented alluvial fans in western Fresno County, California. *U.S. Geol. Survey Prof. Paper 552-F*.

——. 1968. Alluvial fans. *Jour. Geol. Educ.* 16:101–6.

——. 1975. Allometric change of landforms. *Geol. Soc. America Bull.* 86:1489–98.

——. 1980. Geomorphic thresholds as defined by ratios. In *Thresholds in Geomorphology*, pp. 259–63. Ed. by D. Coates and J. Vitek. London: Allen and Unwin Ltd.

——. 1984. Tectonic geomorphology. *Jour. Geol. Educ.* 32:310–24.

——. 1990. Stream-terrace genesis: implications for soil development. In *Soils and Landscape Evolution*, eds. Knuepfer, P. and McFadden, L., *Geomorphology* 3:351–67.

——. 1991. *Geomorphic responses to climatic change*. New York: Oxford University Press, 326 p.

Bull, W. B., and Brandon, M. T. 1998. Lichen dating of earthquake-generated regional rockfall events, southern Alps, New Zealand. *Geol. Society America Bulletin.* 110:60–84.

Bull, W. B., and Cooper, A. 1986. Uplifted marine terraces along the Alpine Fault, New Zealand. *Science* 234:1225–28.

Bull, W. B., and Knuepfer, P. 1987. Adjustments by the Charwell River, New Zealand, to uplift and climatic changes. *Geomorphology* 1:15–32.

Bull, W. B., and McFadden, L. D. 1977. Tectonic geomorphology north and south of the Garlock Fault, California. In *Geomorphology in Arid Regions*, pp. 115–38. Ed. by D. O. Doehring. S.U.N.Y. Binghamton: Proc. 8th Ann. Geomorph. Symposium.

Bullard, F. 1962. *Volcanoes*. Austin: Univ. Texas Press.

Buol, S. W., Hole, F. D., and McCracken, R. J. 1973. *Soil genesis and classification*. Ames, IA: Iowa State Univ. Press.

Burbank, D. W., and Beck, R. A. 1991. Rapid long term rates of denudation. *Geology* 19:1169–72.

Burdon, D. J., and Safadi, C. 1963. Ras-el-ain: The great karst spring of Mesopotamia. *J. Hydrol.* 1:58–95.

Burn, C. R. 1992. Thermokarst lakes. *The Can. Geogr.* 36:81–85.

——. 2000. The thermal regime of a retrogressive thaw slump near Mayo, Yukon Territory. *Canadian Jr. Earth Sci.* 34:967–81.

Burn, C. R., and Smith, M. W. 1990. Development of thermokarst lakes during the Holocene at sites near Mayo, Yukon Territory, Canada. *Permafrost and Periglac. Proc.* 1:161–76.

Burnett, A. W., and Schumm, S. A. 1983. Alluvial river response to neotectonic deformation in Louisiana and Mississippi. *Science* 222:49–50.

Buttner, P. J. R. 1990. Shoreline dynamics of eastern Jones Island, New York; coastal recession survey data processing system. *Northeastern Env. Sci.* 9:2–32.

Butzer, K. W. 1964. *Environment and archeology*. Chicago: Aldine Publishing.

Byrd, T. C., Furbish, D. J., and Warburton, J. 2000. Estimating depth-averaged velocities in rough channels. *Earth Surf. Proc. and Landforms* 25:167–73.

Cahill, T., and 4 others. 1996. Saltating particles, playa crusts and dust aerosols at Owens (Dry) Lake, California. *Earth Surf. Proc. Lndfms.* 21:621–40.

Cailleaux, A., and Tricart, J. 1956. Le probleme de la classification des faits geomorphologiques. *Annales Geog.* 65:162–86.

Caine, N. 1968. *The block fields of northeastern Tasmania*. Australian Natl. Univ., *Dept. Geog. Publ.* G/6.

——. 1980. The rainfall-duration control of shallow landslides and debris flows. *Geografiska Annaler* 62A:23–27.

——. 1981. A source of bias in rates of surface soil movement as estimated from marked particles. *Earth Surf. Proc. and Landforms* 6:69–75.

——. 1982. Toppling failures from Alpine cliffs on Ben Lomond, Tasmania. *Earth Surf. Proc. and Landforms* 7:133–52.

——. 1992. Spatial patterns of geochemical denudation in a Colorado alpine environment. In *Periglacial Geomorphology*, pp. 63–88. Ed. by J. C. Dixon and A. D. Abrahams. Chichester: John Wiley.

Caine, N., and Swanson, F. J. 1990. Geomorphic coupling of hillslope and channel systems in two small mountain basins. *Zeit. F. Geomorph.* 33:189–203.

Caine, T. N. 1986. Sediment movement and storage on alpine slopes in the Colorado Rocky Mountains. In *Hillslope Processes*, pp. 115–37. Ed. by A. D. Abrahams. Boston: Allen and Unwin.

Calkin, P., and Cailleaux, A. 1962. A quantitative study of cavernous weathering (tafonis) and its application to glacial chronology in Victoria Valley, Antarctica. *Zeit. F. Geomorph.* 6:317–24.

Callander, R. A. 1969. Instability and river channels. *Jour. Fluid Mech.* 36:465–80.

Calvache, M. L., and Pulido-Bosch, A. 1991. Saltwater intrusion into a small coastal aquifer. *J. Hydrology* 129:195–213.

Campbell, D., and 5 others. 1995. Processes controlling the chemistry of two snow-melt dominated streams in the Rocky Mountains. *Wat. Resources Res.* 31:2811–21.

Campbell, R. H. 1975. Soil slips, debris flows, and rainstorms in the Santa Monica Mountains, southern California. *U.S. Geol. Survey Prof. Paper* 851.

Campbell, W. J., and Rasmussen, L. A. 1969. Three-dimensional surges and recoveries in a numerical glacier model. *Can. J. Earth Sci.* 6:979–86.

Carling, P. A. 1991. An appraisal of the velocity-reversal hypothesis for stable pool-riffle sequences in the River Severn, England. *Earth Surface Processes and Landforms,* 16:19–31.

Carling, P. A., and Wood, N. 1994. Simulation of flow over pool-riffle topography: A consideration of the velocity reversal hypothesis. *Earth Surf. Proc. and Landforms* 19:319–32.

Carling, P. A., Williams, J. J., Kelsey, A., Glaister, M. S., and Orr, H. G. 1998. Coarse bedload transport in a mountain river. *Earth Surf. Proc. and Landforms* 23:141–57.

Carlston, C. W. 1963. Drainage density and streamflow. *U.S. Geol. Survey Prof. Paper* 422-C.

———. 1969. Downstream variations in the hydraulic geometry of streams: Special emphasis on mean velocity. *Am. Jour. Sci.* 267:499–509.

Carr, M. H. 1981. *The surface of Mars.* New Haven: Yale Univ. Press.

Carr, M. H., and Schaber, G. G. 1977. Martian permafrost features. *Jour. Geophysical Res.* 82:4039–54.

Carr, M. H., Saunders, R. S., Strom, R. G., and Wilhelms, D. E. 1984. *The Geology of the terrestrial planets.* Washington: N.A.S.A. SP-469.

Carrara, A. 1983. Multivariate models for landslide hazard evaluation. *Mathematical Geol.* 15:403–26.

Carson, M. A. 1969. Models of hill-slope development under mass failure. *Geographical Analysis* 1:76–100.

Carson, M. A., and Kirkby, M. 1972. *Hillslope form and process.* London: Cambridge Univ. Press.

Carter, L. D. 1983. Fossil sand wedges on the Alaskan arctic coastal plain and their paleoenvironmental significance. In *Proc. Permafrost 4th Internat. Conf.,* pp. 109–14. Natl. Acad. Sci.

———. 1987. Oriented lakes. In *Geomorphic Systems of North America,* pp. 615–19. Ed. by W. Graf. Boulder: Geol. Soc. Amer.

Carter, R. W. G. 1986. The morphodynamics of beach ridge formation: Magilligan, northern Ireland. *Marine Geol.* 73:191–214.

———. 1988. *Coastal environments.* San Diego: Academic Press.

Casagrande, A. 1948. Classification and identification of soils. *Am. Soc. Civil Engineers Trans.* 113:901–91.

Cenderelli, D., and Kite, J. S. 1998. Geomorphic effects of large debris flows on channel morphology at North Fork Mountain, eastern West Virginia, USA. *Earth Surface Processes and Landforms* 23:1–19.

CERC (Coastal Engineering Research Center). 1973. *Shore Protection Manual.* 3 vols. Washington, DC: U.S. Army Corps Engrs.

Cerling, T. 1990. Dating geomorphic surfaces using cosmogenic ^{3}He. *Quaternary Research* 33:148–56.

Chadbourne, B. D., Cole, R. M., Tootill, S., and Walford, M. E. R. 1975. The movement of melting ice over rough surfaces. *Jour. Glaciol.* 14:287–92.

Chamberlain, R. T. 1928. Instrumental work on the nature of glacial motion. *Jour. Geology* 36:1–30.

Chambers, M. J. G. 1970. Investigations of patterned ground at Signy Island, South Orkney Islands, IV. Longterm experiments. *British Antarctic Surv. Bull.* 23:93–100.

Chandler, R. J. editor. 1991. *Slope stability engineering; developments and applications.* London: Telford.

Chang, H. H. 1986. River channel changes: Adjustments of equilibrium. *Journal of Hydraulic Engineering* 122:43–55.

Chapman, M. G. 1996. Application of Kasei Valles, Mars, maps; hydrologic/geologic history, predicting Pathfinder site geology, and demonstrating digital mapping. *Geological Society Amer Abs., Ann. Mtg.* 28:7:128–29.

Cheel, R. 1982. The depositional history of an esker near Ottawa, Canada. *Can. J. Earth Sci.* 19:1417–27.

Chepil, W. S. 1945. Dynamics of wind erosion, II. Initiation of soil movement. *Soil Sci.* 60:397–411.

———. 1959. Equilibrium of soil grains at the threshold of movement by wind. *Soil. Sci. Soc. Am. Proc.* 23:422–28.

Chepil, W. S., and Woodruff, N. P. 1963. The physics of wind erosion and its control. *Advances in Agron.* 15:211–302.

Chin, A. 1998. On the stability of step-pools mountain streams. *Jour. Geology* 106:59–69.

———. 1999. The morphologic structure of step-pool in mountain streams. *Geomorphology* 27:191–204.

Chorley, R .J. 1957. Illustrating the laws of morphometry. *Geol. Mag.* 94:140–50.

———. 1959. The shape of drumlins. *Jour. Glaciol.* 3:339–44.

———. 1962. Geomorphology and the general systems theory. *U.S. Geol. Survey Prof. Paper* 500-B.

———, ed. 1969b. *Water, earth, and man.* London: Methuen and Co.

Chorley, R. J., and Kennedy, B. A. 1971. *Physical geography.* London: Prentice-Hall International.

Chorley, R. J., and Morley, L. S. D. 1959. A simplified approximation for the hypsometric integral. *Jour. Geology* 67:566–71.

Chorley, R. J., Schumm, S. A., and Sugden, D. E. 1984. *Geomorphology.* New York: Methuen, 605p.

Chose, B., Pandy, S., and Lal, G. 1967. Quantitative geomorphology of the drainage basins in the central Lumi basin in western Rajasthan. *Zeit. F. Geomorph.* 11:146–60.

Chou, L., and Wollast, R. 1984. Study of the weathering of albits at room temperature and pressure with a fluidized bed reactor. *Geochim. Cosmochim. Acta* 48:2205–17.

———. 1985. Study of the weathering of albite at room temperature and pressure with a fluidized bed reactor—a reply. *Geochim. Cosmochim. Acta.* 49:1659–60.

Church, M. 1972. Baffin Island sandurs: A study of arctic fluvial processes. *Canada Geol. Survey Bull.* 216.

———. 1974. Hydrology and permafrost with reference to northern North America. In *Permafrost Hydrology, Proceedings of Workshop Seminar,* pp. 7–20. Ottawa: Canadian National Committee, Internat. Hydrol. Decade.

———. 1978. Paleohydrological reconstructions from a Holocene valley. In *Fluvial Sedimentology,* pp. 743–72. Ed. by A. Miall. *Can. Soc. Petrol. Geol. Mem.* 5.

———. 1988. Floods in cold climates. In *Flood Geomorphology,* pp. 205–30. Ed. by V. R. Baker, R. C. Kochel, and P. C. Patton. New York: John Wiley.

Church, M., and Jones, D. 1982. Channel bars in gravel-bed rivers. In *Gravel-bed rivers,* pp. 291–338. Ed. by R. Hey, J. Bathurst, and C. Thorne. New York: John Wiley & Sons.

Church, M., and Miles, M. J. 1987. Meteorological antecedents to debris flows in southwestern British Columbia; some case studies. In *Debris Flows/Avalanches: Process, Recognition, and Mitigation,* pp. 63–79. Ed. by J. E. Costa and G. F. Wieczorek. Boulder, *Geol. Soc. Amer. Rev. in Engin.* Geol 7.

Church, M., and Slaymaker, O. 1989. Disequilibrium of Holocene sediment yield in glaciated British Columbia. *Nature* 337:452–54.

Chutha, P., and Doodge, J. C. I. 1990. The shape parameters of the geomorphologic unit hydrograph. *J. Hydrology* 117:81–97.

Ciolkosz, E. J., Carter, B. J., Hoover, M. T., Cronce, R. C., Waltman, W. J., and Dobos, R. R. 1990. Genesis of soils and landscapes in the Ridge and Valley province of central Pennsylvania. *Geomorph.* 3:245–61.

Ciolkosz, E. J., Cronce, R. C., and Sevon, W. D. 1986. *Periglacial features in Pennsylvania.* Pa. State Univ. Agron. Ser. 92.

Clague, J. J., Evans, S. G., and Blown, G. 1985. A debris flow triggered by the breaching of a moraine-dammed lake, Klattasine Creek, British Columbia. *Can. J. Ear. Sci.* 22:1492–502.

Clapperton, C. M. 1975. The debris content of surging glaciers in Svalbard and Iceland. *Jour. Glaciol.* 14:395–406.

Claridge, G., and Campbell, I. 1984. Mineral transformation during the weathering of dolerite under cold arid conditions in Antarctica: *New Zealand J. Geophys.* 27:537–46.

Clark, G. K., Schmok, J. P., Ommanney, C. S. L., and Collins, S. G. 1986. Characteristics of surge-type glaciers. *Journal of Geophysical Research* 91:7165–80.

Clark, G. M. 1968. Sorted patterned ground: new Appalachian localities south of the glacial border. *Science* 161:355–56.

Clark, G. M., and Ciolkosz, E. J. 1988. Periglacial geomorphology of the Appalachian Highlands and Interior Highlands south of the glacial border—a review. *Geomorph.* 1:191–220.

Clark, G. M., and Hedges, J. 1992. Origin of certain high elevation local broad uplands in the central Appalachians south of the glacial border—a paleoperiglacial hypothesis. In *Periglacial Geomorphology*, pp. 31–62. Ed. by J. C. Dixon and A. D. Abrahams. Chichester: John Wiley.

Clark, M. J., Gurnell, A. M., and Threlfall, J. L. 1988. Suspended sediment transport in Arctic rivers. In *Permafrost, fifth international conference proceedings*, pp. 558–63. Ed. by K. Senneset. Trondheim: Tapir Publishers.

Clark, S. P., and Jager, E. 1969. Denudation rate in the Alps from geochronological and heat flow data. *Am. Jour. Sci.* 267:1143–60.

Clarke, G. K., Collins, S. G., and Thompson, D. E. 1984. Flow, thermal structure, and subglacial conditions of a surge-type glacier. *Canadian Journal of Earth Science* 21:232–40.

Clayton, L. 1967. Stagnant-glacial features of the Missouri Coteau in North Dakota. *North Dakota Geol. Survey Misc. Series* 30:25–46.

Clayton, T. D. 1989. Artificial beach replenishment on the U.S. Pacific Shore; a brief overview. *Proc. Sympos. on Coastal and Ocean Management* 6:2033–45.

Cleary, W. J., and Hosier, P. E. 1987. North Carolina coastal geologic hazards: an overview. *Bull. Assoc. Engin. Geol.* 24:469–88.

Cleary, W. J., and Pilkey, O. H., 1996. Environmental coastal geology: Cape Lookout to Cape Fear, N.C. (Fieldtrip Guidebook) In *Environmental coastal geology, Cape Lookout to Cape Fear, N.C.*, pp. 87–128. Ed. by W. J. Cleary, Wilmington N.C.: Carolina Geological Society.

Cleaves, E., Fisher, D., and Bricker, O. 1974. Chemical weathering of serpentinite in the eastern piedmont of Maryland. *Geol. Soc. America Bull.* 85:437–44.

Clemmensen, L., Olsen, H., and Blakey, R. 1989. Erg-margin deposits in the Lower Jurassic Moenave Formation and Wingate Sandstone, southern Utah. *Geol. Soc. America Bull.* 101:759–73.

Coates, D. R. 1991. Glacial deposits. In G. A. Kiersch (ed.). *The heritage of engineering geology, the first hundred years.* Boulder, Geological Society of America, Centennial Special Volume 3, pp. 299–322.

Coates, D. R., and Vitek, J. D., eds. 1980. *Thresholds in geomorphology.* London: Allen and Unwin Ltd.

Cochran, M., and Berner, R. 1996. Promotion of chemical weathering by higher plants: Field observations on Hawaiian basalts. *Chem. Geol.* 132:71–77.

Cohn, M. E., and Kochel, R. C. 1993. The relative role of geomorphic processes in storm recovery of washover sites on the Virginia barrier islands. *Geol. Soc. Amer. Abs. w. Prog.* 25:2:9.

Colaprete, A., and Jakosky, B. M. 1998. Ice flow and rock glaciers on Mars. *Jour. Geophysical Res., E. Planets.* 103:5897–5909.

Colbeck, S. C., and Evans, R. J. 1973. A flow law for temperate glacier ice. *Jour. Glaciol.* 12:71–86.

Colby, B. 1963. Fluvial sediments: A summary of source, transportation, deposition, and measurement of sediment discharge. *U.S. Geol. Survey Bull.* 1181–A:21.

———. 1964. Scour and fill in sand bed streams. *U.S. Geol. Survey Prof. Paper* 462-D.

Coleman, J. M. 1968. Deltaic evolution. In *encyclopedia of geomorphology,* pp. 255–60. Ed. by R. W. Fairbridge. New York: Reinhold Book Corp.

———. 1988. Dynamic changes and processes in the Mississippi River delta: *Geological Society of America Bulletin* 100:999–1015.

Coleman, J., Roberts, H., and Stone, G. 1998. Mississippi River Delta: An overview. *J. Coastal Res.* 14:698–716.

Colman, S. 1987. Limits and constraints of the diffusion equation in modeling geological processes of scarp degradation. In Crone, A., and Omdahl, E., (eds.). *Directions in Paleoseismology: U.S. Geological Survey Open File Report,* 87–673, 311–16.

Colman, S. M. 1982. Clay mineralogy of weathering rinds and possible implications concerning the sources of clay minerals in soils. *Geology* 10:370–75.

———. 1983. Progressive changes in the morphology of fluvial terraces and scarps along the Rappahannock River, Virginia. *Earth Surf. Proc. and Landforms* 8:201–12.

———. 1986. Glacial sequence near McCall, Idaho: Weathering rinds, soil development, morphology, and other relative age criteria. *Quaternary Research* 25:25–43.

Colman, S. M., and Watson, K. 1983. Ages estimated from a diffusion model for scarp degradation. *Science* 221:263–65.

Connell, W., and Patrick, W. 1968. Sulfate reduction in soil. Effects of redox potential and pH. *Science* 159:86–87.

Cook, J. H. 1946. Kame complexes and perforation deposits. *Am. Jour. Sci.* 244:573–83.

Cooke, H. J. 1973. Tropical karst in northeast Tanzania. *Zeit. F. Geomorph.* 17:443–59.

Cooke, R. U. 1970. Morphometric analysis of pediments and associated landforms in the western Mojave Desert, California. *Am. Jour. Sci.* 269:26–38.

Cooke, R. U., and Doornkamp, J. 1974. *Geomorphology in environmental management.* London: Clarendon Press.

Cooke, R. U., and Reeves, R. W. 1972. Relations between debris size and the slope of mountain fronts and pediments in the Mojave Desert, California. *Zeit. F. Geomorph.* 16:76–82.

Cooke, R. U., and Warren, A. 1973. *Geomorphology in deserts.* London: Batsford Ltd.

Cooke, R. U., Warren, A., and Goudie, A. 1993. *Desert geomorphology.* London: UCL Press, 526 p.

Cooley, M. E. 1990. Use of paleoflood investigations to improve flood frequency analyses of plains streams in Wyoming. *U.S. Geol. Survey Water Resources Investigations Rept.* 88–4209.

Cooley, R., Fiero, G., Lattman, L., and Mindling, A. 1973. Influence of surface and near-surface caliche distribution on infiltration characteristics and flooding, Las Vegas area, Nevada. *Univ. Nevada, Reno, Desert Research Inst. Proj. Rept.* 21.

Cooper, W. S. 1958. Coastal sand dunes of Oregon and Washington. *Geol. Soc. America Mem.* 72.

Corbel, J. 1957. Karsts hauts-Alpins. *Rev. Geogr. Lyon* 32:135–58.

———. 1959. Vitesse de l'erosion. *Zeit. F. Geomorph.* 3:1–28.

———. 1964. L'erosion terrestre, étude quantitative. *Annales Geog.* 73:385–412.

Corominas, L., and Pasuto, A. 1999. Reconstructing recent landslide activity in relation to rainfall in the Llobregat River basin, eastern Pyrenees, Spain. *Geomorphology* 30:79–93.

Corte, A. E. 1969. Geocryology and Engineering. In *Reviews in Engineering Geology 2*, p. 119–85. Ed. by D. Varnes and G. Kiersch. Boulder, CO: Geol. Soc. America.

Costa, J. E. 1974a. Stratigraphic, morphologic, and pedologic evidence of large floods in humid environments. *Geology* 2:301–3.

———. 1974b. Response and recovery of a piedmont watershed from tropical storm Agnes, June 1972. *Water Resour. Res.* 10:106–12.

———. 1978. Holocene stratigraphy in flood frequency analysis. *Water Resour. Res.* 14:626–32.

———. 1983. Paleohydraulic reconstruction of flash-flood peaks from boulder deposits in the Colorado Front Range. *Geol. Soc. America Bull.* 94:986–1004.

———. 1984. The physical geomorphology of debris flows. In *Developments and Applications of Geomorphology*, pp. 268–317. Ed. by J. E. Costa and P. J. Fleisher. Berlin: Springer-Verlag.

———. 1987. Hydraulics and basin morphometry of the largest flash floods in the conterminous United States. *J. Hydrology* 93:313–38.

———. 1988. Rheologic, geomorphic, and sedimentologic differentiation of water floods, hyperconcentrated flows, and debris flows. In *Flood Geomorphology*, pp. 113–22. Ed. by V. R. Baker, R. C. Kochel, and P. C. Patton. New York: John Wiley.

———. 1997. Hydraulic modeling for lahar hazards at Cascades Volcanoes. *Environmental and Engineering Geoscience* 3:21–30.

Costa, J. E., and Baker, V. R. 1981. *Surficial geology—Building with the Earth.* New York: John Wiley & Sons.

Costa, J. E., and Jarrett, R. D. 1981. Debris flows in small mountain stream channels of Colorado, and their hydrologic implications. *Bull. Assoc. Eng. Geol.* 18:309–22.

Costa, J. E., and O'Connor, J. E., 1995. Geomorphically effective floods. In *Natural and anthropogenic influences in fluvial geomorphology*, pp. 151–64. Ed. by J. E. Costa et al. Geophysical Monograph 89, American Geophysical Union.

Costa, J. E., and Wieczorek, G. F., eds. 1987. Debris flows/avalanches: Process, recognition, and mitigation. Boulder: *Geol. Soc. Amer. Rev. in Engin. Geol 7.*

Costa, J. E., and Williams, G. P. 1984. Debris flow dynamics: *U.S. Geol. Survey Open File Report* 84–606, Video Tape.

Costard, F., Aguirre-Puente, J., Greeley, R., Guillemet, G., and Malchoufi, N. 1996. Martian fluvial erosion; laboratory simulation. *27th Lunar & Planetary Sci. Conf. Abs.*, 261–62.

Coutard, J., and Francou, B. 1989. Rock temperature measurements in two alpine environments: Implications for frost shattering. *Arctic and Alpine Res.* 21:399–416.

Couture, N. J., and Pollard, W. H. 1998. An assessment of ground ice volume near Eureka, Northwest Territories. *Collection Nordicana* 57:195–200.

Cox, E. T. 1874. *Fifth annual report of the Geological Survey of Indiana*, pp. 280–305. Indianapolis: Indiana Geol. Survey.

Cox, T. 1994. Analysis of drainage-basin symmetry as a rapid technique to identify areas of possible Quaternary tilt-block tectonics: An example from the Mississippi Embayment. *Geol. Soc. Amer. Bull.* 106:571–81.

Cramer, H. 1941. Die Systematik der Karstdolinen. *Neues Jb. Miner. Geol. Palaont.* 85:293–382.

Crary, A. P. 1966. Mechanisms for fiord formation indicated by studies of an ice-covered inlet. *Geol. Soc. America Bull.* 77:911–30.

Crittenden, M. 1963. New data on the isostatic deformation of Lake Bonneville. *U.S. Geol. Survey Prof. Paper* 454-E.

Cronin, J. E. 1983. Design and performance of a liquid natural convection subgrade cooling system for construction on ice-rich permafrost. In *Proc. Permafrost 4th Internat. Conf.*, pp. 198–203. Natl. Acad. Sci.

Crosby, W. O. 1902. Origin of eskers. *Am. Geologist* 30:1–38.

Crowley, K. D. 1983. Large-scale bed configurations (macroforms), Platte River basin, Colorado and Nebraska: Primary structures and formative processes. *Geol. Soc. America Bull.* 94:117–33.

Crowson, R., Birkemeier, W., Klein, H., and Miller, H. 1988. SUPERDUCK Nearshore Processes Experiment; summary of studies at CERC Field Research Facility. *U.S. Army Coast Eng. Res. Center. Tech Rpt.* CERC-88-12.

Crozier, M. J. 1973. Techniques for the morphometric analysis of landslips. *Zeit. F. Geomorph.* 17:78–101.

———. 1984. Field assessment of slope instability. In *Slope Instability*, p. 103–42. Ed. by D. Brunsden and W. Prior. New York: John Wiley and Sons.

———. 1986. *Landslides: Causes, Consequences, and Environment.* London: Croom Helm, 252p.

Cruden, D. M., and Eaton, T. M. 1987. Reconnaissance of rockslide hazards in Kananaskis Country, Alberta. *Can. Geotech. J.* 24:414–29.

Cu, P. 1999. Impact of debris flow on river channels in the upper reaches of the Yangtze River. *Internat. Jour. Sediment Res.* 14:201–3.

Cuffey, K. M., Conway, H., Grades, A. M., Hallet, B., Lorrain, R., Severinghaus, J. P., Steig, E. J., Vaughn, B., and White,

J. W. C. 2000. Entrainment at cold glacier beds. *Geology* 28:351–54.

Cunnane, C. 1987. Review of statistical models for flood frequency estimation. In *Hydrologic Frequency Modeling,* pp. 49–95. Ed. by V. J. Singh. Dordrecht: Reidel.

Cunningham, F., and Griba, W. 1973. A model of slope development, and its applications to the Grand Canyon, Arizona. *Zeit. F. Geomorph.* 17:43–77.

Curray, J. R. 1960. Sediments and history of Holocene transgression, continental shelf, northwest Gulf of Mexico. In *Recent Sediments, Northwest Gulf of Mexico,* pp. 221–66. Ed. by Shepard, F. P., et al. Amer. Assoc. Petroleum Geologists.

———. 1961. Late Quaternary sea level; a discussion. *Geol. Soc. America Bull.* 72:1707–12.

———. 1965. Late Quaternary history, continental shelves of the United States. In *The Quaternary of the United States,* pp. 221–66. Ed. by Shepard, F. P., et al. Amer. Assoc. Petroleum Geologists.

Cvijíc, J. 1893. Das Karstphanomen. *Geogr. Abh.* 5:217–329.

Czudek, T., and Demek, J. 1970. Thermokarst in Siberia and its influence on the development of lowland relief. *Quat. Res.* 1:103–20.

Dahl, R. 1965. Plastically sculptured detail forms on rock surfaces in northern Nordland. *Geogr. Annlr.* 47:83–140.

Dahlgren, R., Ugolini, F., and Casey, W. 1999. Field weathering rates of Mt. St. Helens tephra. *Geochim. Cosmochim. Acta* 63:587–98.

Dallimore, S. R., and Wolfe, S. A. 1988. Massive ground ice associated with glaciofluvial sediments, Richards Island, N.W.T., Canada. In *Permafrost, fifth international conference proceedings,* pp. 132–37. Ed. by K. Senneset. Trondheim: Tapir Publishers.

Dalrymple, J. B., Blong, R. J., and Conacher, A. J. 1968. A hypothetical nine-unit landsurface model. *Zeit. F. Geomorph.* 12:60–76.

Dalrymple, T. 1960. Flood frequency analysis. *Manual of hydrology,* part 3, Flood flow techniques. *U.S. Geol. Survey Water Supply Paper* 1543-A.

Daly, R. 1933. *Igneous rock and the depths of the Earth.* New York: McGraw-Hill.

Daniel, J. R. K. 1981. Drainage density as an index of climatic geomorphology. *J. Hydrol.* 50:147–54.

Dansgaard, W., White, J., and Johnson, S. 1989. The abrupt termination of the Younger Dryas climate event. *Nature* 339:532–34.

Darby and Simon. 1999. *Incised river channels.* New York: John Wiley and Sons.

Davies, J. L. 1973. *Geographical variation in coastal development.* New York: Hafner.

Davis, D. G. 1980. Cave development in the Guadalupe Mountains, a critical review of recent hypotheses. *Natl. Speleol. Soc. Bull.* 42:42–48.

Davis, J. L. 1964. A morphogenetic approach to world shorelines. *Zeit. F. Geomorph.* 8:127–42.

Davis, R. A., ed. 1985. *Coastal sedimentary environments.* New York: Springer-Verlag.

Davis, R. A., and Dolan, R. 1993. Nor'easters. *Amer. Sci.* 81:428–39.

Davis, R. A., and Fox, W. T. 1972. Coastal processes and nearshore bars. *J. Sediment. Petr.* 42:401–12.

Day, M. 1978. Morphology and distribution of residual limestone hills (mogotes) in the karst of northern Puerto Rico. *Geol. Soc. America Bull.* 89:426–32.

Dean, R. G. 1997. Models for barrier island restoration. *Jour. Coastal Res.* 13:694–703.

Dean, R. G., and Perlin, M. 1977. Coastal engineering study of Ocean City Inlet, Maryland. In *Coastal sediments 77.* Am. Soc. Civil Engineers, *Symposium Water, Port, Coastal and Ocean Div.* 5:520–42.

Dean, R. G., Healy, T. R., Dommerholt, A. P. 1993. A "blind-folded" test of equilibrium beach profile concepts with New Zealand data. *Marine Geol.* 109:253–66.

Decker, R., Wright, T., and Stauffer, P., eds. 1987. Volcanism in Hawaii. *U.S. Geol. Survey Prof. Paper* 1350, vol. 1 and 2.

Deere, D. U., and Peck, R. B. 1959. Stability of cuts in fine sands and varved clays, Northern Pacific Railroad, Noxon Rapids line change, Montana. *Proc. AREA* 59, pp. 807–15.

Denny, C. S. 1956. Surficial geology and geomorphology of Potter County, Pennsylvania. *U.S. Geol. Survey Prof. Paper* 288.

———. 1967. Fans and pediments. *Am. Jour. Sci.* 265:81–105.

DePolo, C., and Slemmens, D. 1990. Estimation of earthquake size for seismic hazards. In Krintzky, E., and Slemmens, D. (eds.), Neotectonics in earthquake evaluation: *Geological Society of America, Reviews in Engr. Geol. VIII,* 1–28.

deVries, J. J. 1995. Seasonal expansion and contraction of stream networks in shallow groundwater systems. *Jour. Hydrology* 170:15–26.

DeWolf, Y. 1988. Stratified slope deposits. In *Advances in Periglacial Geomorphology,* pp. 91–110. Ed. by M. J. Clark. London: John Wiley.

Dietrich, W. E. 1987. Mechanics of flow and sediment transport in river bends. In *River channels: Environment and process,* pp. 179–227. Ed. by K. Richards. Oxford: Basil Blackwell, Ltd.

Dietrich, W. E., and Dunne, T. 1978. Sediment budget for a small catchment in mountainous terrain. *Zeit. F. Geomorph.,* Suppl. Bd. 29:191–206.

Dietrich, W. E., and Smith, J. D. 1983. Influence of the point bar on flow through curved channels. *Water Resour. Res.* 19:1173–92.

Dingman, S.L. 1978. Drainage density and streamflow: A closer look. *Water Resour. Res.* 14:1183–87.

Dionne, J. C. 1974. Polished and striated mud surfaces in the St. Lawrence tidal flats, Quebec. *Can. J. Earth Sci.* 11:860–66.

Dixon, J. 1994. Aridic soils, patterned ground, and desert pavements. In *Geomorphology of desert environments,* pp. 64–81. Ed. by A. Abraham and A. Parsons. London: Chapman and Hall.

Dixon, J. C. 1986. Solute movement on hillslopes in the alpine environment of the Colorado Front Range. In *Hillslope processes,* pp. 139–59. Ed. by A. D. Abrahams. Boston: Allen and Unwin.

Dixon, J. C., and Abrahams, A. D. 1992. *Periglacial geomorphology.* Chichester: John Wiley.

Dohrenwend, J. 1994. Pediments in arid environments. In *Geomorphology of desert environments,* pp. 321–53. Ed. by A. Abrahams and A. Parsons. London: Chapman & Hall.

Dolan, R., and Ferm, J.C. 1968. Crescentic landforms along the mid-Atlantic coast. *Science* 159:627–29.

Dolan, R., and Lins, H. F. 1986. The Outer Banks of North Carolina: *U.S. Geol. Survey Prof. Paper* 1177–B.

Dolan, R., Hayden, B., and Felder, W. 1979. Shoreline periodicities and edge waves. *Jour. Geology* 87:175–85.

Dolan, R., Hayden, B., and Lins, H. 1980. Barrier islands. *Am. Scientist* 68:16–25.

Dolan, R., Lins, H. F., and Hayden, B. P. 1988. Mid-Atlantic coastal storms. *J. Coastal Res.* 4:417–33.

Dolan, R., Vincent, L., and Hayden, B. 1974. Crescentic coastal landforms. *Zeit. F. Geomorph.* 18:1–12.

Doornkamp, J. C., and King, C. A. M. 1971. *Numerical analysis in geomorphology: An introduction.* London: Edward Arnold Ltd.

Dorn, R., DeNiro, M., and Ajie, H. 1987. Isotopic evidence for climatic influence of alluvial-fan development in Death Valley, California. *Geology* 15:108–10.

Dorn, R. 1988. A rock-varnish interpretation of alluvial-fan development in Death Valley, California. *Natl. Geogr. Res.* 4:56–73.

Dott, R. 1992. An introduction to the ups and downs of eustasy. In Dott, R., ed., Eustasy: The historical ups and downs of a major geological concept. *Geol. Soc. America Memoir* 180:1–16.

Douglas, G. R., McGreevey, J. P., and Whalley, W.B. 1983. Rock weathering by frost shattering processes. In *Proc. permafrost 4th internat. conf.*, pp. 244–48. Natl. Acad. Sci.

———. 1967. Man, vegetation, and sediment yields of rivers. *Nature* 215:925–28.

Drake, J. J., and Harmon, R. S. 1973. Hydrochemical environments of carbonate terrains. *Water Resour. Res.* 9:949–57.

Drake, J. J., and Wigley, T. M. L. 1975. The effect of climate on the chemistry of carbonate groundwater. *Water Resour. Res.* 11:958–62.

Drake, L. D. 1968. *Till studies in New Hampshire.* Ph.D. dissertation, Ohio State University.

———. 1974. Till fabric control by clast shape. Geol. *Soc. America Bull.* 85:247–50.

Drake, L. D., and Shreve, R. L. 1973. Pressure melting and regelation of ice by round wires. *Proc. Royal Soc. London,* ser. A:332:51–83.

Dreimanis, A. 1988. Tills: Their genetic terminology and classification. In *Genetic classification of glacigenic deposits,* pp. 17–83. Ed. by R. P. Goldthwait and C. L. Matsch. Rotterdam: Balkema.

Dreimanis, A., and Vagners, U. J. 1971. Bimodal distribution of rock and mineral fragments in basal till. In *Till: A symposium.* Ed. by R. P. Goldthwait. Columbus: Ohio State Univ. Press.

Drever, J., and Clow, D. 1995 Weathering rates in catchments. In White, A., and Brantley, S., eds., Chemical weathering rates of silicate minerals, pp. 463–83. *Reviews in Mineralogy* 31 Mineral. Soc. Amer., Wash., D.C.

Drew, D. and Hoetzl, H. 1999. *Karst Hydrogeology and Human Activities; Impacts, Consequences and Implication.* Rotterdam, A. A. Balkema.

Drewry, D. 1986. *Glacial geologic processes.* Baltimore: Edward Arnold.

Dreybrodt, W. 1998. Principles of karst evolution from initiation to maturity and their relation to physics and chemistry. In *Global Karst Correlation,* D. X. Yuan and Z. H. Lie (eds.), Beijing, Science Press, pp. 33–49.

Dreybrodt, W. 1981. Kinetics of the dissolution of calcite and its application to karstification. *Chemical Geology* 31:245–269.

———. 1990. The role of dissolution kinetics in the development of karst aquifers in limestone: A model simulation of karst evolution. *Journal of Geology* 98:639–55.

Dreybrodt, W., Gabrovšek, F., and Siemers, J. 1997. Dynamics of the evolution of early karst. *Theoretical and Applied Karstology.* 10:9–27.

Druitt, T., and Sparks, R. 1985. On the formation of calders during ignimbrite eruptions. *Nature* 310:679–81.

Duane, D. B., and James, W. R. 1980. Littoral transport in the surf zone elucidated by an Eulerian sediment tracer experiment. *Jour. Sed. Petrology* 50:929–42.

DuBois, R. N. 1972. Inverse relation between foreshore slope and mean grain size as a function of the heavy mineral content. *Geol. Soc. America Bull.* 83:871–76.

———. 1978. Beach topography and beach cusps. *Geol. Soc. America Bull.* 89:1133–39.

———. 1981. Foreshore topography, tides and beach cusps, Delaware. *Geol. Soc. America Bull.* 92:132–38.

Dunkerley, D. L. 1980. The study of the evolution of slope form over long periods of time: A review of methodologies and some new observational data from Papua, New Guinea. *Zeit. F. Geomorph.* 24:52–67.

Dunne, L. 1988. Fan-delta and braid delta: Varieties of coarse-grained deltas: discussion. *Geol. Soc. American Bull.* 100:1308–09.

Dunne, T. 1978. Field studies of hillslope flow processes. In *Hillslope hydrology,* p. 227–94. Ed. by M. J. Kirby. New York: John Wiley & Sons.

———. 1979. Sediment yield and land use in tropical catchments. *J. Hydrol.* 42:281–300.

———. 1990. Hydrology, mechanics, and geomorphic implications of erosion by subsurface flow. In *Groundwater geomorphology,* pp. 1–28. Ed. by C. G. Higgins and D. R. Coates. Boulder: *Geol. Soc. Amer. Sp. Paper* 252.

Dunne, T., and Black, R. D. 1970a. An experimental investigation of runoff production in permeable soils. *Water Resour. Res.* 6:478–90.

———. 1970b. Partial area contribution to storm runoff in a small New England watershed. *Water Resour. Res.* 6:1296–311.

Dunne, T., and Leopold, L. B. 1978. *Water in environmental planning.* San Francisco: W. H. Freeman.

Dury, G. H. 1965. Theoretical implications of underfit streams. *U.S. Geol. Survey Prof. Paper* 452-C.

———. 1966. Pediment slope and particle size at Middle Pinnacle, near Broken Hill, New South Wales. *Austr. Geog. Studies* 4:1–17.

———. 1973. Magnitude-frequency analysis and channel morphology. In *Fluvial geomorphology,* pp. 91–121. Ed. by M. Morisawa. Binghamton: S.U.N.Y., Pubs. in Geomorphology.

Duval, P. 1981. Creep and fabrics of polycrystalline ice under shear and compression. *Jour. Glaciol.* 27:129–40.

Duval, P., and Hughes, L. G. 1980. Does the permanent creep-rate of polycrystalline ice increase with crystal size? *Jour. Glaciol.* 25:151–57.

Dylik, J. 1964. The essentials of the meaning of the term "Periglacial." *Soc. Sci. et Lettres Lodz Bull.* 15:2:1–19.

Eakin, H. M. 1916. The Yukon-Koyukuk region. Alaska. *U.S. Geol. Survey Bull.* 631.

Eardley, A. G. 1967. Rates of denudation as measured by bristlecone pines, Cedar Breaks, Utah. *Utah Geol. and Mineral Survey Spec. Studies* 21.

Eaton, L. S. 1999. Debris flows and landscape evolution in the Upper Rapidan Basin, Blue Ridge Mountains, Virginia: Dissertation, Univ. Virginia, Charlottesville.

Eaton, L. S., Kochel, R. C., Howard, AD, Sherwood, W. C. 1997. Debris forward stratified slope wash deposits in the central Blue Ridge of Virginia: *Geological Soc Amer Abs Ann. Mtg* 29:7:410.

Echelmeyer, K., and Zhongxiang, W. 1987. Direct observation of basal sliding deformation of basal drift at sub-freezing temperatures. *Journal of Glaciology* 33:83–98.

Ede, D. P. 1975. Limestone drainage systems. *J. Hydrol.* 27:297–318.

Egemeir, S. J. 1981. Cavern development by thermal waters. *Natl. Speleol. Soc. Bull.* 43:31–52.

Eggleton, R. A., Foudoulis, C., and Varkevisser, D. 1987. Weathering of basalt: Changes in rock chemistry and mineralogy. *Clays and Clay Minerals* 35:161–69.

Einstein, H. A. 1950. The bedload function for sediment transportation in open channel flows. *U.S. Dept. Agri. Tech. Bull.* 1026.

Eklund, A., and Hart, J. K. 1996. Glaciotectonic deformation within a flute from the Isfallsglaciären, Sweden. *Journal of Quaternary Science* 11:299–310.

El-Baz, F., and Maxwell, T. A., eds. 1982. *Desert landforms of southwest Egypt: A basis for comparison with Mars.* NASA Sci. and Tech. Information Branch, Washington, DC, CR–3611.

Elliston, G. R. 1963. Catastrophic glacier advances. *Int. Assoc. Sci. Hydrol. Bull.* 8:65–66.

Elson, J. A. 1968. Washboard moraines and other minor moraine types. In *Encyclopedia of geomorphology*, pp. 1213–19. Ed. by R. W. Fairbridge. New York: Reinhold Book Corp.

———. 1988. Comment on glacitectonite, deformation till, and comminution till. In *Genetic classification of glacigenic deposits*, pp. 85–91. Ed by R. P. Goldthwait and C. L. Matsch, Rotterdam: Balkema.

Ely, L., and Baker, V. R. 1990. Large floods and climate change in the southwestern United States. In French, R. H., ed. Hydraulics/Hydrology of Arid Lands. *Amer. Soc. Civil Eng.* 361–66.

Embleton, C., and King, C. A. M. 1968. *Glacial and periglacial geomorphology.* Edinburgh: Edward Arnold Ltd.

———. 1975. *Periglacial Geomorphology.* New York: Halsted Press.

Emery, K. O., and Kuhn, G. G. 1982. Sea cliffs: Their processes, profiles and classifications. *Geol. Soc. America Bull.* 93:644–54.

Engelhardt, H. F., Harrison, W. D., and Kamb, B. 1978. Basal sliding and conditions at the glacier bed as revealed by bore-hole photography. *Jour. Glaciol.* 20:469–508.

England, P., and Molnar, P. 1990. Surface uplift, uplift of rocks, and exhumation of rocks. *Geology* 18:1173–77.

Ensminger, S. L., Evenson, E. B., Alley, R. B., Larson, G. J., Lawson, D. E., and Strasser, J. C. 1999. In *Glacial Processes Past and Present*, D. M. Mickelson and J. W. Attig, eds., pp. 11–22. Geological Society of America Special Paper 337.

Ericksen, G. E., Pflacker, G., and Fernandez, J. V. 1970. Preliminary report on the geological events associated with the May 31, 1970, Peru earthquake. *U.S. Geol. Survey Circ.* 639.

Evans, D. J. A., ed., 1994. *Cold Climate landforms.* New York: John Wiley.

Evans, I. S. 1969. Salt crystallization and rock weathering: A review. *Rev. Geomorph. Dynamique* 19:157–77.

Evans, O. F. 1938. Classification and origin of beach cusps. *Jour. Geology* 46:615–27.

———. 1940. The low and ball of the east shore of Lake Michigan. *Jour. Geology* 48:476–511.

———. 1942. The origin of spits, bars, and related structures. *Jour. Geology* 50:846–63.

Evans, R. 1976. Observations on a stripe pattern. *Biuletyn Perygl.* 25:9–22.

Everett, D. H. 1961. The thermodynamics of frost damage to porous solids. *Trans. Faraday Soc.* 57:1541–51.

Everett, J., Morisawa, M., and Short, N. 1986. Tectonic landforms. In Short, N. and Blair, R. (eds.), *Geomorphology from Space: NASA SP-486*, 27–183 p.

Everitt, B. L. 1968. Use of cottonwood in an investigation of recent history of a flood plain. *Am. Jour. Sci.* 266:417–39.

Eyles, N., and Rogerson, R. J. 1978. Sedimentology of medial moraines in Berendon Glacier, British Columbia, Canada: Implications for debris transport in a glacierized basin. *Geol. Soc. America Bull.* 89:1688–93.

Eynon, G., and Walker, R. G. 1974. Facies relationships in Pleistocene outwash gravels, southern Ontario: A model for bar growth in braided rivers. *Sedimentology* 21:43–70.

Fahey, B. D. 1975. Nonsorted circle development in a Colorado alpine location. *Geogr. Annlr.* 57A:153–64.

Fahnestock, R. K. 1961. Competence of a glacial stream. *U.S. Geol. Survey Prof. Paper* 424-B:211–13.

———. 1963. Morphology and hydrology of a glacial stream—White River, Mount Rainier, Washington. *U.S. Geol. Survey Prof. Paper* 422-A.

———. 1969. Morphology of the Slims River. In *Icefield Ranges Research Project, Scientific Results #1*, pp. 161–72. Ed. by V. C. Bushell and R. H. Ragle. Am. Geog. Soc. and Arctic Inst. N. America.

Fairbridge, R. W. 1968. Land mass and major landform classification. In *Encyclopedia of Geomorphology*, pp. 618–26. Ed. by R. W. Fairbridge. New York: Reinhold Book Corp.

Fairchild, J. C. 1973. Longshore transport of suspended sediment. *Proc. 13th Conf. on Coast. Eng.* 1069–88.

Farres, P. J. 1987. The dynamics of rainsplash erosion and the role of soil aggregate stability. *Catena* 14:119–130.

Feda, J. 1988. Collapse of loess upon wetting. *Engin. Geol.* 25:263–69.

Federal Highway Administration, 1985. *PCSTABLE4, Computer Modeling of Slope Stability*, prepared by Purdue University, School of Civil Engineering, West Lafayette, Indiana.

Federov, A. V. 1996. Effects of recent climatic change on permafrost landscapes in central Sakha *Polar geography* 20:99–108.

Fenster, M. J., and Dolan, R. 1996. Assessing the impact of tidal inlets or adjacent barrier island shorelines. Jour. Coastal Res. 12:294–310.

Ferguson, R. 1987. Hydraulic and sedimentary controls of channel pattern. In *River Channels: Environment and Process.* pp. 129–58. Ed. by K. Richards. London: Basil Blackwell.

Ferguson, R. I. 1975. Meander irregularity and wavelength estimation. *J. Hydrol.* 26:315–33.

Ferguson, R. I. 1993. Understanding braiding processes in gravel-bed rivers: Progress and unsolved problems. In *Braided Rivers,* Ed. by J. L. Best and C. S. Bristow, pp. 73–87. Geo. Soc. America Special Publication No. 75.

Ferguson, R. I., and 5 others. 1998. Downstream fining of river gravels: Integrated field, laboratory, and modeling study. In *Gravel-bed rivers in the environment,* pp. 85–114. Ed. by P. C. Klingeman, R. L. Beschta, P. D. Komar, and J. B. Bradley, Highlands Ranch: Water Resources Publications, LLC.

Fernandes, N. F., and Dietrich, W. E. 1997. Hillslope evolution by diffusive processes; the timescale for equilibrium adjustments. *Water Resources Res.* 33:1307–18.

Ferrians, O. J. 1965. Permafrost map of Alaska. *U.S. Geol. Survey Misc. Geol. Inv. Map* I-445.

———. 1988. Pingos in Alaska: A review. In *Permafrost, Fifth International Conference Proceedings,* pp. 734–39. Ed. by K. Senneset. Trondheim: Tapir Publishers.

Ferrians, O. J., Kachadoorian, R., and Greene, G. W. 1969. Permafrost and related engineering problems in Alaska. *U.S. Geol. Survey Prof. Paper* 678.

Finlayson, B. 1981. Field measurements of soil creep. *Earth Surf. Proc. and Landforms* 6:35–48.

Fisher, D., and 4 others 1998. Effect of subducting sea-floor roughness on fore-arc kinematics, Pacific Coast, Costa Rica. *Geology* 26, 5:467–70.

Fisher, J. J., and Simpson, E. J. 1979. Washover and tidal sedimentation rates as environmental factors in development of a transgressive barrier shoreline. In *Barrier Islands.* Ed. by S. Leatherman. New York: Academic Press.

Fisher, T. G., and Shaw, J. 1992. A depositional model for Rogen moraine, with examples from the Avalon Peninsula, Newfoundland. *Can. J. Earth Sci.* 29:669–86.

Fisk, H. N. 1944. *Geological Investigation of the Alluvial Valley of the Lower Mississippi River.* Vicksburg, MS: Mississippi River Comm.

———. 1951. Loess and Quaternary geology of the lower Mississippi Valley. *Jour. Geology* 59:333–56.

Fiske, R. S., Hopson, C. A., and Waters, A. C. 1963. Geology of Mount Rainier National Park. *U.S. Geol. Survey Prof. Paper* 444.

Fitze, P. 1971. Messungen von Bodenbewegungen auf West-Spitzbergen. *Geog. Helvetica* 26:148–52.

FitzGerald, D. M. 1984. Interactions between the ebb-tidal delta and landward shoreline; Price Inlet, South Carolina. *J. Sedy. Petrol.* 54:1303–18.

Fitzsimons, S. J. 1991. Supraglacial eskers in Antarctica. *Geomorphology* 4:293–99.

Flemal, R. C. 1976. Pingos and pingo scars: Their characteristics, distribution, and utility in reconstructing former permafrost environments. *Quat. Res.* 6:37–53.

Flemal, R. C., Hinkley, K. C., and Hesler, J. L. 1976. DeKalb mounds: A possible Pleistocene (Woodfordian) pingo field in north-central Illinois. *Geol. Soc. America Mem.* 136:229–50.

Fleming, R. W., and Johnson, A. M. 1975. Rates of seasonal creep of silty clay soil. *Quart. Jour. Engr. Geol.* 8:1–29.

Fleming, R. W., Johnson, R. B., and Scuster, R. L. 1987. The reactivation of the Manti Landslide, Utah. *U.S. Geol. Survey Prof. Paper* 1311–A, 22p.

Fletcher, C. H. 1992. Sea level trends and physical consequences. *Earth Sci. Rev.* 33:73–109.

———. 1971. *Glacial and Quaternary geology.* New York: John Wiley & Sons.

Florsheim, J. L., and Keller, E. A. 1987. Influence of fire and channel morphology on fluvial sediment transport and deposition. In *Erosion and sedimentation of the Pacific Rim,* pp. 279–80. Ed. by R. L. Beschta et al. Int. Assoc. Hydr. Sci. Pub. 165.

Florsheim, J. L., Keller, E. A., and Best, D. W. 1991. Fluvial sediment transport in response to moderate storm flows following chaparral wildfire, Ventura County, southern California. *Geol. Soc. Amer. Bull.* 103:504–11.

Foley, M. G. 1978. Scour and fill in steep, sand-bed ephemeral streams. *Geol. Soc. America Bull.* 89:559–70.

———. 1980a. Bed-rock incision by streams. *Geol. Soc. America Bull.* 91 (pt. 2):2189–213.

———. 1980b. Quaternary diversion and incision, Dearborn River, Montana. *Geol. Soc. America Bull.* 91 (pt. 2):2152–88.

———. 1975. Glacial deposits identified by chattermark trails in detrital garnets. *Geology* 3:473–75.

Folk, R. L., and Patton, E. B. 1982. Buttressed expansion of granite and development of grus in central Texas. *Zeit. F. Geomorph.* 26:17–32.

Foose, R. M. 1967. Sinkhole formation by groundwater withdrawal: Far West Rand, South Africa. *Science* 157:3792:1045–48.

Forbes, J. D. 1843. *Travels through the Alps of Savoy.* Edinburgh: Oliver and Boyd.

Ford, D., and Ewers, R. 1978. The development of limestone cave systems in length and depth. *Can. J. Earth Sci.* 15:1783–98.

Ford, D. C. 1987. Effects of glaciations and permafrost upon the development of karst in Canada. *Earth Surface Process and Landforms* 12:507–21.

Ford, D. C., and Williams, P. 1989. *Karst geomorphology and hydrology.* Winchester, MA: Unwin Hyman, Ltd.

Ford, D. C., Palmer, A. N., and White, W. B. 1988. Landform development; Karst. In *Hydrogeology.* Ed. by W. Bach, J. S. Rosenshein, and P. R. Geolen. Boulder, Geological Society of America, The Geology of North America, V. 0–2:401–12 p.

Forland, K. S., Forland, T., and Ratkje, S. K. 1988. Frost heave. In *Permafrost, fifth international conference proceedings,* pp. 344–48. Ed. by K. Senneset. Trondheim: Tapir Publishers.

Forman, S., and Maat, P. 1990. Stratigraphic evidence for late Quaternary dune activity near Hudson on the Piedmont of northern Colorado. *Geology* 18:745–8.

Forman, S., Goetz, A., and Yuhas, R. 1992. Large-scale stabilized dunes on the High plains of Colorado: Understanding the landscape response to Holocene climates with the aid of images from space. *Geology* 20:145–48.

Formento-Trigilio, M., and Pazzaglia, F. 1998. Tectonic geomorphology of the Sierra Nacimento: Traditional and new techniques in assessing long-term landscape evolution in the southern Rocky Mountains. *Jour. Geol.* 106:433–53.

Foster, I. D. L., Dearing, J. A., and Grew, R. 1990. The sedimentary data base; an appraisal of lake and reservoir sediment based on studies of sediment yield. In *Erosion, Transport, and Depositional Processes,* pp. 19–43. Ed. by D. E. Walling, A. Yair, and S. Berkowiczs. IAHS-AISH Pub. 189.

Fournier, M. F. 1960. *Climat et erosion.* Paris: Presses Univ. France.

Fowler, A. C. 1986. Sub-temperate basal sliding. *Jour. Glaciol* 32:3–5.

———. 1987. A theory of glacier surges. *Journal of Geophysical Research,* 92:9111–20.

Francis, P. 1993. *Volcanoes: A planetary perspective.* New York: Oxford Univ. Press.

Francou, B. 1990. Stratification mechanisms in slope deposits in high subequatorial mountains. *Permafrost and Periglac.* Proc. 1:249–63.

Frank, A., and Kocurek, G., 1996a. Towards a model for airflow on the lee side of aeolian dunes. *Sedimentol.* 43:451–58.

———. 1996b. Airflow up the stoss side of sand dunes: Limitations of current understanding. *Geomorph.* 17:47–54.

Franklin, J. A. 1984. Slope instrumentation and monitoring. In *Slope instability,* pp. 143–69. Ed. by D. Brunsden and D. B. Prior. New York: John Wiley.

Free, G. R. 1960. Erosion characteristics of rainfall. *Agri. Engineering* 41:447–49, 455.

Freer, J., McDonnel, J., Beuen, K. J., Brammer, D., Burns, D., Hooper, R. P., and Kendal, C. 1997. Topographic controls on subsurface storm flow at the hillslope scale for two hydrologically distinct small catchments. *Hydrological Processes* 11:1347–52.

Freeze, R. A. 1980. A stochastic-conceptual analysis of rainfall-runoff processes on a hillslope. *Water Resour. Res.* 16:391–408.

French, H. M. 1974. Mass-wasting at Sachs Harbour, Barks Island, N.W.T., Canada. *Arc. Alp. Res.* 6:77–78.

———. 1976. *The periglacial environment.* London: Longman.

French, H. M., and Guglielmin, M. 1999. Observations on the ice-marginal, peroglacial geomorphology of Terra Nova Bay, Northern Victoria Land, Antarctica. *Permafrost and Periglacial Processes* 10:331–47.

Friedkin, J. F. 1945. A laboratory study of the meandering of alluvial rivers. U.S. Army Corps Engrs., *U.S. Waterways Eng. Exp. Sta.*

Frye, J. C., and Leonard, A. R. 1954. Some problems of alluvial terrace mapping. *Am. Jour. Sci.* 252:242–51.

Fucella, J. E., and Dolan, R. 1996. Magnitude of subaerial beach disturbance during northeast storms. *Jour. Coastal Res.* 12:420–29.

Fuller, M. L. 1914. The geology of Long Island, New York. *U.S. Geol. Survey Prof. Paper* 82.

Furuya, G., Sassa, K., Hiura, H., and Fukuoka, H. 1999. Mechanism of creep movement caused by landslide activity and underground erosion in cystalline schist, Shikoku Island, Southwestern Japan. *Engineering Geology.* 53:311–25.

Galvin, C. J. 1967. Longshore current velocity: A review of theory and data. *Revs. in Geophys.* 5:3:287–304.

———. 1968. Breaker type classification on three laboratory beaches. *Jour. Geophys. Research* 73:3651–59.

Galvin, C. J., and Eagleson, P. S. 1965. Experimental study of longshore currents on a plane beach. U.S. Army Corps Engrs., *Coastal Eng. Res. Center Tech. Memo* 10.

Gams, I. 1993. Origin of the term "karst," and the transformation of the Classical Karst (kras). *Environmental Geology,* 21:110–114.

Gams, I. 1969. Some morphological characteristics of the Dinaric karst. *Geogr. Jour.* 135:563–72.

———. 1978. The polje: The problem of definition. *Aeits. Geomorph.* 22:170–81.

Gardiner, V. 1990. Drainage basin morphometry. In *Geomorphological techniques,* pp. 71–81. Ed. by A. S. Goudie. London: Unwin Hyman.

Gardner, J. S. 1979. The movement of material on debris slopes in the Canadian Rocky Mountains. *Zeit. F. Geomorph.* 23:45–57.

Gardner, J. S., and Bajewsky, I. 1987. Hilda rock glacier stream discharge and sediment load characteristics, Sunwapta Pass area, Canadian Rocky Mountains. In *Rock glaciers,* pp. 265–87. Ed. by J. R. Giardino, J. F. Shroder, and J. D. Vitek. Boston: Allen and Unwin.

Gardner, R., and Walsh, N. 1996. Chemical weathering of metamorphic rocks from low elevations in the southern Himalaya. *Chem. Geol.* 127:161–76.

Gardner, T., and 6 others. 1992. Quaternary uplift astride the aseismic Cocos Ridge, Pacific Coast, Costa Rica. *Geol. Soc. Amer. Bull.* 104:219–32.

Gardner, T., Jorgensen, D., Schumm, D., and Lemeieux, C. 1987. Geomorphic and tectonic process rates: Effects of measured time interval, *Geology* 15:259–61.

Gardner, T. W. 1983. Experimental study of knickpoint and longitudinal profile evolution in cohesive, homogeneous material. *Geol. Soc. America Bull.* 94:664–72.

Gardner, T. W., and others. 1991. A periglacial stratified slope deposit in the Valley and Ridge Province of central Pennsylvania, USA: Sedimentology, stratigraphy, and geomorphic evolution. *Permafrost and Periglac.* Proc. 2: 141–62.

Garrels, R. M., and Mackenzie, F. T. 1971. *Evolution of sedimentary rocks.* New York: Norton and Co.

Gates, W. C. B. 1987. The fabric of rockslide avalanche deposits. *Bull. Assoc. Engin. Geol.* 24:389–402.

Gatto, L. W., and Anderson, D. W. 1975. Alaskan thermokarst terrain and possible Martian analog. *Science* 188:255–57.

Geist, D., Naumann, T., and Larson, P. 1998. Evolution of Galapogas magmas: Mantle and crustal level fractionation without assimilation. *Jour. Petrology* 39, 953–71.

Geist, D., White, W., Naumann, T. and Reynolds, R. 1999. Illegitimate magmas of the Galapogas: Insights into mantle mixing and magma transport. *Geology* 27, 1103–06.

Gentilli, J. 1968. Exfoliation. In *Encyclopedia of geomorphology,* pp. 336–39. Ed. by R. W. Fairbridge. New York: Reinhold Book Corp.

Germanoski, D. 1989. The effects of sediment load and gradient on braided river morphology. Unpublished Dissertation. Colorado State University.

———. 1990. Comparison of bar-forming processes and differences in morphology in sand- and gravel-bed braided rivers. *Ecol. Soc. America Abs. with Programs.* Annual Mtg 27:A110.

———. 2000. Bar forming processes in gravel-bed braided rivers, with implications for small-scale gravel mining. In *Applying geomorphology to environmental management,* pp. 1–26. Ed. by D. J. Anthony, M. Harvy, J. Laronne, and P. Mosley.

Gerrard, A. J. 1981. *Soils and landforms.* London: Allen and Unwin Ltd.

Gerrard, J. 1993. Soil geomorphology, present dilemmas and future challenges. *Geomorphology* 7:61–84.

Giardino, J. R. 1983. Movement of ice-cemented rock glaciers by hydrostatic pressure; an example from Mount Mestas, Colorado. *Zeit. F. Geomorph.* 27:297–310.

Giardino, J. R., Shroeder, J. F., and Vitek, J. D., eds. 1987. *Rock glaciers.* Boston: Allen and Unwin.

Giardino, J. R., and Vitek, J. D. 1988. Rock glacier rheology: A preliminary assessment. In *Permafrost, fifth international conference proceedings,* pp. 744–48. Ed. by K. Senneset. Trondheim: Tapir Publishers.

Giardino, J. R., Vitek, J. D., and DeMorett, J. L. 1992. A model of water movement in rock glaciers and associated water characteristics. In *Periglacial geomorphology,* pp. 159–84. Ed. by J. C. Dixon and A. D. Abrahams. Chichester: John Wiley.

Gibbs, R. J. 1967. The geochemistry of the Amazon River system, part I. *Geol. Soc. America Bull.* 78:1203–32.

Gilbert, G. K. 1877. *Geology of the Henry Mountains (Utah).* U.S. Geog. and Geol. Survey of the Rocky Mtn. Region. Washington, D.C.: U.S. Govt. Printing Office.

———. 1890. Lake Bonneville. *U.S. Geog. Surv. Monograph* 1.

———. 1917. Hydraulic-mining debris in the Sierra Nevada. *U.S. Geol. Survey Prof. Paper* 105.

Gile, L., and Grossman, R. 1979. *The desert project soil monograph.* U.S. Dept. Agri., Soil Conserv. Serv.

Gile, L., Hawley, J., and Grossman, R. 1981. Soils and geomorphology in the Basin and Range area of Southern New Mexico—Guidebook to the Desert Project. *New Mexico Bur. Mines and Min. Res. Memo* 39.

Gile, L. H. 1966. Cambic and certain non-cambic horizons in desert soils of southern New Mexico. *Soil Sci. Soc. Am. Proc.* 30:773–81.

———. 1975. Holocene soils and soil-geomorphic relations in an arid region of southern New Mexico. *Quat. Res.* 5:321–60.

Gile, L. H., Peterson, F., and Grossman, R. 1965. The K horizon. A master soil horizon of carbonate accumulation. *Soil Sci.* 99:74–82.

———. 1966. Morphological and genetic sequences of carbonate accumulation in desert soils. *Soil Sci.* 101:347–60.

Gili, J. A., Corominas, J., and Rius, J. 2000. Using global positioning system techniques in landslide monitoring. *Engineering Geology* 55:167–92.

Gill, T., 1996. Eolian sediments generated by anthropogenic disturbance of playas: Human impacts on the geomorphic system, and the geomorphic effects on the human system. *Geomorph.* 17:207–28.

Gillette, D., 1999. Physics of aeolian movement emphasising changing of the aerodynamic roughness height by saltating grains (the Owen Effect). In *Aeolian environments, sediments and landforms,* pp. 129–42. Ed. by A. Goudie, I. Livingstone, and S. Stokes, Chichester: Wiley.

Gillette, D., and 5 others. 1997a. Relation of vertical flux of particles smaller than 10 μm to total aeolian horizontal mass flux at Owens Lake. *J. Geophys. Res.* 102 (D22), 26009–015.

Gillette, D., Hardebeck, E., and Parker, J., 1997b. Large-scale variability of wind erosion mass flux rates at Owens Lake: The role of roughness change, particle limitation, change of threshold friction velocity, and the Owen effect. *J. Geophys. Res.* 25:25989–98.

Gillieson, D. 1996. *Caves: Processes, Development, Management.* Oxford, Blackwell Publishers, p. 324.

Gilluly, J. 1949. The distribution of mountain-building in geologic time. *Geol. Soc. America Bull.* 60:561–90.

———. 1955. Geologic contrasts between continents and ocean basins. *Geol. Soc. America Spec. Paper* 62:7–18.

———. 1964. Atlantic sediments, erosion rates, and the evolution of the Continental Shelf—Some speculations. *Geol. Soc. America Bull.* 75:483–92.

———. 1969. Geological perspectives and the completeness of the geologic record. *Geol. Soc. America Bull.* 80:2303–12.

Gjessing, J. 1967. On plastic scouring and subglacial erosion. *Norsk. Geogr. Tidsskr.* 20:1–37.

Glasser, N. F.; Crawford, K. R.; Hambrey, J. J.; Bennett, M. R.; and Huddart, D. 1998. Lithological and structural controls on the surface wear characteristics of glaciated metamorphic bedrock surfaces; Ossian Sarsfjellet, Svalbard. *Jour. Geology* 106:319–329.

Glen, J. W. 1952. Experiments on the deformation of ice. *Jour. Glaciol.* 2:111–14.

———. 1955. The creep of polycrystalline ice. *Proc. Royal Soc. London,* ser. A: 228:519–38.

———. 1958. Mechanical properties of ice. I. The plastic properties of ice. *Philos. Mag.,* Suppl. 7:254–65.

———. 1987. Fifty years of progress in ice physics. *Journal of Glaciology,* special issue, p. 52–59.

Glen, J. W., and Lewis, W. V. 1961. Measurements of side-slip at Austerdalsbreen, 1959. *Jour. Glaciol.* 3:1121.

Glennie, K. W. 1970. *Desert Sedimentary Environments.* Amsterdam: Elsevier.

Godfrey, A. E. 1997. Mass movement of Mancos Shale near Caineville, Utah; a 30-year record. *Geografesleu Annaler Ser. A.* 79:185–94.

Godfrey, P. J. 1976. Barrier beaches of the East Coast. *Oceanus* 19:27–40.

Gold, L. W., and Lachenbruch, A. H. 1973. Thermal conditions in permafrost—A review of North American literature. In *Proc. Permafrost 2nd Internat. Conf.* pp. 3–23. Natl. Acad. Sci.–Natl. Res. Council, Yakutsk, U.S.S.R., 1973.

Goldich, S. 1938. A study of rock weathering. *Jour. Geology* 46:17–58.

Goldthwait, R. P. 1951. Development of end moraines in east central Baffin Island. *Jour. Geology* 59:567–77.

———. 1969. Patterned soils and permafrost on the Presidential Range (abs). Paris. 8th INQUA Cong. *Resumes des Communications.* 150.

———. 1973. Jerky glacier motion and meltwater. *Int. Assoc. Sci. Hydrol. Bull.* 95:183–88.

———. 1976. Frost sorted patterned ground: A review. *Quat. Res.* 6:27–35.

———. 1979. Giant grooves made by concentrated basal ice streams. *Jour. Glaciol.* 23:297–307.

———, ed. 1971. *Till: A symposium.* Columbus: Ohio State Univ. Press.

Gomez, B., and Church, M. 1989. An assessment of bed load sediment transport formule for gravel bed rivers. *Water Resour. Res.* 25:1161–86.

Gomez, B., and Marron, D. 1991. Neotectonic effects on sinuosity and channel migration, Belle Fourche River, Western South Dakota. *Earth Surf. Proc. Ldfrms.* 16:227–35.

Gomez, B., and 4 others. 1995. Sediment characteristics of an extreme flood: 1993 upper Mississippi River valley. *Geology* 23:963–66.

Gomez, B., and 3 others. 1997. Floodplain sedimentation and sensitivity: Summer 1993 flood, upper Mississippi River valley. *Earth Surf. Proc. Landfms.* 22:923–36.

———. 1998. Floodplain construction by recent, rapid vertical accretion: Waipaoa River, New Zealand. *Earth Surf. Proc. Landfms.* 23:405–13.

Gordon, J. E. 1981. Ice-scoured topography and its relationship to bedrock structure and ice movements in parts of northern Scotland and West Greenland. *Geogr. Annlr.* 63A:55–65.

Gordon, M., Tracey, J., and Ellis, M. 1958. Geology of the Arkansas.

Gornitz, V., and Kanciruk, P. 1989. Assessment of global hazards from sea level rise. *Coastal Zone '89* 1345–59.

Gorrel, G., and Shaw, J. 1991. Deposition in an esker, bead and fan complex, Lanark, Ontario, Canada. *Sedimentary Geology.* 72:285–314.

Gorsline, D. S. 1966. Dynamic characteristics of west Florida Gulf Coast beaches. *Marine Geol.* 4:187–206.

Gottsfield, A. S., and Gottsfield, L. M. 1990. Floodplain dynamics of a wandering river dendrochronology of the Morice River, British Columbia, Canada. *Geomorphology* 3:159–79.

Goudie, A. S. 1997. Weathering processes. In *Arid Zone Geomorphology; process form and change in Drylands.* Ed by D. S. G. Thomas. Chichester: John Wiley.

Goudie, A. 1983. Dust storms in space and time. *Prog. Phys. Geogr.* 7:502–29.

Goudie, A. 1999. Wind erosional landforms: yardangs and pans. In *Aeolian environments, sediments and landforms,* pp. 167–80. Ed. by A. Goudie, I. Livingstone, and S. Stokes, Wiley, Chichester.

Goudie, A., Livingstone, I., and Stokes, S., eds. 1999. Aeolian environments, sediments and landforms. Chichester: Wiley.

Goudie, A., and Middleton, N. 1992. The changing frequency of dust storms through time. *Climatic Change* 20:197–225.

Goudie, A., and Wells, G., 1995. The nature, distribution and formation of pans in arid zones. *Earth Sci. Rev.* 38, 1–69.

Goudie, A. S., Viles, H. A., and Parker, A. G. 1997. Monitoring of rapid salt weathering in the central Namib Desert using limestone blocks. *Jour. Arid Environments* 37:581–98.

Goudie, A. S. 1986. *Environmental change.* Oxford: Clarendon Press.

———. 1989. Weathering processes. In *Arid Zone Geomorphology,* pp. 1–12. Ed. by D. S. G. Thomas, New York: John Wiley.

Gow, A. J., and Williamson, T. 1976. Rheological implications of the internal structure and crystal fabrics of the West Antarctic ice sheet as revealed by deep core drilling at Byrd Station. *Geol. Soc. America Bull.* 87:1665–77.

Graf, J. B.; Webb, R. H.; and Hereford, R. 1991. Relation of sediment load and floodplain formation to climatic variability, Paria River drainage basin, Utah and Arizona. *Geol. Soc. Amer. Bull.* 103:1405–15.

Graf, W. L. 1976. Cirques as glacier locations. *Arc. Alp. Res.* 8:79–90.

———. 1979. The development of montane arroyos and gullies. *Earth Surf. Proc. and Landforms* 4:1–14.

———. 1988. Definition of flood plain along arid-region rivers. In *Flood Geomorphology,* Ed. by V. Baker, R. Kochel, J. B. Graf, R. H. Webb, and R. Hereford. 1991. Relation of sediment load and floodplain formation to climatic variability, Paria River drainage basin, Utah and Arizona. *Geol. Soc. Amer. Bull.* 103:1405–15. Patten, P. New York; Wiley-Interscience. 231–42.

Graham, J. 1984. Methods of stability analysis. In *Slope instability,* pp. 171–215. Ed. by D. Brunsden and D. B. Prior, New York: John Wiley.

Grainger, P., and Kalaugher, P. G. 1987. Intermittent surging movements of a coastal landslide. *Earth Surf. Proc. and Landforms* 12:597–603.

Grant, G., and Swanson, F. 1995. Morphology and processes of valley floors in mountain streams, western Cascades, Oregon. In *Natural and anthropogenic influences in fluvial geomorphology.* Am. Geophys. Un., Geophys. Monograph 89, 83–101.

Grant, G. E., and Wolff, A. L. 1991. Long-term patterns of sediment transport after timber harvest, western Cascade Mountains, Oregon, USA. In *Sedimentation and Stream Water Quality in a Changing Environment: Trends and Explanation,* pp. 31–40. Ed. by W. E. Peters and D. E. Walling. IAHS Pub. 203.

Grant, G. E., Swanson, F. J., and Wolman, M. G. 1990. Pattern and origin of stepped-bed morphology in high-gradient streams, western Cascades, Oregon. *Geol. Soc. America Bull.* 102:340–52.

Gravenor, C. P. 1953. The origin of drumlins. *Am. Jour. Sci.* 251:674–81.

———. 1982. Chattermarked garnets in Pleistocene glacial sediments. *Geol. Soc. America Bull.* 93:751–58.

Gravenor, C. P., and Kupsch, W. O. 1959. Ice disintegration features in western Canada. *Jour. Geology* 67:48–64.

Gravenor, C. P., and Meneley, W. A. 1958. Glacial flutings in central and northern Alberta. *Am. Jour. Sci.* 256:715–28.

Gray, J. M. 1982. Unweathered, glaciated bedrock on an exposed lake bed in Wales. *Jour. Glaciol.* 28:483–97.

———. 1992. Scarisdale: P-forms. In *The south-west Scottish highlands: Field guide,* pp. 85–88. Ed. by M. J. C. Walker, J. M. Gray, and J. J. Lowe. Cambridge: Quaternary Research Association.

Gray, W. M. 1965. Surface spalling by thermal stresses in rocks. In *Rock Mechanics Symposium.* Toronto: Proc. Ottawa, Can. Dept. Mines and Tech. Surveys.

Greeley, R., and Iversen, J. 1985. *Wind as a geological process on Earth, Mars, Venus and Titan.* Cambridge: Cambridge Univ. Press.

Greeley, R., Christensen, P., and Carrasco, R. 1989. Shuttle radar images of wind streaks in the Altimplano, Bolivia. *Geology* 17:665–8.

Greeley, R., and others, 1997. Applications of spaceborne radar laboratory data to the study of aeolian processes. *J. Geophys. Res.,* 102 (E5):10971–83.

Green, J., and Short, N. 1971. *Volcanic Landforms and Surface Features.* New York: Springer-Verlag.

Greenwood, B., and Davidson-Arnott, R. G. D. 1979. Sedimentation and equilibrium in wave formed bars. *Can. J. Earth Sci.* 16:312–22.

Gregory, D. I., and Schumm, S. A. 1987. The effect of active tectonics on alluvial river morphology. In *River channels: Environment and process,* pp. 41–68. Ed. by K. Richards. London: Basil Blackwell.

Gregory, K. J., ed. 1983. *Background to paleohydrology, a perspective.* New York: John Wiley.

Gregory, K. J., and Walling, D. E. 1973. *Drainage basin form and process.* New York: Halsted Press.

Griggs, D. 1936a. The factor of fatigue in rock exfoliation. *Jour. Geology* 44:783–96.

———. 1936b. Deformation of rocks under high confining pressures. *Jour. Geology* 44:541–77.

Grim, R. 1962. *Applied clay mineralogy.* New York: McGraw-Hill.

Gromko, G. J. 1974. Review of expansive soils. *J. Geotech. Div. Amer. Soc. Civ. Engin.* 6:667–87.

Grove, C. G. and Howard, A. D. 1994. Minimum hydrochemical conditions allowing limestone cave development. *Water Res. Research.* 30:607–615.

Grove, J. M. 1960. The bands and layers of Vesl-Skautbreen. In *Norwegian Cirque Glaciers,* pp. 11–23. Ed. by W. V. Lewis. Royal Geog. Soc. Res. Ser. 4.

Guccione, M., Miller, J., and Van Arsdale, R. 1994. Amount and timing of deformation near the St. Francis "Sunklands", northeastern Arkansas. Geol. Soc. Amer. abst. w. Prog. 26, 1, 7.

Guilcher, A. 1958. *Coastal and submarine morphology.* London: Methuen and Co.

———. 1994. Quaternary nivation, cryoplanation, and solifluction in western Brittany and the North Devonshire Hills. In *Cold climate landforms,* pp. 187–204. Ed. by D. J. A. Evans. New York: John Wiley.

Gunn, J. 1986. Solute processes and karst landforms. In *Solute Processes.* S. T. Trudgill (ed.), London: John Wiley and Sons, Ltd., p. 363–437.

Gunn, J. 1981. Hydrological processes in karst depressions. *Zeit. F. Geomorph.* 25:313–31.

———. 1983. Point recharge of limestone aquifers—A model from New Zealand karst. *J. Hydrol.* 61:19–29.

Gupta, A., and Fox, H. 1974. Effects of high-magnitude floods on channel form: A case study in Maryland Piedmont. *Water Resour. Res.* 10:499–509.

Gustavson, T. 1991. Buried vertisols in lacustrine facies of the Pliocene Fort Hancock Formation.

Gustavson, T. C. 1975. Microrelief (gilgai) structures on expansive clays of the Texas coastal plain—their recognition and significance in engineering construction. Austin: Univ. Texas, *Bur. Econ. Geol. Circ.* 75–87.

Gustavson, T. C., and Boothroyd, J. C. 1987. A depositional model for outwash, sediment sources, and hydrologic characteristics, Malaspina Glacier, Alaska: A modern analog of the southeastern margin of the Laurentide Ice Sheet. *Geological Society of America Bulletin* 99:187–200.

Gutenberg, B. 1941. Changes in sea level, postglacial uplift, and mobility of the earth's interior. *Geol. Soc. America Bull.* 52:721–72.

Guza, R. T., and Inman, D. L. 1975. Edge waves and beach cusps. *Jour. Geophys. Research* 80:21:2997–3012.

Haan, C. T., and Johnson, H. P. 1966. Rapid determination of hypsometric curves. *Geol. Soc. America Bull.* 77:123–25.

Habbe, K. A. 1992. On the origin of the drumlins of the South German Alpine Foreland (II): the sediments underneath. *Geomorphology* 6:69–78.

Hack, J. T. 1941. Dunes of the western Navajo country. *Geogr. Rev.* 31:240–63.

———. 1957. Studies of longitudinal stream profiles in Virginia and Maryland. *U.S. Geol. Survey Prof. Paper* 294-B:45–97.

———. 1960a. Relation of solution features to chemical character of water in the Shenandoah Valley, Virginia. *U.S. Geol. Survey Prof. Paper* 400-B:387–90.

———. 1960b. Interpretation of erosional topography in humid temperate regions. *Am. Jour. Sci.* (Bradley Vol.) 258-A:80–97.

———. 1965. Postglacial drainage evolution in the Ontonagan area, Michigan. *U.S. Geol. Survey Prof. Paper* 504-B:1–40.

———. 1966. Circular patterns and exfoliation in crystalline terrane, Grandfather Mountain area, North Carolina. *Geol. Soc. America Bull.* 77:975–86.

———. 1973. Stream-profile analysis and stream-gradient index. *U.S. Geol. Survey Jour. Research* 1:421–29.

Hack, J. T., and Goodlett, J. C. 1960. Geomorphology and forest ecology of a mountain region in the central Appalachians. *U.S. Geol. Survey Prof. Paper* 347.

Hadley, R. F. 1961. Influence of riparian vegetation on channel shape, northeastern Arizona. *U.S. Geol. Survey Prof. Paper* 424-C:30–31.

Hadley, R. F., and Schumm, S. A. 1961. Sediment sources and drainage basin characteristics in upper Cheyenne River basin. *U.S. Geol. Survey Water Supply Paper* 1531-B:137–96.

Haeberli, W. 1985. Creep of mountain permafrost: Internal structure and flow of alpine rock glaciers. Mitteilungen der Versuchsanstalt fur Wasserbau., *Hydrologie und Glaziologie* 77: 142 p.

———. 1992. Construction, environmental problems and natural hazards in periglacial mountain belts. *Permafrost and Periglacial Processes* 3:111–24.

———, ed. 1990. Pilot analysis of permafrost cores from the active rock glacier Murtel I, Piz Corvatsch, Eastern Swiss Alps. Arbeitsheft 9, Versuchsanstalt fur Wasserbau, *Hydrologie und Glaziologie.* Zurich: ETH, 38p.

Hagerty, D. J. 1980. Multifactor analysis of bank caving along a navigable stream. In *Natl. Waterways Roundtable Proc.* U.S. Army Engr. Water Res. Support Ctr., Inst. for Water Resources, IWR-80-1:463–92.

Hagerty, D. J., and Hamel, J. V. 1989. Geotechnical aspects of river bank erosion. *Proceedings of the National Conference on Hydraulic Engineering.* New Orleans, Aug. 14–18.

Haig, M. 1979. Ground retreat and slope evolution on regraded surface-mine dumps, Waunafon, Gwent. *Earth Surf. Proc. and Landforms* 4:183–89.

Haigh, M. J., and Wallace, W. L. 1982. Erosion of strip-mine dumps in LaSalle County, Illinois: Preliminary results. *Earth Surf. Proc. and Landforms* 7:79–84.

Hairsine, P. B., and Rose, C. W. 1991. Rainfall detachment and deposition: Sediment transport in the absence of flow-driven processes. *Soil Sci. Soc. Amer. J.* 55:320–24.

Halfman, J. D., and Johnson, T. C. 1988. High resolution record of cyclic climatic change during the past 4 ka from Lake Turkana, Kenya. *Geology* 16:496–500.

Hall, K. J. 1992. Mechanical weathering in the Antarctic: A maritime perspective. In *Periglacial geomorphology,* pp. 103–24. Ed. by J. C. Dixon and A. D. Abrahams. Chichester: John Wiley.

———. 1997. Observations on cryoplanation benches in Antarctica. *Antarctic Science* 9:181–87.

———. 1998. Rock temperatures and implications for cold region weathering. New data from Rothera, Adelaide Island, Antarctica. *Permafrost and Periglacial Processes* 9:47–55.

Hallberg, G. R. 1979. Wind-aligned drainage in loess in Iowa. *Iowa Acad. Sci. Proc.* 86:4–9.

Hallet, B. 1981. Glacial abrasion and sliding: their dependence on the debris concentration in basal ice. *Annals of Glaciology,* 2:23–28.

Hallet, B. 1976a. Deposits formed by subglacial precipitation of $CaCO_3$. *Geol. Soc. America Bull.* 87:1003–15.

———. 1976b. The effect of subglacial chemical processes on glacier sliding. *Jour. Glaciol.* 17:209–21.

———. 1979a. A theoretical model of glacial abrasion. *Jour. Glaciol.* 23:39–50.

———. 1979b. Subglacial regelation water film. *Jour. Glaciol* 23:321–34.

———. 1981. Glacial abrasion and sliding: their dependence on the debris concentration in basal ice. Annals of Glaciology, 2:23–28.

———. 1996. Glacial quarrying: a simple theoretical model. *Annals of Glaciology* 22:1–8.

———. 1998. Measurement of soil motion in sorted circles, western Spitsbergen. *7th International Permafrost Conf. Proc.,* 415–420.

Hallet, B., Anderson, S. P., Stubbs, C. W., and Gregory, E. C. 1988. Surface soil displacements in sorted circles, western Spitzbergen. In *Permafrost, fifth international conference proceedings,* pp. 779–85. Ed. by K. Senneset. Trondheim: Tapir Publishers.

Hallet, B., and Prestrud, S. 1986. Dynamics of periglacial sorted circles in western Spitzbergen. *Quaternary Res.* 26:81–99.

Hallet, B., and Waddington, E. D. 1992. Buoyancy forces induced by freeze thaw in the active layer: Implications for diapirism and soil circulation. In *Periglacial geomorphology,* pp. 251–80. Ed. by J. C. Dixon and A. D. Abrahams. Chichester: John Wiley.

Hamblin, W. K., and Christiansen, E. H. 1990. *Exploring the planets.* New York: Macmillan Pub.

Hambrey, M., and Alean, J. 1992. *Glaciers.* New York: Cambridge University Press, 208 p.

Hamilton, S. F., and Whalley, B. W. 1995. Rock glacier nomenclature: A reassessment. *Geomorphology* 14:73–80.

Hanks, T., Bucknam, R., Lajoie, K., and Wallace, R. 1984. Modification of wave-cut and faulting-controlled landforms. *Jour. Geophys. Res.* 89:5771–90.

Hansen, A. 1984, Landslide hazard analysis. In *Slope instability,* pp. 523–602. Ed. by D. Brunsden and D. B. Prior. New York: John Wiley.

Hansen, J. E., and others. 1984. Climate sensitivity: Analysis of feedback mechanisms. In *Climate processes and climate sensitivity,* pp. 130–63. Ed. by J. Hansen and T. Takahashi. Wash., DC: Amer. Geophys. Union.

Hanshaw, B., and Back, W. 1980. Chemical mass-wasting of the northern Yucatan Peninsula by groundwater dissolution. *Geology* 8:222–24.

Hanson, K., and 4 others. 1994. Correlation, ages, and uplift rates of Quaternary marine terraces: South-central coastal California. Geol. Soc. Amer. Spec. Paper 292, 45–71.

Hanvey, P. M. 1992. Variable boulder concentrations in drumlins indicating diverse accretionary mechanisms-examples from western Ireland. *Geomorphology* 6:69–78.

Harayama, S. 1992. Youngest exposed granitoid pluton on Earth: Cooling and rapid uplife of the Phocene-Quaternary Takidano Granodiorite in the Japan Alps, central Japan. *Geology* 20:657–60.

Harbor, J. M. 1990. A discussion on Hirano and Aniya's (1988, 1989) explanation of glacial-valley cross profile development. *Earth Surface Processes and Landforms* 15:369–77.

———. 1992. Numerical modeling of the development of U-shaped valleys by glacial erosion. *Geological Society of America Bulletin* 104:1364–75.

———. 1995. Development of glacial-valley cross sections under conditions of spatially variable resistance to erosion. *Geomorphology* 14:99–107.

Harden, J. 1982. A quantitative index of soil development from field descriptions: Examples from a chronosequence in central California. *Geodermo* 28:1–28.

Harden, J. W. 1987. Soils developed in granitic alluvium near Merced, California. *U.S. Geol. Surv. Bull.* 1590-A, 65 p.

Harpstead, M., and Hole, R. 1980. *Soil science simplified.* Ames: Iowa State Univ. Press.

Harris, C. 1973. Some factors affecting the rates and processes of periglacial mass movement. *Geogr. Annlr.* 55A:24–58.

———. 1987. Solifluction and related periglacial deposits in England and Wales. In *Periglacial Processes and Landforms in Britian and Ireland,* pp. 209–23. Ed. by J. Boardman. Cambridge: Cambridge Univ. Press.

———. 1990. Periglacial landforms. In *Natural Landscapes of Britain from the Air,* pp. 43–48. Ed. by N. Stephens. Cambridge: Cambridge Univ. Press.

Harris, S. A. 1983. Comparison of the climatic and geomorphic methods of predicting permafrost distribution in western Yukon Territory. In *Proc. Permafrost 4th Internat. Conf.* pp. 450–55. Natl. Acad. Sci.

———. 1994. Climatic zonality of periglacial landforms in mountain areas. *Arctic* 47:184–192.

———. 1998a. Non-sorted circles on Plateau Mountain, S.W. Alberta, Canada. *7th Internat. Permafrost Conf. Proc. Collection Nordicana* 57:441–48.

———. 1998b. A genetic classification of the palsa-like mounds in western Canada. *Biuleyn Peryglocjalny.* 37:115–29.

Harris, S. A., Cheng, G., Zhao, X., and Yangquin, D. 1998. Nature and dynamics of an active blockstream, KunlunPass, Quinghai Province. *China Geogr. Annaler Ser. A.* 80:123–33.

Harrison, A. E. 1964. Ice surges on the Muldrow Glacier, Alaska. *Jour. Glaciol.* 5:365–68.

Harrison, J. B. J., McFadden, L. D., and Weldon, R. J. III. 1990. Spatial soil variability in the Cajon Pass chronosequence: Implications for the use of soils as a geochronological tool. *Geomorphology* 3:399–416.

Harrison, W. D. 1975. Temperature measurements in a temperate glacier. *Jour. Glaciol.* 14:23–30.

Hart, J. K. 1998. The deforming bed/debris-rich basal ice continuum and its implications for the formation of glacial landforms (flutes) and sediments (melt-out till). *Quaternary Science Reviews* 17:737–54.

Harvey, A. 1984a. Aggradation and dissection sequences on Spanish alluvial fans: influence on morphological development. *Catena* 11:289–304.

———. 1984b. Debris flow and fluvial deposits in Spanish Quaternary alluvial fans: Implications for fan morphology. In *Sedimentology of gravels and conglomerates,* pp. 123–32. Ed. by E. Koster and R. Steel. Calgary, Can. Soc. Petrol. Geol., Mem. 10.

———. 1987. Alluvial fan dissection: relationship between morphology and sedimentation. In *Desert sediments: ancient and modern.* Ed. by L. Frostick and I. Reid. London, Geol. Soc. London Spec. Publ. 35, 87–103.

———. 1988. Controls on alluvial fan development: the alluvial fans of the Sierra de Carrascoy, Murcia, Spain. Catena Suppl. 13:123–37.

———. 1997. The role of alluvial fans in arid zone fluvial systems. In *Arid zone geomorphology: process, form and change in drylands,* pp. 231–59. Ed. by D. Thomas. Chichester: Wiley.

Harvey, A., and Wells, S. 1987. Response of Quaternary fluvial systems to differential epeirogenic uplift: Aguas and Feos river systems, southeast Spain. *Geology* 15:689–93.

———.1994. Late Pleistocene and Holocene changes in sediment supply to alluvial fan systems: Zzyzx, California. In *Environmental changes in drylands,* pp. 67–84. Chichester: Wiley. Ed. by A. Millington and K. Pye.

Harvey, A. M., Hitchcock, D. H., and Hughes, D. J. 1979. Event frequency and morphological adjustment of fluvial systems. In *Adjustments of the Fluvial System,* pp. 139–67. Ed. by D. D. Rhodes and G. P. Williams. Dubuque, IA: Kendall Hunt.

Harvey, M. D., and Schumm, S. A. 1987. Response of Dry Creek, California to land use change, gravel mining, and dam closure. In *Erosion and Sedimentation on the Pacific Rim,* pp. 451–60. IAHS-AISH Pub. 165.

Harwood, T. A. 1969. Some possible problems with pipelines in permafrost regions. *Proc. 3rd Canadian Conf. on Permafrost,* pp. 79–84. *Natl. Res. Council of Canada, Tech. Memo* 96.

Hassan, M. A., and Reid, I. 1990. The influence of microform bed roughness elements on flow and sediment transport in gravel bed rivers. *Earth Surf. Proc. and Landforms* 15:739–50.

Hastenrath, S., and Kruss, P. 1982. On the secular variation of ice flow velocity at Lewis Glacier, Mount Kenya, Kenya. *Jour. Glaciol* 28:333–39.

Hausenbuiller, R. 1972. *Soil science: Principles and practices.* Dubuque, IA: Wm. C. Brown Company Publishers.

Havholm, K., and Kocurek, G. 1988. A preliminary study of the dynamics of modern draa, Algodones, southeastern California. *Sedimentology* 35:649–69.

Hayden, B. P., Santos, M. C. V., Shao, G., and Kochel, R. C. 1995. Geomorphological controls on coastal vegetation at the Virginia Coast Reserve, *Geomorphology* 13:283–300. In *BioGeomorphology, terrestrial and freshwater systems.* Ed. by C. R. Hupp, W. R., Osterkamp, and A. D. Howard.

Hayes, M. O. 1975. Morphology of sand accumulations in estuaries. In *Estuarine research, v. 2: Geology and engineering,* pp. 3–22. Ed. by L. E. Cronin.

Hayley, D. W. 1988. Maintenance of a railway grade over permafrost in Canada. In *Permafrost, fifth international conference proceedings,* pp. 1413–16. Ed. by K. Senneset. Trondheim: Tapir Publishers.

Haynes, C. 1989. Regnold's barchan: A 57-year record of dune movement in the eastern Sahara and implication for dune origin and palaeoclimate since Neolithic times. *Quat. Res.* 32:153–67.

Haynes, V. C. 1982. The Darb El-Arba'in desert: A product of Quaternary climatic change. In *Desert landforms of southwest Egypt: A basis for comparison with Mars,* pp. 91–118. Ed. by F. El-Baz and T. Maxwell. NASA, CR-3611.

Hays, J., Imbrie, J., and Shackleton, N. 1976. Variations in the Earth's orbit: Pacemaker of the ice ages. *Science* 194:1121–32.

Heim, A. 1932. *Bergsturz und Menschenleben.* Zurich: Fretz and Wasmuth Verlag.

Helgeson, J., Murphy, W., and Aagaard, P. 1984. Thermodynamic and kinetic constraints on reaction rates among numerals and aqueous solution. II Rate constants, effective surface area, and the hydrolysis of feldspar. *Geochim. Cosmochim. Acta.* 48:2405–32.

Heller, P. L. 1981. Small landslide types and controls in glacial deposits: Lower Skagit River drainage, northern Cascade Range, Washington. *Environ. Geol.* 3:221–28.

Hennion, F. B., and Lobacz, E. F. 1973. Corps of engineers technology related to design of pavements in areas of permafrost. In *Proc. Permafrost 2nd Internat. Conf.,* pp. 426–29. Natl. Acad. Sci.–Natl. Res. Council, Yakutsk, U.S.S.R., 1973.

Hey, R. D., and Thorne, C. R. 1975. Secondary flows in river channels. *Area* 7:191–95.

Hickin, E. J. 1974. The development of meanders in natural river channels. *Am. Jour. Sci.* 274:414–42.

Hickin, E. J., and Nanson, G. C. 1975. The character of channel migration on the Beatton River, northeast British Columbia, Canada. *Geol. Soc. America Bull.* 86:487–94.

Hicks, S. D., and Crosby, J. E. 1974. Trends and variability of yearly mean sea level: *NOAA/NOS Tech. Memo.* NOS 13. Rockville, 14p.

Hill, A. R. 1971. The internal composition and structure of drumlins in north Down and south Antrim, northern Ireland. *Geogr. Annlr.* 53:14–31.

Hill, C. A. 1981. Speleogenesis of Carlsbad Caverns and other caves of the Guadalupe Mountains. *Proc. 8th Intl. Cong. Speleol.* Bowling Green, KY, pp. 143–44.

Hinkel, K. M. 1988. Frost mounds formed by degradation at Slope Mountain, Alaska, USA. *Arctic and Alpine Research* 20:76–85.

Hinkel, K. M., F. E. Nelson, and S. I. Outcalt. 1987. Frost mounds at Toolik Lake, Alaska, *Physical Geography* 8:148–59.

Hirano, M., and Aniya, M. 1988. A rational explanation of cross-profile morphology for glacial valleys and of glacial valley development. *Earth Surface Processes and Landforms* 13:707–16.

————. 1989. A rational explanation of cross-profile morphology for glacial valleys and of glacial valley development: A further note. *Earth Surface Processes and Landforms* 14:173–74.

Hirschboeck, K. K. 1987. Flood Hydroclimatology. In *Flood Geomorphology*, pp. 27–49. Ed. by V. R. Baker, R. C. Kochel, and P. C. Patton. New York: John Wiley and Sons.

Hjulström, F. 1939. Transportation of detritus by moving water. *In Recent marine sediments: A symposium*, edited by P. Trask. Tulsa, OK: Am. Assoc. Petroleum Geologists.

Hobbs, H. 1999. Origin of driftless area by subglacial drainage—a new hypothesis. In *Glacial processes past and present*, pp. 93–102. Ed. by D. M. Mickelson and J. W. Attig. Geological Society of America Special Paper 337.

Hoch, A., Reddy, M., and Drever, J. 1999. Importance of mechanical disaggregation in chemical weathering in a cold alpine environment, San Juan Mountains, Colorado. *Geol. Soc. Amer. Bull.* 111:304–14.

Hock, R., and Hooke, R. L. 1993. Evolution of the internal drainage system in the lower part of the ablation area of Storglaciaren, Sweden. *Geological Society of America Bulletin* 105:537–46.

Hodge, S. M. 1974. Variations in the sliding of a temperate glacier. *Jour. Glaciol.* 13:349–69.

————. 1976. Direct measurement of basal water pressures: A pilot study. *Jour. Glaciol.* 16:205–17.

Hodgkin, E. P. 1964. Rate of erosion of intertidal limestone. *Zeit. F. Geomorph.* 8:385–92.

Hoey, T. 1992. Temporal variation in bedload transport rates and sediment storage in gravel river beds. *Progress in Physical Geography* 16:319–38.

Hokkaido Tsunami Survey Group, 1993. Tsunami devastates Japanese coastal region. *EOS* 74(37): 417.

Holbrook, W., Mooney, W., and Christensen, N. 1992. The seismic velocity structure of the deep continental crust. In *Continental lower crust*. Ed by D. Fountain et al. pp. 1–43, Elsevier.

Holdren, G., and Speyer, P. 1985. Reaction rate surface area relationships during the early stages of weathering—I. Initial observation. *Geochim. Cosmochim. Acta.* 49:674–81.

Holliday, V. 1989a. Middle Holocene drought on the southern High Plains. *Quaternary Research* 31:74–82.

————. 1989b. The Blackwater Draw Formation (Quaternary): A 1.4 plus-m.y. record of eolian sedimentation and solid formation on the Southern High Plains. *Geol. Soc. America. Bull.* 101:1598–607.

Holland, H. D., Kirsipu, T. V., Huebner, J. S., and Oxburgh, V. M. 1964. On some aspects of the chemical evolution of cave waters. *Jour. Geology* 72:36–67.

Holland, K. T. 1998. Beach cusp formation and spacing at Duck, USA. *Continental Shelf Res.* 18:1081–98.

Holman, R. A. 1983. Edge waves and the configuration of the shoreline. In *CRC Handbook of Coastal Processes and Erosion*, pp. 21–33. Ed. by P. Komar. Boca Raton, FL: CRC Press.

Holmes, C. D. 1947. Kames. *Am. Jour. Sci.* 245:240–49.

————. 1960. Evolution of till-stone shapes, central New York. *Geol. Soc. America Bull.* 71:1645–60.

Holmes, G. W., Hopkins, D. M., and Foster, H. L. 1968. Pingos in central Alaska. *U.S. Geol. Survey Bull.* 1241-H.

Holtedahl, H. 1967. Notes on the formation of fjords and fjord valleys. *Geogr. Annlr.* 49:188–203.

Hooke, J. M. 1987. Changes in meander morphology. In V. Gardiner (ed.), *International Geomorphology*, Part I., 591–609 p.

Hooke, R. L. 1989. Englacial and subglacial hydrology: A qualitative review. *Arctic and Alpine Research* 21:221–33.

————. 1991. Positive feedbacks associated with erosion of glacial cirques and overdeepenings. *Geological Society of America Bulletin*, 103:1104–08.

————. 1998. *Principles of glacier mechanics*. Upper Saddle River: Prentice Hall.

Hooke, R. LeB. 1967. Processes on arid-region alluvial fans. *Jour. Geology* 75:438–60.

————. 1968. Steady-state relationships on arid-region alluvial fans in closed basins. *Am. Jour. Sci.* 266:609–29.

————. 1972. Geomorphic evidence for Late Wisconsin and Holocene tectonic deformation, Death Valley, California. *Geol. Soc. America Bull.* 83:2073–97.

————. 1975. Distribution of sediment transport and shear stress in a meander bend. *Jour. Geology* 83:543–66.

————. 1977. Basal temperatures in polar ice sheets: A qualitative review. *Quat. Res.* 7:1–13.

Hooke, R. LeB., and Hudleston, P. J. 1980. Ice fabrics in a vertical flow plane, Barnes Ice Cap, Canada. *Jour. Glaciol.* 25:195–214.

————. 1981. Ice fabrics from a borehole at the top of the south dome, Barnes Ice Cap, Baffin Island. *Geol. Soc. America Bull.* 92 pt. 1:274–81.

Hooke, R. LeB., and Rohrer, W. L. 1977. Relative erodibility of source-area rock types, as determined from second-order variations in alluvial-fan size. *Geol. Soc. America Bull.* 88:1177–82.

————. 1979. Geometry of alluvial fans: Effects of discharge and sediment size. *Earth Surf. Proc. and Landforms* 4:147–66.

Hopkins, D. M., and Sigafoos, R. S. 1951. Frost action and vegetation patterns on Seward Peninsula, Alaska. *U.S. Geol. Survey Bull.* 974-C:51–100.

Hoppe, G., and Schytt, V. 1953. Some observations on fluted moraine surfaces. *Geogr. Annlr.* 35:105–15.

Horn, D. P. 1992. A review and experimental assessment of equilibrium grain size and the ideal wave-graded profile. *Marine Geol.* 108:161–74.

Hornberger, G. M., Germann, P. F., and Beve, K. J. 1991. Throughflow and solute transport in an isolated sloping soil block in a forested catchment. *J. Hydrology* 124:81–99.

Horta, J. C. 1985. Salt heaving in the Sahara. *Geotechnique* 35:329–37.

Horton, R. E. 1933. The role of infiltration in the hydrological cycle. *Am. Geophys. Union Trans.* 14:446–60.

————. 1945. Erosional development of streams and their drainage basins. Hydrophysical approach to quantitative morphology. *Geol. Soc. America Bull.* 56:275–370.

Horton, T., and 3 others. 1999. Chemical weathering and lithologic controls of water chemistry in a high-elevation river system: Clarks Fork of the Yellowstone River, Wyoming and Montana. *Wat. Resources Res.* 35:1643–55.

Howard, A., and 3 others, 1978. Sand transport model of barchan dune equilibrium. *Sedimentol.* 25:307–38.

Howard, A. D., and Kerby, G. 1983. Channel changes in badlands. *Geol. Soc. America Bull.* 94:739–52.

Howard, A. D., and Kochel, R. C. 1988. Introduction to cuesta landforms and sapping processes on the Colorado Plateau. In *Sapping features of the Colorado Plateau; A comparative planetary geology field guide,* pp. 6–56. Ed. by A. D. Howard, R. C. Kochel, and H. H. Holt. Washington: N.A.S.A. SP-491.

Howard, A. D., Kochel, R. C., and Holt, H. E. 1988. *Sapping features of the Colorado Plateau; A comparative planetary geology field Guide.* Washington: N.A.S.A. SP-491.

Howard, Alan D. 1971. Simulation model of stream capture. *Geol. Soc. America Bull.* 82:1355–76.

Howard, Arthur D. 1967. Drainage analysis in geologic interpretation: A summation. *Am. Assoc. Petroleum Geologists Bull.* 51:2246–59.

Hoyt, J. H., and Henry, V. J. 1971. Origin of capes and shoals along the southeastern coast of the United States. *Geol. Soc. America Bull.* 82:59–66.

Hsu, K. J. 1975. Catastrophic debris streams (sturzstroms) generated by rockfalls. *Geol. Soc. America Bull.* 86:129–40.

Hubbard, B., and Sharp, M. J. 1989. Basal ice formation and deformation: A review. *Progress in physical geography* 13:529–558.

Hubbard, B., and Sharp, M. J. 1993. Weertman regelation, multiple refreezing effects and the isotopic evolution of the basal ice layer. *Jour. Glaciol* 39:275–91.

Hubbert, M. K. 1940. The theory of groundwater motion. *Jour. Geology* 48:785–944.

Huddart, D., and Lister, H. 1981. The origin of ice marginal terraces and contact ridges of East Kangerdluarssuk Glacier, SW Greenland. *Geogr. Annlr.* 63A:31–39.

Hughes, M. G., and Cowell, P. J. 1987. Adjustment of reflective beaches to waves. *J. Coastal Res.* 3:153–67.

Humlum, O. 1985. Genesis of an imbricate push moraine, Hofdabrekkujokull, Iceland. *Journal of Geology* 93:185–95.

———. 1988. Rock glacier appearance level and rock glacier initiation line altitude; a methodological approach to the study of rock glaciers. *Arctic and Alpine Res.* 20:160–78.

———. 1997. Active layer thermal regime of three rock glaciers in Greenland. *Permafrost and Periglacial Processes* 8:383–408.

———. 1998. The climatic significance of rock glaciers. *Permafrost and Peroglacial Processes* 9:375–95.

Hungr, O., Evans, S. G., and Hazzard, J. 1999. Magnitude and frequency of rock slides along the main transportation corridors of southwestern British Columbia. *Geotechnique* 36:226–38.

Hungr, O., Morgan, G. C., Van Dine, D. F., and Lister, D. R. 1987. Debris flow defenses in British Columbia. In *Debris Flows/ Avalanches: Process, Recognition, and Mitigation,* pp. 201–22. Ed. by J. E. Costa, and G. F. Wieczorek, Boulder, Geol. Soc. Amer. Rev. in *Engin. Geol* 7.

Huntley, D. A., and Bowen, A. J. 1973. Field observations of edge waves. *Nature* 243:160–61.

Huntoon, P. W. 1974. The karstic groundwater basins of the Kaibab Plateau, Arizona. *Water Resour. Res.* 10:579–90.

Hupp, C. R. 1983. Geo-botanical evidence of late Quaternary mass wasting in blockfield areas of Virginia. *Earth Surface Proc. and Landforms* 8:439–50.

———. 1984. Dendrogeomorphic evidence of debris flow frequency and magnitude at Mount Shasta, California. *Env. Geol. and Water Sci.* 6:21–28.

———. 1988. Plant ecological aspects of flood geomorphology and paleoflood history. In *Flood Geomorphology,* pp. 335–56. Ed. by. V. R. Baker, R. C. Kochel, and P. C. Patton, New York: John Wiley.

Hursh, C. R. 1936. Storm-water and absorption. *Am. Geophys. Union Trans.* 17:301–2.

Hursh, C. R., and Brater, E. F. 1941. Separating storm hydrographs from small drainage areas into surface and subsurface flow. *Am. Geophys. Union Trans.* 22:863–70.

Hutchinson, J. N. 1968. Mass movement. In *Encyclopedia of Geomorphology,* pp. 688–96. Ed. by R. W. Fairbridge. New York: Reinhold Book Corp.

Hutter, K. 1982. Glacier flow. *Am. Scientist* 70:26–34.

Hyatt, J. A. and Jacobs, P. M. 1996. Distribution and morphology of sinkholes triggered by flooding following Tropical Storm Alberto at Albany, Georgia, USA. *Geomorphology,* 17:305–316.

Hydrologic Engineering Center 1982. *HEC-2, Water Surface Profiles: Program User's Manual.* U.S. Army Corps Engineers, Davis, California.

Ichim, I. 1990. The relationship between sediment delivery ratio and stream order; a Romanian case study. In *Erosion, transport, and deposition processes,* pp. 79–86. Ed. by D. E. Walling, A. Yair, and S. Berkowicz. IAHS-AISH Pub. 189.

Ida, Y. 1995. Magma chamber and eruptive processes at Izu-Oshima volcano, Japan: Buoyancy control of magma migration. *Jour. Volcanol. Geothermal Res.* 66:53–67.

Iida, T., and Okunish, K. 1983. Development of hillslopes due to landslides. *Zeit. F. Geomorph. Supplmentbd.* 46:67–77.

Iken, A., and Bindschadler, R. A. 1986. Combined measurements of subglacial water pressure and surface velocity of Findelenglescher, Switzerland: Conclusions about drainage system and sliding mechanism. *Journal of Glaciology* 32:101–19.

Iken, A., Rothlisberger, H., Flotron, A., and Haeberli, W. 1983. The uplift of Unteraargletscher at the beginning of the melt season—A consequence of water storage at the bed? *Jour. Glaciol.* 19:28–47.

Imbrie, J., and 8 others. 1984. The orbital theory of Pleistocene climate; Support from a revised chronology of the marine 18O record. In *Milankovitch and climate, Part I,* pp. 269–305. Ed. by A. Berger et al. Dordrecht: Reidel Publ. Co.

Inglis, C. C. 1949. The behavior and control of rivers and canals. *Research Pub.* Poona, India, no. 13, 2 vols.

Inman, D. L., and Bagnold, R. A. 1963. Littoral processes. In *The Sea,* edited by M. N. Hill, 3:529–53. New York: Interscience.

Inman, D. L., and Frautschy, J. D. 1966. Littoral processes and the development of shoreline. *Proc. Coast. Eng. Speciality Conf.,* Am. Soc. Civil Engineers (Santa Barbara, CA.), pp. 511–36.

Inman, D. L., and Jenkins, S. A. 1999. Climate change and the episodicity of sediment flux of small California rivers. *Jour. Geology* 107:251–70.

Inman, D. L., and Nordstrom, C. E. 1971. On the tectonic and morphologic classification of coasts. *Jour. Geology* 79:1–21.

Ippen, A. T., ed. 1966. *Estuary and Coastline Hydrodynamics.* New York: McGraw-Hill.

Irfan, T. Y. 1998. Structurally controlled landslides in saprolitic soils in Hong Kong. *Geotechnical and Geological Engineering* 16:215–38.

Isherwood, D., and Street, A. 1976. Biotite-induced grussification of the Boulder Creek Granodiorite, Boulder County, Colorado. *Geol. Soc. America Bull.* 87:366–70.

Iverson, N. R. 1991a. Morphology of glacial striae: Implications for abrasion of glacier beds and fault surfaces. *Geological Society of America Bulletin* 103:1308–16.

———. 1997. The physics of debris flows. *Reviews of Geophysics* 35:245–96.

———. 1991b. Potential effects of subglacial water-pressure fluctuations on quarrying. *Journal of Glaciology* 37:27–36.

Iverson, N. R., Hanson, B., Hooke, R. L., and Jansson, P. 1995. Flow mechanism of glaciers on soft beds. *Science* 267:80–81.

Iverson, R. M., 2000. Landslide triggering by rain infiltration. *Water Resources Research.* V. 36: 1897–1910.

Ives, J. D., and Fahey, B. D. 1971. Permafrost occurrence in the Front Range, Colorado Rocky Mountains, U.S.A. *Jour. Glaciol.* 10:105–11.

Ives, J. D., and Krebs, P. V. 1978. Natural hazards research and land use planning responses in mountainous terrain: The town of Vail, Colorado Rocky Mountains, USA. *Arctic and Alpine Res.* 10:213–22.

Jackson, J. A., Gagnepain, J., Houseman, G., King, G. C. P., Papadimitriou, P., Soufleris, C., and Virieux, J. 1982. Seismicity, normal faulting, and the geomorphological development of the Gulf of Corinth (Greece): The Corinth earthquakes of February and March 1981. *Earth and Planat. Sci. Letters* 57:377–97.

Jackson, M., Hseung, Y., Corey, R., Evans, E., and Heuval, R. 1952. Weathering sequence of clay size minerals in soils and sediments. *Soil Sci. Soc. Am. Proc.* 16:3–6.

Jackson, T., and Keller, W. 1970. A comparative study of the role of lichens and "inorganic" processes in the chemical weathering of recent Hawaiian lava flows. *Am. Jour. Sci.* 269:446–66.

Jacobel, R. W. 1982. Short-term variations in velocity of South Cascade Glacier, Washington, U.S.A. *Jour. Glaciol* 28:325–32.

Jacobson, R. B., and Pomeroy, J. S. 1987. Slope failures in the Appalachian Plateau. In Mills, H. H. et al. 1987. Appalachian mountains and plateaus, pp. 21–29. In *Geomorphic Systems of North America.* Ed. by W. L. Graf. Boulder: *Geol. Soc. Amer., Centennial Spec.* Vol. 2.

Jacobson, R. B., Cron, E. D., and McGeehin, J. P. 1989. Slope movements triggered by heavy rainfall, November 3–5, 1985, in Virginia and West Virginia, USA. In *Landslide processes of the Eastern United States and Puerto Rico,* pp. 1–13. Ed. by A. P. Shultz. Boulder: *Geol. Soc. Amer. Spec. Paper* 236.

Jacobson, R. B., Miller, A. J., and Smith, J. A. 1989. The role of catastrophic geomorphic events in central Appalachian landscape evolution. In *Appalachian geomorphology.* Ed. by T. W. Gardner and W. D. Sevon. *Geomorphology* 2:257–84.

Jahn, A. 1960. Some remarks on evolution of slopes on Spitsbergen. *Zeit. F. Geomorph.* Suppl. 1:49–58.

Jansen, J. M. L., and Painter, R. B. 1974. Predicting sediment yield from climate and topography. *J. Hydrol.* 21:371–80.

Jansson, M. B. 1988. A global survey of sediment yield. *Geografiska Annaler.* 70A:81–98.

Jansson, P., and Hooke, R. L. 1989. Short-term variations in strain and surface tilt on Storglaciaren, Kebnekaise, northern Sweden. *Journal of Glaciology* 35:201–8.

Jarrett, R. D. 1991. Paleohydrology and its value in analyzing floods and droughts. In *National water summary 1988–89; Hydrologic events and floods.* Ed. by R. W. Paulson and others. *U.S. Geol. Survey Water Supply Paper.*

Jarvis, G. T., and Clarke, G. K. C. 1975. The thermal regime of Trapridge Glacier and its relevance to glacier surging. *Jour. Glaciol.* 14:235–49.

Jarvis, R. S., and Sham, C. H. 1981. Drainage network structure and the diameter-magnitude relation. *Water Resour. Res.* 17:1019–27.

Jennings, J. N. 1967. Some karst areas of Australia. In *Landform studies from Australia and New Guinea,* pp. 256–92. Ed. by J. N. Jennings and J. A. Mabbutt. Canberra.

———. 1971. *Karst.* Cambridge, MA: M.I.T. Press.

———. 1983. Karst landforms. *Am. Scientist* 71:578–86.

———. 1985. *Karst geomorphology.* New York: Basil Blackwell Inc.

Jennings, J. N., and Bik, M. J. 1962. Karst morphology in Australian New Guinea. *Nature* 194:1036–38.

Jenny, H. 1941. *Factors of soil formation.* New York: McGraw-Hill.

Jenny, H., and Leonard, C. 1939. Functional relationships between soil properties and rainfall. *Soil Sci.* 38:363–81.

Jensen, J. L., and Sorensen, M. 1986. Estimation of some aeolian saltation transport parameter: a re-analysis of William's data. *Sed.* 33:547–58.

Jeong, G. 1998a. Formation of vermicular kaolinite from halloysite aggregate in the weathering of plagioclase feldspar. *Clays & Clay Min.* 46:270–79.

Jeong, G. 1998b. Vermicular kaolinite epitactic on primary phyllosilicates in the weathering profiles of anorthosite. *Clays and Clay Min.* 46:509–20.

Jetchick, E., and Allard, M. 1990. Soil wedge polygons in northern Quebec: description and paleoclimatic significance. *Boreas* 19:33–67.

Johnson, A. 1970. *Physical process in geology.* San Francisco: Freeman, Cooper and Co.

Johnson, A. M., and Rahn, P. H. 1970. Mobilization of debris flows. *Zeit. F. Geomorph.* Suppl. 9:168–86.

Johnson, A. M., and Rodine, J. R. 1984. Debris flow. In *Slope instability,* pp. 257–361. Ed. by D. Brunsden and D. B. Prior. New York: John Wiley.

Johnson, D., Keller, E., and Rockwell, T. 1990. Dynamic pedogenesis: New views on some key soil concepts, and a model for interpreting Quaternary soils. *Quaternary Research* 33:306–19.

Johnson, D. L., and Watson-Stegner, D. 1987. Evolution model of pedogenesis. *Soil Sci.* 143:349–66.

Johnson, D. W. 1910. Beach cusps. *Geol. Soc. America Bull.* 21:604–21.

———. 1919. *Shore Processes and Shoreline Development.* New York: John Wiley & Sons. Facsimile edition: Hafner: New York, 1965.

———. 1925. *New England-Acadian shoreline.* New York: John Wiley & Sons.

Johnson, J. W. 1956. Dynamics of nearshore sediment movement. *Am. Assoc. Petroleum Geologists Bull.* 40:2211–32.

Johnson, P. J. 1992. Micro-relief on a rock glacier, Dalton Range, Yukon, Canada. *Permafrost and Periglac. Proc.* 3:41–47.

Johnson, W. D. 1904. The profile of maturity in alpine glacial erosion. *Jour. Geology* 12:7:569–78.

Johnson, W. H. 1990. Ice wedge casts and relict patterned ground in central Illinois and their environmental significance. *Quaternary. Res.* 33:51–72.

Johnston, G. H. 1963. Pile construction in permafrost. *Proc. Permafrost Internat. Conf.,* (Lafayette, Ind., 1963). *Natl. Acad. Sci.–Natl. Res. Council Pub.* 1287, pp. 477–81.

———. 1983. Performance of an insulated roadway on permafrost, Inuvik, N. W. T. In *Proc. Permafrost 4th Internat. Conf.,* pp. 548–51. Natl. Acad. Sci.

Jones, D. E., and Holtz, W. G. 1973. Expansive soils—the hidden disaster. *Amer. Soc. Civil Engin.* 43:49–51.

Jopling, A. V. 1966. Some application of theory and experiment to the study of bedding genesis. *Sedimentology* 7:71–102.

Judson, S. 1968a. Erosion rates near Rome, Italy. *Science* 160:1444–46.

———. 1968b. Erosion of the land. *Am. Scientist* 56:356–74.

Judson, S., and Ritter, D. F. 1964. Rates of regional denudation in the United States. *Jour. Geophys. Research* 69:3395–401.

Kachadoorian, R., and Ferrians, O. J., Jr. 1973. Permafrost-related engineering problems posed by the Trans Alaskan Pipeline. *Permafrost 2nd Internat. Conf.,* pp. 684–87. Natl. Acad. Sci.–Natl. Res. Council, Yakutsk, U.S.S.R., 1973.

Kale, V. S., Mishra, S., and Barker, V. R. 1997. A 2000-year paleoflood record from Sakarghat on Narmada, Central India. *Jour. Geol. Soc. of India* 50:283–88.

Kamb, B. 1987. Glacier surge mechanism based on linked cavity configuration of the basal water conduit system. *Journal of Geophysical Research* 92:9083–9100.

———. 1991. Rheological nonlinearity and flow instability in the deforming bed mechanism of ice stream motion. *Jour. Geophys. Research* 96:585–95.

Kamb, B., and LaChapelle, E. 1964. Direct observation of the mechanism of glacier sliding over bedrock. *Jour. Glaciol.* 5:159–72.

———. 1964a. Outline of Pleistocene geology of Martha's Vineyard, Massachusetts. *U.S. Geol. Survey Prof. Paper* 501-C:134–39.

———. 1964b. Illinoian and early Wisconsin moraines of Martha's Vineyard, Massachusetts. *U.S. Geol. Survey Prof. Paper* 501-C:140–43.

Kamb, B., Raymond, C. F., Harrison, W. D., Engelhardt, H., Echelmeyer, K. A., Humphrey, N., Brugman, M. M., and Pfeffer, T. 1985. Glacier surge mechanism: 1982–1983 surge of Variegated Glacier, Alaska. *Science* 227:469–79.

Kastning, E. H. 1990. Sympathetic evolution of conduits and fluvial networks in karst terranes: A synthesis based on geologic structure. *Geological Society of America Abstracts with Programs,* p. 108.

Keefer, D. K. 1999. Earthquake-induced landslides and their effects on alluvial fans. *Jour. Sedimentary Research* 69:84–104.

Keefer, D. K., and Johnson, A. M. 1983. Earth flows: Morphology, mobilization, and movement. *U.S. Geol. Survey Prof. Paper* 1264, 56p.

Kehew, A., and Lord, M. 1986. Origin and large-scale erosional features of glacial-lake spillways in the northern Great Plains: *Geol. Soc. America Bull.* 97:162–77.

Kehew, A. E., and Lord M. L. 1987. Glacial-lake outbursts along the mid-continent margins of the Laurantide ice-sheet. In *Catastrophic Flooding,* pp. 95–120. Ed. by L. Mayer and D. Nash. Binghamton Symposium in Geomorphology. Boston: Allen and Unwin.

Keller, E., and Pinter, N. 1996. Active tectonics: Earthquakes, uplift, and landscape. Prentice-Hall, Inc. 338p.

Keller, E. A. 1971. Areal sorting of bedload material. *Geol. Soc. America Bull.* 82:753–56.

Keller, E. A., and Melhorn, W. 1973. Bedforms and fluvial processes on alluvial stream channels: Selected observations. In *Fluvial geomorphology,* pp. 253–83. Ed. by M. Morisawa. S.U.N.Y., Binghamton: Pubs. in Geomorphology.

———. 1978. Rhythmic spacing and origin of pools and riffles. *Geol. Soc. America Bull.* 89:723–30.

Keller, E. A., and Swanson, F. J. 1979. Effects of large organic material on channel form and fluvial processes. *Earth Surf. Proc. and Landforms* 4:361–80.

Keller, E. A., and Tally, T. 1979. Effects of large organic debris on channel form and fluvial processes in the coastal redwood environment. In *Adjustments of the Fluvial System,* pp. 169–97. Ed. by D. Rhodes and G. Williams. Dubuque, IA: Kendall Hunt Publishing Company.

Keller, E. A., Bonkowski, M. S., Korsch, R. J., and Shlemon, R. J. 1982. Tectonic geomorphology of the San Andreas fault zone in the southern Indio Hills, Coachella Valley, California. *Geol. Soc. America Bull.* 93:46–56.

Keller, W. 1978. Kaolinization of feldspar as displayed in scanning electron micrographs. *Geology* 6:184–88.

———. 1982. Kaolin—A most diverse rock in genesis, texture, physical properties and uses. *Geol. Soc. America Bull.* 93:27–36.

Kellerhals, R. 1967. Stable channels with gravel-paved beds. Am Soc. Civil Engineers Proc., *Jour. Waterways and Harbors* 93:63–84.

Kelsey, H., and Bockheim, J. 1994. Coastal landscape evolution as a function of eustasy and surface uplift rate. *Geol. Soc. Amer. Bull.* 106:840–54.

Kelsey, H., and 4 others. 1996. Quaternary upper plate deformation in coastal Oregon. *Geol. Soc. Amer.* Bull. 108:843–60.

Kelsey, H. M. 1980. A sediment budget and analysis of geomorphic process in the Van Duzen River basin, north coastal California, 1941–1975. *Geol. Soc. America Bull.* 91:190–95.

Kelson, K. I., and Wells, S. G. 1989. Geologic influences on fluvial hydrology and bedload transport in small mountain watersheds, northern New Mexico: *Earth Surface Processes and Landforms* 14:671–90.

Kemmerly, P., and Towe, S. 1978. Karst depressions in a time context. *Earth Surf. Proc. and Landforms* 3:355–61.

Kemmerly, P. R. 1976. Definitive doline characteristics in the Clarksville quadrangle, Tennessee. *Geol. Soc. America Bull.* 87:42–46.

———. 1980a. Sinkhole collapse in Montgomery County, Tennessee. *Tenn. Div. Geol., Environ. Geol. Ser.* no. 6.

———. 1980b. A time-distribution study of doline collapse: Framework for prediction. *Environ. Geol.* 3:123–30.

———. 1982. Spatial analysis of a karst depression population: Clues to genesis. *Geol. Soc. America Bull.* 93:1078–86.

Kennedy, F. E., Phetteplace, G., Humiston, N., and Prabhakar, V. 1988. Thermal performance of a shallow utilidor. In *Permafrost, fifth international conference proceedings,* pp. 1262–67. Ed. by K. Senneset. Trondheim: Tapir Publishers.

Kenney, C. 1984. Properties and behaviours of soils relevant to slope instability. In *Slope Instability,* pp. 27–65. Ed. by D. Brunsden and D. B. Prior. New York: John Wiley and Sons.

Kesel, R. H. 1973. Inselberg landform elements: Definition and synthesis. *Rev. Geomorph. Dynamique* 22:97–108.

———. 1977. Some aspects of the geomorphology of inselbergs in central Arizona, U.S.A. *Zeit. F. Geomorph.* 21:119–46.

Kesel, R. H., Dunne, K., McDonald, R., Allison, K., and Spicer, B. 1974. Lateral erosion and overbank deposition on the Mississippi River in Louisiana caused by 1973 flooding. *Geology* 2:461–64.

Kesseli, J. E. 1941. Rock streams in the Sierra Nevada, California. *Geogr. Rev.* 31:203–27.

Keyes, C. R. 1912. Deflative scheme of the geographic cycle in an arid climate. *Geol. Soc. America Bull.* 23:537–62.

Kiersch, G. A. 1964. Vaiont reservoir disaster. *Civil Engineering* 34:32–39.

Kilpatrick, F. A., and Barnes, H. H. 1964. Channel geometry of piedmont streams as related to frequency of floods. *U.S. Geol. Survey Prof. Paper* 422E:1–10.

King, C. A. M. 1972. *Beaches and coasts.* New York: St. Martin's Press.

King, C. A. M., and Buckley, J. T. 1968. The analysis of stone size and shape in Arctic environments. *Jour. Sed. Petrology* 38:200–214.

King, C. A. M., and Lewis, W. V. 1961. A tentative theory of ogive formation. *Jour. Glaciol.* 3:913–39.

King, C. A. M., and McCullagh, M. J. 1971. A simulation model of a complex recurved spit. *Jour. Geology* 79:22–37.

King, C. A. M., and Williams, W. W. 1949. The formation and movement of sandbars by wave action. *Geogr. Jour.* 107:70–84.

King, G., and Stein, R. 1983. Surface folding, river terrace deformation rate, and earthquake repeat time in a reverse faulting environment: The 1983 Coalinga, California earthquakes. *California Division of Mines and Geology Special Publication* 66:165–76.

King, L. C. 1953. Canons of landscape evolution. *Geol. Soc. America Bull.* 64:751–52.

Kirkby, M. J. 1967. Measurement and theory of soil creep. *Jour. Geology* 75:359–78.

———. 1969. Infiltration, throughflow, and overland flow; and erosion by water on hillslopes. In *Water, earth, and man,* pp. 215–38. Ed. by R. J. Chorley. London: Methuen and Co.

Kirkby, M. J., and Chorley, R. J. 1967. Throughflow, overland flow and erosion. *Int. Assoc. Sci. Hydrol. Bull.* 12:5–21.

Kirkby, M. J., and Kirkby, A. V. 1969. Erosion and deposition on a beach raised by the 1964 earthquake, Montague Island, Alaska. *U.S. Geol. Survey Prof. Paper* 543-H:1–41.

Kirkby, R. P. 1969. Variation in glacial deposition in a subglacial environment: An example from Midlothian. *Scott. J. Geol.* 5:49–53.

Klemes, V. 1987. Hydrological and engineering relevance of flood frequency analysis. In *Hydrologic Frequency Modeling,* pp. 1–18. Ed. by V. J. Singh. Dordrecht: Reidel.

Klimchouk, A., Lowe, David, Cooper, A. H., and Sauro, J. 1996. *Gypsum Karst of the World.* Rome, Societa Speleologica Italiana.

Kling, J. 1997. Observations on sorted circle development Abisko, Northern Sweden: *Permafrost and Periglacial Processes* 8:447–53.

Klingeman, P. C., Beschta, R. L., Komar P. D., and Bradley, J. B. 1998. *Gravel-bed rivers in the environment.* Highlands Ranch: Water Resources Publications, LLC.

Kneale, W. R. 1982. Field measurements of rainfall drop-size distribution, and the relationship between rainfall parameters and soil movement by rainsplash. *Earth Surf. Proc. and Landforms* 7:499–502.

Knighton, A., and Nanson, G. 1977. Alternative derivation of the minimum variance hypothesis. *Geol. Soc. America Bull.* 88:364–66.

———. 1987. River channel adjustment—the downstream dimension. In *River channels: Environment and process,* pp. 95–128. Ed. by K. Richards. Oxford: Basil Blackwell, Ltd.

———. 1993. Anastomosis and the continuum of channel pattern. *Earth Surface Processes landf.* 18:613–25.

Knighton, A. D. 1974. Variation in width-discharge relation and some implications for hydraulic geometry. *Geol. Soc. America Bull.* 85:1069–76.

———. 1980. Longitudinal changes in size and sorting of stream-bed material in four English rivers. *Geol. Soc. America Bull.* 91:55–62.

———. 1998. *Fluvial forms and processes: A new prospective.* London: Arnold.

Knox, J. C. 1976. Concept of the graded stream. In: *Theories of landform development.* W. Melhorn and R. Flemal, eds., Binghamton, New York: State University of New York Publications in Geomorphology, pp. 169–198.

———. 1977. Human impacts on Wisconsin stream channels. *Ann. Assoc. Amer. Geogr.* 67:401–10.

———. 1987. Historical valley floor sedimentation in the upper Mississippi Valley, *Ann. Amer. Assoc. Geogr.* 77:224–44.

———. 1993. Large increases in flood magnitude in response to modest changes in climate. *Nature* 361:430–32.

———. 2000. Sensitivity of modern and Holocene floods to climate change. *Quaternary Science Reviews* 19:439–57.

Knox, J. C., and Kundzewicz, Z. W. 1997. Extreme hydrological events, paleo-information and climate change. *Hydrological Science Journal* 42:765–79.

Knuepfer, P., and McFadden L., eds. 1990. Soils and landscape evolution. *Geomorphology*, 3, 3/4, Proc. 21st Binghamton Symp. in Geomorphology, 378 p.

Kochel, R. C. 1975. *Morphology, structure, and origin of two blockfields and associated deposits, Northern Berks County, Pennsylvania.* Unpub. Senior Thesis, Lancaster: Franklin and Marshall College.

———. 1987. Holocene debris flows in central Virginia. In *Debris flows/avalanches: process, recognition, and mitigation,* pp. 139–55. Ed. by J. E. Costa and G. F. Wieczorek. Boulder: *Geol. Soc. Amer. Rev. in Engin. Geol* 7.

———. 1988. Extending stream records with slackwater paleoflood hydrology: Examples from west Texas. In *Flood geomorphology,* pp. 377–92. Ed. by V. R. Baker, R. C. Kochel, and P. C. Patton. New York: John Wiley.

———. 1988b. Geomorphic impact of large floods: Review and new perspectives on magnitude and frequency. In *Flood Geomorphology,* pp. 169–187. Ed. by V. R. Baker, R. C. Kochel, R. C., and P. C. Patton, New York: John Wiley and Sons.

———. 1990. Humid fans of the Appalachian Mountains. In *Alluvial fans: A field approach,* pp. 109–29. Ed. by A. J. Rachocki and M. Church. New York: John Wiley.

———. 1992. Floods. In *Encyclopedia of earth system science,* pp. 227–40. San Diego: Academic Press.

Kochel, R. C., and Baker, V. R. 1982. Paleoflood hydrology. *Science* 215:353–61.

———. 1988. Paleoflood analysis using slackwater deposits. In *Flood geomorphology,* pp. 357–76. Ed. by V. R. Baker, R. C. Kochel, and P. C. Patton. New York: John Wiley.

Kochel, R. C., and Dolan, R. 1986. The role of overwash on a mid-Atlantic coast barrier island. *J. Geology* 94:902–06.

Kochel, R. C., and Johnson, R. A. 1984. Geomorphology and sedimentology of humid-temperate alluvial fans, central Virginia. In *Gravels and conglomerates,* pp. 109–22. Ed. by E. Koster and R. Steel. *Can. Soc. Petrol. Geol.* Mem. 10.

Kochel, R. C., and Parris, A. 2000. Macroturbulent erosional and depositional evidence for large-scale pleistocene paleofloods in the lower Susquehanna bedrock gorge new Holtwood, Pennsylvania. *Geological Society Amer Abs. Ann. Mt.* 32:7:121.

Kochel, R. C., Baker, V. R., and Patton, P. C. 1982. Paleohydrology of Southwestern Texas. *Water Resour. Res.* 18:1165–83.

Kochel, R. C., Eaton, L. S., Daniels, N., and Howard, A. D. 1997. Impact of 1995 debris flows on geomorphic evolution of Blue Ridge debris fans, Madison County, Virginia: *Geological Society America, Abs. Ann. Mtg.* 29:4:410.

Kochel, R. C., and Miller, J. R. 1993. Complex response of arid fluvial systems to short term cyclic climatic changes, Anza-Borrego Desert, southern California. *Third Intl. Geomorph. Conf. Abs. w. Prog.,* Hamilton, Ontario p. 172.

———, eds. 1997. *Geomorphic responses to short-term climate change.* Special Issue, *Geomorphology* 19:170–368.

Kochel, R. C., and Piper, J. F. 1986. Morphology of large valleys on Hawaii; evidence for groundwater sapping and comparisons with Martian valleys. *J. Geophys. Res.* 91: E175–92.

Kochel, R. C., and Ritter, D. 1987. Implications of flume experiments on the interpretation of slackwater paleoflood sediments. In *Flood frequency and risk analysis* pp. 371–90. Ed. by V. Singh. Boston: Reidel.

Kochel, R. C., and Ritter, D. F. 1990a. Complex geomorphic response to minor climate changes, San Diego County, California. In *Hydraulics/Hydrology of Arid Lands,* pp. 148–53. Ed. by R. H. French.: Amer. Soc. Civil Engr., New York.

———. 1990b. Catastrophic floodplain destruction in southern Illinois: Example of a stream in disequilibrium. *Geol. Soc. Amer. Abs. w. Prog.* 22(7):111.

Kochel, R. C., and Wampfler, L. A. 1989. Relative role of overwash and aeolian processes on washover fans, Assateague Island, Virginia-Maryland. *J. Coastal Res.* 5: 453–75.

Kochel, R. C., Howard, A. D., and McLane, C. 1985. Channel networks developed by groundwater sapping in fine-grained sediments; analogs to some Martian valleys. In *Models in geomorphology,* pp. 313–41. Ed. by M. Woldenberg, Boston: Allen and Unwin.

Kochel, R. C., McGeehan, K. A., Valastro, S., and Carlson, C. R. 1991. Holocene overwash dynamics on the Virginia barrier islands. *Geol. Soc. Amer. Abs. w. Prog.* 23, no. 5:205–06.

Kochel, R. C., Miller, J. R., and Ritter, D. F. 1997. Geomorphic response to minor cyclic climate changes, San Diego County, California. *Geomorphology* 19:277–302.

Kochel, R. C., Ritter, D. F., and Miller, J. R. 1987. Role of tree dams in the construction of pseudo-terraces and variable geomorphic response to floods in Little River Valley, Virginia. *Geology* 15: 718–21.

Kocurek, G. 1981. Erg reconstruction: The Entrada Sandstone (Jurassic) of northern Utah and Colorado. *Paleogeog., Paleoclim. and Paleoecol.* 36:125–53.

Kocurek, G., and Lancaster, N. 1999. Aeolian sediment states: Theory and Mojave Desert Kelso Dunefield example. *Sedimentol.* 46:505–16.

Kocurek, G., and Nielson, J. 1986. Conditions favourable for the formation of warm-climate aeolian sand sheets. *Sedimentology* 33:795–816.

Kocurek, G., and 3 others, 1991. Amalgamated accumulations resulting from climatic and eustatic changes, Akchar Erg, Mauritania. *Sedimentol.* 38:751–72.

Komar, P. D. 1971a. The mechanics of sand transport on beaches. *Jour. Geophys. Research* 76:3:713–21.

———. 1971b. Nearshore cell circulation and the formation of giant cusps. *Geol. Soc. America Bull.* 82:2643–50.

———. 1973. Observations of beach cusps at Mono Lake, California. *Geol. Soc. America Bull.* 84:3593–3600.

———. 1975. Nearshore currents: Generation by obliquely incident waves and longshore variations in breaker height. In *Proc. symposium on nearshore sediment dynamics,* pp. 17–45. Ed. by J. R. Hails and A. Carr. London: John Wiley & Sons.

———. 1976. *Beach processes and sedimentation.* Englewood Cliffs, NJ: Prentice-Hall.

———. 1983a. Beach processes and erosion—an introduction. In *CRC handbook of coastal processes and erosion,* pp. 1–20. Ed. by P. Komar. Boca Raton, FL: CRC Press.

———, ed. 1983b. *CRC handbook of coastal processes and erosion.* Boca Raton, FL: CRC Press.

Komar, P. D., and Good, J. W. 1989. Long-term erosion impacts of the 1982–83 El Niño on the Oregon coast. In *Coastal zone '89., Proc. 6th symp. on coast. and ocean manag.* pp. 3785–93. Ed. by T. O. Magoon and others.

Komar, P. D., and Inman, D. L. 1970. Longshore sand transport on beaches. *Jour. Geophys. Research* 75:30:5914–27.

Komar, P. D., and Shih, S. M. 1992. Equal mobility versus changing bedload grain sizes in gravel-bed streams. In *Dynamics of gravel-bed rivers*, pp. 73–106. Ed. by P. Billi, R. D. Hey, C. R. Thorne, and P. Tacconi. New York: John Wiley and Sons, Ltd.

Konrad, J. M., and Morgenstern, N. R. 1983. Frost susceptibility of soils in terms of their segregation potential. In *Proc. Permafrost 4th Internat. Conf.,* pp. 660–65. Natl. Acad. Sci.

Kooi, H., and Beaumont, C. 1994. Escarpment evolution on high-elevation rifted margins: Insights derived from a surface processes model that combines diffusion, advection, and reaction. *J. Geophysical Res.* 99:12 191–1209.

Koons, P. 1989. The topographic revolution of collisional mountain belts: A numerical look at the Southern Alps, New Zealand. *American Journal of Science* 289:1041–69.

Kor, P. S. G., Shaw, J., and Sharpe, D. R. 1991. Erosion of bedrock by subglacial meltwater, Georgian Bay, Ontario: A regional view. *Can. J. Earth Sci.* 28:623–42.

Kottlowski, F., Cooley, M., and Ruhe, R. 1965. Quaternary geology of the southwest. In *The Quaternary of the United States.* Ed. by H. Wright and D. Frey. Princeton, NJ: Princeton Univ. Press.

Koutaniemi, L. 1999. Twenty-one years of string movements on the Liipasuuuaape mire, Finland, *Boreas.* 28:521–30.

Krabill, W. B., et al. 2000. Airborne laser mapping of Assateague National Seashore Beach. *Photogrammetric Engineering and Remote Sensing.* 66:65–71.

Kraft, J. C., and Chryzatowski, M. J. 1985. Coastal stratigraphic sequences. In *Coastal sedimentary environments*, pp. 625–63. New York: Springer-Verlag.

Kraus, M., and Middleton, L. 1987. Dissected paleotopography and base-level changes in a Triassic fluvial sequence. *Geology* 15:18–23.

Krigstrom, A. 1962. Geomorphological studies of sandar plains and their braided rivers in Iceland. *Geogr. Annlr.* 44:328–46.

Krinsley, D. H., and Donahue, J. 1968. Environmental interpretation of sand grain surface textures of electron microscopy. *Geol. Soc. America Bull.* 79:743–48.

Krumbein, W. C. 1944. Shore currents and sand movement on a model beach. *U.S. Army Corps Engrs., Beach Erosion Board Tech. Memo 7.*

Krumbein, W. C., and Oshiek, L. E. 1950. Pulsation transport of sand by shore agents. *Am. Geophys. Union Trans.* 31:216–20.

Kuenen, P. H. 1948. The formation of beach cusps. *Jour. Geology* 56:34–40.

Kuhn, G. G., and Osborne, R. H. 1989. Historical climatic fluctuations in southern California and their impact on coastal erosion and flooding: 1862 to present. In *Coastal Zone '89., Proc. 6th symp. on coast. and ocean manag.* pp. 4391–405. Ed. by T. O. Magoon and others.

Kuhn, G. G., and Shepard, F. P. 1983. Beach processes and sea cliff erosion in San Diego County, California. In *CRC handbook of coastal processes and erosion*, pp. 267–84. Ed. by P. Komar. Boca Raton, FL: CRC Press.

Kumar, A., and Chandler, S. 1987. Statistical flood frequency analysis—an overview. In *Hydrologic frequency modeling*, pp. 19–35. Ed. by V. J. Singh. Dordrecht: Reidel.

Kuno, H. 1969. Plateau basalts. In *The Earth's crust and upper mantle*, pp. 495–500. Ed. by P. Hart, Am. Geophys. Union, Geophys. Monograph 13.

Kupsch, W. O. 1955. Drumlins with jointed boulders near Dollard, Saskatchewan. *Geol. Soc. America Bull.* 66:327–38.

Kutvitskaya, N. B., and Gokhman, M. R. 1988. Ventilated surface foundations on permafrost soils. In *Permafrost, fifth international conference proceedings*, pp. 1413–16. Ed. by K. Senneset. Trondheim: Tapir Publishers.

Lachenbruch, A. H. 1966. Contraction theory of ice wedge polygons: A qualitative discussion. *Proc. permafrost internat. conf.* (Lafayette, Ind. 1963). Natl. Acad. Sci.–Natl. Res. Council Pub. 1287, pp. 63–71.

———. 1970. Some estimates of the thermal effects of a heated pipeline in permafrost. *U.S. Geol. Survey Circ.* 632.

Lachenbruch, A. H., and Marshall, B. V. 1969. Heat flow in the Arctic. *Arctic* 22:300–311.

Lagache, M. 1976. New data on the kinetics of the dissolution of alkali feldspars at 200° C in CO^2 charged water. *Geochim. Cosmochim. Acta.* 40:157–61.

Lagally, M. 1934. *Mechanik und Thermodynamik des stationaren Gletschers.* Leipzig.

Lagerback, R., and Rodhe, L. 1986. Pingos and palsas in northernmost Sweden; preliminary notes on recent investigations. *Geografiska Annaler* 68A: 149–54.

Laity, J. 1987. Topographic effects on ventifact development, Mojave Desert, California. *Physical Geogr.* 8:113–32.

Laity, J., 1994. Landforms of aeolian erosion. In *Geomorphology of desert environments*, pp. 506–35. Ed. by A. Abrahams and A. Parsons. London: Chapman and Hall.

Laity, J. E. 1983. Diagenetic controls on groundwater sapping and valley formation, Colorado Plateau, as revealed by optical and electron microscopy. *Phys. Geogr.* 4: 103–25.

Laity, J. E., and Malin, M. C. 1985. Sapping processes and the development of theater-headed valley networks in the Colorado Plateau. *Geological Society of America Bull.* 96:203–17.

Lajczak, A. 1995. The impact of river regulation, 1850–1990, on the channel and floodplain of the upper Vistula River, southern Poland. In River geomorphology, pp. 209–33. Ed by E. Hickin, New York: Wiley,

Lambe, T. 1953. The structure of inorganic soils. *Am. Soc. Civil Engineers Proc.* 79:Separate 315.

Lancaster, N. 1978. The pans of the southern Kalahari, Botswana. *Geogr. Journal* 144:81–98.

———. 1980. The formation of seif dunes from barchans—Supporting evidence for Bagnold's model from the Namib Desert. *Zeit. F. Geomorph.* 24:160–67.

———. 1982. Dunes on the Skeleton Coast, Namibia (South West Africa): Geomorphology and grain size relationships. *Earth Surf. Proc. and Landforms* 7:575–87.

———. 1983. Controls on dune morphology in the Namib Sand Sea. In *Eolian Sediments and Processes*, pp. 261–90. Ed. by M. Brookfield and Ahlbrandt. Amsterdam:Elsevier.

———. 1985. Variations in wind velocity and sand transport on the windward flanks of desert and dunes. *Sedimentology* 32:581–93.

———. 1988a. The development of large aeolian bedforms. *Sedimentary Geol.* 55:69–90.

———. 1988. Controls of eolian dune size and spacing. *Geology* 16:972–75.

———. 1989. The dynamics of star dunes: an example from the Gran Desierto, Mexico. *Sedimentology* 36:273–90.

———. 1992. Relations between dune generations in the Gran Desierto, Mexico. *Sedimentol.* 39:631–44.

———. 1995. *Geomorphology of desert dunes.* London: Routledge.

———. 1996. Airflow up the stoss side of sand dunes: limitations of current understanding. *Geomorph.* 17, 47–54.

Lancaster, N., and 3 others. 1996. Sediment flux and airflow on the stoss slope of a barchan dune. *Geomorph.* 17:55–62.

———. 1999. Geomorphology of desert sand seas. *Aeolian environments, sediments, and landforms,* pp. 49–69. Ed. by A. Goudie, I. Livingstone, and S. Stokes. Chichester, Wiley.

Lane, E. W. 1955. Design of stable channels. *Am. Soc. Civil Engineers Trans.* 120:1234–79.

———. 1957. A study of the shape of channels formed by natural streams in erodible material. *M.R.D. Sediments Series no. 9,* U.S. Army Corps Engrs., Eng. Div., Missouri River, Omaha, Neb.

Langbein, W. B., and Leopold, L. B. 1964. Quasi-equilibrium states in channel morphology. *Am. Jour. Sci.* 262:782–94.

———. 1966. River meanders: Theory of minimum variance. *U.S. Geol. Survey Prof. Paper* 422-H.

Langbein, W. B., and others. 1949. Annual runoff in the United States. *U.S. Geol. Survey Circular* 52.

Langbein, W. B., and Schumm, S. A. 1958. Yield of sediment in relation to mean annual precipitation. *Am. Geophys. Union Trans.* 39:1076–84.

Langford-Smith, T., and Dury, G. H. 1964. A pediment at Middle Pinnacle, near Broken Hill, New South Wales. *Jour. Geol. Soc. Australia* 11:79–88.

Larkin, R., and Sharp, J. 1992. On the relationship between river-basin geomorphology, aquifer hydraulics, and ground-water flow direction in alluvial aquifers. *Geological Society of America Bulletin* 104:1608–20.

Larson, M., and Kraus, N. C., 1994. Temporal and spatial scales of beach profile change, Duck, North Carolina. *Marine Geology* 117:75–94.

Lasaga, A., and 4 others 1994. Chemical weathering rate laws and global geochemical cycles. *Geochim. Cosmochim. Acta* 58: 2361–86.

Lattman, L. 1973. Calcium carbonate cementation of alluvial fans in Southern Nevada. *Geol. Soc. America Bull.* 84:3013–28.

Laury, R. L. 1971. Stream bank failure and rotational slumping. Preservation and significance in the geologic record. *Geol. Soc. America Bull.* 82:1251–66.

LaValle, P. 1967. Some aspects of linear karst depression development in south central Kentucky. *Ann. Assoc. Am. Geog.* 57:49–71.

———. 1968. Karst depression morphology in south-central Kentucky. *Geogr. Annlr.* 50A:94–108.

Lawson, A. C. 1915. The epigene profiles of the desert. *Univ. of Calif. Dept. Geol. Bull.* 9:23–48.

Lawson, D. 1981. Mobilization, movement and deposition of active subaerial sediment flows, Matanuska glacier, Alaska. *Jour. Geology* 90:279–300.

Lecce, S. 1997. Spatial patterns of historical sedimentation and floodplain evolution, Blue River, Wisconsin. *Geomorph.* 27:41–60.

Leclerc, R., and Hickin, E. 1997. The internal structure of scrolled floodplain deposits based on ground-penetrating radar. North Thompson River, British Columbia. *Geomorph.* 21:17–38.

Lee, F. T., Odum, J. K., and Lee, J. D. 1997. Slope failures in northern Vermont, USA. *Environmental and Engineering Geoscience.* 3:161–82.

Lee, M., Hodson, M., and Parsons, I. 1998. The role of intergranular microtextures and microstructures in chemical and mechanical weathering: Direct comparisons of experimentally and naturally weathered alkali feldspars. *Geochim. Cosmochim. Acta* 62:2771–88.

Lee, W., and Uyeda, S. 1965. Review of heat flow data. In *Terrestrial heat flow,* edited by W. Lee. *Am. Geophys. Union, Geophys. Monograph* 8.

Legget, R. 1967. Soil: Its geology and use. *Geol. Soc. America Bull.* 78:1433–60.

Lehman, D. 1963. Some principles of chelation chemistry. *Soil Sci. Soc. Am. Proc.* 27:167–70.

Lehmann, H. 1936. Morphologische Studien auf Java. Stuttgart: *Geogr. Abh.* 3:9.

Lehre, A. K. 1982. Sediment budget of a small coast range drainage basin in north-central California. In *Sediment budgets and routing in forested drainage basins,* pp. 67–77. Ed. by F. J. Swanson et al. U.S.D.A., Forest Serv. Genl. Tech. Rpt. PNW-141.

Leigh, C. 1982. Sediment transport by surface wash and throughflow at the Pasoh Forest Reserve, Negri Sembilan, Peninsular Malaysia. *Geogr. Annlr.* 64A:171–80.

Leighton, M. W., and Pendexter, C. 1962. Carbonate rock types. In *Classification of carbonate rocks,* pp. 33–61. Ed. by W. E. Ham. *Am. Soc. Petroleum Geologists Mem.* 1.

Leliavsky, S. L. 1966. *An introduction to fluvial hydraulics.* New York: Dover Publications.

Lemke, R. W. 1958. Narrow linear drumlins near Velva, North Dakota. *Am. Jour. Sci.* 256:270–83.

Leopold, L. B. 1982. Water surface topography in river channels and implications for meander development. In *Gravel-bed rivers,* pp. 859–88. Ed. by R. Hey, J. Bathurst, and C. Thorne. New York: John Wiley & Sons.

Leopold, L. B., and Emmett, W. W. 1976. Bedload measurements, East Fork River, Wyoming. *Natl. Acad. Sci. Proc.* 73:1000–1004.

———. 1977. 1976 bedload measurements, East Fork River, Wyoming. *Natl. Acad. Sci. Proc.* 74:2644–48.

Leopold, L. B., and Langbein, W. B. 1962. The concept of entropy in landscape evolution. *U.S. Geol. Survey Prof. Paper* 500-A:20.

———. 1963. Association and indeterminancy in geomorphology. In *The fabric of geology,* pp. 184–92. Ed. by C. C. Albritton. Reading, MA: Addison-Wesley.

Leopold, L. B., and Maddock, T., Jr. 1953. The hydraulic geometry of stream channels and some physiographic implications. *U.S. Geol. Survey Prof. Paper* 252.

Leopold, L. B., and Wolman, M. G. 1957. River channel patterns; braided, meandering and straight. *U.S. Geol. Survey Prof. Paper* 282-B.

———. 1960. River meanders. *Geol. Soc. America Bull.* 71:769–94.

Leopold, L. B., Wolman, M. G., and Miller, J. P. 1964. *Fluvial processes in geomorphology.* San Francisco: W. H. Freeman.

Lewis, W. V. 1954. Pressure release and glacial erosion. *Jour. Glaciol.* 2:417–22.

Lewkowitz, A. G. 1992. A solifluction meter for permafrost sites. *Permafrost and Periglacial Proc. 3:*11–18.

Li, Y. H. 1976. Denudation of Taiwan Island since the Pliocene Epoch. *Geology* 4:105–7.

Limerinos, J. T. 1970. Determination of the Manning coefficient from measured bed roughness in natural channels. U.S. Geol. Survey Water-Supply Paper 1898B.

Linell, K. A., and Johnston, G. H. 1973. Engineering design and construction in permafrost regions. *Proc. Permafrost 2nd Internat. Conf.,* pp. 553–75. Natl. Acad. Sci.-Natl. Res. Council, Yakutsk, U.S.S.R., 1973.

Lipman, P., and Mullineaux, D., eds. 1981. The 1980 eruptions of Mount St. Helens, Washington. U.S. Geol. Surv. Prof. Paper 1250, 844p.

Lisle, L. D. 1982. Annotated bibliography of sea level changes along the Atlantic and Gulf coasts of North America. *Shore and Beach* July, 50:24–34.

Lisle, T. F. 1987. Channel morphology and sediment transport in steepland streams. In *Erosion and Sedimentation on the Pacific Rim,* pp. 287–98. IAHS-AISH Pub. 165.

Liu, C., and Yu, S. 1989. Fast uplifting along the plate boundary in Taiwan. *EOS* 70:403.

Livingstone, I., and Warren. 1996. Aeolian geomorphology: An introduction. Harlow: Addison Wesley Longman.

Lliboutry, L. 1964. Subglacial "supercavitation" as a cause of the rapid advances of glaciers. *Nature* 202:77.

———. 1968. General theory of subglacial cavitation and sliding of temperate glaciers. *Jour. Glaciol.* 7:21–58.

Lliboutry, L., and Reynaud, L. 1981. "Global dynamics" of a temperature valley glacier. Mer de Glace and past velocities deduced from Forbes bands. *Jour. Glaciol.* 27:207–26.

Lobacz, E. F., and Quinn, W. F. 1963. Thermal regime beneath buildings constructed on permafrost. *Proc. permafrost internat. conf.,* (Lafayette, Ind., 1963). Natl. Acad. Sci.–Natl. Res. Council Pub. 1287, pp. 159–64.

Lohnes, R. A., and Handy, R. L. 1968. Slope angles in friable loess. *Jour. Geology* 76:247–58.

Lombard, R. E., Miles, M. B., Nelson, L. M., Kresch, D. L., and Carpenter, P. J. 1981. Channel conditions in the lower Toutle and Cowlitz Rivers resulting from the mudflows of May 18, 1980. *U.S. Geol. Survey Circ.* 850-C.

Longuet-Higgins, M. S. 1970. Longshore currents generated by obliquely incident sea waves. *Jour. Geophys. Research* 75:6778–801.

Loope, D. 1984. Eolian origin of upper Paleozoic sandstones, southeastern Utah. *Jour. Sediment. Petrology* 54:563–80.

Lord, M., and Kehew, A. 1987. Sedimentology and paleohydrology of glacial-lake outburst deposits in southeastern Saskatchewan and northwestern North Dakota. *Geol. Soc. America Bull.* 99:663–73.

Loughnan, F. 1969. *Chemical weathering of the silicate minerals.* New York: American Elsevier.

Lowe, D. J. 1992. The origin of limestone caverns: an inception horizon hypothesis. Unpublished PhD Thesis, Manchester Metropolitan University/Council for National Academic Awards, 511 pp.

Lowe, D. J. and Gunn, J. 1997. Carbonate speleogenesis: An inception horizon hypothesis. *Acta Carsologica* 38:457–488.

Lozíinski, W. 1912. Die periglaziale Fazies der mechanischen Verwitterung. Internat. Geol. Cong., 11th, Stockholm, 1910, *Compte rendu,* pp. 1039–53.

Lucchitta, B. K. 1978a. Morphology of chasma walls, Mars. *U.S. Geol. Survey Jour. Res.* 6:651–62.

———. 1978a. Landslides in the Valles Marineris, Mars. *NASA Technical Memo* 79729:288–90.

———. 1979. Debris flows on Olympus Mons. *NASA Technical Memo* 80339:34–35.

Lugn, A. L. 1962. *The origin and sources of loess.* Lincoln: Univ. Nebraska Studies, new series, no. 26.

———. 1968. The origin of loesses and their relation to the Great Plains in North America. In *Loess and Related Eolian Deposits of the World,* p. 139. Ed. by C. B. Schultz and J. C. Frye. Lincoln: Univ. Nebraska Press.

Luk, S. H. 1979. Effect of soil properties on erosion by wash and splash. *Earth Surf. Proc. and Landforms* 4:241–55.

Lundberg, N., and Dorsey, R. 1990. Rapid Quaternary emergence, uplift, and denudation of the Coastal Range, Eastern Taiwan. *Geology* 18:638–41.

Lundqvist G. 1951. En palsmyr sydost om Kebnekaise. *Geologiska Föreningen Förhandlingar* 73:209–25.

Luttenegger, A. J., and Hallberg, G. R. 1988. Stability of loess. *Engin. Geol.* 25:247–61.

Lynch, D. K. 1982. Tidal bores. *Sci. Amer.* 247:146–56.

Mabbutt, J. A. 1966. Mantle-controlled planation of pediments. *Am. Jour. Sci.* 264:78–91.

———. 1971. The Australian arid zone as a prehistoric environment. In *Aboriginal man and environment in Australia,* pp. 66–79. Ed. by D. Mulvaney and J. Golson. Canberra: Australian Natl. Univ. Press.

Macdonald, G. 1972. *Volcanoes.* Englewood Cliffs, NJ: Prentice-Hall.

Machette, M. N. 1978. Dating Quarternary faults in the southwestern United States by using buried calcic paleosols. *U.S. Geol. Surv. Journal Res.* 6: 369–81.

———. 1985. Calcic soils and calcretes of the southwestern United States. *Geol. Soc. Am. Spec. Paper* 203, 1–21.

———. 1988. Quarternary movement along the La Jencia fault, central New Mexico. U.S. Geol. Surv. Prof. Paper 1440.

Mack, G., and James, W. 1992. Calcic paleosols of the Plio-Pleistocene Camp Rice and Palomas Formations, southern Rio Grande rift, USA. *Sedimentary Geology* 77:89–109.

Mack, G., James, W., and Monger, H. 1993. Classification of paleosols. *Geol. Soc. America Bull.* 105:129–36.

Mackay, J. R. 1970. Disturbances to the tundra and forest tundra environment of the western Arctic. *Can. Geotechnical Jour.* 7:420–32.

———. 1973. The growth of pingos, western Arctic coast, Canada. *Can. J. Earth Sci.* 10:979–1004.

———. 1974. Ice-wedge cracks, Garry Island, Northwest Territories. *Can. J. Earth Sci.* 11:1366–83.

———. 1975a. The closing of ice-wedge cracks in permafrost, Garry Island, Northwest Territories. *Can. J. Earth Sci.* 12:1668–74.

———. 1975b. The stability of permafrost and recent climatic change in the Mackenzie Valley, NWT. *Canada Geol. Survey Paper* 75-1B:173–76.

———. 1984. The frost heave of stones in the active layer above permafrost with downward and upward freezing. *Arctic and Alpine Res.* 16:439–46.

———. 1986. The first 7 years (1978–85) of ice wedge growth, Illisarvik experimental drained lake, western Arctic coast. *Can. J. Earth Sci.* 23:1782–95.

———. 1990. Seasonal growth bands in pingo ice. *Can J. Earth Sci.* 27:1115–25.

———. 1992. The frequency of ice wedge cracking (1967–87) at Garry Island, western Arctic coast, Canada. *Can. J. Earth Sci.* 29:236–48.

———. 1998. Pingo growth and collapse, Tuktoyaktuk Peninsular area, western Arctic coast, Canada: A long-term field study: *Geographic physique et Quaternaire* 52:271–83.

Mackay, J. R., and Matthews, W. H. 1974. Movement of sorted stripes, the Cinder Cone, Garibaldi Park, B.C., Canada. *Arc. Alp. Res.* 6:347–59.

Mackin, J. H. 1937. Erosional history of the Big Horn Basin, Wyoming. *Geol. Soc. America Bull.* 48:813–93.

———. 1948. Concept of the graded river. *Geol. Soc. America Bull.* 59:463–512.

Maddock, T., Jr. 1969. The behavior of straight open channels with movable beds. *U.S. Geol. Survey Prof. Paper* 622-A:70.

Major, J. J. 1998. Pebble orientation on large, experimental debris-flow deposits. *Sedimentary Geology* 117:151–64.

Major, J. J., and Iverson, R. M. 1999. Debris-flow deposition effects of pore-fluid pressure and friction concentrated at flow margins. *Geological Society America Bull.* 111:1424–34.

Malin, M. C. 1974. Salt weathering on Mars. *J. Geophys. Res.* 79:3888–94.

Malott, C. A. 1939. Karst valleys. *Geol. Soc. America Bull.* 50:1984.

Mammerickx, J. 1964. Quantitative observations on pediments in the Mojave and Sonoran deserts (southwestern United States). *Am. Jour. Sci.* 262:417–35.

Manley, E. P., and Evans, L. J. 1986. Dissolutions of feldspars by low-molecular-weight aliphatic and aromatic acids. *Soil Science* 141:106–112.

Manley, S., and Manley, R. 1968. *Beaches; their lives, legends and lore.* Philadelphia: Chilton.

Marion, G. W., Schlesinger, W. H., and Fonteyn, P. J. 1985. CALDEP: A regional model for soil $CaCO_3$ (caliche) deposition in southwestern deserts. *Soil Science* 139:468–81.

Mark, D. M. 1974. Line intersection method for estimating drainage density. *Geology* 2:235–36.

Mark, R., and Moore, J. 1987. Slopes of the Hawaiian Ridge. *U.S. Geol. Survey. Prof. Paper* 1350:101–07.

Markewich, H., and 4 others. 1990. A guide for using soil and weathering profile data in chronosequence studies of the Coastal Plain of the Eastern United States. *U.S. Geol. Surv. Bull.* 1589.

Markham, A. J., and Thorne, C. R. 1992. Geomorphology of gravel-bed river bends. In *Dynamics of gravel-bed rivers,* pp. 433–56. Ed. by P. Billi, R. D. Hey, C. R. Thorne, and P. Tacconi. New York: John Wiley and Sons, Ltd.

Marion, T., Filion, L., and Hetu, B. 1995. The Holocene development of a debris slope in subarctic Quebec, Canada: *The Holocene* 5:409–19.

Marrs, R. W., and Kolm, K. E., eds. 1982. *Interpretation of windflow characteristics from eolian landforms. Geol. Soc. America Spec. Paper* 192.

Marsh, B. 1987. Pleistocene pingo scars in Pennsylvania. *Geology* 15:945–47.

Marsh, G. P. 1885. *The Earth as modified by human action.* New York: Charles Scribner's Sons.

Marshall, J., and Anderson, R. 1995. Quaternary uplift and seismic cycle deformation, Peninsula de Nicoya, Costa Rica. *Geol. Soc. Amer. Bull.* 107:463–73.

Martel, C. J. 1988. Developing a thawing model for sludge freezing beds. In *Permafrost, Fifth International Conference Proceedings,* pp. 1426–30. Ed. by K. Senneset. Trondheim: Tapir Publishers.

Martel, E. A. 1921. *Nouveau traite des eaux souterraines.* Paris: Delagrave.

Marticorena, B., and 3 others. 1997. Factors controlling threshold friction velocity in semiarid and arid areas of the United States. *J. Geophys. Res.* 102 (D19):23277–87.

Martin, C. W. 1992. Late Holocene allovial chronology and climate change in the central Great Plains. *Quaternary Research* 37:315–22.

Martin, E. H., and Whalley, W. B. 1987. Rock glaciers; Part 1, Rock glacier morphology; classification and distribution. *Prog. in Phys Geogr.* 11:260–82.

Martini, I. P. 1978. Tafoni weathering, with examples from Tuscany, Italy. *Zeit. F. Geomorph.* 22:44–67.

Marutani, T., Kasai, M., Reid, L. M., and Trustum, N. A. 1999. Influence of storm-related sediment storage on the sediment delivery from tributary catchments in the upper Waipaoa River, New Zealand. *Earth Surface Processes and Landforms* 24:881–96.

Mason, J. A., and Knox, J. C. 1997. Age of colluvium indicates accelerated late Wisconsinian hillslope erosion in the Upper Mississippi Valley. *Geology* 25:267–70.

Masselink, G., Hegge, B. J., and Pattiaratch, C. B. 1997. Beach cusp. morphodynamics. *Earth Surface Processes and Landforms* 22:1139–55.

Mathews, J. A., Harris, C., and Ballantyne, C. K. 1986. Studies on a gelifluction lobe, Jotunheimen, Norway; 14C chronology, stratigraphy, sedimentology, and paleoenvironment. *Geografiska Annaler* 68A:345–60.

Mathewson, C. C., and Keaton, J. R. 1986. Role of bedrock groundwater in the initiation of debris flow. *Abstract, Assoc. Engin. Geol,* 29th Ann. Meeting, p. 56.

Mathewson, C. C., Castleberry, J. P., and Lytton, R. L. 1975. Analysis and modeling of the performance of home foundations on expansive soils in central Texas. *Bull. Assoc. Engin. Geol.* 12:275–302.

Matmon, A., Enzel, Y., Zilberman, E., and Heimann, A. 1999. Late Pliocene and Pleistocene reversal of drainage systems in northern Israel: Tectonic implications. *Geomorphology* 28:43–59.

Matschinski, M. 1968. Alignment of dolines northwest of Lake Constance, Germany. *Geol. Mag.* 105(1):56–61.

Matsukura, Y. 1996. The role of the degree of weathering and groundwater fluctuation in landslide movement in a colluvium of weathered hornblende-gabbra. *Catena* 27:63–78.

Matsumota, T. 1967. Fundamental problems in the circum-Pacific orogenesis. *Tectonophysics* 4:595–613.

Matsuoka, N. 1991. A model of the rate of frost shattering: Application to field data from Japan, Svalbard and Antarctica. *Permafrost and Periglac. Proc.* 2:271–81.

———. 1995. Rock weathering processes and landform development in the Sor Rondane Mountains, Antarctica. *Geomorphology* 12:323–39.

———. 1996. The role of the degree of weathering and groundwater fluctuation in landslide movement in a colluvium of weathered hornblende gabbro. *Catena.* 27:63–78.

———. 1998. The relationship between frost heave and downslope movement. Field measurements in the Japanese Alps. *Permafrost and Periglacial Processes* 9:121–33.

Matsuoka, N., and Sakai, H. 1999. Rockfall activity from an alpine cliff during thaw periods. *Geomorphology* 28:309–28.

Matthes, F. E. 1900. Glacial sculpture of the Bighorn Mountains, Wyoming. *U.S. Geol. Survey 21st Ann. Rept., 1899–1900,* pt. 2, pp. 167–90.

———. 1930. Geologic history of the Yosemite Valley. *U.S. Geol. Survey Prof. Paper* 160.

May, J. P., and Tanner, W. F. 1973. The littoral power gradient and shoreline changes. In *Coastal Geomorphology,* pp. 43–60. Ed. by D. R. Coates, S.U.N.Y., Binghamton: 3rd Ann. Geomorph. Symposium.

May, S. K., Dolan, R., and Hayden, B. P. 1983. Erosion of U.S. shorelines. *EOS* 64:521–23.

Mayer, L. 1987. Sources of error in morphologic dating of fault scarps: In Crone, A. and Omdahl, E. (eds.), Directions in Paleoseismology, *U.S. Geol. Survey Open File Report,* 87–673, 302–10.

Mayer, L., McFadden, L. D., and Harden, J. W. 1988. Distribution of calcium carbonate in desert soils: A model. *Geology* 16:303–06.

Mayer, L., and Nash, D. eds. 1987. *Catastrophic flooding.* Boston: Allen and Unwin.

McCalpin, J. P., and Berry, M. E. 1996. Soil catenas to estimate ages of movements on normal faults scarps, with an example from the Wasatch fault zone, Utah, USA. *Catena* 27: 265–86.

McCauley, J., Breed, C., and Grolier, M. 1977. Yardangs. In *Geomorphology in arid regions,* pp. 233–69. Ed. by D. Doehring. Boston:Allen and Unwin.

McComas, M., Hinkley, K., and Kempton, J. 1969. Coordinated mapping of geology and soils for land-use planning. *Illinois Geol. Survey Environ. Geol. Note* 29.

McCoy, R. M. 1971. Rapid measurement of drainage density. *Geol. Soc. America Bull.* 82:757–62.

McDonald, E. V., Pierson, F. B., Flerchinger, G. N., and McFadden, L. D. 1996. Application of soil-water balance model to evaluate the influence of Holocene climate change on calcic soils, Mojave Desert, California, U.S.A. *Geoderma* 74:167–92.

McDonald, R. C. 1975. Observations on hillslope erosion in tower karst topography of Belize. *Geol. Soc. America Bull.* 86:255–56.

———. 1979. Tower karst geomorphology in Belize. *Zeit F. Geomorph.,* Suppl. 32:35–45.

McDowall, I. C. 1960. Particle size reduction of clay minerals by freezing and thawing. *New Zealand J. Geol. and Geophys.* 3:337–43.

McDowell, P. F., Webb, T., and Bartlein, P. J. 1990. Long-term environmental change. In *The earth as transformed by human action,* pp. 143–62. Ed. by I. Turner and others. Cambridge: Cambridge Univ. Press.

McEwan, I., 1993. Bagnold's kink: a physical feature of a wind velocity profile modified by blown sand. *Earth Surf. Proc. Lndfms.* 18:145–56.

McFadden, L., and 5 others. 1998. The vesicular layer and carbonate collars of desert soils and pavements: Formation, age and relation to climate change. *Geomorph.* 24:101–45.

McFadden, L., Wells, S., and Jercinovich, M., 1987. Influences of eolian and pedogenic processes on the origin and evolution of desert pavements. *Geology* 15:504–8.

McFadden, L. D. 1988. Climatic influences on rates and processes of soil development in Quaternary deposits of southern California. In Reinhardt, J. and Sigleo, W. (eds.), Paleosols and weathering through geologic time. *Geol. Soc. Amer. Spec. Paper* 216:153–77.

McFadden, L. D., Amundson, R. G., and Chadwick, O. A. 1991. Numerical modeling, chemical, and isotopic studies of carbonate accumulation in soils of arid regions. *Soil Science Society of America Special Publication No. 26,* 17–35.

McFadden, L. D., and Hendricks, D. M. 1985. Changes in the content and composition of pedogenic iron oxyhydroxides in a chrosequence of soils in southern California. *Quat. Res.* 23:189–204.

McFadden, L. D., and McAuliffe, J. R. 1997. Lithologically influenced geomorphic responses to Holocene climatic changes in the southern Colorado Plateau, Arizona; a soil-geomorphologic and ecologic perspective. In geomorphic response to short-term climate change. Ed. by R. C. Kochel and J. R. Miller. *Geomorphology* 19:303–22.

McFadden, L. D., and Knuepfer, P. L. K. 1990. Soil geomorphology: The linkage of pedology and surficial processes. *Geomorphology* 3:197–205.

McFadden, L. D., and Weldon, R. J. 1987. Rates and processes of soil development on Quaternary terraces in Cajon Pass, California. *Geol. Soc. Amer. Bull.* 98:280–93.

McGee, W. J. 1897. Sheetflood erosion. *Geol. Soc. America Bull.* 8:87–112.

McGreevey, J. P. 1981. Some perspectives on frost shattering. *Proc. in Phys. Geog.* 5:56–75.

McHattie, R. L., and Esch, D. C. 1983. Benefits of a peat underlay used in road construction on permafrost. In *Proc. permafrost 4th internat. conf.,* pp. 826–31. Natl. Acad. Sci.

McKee, E. D. 1966. Structures of dunes at White Sands National Monument, New Mexico, and a comparison with structures of dunes from other selected areas. *Sedimentology* 7:1–69.

———. 1979. Introduction to a study of global sand seas. In *A study of global sand seas,* edited by E. McKee. *U.S. Geol. Survey Prof. Paper* 1052:1–20.

McKee, E. D., and Tibbitts, G. C., Jr. 1964. Primary structures of a seif dune and associated deposits in Libya. *Jour. Sed. Petrology* 34:5–17.

McKenna Neuman, C., Lancaster, N., and Nickling, W. 1997. Relations between dune morphology, air flow, and sediment flux on reversing dunes, Silver Peak, Nevada. *Sedimentol.* 44:1103–14.

McKenna Neuman, C., and Maljaars, M. 1997. Wind tunnel measurements of boundary-layer response to sediment transport. *Boundary-Layer Meteor.* 84:67–83.

McKenna Neuman, C., and Nickling, W. 1989. A theoretical wind tunnel investigation of the effect of capillary water on the entrainment of sediment by wind. *Can. Jour. Soil Sci.* 69:79–96.

McKenzie, G. D. 1969. Observations on a collapsing kame terrace in Glacier Bay National Monument, S. E. Alaska. *Jour. Glaciol.* 8:413–25.

McKeown, F.; Jones-Cecil, M.; Askew, B.; and McGrath, M. 1988. Analysis of stream profile data and inferred tectonic activity, eastern Ozark mountains region, *U.S. Geol. Surv. Bull.* 1807B, 39 p.

McLean, R. F., and Kirk, R. M. 1969. Relationship between grain size, size-sorting and foreshore slope on mixed sand-shingle beaches. *New Zealand J. Geol. and Geophys.* 12:138–55.

McLennan, S. M. 1993. Weathering and global denudation. *J. Geology* 101:295–303.

McPhee, J. 1989. *The Control of nature,* Toronto: Collins Publ.

McPherson, H. J., and Rannie, W. F. 1967. Geomorphic effects of the May, 1967, flood in Graburn watershed, Cypress Hills, Alberta, Canada. *J. Hydrol.* 9:307–21.

McPherson, J., Shanmugam, G., and Moiola, R. 1987. Fan-deltas and braid-deltas: Varieties of coarse-grained deltas. *Geol. Soc. America Bull.* 99:331–40.

McPherson, M. B. 1974. *Hydrological effects of urbanization.* Paris: UNESCO Press.

McQueen, K. C., Vitek, J. D., and Carter, B. J. 1993. Paleoflood analysis of an alluvial channel in the south-central Great Plains; Black Bear Creek, Oklahoma. *Geomorphology* 8:131–46.

Meade, R. H. 1969. Errors in using modern stream-load data to estimate natural rates of denudation. *Geol. Soc. America Bull.* 80:1265–74.

———. 1982. Sources, sinks and storage of river sediment in the Atlantic drainage of the United States. *Jour. Geology* 90:235–52.

Meier, M. F. 1960. Mode of flow of Saskatchewan glacier, Alberta, Canada. *U.S. Geol. Survey Prof. Paper* 351.

Meier, M. F., and Johnson, A. 1962. The kinematic wave on Nisqually Glacier, Washington. *Jour. Geophys. Research* 67:886.

Meier, M. F., and Post, A. S. 1969. What are glacier surges? *Can. J. Earth Sci.* 6:807–17.

Meier, M. F., Kamb, W. B., Allen, C. R., and Sharp, R. P. 1974. Flow of Blue Glacier, Olympic Mountains, Washington, U.S.A. *Jour. Glaciol.* 13:187–212.

Mellor, M. 1970. Phase composition of pore water in cold rocks. *U.S. Army Corps Engrs., Cold Regions Res. and Eng. Lab. Research Rept.* 292.

Melton, F. A. 1940. A tentative classification of sand dunes. *Jour. Geology* 48:113–73.

Melton, M. A. 1958. Correlation structure of morphometric properties of drainage systems and their controlling agents. *Jour. Geology* 66:442–60.

———. 1965a. The geomorphic and paleoclimatic significance of alluvial deposits in southern Arizona. *Jour. Geology* 73:1–38.

———. 1965b. Debris-covered hillslopes of the southern Arizona desert—consideration of their stability and sediment contribution. *Jour. Geology* 73:715–29.

Menard, H. W. 1961. Some rates of regional erosion. *Jour. Geology* 69:155–61.

Menges, C. 1990. Late Quaternary fault scarps, mountain-front landforms, and Pliocene-Quaternary segmentation on the range-bounding fault zone, Sangre de Cristo Mountains, New Mexico. In Neotectonics in Earthquake Evaluation. Geol. Soc. America Reviews, in *Engr. Geol.* VIII, pp. 131–56. Ed. by E. Krinitzky and D. Slemmons.

Menzies, J. 1979. The mechanics of drumlin formation with particular reference to the change in pore-water content of the till. *Jour. Glaciol.* 22:373–84.

———. 1981. Temperatures within subglacial debris—A gap in our knowledge. *Geology* 9:271–73.

———. 1989. Subglacial hydraulic conditions and their possible impact upon subglacial bed formation. In J. Menzies and J. Rose (eds.), *Subglacial Bedforms-Drumlins, Rogen Moraine and Associated Subglacial Bedforms, Sedimentary Geology* 62:125–50.

Menzies, J., and Rose, J. 1989. Subglacial Bedforms—An introduction. *Sedimentary Geology* 62:117–22.

Merritts, D., and Bull, W. 1989. Interpreting Quaternary uplift rates at the Mendocino triple junction, northern California, from uplifted marine terraces. *Geology* 17:1020–24.

Merritts, D., and Hesterberg, T. 1994. Stream networks and long-term surface uplift in the New Madrid Seismic zone, *Science* 265:1081–84.

Merritts, D., and Vincent, K. 1989. Geomorphic response of coastal streams to low, intermediate, and high rates of uplift, Mendocino triple junction region, northern California, *Geol. Soc. Amer. Bull.* 101:1378–88.

Merritts, D., Chadwick, O., Hendricks, D., Brimhall, G., and Lewis, C. 1992. The mass balance of soil evolution on late Quaternary marine terraces, northern California. *Geol. Soc. Amer. Bull.* 104:1456–70.

Merritts, D., Vincent, K., and Wohl, E. 1994. Long river profiles, tectonism, and eustasy: A guide to interpreting fluvial terraces. *J. Geophys. Res.* 99(B7):14031–50.

Mertes, L. 1994. Rates of floodplain sedimentation on the central Amazon River. *Geology* 22:171–74.

Mertes, L., Dunne, T., and Martinelli, L. 1996. Channel-floodplain geomorphology along the Solimoes-Amazon River, Brazil. *Geol. Soc. Amer. Bull.* 108:1089–1107.

Meyer, G. A., and Wells, S. G. 1997. Five related sedimentation events on alluvial fans, Yellowstone National Park, USA. *Jour. Sedimentary Research* 67:776–91.

Meyer-Peter, E., and Muller, R. 1948. Formulas for bed-load transport. *Intnal. Assoc. for Hydr. Structures Res. Proc.,* 2nd Meeting, Stockholm, pp. 39–65.

Michaels, P. J. 1985. *Virginia climate advisory.* Charlottesville: Univ. Virginia, 9:30.

Michaud, Y., Dionne, J. C., and Dyla, L. D. 1989. Frost bursting: A violent expression of frost action in rock. *Can. J. Ear. Sci.* 26:2075–80.

Mickelson, D. M., and Berkson, J. M. 1974. Till ridges presently forming above and below sea level in Wachusett Inlet, Glacier Bay, Alaska. *Geogr. Annlr.* 56 A:111–19.

Middelkoop, H. and Asselman, N. 1998. Spatial variability of floodplain sedimentation at the event scale in the Rhine-Meuse Delta, The Netherlands. *Earth Surf. Proc. Landfms.* 23:561–73.

Mielenz, R., and King, M. 1955. Physical-chemical properties and engineering performance of clays. *California Div. of Mines Bull.* 169:196–254.

Milana, J., and Ruzycki, L. 1999. Alluvial-fan slope as a function of sediment transport efficiency. *Jour. Sed. Res.* 69:553–62.

Miller, D. J., and Sias, J. 1998. Deciphering large landslides. Linking hydrological, groundwater, and slope stability modes through GIS. *Hydrological Processes* 12:923–41.

Miller, J. and 6 others. 1999. Effects of the 1997 flood on the transport and storage of sediment and Mercury within the Carson River valley, west-central Nevada. *Jour. Geol.* 107:313–27.

Miller, J. P. 1958. High mountain streams; effects of geology on channel characteristics and bed material. *New Mexico State Bur. Mines and Min. Res. Memo* 4.

Miller, J. R. 1990. The influence of bedrock geology on knickpoint development and channel-bed degradation along downcutting streams in south-central Indiana. *Journal of Geology* 9:591–605.

———. 1991. Controls on channel form along bedrock-influenced alluvial streams in south-central Indiana. *Physical Geography* 12:167–86.

Miller, J. R., Barr, R., Grow, D., Lechler, P., Richardson, D., Waltman, K., and Warwick, J. 1999. Effects of the 1997 flood on the transport and storage of sediment and mercury within the Carson River Valley, west-central Nevada. *Jour. Geology* 107:313–27.

Miller, J. R., Orbock Miller, S. M., Torzynski, C. A., and Kochel, R. C. 1989. Beach cusp destruction, formation, and evolution during and subsequent to an extratropical storm, Duck, North Carolina: *J. Geol.* 97:747–60.

Miller, J. R., Ritter, D. F., and Kochel, R. C. 1990. Morphometric assessment of lithologic controls on drainage basin evolution in the Crawford Uplands, south-central Indiana. *Amer. J. Science* 290:569–99.

Miller, R. D. 1966. Phase equilibria and soil freezing. *Proc. permafrost internat. conf.* (Lafayette, Ind., 1963). Natl. Acad. Sci.-Natl. Res. Council Pub. 1287, pp. 193–97.

Milliman, J. D. 1997. Fluvial sediment discharge to the sea and the importance of regional tectonics. In Tectonic uplift and climate change, pp. 239–57. Ed. by W. F. Ruddiman. New York: Plenum Press.

Milliman, J. D., and Meade, R. H. 1983. World-wide delivery of river sediment to the oceans. *Jour. Geology* 91:1–22.

Mills, H., and Starnes, D. 1983. Sinkhole morphometry in a fluviokarst region: Eastern Highland Rim, Tennessee, U.S.A. *Zeit F. Geomorph.* 27:39–54.

Mills, H. C., and Wells, P. D. 1974. Ice-shove deformation and glacial stratigraphy of Port Washington, Long Island, New York. *Geol. Soc. America Bull.* 85:357–64.

Mills, H. H. 1977. Textural characteristics of drift from some representative Cordilleran glaciers. *Geol. Soc. America Bull.* 88:1135–48.

———. 1987. Variation in sedimentary properties of colluvium as a function of topographic setting, Valley and Ridge Province, Virginia. *Zeit. F. Geomorph.* 31:277–2.

———. 1988. Surficial geology and geomorphology of the Mountain Lake area, Giles County, Virginia, including sedimentological studies of colluvium and boulder streams. *U.S. Geol. Survey Prof. Paper* 1469.

Mills, H. H., Brakenridge, G. R., Jacobson, R. B., Newell, W. L., Pavitch, M. J., and Pomeroy, J. S. 1987. Appalachian mountains and plateaus, pp. 5–50. In *Geomorphic systems of North America.* Ed. by W. L. Graf. Boulder: *Geol. Soc. Amer.,* Centennial Spec. Vol. 2.

Miotke, F. D. 1974. Carbon dioxide and the soil atmosphere: Munich, Abhandlungen Karst and Hohlenkunde, Reihe A., *Heft* 9, 49p.

Mitha, S., Tran, B., Werner, B., and Haff, P. 1986. The grain-bed impact process in aeolian saltation. *Acta Mechanica* 63:267–78.

Montgomery, D. R., Abbe, T. B., Buffington, J. M., Peterson, N. P., Schmidt, K. M., and Stock, J. D. 1996. Distribution of bedrock and alluvial channels in forested mountain drainage basins. *Nature* 381:587–89.

Montgomery, D. R., Schmidt, K. M., Greenberg, N. M., and Dietrich, W. E. 2000. Forest clearing and regional landsliding. *Geology* 28:311–14.

Moody, J., Pizzuto, J., and Meade, R. 1999. Ontogeny of a flood plain. *Geol. Soc. Amer. Bull.* 111:291–303.

Moon, J. W. 1980. On the expected diameter of random channel networks. *Water Resour. Res.* 16:1119–20.

Moore, J. 1970. Relationship between subsidence and volcanic load, Hawaii. *Bull. Volcanol.* 34:562–76.

Moore, J., and Clague, D. 1992. Volcano growth and evolution of the island of Hawaii. *Geol. Soc. Amer. Bull.* 104:1471–84.

Moore, J. G., and others. 1984. Age of debris from a huge Pleistocene wave on Lanai, Hawaii. *EOS* 65: 1082.

Moore, T. R. 1979. Land use and erosion in the Machakos Hills. *Ann. Assoc. Am. Geographers* 69:419–31.

Moran, S., Clayton, L., Hooke, R. LeB., Fenton, M., and Andriashek, L. 1980. Glacier-bed landforms of the prairie region of North America. *Jour. Glaciol.* 25:457–76.

Morehouse, D. F. 1968. Cave development via the sulfuric acid reactions. *Natl. Speleol. Soc. Bull.* 30:1–10.

Morey, G., Fournier, R., and Rowe, J. 1962. The solubility of quartz in water in the temperature interval from 25° C to 300°C. *Geochim. et Cosmochim. Acta* 26:1029–43.

Morgan, B. A., Weiczorek, G. F., and Campbell, R. H. 1999. Historical and potential debris flow and flood hazard map of the area affected by the June 27, 1995, storm in Madison

County, Virginia. *U.S. Geological Survey, Geol. Investigations Series.* I-2623-B.

Morgan, B. A., Weiczorek, G. J., Campbell, R. H., and Gori, P. L. 1997. Debris-flow hazards in areas affected by the June 27, 1995, storm in Madison County, Virginia. *U.S. Geological Survey Open-File Report OF-97-0438.*

Morgan, J. P. 1970. Deltas—A résumé. *Jour. Geol. Educ.* 18:107–17.

Morgan, P., and Swanberg, C. 1985. On the Cenozoic uplift and tectonic stability of the Colorado Plateau. *J. Geodynamics* 3:39–63.

Morin, H., and Payette, S. 1988. Holocene gelifluction in a snow-patch environment at the Forest-Tundra Transition along the eastern Hudson Bay Coast, Canada. *Boreas* 17:79–88.

Morisawa, M. 1985. *Rivers.* New York: Longman Group Ltd.

Morisawa, M., and Hack, J. eds. 1985. Tectonic geomorphology. *Proc. 15th Binghamton geomorph. symposium.* Boston: Allen and Unwin.

Morisawa, M. E. 1962. Quantitative geomorphology of some watersheds in the Appalachian Plateau. *Geol. Soc. America Bull.* 73:1025–46.

———. 1964. Development of drainage systems on an upraised lake floor. *Am. Jour. Sci.* 262:340–54.

Morner, N. A. 1980. *Earth rheology, isostasy, and eustasy.* New York: John Wiley & Sons.

Morris, E. M. 1976. An experimental study of the motion of ice past obstacles by the process of regelation. *Jour. Glaciol.* 17:79–98.

Morris, S. 1986. The significance of rainsplash in the surficial debris cascade of the Colorado Front Range foothills. *Earth Surf. Proc. Ldfms.* 11:11–22.

Morton, R. A., and Donaldson, A. C. 1978. Hydrology, morphology, and sedimentology of the Guadalupe fluvial-deltaic system. *Geol. Soc. America Bull.* 89:1030–36.

Morton, R. A., Gibeaut, J. C., and Paine, J. G. 1995. Meso scale transfer of sand during and after storms; implications for predicting shoreline movement. *Marine Geology* 126:161–79.

Mosley, M. P. 1979. Streamflow generation in a forested watershed, New Zealand. *Water Resour. Res.* 15:795.

———. 1982. The effect of a New Zealand beech forest canopy on the kinetic energy of water drops and on surface erosion. *Earth Surf. Proc. and Landforms* 7:103–7.

Moser, M., and Hohensinn, F. 1983. Geotechnical aspects of soil slips in alpine regions. *Engin. Geol.* 19:185–211.

Moss, J. H., and Bonini, W. 1961. Seismic evidence supporting a new interpretation of the Cody terrace near Cody, Wyo. *Geol. Soc. America Bull.* 72:547–56.

Moss, J.H. and Kochel, R. C. 1978.Unexpected geomorphic effects of the Hurricane Agnes storm and flood, Conestoga drainage basin, south-eastern Pennsylvania. *Journal of Geology,* 86:1–11.

Mothershead, D. N., and Pye, K. 1994. Tafoni on coastal slopes, South Devon, U.K. *Earth Surface Processes and Landforms.* 19:543–63.

Moyersons, J. 1988. The complex nature of creep movements on steeply sloping ground in southern Rwanda. *Earth Surf. Proc. and Landforms* 13:511–24.

Muhs, D., Rockwell, T., and Kennedy, G. 1992. Late Quaternary uplift rates of marine terraces on the Pacific Coast of North America, southern Oregon to Baja California Sur. *Quatern. Intl.* 15/16:121–33.

Muir, Wood, A. M. 1969. *Coastal hydraulics.* London: Macmillan.

Mukerji, A. B. 1976. Terminal fans of inland streams in Suglej-Yamauna plain, India. *Zeit. F. Geomorph.* 20:190–204.

Mullenders, W., and Gullentops, F. 1969. The age of the pingos of Belgium. In *The Periglacial Environment.* Ed. by T. Péwé. Montreal: McGill-Queens Univ. Press.

Muller, S. W. 1947. *Permafrost or permanently frozen ground and related engineering problems.* Ann Arbor, MI: J. W. Edwards.

Mulligan, K. 1988. Velocity profiles measured on the windward slope of a transverse dune. *Ea. Surf. Proc. Landfms.* 13:573–82.

Munk, W. H. 1949. Surf beats. *Am. Geophys. Union Trans.* 30:849–54.

Murphey, J. B., Wallace, D. E., and Lane, L. J. 1977. Geomorphic parameters predict hydrograph characteristics in the southwest. *Water Resour. Res. Bull.* 13:25–38.

Murphy, S., and 4 others. 1998. Chemical weathering in a tropical watershed, Luquillo Mountains, Puerto Rico: II. Rate and mechanism of biotite weathering. *Geochim. Cosmochim. Acta* 62:227–43.

Nahon, D. 1991. *Introduction to the petrology of soils and chemical weathering.* New York: John Wiley & Sons.

Nanson, G., and Croke, J. 1992. A genetic classification of floodplains. *Geomorph* 4:459–86.

Nanson, G. C. 1980. Point bar and floodplain formation of the meandering Beatton River, northeastern British Columbia, Canada. *Sedimentology* 27:3–29.

———. 1993. Anabranching rivers, and explanation and classification. *Third International Geomorphology Conference, Programme with Abstracts.* Hamilton, Canada, p. 204.

Nanson, G. C., and Hickin, E. J. 1986. A statistical analysis of bank erosion and channel migration in western Canada. *Geological Society of America Bulletin* 97:497–504.

Nanson, G. C. and Hickin, E. J. 1983. Channel migration and incision on the Beatton River. *Journal of Hydraulic Engineering,* 109:327–337.

Nanson, G. C., and Knighton, A. D. 1996. Anabranching rivers: Their cause, character and classification. *Earth Surf. Proc. and Landforms* 21:217–39.

Nanson, G. C., and Young, R. W. 1981. Overbank deposition and floodplain formation on small coastal streams of New South Wales. *Zeit. F. Geomorph.* 25:332–45.

Nash, D. 1980a. Forms of bluffs degraded for different lengths of time in Emmet County, Michigan. U.S.A. *Earth Surf. Proc. and Landforms* 5:331–45.

———. 1980b. Morphologic dating of degraded normal fault scarps. *Jour. Geology* 88:353–60.

———. 1987. Reevaluation of the linear diffusion model for morphologic dating of scarps. In *Directions in Paleoseismology.* Ed. by A. Crone and E. Omdahl. *U.S. Geol. Survey Open File Report,* 87–673, 325–38.

Nash, D. B. 1994. Effective sediment-transporting discharge from magnitude-frequency analysis. *Jour. Geology* 102:79–95.

National Academy of Sciences. 1983. *Proc. permafrost 4th internat. conf.* Washington, DC: National Academy Press.

National Research Council of Canada. 1978. *Proc. permafrost 3rd internat. conf.* Ottawa, Ont.: National Research Council of Canada.

———. 1988. Glossary of permafrost and related ground-ice terms. *Permafrost Subcommittee, Technical Memo.* 142.

National Research Council 1984. *The Utah landslides, debris flows, and floods of May and June 1983.* Washington, National Acad. Press.

Nelson, F. E., Hinkel, K. M., and Outcalt, S. I. 1992. Palsa-scale frost mounds. In *Periglacial geomorphology,* pp. 305–26. Ed. by J. C. Dixon and A. D. Abrahams. Chichester: John Wiley.

Nesbitt, H. W., and Young, G. M. 1984. Prediction of some weathering trends of plutonic and volcanic rocks based on thermodynamic and kinetic considerations. *Geochim. Cosmochim. Acta.* 48:1523–34.

———. 1989. Formation and diagenesis of weathering profiles. *Jour. Geology* 97:129–47.

Nesje, A., Blikra, L. H., and Anda, E. 1994. Dating rockfall-avalanche deposits from degree of rock-surface weathering by Schmidt-hammer tests; a study from Norangsdalen, Sunnmore, Norway. *Norsk Geologisk Tidsskrift* 74:108–13.

Newson, M. D. 1980. The geomorphic effectiveness of floods—a contribution stimulated by two recent events in mid-Wales. *Earth Surface Processes* 5:106–12.

Nickling, W., ed. 1986. Aeolian geomorphology. *Proc. 17th ann. Binghamton geomorph. symp.* Boston: Allen and Unwin.

———. 1988. The initiation of particle movement by wind. *Sedimentology* 35:499–511.

Nickling, W. G., and Bennett, L. 1984. The shear strength characteristics of frozen coarse granular debris. *Jour. Glaciol* 30:348–57.

Nilson, T. 1982. Alluvial fan deposits. In *Sandstone depositional environments,* pp. 49–86. Ed. by P. Scholle and P. Spearing. *Am. Asso. Petroleum Geol. Memoir* 31.

Nixon, J. F. 1989. Ground freezing and frost heave—a review. *The Northern Engineer* 19:8–18.

———. 1991. Discrete ice lens theory for frost heave in soils. *Can. Geotech. J.* 28:843–59.

Nordstrom, K. 1987. Predicting shoreline changes at tidal inlets on a developed coast. *The Prof. Geogr.* 39:457–65.

Norrman, J. O. 1980. Coastal erosion and slope development in Surtsey Island. *Zeit. F. Geomorph.* Suppl. 34:20–38.

Nunn, K. R., and Rowell, D. M. 1967. Regelation experiments with wires. *Philos. Mag.* 16:1281–83.

Nye, J. F. 1952a. The mechanics of glacier flow. *Jour. Glaciol.* 2:82–93.

———. 1952b. A comparison between the theoretical and the measured long profiles of the Unteraar glacier. *Jour. Glaciol.* 2:103–7.

———. 1957. The distribution of stress and velocity in glaciers and ice-sheets. *Proc. Royal Soc. London,* ser. A:239:113–33.

———. 1960. The response of glaciers and ice-sheets to seasonal and climatic changes. *Proc. Royal Soc. London,* ser. A:256:559–84.

Nye, J. F., and Martin, P. C. S. 1967. *Glacial Erosion,* pp. 78–83. Int. Assoc. Sci. Hydrol., Comm. Snow and Ice, Bern.

Oberlander, T. 1989. Slope and pediment systems. In *Arid zone geomorphology,* pp. 56–84. Ed. by D. Thomas. New York: Halsted Press.

Oberlander, T. 1997. Slope and pediment systems. In *Arid zone geomorphology,* pp. 135–64. Ed. by D. Thomas. Chichester: Wiley.

Oberlander, T. M. 1972. Morphogenesis of granitic boulder slopes in the Mojave Desert, California. *Jour. Geology* 80:1–20.

———. 1974. Landscape inheritance and the pediment problem in the Mojave Desert of southern California. *Am. Jour. Sci.* 274:849–75.

———. 1977. Origin of segmented cliffs in massive sandstones of southeastern Utah. In *Geomorphology in arid regions,* pp. 79–114. Ed. by D. O. Doehring. Boston: Allen and Unwin.

O'Connor, J., Ely, L., Wohl, E., Stevens, L., Melis, T., Kale, V., and Baker, V. 1994. A 4500-year record of large floods on the Colorado River in the Grand Canyon, Arizona. *Jour. Geology* 102:1–9.

O'Connor, J. E., and Baker, V. R. 1992. Magnitudes and implications of peak discharges from glacial Lake Missoula. *Geol. Soc. Amer. Bull.* 104:267–79.

O'Connor, J. E., and Webb, R. H. 1988. Hydraulic modeling for paleoflood analysis. In *Flood geomorphology,* pp. 393–402. Ed. by V. R. Baker, R. C. Kochel, and P. C. Patton. New York: John Wiley.

Odgaard, A. J.; Jain, S. C.; and Luzbetak, D. J. 1989. Hydraulic mechanisms of riverbank erosion. *Proceedings of the National Conference on Hydraulic Engineering.* New Orleans, Aug. 14–18.

Odegard, R.; Liestol, O.; and Sollid, J. L. 1988. Periglacial forms related to terrain parameters in Jotunheimen, southern Norway. In *Permafrost, Fifth International Conference Proceedings,* p. 59–61. Ed. by K. Senneset. Trondheim: Tapir Publishers.

Oertel, G. F., and Kraft, J. C. 1994. New Jersey and Delmarva barrier islands. *Geology of Holocene Barrier Island Systems,* pp. 207–32. Ed. by R. A. Davis. Berlin: Springer-Verlag.

Oertel, G. F.; Kraft, J. C.; Kearney, M. S.; and Woo, H. J. 1992. A rational theory for barrier lagoon development. *SEPM Spec. Pub.* 408, p. 77–87, Tulsa.

Ollier, C. 1988, *Volcanoes.* New York: Blackwell.

Ollier, C. D. 1963. Insolation weathering: Examples from central Australia. *Am. Jour. Sci.* 261:376–81.

———. 1969. *Weathering.* Edinburgh: Oliver and Boyd.

———. 1976. Catenas in different climates. In *Geomorphology and climate.* Ed. by E. Derbyshire. London: John Wiley & Sons.

Ollier, C. D., and Ash, J. E. 1983. Fire and rock breakdown. *Zeit. F. Geomorph.* 27:363–74.

Olyphant, G. 1981a. Allometry and cirque evolution. *Geol. Soc. America Bull.* 92:697–85.

Orbock-Miller, S., Ritter, D., Kochel, R., and Miller, J. 1993. Fluvial response to land-use changes and climatic variations within the Drury Creek watershed, southern Illinois. *Geomorphology* 6:309–29.

Orme, A. 1998. Late Quaternary tectonism along the Pacific coast of the Californias: A contrast in style. In *Coastal tectonics.* Ed. by I. Stewart and C. Vita-Finzi. Geol Soc., London, Spec. Publ. 146, 179–97.

Orme, A. R. 1973. Barrier and lagoon systems along the Zululand coast, South Africa. In *Coastal geomorphology,* pp. 181–217. Ed. by D. R. Coates. S.U.N.Y., Binghamton: 3rd Ann. Geomorph. Symposium.

———. 1974. Quaternary deformation of marine terraces between Ensenada and El Rosario, Baja California, Mexico. In *Geology of Peninsular California, Pacific Sections,* pp. 67–79. AAPG, SEPM, and SEG.

Osborn, G. D. 1975. Advancing rock glaciers in the Lake Louise area, Banff National Park, Alberta. *Can. J. Earth Sci.* 12:1060–62.

Ostrem, G. 1964. Ice-cored moraines in Scandinavia. *Geogr. Annlr.* 46:282–337.

Ouchi, S. 1985. Response of alluvial rivers to slow active tectonics. *Geol. Soc. America Bull.* 96:504–15.

Outcalt, S. I., and Benedict, J. B. 1965. Photo-interpretation of two types of rock glaciers in the Colorado Front Range, U.S.A. *Jour. Glaciol.* 5:849–56.

Outcalt, S. I., and Nelson, F. 1984. Computer simulation of buoyancy and snow-cover effects in palsa dynamics. *Arctic and Alpine Research* 16:259–63.

Overpeck, J., Webb, R., and Webb, T. 1992. Mapping eastern North America vegetation change of the past 18 ka: No-analogs and the future. *Geology* 20:1071–74.

Owen, P. 1964. Saltation of uniform grains in air. *Jour. Fluid Mech.* 20:225–42.

Owens, E. H., and Harper, J. R. 1977. Frost-table and thaw depths in the littoral zone near Pearl Bay, Alaska. *Arctic* 30:155–68.

Owens, L., and Watson, J. 1986. Weathering rates of gneiss and depletion of exchangeable cations in soils under environmental acidification. *Jour. Geol. Soc.,* London, 143:673–77.

Ozouf, J. C., Texier, J. B., Bertran, P., and Coutard, J. P. 1995. Quelqoes Coups Caracter istiques dans les depos de Versant Aquitaine Septentriionale. *Permafrost and Periglacial Processes* 6:89–106.

Paige, S. 1912. Rock-cut surfaces in the desert ranges. *Jour. Geology* 20:442–50.

Pakiser, L. C., and Robinson, R. 1966. Composition of the continental crust as estimated from seismic observations. In *The earth beneath the continents,* pp. 620–26. Ed. by J. Steinhart and T. Smith. *Am. Geophys. Union, Geophys.* Monograph 10.

Palmer, A. C. 1972. A kinematic wave model of glacier surges. *Jour. Glaciol.* 11:65–72.

Palmer, A. N. 1975. Origin of maze caves. *Natl. Speleol. Soc. Bull.* 37:57–76.

———. 1981. *A geological guide to Mammoth Cave National Park.* Teaneck, NJ: Zephyrus Press.

———. 1984. Recent trends in karst geomorphology. *Jour. Geol. Educ.* 32:247–53.

———. 1990. Groundwater processes in karst terranes. In Groundwater geomorphology; the role of subsurface water in earth-surface processes and landforms. Boulder, *Geological Society of America Special Paper 252,* pp. 177–209. Ed. by C. G. Higgens and D. R. Coates.

———. 1991. Origin and morphology of limestone caves. *Geological Society of America Bulletin.* 103:1–21.

Palmer, V. E., and Palmer, A. N. 1975. Landform development of the Mitchell Plain of southern Indiana: Origin of a partially karstic plain. *Zeit. F. Geomorph.* 19:1–39.

Palmquist, R. C. 1979. Geological controls on doline characteristics in mantled karst. *Zeit. F. Geomorph.* Suppl. 32:90–106.

Panoś, V., and Ŝtelcl, O. 1968. Physiographic and geologic control in development of Cuban mogotes. *Zeit. F. Geomorph.* 12:117–65.

Paola, C., and Seal, R. 1995. Grain size patchiness as a cause of selective deposition and downstream fining. *Water Resour. Res.* 31:1395–1407.

Paredes, J. R., and Buol, S. W. 1981. Soils in an aridic, ustic, udic, climosequence in the Maracaibo Lake Basin, Venezuela. *Soil Sci. Soc. Am. Proc.* 45:385–91.

Parizek, R. 1969. *Glacial ice-contact rings and ridges. Geol. Soc. America Spec. Paper* 123:49–102.

Park, C. 1977. World-wide variations in hydraulic geometry exponents of streams channels: An analysis and some observations. *J. Hydrol.* 33:133–46.

Parker, G. 1976. On the cause and characteristic scale of meandering and braiding in rivers. *Jour. Fluid Mech.* 76:459–80.

Parker, G., Klingeman, P. C., and McLean, D. G. 1982. Bedload and size distribution in paved gravel-bed streams. *Proceedings of the American Society of Civil Engineers, Journal Hydraulics Division.* HY4, 108:544–571.

Parrett, C. 1985. Fire-related debris flows in the Beaver Creek drainage, Lewis and Clark Co., Montana: *U.S. Geol. Survey Selected Papers in the Hydrologic Sciences.*

Parsons, A. J. 1988. *Hillslope Form.* London: Routledge.

Parsons, A. J., and Abrahams, A. D. 1987. Gradient-particle size relations on quartz monzonite debris slopes in the Mojave Desert, *Jour. Geology* 95:423–52.

Parsons, M. L. 1998. Salt marsh sedimentary record of the landfall of Hurricane Andrew on the Louisiana coast; diatoms and other paleoindicators. *Jour Coastal Res.* 14:939–50.

Parsons, T., and McCarthy, J. 1995. The active southwest margin of the Colorado Plateau: Uplift of mantle origin. *Geol. Soc. Amer. Bull.* 107:139–47.

Parsons, T., Thompson, G., and Sleep, N. 1994. Mantle plume influence on the Neogene uplift and extension of the U.S. western Cordillera. *Geology* 22:83–86.

Paterson, W. S. B. 1964. Variations in velocity of Athabasca Glacier with time. *Jour. Glaciol.* 5:277–85.

———. 1969. *The physics of glaciers.* Oxford: Pergamon Press.

———. 1994. *The physics of glaciers,* 3d ed. Oxford: Pergamon.

Patterson, C. J., and Hooke, R. Le B. 1996. Observations on the internal structure and origin of some flutes in glaciofluvial sediments, Blomstrandbreen, north-west Spitsbergen. *Jour. Glaciol.* 13:393–400.

Patton, H. B. 1910. Rockstreams of Veta Park, Colorado. *Geol. Soc. America Bull.* 22:663–76.

Patton, P., and Boison, P. 1986. Processes and rates of Holocene alluvial terraces in Harris Wash, Escalante River basin, south-central Utah. *Geol. Soc. Amer. Bull.* 97:369–78.

Patton, P. C. 1988. Drainage basin morphometry and floods. In *Flood Geomorphology,* pp. 51–64. Ed. by V. R. Baker, R. C. Kochel, and P. C. Patton. New York: John Wiley.

Patton, P. C., and Baker, V. R. 1976. Morphometry and floods in small drainage basins subject to diverse hydrogeomorphic controls. *Water Resour. Res.* 12:941–52.

———. 1977. Geomorphic response of central Texas stream channels to catastrophic rainfall and runoff. In *Geomorphology of Arid and Semiarid Regions,* pp. 189–217. Ed. by D. O. Doehring. S.U.N.Y., Binghamton: Pubs. in Geomorphology.

Patton, P. C., and Dibble, D. S. 1982. Archeologic and geomorphic evidence for the paleohydrologic record of the Pecos River in west Texas. *Am. Jour. Sci.* 282:97–121.

Patton, P. C., and Schumm, S. A. 1975. Gulley erosion, northwestern Colorado: A threshold phenomenon. *Geology* 3:88–90.

———. 1981. Ephemeral-stream processes: Implications for studies of Quaternary Valley fills. *Quat. Res.* 15:24–43.

Patton, P. C., Baker, V. R., and Kochel, R. C. 1979. Slack-water deposits: A geomorphic technique for the interpretation of fluvial paleohydrology. In *Adjustments of the Fluvial System,* pp. 225–53. Ed. by D. D. Rhodes and G. P. Williams. Dubuque, IA: Kendall Hunt.

Pavich, M. 1986. Processes and rates of sapprolite production and erosion on a foliated granite rock of the Virginia Piedmont. In *Rates of chemical weathering of rocks and minerals,* pp. 551–90. Ed. by S. Colman and D. Dethies. Orlando, FL: Academic Press.

———. 1989. Regolith residence time and the concept of surface age of the Piedmont "peneplain." *Geomorphology* 2:181–96.

Pavich, M., Brown, L., Valette-Silver, J., Klein, J., and Middleton, R. 1985. ^{10}Be analysis of a Quaternary weathering profile in the Virginia Piedmont. *Geology* 13:39–41.

Pavich, M., Leo, G., Obermeier, S., and Estabrook, J. 1989. Investigations of the characteristics, origin and residence time of the upland residual mantle of the Piedmont of Fairfax County, VA. *U.S. Geol. Survey Prof. Paper* 1352.

Pazzaglia, F., and Gardner, T. 1993. Fluvial terraces of the lower Susquehanna River. *Geomorph.* 8:83–113.

———. 1994. Late Cenozoic flexural deformation of the middle U.S. Atlantic margin. *J. Geophys. Res.* 99(B6):12143–57.

Pazzaglia, F., Gardner, T., and Merritts, D. 1998. Bedrock fluvial incision and longitudinal profile development over geologic time scales determined by fluvial terraces. In *Rivers over rock: fluvial processes in bedrock channels,* pp. 207–35. Ed by K. Tinkler and E. Wohl. Am. Geophys. Un., Wash., D.C.

Peltier, L. 1950. The geographical cycle in periglacial regions as it is related to climatic geomorphology. *Ann. Assoc. Am. Geog.* 40:214–36.

Penland, S., and others. 1989. Morphodynamic signature of the 1985 hurricane impacts on the northern Gulf of Mexico. In Representative publications from the Louisiana Barrier Island erosion study, pp. 439–53. Ed. by W. S. Jeffress and others. Reston: *U.S. Geol. Survey Open File Report.*

Percival, J., and Berry, M. 1987. The lower crust of the continents. In *Composition, structure and dynamics of the lithosphere-asthenosphere system.* Ed. by K. Fuchs and C. Froidevaux, Am. Geophys. Un., Wash. D.C., Geodynamics Series 16:33–59.

Perez, F. L. 1992. Miniature sorted stripes in the Paramo de Piedras Blancas (Venezuelan Andes). In *Periglacial geomorphology,* pp. 125–58. Ed. by J. C. Dixon and A. D. Abrahams. Chichester: John Wiley.

———. 1998. Talus fabric, clast morphology, and botanical indicators of slope processes on the Chaos Crags, (California Cascades), USA. *Geographie Physique et Quaternaire* 52:47–68.

Perutz, M. F. 1940. Mechanism of glacier flow. *Proc. Royal Soc. London,* ser. A:52:132–35.

Pesci, M. 1968. Loess. In *Encyclopedia of geomorphology,* pp. 674–78. Ed. by R. W. Fairbridge. New York: Reinhold Book Corp.

Peterson, J., and Junge, C. 1971. Source of particulate matter in the atmosphere. In *Man's Impact on Climate,* p. 310–320. Ed. by W. Matthews, W. Kellog, and G. Robinson. Cambridge: MIT Press.

Pethick, J. S. 1984. *An introduction to coastal geomorphology.* London: Arnold.

Petrie, G., and Price, R. J. 1966. Photogrammetric measurements of the ice wastage and morphological changes near the Casement Glacier, Alaska. *Can. J. Earth Sci.* 3:827–40.

Péwé, T. L. 1955. Origin of the upland silt near Fairbanks, Alaska. *Geol. Soc. America Bull.* 66:699–724.

———. 1959. Sand-wedge polygons (Tesselations) in the McMurdo Sound region, Antarctica—A progress report. *Am. Jour. Sci.* 257:545–52.

———. 1966. Ice-wedges in Alaska—Classification, distribution and climatic significance. In *Proc. Permafrost Internat. Conf.* (Lafayette, Ind., 1963). Natl. Acad. Sci.-Natl. Res. Council Pub. 1287, pp. 76–81.

———. 1969. *The periglacial environment.* Montreal: McGill-Queens Univ. Press.

———. 1973. Ice wedge casts and past permafrost distribution in North America. *Geoform* 15:15–26.

———. 1981. Desert dust: Origin, characteristics, and effect on man. *Geol. Soc. America Spec. Paper* 186.

Péwé, T. L., and Journaux, A. 1983. Origin and character of loesslike silt in unglaciated south-central Yakutia, Siberia, U.S.S.R. *U.S. Geol. Surv. Prof. Paper* 1262.

Peyronnin, C. A., Jr. 1962. Erosion of Isles Dernieres and Timbalier Islands. Am. Soc. Civil Engineers. *Jour. Waterways and Harbors.* 1:57–69.

Phillip, H. 1920. Geologische Untersuchungen über den Mechanismus der Gletscher Bewegung und die Entstehung der gletschertextur. *Neuer Jb. Miner. Geol. Palaont.* 43:439–556.

Phillips, J., and Renwick, W., eds. 1992. *Geomorphic systems.* Proc. 23rd Binghamton symp. Elsevier, 487p.

Phillips, J. D. 1990. The instability of hydraulic geometry. *Water Resources Research* 26:739–44.

———. 1991. Fluvial sediment delivery to a coastal plain estuary in the Atlantic drainage of the United States. *Marine Geology* 98:121–34.

Phillips, L. F., and Schumm, S. A. 1987. Effect of regional slope on drainage networks. *Geology* 15:813–16.

Phipps, R. L. 1985. Collecting, preparing, cross-dating, and measuring tree increment cores. *U.S. Geol. Survey Water Res. Invest.* Rep. 85-4148. Wash., DC.

Picknett, R. G., Bray, L. F., and Stenner, R. D. 1976. The chemistry of cave waters. In *The Science of Speleology.* T. D. Ford and C. H. D. Cullingford (eds.), London: Academic Press, pp. 213–266.

Pickup, G. 1976. Adjustment of stream-channel shape to hydrologic regime. *J. Hydrology* 29:51–75.

———. 1977. Simulation modelling of river channel erosion. In *River channel changes,* pp. 47–60. London: John Wiley & Sons.

Pierce, J. W. 1970. Tidal inlets and washover fans. *Jour. Geology* 78:230–34.

Pierce, K., and Colman, S. 1986. Effect of height and orientation (microclimate) on geomorphic degradation rates and processes, late-glacial terrace scarps in central Idaho. *Geol. Soc. America Bull.* 97:869–85.

Pierson, T. C. 1980. Erosion and deposition by debris flows at Mount Thomas, New Zealand. *Earth Surf. Proc.* 5:227–47.

Pierson, T. C., and Costa, J. E. 1987. A rheologic classification of subaerial sediment-water flows. In *Debris flows/avalanches: process, recognition, and mitigation*, pp. 1–12. Ed. by J. E. Costa and G. F. Wieczorek. Boulder: *Geol. Soc. Amer. Rev. in Engin. Geol 7*.

Pike, R. J., and Wilson, S. E. 1971. Elevation-relief ratio, hypsometric integral, and geomorphic area-altitude analysis. *Geol. Soc. America Bull.* 82:1079–84.

Pillans, B. 1990. Pleistocene marine terraces in New Zealand: A review. *New Zealand Jour. Geol. Geophys.* 33:219–31.

Pissart, A. 1970. *The pingos of Prince Patrick Island (76°N–120°W)*. Natl. Res. Council of Canada, Tech. Trans. 1401.

Pitlick, J. 1992. Flow resistance under conditions of intense gravel transport. *Water Resour. Res.* 28:891–903.

———. 1993. Response and recovery of a subalpine stream following a catastrophic flood. *Geol. Soc. America Bull.* 105:657–70.

Pizzuto, J. E. 1984. Bank erodibility of shallow sandbed streams. *Earth Surface Processes and Landforms* 9:113–24.

———. 1986. Barrier island migration and onshore sediment transport, southwestern Delaware Bay, Delaware USA. *Marine Geol.* 71:299–325.

———. 1995. Downstream fining in a network of gravel-bedded rivers. *Water Resour. Res.* 31:753–59.

Plummer, L. 1975. Mixing of seawater with calcium carbonate groundwater. In *Quantitative studies in the geological sciences*, pp. 219–36. Ed. by E. Whitten. Geol. Soc. America Mem. 142.

Plummer, L., Vacher, H., Mackenzie, F., Bricker, O., and Land, L. 1976. Hydrochemistry of Bermuda: A case history of groundwater diagenesis of biocalcarenites. *Geol. Soc. America Bull.* 87:301–16.

Pohl, E. R. 1955. *Vertical shafts in limestone caves*. Natl. Speleol. Soc. Occasional Paper 2.

Pohl, E. R., and White, W. B. 1965. Sulfate minerals: Their origin in the central Kentucky karst. *Am. Mineralogist* 50:1461–65.

Poldervaart, A., ed. 1955. Chemistry of the Earth's crust. In *Crust of earth, Geol. Soc. America Spec. Paper* 62:119–44.

———. 1982. Landslides in the greater Pittsburgh region, Pennsylvania. *U.S. Geol. Survey Prof. Paper* 1229.

Pollard, W. H., and H. M. French. 1983. Seasonal frost mound occurrence. North Fork Pass, Ogilvie Mountains, northern Yukon, Canada. *Proceedings of the Fourth International Conference of Permafrost* 1:1000–1004.

———. 1984. The groundwater hydraulics of seasonal frost mounds, North Fork Pass, Yukon Territory. *Canadian Journal of Earth Sciences* 21:1073–81.

Pollard, W. H., and H. M. French. 1985. The internal structure and ice crystallography of seasonal frost mounds. *Journal of Glaciology* 31:157–62.

Pope, G., Dorn, R., and Dixon, J. 1995. A new conceptual model for understanding geographical variations in weathering. *An Amer. Asso. Geogr.* 85:38–64.

Porslid, A. E. 1938. Earth mounds in unglaciated Arctic northwestern America. *Geogr. Rev.* 28:46–58.

Porter, S. C. 1977. Present and past glaciation threshold in the Cascade Range, Washington, U.S.A. Topographic and climatic controls, and paleoclimatic implications. *Jour. Glaciol.* 18:101–16.

Porter, S. C., and Orombelli, G. 1980. Catastrophic rockfall of September 12, 1717, on the Italian flank of the Mont Blanc Massif. *Zeit f. Geomorph.* 24:200–18.

———. 1981. Alpine rockfall hazards. *American Scientist* 69:67–77.

Post, A. S. 1960. The exceptional advances of the Muldrow, Black Rapids and Sustina glaciers. *Jour. Geophys. Research* 65:3703–12.

———. 1966. The recent surge of Walsh glacier, Yukon and Alaska. *Jour. Glaciol.* 6:375–81.

———. 1967. Effects of the March 1964 Alaskan earthquake on glaciers. *U.S. Geol. Survey Prof. Paper* 544-D.

———. 1969. Distribution of surging glaciers in western North America. *Jour. Glaciol.* 8:229–40.

Potter, N., Jr. 1972. Ice-cored rock glacier, Galena Creek, northern Absaroka Mountains, Wyoming. *Geol. Soc. America Bull.* 83:3025–57.

Potter, N., Jr., and Moss, J. H. 1968. Origin of the Blue Rocks block field deposits, Berks County, Pennsylvania. *Geol. Soc. America Bull.* 79:255–62.

Potts, A. S. 1970. Frost action in rocks: Some experimental data. *Inst. Brit. Geog. Trans.* 49:109–24.

Powers, R. W. 1962. Arabian Upper Jurassic carbonate reservoir rocks. In *Classification of Carbonate Rocks*. Ed. by W. E. Ham. *Am. Assoc. Petroleum Geologists Mem.* 1:122–92.

Prestegaard, K. L. 1983a. Bar resistance in gravel bed streams at bankfull stage. *Water Resour. Res.* 19:472–76.

———. 1983b. Variables influencing water-surface slopes in gravel-bed streams at bankfull stage. *Geol. Soc. America Bull.* 94:673–78.

Price, L. W. 1972. *The periglacial environment, permafrost, and man*. Assoc. Am. Geog., Comm. on College Geog. Resource Paper 14.

———. 1991. Subsurface movement on solifluction slopes in the Ruby Range, Yukon Territory, Canada. A 20-year Study. *Arctic and Alpine Res.* 23: 200–05.

Price, R. J. 1966. Eskers near the Casement glacier, Alaska. *Geogr. Annlr.* 48:111–25.

———. 1969. Moraines, sandar, kames, and eskers near Breidamerkùrjökull, Iceland. *Inst. Brit. Geog. Trans.* 46:17–43.

———. 1970. Moraines at Fjallsjökull, Iceland. *Arc. Alp. Res.* 2:27–42.

———. 1973. *Glacial and fluvioglacial landforms*. New York: Hafner.

Priesnitz, K. 1990. Geomorphic activity of rivers during snow melt and breakup, Richardson Mountains, Yukon and Northwest Territories, Canada. *Permafrost and Periglacial Proc.* 1:295–99.

Pritchard, G. B. 1962. Inuvik, Canada's new Arctic town. *Polar Record* 11:71:145–54.

Prowse, T. D., and Ommanney, C. S. L., eds. 1990. Northern hydrology: Canadian perspectives. Ottawa: Environment Canada, *NHRI Science Report* 1.

Pubellier, M., Deffontaines, B., Quebral, R., and Rangin, C. 1994. Drainage network analysis and tectonics of Mindanao, southern Philippines. *Geomorphology* 9:325–42.

Purdue University, 1985. *PCSTABLE4,* Computer Modeling of Slope Stability, prepared for Federal Highway Administration (1985). School of Civil Engineering, West Lafayette, Indiana.

Pye, K., 1987. *Aeolian dust and dust deposits.* Academic Press, London, 334pp.

Pye, K., and Lancaster, N. (eds.), 1993. Aeolian sediments, Spec. Publ. 16, *Intl. Asso. Sedimentologists,* Oxford: Blackwell Scientific Publ. 167 p.

Pye, K., and Lancaster, N. (eds.). 1993. Aeolian sediments: Ancient and Modern. IAS Spec. Publ. 16, Blackwell Science, Oxford.

Pye, K., and Tsoar, H. 1990. *Aeolian Sand and Sand Dunes.* London: Unwin Hyman, 396 p.

Quinlan, J. F. 1982. Groundwater basin delineation with dye-tracing potentiometric surface mapping, nd cve mapping, Mammoth Cave Region, Kentucky. U.S.A. *Beitraege zur Geologie der Schweiz-Hydrologie,* 28:177–190.

Ragan, R. M. 1968. An experimental investigation of partial area contributions. *Intl. Assoc. Sci. Hydrol.* Pub. 76:241–51.

———. 1967. Sheetfloods, streamfloods, and the formation of pediments. *Ann. Assoc. Am. Geog.* 57:593–604.

———. 1976. Coulee alignment and the wind in southern Alberta, Canada: Discussion. *Geol. Soc. America Bull.* 87:157.

Rahn, P. H. 1966. Inselbergs and nickpoints in southwestern Arizona. *Zeit. F. Geomorph.* 10:217–25.

———. 1967. Sheetfloods, streamfloods, and the formation of pediments. *Ann. Assoc. Am. Geog.* 57:593–604.

Rains, B., Shaw, J., Skoye, R., Sjogren, D., and Kvill, D. 1993. Lake Wisconsin subglacial megaflood paths in Alberta. *Geology* 21:323–26.

Rains, B., and Welch, J. 1988. Out-of-phase Holocene terraces in part of the North Saskatchewan River basin, Alberta. *Can. J. Earth Sci.* 25:454–61.

Rains, R., and Shaw, J. 1981. Some mechanisms of controlled moraine development, Antarctica. *Jour. Glaciol* 27:113–28.

Ramanathan, V. 1988. The greenhouse theory of climate change: A test by an inadvertent global experiment. *Science* 240:293–99.

Ramirez-Herrera, M-T., and Urrutia-Fucugauchi, J. 1999. Morphotectonic zones along the coast of Pacific continental margin, southern Mexico. *Geomorphology* 28:237–50.

Rampino, M., and Self, S. 1993. Climate-volcanism feedback and the Toba eruption of 74,000 years ago. *Quaternary Res.* 40:269–80.

Rampton, V. N. 2000. Large-scale effects of subglacial meltwater flow in the southern Slave Province, Northwest Territories, Canada. *Can. J. Earth. Sci.* 37:81–93.

Ramsey, M., and 3 others. 1999. Identification of sand sources and transport pathways at the Kelso Dunes, California using thermal infared remote sensing. *Geol. Soc. Amer. Bull.* 111:646–62.

Rankin, J. K. 1952. Development of the New Jersey shore. In *3rd coastal engr. conf. proc.,* pp. 306–17. Ed. by J. W. Johnson. Cambridge, MA: Council of Wave Research.

Rapp, A. 1960. Recent development of mountain slopes in Karkevagge and surroundings, northern Scandinavia. *Geogr. Annlr.* 42:65–206.

Ray, L. L. 1951. Permafrost. *Arctic* 4:196–203.

Ray, R. J., Krantz, W. B., Caine, T. N., and Gunn, R. D. 1983a. A mathematical model for patterned ground: Sorted polygons and stripes, and underwater polygons. In *Proc. permafrost 4th internat. conf.,* pp. 1036–41. Natl. Acad. Sci.

———. 1983b. A model for sorted patterned ground regularity. *Jour. Glaciol.* 29:317–37.

Raymond, C. F. 1971. Flow in a transverse section of Athabasca Glacier, Alberta, Canada. *Jour. Glaciol.* 10:55–84.

———. 1987. How do glaciers surge: A review. *Journal of Geophysical Research* 92:9121–34.

Rea, B. R., Whalley, W. B., and Porter, E. M. 1996. Rock weathering and the formation of summit block-field slopes in Norway. Examples and implications. In *Advances in hillslope processes,* v. 2, pp. 1257–75. Ed. by M. G. Anderson and S. M. Brooks. New York: John Wiley.

Reed, B., Galvin, C. J., and Miller, J. P. 1962. Some aspects of drumlin geometry. *Am. Jour. Sci.* 260:200–210.

Reeve, I. J. 1982. A splash transport model and its application to geomorphic measurement. *Zeit. F. Geomorph.* 26:55–71.

Reheis, M., 1987a. Gypsic soils on the Kane alluvial fans, Big Horn County, Wyoming, *U.S. Geol. Surv. Bull.* 159-C.

———. 1987b. Climate implications of alternating clay and carbonate formation in semi arid soils of south-central Montana. *Quat. Res.* 29:270–82.

Reheis, M., and Kihl, R. 1995. Dust deposition in southern Nevada and California, 1984–1989: Relations to climate, source area, and source lithology. *J. Geophys. Res.* 100 (D5):8893–918.

Reheis, M., and 7 others. 1995. Quaternary soils and dust deposition in southern Nevada and California. *Geol. Soc. Amer. Bull.* 107:1003–22.

Reheis, M. C. 1987. Climatic supplication of alternating clay and carbonate formation in semiarid soils of south-central Montana. *Quaternary Res.* 27:270–82.

Reidel, S., and Tolan, T. 1992. Eruption and emplacement of flood basalt: An example from the large-volume Teepee Butte Member, Columbia River Basalt Group. *Geol. Soc. America Bull.* 104:1650–71.

Reinhardt, J., and Sigleo, W., eds. 1988. Paleosols and weathering through geologic time. *Geol. Soc. Amer. Sepc. Paper.*

Rendell, H. 1982. Clay hillslope erosion rates in Basento Valley, S. Italy, *Geogr. Annlr.* 64A:141–47.

Reneau, S. L., and Dietrich, W. E. 1987. The importance of hollows in debris flow studies; Examples from Marin County, California. In *Debris flows/avalanches: process, recognition, and mitigation,* pp. 165–80. Ed. by J. E. Costa and G. F. Wieczorek. Boulder, *Geol. Soc. Amer. Rev. in Engin. Geol* 7.

Reneau, S. L., Dietrich, W. E., Donohue, D. J., Jull, A. J. T., and Rubin, M. 1990. Late Quaternary history of colluvial deposition and erosion in hollows, central California Coast Ranges. *Geol. Soc. Amer. Bull.* 102:969–82.

Reneau, S. L., Dietrich, W. E., Rubin, M., Donahue, D. J., and Jull, A. J. T. 1989. Analysis of hillslope erosion rates using colluvial deposits. *J. Geology* 97:45–63.

Repelewska-Pekabwa, J., and Gluza, A. 1988. Dynamics of permafrost active layer—Spitsbergen. In *Permafrost, Fifth International Conference Proceedings*, pp. 448–53. Ed. by K. Senneset. Trondheim: Tapir Publishers.

Retallack, G. J. 1983. Late Eocene and Oligocene paleosols from Badlands National Park, South Dakota. *Geol. Soc. Amer. Spec. Paper* 193.

———. 1988. Field recognition of paleosols: In Paleosols and weathering through geologic time. Ed. by J. Rheinhardt and W. Sigleo. *Geol. Soc. Amer. Spec. Paper,* 216:1–20.

———. 1990. *Soils of the past: An introduction to paleopedology.* Boston: Unwin Hyman.

Revue de géomorphologie dynamique. 1967. Field methods for the study of slope and fluvial processes. *Rev. geomorph. dynamique* 17:145–88.

Rhea, S. 1989. Evidence of uplift near Charleston, South Carolina. *Geology* 17:311–15.

Rhoads, B. L., and Welford, M. R. 1991. Initiation of river meandering. *Progress in Physical Geography* 15:127–56.

Rhodes, B. 1992. Fluvial geomorphology. *Progr. in Phys. Geogr.* 16:489–96.

Rhodes, D. D. 1977. The b-f-m diagram: Graphical representation and interpretation of at-a-station hydraulic geometry. *Am. Jour. Sci.* 277:73–96.

Rhodes, J., and 4 others. 1989. Geochemical evidence for invasion of Kilauea's plumbing system by Mauna Loa magma. *Nature* 337:257–60.

Rice, A. 1976, Insolation warmed over. *Geology* pp. 61–62.

Rice, T. E., Niedoroda, A. W., and Pratt, A. P. 1976. The coastal processes and geology: Virginia barrier islands. In *Virginia coast reserve study: Ecosystem description,* Ed. by R. D. Dueser. *The Nature Conservancy.*

Richards, J. F. 1990. Land transformation. In *The earth as transformed by human action,* pp. 163–178. Ed. by. I. Turner and others. Cambridge: Cambridge Univ. Press.

Richards, K. S. 1976a. Channel width and the riffle-pool sequence. *Geol. Soc. America Bull.* 87:883–90.

———. 1976b. The morphology of riffle-pool sequences. *Earth Surf. Proc. and Landforms* 1:71–88.

———. 1979. Prediction of drainage density from surrogate measures. *Water Resour. Res.* 15:435–42.

———. 1982. *Rivers.* London: Methuen and Co.

Richardson, H. W. 1942. Alcan—America's glory road, parts I and II. *Engr. News Record* 129:25:81–96 and 27:35–42.

———. 1943. Alcan-America's glory road, part III. *Engr. News Record* 130:1:131–38.

———. 1944. Controversial Canol. *Engr. News Record* 132:2:78–84.

Richardson, J. L., Wilding, L. P., and Daniels, R. B. 1992. Recharge and discharge of groundwater in aquic conditions illustrated with flownet analysis. *Geoderma* 53:65–78.

Rickenmann, D., and Zimmermann, M. 1993. The 1987 debris flows in Switzerland: Documentation and analysis. *Geomorph.* 8:175–89.

Riddle, C. H., Rooney, J. W., and Bredthauer, S. R. 1988. Yukon River bank stabilization: A case study. In *Permafrost, fifth international conference proceedings,* pp. 1312–16. Ed. by K. Senneset. Trondheim: Tapir Publishers.

Rieke, R. D., Vinson, T. S., and Mageau, D. W. 1983. The role of specific surface area and related index properties in the frost heave susceptibility of soils. In *Proc. permafrost 4th internat. conf.,* pp. 1066–71. Natl. Acad. Sci.

Rigsby, G. P. 1960. Crystal orientation in glacier and in experimentally deformed ice. *Jour. Glaciol.* 3:589–606.

Rimstidt, J. 1997. Quartz solubility at low temperatures. *Geochim. Cosmochim. Acta* 61, 2553–58.

Ritchie, W., and Penland, S. 1988. Rapid dune changes associated with overwash processes on the deltaic coast of south Louisiana. *Marine Geol.* 81: 97–122.

Ritter, D. F. 1967. Rates of denudation. *Jour. Geol. Educ.* 15, C.E.G.S. short rev. 6:154–59.

———. 1972. The significance of stream capture in the evolution of a piedmont region, southern Montana. *Zeit. F. Geomorph.* 16:83–92.

———. 1975. Stratigraphic implications of coarse-grained gravel deposited as overbank sediment, southern Illinois. *Jour. Geology* 83:645–50.

Ritter, D. 1988a. Landscape analysis and the search for geomorphic unity. *Geol. Soc. America Bull.* 100:160–71.

———. 1988b. Floodplain erosion and deposition during the December 1982 floods in southeast Missouri, in *Flood Geomorphology* pp. 243–59. Ed. by V. Baker, R. Kochel, and P. Patton. New York: Wiley-Interscience.

Ritter, D., and Dutcher, R. 1990. Geomorphic controls on the origin and location of the Tolman Ranch ventifact site, Park County, Wyoming, USA, *Jour. Geology* 98:943–54.

Ritter, D. F., Kinsey, W. F., and Kauffman, M. E. 1973. Overbank sedimentation in the Delaware River valley during the last 6,000 years. *Science* 179:374–75.

Ritter, D., Kochel, R., and Miller, J. 1999. The disruption of Grassy Creek: Implications concerning catastrophic events and thresholds. *Geomorphology,* 29:323–38.

Ritter, D., and TenBrink, N. 1986. Alluvial fan development and the glacial-glaciofluvial cycle. Nenana Valley, Alaska. *Jour. Geology* 94:613–25.

Ritter, J., and 16 others. 1993. Quaternary evolution of Cedar Creek alluvial fan, Montana. *Geomorphology* 8:287–304.

Ritter, J., Miller, J. Enzel, Y. and Wells, S. 1995. Reconciling the roles of tectonism and climate in Quaternary alluvial fan evolution. *Geology* 23:245–48.

Roberge, J., and Plamondon, A. P. 1987. Snowmelt runoff pathways in a boreal forest hillslope, the role of pipe throughflow. *J. Hydrology* 95:39–54.

Roberts, H. 1997. Dynamic changes of the Holocene Mississippi River delta plain: The delta cycle. *J. Coastal Res.* 13:605–27.

Robin, G. deQ., and Weertman, J. 1973. Cyclic surging of glaciers. *Jour. Glaciol.* 12:3–18.

Robinson, G. 1966. Some residual hillslopes in the Great Fish River Basin, South Africa. *Geogr. Jour.* 132:386–90.

Robinson, L.A. 1977. Marine erosive processes at the cliff foot. *Marine Geol.* 23:257–71.

Rodriquez-Itrube, I., and Valdes, J. B. 1979. The geomorphologic structure of hydrologic response. *Water Resources Res.* 15:1409–20.

Rodriguez-Navarro, C., Doehne, E., and Sebastian, E. 1999. Origins of honeycomb weathering; the role of salts and wind. *Geological Society America Bull.* III: 1250–55.

Roman, C. T., and Nordstrom, K. F. 1988. The effect of erosion rate on vegetation patterns of an east coast barrier island. *Estuarine, Coastal and Shelf Sci.* 26:233–42.

Romero-Diaz, M. A., and others. 1988. Variability of overland flow erosion rates in a semi-arid Mediterranean environment under natural cover, Murcia, Spain. *Catena Suppl.* 13:1–11.

Ronov, A., and Yaroshevsky, A. 1969. Chemical composition of the Earth's crust. In *The earth's crust and upper mantle*, pp. 37–57. Ed. by P. Hart. *Am. Geophys. Union, Geophys. Monograph* 13.

Rossby, C. G. 1941. The scientific basis of modern meteorology. In *Climate and man.* U.S. Dept. Agri. Yearbook, pp. 599–655.

Röthlisberger, H. 1972. Water pressure in intra- and sub-glacial channels. *Jour. Glaciol.* 11:177–203.

Rouse, L. J., Jr., Roberts, H. H., and Cunningham, R. 1978. Satellite observation of subaerial growth of the Atchafalaya Delta, Louisiana. *Geology* 6:405–8.

Rubey, W. W. 1938. The force required to move particles on a stream bed. *U.S. Geol. Survey Prof. Paper* 189-E.

———. 1952. Geology and mineral resources of the Hardin and Brussels quadrangles (in Illinois). *U.S. Geol. Survey Prof. Paper* 218.

Rubin, D. M., and Hunter, H. E. 1985. Why deposits of longitudinal dunes are rarely recognized in the geologic record. *Sedimentology* 32:147–57.

Ruddiman, W., ed. 1997. *Tectonic uplift and climate change.* New York: Plenum Press.

Rudnick, R. 1992. Xenoliths-samples of the lower continental crust. In *Continental lower crust,* pp. 269–308. Ed. by D. Fountain et al. Elsevier,

Ruessink, B. G., and Terwandt, J. H. J. 2000. The behavior of nearshore bars or the time scale of years; a conceptual model. *Marine Geology* 163:289–302.

Ruhe, R. V. 1952. Topographic discontinuities of the Des Moines lobe. *Am. Jour. Sci.* 250:46–56.

———. 1964. Landscape morphology and alluvial deposits in southern New Mexico. *Ann. Assoc. Am. Geog.* 54:147–59.

———. 1965. Quaternary paleopedology. In *The Quaternary of the United States,* pp. 735–64. Ed. by H. Wright and D. Frey. Princeton, NJ: Princeton Univ. Press.

———. 1967. Geomorphic surfaces and surficial deposits in southern New Mexico. *New Mexico State Bur. Mines and Min. Res. Memo* 18.

———. 1969. *Quaternary landscapes in Iowa.* Ames: Iowa State Univ. Press.

———. 1975. *Geomorphology.* Boston: Houghton Mifflin Co.

Rumpel, D. A. 1985. Successive aeolian saltation: Studies of idealized collisions. *Sedimentology* 32:267–80.

Rundle, A. S. 1985. Braid morphology and the formation of multiple channels, the Rakaia, New Zealand. *Zeit. F. Geomorph., Supplement Band* 55:15–37.

Runnells, D. D. 1969. Diagenesis, chemical sediments, and the mixing of natural waters. *Jour. Sed. Petrology* 39:1188–1201.

Russ, D. 1982. Style and significance of surface deformation in the vicinity of New Madrid, Missouri. In *Investigation of the New Madrid, Missouri earthquake region,* pp. 95–114. Ed. by F. McKeown and L. Pakiser, U.S. Geol. Surv. Prof. Paper 1236. 93–114.

Russell, R. 1943. Freeze-thaw frequencies in the United States. *Am. Geophys. Union Trans.* 24:125–33.

———. 1967. *River and delta morphology.* Louisiana State Univ., Coastal Studies Inst. Tech. Rept. 52.

Russell, R. J. 1967. Aspects of coastal morphology. *Geogr. Annlr.* 49A:299–309.

Russell, R. J., and McIntire, W. G. 1965. Beach cusps. *Geol. Soc. America Bull.* 76:307–20.

Russell-Head, D. S., and Budd, W. F. 1979. Ice-sheet flow properties derived from bore-hole shear measurements combined with ice-core studies. *Jour. Glaciol.* 24:117–30.

Rust, B. R. 1972. Structure and process in a braided river. *Sedimentology* 18:221–45.

———. 1977. Mass flow deposits in a Quaternary succession near Ottawa, Canada: Diagnostic criteria for subaqueous outwash. *Can. J. Earth Sci.* 14:175–84.

Ruxton, B. P., and Berry, L. 1961. Weathering profiles and geomorphic position on granite in two tropical regions. *Rev. geomorph. dynamique* 12:16–31.

Ruxton, B. P., and McDougall, I. 1967. Denudation rates in northeast Papua from potassium-argon dating of lavas. *Am. Jour. Sci.* 265:545–61.

Ryckborst, H. 1975. On the origin of pingos. *J. Hydrol.* 26:303–14.

Ryder, J. M. 1971a. The stratigraphy and morphology of paraglacial alluvial fans in south-central British Columbia. *Can. J. Earth Sci.* 8:279–98.

———. 1971b. Some aspects of the morphometry of paraglacial alluvial fans in south-central British Columbia. *Can. J. Earth Sci.* 8:1252–64.

Sabin, T., and Holliday, V. 1995. Playas and lunettes on the southern high plains: Morphometric and spatial relationships. *Ann. Amer. Asso. Geogr.* 85:286–305.

Sakamoto-Arnold, C. M. 1981. Eolian features produced by the December 1977 windstorm in southern San Joaquin Valley, California. *Jour. Geology* 89:129–37.

Sala, M. 1988. Slope runoff and sediment production in two Mediterranean mountain environments. *Catena Supplement* 12:13–19.

Sallenger, A. D. 1979. Beach-cusp formation. *Marine Geol.* 29:23–37.

Sanderson, P. G., Eliot, I., Hegge, B., and Maxwell, S. 2000. Regional variation of coastal morphology in SW Australia: A synthesis. *Geomorphology* 34:73–88.

Sasaki, Y., Fujii, A., and Asai, K. 2000. Soil creep process and its role in debris slide generation; field measurements on the north side of Tsukuba Mountain in Japan. *Engineering Geology* 56:163–83.

Saucier, R. 1987. Geomorphological interpretation of late Quaternary terraces in western Tennessee and their regional tectonic implications. U.S. Geol. Surv. Prof. Paper 1336A.

Saucier, R. T., and Fleetwood, A. R. 1970. Origin and chronologic significance of Late Quaternary terraces, Ouachita River, Arkansas and Louisiana. *Geol. Soc. America Bull.* 81:869–90.

Savat, J. 1981. Work done by splash: Laboratory experiments. *Earth Surf. Proc. and Landforms* 6:275–83.

Savigear, R. 1952. Some observations on slope development in South Wales. *Inst. Brit. Geog. Trans.* 18:31–51.

Sawkins, J. 1869. Report on the geology of America. *Mem. Geol. Survey.*

Schalscha, E., Appelt, H., and Schatz, A. 1967. Chelation as a weathering mechanism, I. Effect of complexing agents on the solubilization of iron from minerals and granodiorite. *Geochim. et Cosmochim. Acta* 31:587–96.

Schatz, A. 1963. Chelation in nutrition, soil microorganisms and soil chelation. The pedogenic action of lichens and lichen acids. *Jour. Agri. and Food Chem.* 11:112–18.

Schatz, A., Cheronis, N., Schatz, V., and Trelawney, G. 1954. Chelation (sequestration) as a biological weathering factor in pedogenesis. *Pennsylvania Acad. Sci. Proc.* 28:44–57.

Schlesinger, W. H. 1985. The formation of caliche in soils of the Mojave Desert, California. *Geochim Cosmochim. Acta* 49:57–66.

Schmidlin, T. W. 1988. Alpine permafrost in eastern North America: A Review. In *Permafrost, Fifth International Conference Proceedings,* pp. 241–45. Ed. by K. Senneset. Trondheim: Tapir Publishers.

Schimmelmann, A., Zhao, M., Harvey, C. C., and Lange, C. B. 1998. A large California flood and correlative global climatic events 400 years ago. *Quaternary Research* 49:51–61.

Schowengerdt, R. A., and Glass, C. E. 1983. Digitally processed topographic data for regional tectonic evaluation. *Geol. Soc. America Bull.* 94:549–56.

Schrott, L., and Pasuto, A. 1999. Temporal stability and activity of landslides in Europe with respect to climate change. *Geomorphology* 30:1–2.

Schulz, M., and White, A. 1999. Chemical weathering in a tropical watershed, Luquillo Mountains, Puerto Rico III: Quartz dissolution rates. *Geochim. Cosmochim. Acta, 63,* 337–50.

Schumm, S. A. 1956. Evolution of drainage systems and slopes in badlands at Perth Amboy, New Jersey. *Geol. Soc. America Bull.* 67:597–646.

———. 1960. The shape of alluvial channels in relation to sediment type. *U.S. Geol. Survey Prof. Paper* 352-B.

———. 1962. Erosion of miniature pediments in Badlands National Monument, South Dakota. *Geol. Soc. America Bull.* 73:719–24.

———. 1963a. A tentative classification of alluvial river channels. *U.S. Geol. Survey Circ.* 477.

———. 1963b. Sinuosity of alluvial channels on the Great Plains. *Geol. Soc. America Bull.* 74:1089–1100.

———. 1963c. Disparity between present rates of denudation and orogeny. *U.S. Geol. Survey Prof. Paper* 454-H.

———. 1965. Quaternary paleohydrology. In *The Quaternary of the United States,* pp. 783–94. Ed. by H. E. Wright and D. G. Frey. Princeton, NJ: Princeton Univ. Press.

———. 1967a. Rates of surficial rock creep on hillslopes in western Colorado. *Science* 155:560–61.

———. 1967b. Meander wavelength of alluvial rivers. *Science* 157:1549–50.

———. 1968. River adjustment to altered hydrologic regimen, Murrumbidgee River and paleochannels, Australia. *U.S. Geol. Survey Prof. Paper* 598.

———. 1969. River metamorphosis. Am. Soc. Civil Engineers, *Jour. Hydraulics Div.* HY1:255–73.

———. 1971. Fluvial geomorphology. In *River Mechanics,* edited by H. W. Shen. chs. 4 and 5. Ft. Collins, CO: Colorado State Univ.

———. 1973. Geomorphic thresholds and complex response of drainage systems. In *Fluvial Geomorphology,* pp. 299–310. Ed. by M. Morisawa, S.U.N.Y., Binghamton: Pubs. in Geomorphology, 4th Ann. Mtg.

———. 1977. *The fluvial system.* New York: John Wiley & Sons.

———. 1979. Geomorphic thresholds: The concept and its applications. *Progress in Phys. Geogr. Trans.* 4:485–515.

———. 1980. Some applications of the concept of geomorphic thresholds. In *Thresholds in Geomorphology,* pp. 473–86. Ed. by D. Coates and J. Vitek. London: Allen and Unwin Ltd.

———. 1981. Evolution and response of the fluvial system; sedimentologic implications. In *Recent and ancient nonmarine depositional environments: Models for exploration.* Ed. by F. Ethridge and R. Flores. Tulsa, Soc. Econ. Paleontologists and Mineralogists Spec. Publ. 31:19–39.

Schumm, S. A., and Chorley, R. J. 1964. The fall of Threatening Rock. *Am. Jour. Sci.* 262:1041–54.

———. 1966. Talus weathering and scarp recession in the Colorado Plateaus. *Zeit. F. Geomorph.* 10:11–36.

Schumm, S. A., and Brakenridge, G. R. 1987. River response. In *North America and adjacent oceans during the last deglaciation.* Ed. by W. F. Ruddiman and H. E. Wright, Jr. Boulder, Colorado, Geol. Soc. America, v. K-3.

Schumm, S. A., and Khan, H. R. 1972. Experimental study of channel patterns. *Geol. Soc. America Bull.* 83:1755–70.

———. 1965. Time, space and causality in geomorphology. *Am. Jour. Sci.* 263:110–19.

Schumm, S. A., and Parker, R. S. 1973. Implications of complex response of drainage systems for Quaternary alluvial stratigraphy. *Nat. Phys. Sci.* 243:99–100.

Schumm, S. A., and Phillips, L. 1986. Composite channels of the Canterbury Plain, New Zealand; a Martian analog? *Geology* 14:326–29.

Schumm, S. A., Bean, D. W., and Harvey, M. D. 1982. Bed-form-dependent pulsating flow in Medano Creek, Southern Colorado. *Earth Surf. Proc. and Landforms* 7:17–28.

Schumm, S. A., Mosley, M. P., and Weaver, W. E. 1987. *Experimental fluvial geomorphology.* New York: John Wiley.

Schumm, S. A., and Rea, D. K. 1995. Sediment yield from disturbed earth systems. *Geology* 23:391–94.

Schweizer, J., and Iken, A. 1992. The role of bed separation and friction in sliding over an undeformable bed. *Journal of Glaciology* 38:77–92.

Scott, A. J., and Fisher, W. L. 1969. Delta systems and deltaic deposition. In *Delta systems in the exploration for oil and gas.* Ed. by W. Fisher, L. Brown, A. Scott, and J. McGowen. Univ. Texas, Austin, Bureau Econ. Geology.

Scott, K. M. 1988. Origins, behavior, and sedimentology of lahars and lahar-runout flows in the Toutle-Cowlitz River system. *U.S. Geol. Survey Prof. Paper* 1447-A, 74p.

Scott, P., and Erskine, W. 1994. Geomorphic effects of a large flood on fluvial fans. *Earth Surf. Proc. Landfms.* 19:95–108.

Seeber, L., and Gornitz, V. 1983. River profiles along the Himalayan arc as indicators of active tectonics. *Tectonophysics* 92:335–67.

Seed, H., Woodward, R., and Lundgren, R. 1964. Clay mineralogical aspects of the Atterberg limits. *Am. Soc. Civil Engineers Proc.* 90:SM4:107–31.

Selby, M. J. 1966. Methods of measuring soil creep. *J. Hydrol.* 5:54–63.

———. 1967. Aspects of the geomorphology of the Greywacke ranges bordering the lower and middle Waikato basins. *Earth Sci. Jour.* 1:1–22.

———. 1980. A rock mass strength classification for geomorphic purposes: With tests from Antarctica and New Zealand. *Zeit. F. Geomorph.* 24:31–51.

———. 1982. *Hillslope Materials and Processes.* New York: Oxford Univ. Press.

Selby, M. J., and Wilson, A. T. 1971. The origin of the Labyrinth, Wright Valley, Antarctica. *Geol. Soc. Amer. Bull.* 82:471–76.

Senneset, K., ed. 1988. *Permafrost, fifth international conference proceedings,* Trondheim: Tapir Publishers.

Senstius, M. 1958. Climax forms of chemical rock-weathering. *Am. Scientist* 46:355–67.

Seppälä, M. 1982. An experimental study of the formation of palsas. *Proceedings of the fourth Canadian permafrost conference* 36–42.

Seppälä, M. 1986. The origin of palsas. *Geografiska Annaler* 68A:141–47.

Sevon, W. D. 1969. Sedimentology of some Mississippian and Pleistocene deposits of northeastern Pennsylvania. In *Geology of Selected Areas in New Jersey and Eastern Pennsylvania.* New Brunswick, NJ: Rutgers Univ. Press.

———. 1951. Features of the firn on upper Seward Glacier, St. Elias Mountains, Canada. *Jour. Geology* 59:599–621.

———. 1953. Deformation of a vertical bore hole in a piedmont glacier. *Jour. Glaciol.* 2:182–84.

———. 1963. Wind ripples. *Jour. Geology* 71:617–36.

———. 1964. Wind-driven sand in Coachella Valley, California. *Geol. Soc. America Bull.* 75:785–804.

———. 1966. Kelso Dunes, Mojave Desert, California. *Geol. Soc. America Bull.* 77:1045–74.

———. 1979. Intradune flats of the Algodones chain, Imperial Valley, California. *Geol. Soc. America Bull.* 90:908–16.

———. 1980. Wind-driven sand in Coachella Valley, California: Further data. *Geol. Soc. America Bull.* 91:724–30.

———. 1989. Erosion in the Juniata River drainage basin, Pennsylvania. In *Appalachian Geomorphology.* Ed. by T. W. Gardner et al. *Geomorph.* 2: 3:3–18.

Sexton, W. J. 1995. The post-storm Hurricane Hugo recovery of the undeveloped beaches along the South Carolina coast, Cape Isle to Santee Delta. *Jour. Coastal Res.* 11:1020–25.

Shao, G., Young, D. R., Porter, J. H., and Hayden, B. P. 1998. An interpretation of remote sensing and GIS to examine the responses of shrub thicket distributions and shoreline changes on the Marthas Vineyard Islands. *Jour. Coastal Res.* 14:299–307.

Shao, Y., Raupach, M., and Findlater, P. 1993. Effect of saltation bombardment on the entrainment of dust by wind. *J. Geophys. Res.* 98 (D7):12719–26.

Sharp, D. R., and Shaw, J. 1989. Erosion of bedrock by subglacial meltwater, Cantley, Quebec. *Geological Society of America Bulletin* 101:1011–20.

Sharp, M. 1988. Surging glaciers: Geomorphic effects. *Progress in Physical Geography* 12:533–59.

Sharp, M. J., Dowdeswell, J. A., and Gemmel, J. C. 1989. Reconstructing past glacier dynamics and erosion from glacial geomorphic evidence: Snowdon, North Wales. *Journal of Quaternary Science* 4:115–30.

Sharp, R. 1949. Pleistocene ventifacts east of the Bighorn Mountains, Wyoming. *Jour. Geology* 57:175–78.

———. 1953. Deformation of a vertical bore hole in a piedmont glacier. *Jour. Glaciol.* 2:182–84.

———. 1960. *Glaciers.* Eugene, OR: Univ. Oregon Press.

———. 1964. Wind-driven sand in Coachella Valley, California. *Geol. Soc. America Bull.* 75:785–804.

———. 1966. Kelso Dunes, Mojave Desert, California. *Geol. Soc. America Bull.* 77:1045–74.

———. 1973. Mars: Fretted and chaotic terrains. *J. Geophys. Res.* 78:4222–30.

———. 1980. Wind-driven sand in Coachella Valley, California: Further data. *Geol. Soc. America Bull.* 91:724–30.

———. 1985. *Living ice: Understanding glaciers and glaciation.* New York: Cambridge University Press.

———. 1988. *Living ice: Understanding glaciers and glaciation.* New York: Cambridge University Press.

Sharp, R. P., and Noble, L. H. 1953. Mudflow of 1941 at Wrightwood, southern California. *Geol. Soc. America Bull.* 64:547–60.

Sharpe, C. F. S. 1938. *Landslides and related phenomena.* New York: Columbia Univ. Press.

Sharpe, D. R. 1987. Significance of meltwater erosion beneath an ice sheet. INQUA Congress, Ottawa, Program with abstracts, p. 263.

Shaw, J. 1972. Sedimentation in the ice-contact environment, with examples from Shropshire (England). *Sedimentology* 18:23–62.

———. 1979. Genesis of the Sveg tills and Rogen moraines of central Sweden: a model of basal melt-out. *Boreas* 8:409–26.

———. 1980. Drumlins and large-scale flutings related to glacier folds. *Arc. Alp. Res.* 12:287–98.

———. 1983. Drumlin formation related to inverted melt-water erosional marks. *Journal of Glaciology* 29:461–79.

———. 1985. Subglacial and ice marginal environments. In *Glacial sedimentary environments.* Ed. by G. M. Ashley, J. Shaw, and N. D. Smith. Society of Econ. *Peleontol. Mineral.,* Short Course, 16:7–84.

———. 1989. Drumlins, subglacial meltwater floods, and ocean responses. *Geology* 17:853–56.

———. 1993. Geomorphology. In *Edmonton Beneath our Feet: A Guide to the Geology of the Edmonton Region,* pp. 21–31. Ed. by J. D. Godfrey. Edmonton Geological Society, University of Alberta.

Shaw, J., and Kvill, D. 1984. A glaciofluvial origin for drumlins of the Livingstone Lake area, Saskatchewan. *Canadian Journal of Earth Science* 21:1442–59.

Shaw, J., Kvill, D., and Rains, B. 1989. Drumlins and catastrophic subglacial floods. In Subglacial Bedforms. Ed. by J. Menzies and J. Rose. *Sedimentary Geology* 62:177–202.

Shaw, J., and Sharpe, D. R. 1987. Drumlin formation by subglacial meltwater erosion. *Canadian Journal of Earth Science* 24:2316–22.

Shepard, F. P. 1950. Beach cycles in southern California. U.S. Army Corps Engrs., *Beach Erosion Board Tech. Memo* 20.

———. 1952. Revised nomenclature for depositional coastal features. *Am. Assoc. Petroleum Geologists Bull.* 36:1902–12.

———. 1963. *Submarine geology.* New York: Harper & Row.

Shepard, F. P., Emery, K. O., and LaFond, E. C. 1941. Rip currents: A process of geological importance. *Jour. Geology* 49:337–69.

Shepard, F. P., and Grant, U. S., IV. 1947. Wave erosion along the southern California coast. *Geol. Soc. America Bull.* 58:919–26.

Shepard, F. P., and Inman, D. L. 1950. Nearshore circulation related to bottom topography and wave refraction. *Am. Geophys. Union Trans.* 31:2:196–212.

Shepard, F. P., and Wanless, H. R. 1971. *Our changing coastlines.* New York: McGraw-Hill.

Shepherd, R. G., and Schumm, S. A. 1974. Experimental study of river incision. *Geol. Soc. America Bull.* 85:257–68.

Sherman, D. J., and Bauer, B. O. 1993. Coastal geomorphology through the looking glass. *Geomorph.* 7:225–49.

Shimokawa, E., and Jitousono, T. 1998. A study of the change from landslide to debris flow at Harihara, Southern Kyu-sh. *Jour. Natural Disaster Science* 20:75–81.

Shlemon, R. J. 1975. Subaqueous delta formation—Atchafalaya Bay, Louisiana. In *Deltas,* pp. 209–21. Ed. by M. Brousard. Houston: Houston Geologic Society.

Shoemaker, E. M. 1988. On the formulation of basal debris drag for the case of sparse debris. *Journal of Glaciology* 34:259–64.

Shoji, S., Yamada, I., and Kurashima, K. 1981. Mobilities and related factors of chemical elements in the topsoils of andosols in Tohuku, Japan: 2. Chemical and mineralogical compositions of size fractions and factors influencing the mobilities of major chemical elements. *Soil Sci.* 132:330–46.

Short, A. D. 1975. Multiple offshore bars and standing waves: *J. Geophys. Res.* 80:3838–40.

———. 1979. Three-dimensional beach stage model. *Jour. Geology* 87:553–71.

Short, A. D., and Wright, L. D. 1981. Beach systems of the Sydney Region. *Aust. Geog.* 15:8–16.

Short, N. 1986. Volcanic landforms. In Short, N. and Blair, R. eds. *Geomorphology from space, NASA SP-486,* pp. 185–253. Ed. by N. Short and R. Blair.

Shreve, R. L. 1966a. Statistical law of stream numbers. *Jour. Geology* 74:17–37.

———. 1966b. Sherman landslide, Alaska. *Science* 154:1639–43.

———. 1967. Infinite topologically random channel networks. *Jour. Geology* 75:178–86.

———. 1968. *The Blackhawk landslide. Geol. Soc. America Spec. Paper* 108.

———. 1974. Variation of mainstream length with basin area in river networks. *Water Resour. Res.* 10:1167–77.

———. 1984. Glacier sliding at subfreezing temperatures. *Jour. Glaciol.,* 30:341–47.

———. 1985. Esker characteristics in terms of glacier physics, Katahdin esker system, Maine. *Geological Society of America Bulletin.* 96:639–46.

Shulmeister, J. 1989. Flood deposits in the Tweed Esker (southern Ontario, Canada). *Sedimentary Geology* 65:153–63.

Shuster, E. T., and White, W. B. 1971. Seasonal fluctuations in the chemistry of limestone springs: A possible means for characterizing carbonate aquifers. *J. Hydrol.* 14:93–128.

Shuster, R. L., Nieto, A. S., O'Rourke, T. D., Crespo, E., Plaza-Nieto, G. 1996. Mass wasting triggered by the 5 March 1987 Ecuador earthquakes. *Engineering Geology* 42:1–23.

Sidle, R., and Swanston, D. 1982. Analysis of a small debris slide in coastal Alaska. *Can. Geotechnical Jour.* 19:167–74.

Sigafoos, R. S. 1964. Botanical evidence of floods and floodplain deposition. *U.S. Geol. Survey Prof. Paper* 485-A.

———. 1976. Botanical evidence of floods and floodplain deposition. *U.S. Geol. Survey Prof. Paper* 424-C.

Silar, J. 1965. Development of tower karst of China and North Vietnam. *Natl. Speleol. Soc. Bull.* 27(2):35–46.

Silvester, R. 1966. Wave refraction. In *Encyclopedia of oceanography,* pp. 975–76. Ed. by R. W. Fairbridge. New York: Reinhold Book Corp.

Simmons, G., and Richter, D. 1976. Microcracks in rocks. In *The physics and chemistry of minerals and rocks,* pp. 105–37. Ed. by R. G. J. Strens. London: John Wiley & Sons.

Simon, A. 1989. The discharge of sediment in channelized alluvial streams. *Water Resources Res. Bull.* 25:1177–88.

Simon, A., and Curini, A. 1998. Pore pressure and bank stability: The influence of matric suction. In *Water Resources Engineering 98,* pp. 358–363. Ed. by S. R. Abt, volume 1. New York: American Society of Civil Engineers.

Simon, A., Curini, A., Darby, S., and Langendoen, E. J. 1999. Streambank mechanics and the role of bank and near-bank processes in incised channels. In *Incised river channels,* pp. 123–51. Ed. by S. E. Darby and A. Simon. New York: John Wiley and Sons, Ltd.

Simons, D. B., and Richardson, E. V. 1962. Resistance to flow in alluvial channels. *Am. Soc. Civil Engineers Trans.* 127:927–52.

———. 1963. Forms of bed roughness in alluvial channels. *Am. Soc. Civil Engineers Trans.* 128:284–302.

———. 1966. Resistance to flow in alluvial channels. *U.S. Geol. Survey Prof. Paper* 422-J.

Simpson, D. 1964. Exfoliation in the upper Pocahontas Sandstone, Mercer County, West Virginia. *Am. Jour. Sci.* 262:545–51.

Singer, A. 1980. The paleoclimate interpretation of clay minerals in soils and weathering profiles. *Earth Sci. Rev.* 15:303–26.

Singh, V. J., ed. 1987. *Regional flood frequency analysis.* Dordrecht: Reidel Pub.

Sinha, S. K., and Parker, G. 1996. Causes of concavity in longitudinal profiles of rivers. *Water Resour. Res.* 32:1417–28.

Skempton, A. W. 1953. Soils mechanics in relation to geology. *Yorkshire Geol. Soc. Proc.* 29:33–62.

———. 1964. The long-term stability of clay slopes. *Geotechnique* 2:75–102.

Sklar, L., and Dietrich, W. 1998. River longitudinal studies and bedrock incision models: stream power and the influence of sediment supply. In *Rivers over rock: fluvial processes in bedrock channels,* pp. 237–60. Ed. by K. Tinkler and E. Wohl. Am. Geophys. Un., Wash., D.C.

Sleep, N. 1990. Hotspots and mantle plumes: Some phenomenology. *J. Geophys. Res.* 96:6715–36.

Small, R., Clark, M., and Cawse, T. J. 1979. The formation of medial moraines on Alpine glaciers. *Jour. Glaciol.* 22:43–52.

Smalley, I. J. 1966. Drumlin formation. A rheological model. *Science* 151:1379.

———. 1970. Cohesion of soil particles and the intrinsic resistance of simple soil systems to wind erosion. *Jour. Soil Sci.* 21:154–61.

———. 1981. Conjectures, hypotheses, and theories of drumlin formation. *Jour. Glaciol.* 27:503–5.

Smalley, I. J., and Unwin, D. J. 1968. The formation and shape of drumlins and their distribution and orientation in drumlin fields. *Jour. Glaciol.* 7:377–90.

Smart, J. S., and Wallis, J. R. 1971. Cis and trans links in natural channel networks. *Water Resour. Res.* 7:1346–48.

Smart, P. L., and Hobbs, S. L., 1986. Characterization of carbonate aquifers: a conceptual base. Proceedings of the Environmental Problems in Karst Terranes and Their Solutions Conference. *National Water Well Association,* Dublin, Ohio, pp. 1–13.

Smith, D. D., and Wischmeier, W. H. 1962. Rainfall erosion. *Advances in Agron.* 14:109–48.

Smith, D. G. 1976. Effect of vegetation on lateral migration of anastomosed channels of a glacier meltwater river. *Geol. Soc. America Bull.* 87:857–60.

———. 1980. River ice processes: Thresholds and geomorphological effects in northern and mountain rivers. In *Thresholds in Geomorphology,* pp. 323–43. Ed. by D. R. Coates and J. D. Vitek. London: Allen and Unwin.

———. 1986. Anastomosing river deposits, sedimentation rates and basin subsidence, Magdalena River, northwestern Colombia, South America. *Sedimentary Geology* 46:177–96.

Smith, D. G., and Smith, N. D. 1976. Sedimentation in anastamosed river systems: Examples from alluvial valleys near Banff, Alberta. *Jour. Sed. Petrology* 50:157–64.

Smith, D. I. 1969. The solution erosion of limestone in an arctic morphogenetic region. In *Problems of the karst denudation,* pp. 99–110. Ed. by O. Stelcl. Brno.

Smith, D. I., and Atkinson, T. 1976. Process landforms and climate in limestone regions. In *Geomorphology and climate.* Ed. by E. Derbyshire. London: John Wiley & Sons.

Smith, D. J. 1992. Long-term rates of contemporary solifluction in the Canadian Rocky Mountains. In *Periglacial geomorphology,* pp. 203–22. Ed. by J. C. Dixon and A. D. Abrahams. Chichester: John Wiley.

Smith, H. T. U. 1953. The Hickory Run boulder field, Carbon County, Pennsylvania. *Am. Jour. Sci.* 251:625–42.

Smith, M. W., and Patterson, D. E. 1989. Detailed observations on the nature of frost heaving at a field scale. *Can Geotech. J.* 26:306–12.

Smith, N. D. 1970. The braided stream depositional environment: Comparison of the Platte River with some Silurian clastic rocks, north-central Appalachians. *Geol. Soc. America Bull.* 81:2993–3014.

———. 1971. Transverse bars and braiding in the Lower Platte River, Nebraska. *Geol. Soc. America Bull.* 82:3407–20.

———. 1974. Sedimentology and bar formation in the upper Kicking Horse River, a braided outwash stream. *Jour. Geology* 82:205–23.

———. 1985. Proglacial fluvial environment. In *Glacial sedimentary environments,* pp. 85–134. Ed. by G.M. Ashley, J. Shaw, and N. D. Smith. *Society of Paleontologists and Mineralogists Short Course No. 16.*

Snodgrass, D., Groves, G., Hasselmann, K., Miller, G., Munk, W., and Powers, W. 1966. Propagation of ocean swell across the Pacific. *Phil. Trans. Royal Soc. London,* ser. A:259:431–97.

Soil Survey Division Staff. 1993. *Soil survey manual.* U.S. Dept. of Agriculture Handbook No. 18.

Soil Survey Staff. 1951. *Soil survey manual.* U.S. Dept. Agri. Handbook 18, Soil Conserv. Serv.

———. 1960. *Soil classification, a comprehensive system—7th approximation.* U.S. Dept. Agri., Soil Conserv. Serv.

———. 1975. *Soil taxonomy.* U.S. Dept. Agri. Handbook 436, Soil Conserv. Serv.

———. 1981. Replacement chapter to Handbook 18, *Soil survey manual,* released May 1981. U.S. Dept. Agri., Soil Conserv. Serv.

———. 1998. *Dominant soil orders and suborders—Soil taxonomy 1998, United States of America.* Maps and Soil Photographs, USDA Natural Resources Conservation Service, NSSC 5502-0898-01.

———. 1999. *Soil taxonomy,* U.S. Dept. Handbook 436, 2nd ed. Natural Resources Conservation Service.

Soloviev, P. A. 1962. Alas relief and its origin in central Yakutia. In *Mnoholetnemerlyye Porody i Sopotstvuyuhchiye yim Yavleniya na territorii JASSR:* Moscow p. 38–53.

———. 1973. *Alas thermokarst relief of central Yakutia.* Guidebook, Second International Permafrost Conference, Yakutsk, USSR.

Sonu, C. J. 1972. Field observation of nearshore circulation and meandering currents. *Jour. Geophys. Research* 77:18:3232–47.

Sophocleous, M. A. 1991. Stream floodwave propagation through the Great Bend alluvial aquifer, Kansas; field measurements and numerical simulations. *J. Hydrology* 124:207–28.

Soucie, G. 1973. Where beaches have been going: Into the ocean. *Smithsonian* 4:3:55–61.

Southworth, C. S. 1988. Large Quaternary landslides in the central Appalachian Valley and Ridge Province near Petersburg, West Virginia. *Geomorph.* 1:317–29.

Sperling, C. H. B., and Cooke, R. U. 1985. Laboratory simulation of rock weathering by salt crystallization and hydration processes in hot, arid environments. *Earth Surf. Proc. and Landforms* 10:541–55.

Spitz, W., and Schumm, S. 1997. Tectonic geomorphology of the Mississippi Valley between Osceola, Arkansas, and Friars Point, Mississippi. *Engr. Geol.* 46:259–80.

Springer, M. E. 1958. Desert pavement and vesicular layer of some desert soils in the desert of the Lahontan Basin, Nevada. *Soil Sci. Am. Proc.* 22:63–66.

Squyres, S. W. 1979. The distribution of lobate debris flow aprons and similar flows on Mars. *Jour. Geophys. Res.* 84:8087–96.

Stalker, A. MacS. 1960. Ice-pressed drift forms and associated deposits in Alberta. *Canada Geol. Survey Bull.* 57.

Stanley, D., and Warne, A. 1994. Worldwide initiation of Holocene marine deltas by deceleration of sea-level rise. *Science* 265:228–31.

Stanley, D. J., Krinitzsky, E. L., and Compton, J. R. 1966. Mississippi River bank failure, Fort Jackson, Louisiana. *Geol. Soc. America Bull.* 77:850–66.

Stanley, J. M., and Cronin, J. E. 1983. Investigations and implications of subsurface conditions beneath the Trans-Alaska Pipeline in Atigun Pass. In *Proc. permafrost 4th internat. conf.,* pp. 1188–93. Natl. Acad. Sci.

Stanley, S. R., and Ciolkosz, E. J. 1981. Classification and genesis of spodosols in the central Appalachians. *Soil Sci. Soc. Am. Proc.* 45:912–17.

Starkel, L., Gregork, K. J., and Thornes, J. B., eds. 1991. *Temperate paleohydrology.* London: John Wiley and Sons.

Statham, I., and Francis, S. C. 1986. Influence of scree accumulation and weathering on the development of steep mountain slopes. In *Hillslope processes,* pp. 245–67. Ed. by Abrahams. Boston: Allen and Unwin.

Stearns, S. R. 1966. *Permafrost (perenially frozen ground).* U.S. Army Corps Engrs., Cold Regions Res. and Eng. Lab, Cold Regions Sci. and Eng. 1 (A2).

Stedinger, J. R., and Baker, V. R. 1987. Surface water hydrology: Historical and paleoflood information. *Rev. of Geophys.* 25:119–24.

Stedinger, J. R., and Cohn, T. A. 1986. Flood frequency analysis with historical and paleoflood information. *Water Resources Res.* 22:785–93.

Steig, E. J., Fitzpatrick, J. J., Potter, N. Jr., and Clark, D. H. 1998. The geochemical record in rock glaciers. *Geografiska Annaler Ser A.* 80:3–4.

Steers, J. A. 1962. *The sea coast.* London: Collins.

Stein, R., and King, G. 1984. Seismic potential as revealed by surface folding. 1983 Coaling, California, earthquake. *Science* 224:869–72.

Stene, L. P. 1980. Observations on lateral and overbank deposition—Evidence from Holocene terraces, southwestern Alberta.

Stewart, I., and Vita-Finzi, C., eds. 1998. *Coastal tectonics.* Geological Society, London. Spec. Publ. 146.

Stokes, D., and Swinehart, J. 1997. Middle- and late-Holocene dune reactivation in the Nebraska Sand Hills, USA. *The Holocene* 7:273–82.

Stone, G., and Donley, J., eds., 1998. The worlds deltas conference: A tribute to the late Professor James Plummer Morgan: 1919–1995. Thematic Section I. *J. Coastal Res.* 14:695–858.

———. A tribute to James Plummer Morgan (1919–1995) and review of his scientific contributions to the studies of coastal and deltaic systems. Thematic Section II. *J. Coastal Res.* 14:859–915.

Stone, G. W., and Stapor, F. W. Jr. 1996. A nearshore sediment transport model for the northeast Gulf of Mexico coast, USA. *Jour. Coastal Res.,* 12:786–793.

Stone, R. 1968. Deserts and desert landforms. In *Encyclopedia of geomorphology,* pp. 271–79. Ed. by R. W. Fairbridge. New York: Reinhold Book Corp.

Stone, G. W. and Stapor, F. W. Jr., 1996. A nearshore sediment transport model for the northeast Gulf of Mexico coast, USA. *Journal of Coastal Research,* v. 12:786–93.

Stonestrom, D., White, A., and Akstin, C. 1998. Determining rates of chemical weathering in soils—Solute transport versus profile evaluation. *Jour. Hydrol.* 209:331–45.

Strahler, A. N. 1950. Equilibrium theory of slopes approached by frequency distribution analysis. *Am. Jour. Sci.* 248:800–814.

———. 1952a. Dynamic basis of geomorphology. *Geol. Soc. America Bull.* 63:923–38.

———. 1952b. Hypsometric (area-altitude) analysis of erosional topography. *Geol. Soc. America Bull.* 63:1117–42.

———. 1957. Quantitative analysis of watershed geomorphology. *Am. Geophys. Union Trans.* 38:913–20.

———. 1958. Dimensional analysis applied to fluvially eroded landforms. *Geol. Soc. America Bull.* 69:279–99.

———. 1964. Quantitative geomorphology of drainage basins and channel networks. In *Handbook of Applied Hydrology,* pp. 439–76. Ed. by V. T. Chow. New York: McGraw-Hill.

———. 1966. Tidal cycle of changes on an equilibrium beach. *Jour. Geology* 74:247–68.

———. 1968. Quantitative geomorphology. In *Encyclopedia of geomorphology,* pp. 898–912. Ed. by R. W. Fairbridge. New York: Reinhold Book Corp.

Strakhov, N. 1967. *Principles of lithogenesis.* London: Oliver and Boyd.

Stringfield, V. T., and LeGrand, H. E. 1969. Hydrology of carbonate rock terranes—A review. *J. Hydrol.* 8:349–413.

Strunk, H. 1997. Dating of geomorphological processes using dendrogeomorphological methods. *Catena* 31:137–51.

Sugden, D. E., and John, B. S. 1976. *Glaciers and landscape.* London: Edward Arnold Ltd.

Summerfield, M. 1991. Tectonic geomorphology. *Prog. Phys. Geogr.* 15:193–206.

Sunamura, T. 1975. A laboratory study of wave-cut platform formation. *Jour. Geology* 83:389–97.

———. 1976. Feedback relationship in wave erosion of laboratory rocky coast. *Jour. Geology* 84:427–37.

———. 1977. A relationship between wave-induced cliff erosion and erosive force of waves. *Jour. Geology* 85:613–18.

———. 1978. Mechanisms of shore platform formation on the southeast coast of the Izu peninsula, Japan. *Jour. Geology* 86:211–22.

———. 1982. A predictive model for wave-induced erosion, with application to Pacific coasts of Japan. *Jour. Geology* 90:167–78.

———. 1983. Processes of sea cliff and platform erosion. In *CRC Handbook of Coastal Processes and Erosion,* pp. 233–65. Ed. by P. Koma. Boca Raton, FL: CRC Press.

Sunamura T., and Auki, H. 2000. A field experiment of cusp formation on a coarse clastic beach using a suspended video camera system. *Earth Surface Processes and Landforms* 25:329–33.

Suzuki, T., and Takahashi, K. 1981. An experimental study of wind abrasion. *Jour. Geology* 89:509–22.

Swanson, D. 1972. Magma supply rate at Kilauea Volcano, 1925–1971. *Science* 175:169–70.

———. 1985. Soil catenas on Pinedale and Bull Lake moraines, Wind River Mountains, Wyoming. *Catena* 12:329–42.

———. 1996. Soil geomorphology on bedrock and colluvial terrain with permafrost in central Alaska, USA. *Geoderma* 71:157–72.

Sweeney, S. J., and Smalley, I. J. 1988. Occurrence and geotechnical properties of loess in Canada. *Engin. Geol.* 25:123–34.

Sweet, M. 1992. Lee-face airflow, surface processes, and stratification types: Their significance for refining the use of eolian cross-strata as paleocurrent indicators. *Geol. Soc. America Bull.* 104:1528–38.

Sweeting, M., and Lancaster, N. 1982. Solutional and wind erosion forms on limestone in the central Namib Desert. *Zeit. F. Geomorph.* 26:197–207.

Sweeting, M. M. 1958. The karstlands of Jamaica. *Geogr. Jour.* 124:184–99.

———. 1973. *Karst landforms.* New York: Columbia Univ. Press.

Swineford, A., and Frye, J. C. 1951. Petrography of the Peorian loess in Kansas. *Jour. Geology* 59:306–22.

Swinzow, G. K. 1969. Certain aspects of engineering geology in permafrost. *Eng. Geol.* 3:177–215.

Taber, S. 1929. Frost heaving. *Jour. Geology* 37:428–61.

———. 1930. The mechanics of frost heaving. *Jour. Geology* 38:303–17.

———. 1943. Perennially frozen ground in Alaska: Its origin and history. *Geol. Soc. America Bull.* 54:1433–1548.

———. 1953. Origin of Alaska silts. *Am. Jour. Sci.* 251:321–36.

Takabatake, H., Hirano, M., Moriyama, T., and Kawahara, K. 1998. Estimation of the occurrence condition of a debris flow. *Jour. Hydroscience* and *Hydraulic Engineering* 16:63–70.

Tanner, W. F. 1958. The equilibrium beach. *Am. Geophys. Union Trans.* 39:889–91.

Tator, B. A. 1952. Pediment characteristics and terminology. *Ann. Assoc. Am. Geog.* 42:295–317.

———. 1953. Pediment characteristics and terminology. *Ann. Assoc. Am. Geog.* 43:47–53.

Taylor, F., Jouannic, C., and Bloom, A. 1985. Quaternary uplift of the Torres Islands, northern New Hebrides frontal arc: Comparison with Santo and Malekula islands, central New Hebrides, *Jour. Geology* 93:419–38.

———. 1984. Recent trends in karst geomorphology. Jour. Geol. Educ. 32:247–53.

———. 1990. Groundwater processes in karst terranes. In Groundwater geomorphology; The role of subsurface water in earth-surface processes and landforms, pp. 177–209. Ed. by C. G. Higgens and D. R. Coates. Boulder, *Geological Society of America Special Paper* 252.

———. 1991. Origin and morphology of limestone caves. *Geological Society of America Bulletin* 103:1–21.

Taylor, K., and 7 others. 1993. The "flickering switch" of late Pleistocene climate change. *Nature* 361:432–36.

Taylor, M., and Lewin, J. 1997. Non-synchronous response of adjacent foodplain systems to Holocene environmental change. *Geomorph* 18:251–64.

Tchakerian, V., ed. 1995. *Desert aeolian processes.* New York: Chapman and Hall.

Teisseyre, A. K. 1978. Physiography of bedload meandering streams: Imbricated gravels in fine-grained overbank deposits, *Geol. Sudetica* 13:87–92.

Teller, J. T. 1987. Proglacial lakes and the southern margin of the Laurentide Ice Sheet. In North America and adjacent oceans during the last deglaciation, pp. 36–69. Ed. by A. R. Palmer and J. O. Wheeler. *The Geological Society of America.*

Ten Brink, N. W. 1974. Glacio-isostasy: New data from West Greenland and geophysical implications. *Geol. Soc. America Bull.* 85:219–28.

Terranova, T., and Kochel, R. C. 1987. Multivariate analysis of factors related to debris avalanching in Nelson County, central Virginia. *Geol. Soc. Amer. Abs. w. Prog.* 9 no. 7, p. 866.

Terzaghi, K. 1936. The shearing resistance of saturated soils. *Proc. 1st internat. conf. on soils mech. and foundation eng.* 1:54–66.

———. 1943. *Theoretical soils mechanics.* New York: John Wiley & Sons.

———. 1950. Mechanism of landslides. In *Application of geology to engineering practice,* pp. 83–123. Ed. by S. Paige. Geol. Soc. America Berkey Vol.

———. 1962. Stability of steep slopes on hard unweathered rock. *Geotechnique* 12:251–70.

Theilig, E., and Greeley, R. 1978. Episodic channeling and layered terrain on Mars: Implications for ground ice. In *Proc. second colloquium of planetary water and polar processes,* pp. 151–57. Wash., DC, NASA Planetary Geology Program.

Thie, J. 1974. Distribution and thawing of permafrost in the southern part of the discontinuous permafrost zone in Manitoba. *Arctic* 27:189–200.

Thomas, D., ed. 1997. *Arid zone geomorphology.* London: Bellhaven/Halsted Press.

Thomas, H. P., and Ferrell, J. E. 1983. Thermokarst features associated with buried sections of the Trans Alaska pipeline. In *Proc. permafrost 4th internat. conf.,* pp. 1245–50. Natl. Acad. Sci.

Thomas, R. H., MacAyeal, D. R., Bentley, C. R., and Clapp, J. L. 1980. The creep of ice, geothermal heat flow, and Roosevelt Island, Antarctica. *Jour. Glaciol.* 25:47–60.

Thompson, A. 1986. Secondary flows and the pool-riffle unit: a case study of the processes of meander development. *Earth Surf. Proc. and Landforms* 11:631–41.

Thompson, D. M., Wohl, E. E., and Jarrett, R. D. 1996. A revised velocity-reversal and sediment-sorting model for a high-gradient, pool-riffle stream. *Physical Geogr.* 17:142–56.

Thompson, D. M., Nelson, J. M., and Wohl, E. E. 1998. Interactions between pool geometry and hydraulics. *Water Resour. Res.* 12:3673–81.

Thorn, C., and Welford, M. 1994. The equilibrium concept in geomorphology. *An. Asso. Amer. Geogr.* 84:666–96.

Thorn, C. E. 1976. Quantitative evaluation of nivation in the Colorado Front Range. *Geol. Soc. America Bull.* 87:1169–78.

———. 1979. Bedrock freeze-thaw weathering regime in an alpine environment, Colorado Front Range. *Earth Surf. Proc. and Landforms* 4:211–28.

———. 1992. Periglacial geomorphology: What? Where? When? In *Periglacial Geomorphology,* pp. 1–30. Ed. by J. C. Dixon and A. D. Abrahams. Chichester: John Wiley.

Thorn, C. E., and Hall, K. 1980. Nivation: An arctic-alpine comparison and reappraisal. *Jour. Glaciol.* 25:109–24.

Thorne, C. R. 1982. Processes and mechanisms of river bank erosion. In *Gravel-Bed Rivers,* pp. 227–72. Ed. by R. Hey, J. Bathurst, and C. Thorne. New York: John Wiley & Sons.

Thorne, C. R., and Lewin, J. 1979. Bank processes, bed material movement, and platform development in a meandering river. In *Adjustments of the fluvial system,* pp. 117–37. Ed. by D. D. Rhodes and G. P. Williams. Dubuque, IA: Kendall Hunt.

Thorne, C. R., and Tovey, N. K. 1981. Stability of composite river banks. *Earth Surf. Proc. and Landforms* 6:469–84.

Thornton, E. B. 1973. Distribution of sediment transport across the surf zone. *Proc. 13th conf. on coast. eng.,* pp. 1049–68.

Thorp, J., and Smith, G. 1949. Higher categories of soil classifications, Order, suborder, and great soil groups. *Soil Sci.* 67:117–26.

Thouret, J-C. 1999. Volcanic geomorphology—an overview. *Earth Sci. Reviews* 47:95–131.

Thrailkill, J. 1968. Chemical and hydrologic factors in the excavation of limestone caves. *Geol. Soc. America Bull.* 79:19–45.

———. 1972. Carbonate chemistry of aquifer and stream water in Kentucky. *J. Hydrol.* 16:93–104.

Thrailkill, J., and Robl, T. 1981. Carbonate geochemistry of vadose water recharging limestone aquifers. *J. Hydrol.* 54:195–208.

Tinkler, K., and Wohl, E. 1998. A primer on bedrock channels. *In Rivers over rock: Fluvial processes in bedrock channels,* pp. 1–18. Ed. by K. Tinkler and E. Wohl. American Geophysical Union: Geophysical Monograph 107.

Tinkler, K. J. 1982. Avoiding error when using the Manning equation. *Jour. Geology* 90:326–28.

Todd, D. K. 1959. *Groundwater hydrology.* New York: John Wiley & Sons.

———, ed. 1970. *The water encyclopedia.* Port Washington, NY: Water Information Center.

Torres, R., Dietrich, W. E., Montgomery, D. R., Anderson, S. P., and Loague, K. 1998. Unsaturated zone processes and the hydrologic response of a steep, unchanneled catchment, *Water Resources Res.* 34:1865–79.

Tosdal, T., Clark, A., and Farrar, E. 1984. Cenozoic polyphase landscape and tectonic evolution of the Cordillera Occidental, southernmost Peru. *Geol. Soc. America Bull.* 95:1318–32.

Toulmin, P., Baird, A. K., Clark, B. C., Keil, K., Rose, H. J., Christian, R. P., Evans, P. H., and Kelliher, W. C. 1977. Geochemical and mineralogical interpretation of the Viking inorganic chemical results. *J. Geophys. Res.* 82:4625–34.

Townsend, D. W., and Vickery, R. P. 1967. An experiment in regelation. *Philos. Mag.* 16:1275–80.

Toy, T. J. 1977. Hillslope form and climate. *Geol. Soc. America Bull.* 88:16–22.

———. 1982. Accelerated erosion: Process, problems, and prognosis. *Geology* 10:524–29.

Trauth, M. H., Alonso, R. A., Haselton, K. R., Hermanns, R. L., and Strecker, M. R. 2000. Climate change and mass movements in the N.W. Argentine Andes. *Earth and Planetary Science Letters* 179:243–56.

Trenhaile, A. S. 1971. Drumlins: Their distribution, orientation and morphology. *Can. Geog.* 15:113–26.

———. 1977. Cirque elevation and Pleistocene snowlines. *Zeit. F. Geomorph.* 21:445–59.

———. 1979. The morphology of valley steps in the Canadian Cordillera. *Zeit. F. Geomorph.* 23:27–44.

Tricart, J. 1967. Le modelé des régions périglaciaires. In *Traite de geomorphologie 2.* Ed. by J. Tricart and A. Cailleux. Paris: SEDES.

———. 1969. *Geomorphology of cold environments.* Trans. by Edward Watson. New York: St. Martin's Press.

Tricart, J., and Cailleux, A. 1972. *Introduction to climatic geomorphology.* London: Longman.

Trimble, S. W. 1977. The fallacy of stream equilibrium in contemporary denudation studies. *Am. Jour. Sci.* 277:876–87.

Trimble, S. W., and Lund, S. W. 1982. Soil conservation and the reduction of erosion and sedimentation in the Coon Creek Basin, Wisconsin. *U.S. Geol. Survey Prof. Paper* 1234.

Troester, J. W., White, E. L., and White, W. B. 1984. A comparison of sinkhole depth frequency distributions in temperate and tropical karst regions. In *Sinkholes, their geology, engineering, and environmental impact,* pp. 65–73. Ed. by B. F. Beck. Rotterdam: A. A. Balkema.

Trudgill, S. 1976. The marine erosion of limestones on Aldabra Atoll, Indian Ocean. *Zeit. F. Geomorph.,* Suppl. 26:164–200.

Truman, C. C., and Bradford, J. M. 1990. Effect of antecedent soil moisture on splash detachment under simulated rainfall. *Soil Sci.* 150:787–98.

Tsoar, H. 1983. Dynamic processes acting on a longitudinal (seif) sand dune. *Sedimentology* 30:567–78.

Tsoar, H., and Pye, K. 1987. Dust transport and the question of desert loess formation. *Sedimentology* 34:139–53.

Tuan, Ti-Fu. 1959. *Pediments in southeastern Arizona.* Berkeley, CA: Univ. Calif. Pubs. in Geog. 13.

———. 1962. Structure, climate and basin landforms in Arizona and New Mexico. *Ann. Assoc. Am. Geog.* 52:51–68.

Tucker, G. E., and Slingerland, R. 1997. Drainage basin responses to climate change. *Water Resources Res.* 33:2031–47.

Tucker, M. J. 1950. Surf beats: Sea waves of 1 to 5 minute periods. *Proc. Royal Soc. London,* ser. A, 202:565–73.

Turko, J., and Knuepfer, P. 1991. Late Quaternary fault segmentation from analysis of scarp morphology. *Geology* 19:718–21.

Turner, B. L., Clark, W. C., Kates, R. W., Richards, J. F., Mathews, J. T., and Meyer, W. B., eds. 1990. *The earth transformed by human action.* Cambridge: Cambridge Univ. Press.

Twidale, C. R. 1962. Steepened margins of inselbergs from northwestern Eyre Peninsula, South Australia. *Zeit. F. Geomorph.* 6:51–69.

———. 1964. Erosion of an alluvial bank at Birdwood, South Australia. *Zeit. F. Geomorph.* 8:189–211.

———. 1967. Origin of the piedmont angle as evidenced in South Australia. *Jour. Geology* 75:393–411.

———. 1968. Weathering. In *Encyclopedia of geomorphology,* pp. 1228–32. Ed. by R. W. Fairbridge. New York: Reinhold Book Corp.

———. 1971. *Structural landforms.* Canberra: Australian National Univ. Press.

———. 1972. Evolution of sand dunes in the Simpson Desert, central Australia. *Inst. Brit. Geog. Trans.* 56:77–110.

———. 1973. On the origin of sheet jointing. *Rock Mechanics* 3:163–87.

———. 1978. On the origin of pediments in different structural settings. *Am. Jour. Sci.* 278:1138–76.

———. 1982. *Granite landforms.* Amsterdam: Elsevier.

———. 1990. The origin and implications of some erosional landforms. *Jour. Geology,* 98:343–64.

Twidale, C. R., and Bourne, J. A. 1975. Episodic exposure of inselbergs. *Geol. Soc. America Bull.* 86:1473–81.

Twidale, C. R., and Campbell, E. M. 1986. Localised inversion on steep hillslopes: Gully gravure in weak and in resistant rocks. *Zeit. F. Geomorph.* 30:35–46.

Tyson, P., and Seely, M., 1980. Local winds over the central Namib. *South Afr. Geogr. Jour.* 62, 135–50.

Uchupi, E., and Oldale, R. N. 1994. Spring sapping origin of the enigmatic relief valleys of Cape Cod and Martha's Vineyard and Nantucket Islands, Massachusetts. *Geomorphology* 9:83–95.

Ueta, H. T., and Garfield, D. E. 1968. Deep core drilling program at Byrd Station 1967–68. *U.S. Arctic Journal* 3:111–12.

Ullrich, C. R., Hagerty, D. J., and Holmberg, R.W. 1986. Surficial failures of alluvial stream banks. *Canadian Geotechnical Journal.* 23:304–316.

Ungar, J. E., and Haff, P. K. 1987. Steady state saltation in air. *Sed.* 34:289–99.

Ursic, S. J., and Dendy, F. E. 1965. Sediment yields from small watersheds under various land uses and forest covers. *Proc. Fed. Inter-Agency Sedimentation Conf.* (1963). U.S. Dept. Agri. Misc. Publ. 970:47–52.

U.S. Army Corps of Engineers. 1971. National shoreline study. Washington, DC: U.S. Army Corps Engr.

———. 1985. *Flood hydrograph package and users manual— HEC-1.* Davis, CA: The Hydrologic Engineering Center.

U.S. Boundary Water Commission. 1900–1977. International Boundary and Water Commission (1930–1977). Flow of the Rio Grande and related data. *Water Bulletins* Nos. 1–47.

U.S. Geological Survey. 1982. Goals and tasks of the landslide part of a ground failure hazards reduction program. *Circular 880.*

U.S. Water Resources Council. 1981. Guidelines for determining flood flow frequency. *Bull No. 17B,* U.S. Water Res. Council. Washington, DC.

Vagners, V. J. 1966. *Lithologic relationship of till to carbonate bedrock in southern Ontario.* M.S. thesis, Geology Dept., Univ. Western Ontario.

Valeton, I. 1994. Element concentration and formation of ore deposits by weathering. *Catena* 21:99–129.

Vallance, J. W. 1999. Postglacial lahars and potential hazards in the White Salmon River system on the southwest flank of Mount Adams, Washington. *U.S. Geological Survey Bulletin,* 2161.

Van Arsdale, R. 1982. Influence of calcrete on the geometry of arroyos near Buckeye, Arizona. *Geol. Soc. America Bull.* 93:20–26.

Van Asch, T. W. J., Buma, J., and Van Beek, L. P. H. 1999. A view on some hydrological triggering systems in landslides. *Geomorphology* 30:25–32.

VanDine, D. F. 1985. Debris flows and debris torrents in the southern Canadian Cordillera. *Can. Geotech. J.* 22:44–68.

Van Dorn, W. G. 1965. Tsunamis. In *Hydroscience advances 2.* New York: Academic Press.

———. 1966. Tsunamis. In *Encyclopedia of oceanography,* pp. 941–43. Ed. by R. W. Fairbridge. New York: Reinhold Book Corp.

Van Everdingen, R. O. 1978. Frost mounds at Bear Rock, near Fort Norman, Northwest Territories 1975–1976. *Canadian Journal of Earth Sciences* 15:263–76.

———. 1982. Frost blisters of the Bear Rock Spring area near Fort Norman, N.W.T. *Artic* 35:243–65.

———. 1990. Groundwater hydrology. In *Northern hydrology: Canadian perspectives,* pp. 77–103. Ed. by T. D. Prowse and C. S. L. Ommanney. Ottawa: Environment Canada, NHRI Science Report 1.

Van Heukon, T. K. 1977. Distant source of 1976 dustfall in Illinois and Pleistocene weather models. *Geology* 5:693–95.

Van Steijn, H., Bertran, P., Francou, B., Hetu, B., and Texier, J. P. 1995. Models for the genetic and environmental interpretation of stratified slope deposits. Review. *Permafrost and Periglacial Processes* 6:125–46.

Van Vilet-Lanoe, B. V. 1998a. Frost and soils: Implication for paleosols, paleoclimate and stratigraphy, 34:157–83.

———. 1998b. Patterned ground, hummock, and Holocene climate changes. *Eurasian Soil Science* 31:507–13.

Vanoni, V. A. 1941. Some experiments on the transportation of suspended load. *Am. Geophys. Union Trans.* 22nd Ann. Mtg., pt. 3:608–20.

———. 1946. Transportation of suspended sediment by water. *Am. Soc. Civil Engineers Trans.* 3:67–133.

Vanstone, S. 1991. Early Carboniferous (Mississippian) paleosols from southwest Britain: Influence of climatic change on soil development. *Jour. Sed. Petrology* 61:445–57.

Varnes, D. J. 1958. Landslide types and processes. In *Landslides and engineering practice,* pp. 20–47. Ed. by E. Eckel. Washington, DC: Highway Research Board Spec. Rept. 29.

———. 1978. Slope movement types and processes. In *Landslides,* pp. 11–33. Ed. by R. Schuster and R. Krizak. Washington, DC: Trans. Res. Board, Natl. Acad. Sci.

Veillette, J. J. 1986. Former southwesterly ice flows in Abitibi-Timiskaming region: implications for the configuration of the Late Wisconsinan ice sheet. *Can. J. Earth Sci.* 23:1724–41.

Vernon, P. 1966. Drumlins and Pleistocene ice flow over the Ards Peninsula. *Jour. Glaciol.* 6:401–9.

Verstappen, H. Th. 1964. Karst morphology of the Star Mountains (central New Guinea) and its relation to lithology and climate. *Zeit. F. Geomorph.* 8:40–49.

Viers, J., and 5 others. 1997. Chemical weathering in the drainage basin of a tropical watershed (Nsimi-Zoetele site, Cameroon): Comparison between organic-poor and organic-rich waters. *Chem. Geol.* 140:181–206.

Violante, P., and Wilsen M. J. 1983. Mineralogy of some Italian andesols with specific reference to the origin of the clay fraction. *Geoderma* 29:157–74.

Vitek, J. 1989. A perspective of geomorphology in the twentieth century: links to the past and future. In *History of Geomorphology from Hutton to Hack,* pp. 293–324. Ed. by K. Tinkler. London: Unwin Hyman.

Vitek, J. D. 1983. Stone polygons: Observations of surficial activity. In *Proc. permafrost 4th internat. conf.,* pp. 1326–31. Natl. Acad. Sci.

Vivian, R. 1970. Hydrologie et erosion sous-glaciaires. *Rev. Géog. Alp.* 58:241–64.

———. 1980. The nature of the ice-rock interface: The results of investigation on 20,000 m^2 of the rock bed of temperate glaciers. *Jour. Glaciol.* 25:267–77.

Wade, F. A., and deWys, J. N. 1968. Permafrost features on the Martian surface. *Icarus* 9:175–85.

Wahrhaftig, C. 1965. Stepped topography of the southern Sierra Nevada. *Geol. Soc. America Bull.* 76:1165–90.

Wahrhaftig, C., and Cox, A. 1959. Rock glaciers in the Alaska Range. *Geol. Soc. America Bull.* 70:383–436.

Wainwright, J. 1996. Hillslope response to extreme storm events: Example from the Vaison-La-Romaine event. In *Advances in Hillslope Processes,* pp. 997–1026. Ed. by M. G. Anderson and S. M. Brooks. New York: John Wiley.

Waitt, R. B., Jr. 1980. About forty last-glacial Lake Missoula jökulhlaups through southern Washington. *Jour. Geology* 88:653–79.

Walder, J., and Hallet, B. 1985. A theoretical model of the fracture of rock during freezing. *Geological Society of America Bulletin* 96:336–46.

Walker, H. J. 1988. Permafrost and coastal processes. In *Permafrost, fifth international conference proceedings*, pp. 35–41. Ed. by K. Senneset. Trondheim: Tapir Publishers.

Wallace, R., ed. 1986. *Active tectonics*. Washington, DC: National Academic Press.

Wallace, R. E. 1977. Profiles and ages of young fault scarps, north-central Nevada. *Geol. Soc. America Bull.* 88:1267–81.

———. 1978. Geometry and rates of change of fault-generated range fronts, north-central Nevada. *U.S. Geol. Survey Jour. of Res.* 6:637–49.

Walling, D. E. 1987. Rainfall, runoff and erosion of the land; a global view. In Energetics of the physical environment, pp. 89–117. Ed. by K. J. Gregory. Chichester: John Wiley.

Walling, D. E., and Quinne, T. A. 1990. Use of Cesium-137 to investigate patterns and rates of soil erosion on arable fields. In *Soil Erosion on Agricultural Land*, pp. 33–53. Ed. by J. Boardman, D. L. Foster, and J. A. Dearing. Chichester: John Wiley.

Walling, D. E., and Webb, B. W. 1996. Erosion and sediment yield: A global review. In *Erosion and Sediment Yield: Global and Regional Perspectives*, pp. 3–19. Ed. by D. E. Walling and B. W. Webb. Int. Assoc. Hyd. Sciences Pub. 236.

Walters, J. C. 1978. Polygonal patterned ground in central New Jersey. *Quat. Res.* 10:42–54.

Warren, A., and Allison, D. 1998. The palaeoenvironmental significance of dune size hierarchies. *Palaeogeog., Palaclim., Palaeoecol.* 137, 289–303.

Warren, W. P. and Ashley, G. M. 1994. Origins of the ice-contact stratified ridges (eskers) of Ireland. *J. Sed. Res.* A64:433–449.

Washburn, A. L., 1980, *Geocryology*. New York: John Wiley & Sons.

Waters, M. R. 1988. Holocene alluvial geology and geoarchaeology of the San Xavier reach of the Santa Cruz River, Arizona. *Geol. Soc. Amer. Bull.* 100:479–91.

Waters, M. and Nordt, L. 1995. Late Quaternary floodplain history of the Brazos River in east-central Texas. *Quat. Res.* 43, 311–19.

Wayne, W. J. 1967. Periglacial features and climatic gradient in Illinois, Indiana and western Ohio, east-central United States. In *Quaternary Paleoecology*, pp. 393–414. Ed. by E. Cushing and H. Wright. New Haven, CT: Yale Univ. Press.

———. 1981. Ice segregation as an origin for lenses of nonglacial ice in "ice-cemented" rock glaciers. *Jour. Glaciol.* 27:506–10.

Wayne, W. J. 1991. Ice wedge casts of Wisconsinan age in eastern Nebraska. *Permafrost and Periglac. Proc.* 2:211–23.

Waythomas, C. F., and Jarrett, R. D. 1994. Flood geomorphology of Arthurs Rock Gulch, Colorado paleoflood history. *Geomorphology* 11:15–40.

Waythomas, C. F., and Williams, G. P. 1988. Sediment yield and spurious correlation; toward a better portrayal of the annual suspended sediment load of rivers. *Geomorph.* 1:309–16.

Wear, J., and White, J. 1951. Potassium fixation in clay mineral studies as related to crystal structure. *Soil Sci.* 71:1–14.

Webb, J. A., and Fielding, C. R. 1999. Debris flow and sheet flood fans on the northern Prince Charles Mountains, East Antarctica. In *Varieties of fluvial form,* pp. 317–41. Ed. by A. J. Miller and A. Gupta. New York: John Wiley.

Weertman, J. 1957. On the sliding of glaciers. *Jour. Glaciol.* 3:33–38.

———. 1964. The theory of glacier sliding. *Jour. Glaciol.* 5:287–303.

———. 1967. An examination of the Lliboutry theory of glacier sliding. *Jour. Glaciol.* 6:489–94.

———. 1972. General theory of water flow at the base of a glacier or ice sheet. *Reviews of Geophysics and Space Physics,* 10:287–333.

———. 1979. The unsolved general glacier sliding problem. *Jour. Glaciol.* 23:97–111.

———. 1986. Basal water and high-pressure ice. *Journal of Glaciology* 32:455–63.

Weertman, J., and Weertman, J. B. 1964. *Elementary dislocation theory.* New York: Macmillan.

Weinert, H. 1961. Climate and weathered Karroo dolerites. *Nature* 191:325–29.

———. 1965. Climatic factors affecting the weathering of igneous rocks. *Agri. Meterol.* 2:27–42.

Welder, F. A. 1959. Processes of deltaic sedimentation in the lower Mississippi River. Louisiana State Univ., *Coastal Studies Inst. Tech. Rept.* 12.

Wellman, P. 1982. Surging of Fisher Glacier, eastern Antarctica: Evidence from geomorphology. *Jour. Glaciol.* 28:23–28.

Wells, J. T., Prior, D. B., and Coleman, J. M. 1980. Flowslides in muds on extremely low angle tidal flats, northeast South America. *Geology* 8:272–75.

Wells, S. 1976. Sinkhole plain evolution in the central Kentucky karst. *Natl. Speleol. Soc. Bull.* 38:103–6.

Wells, S., and Gardner, T. 1985. Geomorphic criteria for selecting stable uranium tailings disposal sites in New Mexico: Santa Fe, New Mexico. *New Mexico Energy Res. and Develop. Inst. Tech. Rpt.,* 2-69-112, v. 1.

Wells, S., and Harvey, A. 1987. Sedimentologic and geomorphic variations in storm-generated alluvial fans, Howgill Fells, northwest England. *Geol. Soc. America Bull.* 98:182–98.

Wells, S., and 3 others, 1995. Cosmogenic 3He surface exposure dating of stone pavements: Implications for landscape development. *Geology* 23:613–16.

Wells, S. G., and Meyer, G. A. 1993. Valley floor responses to Holocene climatic changes in semiarid and temperate ecosystems of the western USA. Third Intl. Geomorph. Conf. Abs. w. Prog., Hamilton, Ontario, p. 270.

Wentworth, C. K. 1938. Marine beach formation: Water level weathering. *Jour. Geomorphology* 1:6–32.

Werner, B., 1995. Eolian dunes: computer simulations and attractor interpretation. *Geology* 23:1107–10.

Werner, B., and Kocurek, G., 1999. Bedform spacing from defect dynamics. *Geology* 27:727–30.

Werner, B. T., and Haff, P. K. 1988. The impact process in aeolian saltation: Two-dimensional simulations. *Sedimental* 35:189–96.

Werritty, A. 1992. Downstream fining in a gravel-bed river in southern Poland: Litholic controls and the role of abrasion. In *Dynamics of gravel-bed rivers*, pp. 333–50. Ed. by P. Billi,

R. D. Hey, C. R. Thorne, and P. Tacconi. New York: John Wiley and Sons, Ltd.

Wescott, W. A., and Ethridge, F. G. 1980. Fan-delta sedimentology and tectonic-Hallahs Fan delta, southeast Jamaica. *Am. Assoc. Petroleum Geologists Bull.* 64:374–99.

Wesnousky, S. 1986. Earthquakes, Quaternary faults, and seismic hazard in California. *Jour. Geophys. Res.* 9:12:587–631.

Wesnousky, S., Prentics, C., and Sieh, K. 1991. An offset Holocene stream channel and the rate of slip along the northern reach of the San Jacinto fault zone, San Bernardino Valley, California, *Geol. Soc. America Bull.* 103:700–709.

Wesson, R., Helley, E., LaJoie, K., and Wentworth, C. 1975. Faults and future earthquakes. In *Studies for seismic zonation of the San Francisco Bay region,* pp. A5–A30. Ed. by R. Borcherdt, U.S. Geol. Surv. Prof. Paper 941 A.

West, D., and McCrumb, D. 1988. Coastline uplift in Oregon and Washington and the nature of Cascadia subduction-zone tectonics. *Geology* 16:169–72.

Westgate, J. A. 1968. Linear sole markings in Pleistocene till. Geol. Mag. 105:501–5.

Whalley, W. B. 1983. Rock glaciers—Permafrost features or glacial relics. In *Proc. permafrost 4th internat. conf.,* pp. 1396–1401. Natl. Acad. Sci.

Whalley, W. B.; Douglas, G. R.; and McGreevey, J. P. 1982. Crack propogation and associated weathering in igneous rocks. *Zeit. F. Geomorph.* 26:33–54.

Whipple, K. X., and Tucker, G. E. 1999. Dynamics of the stream-power river incision model: Implications for height limits of mountain ranges, landscape response timescales, and research needs. *Jour. Geophys. Research* 104:17661–17, 674.

Whipple, K. X., Hancock, G. S., and Anderson, R. S. 2000. River incision into bedrock: Mechanics and relative efficacy of plucking, abrasion, and cavitation. *Geol. Soc. America Bull.* 112:490–503.

White, A., and Blum, A. 1995. Effects of climate on chemical weathering in watersheds. *Geochim. Cosmochim. Acta* 59: 1729–47.

White, A., and Brantley, S., eds. 1995. Chemical weathering rates of silicate minerals. *Reviews in Mineralogy* 31, Mineral. Soc. Amer., Wash., D.C.

White, A., and 5 others. 1996. Chemical weathering rates of a soil chronosequence on granitic alluvium: I. Quantification of mineralogical and surface area changes and calculation of primary silicate reaction rates. *Geochim. Cosmochim. Acta* 60:2533–50.

White, A., and 7 others, 1998. Chemical weathering in a tropical watershed, Luquillo Mountains, Puerto Rico: I. Long-term versus short-term weathering fluxes. *Geochim. Cosmochim. Acta* 62:209–26.

White, A. F. 1995. Chemical weathering rates of silicate minerals in soils. In Chemical Weathering Rates of Silicate Minerals (ed. A. F. White and S. L. Branley); Rev. Mineral. 31:407–459.

White, B. R., and Schultz, J. C. 1977. Magnus effect in saltation, *Jour. Fluid Mech.* 81:497–512.

White, E., and White, W. 1979. Quantitative morphology of landforms in carbonate rock basins in the Appalachian Highlands. *Geol. Soc. America Bull.* 90:385–96.

———. 1983. Karst landforms and drainage basin evolution in the Obey River basin, north-central Tennessee, U.S.A. *J. Hydrol.* 61:69–82.

White, E. L., and Reich, B. M. 1970. Behaviour of annual floods in limestone basins in Pennsylvania. *J. Hydrol.* 10:193–98.

White, S. E. 1976. Rock glaciers and block fields, review and new data. *Quat. Res.* 6:77–98.

White, W. A. 1966. Drainage asymmetry and the Carolina capes. *Geol. Soc. America Bull.* 77:223–40.

White, W. B. 1969. Conceptual models for carbonate aquifers. *Ground Water* 7:15–21.

———. 1976. Geology and biology of Pennsylvania caves. *Pennsylvania Geol. Survey Gen. Rept.* 66:1–71.

———. 1978. Theory of cave origin and karst aquifer development. Yellow Springs, Ohio, *Cave Research Foundation Annual Report,* pp. 36–37.

———. 1984. Rate processes; Chemical kinetics and karst landform development. In *Groundwater as a geomorphic agent,* pp. 227–48. Ed. by R. G. LaFleur. Boston: Allen and Unwin, Inc.

———. 1988. *Geomorphology and hydrology of karst terrains.* New York, Oxford University Press.

———. 1990. Surface and near-surface karst landforms. In *Groundwater geomorphology; The role of subsurface water in earth-surface processes and landforms,* pp. 157–75. Ed. by G. G. Higgins and D. R. Coates. Boulder: *Geological Society of America Special Paper* 252.

———. 1999. Karst Hydrology: Recent developments and open questions. In Beck, Pettit and Herring (eds.), *Hydrogeology and Engineering Geology of Sinkholes and Karst,* Rotterdam, A. A. Balkema, pp. 3–20.

Whitehouse, I. 1992. Tectonic geomorphology: Recent studies of faulting and tectonic landforms, *Prog. Phys. Geogr.* 16:361–69.

Whitney, M. I., and Dietrich, R. V. 1973. Ventifact sculpture by windblown dust. *Geol. Soc. America Bull.* 84:2561–82.

Whittaker, J. G., and Jaegg, M. N. R. 1982. Origin of step-pool systems in mountain streams. Am. Soc. Civil Engineers, *Jour. Hydraulics Div.* HY6. 108:758–70.

Whittecar, G. R., and Mickelson, D. 1979. Composition, internal structures, and a hypothesis of formation for drumlins, Waukesha County, Wisconsin, U.S.A. *Jour. Glaciol.* 22:357–71.

Wiberg, P. L., and Smith, J. D. 1991. Velocity distribution and bed roughness in high-gradient streams. *Water Resour. Res.* 27:825–38.

Wieczorek, G. F. 1987. Effect of rainfall intensity and duration on debris flows in central Santa Cruz Mountains, California. In *Debris flows/avalanches: process, recognition, and mitigation,* pp. 93–104. Ed. by J. E. Costa and G. F. Wieczorek, Boulder: *Geol. Soc. Amer. Rev. in Engin. Geol* 7.

Wieczorek, G. F., and Sarmiento, J. 1983. Significance of storm intensity-duration for triggering debris flows near La Honda, California. *Geol. Soc. Amer. Abs. w. Prog.* 15(5):289.

Wieczorek, G. F., and Snyder, J. B. 1999. Rockfalls from Glacier Point above Camp Curry, Yosemite National Park, California. *U.S. Geological Survey Open File Report,* 99–385.

Wieczorek, G. F., Mandrane, G, and DeCola, L. 1997. The influence of hillslope shape on debris-flow initiation. *Amer.*

Society Civil Engineers. First Int. Conf. on Debris-flow Hazards, pp. 21–31.

Wieczorek, G. F., Lips, E. W., and Ellen, S. D. 1989. Debris flows and hyperconcentrated floods along the Wasatch Front, Utah, 1983 and 1984. *Bull. Assoc. Engin. Geol.* 26:191–208.

Wieczorek, G. F., Snyder, J. B., Waitt, R. B., Morissey, M. M., Uhrhammer, R. A., Harp, E. L., Norris, R. O., Bursik, M. I., and Finewood, L. G. 2000. Unusual July 10, 1996, rockfall at Happy Isles, Yosemite National Park, California. *Geological Society America Bull.* 112:75–85.

Wiegel, R. L. 1964. *Oceanographical Engineering.* Englewood Cliffs, NJ: Prentice-Hall.

Wigley, T. M. L., and Plummer, L. N. 1976. Mixing of carbonate waters: *Geochimica et Cosmochimica Acta* 40:989–95.

Wilcock, D. N. 1971. Investigation into the relations between bedload transport and channel shape. *Geol. Soc. America Bull.* 82:2159–76.

———. 1975. Relations between planimetric and hypsometric variables in third- and fourth-order drainage basins. *Geol. Soc. America Bull.* 86:47–50.

Wilcock, P. R., and Southard, J. B. 1988. Experimental study of incipient motion in mixed-size sediment. *Water Resources Research* 24:1137–51.

Wilcock, P. R., Muller, D. S., Shea, H. R., and Kerkin, R. T. 1998. Frequency of effective wave activity and the recession of coastal cliffs. Calvert Cliffs, Maryland. *Jour. Coastal Res.* 14:256–68.

Wilford, G. E., and Wall, J. R. D. 1965. Karst topography in Sarawak. *J. Trop. Geogr.* 21:44–70.

Wilgoose, G., Bras, R. L., and Rodriguez-Iturbe, I. 1990. Results from a new model of river basin evolution. *Earth Surf. Proc. and Landforms* 16:237–54.

Williams, A. T. 1973. The problem of beach cusp development. *Jour. Sed. Petrology* 43:857–66.

Williams, D. T., and Julien, P. Y. 1989. Applicability index for sand transport equations. *Jour. Hydraulic Engineering* 115:1578–81.

Williams, G. 1964. Some aspects of the eolian saltation load. *Sedimentology* 3:257–87.

Williams, G. P. 1978. Hydraulic geometry of river cross-sections—Theory of minimum variance. *U.S. Geol. Survey Prof. Paper* 1029.

———. 1989. Sediment concentration versus water discharge during single hydrologic events in rivers. *J. Hydrology,* 111:89–106.

Williams, G. P., and Guy, H. P. 1973. Erosional and depositional aspects of Hurricane Camille in Virginia, 1969. *U.S. Geol. Survey Prof. Paper* 804.

Williams, G. P., and Wolman, M. G. 1984. Downstream effects of dams on alluvial rivers. *U.S. Geol. Survey Prof. Paper* 1286.

Williams, J. R. 1970. Groundwater in the permafrost region of Alaska. *U.S. Geol. Survey Prof. Paper* 696.

Williams, L. 1964. Regionalization of freeze-thaw activity. *Ann. Assoc. Am. Geog.* 54:597–611.

Williams, P. J. 1966. Downslope soil movement at a sub-Arctic location with regard to variations with depth. *Canadian Geotech. Jour.* 3:191–203.

———. 1966. Morphometric analysis of temperate karst landforms. *Ir. Speleol.* 1:23–31.

———. 1971. Illustrating morphometric analyses of karst with examples from New Guinea. *Zeit. F. Geomorph.* 15(1):40–61.

———. 1972a. The analysis of spatial characteristics of karst terrains. In *Spatial analysis in geomorphology,* pp. 135–63. Ed. by R. J. Chorley. London: Methuen and Co.

———. 1972b. Morphometric analysis of polygonal karst in New Guinea. *Geol. Soc. America Bull.* 83:761–96.

———. 1983. The role of the subcutaneous zone in karst hydrology. *J. Hydrol.* 61:45–67.

Williams, P. J., and Smith, M. W. 1989. *The frozen earth: Fundamentals of geocryology.* Cambridge: Cambridge Univ. Press.

Williams, P. W. 1966. Morphometric analysis of temperate karst landforms. *Ir. Speleol.* 1:23–31.

———. 1971. Illustrating morphometric analyses of karst with examples from New Guinea. *Zeit. F. Geomorph.* 15(1):40–61.

———. 1972a. The analysis of spatial characteristics of karst terrains. In *Spatial analysis in geomorphology.* Ed. by R. J. Chorley. London: Methuen and Co.

———. 1972b. Morphometric analysis of polygonal karst in New Guinea. *Geol. Soc. America Bull.* 83:761–96.

———. 1985. Subcutaneous hydrology and the development of doline and cockpit karst. *Zeit. F. Geomorph.* 29:463–82.

———. 1987. Geomorphic inheritance and the development of tower karst. *Earth Surface Processes and Landforms.* 12:453–65.

———. 1990. Hydrological control and the development of cockpit and tower karst. In *Karst hydrogeology and karst environmental protection. International Association of Hydrological Sciences* Publication No. 176, pp. 281–87.

Williams, R. B. G., and Robinson, D. A. 1991. Frost weathering of rocks in the presence of salt; a review. *Permafrost and Periglac. Proc.* 2: 347–53.

Williams, R. P. 1987. Unit hydraulic geometry: An indicator of channel changes. *U.S.G.S. Selected Papers in Hydrological Science,* Water Supply Paper W2330, p. 77–89.

Willis, I. C. 1995. Intra-annual variations in glacier motion: a review. *Progress in Physical Geography* 19:61–106.

Willman, H. B., and Frye, J. C. 1970. Pleistocene stratigraphy of Illinois. *Illinois Geol. Survey Bull.* 94.

Wilson, B. W. 1966. Seiche. In *Encyclopedia of oceanography,* pp. 804–11. Ed. by R. W. Fairbridge. New York: Reinhold Book Corp.

Wilson, I. G. 1972. Aeolian bedforms—their development and origins. *Sedimentology* 19:173–210.

Wilson, L. 1972. Seasonal sediment yield patterns of United States rivers. *Water Resour. Res.* 8:1470–79.

———. 1973. Variations in mean annual sediment yield as a function of mean annual precipitation. *Am. Jour. Sci.* 273:335–49.

———. 1968. Morphogenetic classification. In *Encyclopedia of geomorphology,* pp. 717–28. Ed. by R. W. Fairbridge. New York: Reinhold Book Corp.

Wilson, P. 1992. Small scale patterned ground, Comeragh Mountains, southeast Ireland. *Permafrost and periglac. proc.* 3: 63–70.

Winkler, E. M. 1965. Weathering rates as exemplified by Cleopatra's Needle in New York City. *Jour. Geol. Educ.* 13:50–52.

————. 1975. *Stone: Properties, durability in man's environment,* 2d ed. New York: Springer-Verlag.

Winkler, E. M., and Wilhelm, E. J. 1970. Salt bursts by hydration pressures in architectural stone in urban atmosphere. *Geol. Soc. America Bull.* 81:567–72.

Witkind, I. J. 1988. Potential geologic hazards near the Thistle Landslide, Utah County, Utah. *Bull. Assoc. Engin. Geol.* 25:83–94.

Wohl, E. E. 1993. Becrock channel incision along Picanniny Creek, Australia. *J. Geol.* 101:749–61.

————. 1998. Bedrock channel morphology in relation to erosional processes. In *Rivers over rock: Fluvial processes in bedrock channels,* pp. 133–51. Ed. by K. Tinkler and E. Wohl. American Geophysical Union: Geophysical Monograph 107.

————. 1999. Incised bedrock channels. In *Incised river channels,* pp. 187–218. Ed. by S. E. Darby and A. Simon. New York: John Wiley and Sons Ltd.

Wohl, E. E., and Grodek, T. 1994. Channel bed-steps along Nahal Yael, Negev desert, Israel. *Geomorphology* 9:117–26.

Wohl, E. E., and Ikeda, H. 1998. Patterns of bedrock channel erosion on the Boso Peninsula, Japan. *Jour. Geology* 106:331–45.

Wohl, E. E., Madsen, S., and MacDonald, L. 1997. Characteristics of log and clast bed-steps in step-pool streams of northwestern Montana. USA. *Geomorphology* 20:1–10.

Woida, K., and Thompson, M. 1993. Polygenesis of a Pleistocene paleosol in southern Iowa. *Geol. Soc. Amer. Bull.* 105:1445–61.

Woldenberg, M. J. 1969. Spatial order in fluvial systems. Horton's laws derived from mixed hexagonal hierarchies of drainage basin areas. *Geol. Soc. America Bull.* 80:97–112.

Wolfe, D., and David, P. 1997. Parabolic dunes: examples from the Great Sand Hills, southwestern Saskatchewan. *Can. Geogr.* 41:1103–14.

Wolfe, S., and Nickling, W. 1993. The protective role of sparse vegetation in wind erosion. *Prog. in Phys. Geog.* 17:50–68.

Wolfe, S. A. 1998. Living with frozen ground. A field guide to permafrost in Yellowknife, Northwest Territories. *Geol. Survey of Canada Misc. Rept.* 64.

Wolman, M. G. 1955. The natural channel of Brandywine Creek, Pennsylvania. *U.S. Geol. Survey Prof. Paper* 271.

————. 1959. Factors influencing erosion of a cohesive river bank. *Am. Jour. Sci.* 257:204–16.

————. 1967. A cycle of sedimentation and erosion in urban river channels. *Geogr. Annlr.* 49-A:385–95.

Wolman, M. G., and Gerson, R. 1978. Relative scales of time and effectiveness of climate in watershed geomorphology. *Earth Surf. Proc. and Landforms* 3:189–208.

Wolman, M. G., and Leopold, L. B. 1957. River flood plains; some observations on their formation. *U.S. Geol. Survey Prof. Paper* 282-C.

Wolman, M. G., and Miller, J. P. 1960. Magnitude and frequency of forces in geomorphic processes. *Jour. Geology* 68:54–74.

Womack, W. R., and Schumm, S. A. 1977. Terraces of Douglas Creek, northwestern Colorado: An example of episodic erosion. *Geology* 5:72–76.

Woo, M., and Heron, R. 1981. Occurrence of ice layers at the base of High Arctic snowpacks. *Arc. Alp. Res.* 13:225–30.

Woo, M., Yang, Z., Xia, Z., and Yang, D. 1994. Streamflow processes in an alpine permafrost catchment, Tianshan, China. *Permafrost and Periglacial Processes* 5:71–85.

Wood, A. 1942. The development of hillside slopes. *Proc. Geol. Assoc.* 53:128–40.

Wood, F. J. 1978. *The strategic role of perigean spring tides.* Washington, DC: U.S. Dept. of Commerce.

Woodcock, A. H. 1974. Permafrost and climatology of a Hawaii volcano crater. *Arc. Alp. Res.* 6:49–62.

Woodcock, A. H., Furumoto, A. S., and Woollard, G. P. 1970. Fossil ice in Hawaii. *Nature* 226:873.

Worsley, P., and Gurney, S. 1995. Geomorphology and hydrogeological significance of the Holocene pingos in the Karup Valley area, Traill Island, northern east Greenland. *Jour. Quaternary Sci.* 14: 249–62.

Wright, H. E., and Frey, D. G., eds. 1965. *The Quaternary of the United States.* Princeton, NJ: Princeton Univ. Press.

Wright, J., and Schnitzer, M. 1963. Metallo-organic interactions associated with podsolisation. *Soil Sci. Soc. Am. Proc.* 27:171–76.

Wright, L. D. 1977. Sediment transport and deposition at river mouths: A synthesis. *Geol. Soc. America Bull.* 88:857–68.

Wright, L. D., and Pilkey, O. H. 1989. The effect of hard stabilization upon dry beach width. In *Coastal Zone '89., Proc. 6th symp. on coast. and ocean manag.,* pp. 776–90. Ed. by T. O. Magoon and others.

Wright, L. D., and Short, A. D. 1983. Morphodynamics of beaches and surf zones in Australia. In *CRC handbook of coastal processes and erosion,* pp. 35–64. Ed. by P. Komar. Boca Raton, FL: CRC Press.

Wright, L. D., Chappell, J., Thom, B. G., Bradshaw, M. P., and Cowell, P. J. 1979. Morphodynamics of reflective and dissipative beach and inshore systems, southeastern Australia. *Marine Geology* 32:105–40.

Wright, L. D., Guza, R. T., and Short, A. D. 1982a. Dynamics of a high-energy dissipative surf zone. *Marine Geol.* 45:41–62.

Wright, L. D., Nielson, P. N., Short, A. D., and Green, M. O. 1982b. Morphodynamics of a macrotidal beach. *Marine Geol.* 50:97–128.

Wright, P. V., and Tucker, M. E. 1991. Calcretes: An introduction. In *Calcretes,* pp. 1–22. Ed. by V. P. Wright and M. E. Tucker. Reprinted series vol. 2, Internat. Assoc. Sedimentologists. Oxford: Blackwell Scientific.

Wyllie, P. 1971. *The dynamic earth.* New York: John Wiley & Sons.

Wyzga, B. 1999. Estimating mean flow velocity in channel and floodplain areas and its use for explaining the pattern of overbank deposition and floodplain retention. *Geomorph.* 28:281–97.

Yaalon, D. H. 1975. Conceptual models in pedogenesis. Can soil-forming functions be solved? *Geoderma* 14:189–205.

Yair, A., Lavee, H., Bryan, R. B., and Adar, E. 1980. Runoff and erosion processes and rates in the Zin Valley badlands, northern Negev, Israel. *Earth Surf. Proc. and Landforms* 5:205–25.

Yang, C. T. 1973. Incipient motion and sediment transport. Am. Soc. Civil Engineers, *Jour. Hydraulics Div.* 99:HY10:1679–1704.

Yang, D., and Xie, Y. 1997. Paleoflood slackwater deposits. *Acta Sedimentologia Sinica* 15:29–32.

Yanosky, T. M. 1982. Effects of flooding upon woody vegetation along parts of the Potomac River floodplain. *U.S. Geol. Survey Prof. Paper* 1206.

Yefimov, A. I., and Dukhin, I. E. 1968. Some permafrost thicknesses in the Arctic. *Polar Record* 14:68.

Yoshikawa, K. 1998. The groundwater hydraulics of open system pingos. *7th Internat. Permafrost Conf. Proc. Collection Nordicana* 57:1177–84.

Young, A. 1960. Soil movement by denudational processes on slopes. *Nature* 188:120–22.

———. 1961. Characteristic and limiting slope angles. *Zeit. F. Geomorph.* 5:126–31.

———. 1972a. *Slopes.* Edinburgh: Oliver and Boyd.

———. 1972b. The soil catena: A systematic approach. In *International geography 1972, Congr. Int. Geogr. Commun.,* n. 22, v. 1:287–89.

Young, R. W. 1983. The tempo of geomorphological change: Evidence from southeastern Australia. *Jour. Geology* 91:221–30.

Yuan, D. 1987. New observations on tower karst. In *International Geomorphology, Part II,* pp. 1109–23. Ed. by V. Gardiner.

Zeigler, J. M., Hayes, C. R., and Tuttle, S. D. 1959. Beach changes during storms on outer Cape Cod, Massachusetts. *Jour. Geology* 67:318–36.

Zenkovich, V. P. 1967. *Processes of coastal development.* Translated by D. G. Fry. Ed. by J. A. Steers. Edinburgh: Oliver and Boyd.

Zhang, T., and Stamnes, K, 1998. Impact of climatic factors on the active layer and permafrost at Barrow, Alaska. *Permafrost and Periglacial Processes* 9:229–46.

Zicheng, K., and Jing, L. 1987. Erosion processes and effects of debris flow. In *Erosion and sedimentation of the Pacific Rim,* pp. 233–42. Ed. by R. L. Beschta et al., Int. Assoc. Hydr. Sci. Pub. 165.

Zimpfer, G. L. 1982. Hydrology and geomorphology of an experimental drainage basin. PhD. dissertation. Fort Collins: Colo State Univ.

Zirjacks, W. L., and Hwang, C. T. 1983. Underground utilidors at Barrow, Alaska: A two-year history. In *Proc. permafrost 4th internat. conf.,* pp. 1513–17. Natl. Acad. Sci.

Zolitschka, B. 1998. A 14,000-year sediment yield record from western Germany based on annually laminated lake sediments. *Geomorphology* 22:1–17.

Zoltai, S. C. 1972. Palsas and peat plateaus in central Manitoba and Saskatchewan. *Canadian Journal of Forest Research* 2:291–302.

Zwingle, E. 1993. Ogallala Aquifer: Wellspring of the High Plains. *National Geogr.* 183:81–109.

CREDITS

LINE ART/TABLES

Chapter 1

Table 1.3: Adapted from W.H.K. Lee and S. Uyeda, GEOPHYSICAL MONOGRAPH 8, p. 147, 1965, copyright by the American Geophysical Union; **Table 1.5:** From A. Ronov and A. Yaroshevsky, GEOPHYSICAL MONOGRAPH 18, ed. by P. Hart, pp. 37–57, 1969, copyright by the American Geophysical Union; **Table 1.6:** From L.C. Pakiser and R. Robinson, GEOPHYSICAL MONOGRAPH 10, pp. 620–26, 1966, copyright by the American Geophysical Union.

Chapter 2

Table 2.1: Source: W.B. Bull, "Tectonic Geomorphology," JOURNAL OF GEOLOGIC EDUCATION 32:310–24, 1984; **Table 2.2:** Source: L. Wilson, "Morphogenetic Classification" in ENCYCLOPEDIA OF GEOMORPHOLOGY, ed. by R.W. Fairbridge, copyright 1968 Dowden, Hutchinson & Ross, Inc.

Chapter 3

Table 3.1: After Goldich 1938; **Table 3.2:** Redrawn from "Chemical Weathering and Global Chemical Cycles" by Lasaga et al. in GEOCHEMICA ET COSMOCHIMICA ACTA 58, Table 1, p. 2362, copyright 1994; **Table 3.3:** Redrawn from THE NATURE AND PROPERTIES OF SOILS, 6th Edition by Harry O. Buckman and Nyle C. Brady. Copyright © 1960 by Macmillan Publishing Company; **Table 3.5:** Adapted from the Soil Survey Staff, 1960, 1975, 1981; **Table 3.6:** From Soil Survey Division Staff 1993, SOIL SURVEY MANUAL; **Table 3.7:** From Soil Survey Staff 1999, SOIL TAXONOMY, p. 126; **Table 3.8:** From Soil Survey Staff 1999, SOIL TAXONOMY, p. 127; **Table 3.9:** Source: Gile 1975; **Figure 3.4:** From INTRODUCTION TO THE PETROLOGY OF SOILS AND CHEMICAL WEATHERING by Daniel B. Nahon, Figure 1.4, p. 13. Copyright © 1991 by John Wiley & Sons, Inc. Reprinted by permission; **Figure 3.10:** Figure 2, p. 2535 by White et al. from GEOCHIMICA ET COSMOCHIMICA ACTA. Copyright © 1996. Reprinted with permission from Elsevier Science; **Figure 3.11:** Figure 6, p. 2539 by White et al. from GEOCHIMICA ET COSMOCHIMICA ACTA. Copyright © 1996. Reprinted with permission from Elsevier Science; **Figure 3.25:** Figure 5 by McCalpin and Berry from CATENA, Vol. 27. Copyright © 1996. Reprinted by permission from Elsevier Science.

Chapter 4

Table 4.4: From M. Selby, 1980 in ZEITSCHRIFT FUR GEOMORPHOLOGIE. Used by permission of Gebruder Borntraeger Verlagbuchhandlung, Stuttgart; **Table 4.6:** From D.J. Varnes, 1978, "Landslides: Analysis and Control," TRB SPECIAL REPORTS 176: LANDSLIDES, Transportation Research Board, National Research Council, Washington, D.C. Used by permission; **Figure 4.24:** From "Urban Geology of Boulder, Colorado: A Progress Report" by Victor R. Baker in ENVIRONMENTAL GEOLOGY, Vol. 1, pp. 75–88. Reprinted by permission of Springer-Verlag GmbH & Co. and the author; **Figure 4.25:** From "Urban Geology of Boulder, Colorado: A Progress Report" by Victor R. Baker in ENVIRONMENTAL GEOLOGY, Vol. 1, pp. 75–88. Reprinted by permission of Springer-Verlag GmbH & Co. and the author.

Chapter 5

Table 5.1: From Howard, 1967, reprinted by permission; **Figure 5.2a:** From Howard, 1967, reprinted by permission.

Chapter 6

Figure 6.15: From "Geomorphically Effective Floods" by J.E. Costa and J.E. O'Connor in GEOPHYSICAL MONOGRAPH 89, 1995, Figure 11, p. 54. Copyright © 1995 by the American Geophysical Union; **Figure 6.23:** Figure 6, p. 25 by Stanley A. Schumm from SEPM SPECIAL PUBLICATION, No. 31. Copyright © 1981. Reprinted by permission of SEPM (Society for Sedimentary

INDEX

A

ablation, 301, 306–7
abrasion
    of bedrock river channels, 203
    glacial, 322–23
    by wind-borne particles, 280–82
Absaroka Mountain glaciers, 358
accumulation of water by glaciers, 301,
    306–7
active depositional lobes, 251
active glaciers, 299
active layer, of permafrost, 361
active sea cliffs, 467
activity of clays, 97–98
adsorption, 48
adverse slope, 333–34
aggradation, 205, 356
aggressive water, 409
agriculture, impact on sediment
    yield, 182
A horizon, 61, 72
airfields, on permafrost, 397–98
air-lubrication hypothesis, 113
alas, 381, 382
Alaska
    circular patterned ground, 384
    Gilkey Glacier, 319
    ice-wedge polygons, 378
    La Perouse Glacier, 300
    Little Yanert Glacier, 318
    Matanuska Glacier velocities, 314
    permafrost, 362
    pingos, 379
    rock glaciers, 388
    Schwan Glacier, 330
    Scott Glacier sandar, 355
    Trans-Alaska pipeline, 399–400, 401
    Variegated Glacier surge, 315
    Woodworth Glacier, 340
    Yanert Glacier, 345
Alcan Highway, 397
Alfisols, 69
allogenic recharge, 415, 416

allogenic valleys, 424
alluvial fans
    deltas versus, 264
    deposits and origins, 255–59
    morphology, 251–54
    system components, 248–51, 252, 253
alluvial plain, 260
alternate bars, 215
aluminum oxide, 48, 56
alveolar weathering, 88, 89
anabranching systems, 223–24
anastomosing and anabranching
    channels, 214, 223–24
anastomosing river, inorganic
    floodplains, 234
anastomosing river, organic-rich
    floodplains, 234
anastomotic mazes, 429, 430
angulate passages, 428, 429
annual series, for discharge, 167
annular drainage patterns, 136
Antarctic permafrost, 364–65
apexes, of alluvial fans, 251, 252
Appalachian Mountains, periglacial
    processes in, 395
Appalachian Plateau, slope processes
    in, 110
aquifers, 162–64, 414–18
arches, 87, 469
areal morphometric relationships,
    150–53
arêtes, 328, 330
arid fans, 249
Aridisols, 71
arid morphogenetic systems, 36
Arizona, Kaibab Plateau springs, 417
arroyos, 183
artesian flow, 162
Assateague Island
    erosion rates, 474
    landward migration, 488
    nearshore bars, 454
    recurved spit, 473
    shoreline features, 466

    vegetation and landforms, 482
    washover terrace, 484
astronomical motions, climate and, 37
at-a-station geometry, 207–9
Atchafalaya River, 266–67
atmospheric composition, 37
Atterberg limits, 97–98
attrition of sea cliffs, 468
Australia
    arid regions in, 273
    beach classification studies, 455–57
    exfoliation in, 88
autogenic recharge, 416
available water capacity, 59
avalanches, 113
avulsion, 224
Ayers Rock, 88

B

Bagnold's hypothesis, 291
bajadas, 251
bank erosion
    basic processes, 200–202
    in braided channels, 221
    meandering and, 220
    resistance and channel shape, 225
barchan dunes, 288, 289, 290, 291
barchanoid ridges, 288, 289
barrier islands
    defined, 470, 481
    distribution and characteristics,
    481–83
    geomorphic processes, 483–90
bars
    effects on rip currents, 450
    formation in braided channels,
    221–22
    nearshore, 453–55
basal sliding model, 388
base flow, 141
basin lag, 143
basin morphometry
    evolution of, 156–60

flood hydrographs and, 154–56
    variables of, 147–53
basin size, 177–78
bathyphreatic caves, 431
baymouth bars, 470–73
Bay of Fundy, 446
beach cusps, 461–64
beaches. *See also* coastal processes
    basic features, 451–53
    landforms and shoreline
        configurations, 460–66
    morphodynamic states, 455–60
    nearshore bars, 453–55
    overview of, 451
    pocket, 472
beach face, 452, 453
beach ridges, 473, 487
beach step, 457
beam type cantilever failure, 200
Beartooth Mountains
    cryoplanation, 394
    gelifluction lobes, 387
    stream piracy in, 159
    terminal moraines, 346
bed forms, 192, 194
bedload, 195–96, 221
bed material load, 195
bedrock channels
    erosion in, 202–3, 204
    longitudinal profiles, 211–13
bergschrunds, 331–32
berms, 452–53
BET surface area, 58
B horizon, 61, 62, 72
bifurcation
    of channels, 146
    in deltas, 266, 267
bifurcation ratio, 150
Bingham material, 114
biomass, erosion resistance, 145.
      See also vegetation
Black Mountains, 27
blind valleys, 409, 424
blockfields, 389–91
block-like soil structures, 60
blowouts, 288, 289
Bonner Bridge, 460, 488
bottomset deposits, 266
Bouguer anomalies, 23
boundary resistance, 194
bounding surfaces, 274
braid bars, 221
braid-deltas, 264
braided channels
    appearance, 214, 215
    formation of, 175, 221–23
    transition with meandering, 225
braided river floodplains, 234
branchwork patterns, 429, 430
breakers, 452, 453
breaking strength, 99

breakthrough, in karstification, 413
brittle deformation of glacial
      sediment, 344
building foundations, on permafrost, 397
bulking, 116
buried soils, 75
Byobugaura sea cliff erosion, 470, 471

C

calcification, soil formation by, 70–71
calcite
    dissolution, 410–12, 413
    in limestone, 408
    in speleothems, 428–29
calderas, 33, 35
California
    baymouth bars, 472
    coastal erosion, 3, 4
    composite cones in, 34
    early mining sedimentation effects,
      15–16
    exfoliation in, 87
    expansive soils in, 85
    fault-bounded mountains, 27
    high-relief shorelines, 467
    rip currents, 449
    rockfalls in, 106
    sample basin morphometry data, 152
    San Andreas fault, 30
    sediment yield in, 179–80
    wave-cut platforms, 28
calving of glaciers, 301
Canol pipeline, 399
cantilever failure, 200
canyon passages, 428, 430
Cape Hatteras, 459
capes, 464–66
capillary fringe, 161
carbon dioxide, 410–13
carbonic acid formation, 410
catenas, 67–68
cation exchange capacity, 48
cave patterns, 429
caves, 409, 428–32
cavettos, 326
cavitation, 203
c-axis orientation, 298–99
Cedar Creek alluvial fans, 250
Cedar Island overwash, 480
central bar theory, 222
chambers, defined, 409
channel bars, 194
channel fill, 235
channel initiation on slopes, 144–47.
      See also river channels
channelized flow, 309
channel lag, 235
channel patterns. *See* river channels
channel resistance, 193
channel shape, 213–14

chatter marks, 323, 325
chelation, 51–52
chemical composition of lithosphere,
    10, 11, 12
chemical weathering. *See*
    decomposition
chevron crevasses, 317, 318
Chezy equation, 193
China, tower karst in, 427, 428
chlorite, 54
C horizon, 61
chronofunctions, 73
chronosequences, 56
chutes, 235
circular patterned ground, 383, 384
cirque glaciers, 299
cirques, 328–32
clay plugs, 236
clays
    accumulation in B horizon, 70
    expansion effects in, 84, 85
    ion exchange in, 48–49
    mineral structures and stability,
      53–56
    plasticity of, 97–98
Cleopatra's Needle, 87–88
cliff faces, 125, 126
climate. *See also* storms
    driving forces of, 7–8
    effect of changes on periglacial
      regions, 400–401
    effects of changes on rivers, 228
    effects on soil formation, 68
    fan formation and, 256–58
    forces causes change in, 36–40
    influence on slope profile, 130–31,
      132–33
    permafrost and, 366
    role in debris flows, 118–20
    sediment yield and, 176–77, 183
climate-process systems, 34, 35
climatic geomorphology, 34–40
closed depressions, 409, 419–23
closed-system pingos, 379
Coastal Engineering Research Center-
    Field Research Facility,
    436–37, 453
coastal processes
    barrier islands, 470, 481–90
    beaches (*see* beaches)
    currents, 448–51
    high-relief shoreline topography,
      466, 467–70
    low-relief shoreline topography, 466,
      470–73
    overview of, 436–37
    shoreline change, 473–81
    tides, 445–48
    waves, 437–45
Coastal Research Amphibious Buggy
    (CRAB), 436, 438, 453

coastal storms. *See* storms
coastal zones, 434–36
cockpit karst, 409, 425–26
cockpits, 409
cohesion, shear strength and, 96–99
collapse dolines, 421–22
collapsing breakers, 440–43
colluvium
    debris flow vulnerability of, 118, 119
    defined, 80
    in floodplain sequence, 235
Colorado
    mountain formations, 13
    rockfalls in, 106, 108
Colorado Plateau
    groundwater sapping along, 149
    isostatic compensation in, 23
Colorado River, 149
color of soils, 58–59
cols, 329
Columbia River, 16, 32
comma forms, 326
competence, 197
complex responses, 15
composite cones, 33, 34
composite ridges, 344
compressive flow, 306
computer programs, 102
concave segments, 125, 126, 130
concentrated recharge, 416
conchoidal fractures, 325
cone karst, 409, 427
cones of depression, 163
cones of volcanoes, 32–33
confined aquifers, 162, 163
confined coarse-textured
        floodplains, 234
constant of channel maintenance,
        151–53
construction
    impact on sediment yield, 182
    on permafrost, 396–98
continents
    average elevation, 21–22
    denudation estimates for, 187
continuous creep, 105
continuous permafrost, 365
contorted drainage patterns, 136
controlling obstacle size, in glaciers, 312
convection
    frost sorting by, 371–72
    patterned ground from, 385
convergent flow, 216–17
convex segments, 125, 126, 130
coppice dunes, 289
corrasion, 200, 468
corrosion, 467–68
Coulomb equation, 95
Cowlitz River, 16, 17
Crater Lake, 33, 35

creep
    effects of weathering on, 112
    in glaciers, 302, 303
    overview of, 102–5
    saltation and, 280
crescentic fractures, 325
crescentic landforms, 461–66
crescentic nearshore bars, 454
crescentic scars, 325
crevasses, 266, 316–18
critical bed velocity, 197
critical flow, 191
critical friction velocity, 277
critical length, 145
critical shear stress, 197–98
cross-grading, 146
cross-sectional shape, of river channels,
        213–14
cross-valley moraines, 350
crust
    contrasts between continents and
            ocean floor, 21–22
    mineral composition, 10, 11, 12
cryoplanation terraces, 391–95
cryoturbation, 373
cuestas, 14
cusplets, 462
cusps, 461–64
cut and fill floodplains, 234
Cvijíc, Jovan, 407
cyclic time, 17–18
cyclones, 478

D
dams, 229
Darcy's Law, 161–62
Darcy-Weisbach resistance
        coefficient, 194
dating techniques, 30–31
Davis, W. M., 4–5
dead glaciers, 299
debris, unconsolidated, 92–99, 312–13
debris avalanches, 113
debris flows
    in fan deposition, 255–56
    field criteria for, 257
    on Mars, 402, 403, 404
    properties and causes, 113–21
debulking, 116
Deccan Plain, 32
decomposition
    in deserts, 274
    estimating degree of, 52–56
    ion mobility and, 45, 49–52
    overview of, 44–46
    processes of, 46–49
    rates of, 56–58
    relation to physical weathering, 43,
            92, 93
deflation, 275, 282

deformation
    in glaciers, 303
    rock glacier movement by, 388
    of subglacial sediments, 312–13
deformation till, 337, 349
DeKalb mounds, 380–81
deltas
    classification, morphology, and
            deposits, 264–66
    described, 264
    evolution of, 266–69
    in tidal inlets, 482, 486, 487
delta switching, 266–67
dendritic drainage patterns, 136, 157
denudation. *See also* erosion
    compared to uplift rates, 24
    erosion versus, 173
    of karst topography, 412–13
    measuring, 173, 186–87
    sediment budgets and, 184–85
    sediment yield factors, 176–83
    Universal Soil Loss Equation,
            175–76
    weathering rates and, 173–74
deposition
    in alluvial fans, 255–56 (*see also*
            alluvial fans)
    in anabranching channels, 224
    basic factors in, 204–5
    in braided channels, 221–23
    floodplain origins, 237–42 (*see also*
            floodplains)
    glacial (*see* glacial deposition)
    in low-relief shorelines, 470–73 (*see
            also* coastal processes)
    slope changes from, 229
    from storm overwash, 485–86
depositional terraces, 243–44, 247
depth of compensation, 23
desert basins, 282
desertification, 272
deserts, 264, 272–75
Desert Soil Geomorphology Project,
        75–76, 77
Deuteronilus Mensae region
        (Mars), 404
Devil's Potato Patch, 393
differential subsidence, 398
diffusion-layer model, 45
dilatation joints, 331
dilute debris flows, 257, 258
dimensionless critical shear stress,
        197–98
dimensionless numbers, 150
direct runoff, 138, 141. *See also* runoff
discontinuous permafrost, 365
Discovery Pingo, 379
disintegration
    in deserts, 273–74
    expansion processes, 81–89

growth in voids, 89–92
overview of, 80–81
relation to chemical weathering, 43, 92, 93
role of rain water, 92, 93
disintegration moraines, 343–44
dispersed recharge, 416
dispersion waves, 439
disruptive events, 3
dissipative beaches, 456, 457
dissolution
chemical processes, 45–46, 47–48
in karstification, 410–13
of sea cliffs, 467–68
distal zone, of sandar, 356
distributaries, 264
divergent flow, 217
divide-averaged relief, 153
divides, 135
Dixon Cave, 430
D/L value, 123–25
Dokuchaiev classification scheme, 62–63
dolines, 409, 419–23
dolomite, 408
dome-shaped dunes, 289
dominant discharge, 205
draa, 284, 288, 291
drag velocity, 275–76, 278–79
drainage, in glaciers, 309
drainage basins. *See also* floodplains
basic features, 135–36
channel initiation, 144–47
denudation (*see* denudation)
evolution of, 156–60
morphometric variables, 147–53
slope hydrology and runoff, 137–44
subsurface water sources, 160–64
surface water discharge, 164–73
drainage composition, 148–50
drainage density
in drainage basins, 150–51
as gage of peak flow, 154–55
relation to stream frequency, 158
drainage patterns, 135–36, 137
drawdown, of groundwater, 163
drift, glacial, 335–38
driving forces
defined, 6
types, 7–9
in unconsolidated debris, 92–95
drumlins, 352–54
dry valleys, 424, 425
ductile deformation, 344
Dump fault movement studies, 74
dumping, 341–43
dune patterns, 288–91
dunes, 286–92, 483
duricrusts, 275
dust, loess development from, 293
dust storms, 293

dynamic classification of glaciers, 299
dynamic equilibrium concept, 5, 6

E

earth fall, 201
earthflows, 121–22
earthquakes, 107
East Fork River, 196
ebb tidal deltas, 482, 486, 487
eddy viscosity, 191
edge waves, 450–51, 463
EDTA, 51
effective normal stress, 96–97
effective precipitation, 176
Eh factor, 47, 50–51
E horizon, 61
elevation and relief, sediment yield and, 178–79
elevation averages, 21–22
elliptical tubes, 428
elongate deltas, 265
El Salvador earthquake, 107
Emperor Chain, 32
Enchanted Rock, 83
end moraines, 341–43
endogenic processes
epeirogeny, 22–24
orogeny, 25–31
overview of, 6–7, 21–22
volcanism, 31–34
en echelon crevasses, 318
englacial load, 336
enhanced creep, 312
entrainment
fluvial, 200–201
of sediment in channels, 196–200
by wind, 276–80
entrenchment, 259, 260
environmental science, 3
eolian bed forms, 285. *See also* wind action
epeirogeny, 22–24
equal mobility hypothesis, 199–200
equilibrium, 4–6
equilibrium beach profile, 452
equilibrium lines of glaciers, 301, 302
equipotential surfaces, 161
ergodic hypothesis, 132
ergs, 282–84
erosion. *See also* denudation
of bedrock river channels, 202–3
channel initiation, 144–47
climatic influences, 38–39
coastal, 3, 473–81 (*see also* coastal processes)
from debris flows, 118, 119
glacial (*see* glacial erosion)
reduction by vegetal interception, 137–38

of river banks, 200–202, 220, 221
of sea cliffs, 467–70
sediment yield factors, 176–83
wind action, 280–82
erosional terraces, 243, 244–46, 247
eskers, 348–49
etch forms, 263
etch pits, 45, 46
eustatic change, 37–38, 246
evapotranspiration, 138
exfoliation, 86, 87
exhumed soils, 75
exogenic processes, 6
expansion joints, 82
expansion processes of disintegration, 81–89
expansive soils, 84, 85
experimental weathering rates, 57–58
exsurgences, 409, 417
extending flow, 306
extratropical storms, 478–80
extrinsic thresholds, 14

F

fabric, analyzing rockfalls from, 107
facets, wind abrasion, 280–81
facies types, 257, 258
failure mechanisms of river banks, 200–201
falls, 105–7
fan-deltas, 264
fanhead trenches, 251
fans
alluvial, 248–59
in Desert Project area, 77
role in slope formation, 132–33
fault-bounded mountains, 25–26, 27
faulting, 73, 74
fault-scarp analyses, 30–31
feeder channels, 251, 252
Fenwick Island, 488
fetch, 437
field capacity, 61
field work, importance of, 2
fill, in river channels, 204–5
fine-grained deposits, 292–94
fiords, 335
firn, formation of, 297–98
firn line, 297, 301, 302
fixation of ions, 51
flashy streams, 143, 240–41
flood hydrographs, 141–44, 154–56
flooding
basin morphometry and, 154–56
effect of physical basin characteristics, 141–44
frequency and magnitude, 165–68
impact on channel morphology, 206, 207

flooding—*Cont.*
 paleoflood data, 168–73
 runoff and base flow effects, 141,
  142, 155
floodplains
 classifying, 233–34
 deposits and topography, 235–37
 origins, 237–42
 water retention by, 144
flood recurrence interval, 167–68
flood tidal deltas, 482, 486, 487
flow duration curves, 166
flows
 defined, 99
 in river channels (*see* river channels)
 slides versus, 122–23
 types and causes, 113–23
flow-tills, 337–38
flumes, 190, 199
fluted surfaces, 350–52
flutes, from wind abrasion, 281
fluvial bars and lobes, 257, 258
fluvial entrainment, 200–201
fluvial landforms
 alluvial fans, 248–59
 deltas, 264–69
 floodplains, 233–42
 pediments, 259–64
 terraces, 242–48
fluvial processes. *See* river channels
fluvial sheet deposits, 257, 258
fluvioglacial drift, 338
fluviokarst terrains, 413
foliation, 316
force, defined, 6
foreset deposits, 266
form drag, 194
former sea cliffs, 467
foundations, on permafrost, 397
fracturing, 413
France, Tarn Gorge, 424
free surface resistance, 193
freeze-thaw cycles, 370, 374
friction, internal, 95
friction cracks, 323, 325
friction velocity. *See* drag velocity
fronts, 8
frost action
 cracking, 372–73
 creep, 104, 105, 374, 376
 heave, 104, 105, 370–71, 385
 patterned ground from, 385
 role in disintegration, 90–92
 shattering, 331–32, 368, 369
 sorting, 371–72, 373, 385
 thrusting, 370–71
 wedging, 369–70
frost-pull and frost-push
  hypotheses, 371
Froude number, 191
fulvic acid, 51
furrows, 326

G

gage height, 165
gaging stations, 165
Galera Creek Rock Glacier, 358
Gasconade River, 237
gelifluction
 in block slope development,
  390–91, 392
 common landforms, 386, 387
 process overview, 374–76
 in rock glacier development, 389, 390
gelifluction lobes, 386, 387
genetic floodplains, 233, 234
geochemical weathering, 44–45
geoid, 22
geologic screens, 38–40
geometric surface area, 58
geomorphic effectiveness, 205
geomorphic surfaces, 26–29
geomorphic thresholds, 100
geomorphic unit hydrographs, 155
geomorphic work, in river channels,
  205–6
geomorphology, 2–3. *See also* process
  geomorphology
geothermal gradient, 361
gibbsite, 56
Gilbert, G. K., 4, 126, 244
gilgai, 84
Gilkey Glacier, 319
glacial budget, 301
glacial deposition
 drift types, 335–38
 framework for variations, 338–41
 interior ice-contact features, 340,
  349–54
 marginal ice-contact features, 340,
  341–49
 proglacial features, 354–56
glacial erosion
 basic processes, 322–28
 cirques, 328–32
 troughs, 332–35
glacial grooves, 325
glacial morphogenetic systems, 36
glacial polish, 324
glacial slippage, 308–9
glacial striae, 325
glacial surges, 315
glaciers
 as argument for eustasy, 37–38
 basic movement processes,
  302–3
 deposition (*see* glacial deposition)
 effects on sea level, 476, 481
 erosion processes (*see* glacial
  erosion)
 ice structures, 315–18
 internal motions, 303–8
 isostatic effects, 23, 24
 mass balance, 301–2
 motion with deforming subglacial
  sediments, 312–13
 origins and types, 297–301
 sedimentation from, 185
 sliding, 308–12
 velocity variations with time, 313–15
glacio-isostasy, 23, 24
glacitectonics, 344
global warming, 477
gnammas, 92, 94
Goldich stability series, 52
graded river concept, 227
graded time, 17
gradient. *See* slope
grain roughness, 194
granite
 expansion in, 84–86
 physical and chemical weathering,
  92, 94
gravity, 8–9, 22–23
gravity anomalies, 22
great groups, 64–65
Greenland, isostatic recovery in, 24
Greenland Ice Sheet Program, 37
groins, 459
grooves, from wind abrasion, 281
ground ice, 368–69
ground moraine, 342, 349–50
groundwater
 base flow from, 141
 karstification and, 414–18
 movement patterns, 160–62
 in permafrost regions, 366
 problems with, 162–64
groundwater profile, 160, 161
groundwater sapping, 147, 148, 149
grus, 84–86, 87, 263
gully gravure, 131–32
gypsum, 408–9

H

Half Dome, 87
Hawaii
 groundwater sapping in, 149
 high-relief shorelines, 468
 shields forming, 32
head, of groundwater, 161
headwalls, of glaciated troughs, 334
headwater areas, 227
heat
 internal, 9
 role in climate, 7–8
 surface uplifts from, 23
 thermal expansion, 81–82
 units of, 7
heave
 creep and, 102–5
 defined, 99
 frost action, 370–71

in patterned ground, 385
role in rockfalls, 105
helical flow, 217
hematite, 56
high-constructive deltas, 265
high-destructive deltas, 265–66
high-energy floodplains, 234
high-polar glaciers, 299, 300
high-relief shoreline topography, 466,
467–70
Hofdabrekkujökull
Glacier, 346
Hog Island shoreline changes, 487
Holocene climate, 38
holokarst terrains, 413
hooks, 473
horns, 328, 330, 461
Horton, R. E., 136, 148
Hortonian overland flow, 139,
144–45
human activities
coastal hazards from, 467
contributions to dust storms, 293
as factor in collapse dolines, 422
floodplain formation and, 240
global warming, 477
on permafrost, 396–400
recognizing contributions of, 3
sediment yield and, 180–83
humid-glacial fans, 249
humid-temperate model of floodplain
origins, 237–41
humid temperate morphogenetic
systems, 36
humid-tropical fans, 249
humus, 59
Hurricane Agnes, 142–44
hurricanes, 479
hydration
expansion effects of, 83–87
of salts, 87–89
hydraulic action, on sea
cliffs, 468
hydraulic conductivity, 162
hydraulic geometry of river channels,
207–10
hydraulic gradient, 162
hydrogeology, 161
hydrologic budget, 160
hydrologic cycle
of drainage basins, 160
of slopes, 137, 138
hydrologic floodplains, 233
hydrolysis, 48
hydrostatic pressure, in glaciers,
303, 309
hyperconcentrated flow, 116
hypsometric analysis, 153
hypsometric integral, 153
hysteresis, 210

I

ice-cemented rock glaciers, 387
ice-contact environment, 338–41
ice-cored moraines, 345
ice-cored rock glaciers, 387
ice fabric transition, 298–99
Iceland
cryoplanation, 394
debris flows on talus slopes, 115
push moraines, 346
shields forming, 33
ice lenses, 368, 369
ice sheets, 299
ice streaming, 328
ice-wedge polygons, 377–78
ice wedges, 362, 377
icing, of roads on permafrost, 398
ideal water table caves, 431
igneous rocks, 10, 12
Illinois
DeKalb mounds, 380–81
denudation in, 178, 179
dolines, 420
earthflows in, 123
loess soils, 294, 350
scroll floodplain topography, 236
illite, 49, 54
illuviated zone, 62
impact craters, on Mars, 402
impact threshold, 277
inactive sea cliffs, 467
inception horizons, 431
incised channels, 251, 252
incision rates, 212, 213
inclinometers, 306
India, volcanic plains in, 32
Indiana
bedrock channel erosion, 204
blind valleys, 425
sample basin morphometry data, 152
Texas Creek knickpoints, 229
infiltration, 138–39, 155
infiltration capacity, 138, 179, 181
inlets, tidal, 486–88
inselbergs, 261, 263
insolation, 7
interception, 137–38
interflow, 139
interfluves, 135
interior ice-contact features, 340,
348–54
interior moraines, 349–50
interlobate moraines, 341, 342
interlocking friction, 95
intermediate states of beaches,
456, 457
intermediate zones, 161, 355–56
internal friction, 95
internal heat, 9

intersection point, of alluvial fans,
251, 252
intrinsic thresholds, 14–15
ion exchange, 45–46, 48–49
ionic potential, 49
iron, 47, 48, 51–52
islands, in braided channels, 221–22
isostasy, 23
isostatic anomalies, 23
Italy, expansion joints in, 82

J

Japan
sea cliff erosion, 470, 471
tsunamis, 445
jetties, 487–88
Jockey's Ridge, 483
jointing
dissolution and, 413
permeability from, 410
rockslides and, 110–12, 113
jokulhlaups, 356
Juneau Icefield, 319

K

Kaibab Plateau, 417
kame deposits, 339, 347
kame terraces, 347
kaolinite, 53, 54, 55
Karoo dolerite, 92
karren, 409
karst
caves, 428–32
definitions and characteristics,
407–8, 409
hydrology and drainage, 413–18
processes and controls, 408–13
surficial landforms, 419–28
karst denudation, 412
karst plains, 409
karst valleys, 409, 424–25
karst windows, 409
Kasei Vallis, 81
Kentucky
Dixon Cave, 430
dolines, 423
rockfalls in, 106
kettled sandar, 355
kettle holes, 348
K horizon, 62
kinematic viscosity, 191
kinematic waves, 315
Kinsey Run debris flow, 115, 122
Klamath River baymouth bar, 472
knickpoint inclination, 230
knickpoint replacement, 230
knickpoints, 229–30

L

lag time, 141, 142
La/Le ratio, 423
laminar flow, 190–91
landform/process balance theories, 4–6
landmass movements, effects on climate, 37
landslides
    active versus inactive, 125
    calculating causative factors, 110
    classification, 109, 110
    creep and, 105
    distinguishing deposits from glacial and fluvial sediments, 126
    human factors in, 3
landward migration, of barrier islands, 488–90
Langbein-Schumm analysis, 38–40
La Perouse Glacier, 300
Laramie Range grus development, 84–86
large-scale rhythmic topography, 464–66
lateral accretion, 235, 237–40
lateral migration
    river channels, 236–37, 238, 239–40
    scrolled floodplains, 234
lateral moraines, 341, 342
lateral spreads, 109
laterization, 70
latitude, solar heat and, 7
lava plains, 32
leaching, 49–50
length ratio, 150
levees, 235–36, 266
lichens, 52, 107
lift, 198–99
liftoff speed, 280
limestone caves, 428–32
limestone karstification, 408–13
limonite, 56
linear dunes, 288, 289, 291
linear morphometric measurements, 150, 151
linear passages, 428, 429
linked-cavity systems
    in glaciers, 309, 311, 313
    surges and, 315
liquid limit, 97
lithology
    effects on runoff, 138
    resisting forces of, 9–12
    sediment yield and, 179–80, 181
lithosphere, chemical composition, 10, 11, 12
litter, 59
Little Yanert Glacier, 318
littoral drift
    attempts to model, 458–59

challenges to human activities, 459–60
    representative rates, 461
    spit and baymouth bar formation, 470, 471, 473
littoral zone, 451
lobate bars, 222, 223
lobate deltas, 265
lodgement till, 336–37, 349
loess soils, 99, 292–94
Log Pearson Type III, 168
longitudinal profile of river channels, 211–13
longshore currents, 451, 458
lower delta plain, 266
low-relief shoreline topography, 466, 470–73
lunate fractures, 325

M

macropores, 139–40
macrotidal tides, 447
Madison County flood event, 114, 115, 122
mafic rocks, 10, 11, 12
Magnus effect, 280
Maine, pocket beaches, 472
Manning equation, 193
Manning's *n*, 193, 194
Marbut, C. F., 63
marginal ice-contact features, 340, 341–49
Mars
    groundwater sapping on, 148
    mass movements, 81
    periglacial landforms, 401–4
Marsh, George Perkins, 180
Maryland
    Assateague Island, 466, 474
    creep in vertical shale, 103
    Ocean City Inlet, 488
mass, relation to surface topography, 23
mass balance of glaciers, 301–2
mass movements
    falls, 105–7
    flows, 113–23
    heave and creep, 102–5
    on Mars, 81, 404
    morphology, 123–25
    overview of, 99–100
    periglacial landforms from, 386–95
    in permafrost regions, 374–76
    slides, 107–13
    slope stability calculations, 100–102
mass wasting, 373
master horizons, 61, 62
Matanuska Glacier velocities, 314
matrix suction, 201
Mauna Loa, 33
maximum basin relief, 153

maze patterns, 429, 430
mean annual discharge, 165
mean annual flood, 168
mean daily discharge, 165
meandering channels
    appearance, 214, 215
    formation of, 217–20
    transition with braided, 225
meander scrolls, 235, 236
mechanical weathering. *See* physical weathering
medial moraines, 342, 349
medium-energy floodplains, 234
melt-out till, 337, 349
meltwater, 114, 120
Merced chronosequence, 56, 57, 58
mesotidal tides, 447
mica, 53–54
Michigan, lake bluff slope evolution, 132
micropiracy, 146
microtidal tides, 447
migration, of dunes, 288, 291
minerals, rock bursts from, 90–92
mineral stability, 52–53
mini crag and tail, 325
mining, 15–16
Mississippi River
    paleoflood data, 172
    scroll floodplain topography, 236
    subdeltas, 266–67, 268
Missouri, lobe sediment deposits, 237
mixing corrosion, 411–12
mobility of ions, 49–52
modal state, 457
mogotes, 409, 427
moisture conditions, 68. *See also* water
Mojave Desert, pediment formation, 264
molecular viscosity, 191
Mollisols, 71
momentum analysis, 451
Montana
    alluvial fans in, 250
    cryoplanation, 394
    frost sorting, 372
    gelifluction lobes, 387
    Late Wisconsinan till, 336
    stone stripe, 383
    terminal moraines, 343, 346
    terrace deposits, 159
montmorillonite
    atomic structure, 53–54
    plastic and liquid limits, 97
    swelling capacity, 83–84
moraine ridges, 350
moraines
    interior, 349–50
    marginal, 341–44, 345, 346
morphogenetic systems, 34, 36
morphological classification
    of glaciers, 299

morphometric indices, 123–25, 150–53
mountain formations
  processes creating, 22, 25–31
  sediment yield in, 178–79
  tectonic effects on, 25–26
mountain fronts, 26, 28, 29
mountain-front sinuosity, 25, 26
mountain ice sheets, 299
Mount Shasta, 34
Mount St. Helens
  debris flows from, 116, 117
  linkages to eruption, 16, 17
mudflows, 113, 255
multibasinal drainage patterns, 136
multiple parallel bars, 454
multiwind hypothesis, 291
muschelbruch, 326

N

nailhead striae, 325
Namib sand sea, 284
natural levees, 235–36
neap tides, 446–47
nearshore bars, 453–55
nearshore currents, 448–51
Nebraska Sand Hills, 282
Nelson County debris flows, 118
nets (patterned ground), 383
net specific budget, 301
network mazes, 429, 430
Nevada, fault-bounded mountains, 27
New Guinea, karst morphometry and hydrology, 418
New Mexico
  Desert Soil Geomorphology Project, 75–76, 77
  Dump fault movement studies, 74
  lithological variations, 14
New York
  drumlins, 353
  stratified drift, 339
Nirgal Vallis, 148
nivation, 329, 331, 373
Noachis Terra, 148
nonstratified drift, 335–38
normal stress, 95, 96–97
North Carolina
  barrier island dunes, 483
  beach cusps, 462, 463, 464
  littoral drift, 459, 460, 488
  piedmont soils, 67
  wave runup, 444
northeasters, 480, 484

O

Ocean City Inlet, 488
ocean currents, 8

ocean water, infiltration of aquifers, 164
Ogallala Aquifer, 164
ogives, 318, 319
O horizon, 61
oil pipelines, 399–400
Olympic Peninsula sea cliffs, 466
open-system pingos, 379–80
orders, in Soil Taxonomy, 64
Oregon
  calderas in, 33, 35
  volcanic plains in, 32
Oregon Inlet, 460, 487, 488
organic matter, 59, 60
orogenic processes, 22, 25–31
orthoclase feldspar, 45, 48, 49
oscillation, waves of, 439
outwash debris, 293, 338
outwash plains, 354
overbank flow, 239
overdeepenings, 333, 334
overdraft, 163–64
overwash, 480, 483–86
oxidation and reduction, 46
oxides, 56
Oxisols, 70

P

paired terraces, 243, 244
paleoflood hydrology, 168–73
paleosols, 66–67, 75
palsas, 380
pans, 282
parabolic dunes, 288, 289
parabolic forms, for glaciers, 332
parallel drainage patterns, 136, 157
parallel nearshore bars, 454
parallel retreat, 131, 230
Parramore Island rip currents, 449
partial area concept, 140
partial duration series, 167
particle-hiding, 199
passages, 428–29
passive glaciers, 299
paternoster lakes, 333
patterned ground, 381–86, 404
pavements, 274
PC-STABL4, 102
peak flow, 154–55
Pecos River
  annual discharge data, 167
  flood frequency curve, 168
  flooding's effects on channel, 206
  paleoflood data, 169, 171
  slackwater deposits, 170
pediment association, 262
pediment formation, 259–64
pedochemical weathering, 44
pedogenesis
  controls and regimes, 65–71
  in permafrost regions, 373–74

pedogenic threshold, 74
peds, 59
Pennsylvania
  blockfields in, 391–93
  sample basin morphometry data, 152
  sediment yield from human activities, 181
  trellis drainage in, 137
percentage of base saturation, 48–49
perched water table, 416
perigean spring tide, 448
perigee, 448
periglacial environments
  climate changes and, 400–401
  human activities in, 396–400
  hydrology, 366–67
  on other planets, 401–4
  overview of, 359–60
  permafrost and ground ice, 360–67
periglacial landforms
  environmental and engineering issues, 396–401
  mass movement related, 386–95
  on other planets, 401–4
  patterned ground, 381–86
  permafrost related, 377–81
  relict features, 395
periglacial morphogenetic systems, 36
periglacial processes
  frost action, 368–73
  mass movements, 374–76
  nivation, 373
  pedogenesis, 373–74
period, of waves, 437
permafrost
  distribution, thickness, and origin, 361–66
  evidence on Mars, 402
  features of, 360–61
  human activities on, 396–400
  periglacial landforms, 377–81
permafrost table, 361
permanent wilting point, 61
permeability of rock, 409–10
P-forms, 323, 326
pH, 50
phreatic caves, 431
phreatic zone, 161
physical weathering. *See* disintegration
piedmont angle, 261
piedmont glaciers, 299
piedmont soils
  alluvial fans, 248–59
  in Desert Project area, 75–76
  drainage basin evolution in, 159
  pediments, 259–64
  variations, 67
pingos, 379–81
pipelines, 399–400
piping, 139–40
pits, 281

plain sandar, 354
planar slides, 201, 202
plane friction, 95
plants. *See* vegetation
plastically moulded forms, 323, 326
plastic deformation, in glaciers, 303–4
plasticity index, 97
plastic limit, 97
plateaus, volcanic, 32
plate-like soil structures, 60
plate-margin analyses, 31
playas, 275
Pleistocene climate, 38
plucking
    of bedrock river channels, 203
    glacial, 322, 327–28
plume concept, 23
plunging breakers, 440, 442, 443
pocket beaches, 472
pocket valleys, 425
podzolization, 68–70
point bar deposition, 237–39
polar glaciers, 299–300
poljes, 409, 414, 423
polygenetic soils, 73–74
polygonal karst, 423
polygons, in patterned ground, 383, 384
ponding, in alluvial channels, 194
pool-riffle sequences, 194, 212, 215–17
pore pressure
    calculating, 101
    normal stress and, 96
    role in slides, 110
    talus slope stability and, 129
porosity of rock, 409–10
potassium, 51
Potato Run, 204
potential of groundwater, 161
potentiometric surface, 162
potholes, 326
Powder River, 91, 166
power flow law, 303, 304
precipitation. *See also* rainwater; runoff
    forces controlling, 8
    sediment yield and, 176–77
primary capes, 462
primary minerals, 53
primary porosity, 409
primary structures of glaciers, 315–16
prism-like soil structures, 60
probabilities, computing for
    floods, 168
processes, 2, 6–7
process geomorphology
    elements of, 3
    Horton's contributions, 136
    importance of soils data, 73–77
    landform/process balance theories,
        4–6
    mechanical forces in, 6–14
    process linkage in, 16

threshold concept, 14–16
    time frameworks in, 16–19
process linkage, 16
profile development index, 72–73
proglacial environment, 341
Protonilus Mensae region (Mars), 403
proximal zone of sandar, 355
pseudoplastic movement, 303
Puerto Rico, sea arches, 469
push moraines, 344, 346
pyrite, 51

Q

quantitative geomorphology, 136
quarrying, 322, 326–27
quartz dissolution, 47
quasi-equilibrium condition,
    206–14, 227
Quaternary geomorphology, 21, 73–74

R

radial crevasses, 317
radial drainage patterns, 136
radiation from sun, 7
radioactivity, 9
rain, forces controlling, 8
Rainfall Erosion Facility, 157
rainfall intensity, 139
rainfall velocity, 173–74
rainwater
    infiltration, 138–39
    interception by vegetation, 137–38
    leaching, 49–50
    role in chemical weathering, 44–45
    role in debris flows, 114, 115, 118
    role in slides, 110, 112
    runoff (*see* runoff)
    soil splash factors, 174
rapid flow, 191
rating curves, for stream discharge,
    165, 166
reaction time of rivers, 227–28
recessional moraines, 342
recession limb, on hydrographs, 141, 142
recovery time, 18, 206
recrystallization, 298, 304
rectangular drainage patterns, 136
recurved spits, 473
redox potential, 47, 50–51
reflection of waves, 440
reflective beaches, 456, 457
refraction of waves, 440, 441
regelation, 311–12
regolith arches, 421, 422
relative density, 158
relict hillslope forms, 132–33
relict periglacial features, 395
relict soils, 75

relief-length ratios, 179, 180
relief morphometric relationships,
    151, 153
relief ratio, 153
reptation, 277
reptation population, 285
resisting elements, 9–14
response time, 37, 315
resurgences, 409, 417
retrograde movement, 374
return flow, 140
reversing dunes, 289
Reynolds number, 191
R horizon, 61
rhythmic topography, 464–66
ribbed moraines, 349–50
ridge and runnel bars, 454
riffles, 194, 212, 215–17
rift zones, 32
rillen, 94
rills
    destruction and reformation, 175
    in overland flow, 139, 145–46
Rio Grande, 75–76
rip currents, 449–51
ripples, 285
rising limb, on hydrographs, 141, 142
river channels
    anastomosing and anabranching, 214,
        223–24
    bank erosion, 200–202
    basic flow mechanics of, 190–93
    bedrock erosion, 202–3, 204
    braided, 214, 215, 221–23
    deposition, 204–5
    entrainment of sediment in, 196–200
    equilibrium conditions, 225–31
    flow equations and resistance factors,
        193–95
    frequency and magnitude of
        processes, 205–6
    karstification and, 413–14
    lateral migration, 236–37, 238
    meandering, 214, 215, 217–20
    pattern classification overview,
        214–15
    pattern continuity, 224–25
    quasi-equilibrium condition, 206–14
    sediment transport concepts and
        measurement, 195–96
    straight, 214, 215–17
    tectonic effects on, 29
river metamorphosis, 230–31
River of Rocks, 391, 392
rivers
    climatic influences, 38–40
    effects of debris flows on, 121
    karstification and, 413–14
    in permafrost regions, 366–67
river stage, 165
river terraces, 29, 38

roads on permafrost, 397–98
rock avalanches, 113
rock bursts, 82, 90
rock-cut terraces, 245
rockfalls, 105–7, 108
rock glaciers, 387–89, 390, 391
rockslides, 110–13
rock strength, 99, 100
rock type. *See* lithology
Rogen moraines, 342, 349–50, 351
rooms, 409
rotational sliding, 110, 111,
    329–31
rotational slip failure, 200
rouche moutonnée, 327, 328
ruggedness number, 155
runoff. *See also* erosion
    basin morphometric variables
        predicting, 149–53, 154–56
    channel initiation and, 144–47
    effect of lithology on, 179–80
    as factor in discharge, 141
    infiltration effects on, 138–39
    karst denudation and, 412–13
    from subsurface stormflow, 139–40
runoff intensity, 145

S

Sacramento River, mining effects on, 16
saltation, 277–80
salt water, 164, 370
saltwater intrusion, 164
salt weathering, 87–89, 90
sampling intervals, 57
San Andreas fault, 30
sand, supplies for beaches, 451. *See*
    also beaches
sandar, 354–56
sand dunes, 286–92, 483
sand seas, 282–84
sand sheets, 282
sand waves, 462
sapping, 201
saprolites, 44
Saskatchewan Glacier, 316, 317
saturated zone, 161
saturation overland flow, 140
scarps
    of fluvial terraces, 242
    of karst valleys, 424
    tectonic effects on, 30–31
Schmidt hammers, 99
Schwan Glacier, 330
Scott Glacier sandar, 355
scouring, 204–5, 209–10, 237–39
scree slopes, 108
Scripps Beach, 449
sea arches, 469
sea caves, 469

sea cliffs
    formation of, 466, 467–70
    tracking erosion, 471, 474–76, 477
sea floor, average elevation, 21–22
sea level
    effect of changes on climate, 37–38
    effects of glaciers on, 476, 481
    effects on coasts and rivers, 38,
        476–77, 478
    effects on terrace development, 246
    role in epeirogeny, 22–23
seasonal creep, 102–5
sea stacks, 466, 469
secondary capes, 462
secondary flow, 217–18, 219
secondary minerals, 53–56
secondary porosity, 410
secondary structures, of glaciers,
    316–18
sedimentary rocks, 10–11
sedimentation. *See also* deposition;
    sediment in channels
    in alluvial fans, 255–56 (*see also*
        alluvial fans)
    beneath glaciers, 312–13 (*see also*
        glacial deposition)
    climatic influences, 38–40
    complex response examples, 15–16
    in drainage basin formation, 135, 136
    effect on streamflow, 195
    paleoflood data from slackwater
        deposits, 169
    process linkage examples, 16, 17
sediment budgets, 184–85
sediment delivery ratios, 182, 185
sediment in channels. *See also*
    floodplains
    bank erosion, 200–202
    bedrock erosion, 202–3, 204
    channel shape and, 213–14
    deposition, 204–5
    entrainment, 196–200
    longitudinal profile and, 211–13
    relation to discharge, 209–10, 211
    transport concepts and measurement,
        195–96
sediment layering, 199
sediment mobilization and
    production, 184
sedimentology, 3
sediment yield
    factors affecting, 176–83
    as symptom of system changes, 185
    Universal Soil Loss Equation,
        175–76
segregation potential, 369
seiches, 445
seif dunes, 288–89, 291
seismic activity, cryoturbation
    versus, 373
Selby classification system, 99, 100

selective transport, 200
selva morphogenetic systems, 36
semiarid morphogenetic systems, 36
sensitive soils, 98–99
sensitivity of rivers, 227
separation pressure, 309
sewage lines, in permafrost
    regions, 399
Sexton Creek, 178, 179
shafts, in karst terrains, 416, 428
shallow slip, 201
shear strength
    elements of, 95–99
    factors affecting, 101
    measuring, 99, 110
    of subglacial sediments, 312–13
shear stress
    defined, 95
    factors affecting, 101
    in glaciers, 303, 305
shear type cantilever failure, 200
sheet flow, 174, 262
sheeting joints, 83
Shields diagrams, 197–98
shield volcanoes, 32
Ship Shoal Island, 484
shoaling of waves, 440–45
shoreline change, 473–81. *See also*
    coastal processes
shoreline configurations, 460–66
Shreve Magnitude, 149–50
sichelwannes, 325, 326
silica, 47–48, 53, 54
silicic rocks
    factors affecting dissolution, 47–48
    as percentage of Earth's crust, 10,
        11, 12
sinkholes, 419–23
sinuosity, 215
sinuous passages, 428, 429
sixth-power law, 197
slab failure, 112–13, 201
slackwater deposits, paleoflood data
    from, 169, 170
slides
    causes of, 107–13
    defined, 99
    flows versus, 122–23
sliding, of glaciers, 303, 308–12
slip planes, 306–7, 308
slip-stick movement, 323
slope
    alluvial fan morphology and, 251–54
    alteration in channels, 228–30
    beach formation and, 453
    effects on patterned ground, 381, 383
    effects on soil formation, 68
    frost creep and, 374
    influence on river channel
        mechanics, 211–13
    of pediments, 261–62

slope—*Cont.*
  relation to river channel discharge, 225, 226
  solution doline occurrence and, 420
slope angles, creep rates and, 104, 105
slope decline, 131
slope hazard regions, 117
slope hydrology, 137–40
slope processes
  applications of, 80
  driving forces, 93–95
  Gilbert's theories, 4
  mass movement types (*see* mass movements)
slope profiles
  determining causes, 125–29
  evolution of, 131–32
  influence of rock and climate, 129–31
  relict forms, 132–33
slope replacement, 131
slope stability, 100–102
smectite group, 53–54
snowfields, nivation in, 373
snowline, 297
snowpacks, 57, 297, 298
soil catenas, 67–68
soil chronosequences, 73
soil convection, 371–72
soil fall, 201
soil horizons, 61–62, 63, 64
soil profiles, 43, 58
soils
  basic properties, 58–61
  classification, 62–65
  defined, 43
  factors in formation of, 58, 65–73
  geomorphic significance, 73–77
  heave and creep, 102–5
  horizon nomenclature and descriptions, 61–62
Soil Taxonomy, 64–65, 66
soil wedges, 378
solar constant, 7
solar radiation, 7, 37
solifluction, 105, 374–76. *See also* gelifluction
solution. *See* dissolution
solution chimneys, 428
solution dolines, 419–21
Sonoma Range, 27
sorting, 371–72
Sourdough Mountain rock glaciers, 388
South Africa, sinkholes, 422
South Cascade Glacier, 307–8
South Dakota, expansive soils in, 85
spalling, 81, 82
speleothems, 409, 428–29
spheroid, 22
spheroidal soil structures, 60
spheroidal weathering, 86–87
spilling breakers, 440, 442

spindle flutes, 326
spits, 470–73, 487
splash, erosion and, 174
splay crevasses, 317
splay deposits, 235
Spodosols, 69
spongework mazes, 429, 430
sporadic permafrost, 365–66
springs, 417–18
spring tides, 446–47
squeezing, 344
stacks, 466, 469
staircase profile, 333–34
stalactites, 428–29
stalagmites, 428–29
star dunes, 289
static equilibrium, 5, 6
steady-state equilibrium, 5, 6
steady time, 16–17
step-pool sequences, 212, 217
steps, in patterned ground, 383
steptoes, 32
sticky spots, 313
stone stripe, 383
Storglaciaren, 332
storm runoff, 138
storms. *See also* climate
  effects on beach cusps, 463, 464
  effects on nearshore bars, 452–54
  shoreline erosion and, 476, 478–81
  threshold concept and, 18
storm surges, 479–80
Strahler method, 148–49
straight channels, 214, 215–17
straight segments, 125, 126, 130
strain rates, 303–4
stratification, of primary glacial features, 316
stratified drift, 338, 339
stratified marginal features, 344–49
stratified slope deposits, 386
stratigraphic debris fan studies, 119, 120
streambed armor, 246
stream discharge. *See also* river channels; rivers
  equilibrium changes and, 227
  flood frequency studies of, 165–68
  hydraulic geometry and, 207–10, 211
  measuring, 164–65
  relation to slope, 225, 226
  runoff as factor, 141
stream frequency, 158
stream hydrology, 154
stream ordering
  in drainage basins, 148–50, 152
  relation to stream length and number, 158
stream power, 199
streams, in permafrost regions, 366–67
strength equilibrium envelope, 127
strength equilibrium slopes, 127–28

stress, defined, 6
stress concentrations, 326
striations, glacial, 323, 324
stripes, in patterned ground, 383, 384–85
structure
  resisting forces of, 12–14
  soil types, 59, 60
sturzstrom, 113
suballuvial bench, 260
subaqueous delta plain, 266
subglacial cavities, 326, 327
subglacial drainage systems, 309–11
subglacial melt-out till, 337, 349
subgroups, in Soil Taxonomy, 64–65
suborders, in Soil Taxonomy, 64
subpolar glaciers, 299–300
subsurface stormflow, 139–40
sulfuric acid, 51, 410
summer profiles, 455
SUPERDUCK, 437
superimposed ice, 302
supraglacial till, 336
surface area, weathering and, 58
surface creep, 277
surface friction speed, 275–76
surf beat, 445
surf-scaling parameter, 457
surf zone, 443
surging breakers, 440, 442, 443
surging glaciers, 315
suspended load, 195, 227
suspension, by wind, 277
Susquehanna River
  basin stream ordering, 152
  Hurricane Agnes flood, 143–44
  sedimentation in, 186
  trellis drainage, 137
swallets, 409, 414
swallow holes, 409, 414
swash bars, 454
swash transport, 458, 462–63
swash zone, 443, 444
Sweden, Storglaciaren, 332
swell, of waves, 439
swelling, from hydration, 83–87
systems approach, 5

T

tafoni, 88, 89
taliks, 363, 365, 399
talus slopes
  debris flows on, 115
  examples, 108
  formation of, 128–29
  relict, 132
Tarn Gorge, 424
tarns, 328
tectonic activity, 246

tectonic geomorphology, 25–31
Teepee Butte Member, 32
temperate glaciers, 299, 300–301
temperature
    effects on soil formation, 68
    effects on wind direction, 275
    erosion and, 38–39
    within glaciers, 300
    of permafrost, 361, 363
Tennessee, dolines, 423
tensile type cantilever failure, 200
tension cracks, 200–201
tephra, 32, 33
terminal moraines
    defined, 341
    hummocky topography, 346
    Pinedale glaciation, 343
    Whitewater, Wis. quadrangle, 342
terminal velocity of raindrops, 173–74
terraces
    in Desert Project area, 76, 77
    fluvial, 242–47, 248
terra rossa, 409
Texas
    expansive soils in, 85, 87
    granite weathering, 94
    sample basin morphometry data, 152
    sheeting joints in, 83
Texas Creek knickpoints, 229
texture, soil types, 59
thalwegs, 215
thermal classification, 299–300
thermal erosion, 366–67
thermal expansion, 81–82
thermal spalling, 81, 82
thermokarst features
    formation of, 381, 382
    on Mars, 402, 403, 404
Thistle landslide flow, 116–18
threshold friction velocity, 277
thresholds, 3, 14–16, 18–19
throughflow, 139, 140
thrustbook moraine, 344
tidal bores, 446
tidal inlets, 486–88
tidal prism, 486
tide-dominated deltas, 265, 266
tides, 445–48
till
    in fluted surfaces, 350–52
    glacial, 335–38
    in ground moraine, 349
time
    effects on slope profiles, 131–32
    effects on soil formation, 71–73
    intervals, 5, 16–19
toe, of alluvial fans, 251
tombolos, 473
topples, 107, 109
toppling slab failure, 200
topset deposits, 266

total resistance, 193–95
Toutle River, 16, 17
tower karst, 409, 426–28
tractive force, 145
traditional-flow deposits, 257
tranquil flow, 191
Trans-Alaska pipeline, 399–400, 401
transitional deposits, 258
transition zones, 62
translational slides, 107
translational waves, 440, 442
translocation, 62
transport-limited slopes, 127, 130
transverse bars, 454
transverse crevasses, 318
transverse dunes, 288, 289
transverse troughs, 326
travertine, 429
treads, of fluvial terraces, 242
tree throw, 373
trellis drainage patterns, 136, 137
tributaries, complex responses in, 15
tropical storms, 479–80
troughs, 332–35, 450
tsunamis, 445
turbulent flow, 191, 276
Types I and II alluvial fans, 251, 253

U

Ultisols, 69
unconfined aquifers, 162
unconsolidated debris, 92–99, 312–13
undulating surfaces, 326
uniform flow, 190
unit hydrographs, 142
unit stream power, 185
universal gravitation law, 8
Universal Soil Loss Equation, 175–76
unloading, 82–83
unpaired terraces, 243, 244
unsaturated zone, 161
uplifts, 23–24
upper delta plain, 266
urbanization, impact on sediment
        yield, 182
Utah
    debris flows in, 116–18
    sample basin morphometry data, 152
    utilidors, 399
    utilities, in permafrost regions, 399
    uvalas, 409, 423

V

vadose caves, 431
vadose flow, 416
vadose seepage, 415, 416
Vaiont River valley expansion
        joints, 82

Valles Marineris, 81, 148
valley glaciers, 299, 332–35
valley sandar, 354, 355
valley trains, 354
variable source concept, 140
Variegated Glacier surge, 315
Varnes classification system, 107,
        109, 110
Vaughan Lewis Icefall, 319
vegetal screens, 38–40
vegetation
    on barrier islands, 482
    effects of changes on rivers, 228
    effects on river channel
        resistance, 194
    effects on soil formation, 68
    erosion resistance, 145
    interception of rainfall, 137–38, 176
    role in disintegration, 90
    shoreline erosion effects on, 474
    solution doline occurrence and, 421
    wind resistance, 272
velocity
    of glaciers, 305–8, 313–15
    of river channel flows, 208–9
    of stream discharge, 164–65
    of wind with height, 275–76
velocity profiles, 191–93
velocity reversal hypothesis, 216
ventifacts, 280, 281
vermiculite, 54
Vermont, rockfalls in, 107
vertical accretion, 235, 239–40
Vertisols, 84
vesicular horizon, 274
Virginia
    Assateague Island, 454, 473, 474
    Cedar Island overwash, 480
    debris flows in, 114, 115, 118, 122
    denudation in, 182
    Hog Island shoreline changes, 487
    nearshore bars, 454
    rip currents, 449
    rockfalls in, 108
    rotational slides in, 111
    sample basin morphometry data, 152
    saprolite formations, 44
    Ship Shoal Island, 484
    undeveloped barrier islands, 483
    Wreck Island landward
        migration, 489
Virginia humid-temperate fans, 249
viscous debris flows, 258
volcanoes, 31–34, 35, 37
Volney Creek, 159

W

Wasatch Front debris flows, 116–18
wash, 174–75

washboard moraines, 342, 350
Washington
  Mount St. Helens eruption, 16, 17
  sea cliffs, 466
  South Cascade Glacier, 307–8
  volcanic plains in, 32
wash load, 195
washover morphology, 483–86
water
  beneath glaciers, 308–11, 314
  capacity of soils to hold, 59
  leaching, 49–50
  role in chemical weathering, 44–45
  role in debris flows, 113–14, 115
  role in disintegration, 92, 93
  role in slides, 110, 112
  role of frost in disintegration, 90–92
water balance, 160
watersheds. *See* drainage basins
water supply lines, in permafrost
    regions, 399
water table
  in caves, 431
  defined, 161
  effects of infiltration, 139–40
  as factor in slope stability, 101–2
  in karst terrains, 414–15, 416
  role in slides, 110
wave-cut platforms, 28, 468–69
wave diffraction, 440
wave dispersion, 439
wave-dominated deltas, 265, 266
wavelength, of meanders, 218–19

wave runup, 444
waves, 437–45. *See also* beaches;
    coastal processes
wave setup, 450
wave shoaling, 440–45
waves of oscillation, 439
waves of translation, 440, 442
wave steepness, 439
wear law, 322
weather, role of solar energy in, 7
weathering
  in deserts, 273–74
  differing scales of, 11–12
  measuring denudation, 173–74
  overview of, 43
  of pediments, 262–63
  soil creep and, 112
weathering front, 263
weathering history, 187
weathering-limited slopes,
    126–28, 130
wedge striae, 325
Weibull Method, 167
wells, effects on hydrostatic level,
    162–64
Western Highland Rim dolines, 423
West Spanish Peak, 13
whaleback forms, 327
width-depth ratio, 214
wind action
  basic properties, 275–76
  desert soil resistance, 274–75
  dunes, 286–92

entrainment and transport,
    276–82
fine-grained deposits, 292–94
ripples, 285
role in wave generation, 437–39
sand seas and, 282–84
winter profiles, 455
Wisconsin, terminal moraines, 342
Wolman-Miller hypothesis, 205
Woodworth Glacier, 340
Wrangell Mountain rock
    glaciers, 388
Wreck Island, 485, 489
Wyoming
  East Fork River measurement
    station, 196
  frost shattered formations, 91
  Galera Creek Rock Glacier, 358

Y

Yanert Glacier, 345
yardangs, 282, 283
Yarmouth-Sangamon Paleosol, 66–67
yield stress, 303
Yosemite National Park, 106, 333
Yuba River, 16

Z

zero annual amplitude, 361
zone of soil moisture, 161